ROWAN UNIVERSITY
CAMPBELL LIBRARY
201 MULLICA HILL RD.
GLASSBORO, NJ 08028-1701

Standard Potentials
in Aqueous Solution

COMMISSIONS ON ELECTROCHEMISTRY AND ELECTROANALYTICAL CHEMISTRY

Memberships of the Commissions during the period (1979-83) this monograph was prepared were as follows:

ELECTROCHEMISTRY

Chairman: 1979-81 R. Parsons (France); 1981-83 K. E. Heusler (FRG); *Vice-Chairman*: 1979-81 A. J. Bard (USA); 1981-83 S. Trasatti (Italy); *Secretary*: 1979-83 J. C. Justice (France); *Titular and Associate Members*: J. N. Agar (UK; Associate 1979-83); A. J. Bard (USA: Titular 1981-83; E. Budevski (Bulgaria; Associate 1979-83); G. Gritzner (Austria; Associate 1979-83); K. E. Heusler (FRG; Titular 1979-81); H. Holtan (Norway; Associate 1979-83); N. Ibl‡ (Switzerland; Associate 1979-81); M. Keddam (France; Associate 1979-83); J. Kuta‡ (Czechoslovakia; Titular 1979-81); R. Memming (FRG; Associate 1979-83); B. Miller (USA; Associate 1979-83); K. Niki (Japan; Titular 1979-83); R. Parsons (France; Associate 1981-83); J. A. Plambeck (Canada; Associate 1979-83); Y. Sato (Japan; Associate 1981-83); R. Tamamushi (Japan; Associate 1979-81); S. Trasatti (Italy; Titular 1979-81); *National Representatives*: A. J. Arvía (Argentina; 1979-83); B. E. Conway (Canada; 1979-83); G. Horányi (Hungary; 1979-83); S. K. Rangarajan (India; 1981-83); J. W. Tomlinson (New Zealand; 1979-83); S. Minč (Poland; 1979-83); E. Mattsson (Sweden; 1981-83); A. K. Covington (UK; 1979-83); M. Karsulin (Yugoslavia; 1979-81); D. Dražić (Yugoslavia; 1981-83).

ELECTROANALYTICAL CHEMISTRY

Chairman: 1979-81 J. F. Coetzee (USA); 1981-83 J. Jordan (USA); *Secretary*: 1979-81 J. Jordan (USA); 1981-83 K. Izutsu (Japan); *Titular and Associate Members*: J. F. Coetzee (USA; Associate 1981-83); A. K. Covington (UK; Titular 1979-83); W. Davison (UK; Associate 1981-83); R. A. Durst (USA; Associate 1979-83); M. Gross (France; Associate 1979-83); K. Izutsu (Japan; Titular 1979-83); J. Juillard (France; Titular 1979-83); K. M. Kadish (USA; Associate 1979-83); R. Kalvoda (Czechoslovakia; Associate 1979-83); H. Kao (China; Associate 1981-83); R. C. Kapoor (India; Titular 1979-83); Y. Marcus (Israel; Associate 1979-83); L. Meites (USA; Associate 1979-81); T. Mussini (Italy; Associate 1979-83); H. W. Nürnberg ‡ (FRG; Associate 1979-83); P. Papoff (Italy; Associate 1979-81); E. Pungor (Hungary; Titular 1979-83); M. Senda (Japan; Associate 1979-83); D. E. Smith‡ (USA; Associate 1979-83); N. Tanaka (Japan; Associate 1979-83); *National Representatives*: D. D. Perrin (Australia; 1979-83); B. Gilbert (Belgium; 1981-83); W. C. Purdy (Canada, 1979-83); R. Neeb (FRG; 1979-83); K. Tóth (Hungary; 1979-83); H. V. K. Udupa (India; 1979-81); S. K. Rangarajan (India; 1981-83); W. F. Smyth (Ireland; 1981-83); E. Grushka (Israel; 1981-83); Z. Galus (Poland; 1979-83); G. Johansson (Sweden; 1981-83); J. Buffle (Switzerland; 1981-83); B. Birch (UK; 1979-83); J. Osteryoung (USA; 1981-83); M. Branica (Yugoslavia; 1979-83).

---

‡Deceased.

INTERNATIONAL UNION OF PURE AND APPLIED CHEMISTRY

IUPAC Secretariat: Bank Court Chambers, 2-3 Pound Way, Cowley Centre, Oxford OX4 3YF, UK

Standard Potentials in Aqueous Solution

Edited by

Allen J. Bard
University of Texas
Austin, Texas

Roger Parsons
Laboratoire d'Electrochimie Interfaciale du CNRS
Meudon, France

and

The University
Southampton, England

Joseph Jordan
The Pennsylvania State University
University Park, Pennsylvania

IUPAC
INTERNATIONAL UNION OF PURE AND APPLIED CHEMISTRY
Physical and Analytical Chemistry Divisions
Commissions on Electrochemistry and Electroanalytical Chemistry

MARCEL DEKKER, INC.   New York and Basel

MONOGRAPHS IN ELECTROANALYTICAL CHEMISTRY AND ELECTROCHEMISTRY

consulting editor
Allen J. Bard
Department of Chemistry
University of Texas
Austin, Texas

Electrochemistry at Solid Electrodes, *Ralph N. Adams*

Electrochemical Reactions in Nonaqueous Systems, *Charles K. Mann and Karen Barnes*

Electrochemistry of Metals and Semiconductors: The Application of Solid State Science to Electrochemical Phenomena, *Ashok K. Vijh*

Modern Polarographic Methods in Analytical Chemistry, *A. M. Bond*

Laboratory Techniques in Electroanalytical Chemistry, *edited by Peter T. Kissinger and William R. Heineman*

Standard Potentials in Aqueous Solution, *edited by Allen J. Bard, Roger Parsons, and Joseph Jordan*

Other Volumes in Preparation

Library of Congress Cataloging in Publication Data
Main entry under title:

Standard potentials in aqueous solution.

(Monographs in electroanalytical chemistry and electrochemistry)
At head of title: International Union of Pure and Applied Chemistry.
Includes indexes.
1. Electrodes–Tables. 2. Electrochemistry–Tables.
I. Bard, Allen J. II. Parsons, Roger. III. Jordan, Joseph, [date]. IV. International Union of Pure and Applied Chemistry. V. Series
QD571.S74 1985   541.3'72'0212   85-10307
ISBN: 0-8247-7291-1

Copyright © 1985 International Union of Pure and Applied Chemistry

All Rights Reserved. No part of this publication may be reproduced, stored in a retrieval system or transmitted in any form or by any means: electronic, electrostatic, magnetic tape, mechanical, photocopying, recording or otherwise, without permission in writing from the copyright holders.

Marcel Dekker, Inc., 270 Madison Avenue, New York, New York 10016
10 9 8 7 6 5 4 3 2 1

PRINTED IN THE UNITED STATES OF AMERICA

Preface

Standard electrode potentials (variously called redox potentials, oxidation-reduction potentials, etc.) are widely used by the scientific community because they provide predictive power about the course of chemical reactions and the behavior of metal/solution interfaces. They represent crucial reference data for diversified disciplines, ranging from the physical sciences to geology, oceanography and biology. Thus, standard potentials are of general interest to all practicing chemists and of special relevance in such widely disparate areas as corrosion research, biochemistry, chemical engineering, physical chemistry, analytical chemistry, electrochemistry, and inorganic synthesis. The need for a reliable compilation has been admirably filled by the classic monograph of Wendell M. Latimer, *Oxidation Potentials* (or *The Oxidation States of the Elements and Their Potentials in Aqueous Solutions*) first published in 1938 and last revised in 1952. Although this widely used work was later supplemented by numerous other compilations, Latimer has remained the authoritative reference in this field. A desire to incorporate a wealth of new data and to bring notations and conventions into accord with current usage has led Commissions I.3 (Electrochemistry) and V.5 (Electroanalytical Chemistry) of the *International Union of Pure and Applied Chemistry* (IUPAC) to undertake the preparation of this book.

Latimer had admirable features of consistency throughout the book. This is an unmatched advantage of single authorship. However, Latimer himself recognized the corollary limitation: "Much...should have been recalculated by modern methods *but the labor involved is beyond the capacity of a single author*" (quoted from the Preface of the first edition, Prentice-Hall, Inc., New York, 1938; emphasis added). That constraint has since become genuinely insurmountable: no single individual could currently undertake such a Herculean task. In view of this situation we have opted for multiple authorship. Chapters were prepared by knowledgeable specialized experts and reviewed by referees whose names are included in this book. It is hoped that the potentials given here represent the *critically selected* values or at least the best estimates now available. To keep this book to a manageable size, discretion was exercised as to which half-reactions would be included for a given element. For similar reasons, data have been limited to a single temperature, 25°C. Furthermore, no attempt was made to give complete citations to all pertinent original publications. The bibliography has been restricted to previous important compilations and "key papers." This monograph is intended as a convenient aid to the reader and the bibliography is *not* meant to be a pantheon of all contributions in the field.

The book is organized in the following way. The first two chapters deal with units, conventions, methods of determination, and the concept of the single electrode potential. These are followed by standard potentials and thermodynamic data for the elements grouped according to their positions in the periodic table.

The editors gratefully acknowledge the assistance and support of the members of the IUPAC Commissions I.3 and V.5, and sincerely thank all those involved (authors and reviewers), who contributed to this effort without financial remuneration to ensure that the book could be produced at a price affordable to students and the rest of the scientific community.

Allen J. Bard
Joseph Jordan
Roger Parsons

Contents

Preface	iii
Contributors	ix

1. Standard Electrode Potentials: Units, Conventions, and Methods of Determination, *Roger Parsons* — 1

2. The Single Electrode Potential: Its Significance and Calculation, *Roger Parsons* — 13

3. Hydrogen, *Philip N. Ross* — 39

4. Oxygen, *James P. Hoare* — 49

5. The Halogens — 67
   - I. Fluorine, *Anselm T. Kuhn and Colin L. Rice* — 67
   - II. Chlorine, *Torquato Mussini and Paolo Longhi* — 70
   - III. Bromine, *Torquato Mussini and Paolo Longhi* — 78
   - IV. Iodine and Astatine, *Pier Giorgio Desideri, Luciano Lepri, and Daniela Heimler* — 84

6. Sulfur, Selenium, Tellurium, and Polonium, *Stephan I. Zhdanov* — 93

7. Nitrogen, Phosphorus, Arsenic, Antimony, and Bismuth — 127
   - I. Nitrogen, *Joseph T. Maloy* — 127
   - II. Phosphorus, *K. S. V. Santhanam* — 139
   - III. Arsenic, *K. S. V. Santhanam and N. S. Sundaresan* — 162
   - IV. Antimony, *Vello Past* — 172
   - V. Bismuth, *B. Lovreček, I. Mekjavić, and M. Metikoš-Huković* — 180

8. Carbon, Silicon, Germanium, Tin, and Lead, *Zbigniew Galus* — 189

9. Gallium, Indium, and Thallium, *V. V. Losev* — 237

10	Zinc, Cadmium, and Mercury	249
	I. Zinc, *Ralph J. Brodd and John Werth*	249
	II. Cadmium, *Yutaka Okinaka*	257
	III. Mercury, *Jan Balej*	265
11	Copper, Silver, and Gold	287
	I. Copper, *Ugo Bertocci and Donald D. Wagman*	287
	II. Silver, *G. V. Zhutaeva and N. A. Shumilova*	294
	III. Gold, *Gerhard M. Schmid*	313
12	Nickel, Palladium, and Platinum	321
	I. Nickel, *Alejandro J. Arvia and Dionisio Posadas*	321
	II. Palladium and Platinum, *Francisco Colom*	339
13	Cobalt, Rhodium, and Iridium	367
	I. Cobalt, *Nobufumi Maki and Nobuyuki Tanaka*	367
	II. Rhodium and Iridium, *Francisco Colom*	382
14	Iron, Ruthenium, and Osmium	391
	I. Iron, *Konrad E. Heusler and W. J. Lorenz*	391
	II. Ruthenium and Osmium, *Francisco Colom*	413
15	Manganese, Technetium, and Rhenium	429
	I. Manganese, *James C. Hunter and Akiya Kozawa*	429
	II. Technetium and Rhenium, *Robert J. Magee and Harry Blutstein*	439
16	Chromium, Molybdenum, and Tungsten	453
	I. Chromium, *Katsumi Niki*	453
	II. Molybdenum, *Nicholas Stolica*	462
	III. Tungsten, *Alexander T. Vas'ko*	484
17	Vanadium, Niobium, and Tantalum	507
	I. Vanadium, *Yecheskel Israel and Louis Meites*	507
	II. Niobium and Tantalum, *H. V. K. Udupa, V. K. Venkatesan, and M. Krishnan*	526
18	Titanium, Zirconium, and Hafnium	539
	I. Titanium, *William J. James and James W. Johnson*	539
	II. Zirconium and Hafnium, *Dennis G. Tuck*	547
19	Boron, Aluminum, and Scandium	555
	I. Boron, *Su-Moon Park*	555
	II. Aluminum, *Georges G. Perrault*	566
	III. Scandium, *Su-Moon Park*	580
20	Yttrium, Lanthanum, and the Lanthanide Elements, *Lester R. Morss*	587

Contents

21	The Actinides, *L. Martinot and J. Fuger*	631
22	Beryllium, Magnesium, Calcium, Strontium, Barium, and Radium	675
	I. Beryllium, *Richard E. Panzer*	675
	II. Magnesium, *Georges G. Perrault*	687
	III. Calcium, Strontium, Barium, and Radium, *Shinobu Toshima*	700
23	Lithium, Sodium, Potassium, Rubidium, Cesium, and Francium, *Hiroyasu Tachikawa*	727
24	Inert Gases, *Bruno Jaselskis*	763

Appendix: Synopsis of Standard Potentials   787
Author Index   803
Subject Index   829

Contributors

Alejandro J. Arvia Instituto de Investigaciones Fisicoquimicas Teoricas y Aplicadas, Universidad Nacional de La Plata, La Plata, Argentina

Jan Balej Institute of Inorganic Chemistry, Czechoslovak Academy of Sciences, Prague, Czechoslovakia

Ugo Bertocci Center for Material Science, National Bureau of Standards, Gaithersburg, Maryland

Harry Blutstein[*] Department of Chemistry, La Trobe University, Melbourne, Victoria, Australia

Ralph J. Brodd[†] Inco Research and Development Center, Inc., Sterling Forest, New York

Francisco Colom Institute of Physical Chemistry, CSIC, Madrid, Spain

Pier Giorgio Desideri Institute of Analytical Chemistry, University of Florence, Florence, Italy

J. Fuger Laboratory of Analytical Chemistry and Radiochemistry, University of Liège, Liège, Belgium

Zbigniew Galus Department of Chemistry, University of Warsaw, Warsaw, Poland

Daniela Heimler Institute of Analytical Chemistry, University of Florence, Florence, Italy

Konrad E. Heusler Abteilung Korrosion und Korrosionsschutz, Institut für Metallkunde und Metallphysik, Technical University of Clausthal, Clausthal, Federal Republic of Germany

---

Current affiliation

[*]Water Branch, Environment Protection Authority of Victoria, Melbourne, Victoria, Australia
[†]Amoco Research Center, Naperville, Illinois

James P. Hoare Electrochemistry Department, General Motors Research Laboratories, Warren, Michigan

James C. Hunter Battery Products Division, Union Carbide Corporation, Westlake, Ohio

Yecheskel Israel Analytical and Electrochemical Department, IMI Institute for Research and Development, Haifa, Israel

William J. James Department of Chemistry and Graduate Center for Materials Research, University of Missouri-Rolla, Rolla, Missouri

Bruno Jaselskis Department of Chemistry, Loyola University of Chicago, Chicago, Illinois

James W. Johnson Department of Chemical Engineering, University of Missouri-Rolla, Rolla, Missouri

Akiya Kozawa Battery Products Division, Union Carbide Corporation, Westlake, Ohio

M. Krishnan Thomas J. Watson Research Center, IBM Corporation, Yorktown Heights, New York

Anselm T. Kuhn* Department of Biomaterials Science, Institute of Dental Surgery, University of London, London, England

Luciano Lepri Institute of Analytical Chemistry, University of Florence, Florence, Italy

Paolo Longhi Department of Physical Chemistry and Electrochemistry, University of Milan, Milan, Italy

W. J. Lorenz Institute of Physical Chemistry and Electrochemistry, Karlsruhe University, Karlsruhe, Federal Republic of Germany

V. V. Losev Karpov Institute of Physical Chemistry, Moscow, USSR

B. Lovreček Institute of Electrochemistry, University of Zagreb, Zagreb, Yugoslavia

Robert J. Magee Department of Chemistry, La Trobe University, Melbourne, Victoria, Australia

Nobufumi Maki Department of Chemistry, Shizuoka University, Hamamatsu City, Japan

Joseph T. Maloy Department of Chemistry, Seton Hall University, South Orange, New Jersey

---

Current affiliation

*Faculty of Science and Technology, Harrow College of Higher Education, Harrow, England

Contributors

L. Martinot Inter-University Institute for Nuclear Sciences and Laboratory of Analytical Chemistry and Radiochemistry, University of Liège, Belgium

Louis Meites* Department of Chemistry, Clarkson College of Technology, Potsdam, New York

I. Mekjavić Laboratory of Physical Chemistry, University of Split, Split, Yugoslavia

M. Metikoš-Huković Institute of Electrochemistry, University of Zagreb, Zagreb, Yugoslavia

Lester R. Morss Chemistry Division, Argonne National Laboratory, Argonne, Illinois

Torquato Mussini Department of Physical Chemistry and Electrochemistry, University of Milan, Milan, Italy

Katsumi Niki Department of Electrochemistry, Yokohama National University, Yokohama, Japan

Yutaka Okinaka[†] Chemical Process Department, Bell Telephone Laboratories, Murray Hill, New Jersey

Richard E. Panzer Consultant, El Sobrante, California

Su-Moon Park Department of Chemistry, University of New Mexico, Albuquerque, New Mexico

Roger Parsons[‡] Laboratoire d'Electrochimie Interfaciale du CNRS, Meudon, France

Vello Past Department of Chemistry, Tartu State University, Tartu, USSR

Georges G. Perrault Laboratoire d'Electrochimie Interfaciale du CNRS, Meudon, France

Dionisio Posadas Universidad Nacional de La Plata, La Plata, Argentina

Colin L. Rice Department of Chemistry, University of Salford, Salford, England

Philip N. Ross Materials and Molecular Research Division, Lawrence Berkeley Laboratory, Berkeley, California

K. S. V. Santhanam Chemical Physics Group, Tata Institute of Fundamental Research, Bombay, India

---

Current affiliation

*Department of Chemistry, George Mason University, Fairfax, Virginia
[†]Materials Science Research Division, Bell Communications Research, Inc. Murray Hill, New Jersey
[‡]Department of Chemistry, The University, Southampton, England

Gerhard M. Schmid Department of Chemistry, University of Florida, Gainesville, Florida

N. A. Shumilova Institute of Electrochemistry, Academy of Sciences of the USSR, Moscow, USSR

Nicholas Stolica Institut für Metallforschung, Westfälische Wilhelms Universität, Münster, Westfalen, Federal Republic of Germany

N. S. Sundaresan* Chemical Physics Group, Tata Institute of Fundamental Research, Bombay, India

Hiroyasu Tachikawa Department of Chemistry, Jackson State University, Jackson, Mississippi

Nobuyuki Tanaka Department of Chemistry, Tohoku University, Sendai, Miyagi, Japan

Shinobu Toshima† Department of Applied Chemistry, Tohoku University, Sendai, Miyagi, Japan

Dennis G. Tuck Department of Chemistry, University of Windsor, Windsor, Ontario, Canada

H. V. K. Udupa‡ Central Electrochemical Research Institute, Karaikudi, Tamil Nadu, India

Alexander T. Vas'ko Institute of General and Inorganic Chemistry, Ukrainian SSR Academy of Sciences, Kiev, USSR

V. K. Venkatesan Central Electrochemical Research Institute, Karaikudi, Tamil Nadu, India

Donald D. Wagman§ Chemical Thermodynamics Data Center, National Bureau of Standards, Gaithersburg, Maryland

John Werth Engelhard Industries, Menlo Park, New Jersey

Stephan I. Zhdanov Institute of the Ministry of Chemical Industry, Moscow, USSR

G. V. Zhutaeva Institute of Electrochemistry, Academy of Sciences of the USSR, Moscow, USSR

Current affiliation

*Department of Chemistry, Sophia College, Bombay, India
†Department of Fuel Chemistry, Akita University, Akita, Japan
‡Titanium Equipment and Anode Manufacturing Company, Ltd., Vandalur, Madras, India
§Retired

1

Standard Electrode Potentials: Units, Conventions, and Methods of Determination*

ROGER PARSONS† *Laboratoire d'Electrochimie Interfaciale du CNRS, Meudon, France*

I. SYMBOLS

The standard electrode potential is a physical quantity and, like all physical quantities, it is the product of a numerical value (a pure number) and a unit (volt). It is represented by the symbol $E°$, the superscript indicating the standard nature of the quantity. More precisely, $E°$ in the context of this book is the standard potential of a cell reaction when that reaction involves the oxidation of molecular hydrogen to solvated protons. When necessary the reaction will be indicated in parentheses or brackets after the symbol $E°$, for example,

$$E°(Zn^{2+}(aq) + H_2(g) \to 2H^+(aq) + Zn(s)) \tag{1.1}$$

which may be abbreviated [assuming that the reaction $H_2(g) \to 2H^+(aq) + 2e^-$ is always present]

$$E°(Zn^{2+}(aq) + 2e^- \to Zn) \tag{1.2}$$

or

$$E°(Zn^{2+}/Zn) \tag{1.3}$$

Other important physical quantities to be used in this work are:

T	Thermodynamic temperature
G	Molar Gibbs energy
S	Molar entropy
C_p	Molar heat capacity
H	Molar enthalpy
x_A	Mole fraction of species A

---

*Reviewers: M. Gross, University of Strasbourg, Strasbourg, France; L. G. Hepler, University of Lethbridge, Lethbridge, Alberta, Canada.
†*Current affiliation*: The University, Southampton, England.

m_A Molality of species A (number of moles of A dissolved in 1 kg of solvent
c_A Concentration of species A, also indicated as [A] (number of moles of A contained in 1 liter of solution, formerly called molarity)
p_A Partial pressure of species A
p_A^* Fugacity of species A
a_A Activity of species A
f_A
γ_A Activity coefficients $\begin{cases} \text{Mole fraction basis} \\ \text{Molality basis} \\ \text{Concentration basis} \end{cases}$
y_A
K Thermodynamic equilibrium constant of a reaction
n Charge number of a cell reaction [see Section III, following (1.8)]

II. UNITS AND CONSTANTS

The units of energy and potential employed are those adopted by the Conférence Générale des Poids et Mesures and known as SI (Système International) units. The primary energy unit is the joule, defined in terms of the SI base units, the kilogram, meter, and second:

$$1 \text{ joule} \equiv 1 \text{ J} \equiv 1 \text{ kg m}^2 \text{ s}^{-2} \tag{1.4}$$

Many data in the literature are quoted in calories. They have been converted into joules using the conversion factor $4.1840 \text{ J cal}^{-1}$. Also, $1 \text{ eV} = 1.6021892 \times 10^{-19}$ J. The unit of electrical potential difference is the volt, defined in terms of these SI base units together with the ampere:

$$1 \text{ volt} \equiv 1 \text{ V} \equiv 1 \text{ kg m}^2 \text{ s}^{-3} \text{ A}^{-1} \equiv 1 \text{ J A}^{-1} \text{ s}^{-1} \tag{1.5}$$

Pressure, molality, and concentration units have most usually been

$$1 \text{ atmosphere} \equiv 101\,325 \text{ N m}^{-2} \text{ (Pa)}$$

$$1 \text{ mol kg}^{-1} \text{ (mass of solvent)}$$

$$1 \text{ mol L}^{-1} = 1 \text{ mol dm}^{-3} \text{ (volume of solution)}$$

and these will continue to be used here. Conversion of pressure or concentration units to the coherent SI units of 1 N m^{-2} and 1 mol m^{-3} involves recalculation of the values of standard thermodynamic quantities [see (1.4) and (1.5)]. Since no measure of international agreement has been reached on this awkward problem, the traditional units have been retained here.

Important fundamental constants used in this work are [1]:
Celsius temperature defined by

$$t = T - T_0 \quad \text{where } T_0 = 273.15 \text{ K}$$

Faraday constant: $F = 9.648456 \times 10^4$ C mol^{-1}

Gas constant: $R = 8.31441$ J mol^{-1} K^{-1}

The combination of fundamental constants RT/F is frequently used. At 25°C it has the value

$RT/F = 0.0256926$ V

$(RT/F) \ln 10 = 0.0591594$ V

III. CONVENTIONS

The galvanic cell is represented by a diagram. For example, the chemical cell consisting of an aqueous solution of hydrogen chloride of concentration 1 mol L^{-1}, a platinum-hydrogen electrode with hydrogen at a partial pressure of 1 atm, and a silver-silver chloride electrode is represented by the diagram

$$\text{Cu(s)} \mid \text{Pt(s)} \mid \text{H}_2\text{(g)} \mid \text{HCl(aq)} \mid \text{AgCl(s)} \mid \text{Ag(s)} \mid \text{Cu(s)}$$
$$p = 1 \text{ atm} \quad c = 1 \text{ mol L}^{-1} \tag{1.6}$$

in which the Cu(s) at either end of the cell represent the terminals of the cell, which must be of the same material (although not necessarily copper). A single vertical bar represents a phase boundary; a dashed vertical bar (¦) represents a junction between miscible liquids, and double dashed vertical bars (¦¦) represent a liquid junction in which the liquid junction potential has been eliminated by some more or less accurate extrapolation or other procedure.

The electric potential difference E of a cell written in this way is always taken as the potential of the right-hand terminal with respect to that of the left-hand terminal. Thus if the cell is written down in the opposite direction, the sign of E must be reversed.

The reaction occurring in the cell is written down in the direction that corresponds to the flow of electrons in the external circuit from the left-hand terminal to the right-hand terminal. Thus for the cell (1.6), the reaction is

$$\text{H}_2(\text{g}) + 2\text{AgCl(s)} \rightarrow 2\text{H}^+(\text{aq}) + 2\text{Cl}^-(\text{aq}) + 2\text{ Ag(s)} \tag{1.7}$$

Again, if the cell is written in the opposite direction, the reaction must also be reversed.

The potential difference measured becomes of thermodynamic significance when the current through the external circuit becomes zero, all local charge transfer equilibria across phase boundaries represented in the cell diagram (except at electrolyte-electrolyte junctions) and local chemical equilibria being established. The resulting value of E is termed the electromotive force (emf) of the cell, a name that no longer represents the reality of the concept but which is so well established that it seems impossible to replace (the same is true for the equivalent terms in other languages).

When all species in the cell reaction are in their standard states (pure solids, unit standard concentrations, etc.), the measured emf of the cell is, to a good approximation, equal to the standard potential of the reaction in the cell $E°$, which is also given by

$$E° = -\Delta G°/nF = (RT/nF) \ln K \qquad (1.8)$$

where $\Delta G°$ is the standard Gibbs energy change in the cell reaction and K is the thermodynamic equilibrium constant of this reaction. n, the charge number of the reaction, is the number of electrons passing around the external circuit when one unit of the cell reaction occurs as written [e.g., for (1.7), $n = 2$].

The *standard electrode potential* of an electrode reaction $E°(Zn^{2+}/Zn)$ is the standard potential of a reaction in a cell whose left-hand electrode is a hydrogen electrode. [Note that this is the point at which the present (IUPAC) convention differs from that used in Latimer's book, where the hydrogen electrode is on the right of the cell in its standard format. Clearly, the standard electrode potentials consequently have the opposite sign.] For example, in the IUPAC convention

$$E°(Zn^{2+}/Zn) = -0.763 \text{ V}$$

At the end of the section on each element a summary of the principal standard electrode potentials for that element at 298.15 K is given in the form of a potential diagram. This summarizes the oxidation states of the element with the highest oxidation state on the left. The numbers between each state give $E°/V$.

IV. STANDARD STATES IN SOLUTION

For many phases in galvanic cells the standard state is the pure element or pure compound at unit pressure and at the temperature of the experiment, frequently 25°C. For mixtures that are miscible in all proportions these standard states may be retained by expressing the composition of the mixture in terms of mole fraction (= pure component). This means, of course, that all components in a mixture miscible in all proportions cannot be simultaneously in their standard states.

The chemical potential of a species A is then written

$$\mu_A = \mu°(p,T) + RT \ln a_A \qquad (1.9)$$

where the activity

$$a_A = f_A x_A \qquad (1.10)$$

Note that $\mu°$ is the chemical potential in the standard state when $x_A \to 1$, $f_A \to 1$, so that the standard state is also the reference state (the state where the activity coefficient goes to a defined limit, i.e., unity).

For many electrolytes the symmetrical system described above is inconvenient, partly because salts and solvents are often not completely miscible. It is therefore more common to use systems based on molality or concentration. The former has the advantage that it is independent of temperature. Equation (1.9) remains valid for the solute but the activity is given by

$$a_A = \gamma_A (m_A/m°) \qquad (1.11)$$

or

$$a_A = y_A(c_A/c^\circ) \qquad (1.12)$$

where $m^\circ(c^\circ)$ is the standard molality (concentration) 1 mol kg^{-1} (1 mol dm^{-3}). Thus the standard state is at a composition in the region of 1 mol kg^{-1} or 1 mol L^{-1} but is significantly different from unit molality or unit concentration because electrolytes are substantially nonideal. This is sometimes expressed by saying that the standard state is "a hypothetical ideal state of unit molality or unit concentration."

In this book the standard state in solution is the hypothetical ideal state of unit concentration measured in mol L^{-1} (unless otherwise stated). Thus equilibrium constants are dimensionless, being products of activities, and c° is taken as 1 mol L^{-1} for all species in solution.

As the solute becomes more and more dilute its mutual interactions decrease and vanish as m_A or $c_A \to 0$. In this reference state γ_A or $y_A = 1$. Thus the reference state is defined as

$$m_A \to 0 \qquad \gamma_A \to 1 \qquad (1.13)$$

or

$$c_A \to 0 \qquad y_A \to 1 \qquad (1.14)$$

while the standard state is defined by

$$a_A^\circ = 1 \qquad (1.15)$$

In this standard state the Gibbs energy of the solvent in aqueous solutions is not very different from that in the pure state. Hence the solvent may be regarded as being approximately in its standard state, too (if that is taken as the pure substance). However, the approximation may in fact be rather poor, and this difficulty is avoided by omitting the solvent from the cell reaction in most of the examples given in this book.

V. STANDARD STATE IN THE GAS PHASE

Up to the present time the generally used standard state pressure (p°) has been 1 standard atmosphere (101 325 Pa). During the time that this book was prepared, a recommendation for a change in this standard state was proposed [2]. The new recommendation for the standard state pressure made by IUPAC is 10^5 Pa (1 bar). In view of the fact that all the data collected in this book are referred to the standard pressure of 1 atm, it has been decided to retain this as the standard state pressure, at which the standard electrode potential is specified.

At the same time, it should be noted that the change to the standard state of 10^5 Pa would affect very few of the data presented here. The major effect for most standard potentials would be that the standard hydrogen electrode potential shifts 0.169 mV in the positive direction, that is, 0.169 mV would have to be subtracted from the quoted standard electrode potentials on this account. For cell reactions that otherwise involve only condensed phases, this is the only significant correction. For cell reactions in which a gas other than hydrogen takes part, there would be a further change that would be of the same order of magnitude. Since the large majority of the standard potentials quoted in this book have an uncertainty of 1 mV or greater, this correction can be neglected.

If a correction to 1 bar is necessary, it may be calculated from the Nernst equation [see (1.22)] that the change in $E°$ is

$$\Delta \nu (RT/nF) \ln(1.01325) = \Delta \nu \times 0.169 \text{ mV}$$

where $\Delta \nu$ is the increase in the number of gaseous molecules as the unit reaction proceeds as written [e.g., $\Delta \nu = -1$ for reaction (1.7)].

VI. COMBINATION OF STANDARD ELECTRODE POTENTIALS

In view of the relation between $E°$ and the change in Gibbs energy given by equation (1.8), it is evident that the former, like the latter, is a function of state and therefore an additive property. This is the basis of the utility of a table of standard electrode potentials. If there are n possible electrodes, the total number of cells that can be made from pairs of them is clearly $\frac{1}{2}n(n-1)$. The standard emfs of all of these may be readily calculated from a tabulation of the values for $n-1$ cells in which $n-1$ electrodes are in turn combined with one particular electrode. The latter is termed the *standard reference electrode* and in aqueous solution (and in other protic solvents) this is taken to be the hydrogen electrode.

For example, from the standard electrode potential of the cell

$$\text{Cu} \mid \text{Pt} \mid \text{H}_2 \mid \text{H}^+ \vdots \text{Zn}^{2+} \mid \text{Zn} \mid \text{Cu} \quad E° = -0.763 \text{ V} \quad (1.16)$$

that of the cell

$$\text{Cu} \mid \text{Pt} \mid \text{H}_2 \mid \text{H}^+ \vdots \text{Ag}^+ \mid \text{Ag} \mid \text{Cu} \quad E° = 0.799 \text{ V} \quad (1.17)$$

may be subtracted to obtain

$$\text{Cu} \mid \text{Ag} \mid \text{Ag}^+ \vdots \text{Zn}^{2+} \mid \text{Zn} \mid \text{Cu} \quad E° = -1.562 \text{ V} \quad (1.18)$$

where it should be noted that in subtracting the cells the cell being subtracted is reversed and added, the common hydrogen electrode being canceled out.

It is, in fact, more precise to note that standard electrode potentials represent the Gibbs energy change in the cell reaction and to write this combination in the form

$$\text{H}_2(g) + \text{Zn}^{2+}(aq) \rightarrow \text{Zn}(s) + 2\text{H}^+(aq) \quad E° = -0.763 \text{ V} \quad (1.19)$$

$$\text{H}_2(g) + 2\text{Ag}^+(aq) \rightarrow 2\text{Ag}(s) + 2\text{H}^+(aq) \quad E° = 0.799 \text{ V} \quad (1.20)$$

$$2\text{Ag}(s) + \text{Zn}^{2+}(aq) \rightarrow 2\text{Ag}^+(aq) + \text{Zn}(s) \quad E° = -1.562 \text{ V} \quad (1.21)$$

Standard Electrode Potentials

However, in adding the reactions it must be noted that they must be written in such a format that the H_2 and H^+ have the same stoichiometric coefficients in the two reactions being subtracted. This is equivalent to saying that the charge number n is the same for both reactions. In the example given, $n = 2$. The Gibbs energy changes calculated from these values of $E°$ with $n = 2$, using (1.8) are consequently the values associated with a mole of the reaction as written.

VII. STANDARD ELECTRODE POTENTIALS FROM GALVANIC CELLS

It is sometimes possible to derive values of $E°$ from electrochemical experiments. A good example is the cell (1.6), which may be set up experimentally and the reactions at the electrodes brought to a stable equilibrium state. This type of cell is frequently called a "cell without liquid junction." This is only approximately correct. In fact, the electrolyte near the Pt electrode is saturated with hydrogen and that near the AgCl electrode is saturated with AgCl. The direct reaction between these two solutions must be prevented and this is usually done by providing a long path between the two electrodes. In principle this leads to a junction between the two electrolytes, but in practice the effect of this is negligible.

Thus it is possible to write the emf of this cell according to the Nernst equation:

$$E = E° - (RT/2F) \ln (a_{H^+}^2 a_{Cl^-}^2 \cdot p°/p_{H_2}^*) \tag{1.22}$$

where $p°$ is the standard pressure. The quotient in the logarithmic term is of the form of an equilibrium constant with the activities or fugacities of the species appearing on the right-hand side of the reaction equation, raised to the power of the appropriate stoichiometric number, in the numerator and the corresponding factors for the species on the left-hand side of the reaction equation in the denominator.

In view of (1.11), the Nernst equation may be written

$$E = E° - (RT/F) \ln[\gamma_\pm^2 m^2 (p°)^{1/2} / (m°)^2 p^{*1/2}] \tag{1.23}$$

where $\gamma_\pm$ is the mean molar stoichiometric activity coefficient defined in terms of the individual ionic activity coefficients:

$$\gamma_\pm = (\gamma_+^{\nu_+} \gamma_-^{\nu_-})^{1/(\nu_+ + \nu_-)} \tag{1.24}$$

for an electrolyte $A_{\nu_+} B_{\nu_-}$.

The standard value $E°$ may be determined from experimental values of E at various values of m, the molality of HCl in the cell, by noting that equation (1.23) may be written

$$E°{'} = E + (RT/F) \ln[m^2(p°)^{1/2}/(m°)^2 p_{H_2}^{*1/2}] = E° - (RT/F) \ln \gamma_\pm^2 \tag{1.25}$$

It is evident that the limit of the quantity $E^{\circ\prime}$ (calculable from experiment) as $m \to 0$ is E° because $\gamma_\pm \to 1$. The extrapolation depends on a judicious plot of $E^{\circ\prime}$ against and appropriate function of m. In favorable cases such as the cell (1.6), very precise values of E° may be obtained.

The quantity $E^{\circ\prime}$ is called the *conditional or formal potential*. It may be regarded as a standard potential for a particular medium in which the activity coefficients are independent (or approximately so) of the reactant concentrations. If activity coefficients from another source are available, E° can be calculated directly from the Nernst equation in a form like (1.23).

One or other of these procedures may be adapted for different types of cell without liquid junction. Clearly, it is not essential to use the hydrogen electrode as one electrode of the experimental cell provided that one of the electrodes being used has a standard electrode potential which is already known.

Many measurements have also been made in cells which have an explicit liquid junction; for example:

$$\text{Cu} \mid \text{Hg}(\ell) \mid \text{Hg}_2\text{Cl}_2(s) \mid \text{KCl(sat)} \vdots \text{CuSO}_4(\text{aq}) \mid \text{Cu} \tag{1.26}$$

The Nernst equation for such a cell may be written in the form

$$E = E^\circ - (RT/2F) \ln (1/a_{\text{Cu}^{2+}} a^2_{\text{Cl}^-}) + E_{\text{LJ}} \tag{1.27}$$

where it must be noted that the activities of Cu^{2+} and Cl^- are in different solutions. Measurements in a cell of this type are usually done by varying the electrolyte concentration in the right-hand half of the cell while keeping that in the left-hand half constant. In such measurements $a_{\text{Cu}^{2+}}$ and E_{LJ} (the liquid junction potential) vary. Neither of these quantities is independently variable, but when one side of the junction consists of saturated KCl and the other an electrolyte of substantially lower concentration, the theory of liquid junctions suggests that E_{LJ} is small and reasonably constant. This argument depends on the equal mobilities of the ions K^+ and Cl^-. If so, cell (1.26) may be used to obtain a value of the standard potential of the right-hand electrode to a reasonable approximation either by an extrapolation procedure as described for cell (1.6) or by calculation from the approximate value of $a_{\text{Cu}^{2+}}$ using the assumption that

$$\gamma_{\text{Cu}^{2+}} = \gamma_\pm(\text{CuSO}_4) \tag{1.28}$$

Note that this type of procedure is best worked out in the procedure for measurement of pH where assumptions equivalent to (1.28) are used.

Finally, it may be remarked that with some electrodes the experimental difficulties in obtaining reproducible results may lead to errors which are greater than those resulting from the uncertainties in liquid junction potentials or activity coefficients. Hence the use of cells with liquid junction potentials is often quite acceptable.

VIII. STANDARD ELECTRODE POTENTIALS FROM THERMODYNAMIC DATA

As indicated by equation (1.8), values of $E°$ may be obtained from nonelectrochemical measurements of the standard Gibbs energy change in a reaction or the equilibrium constant of a reaction. In this way values may be obtained for electrodes which are difficult or impossible to set up experimentally because the electrode reaction is slow or because of other difficulties, such as side reactions, surface film formation, and so on.

Standard Gibbs energies may be obtained by purely thermal methods with the aid of the relation

$$\Delta G° = \Delta H° - T \Delta S° \tag{1.29}$$

The enthalpy change in the reaction may be measured in a calorimeter. The individual standard entropies of the substances taking part in the reaction may also be determined by measuring their heat capacities as a function of temperature, also in a calorimeter. The importance of this route is that it does not depend on the establishment of an equilibrium state for the reaction. However, in the case of ions in solution it is necessary to set up one equilibrium between the ions and a substance whose standard entropy is known. This is most often a solubility equilibrium. For example, the equilibrium between an ionic crystal and its ions,

$$AgCl(s) \rightarrow Ag^+(aq) + Cl^-(aq) \tag{1.30}$$

can be studied in solutions of various concentrations of a salt whose ions do not take part in the equilibrium (e.g., KNO_3 for this case). By extrapolation to zero ionic concentration the equilibrium constant of (1.30) can be found (the solubility product K_S). This yields the standard Gibbs energy of reaction (1.30),

$$\Delta G° = -RT \ln K_S \tag{1.31}$$

and the entropy change can be obtained either by measuring the enthalpy of precipitation [i.e., the reverse of reaction (1.30)] in a calorimeter or by measuring K_S at several temperatures.

$$\Delta S° = -(\partial \Delta G°/\partial T)_p = R[\partial (T \ln K_S)/\partial T]_p \tag{1.32}$$

Entropy changes in reactions involving ions may also be obtained from measurements of the temperature coefficient of the standard emf of a galvanic cell. This follows directly from equations (1.8) and (1.32):

$$\Delta S° = -(\partial \Delta G°/\partial T)_p = nF(\partial E°/\partial T)_p \tag{1.33}$$

Temperature coefficients quoted in this book are those measured (or calculated) for a cell in which the temperature of the whole cell is varied (i.e., with both electrodes always at the same temperature). These are sometimes called "isothermal" temperature coefficients and they satisfy equation (1.33).

IX. CONVENTIONS FOR THERMODYNAMIC PROPERTIES OF IONS

It is important to note that because all experimental measurements concern reactions in which electric charge is conserved, it is not possible to obtain the thermodynamic properties of single ionic species. The standard thermodynamic properties conventionally quoted for aqueous solutions are therefore those of a reactions such as

$$Na(s) + H^+(aq) \to Na^+(aq) + (1/2)H_2(g) \tag{1.34}$$

or

$$Al(s) + 3H^+(aq) \to Al^{3+}(aq) + (3/2)H_2(g) \tag{1.35}$$

in which the ion is produced from the element (or elements) in its standard state while the hydrogen ion is removed from solution and transformed to molecular hydrogen under standard conditions, such that the total charge of the system remains unchanged.

In view of the convention that the enthalpy and Gibbs energy of an element in its standard state are taken to be zero, it follows that the standard enthalpy change in reaction (1.34) at 298 K is [here and in subsequent equations in this section the ions are hydrated but the (aq) subscript is omitted for clarity]

$$\Delta H°(298) = H°_{Na^+}(298) - H°_{H^+}(298) \tag{1.36}$$

and that in reaction (1.35) is

$$\Delta H°(298) = H°_{Al^{3+}}(298) - 3H°_{H^+}(298) \tag{1.37}$$

or in general the standard enthalpy of an ionic species B having charge number z_B is

$$\Delta H°_B(298) = H°_B(298) - z_B H°_{H^+}(298) \tag{1.38}$$

Similarly, the standard Gibbs energy of ionic species B is

$$\Delta G°_B(298) = G°_B(298) - z_B G°_{H^+}(298) \tag{1.39}$$

On the other hand, in view of the convention for standard entropies, that the entropy is zero for a perfectly ordered pure solid at 0 K, the corresponding standard ionic entropy must be calculated from the entropy change in reactions like (1.34) and (1.35), taking account of the entropies the elements in their standard states. For example,

$$S°_{Na^+}(298) = S°'_{Na^+} - S°'_{H^+} \tag{1.40}$$

$$= \Delta S°(298) - (1/2)S°_{H_2}(298) + S°_{Na}(298) \tag{1.41}$$

where the prime indicates the "absolute" entropy of these hydrated ions and $\Delta S°(298)$ is the standard entropy change in reaction (1.34).

Thus the general definition of the standard entropy of species B is

$$S°_B(298) = S°'_B(298) - z_B S°'_{H^+}(298) \tag{1.42}$$

All the thermodynamic properties of ions thus are the difference between the property of a given cation and that of z_+ hydrogen ions or the sum of the property of a given anion and that of z_- hydrogen ions. Owing to the conventional definition of the standard enthalpy and energy of elements, these quantities are also the change in enthalpy (energy) accompanying a real process like (1.34) or (1.35). However, in the case of the entropy a different reference state for the element is used and the entropy change in these real processes must be calculated from

$$\Delta S°(298) = S°_B(298) + (z/2)S°_{H_2}(298) - S°_{eB}(298) \tag{1.43}$$

where $S°_{eB}(298)$ is the entropy of element B at 298 K.

The same definitions of standard thermodynamic properties of ions may be used at other temperatures. This convention is consistent with the use of the hydrogen electrode as a reference electrode at all temperatures, as follows from (1.33).

The conventions defined here are often described as "taking the enthalpy, Gibbs energy, and entropy of the hydrogen ion as zero" or "taking the potential of the hydrogen electrode and its temperature coefficient as zero." Although this description is not incorrect, it seems that less confusion arises if the full nature of the convention is given explicitly, as in equations (1.38), (1.39), and (1.42).

REFERENCES

1. Manual of Symbols and Terminology for Physicochemical Quantities and Units. Pure Appl. Chem., *51*, 1 (1979).
2. Notation for states and processes. Pure Appl. Chem., *54*, 1239 (1982); J. Chem. Thermodyn., *14*, 805 (1982).

2

The Single Electrode Potential: Its Significance and Calculation*

ROGER PARSONS[†] Laboratoire d'Electrochimie Interfaciale du CNRS, Meudon, France

In Chapter 1 the relation of standard electrode potentials to standard thermodynamic quantities was considered. It was emphasized there that by considering whole cell reactions there is no difficulty in principle in relating these quantities, and also that these relationships provide all that is required for the solution of practical problems. Nevertheless, the question of single electrode potentials and their relation to other single ionic properties has existed since the early days of electrochemistry and interest in this question still remains active. This chapter attempts to clarify some of these problems.

I. ELECTRICAL POTENTIALS IN REAL SYSTEMS

Many of the problems in real systems arise because of the attempt to adapt the classical concept of an electrical potential to a system of real condensed phases. In classical electrostatics the potential is defined in terms of the energy of a test charge and the nature of this test charge is unimportant provided that the potential difference between two points in a uniform medium is considered. This occurs in a number of real situations: for example, two points in vacuum or in a dilute gas or, in the most important example, two pieces of metal of the same composition which make up the terminals of a potentiometer or digital voltmeter. It is this last example which is used every time a cell potential difference is measured as described in Chapter 1.

However, the classical definition of potential causes difficulty as soon as an attempt is made to consider the potential difference between two points in different media: for example, between a point in vacuum and a point in a condensed phase. This may be made clear by noting that the energy change in taking a test charge from one point to the other depends on the nature of the test charge because of the short-range "chemical" interaction between the test charge and the condensed phase. There is no complete solution to this problem because there is no method for the separation of "chemical" and "electrical" forces since they are both essentially electrical in nature.

Reviewers: M. Gross, University of Strasbourg, Strasbourg, France;
L. Hepler, University of Lethbridge, Lethbridge, Alberta, Canada;
S. Trasatti, University of Milan, Milan, Italy.
[†]*Current affiliation*: The University, Southampton, England.

Nevertheless, some useful separations can be carried out on the basis of well-defined operations either on the real system or on a simple model of it. In the former case the quantities obtained are measurable, in the latter they are calculable in principle. These two cases can be illustrated first using the simple system of a piece of conducting material β of uniform bulk composition surrounded by free space. The work of inserting a charged particle of species i from a point distant from the conductor into a point in the bulk of the conductor is the electrochemical potential $\tilde{\mu}_i^\beta$. This is a measurable quantity that depends on the nature of the phase β, on the nature of the charged particle i, and on the state of charge of β. Experimentally, it is possible to reduce the electrostatic charge on β to zero. This may be verified by the absence of a field in the free space around β. Under these conditions the electrochemical potential takes a particular value which has become known as the *real potential* α_i^β. The best known example of a real potential is the electronic work function. (Since the latter is a work of extraction, not a work of insertion, it is $-\alpha_e^\beta$.)

If the more general case of a charged phase β is considered again, it is evident that the difference $\tilde{\mu}_i^\beta - \alpha_i^\beta$ arises from the presence of the charge. The charge on a conductor resides on its surface; hence virtually all this energy difference $\tilde{\mu}_i^\beta - \alpha_i^\beta$ arises from the traverse of the particle i from the reference point in free space to a point close to, but not in, the surface of the phase β, that is, a point at which the particle has no short-range interaction with the metal. Consequently, this energy does not depend on the nature of the test charge i and may be represented by a potential in the classical electrostatic sense:

$$\tilde{\mu}_i^\beta - \alpha_i^\beta = z_i e \psi^\beta \tag{2.1}$$

where $z_i e$ is the charge on the particle i, e is the magnitude of the electronic charge, and ψ^β is called the outer potential of phase α.

The real potential may be analyzed further, but only by using a model. The uniform nature of the bulk of phase β means that the charge density averaged over a volume large compared with that of an atom is zero. However, at the surface, even in the absence of a net charge on β, there is a redistribution of charge which may be represented by an electrical double layer, that is, two sheets of charge parallel to the surface having equal and opposite charge densities. This surface double layer contributes a term to the real potential which is independent of the nature of the test charge because the energy of passing right across a double layer is independent of the details of the energy changes while the particle is within the double layer. Hence this energy may be expressed in terms of a potential χ^β, the surface potential. The remaining part of the real potential is called the chemical potential μ_i^β and depends on the nature of i as well as that of β. Clearly, it includes all the interaction of i with the bulk of β.

$$\alpha_i^\beta = \mu_i^\beta + z_i e \chi^\beta \tag{2.2}$$

Finally, it is convenient to define the inner potential ϕ^β by

$$\phi^\beta = \psi^\beta + \chi^\beta \tag{2.3}$$

The Single Electrode Potential

because this includes the two contributions to $\tilde{\mu}_i^\beta$ which are independent of the nature of the test particle, although it is not a measurable quantity.

This system of definitions was introduced by Lange and Mischenko [1] and it seems to be the simplest and most useful, although several others have been proposed. It is consistent with strict thermodynamics in that the electrochemical potential

$$\tilde{\mu}_i^\beta = \mu_i^\beta + z_i e \phi^\beta \tag{2.4}$$

reduces to the chemical potential when the species i is uncharged ($z_i = 0$) and also that the sum of the electrochemical potentials of a group of ions whose total charge is zero (e.g., a salt) is equal to the sum of their chemical potentials ($\Sigma_i z_i = 0$).

These definitions do not require that the surface potential χ^β be independent of the free charge on the surface, but they are simpler if this condition is satisfied. Experimental indications about the variation of α_i^β with charge on the surface would give some indication of whether it is actually satisfied. Not much evidence is available, but the fact that for a macroscopic sample in contact with vacuum, only small charges are built up for quite high potentials suggests that χ^β may vary very little with ψ^β.

II. EQUILIBRIUM AT AN ELECTRODE

The simplest form of electrode consists of a metal in contact with a solution containing a salt of its own ions. The thermodynamic condition for the electrode equilibrium

$$M^{z+} + z_+ e^- \to M \tag{2.5}$$

is

$$\tilde{\mu}_{M^{z+}}^S + z_+ \tilde{\mu}_e^M = \mu_M^M \tag{2.6}$$

where the superscripts S and M represent the solution and metal phases, respectively. Two types of single electrode potentials may be obtained immediately by the application of (2.1) and (2.4), respectively, to (2.6).

1. From (2.1) and (2.6),

$$\alpha_{M^{z+}}^S + z_+ e \psi^S + z_+ (\alpha_e^M - e\psi^M) = \mu_M^M \tag{2.7}$$

or

$$\psi^M - \psi^S = -\frac{\mu_M^M - \alpha_{M^{z+}}^S}{z_+ e} + \frac{\alpha_e^M}{e} \tag{2.8}$$

This difference,

$$\Delta_S^M \psi = \psi^M - \psi^S \tag{2.9}$$

is called the *Volta potential difference*. It has the advantage of being measurable but the disadvantage that it is not determined only by the chemical nature of the phases M and S composing the electrode but also by the state of their free surfaces because the values of $\alpha_{M^{z+}}^S$ and α_e^M depend on the composition of the surface.

2. From (2.4) and (2.6),

$$\mu_{M^{z+}}^S + z_+ e \phi^S + z_+(\mu_e^M - e\phi^M) = \mu_M^M \tag{2.10}$$

$$\phi^M - \phi^S = -\frac{\mu_M^M - \mu_{M^{z+}}^S}{z_+ e} + \frac{\mu_e^M}{e} \tag{2.11}$$

This difference,

$$\Delta_S^M \phi = \phi^M - \phi^S \tag{2.12}$$

is called the *Galvani potential difference*. It is a function only of the chemical nature of the phases in equilibrium but cannot be measured. (Trasatti [2] terms this the *operative* potential of the electrode.)

III. SINGLE ELECTRODE POTENTIALS

When an electrochemical cell consisting of two simple metal electrodes, of the type described in the preceding section, is studied, the measurement of the equilibrium potential necessarily involves the introduction of three boundaries,

$$M \mid M^{z+} \quad M'^{z'+} \mid M' \mid M'' \tag{2.13}$$

if the liquid junction potential is ignored. Thus the potential difference which is measured between two pieces of metal of the same nature M and M'' is

$$E = \phi^{M''} - \phi^M \tag{2.14}$$

and may be analyzed into three Galvani potential differences:

$$E = (\phi^{M''} - \phi^{M'}) + (\phi^{M'} - \phi^S) + (\phi^S - \phi^M) \tag{2.15}$$

$$= \Delta_{M'}^{M''} \phi + \Delta_S^{M'} \phi - \Delta_S^M \phi \tag{2.16}$$

The Single Electrode Potential

This means that the interpretation of the potential difference of a cell of *two* electrodes cannot be made in terms of two galvanic potential differences, so the Galvani potential difference cannot be considered as a single electrode potential in the sense that two of these quantities add up to the total cell potential difference.

A solution that has been proposed to this problem is achieved by expressing the metal-metal Galvani potential in terms of the chemical potential of the electrons. Since there is electronic equilibrium at this interface,

$$\tilde{\mu}_e^{M'} = \tilde{\mu}_e^{M''}$$

and with the use of (2.4) it follows that

$$\Delta_{M'}^{M''}\phi = (\mu_e^M - \mu_e^{M'})/e \qquad (2.17)$$

where the superscript M replaces M" for the chemical potential because M and M" differ only in their electrical state.

If (2.17) is introduced into (2.16), the whole cell potential can be written

$$E = (\Delta_S^{M'}\phi - \mu_e^{M'}/e) - (\Delta_S^M\phi - \mu_e^M/e) \qquad (2.18)$$

so that there are now two terms, each characteristic of one of the two electrodes. It has been suggested therefore (particularly by Trasatii [3]) that the quantity $(\Delta_S^M\phi - \mu_e^M/e)$ would be more reasonably identified with a single electrode potential. Trasatti gives this the symbol

$$_rE^M = \Delta_S^M\phi - \mu_e^M/e \qquad (2.19)$$

and calls it the reduced absolute potential.

An alternative expression for $_rE^M$ may be obtained by inserting (2.11) into (2.19):

$$_rE^M = (\mu_{M^{z+}}^S - \mu_M^M)/z_+e \qquad (2.20)$$

Since $_rE^M$ is not a measurable quantity it has been suggested (particularly by Kanevski [4]) that a better choice is the measurable quantity obtained by replacing $\mu_{M^{z+}}^S$ by the corresponding real potential for this ion in the same solution having a *clean* surface:

$$_kE^M = (\alpha_{M^{z+}}^S - \mu_M^M)/z_+e \qquad (2.21)$$

It follows from (2.2) that these two potentials differ by the surface potential of the solution with a clean surface:

$$_kE^M = {_rE^M} + \chi^S \tag{2.22}$$

For dilute aqueous solutions χ^S is probably small (≤ 0.1 V), so that these two potentials do not differ much.

Finally, Trasatti has suggested [2] that the absolute electrode potential should be defined by an equation like (2.19) but with μ_e^S replacing μ_e^M:

$$_sE^M = -\frac{\mu_M^M - \mu_{M^+}^S}{z_+ e} + \frac{\mu_e^S}{e} \tag{2.23}$$

or

$$_sE^M = {_rE^M} + \mu_e^S/e \tag{2.24}$$

This means that it is a potential related to the reaction in which a metal dissolves to form ions, the electron remaining in solution. It is not evident that this is more closely an "absolute" potential and it is certainly less close to experiment. It is clear that each of these potentials $_kE^M$, $_rE^M$, and $_sE^M$ satisfies the condition that the addition of two such potentials for the two electrodes of a cell gives the measurable cell potential.

IV. CALCULATION OF THE STANDARD SINGLE ELECTRODE POTENTIAL

Since a cell potential difference is equal to the sum of two potentials defined by (2.19), (2.21), or (2.23), it is sufficient to calculate each of these quantities for one electrode. As all potentials in aqueous solution are referred to the hydrogen electrode, it is convenient to calculate each of these quantities for this electrode. The calculation will be described for $_kE^M$ since this may be made exactly, in principle.

The equilibrium at the hydrogen electrode is

$$H^+ + e^- \rightarrow (1/2)H_2 \tag{2.25}$$

which may be expressed as

$$\tilde{\mu}_{H^+}^S + \tilde{\mu}_e^M = (1/2)\mu_{H_2}^G \tag{2.26}$$

which leads to the analog of (2.11):

$$\phi^M - \phi^S = [\mu_{H^+}^S - (1/2)\mu_{H_2}^G]/e + \mu_e^M/e \tag{2.27}$$

so that the equivalent of (2.21) is

$$_kE^H = [\alpha_{H^+}^S - (1/2)\mu_{H_2}^G]/e \tag{2.28}$$

The Single Electrode Potential

In the calculation of $_kE^H$ from this equation it is, of course, important that both energies are referred to the same reference state. For the present purpose, this is conveniently that of the protons and electrons in field-free space. Thus $(1/2)\mu_{H_2}^G$ may be calculated from the sum of the standard Gibbs energies of the reactions

$$H^+(g) + e^-(g) \to H(g) \qquad (2.29)$$

$$H(g) \to (1/2)H_2(g) \qquad (2.30)$$

for which the standard Gibbs energy changes are -13.613 eV and -2.107 eV, respectively. On the basis of Randles' measurements [5] of real energies, Trasatti [3] has suggested that the value for the proton in water is -11.279 eV. More recent work by McTigue and Farrell [6] leads to a value of 11.276 eV. Hence

$$_kE^H = [-11.276 - (-13.613 - 2.107)]/e$$

$$= 4.44 \text{ V} \qquad (2.31)$$

The accuracy of this value is probably in the region of 50 mV.

With the estimate of χ^S for pure water of 0.13 V [3,7], it follows immediately from (2.22) that

$$_rE^H = 4.35 \text{ V} \qquad (2.32)$$

but the uncertainty of this value is difficult to determine because the value of χ^S can only be obtained indirectly.

In these calculations the values given by Trasatti have been used. The major uncertainty in the value of $_kE^H$ arises from the uncertainty in the value of $\alpha_{H^+}^{H_2O}$, which is based on the Volta potential measurements by Randles [5] and by McTigue and Farrell [6].

These are valid within the limits stated provided that Randles was correct in assuming that the dynamic surfaces of the mercury and the solution in his experiment were clean. He provides evidence that this was so. More recent experiments by Gomer and Tryson [8] using static surfaces have given slightly different results. They paid close attention to the metallic electrode but gave no evidence about the state of the solution surface. It is for this reason that the result based on the work of Randles and McTigue and Farrell is preferred here.

V. PHYSICAL SIGNIFICANCE OF THE SINGLE ELECTRODE POTENTIAL

The potential $_kE^M$ may be understood by combining (2.19) and (2.22) to obtain

$$_kE^M = \Delta_S^M\phi - \mu_e^M/e + \chi^S \qquad (2.33)$$

$$= \Delta_S^M \psi - \mu_e^M/e + \chi^M \tag{2.34}$$

in view of (2.3) and consequently with (2.2):

$$_kE^M = \Delta_S^M \psi - \alpha_e^M/e \tag{2.35}$$

Since

$$-\alpha_e^M = \phi^M \tag{2.36}$$

the electronic work function of the metal M, it is clear that $_kE^M$ is the sum of two measurable quantities, the work function and the Volta potential difference. Further, it can be seen that $-e_k E^M$ measures the work, of removing an electron from the Fermi level of M and transferring it to a point near the surface of the electrolyte containing the M^{z+} ions with which this metal is in equilibrium (see Fig. 1).

The significance of $_rE^M$ then follows directly from (2.22) since $-e\chi^S$ is the work done in taking an electron across the surface of the solution. Hence $-e_r E^M$ measures the work done in taking an electron from the Fermi level of the electrode metal and putting it into the solution in equilibrium with the metal without any bulk interaction between the electron and the solution (see Fig. 2). Since χ^S is not measurable, this is not a measurable quantity, although it may be deduced approximately from a model.

From (2.24) and (2.19) it follows that

$$_SE^M = \Delta_S^M \phi - \mu_e^M/e + \mu_e^S/e \tag{2.37}$$

$$= -(\tilde{\mu}_e^M - \tilde{\mu}_e^S)/e \tag{2.38}$$

so that $-e_S E^M$ measures the work of transferring an electron from the Fermi level in a metal to the solvated state at rest in the solution. This is the equilibrium work function of a metal in solution (see Fig. 3) and is a measurable quantity, as can be seen from the fact that it is the difference of two electrochemical potentials.

Since $_SE^M$ is not zero for the majority of normal electrodes, it follows that there is no electronic equilibrium between the metal and the solution. (Electronic equilibrium can occur in certain special cases, for example in sol-

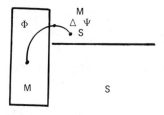

FIG. 1

The Single Electrode Potential

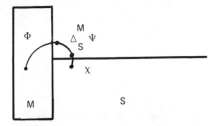

FIG. 2

vents that permit high concentrations of solvated electrons, like liquid ammonia.) However, for some purposes, notably the construction of electron energy diagrams, it is useful to represent the species in solution in terms of an equivalent electron energy. For example, when there is a redox couple in solution such as Cr^{2+}/Cr^{3+}, the reduced species can be regarded as an occupied electron level and the oxidized species as the corresponding unoccupied level. Then an electrochemical potential of electrons in the solution can be *defined* as

$$\tilde{\mu}_e^{S*} \equiv \tilde{\mu}_{Cr^{2+}}^S - \tilde{\mu}_{Cr^{3+}}^S \tag{2.39}$$

where an asterisk is used to indicate that this is not the actual electrochemical potential of electrons in the solution, but shorthand notation for the right-hand side of (2.39). $\tilde{\mu}_e^{S*}$ has the same formal properties as the true electrochemical potential in that it may be separated into a bulk contribution μ_e^{S*} and a surface contribution $\chi^S + \psi^S$, where the latter are the normal surface and outer potentials of the solution. With this definition it is possible to write the equilibrium across the interface of an electrode in the form

$$\tilde{\mu}_e^M = \tilde{\mu}_e^{S*} \tag{2.40}$$

even though the electronic equilibrium implied by such a relation does not actually exist.

The use of (2.40) permits another interpretation of $_kE^M$ since it follows from this equation with (2.1) that, at equilibrium,

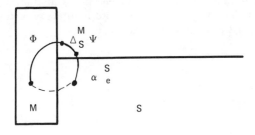

FIG. 3

$$\Delta_S^M \psi = (\alpha_e^M - \alpha_e^{S*})/e \tag{2.41}$$

which with (2.35) yields

$$_k E^M = -\alpha_e^{S*}/e \tag{2.42}$$

This means that $_k E^M$ is a measure of the real potential of an electron in solution, defined as being in equilibrium with the redox couple as described above. In particular, the value of $_k E^M$ calculated in (2.31) is this quantity calculated for a standard hydrogen electrode in aqueous solution at 25°C. Thus it represents the standard energy required to take an electron from an occupied level (H_2) to field-free vacuum leaving an unoccupied level (solvated H^+). Since $e_k E^M$ plays the role of an electronic work function for the solution, this quantity can be used to align the electron levels in the metal with those in solution in the same way as is normally done between two metals. This effectively relates the electrochemist's scale of potentials (with the hydrogen electrode as reference) to the physicist's scale of energies (with the vacuum as reference). Since the former is a scale of potentials whereas the latter is a scale of electron energies, there is of course a difference of sign because the electron has a negative charge.

VI. RELATION BETWEEN ELECTRODE POTENTIALS AND OTHER PHYSICOCHEMICAL PROPERTIES

It has been pointed out above (Section IV) that once one standard single electrode potential such as $_k E^H$ is known, those for all other electrodes in the same solvent and at the same temperature may be obtained from the standard electrode potentials $E°$. The resulting single standard electrode potentials may then be analyzed to demonstrate the relation between these electrochemical quantities and other physicochemical properties.

Since the aim of this section is to show some qualitative tendencies, the data provided in the tables are suitable for this type of illustration rather than for precise calculation. Thus energies at 0 K are given in some cases rather than Gibbs energies at 298 K.

Electrodes consisting of metals in equilibrium with their aquated ions will be considered first. Equation (2.21) shows that the potential of such electrodes is given by the difference of the real solvation energy of the metal ion and the energy of the metal atom in the metal lattice. As in the calculation of Section IV, it is important that these two quantities be referred to the same reference state and again this is conveniently the ions and electrons in field-free space. The quantity μ_M^M is then the Gibbs energy change in the process made up of the formation of atoms from the ions and electrons:

$$M^{z+}(g) + z_+ e^- \rightarrow M(g) \tag{2.43}$$

and the formation of the metal crystal from the dispersed atoms:

$$M(g) \rightarrow M(c) \tag{2.44}$$

The Single Electrode Potential

The Gibbs energy accompanying (2.49) is the Gibbs energy of ionization with the sign reversed, while that for (2.44) is the Gibbs energy of vaporization of the metal with sign reversed.

These three quantities are tabulated in Tables 1, 2, and 3. It may be noted that the solvation energy and the ionization energy are generally large in comparison with the vaporization energy and are of comparable magnitude. Thus the electrode potential is essentially determined by the difference of two large quantities (rather like the solubility of a solid; see below). This means that only rather general correlations of the electrode potential can be made with these physicochemical properties.

Electrodes in equilibrium with an anion in solution are either electrodes like the halogens or electrodes of the second kind like the silver/silver chloride electrode. For the former type the analog of (2.21) is

$$_k E^A = -[\alpha_{A^-}^S - (1/2)\mu_{A_2}]/e \tag{2.45}$$

if the molecule is diatomic. The chemical potential of this compound is then expressed in terms of the energy of removal of the electron from the gaseous ion:

$$A^-(g) \rightarrow A(g) + e^- \tag{2.46}$$

and the energy of formation of the molecule:

$$A(g) \rightarrow (1/2)A_2 \tag{2.47}$$

where the final state may be gas, liquid, or solid. The energy involved in process (2.46) is normally called the electron affinity, defined as the energy *released* when the process (2.46) occurs from right to left at 0 K, while the energy change accompanying (2.47) is the energy of dissociation of the molecule from its standard state with the sign reversed. These quantities are tabulated in Tables 4 and 5. When they are compared with the solvation energies it can be seen that the latter are similar in magnitude to the electron affinities. Again there is a general correlation of the electrode potential with these physicochemical properties.

The standard potential of an electrode of the second kind differs from that of the corresponding metal electrode of the first kind by a term that depends on the standard Gibbs energy of solution of an ionic crystal [as shown in (1.30) and (1.31)]. This energy may be expressed as the difference between the energy of the ions in the crystal lattice (the lattice energy) and the solvation energy of the constituent ions. These two quantities are of comparable magnitude (see Tables 2 and 6) and generally much larger than the standard Gibbs energy of equation (1.31). Thus again this Gibbs energy is correlated with the simpler physicochemical properties only in a rather general way.

The analog of (2.21) for an inert electrode at which a redox reaction of the type

$$M^{(z+n)+} + ne^- \rightarrow M^{z+} \tag{2.48}$$

occurs is

TABLE 1 Ionization Potentials of the Elements, I/eV[a]

		I	II	III	IV	V	VI	VII	VIII	IX	X
1.	H	13.59									
2.	He	24.56	54.1								
3.	Li	5.40	75.7	121.8							
4.	Be	9.32	18.2	153.9	216.6						
5.	B	8.28	25.1	37.9	259.3	338.5					
6.	C	11.27	24.8	47.9	64.5	392.0	(487)				
7.	N	14.55	29.6	47.4	77.4	97.9	551.9	(663)			
8.	O	13.62	35.2	54.9	77.4	113.9	138.1	739.1	(867)		
9.	F	17.43	34.9	62.7	87.3	114.3	157.2	185.2	953.0	(1100)	
10.	Ne	21.56	40.9	63.9	96.4	125.8	~157	(207)	(238)	(1190)	(1350)
11.	Na	5.14	47.3	71.7	98.9	138.6	172.5	209.1	264.3	299.9	(1460)
12.	Mg	7.64	15.0	80.2	109.3	141.2	186.9	225.6	266.8	328.2	367
13.	Al	5.97	18.8	28.5	120.0	153.6	190.3	243.2	285.8	331.6	399.2
14.	Si	8.15	16.4	33.5	45.2	165.9	204.1	245.2	301.7	~349	(407)
15.	P	10.9	19.7	30.2	51.4	65.0	(223)	(268)	(316)	(380)	(433)
16.	S	10.36	23.4	35.1	47.1	(72)	88.1	281	328.9	379.1	(459)
17.	Cl	12.90	23.7	39.9	53.5	67.8	(97)	114.3	346.6	398.8	~453
18.	Ar	15.76	27.5	40.7	~61	~78	89.0	124.0	142.8	(434)	(494)
19.	K	4.34	31.7	45.5	60.6	(83)	(101)	(120)	(155)	176.0	501.4
20.	Ca	6.11	11.9	51.0	67	84.0	(111)	~127	(151)	(189)	211.4
21.	Sc	6.7	12.8	24.8	73.6	91	110.5	(141)	~158	(185)	(227)
22.	Ti	6.84	13.6	27.6	43.3	99.4	119	140.1	(176)	(200)	(223)
23.	V	6.71	14.1	26.5	(48)	64.9[2)]	128.3	150	172.8	(214)	(239)
24.	Cr	6.74	16.7	(32)	(51)	(72)	~90	160.3	184	208.6	(255)
25.	Mn	7.43	15.64	(34)	(53)	76	(101)	119.3	~195	221	247.2
26.	Fe	7.83	16.5	~30	(56)	(79)	(105)	(133)	151.2	233.5	261
27.	Co	7.84	17.4	(34)	(53)	(82)	(109)	(138)	(170)	(202)	(295)
28.	Ni	7.63	18.2	(36)	(56)	(79)	(113)	(143)	(176)	(210)	(246)

The Single Electrode Potential

29.	Cu	7.72	20.2	(38)	(59)	(83)	(109)	(148)	(182)	(217)	(255)
30.	Zn	9.39	18.0	40	(62)	(86)	(114)	(144)	(188)	(244)	(263)
31.	Ga	5.97	20.5	30.8	63.9	(90)	(118)	(149)	(183)	(231)	(271)
32.	Ge	8.13	16.0	34.2	45.7	93.5	(123)	(155)	(189)	(226)	(280)
33.	As	10.5	~20	28.3	50.1	62.6	127.6	(160)	(196)	(234)	(274)
34.	Se	9.73	(21)	34	42.9	72.8	81.7	166	(202)	(241)	(282)
35.	Br	11.76	19.2	35.6	50.2	(60)	(87)	(104)	210	(248)	(291)
36.	Kr	14.0	24.5	36.8	(52)	(66)	(80)	(110)	(127)	256.6	(300)
37.	Rb	4.17	27.3	39.7	(53)	(71)	(86)	(102)	(134)	(153)	(308)
38.	Sr	5.69	11.0	(43)	57.0	(72)	(93)	(109)	(126)	(161)	(182)
39.	Y	6.5	12.3	20.4	(62)	76.9	(94)	(117)	(135)	(152)	(191)
40.	Zr	6.95	14	24.1	34.0	(83)	98.9	(118)	(143)	(163)	(181)
41.	Nb	–	(13)	(25)	(38)	~50	(106)	(125)	(145)	(172)	(193)
42.	Mo	7.06	–	(27)	46.0	~61	~67	(133)	(153)	(174)	(204)
43.	–	–	–	(29)	(43)	(59)	(76)	(94)	(162)	(184)	(206)
44.	Ru	~7.5[3]	(16)	(29)	(47)	(63)	(81)	(100)	(119)	(194)	(217)
45.	Rh	7.7	(18)	(31)	(46)	(67)	(85)	(105)	(126)	(147)	(228)
46.	Pd	8.1	~19.8	(33)	(49)	(66)	(90)	(111)	(132)	(155)	(178)
47.	Ag	7.58	21.4	35.9	(52)	(70)	(89)	(116)	(139)	(162)	(187)
48.	Cd	8.99	16.9	38.1	(55)	(73)	(94)	(115)	(146)	(170)	(195)
49.	In	5.79	18.9	27.9	57.9	(77)	(98)	(121)	(144)	(178)	(204)
50.	Sn	7.30	14.6	30.7	39.4	80.9	(103)	(126)	(151)	(176)	(213)
51.	Sb	8.64	(17)	24.9	44.2	55.7	107.7	(132)	(157)	(184)	(211)
52.	Te	8.96	(19)	30.6	37.8	60.3	72.4	137.3	(164)	(192)	(220)
53.	I	10.44	19.0	(31)	(42)	(52)	(77)	(90)	169.9	(200)	(229)
54.	X	12.13	21.2	32.1	(45)	(57)	(68)	(96)	(110)	204.7	(238)
55.	Cs	3.89	23.4	(34)	(46)	(62)	(74)	(86)	(117)	(132)	(246)
56.	Ba	5.21	10.0	(37)	(49)	(62)	(80)	(93)	(106)	(140)	(156)
57.	L	5.6	11.4	19.1	(52)	(66)	(80)	(100)	(114)	(128)	(165)
58.	Ce	6.54[4]	–	–	36.55)	(70)	(85)	(100)	(122)	(137)	(152)
59.	Pr	~5,86)	–	–	–	–	(89)	(106)	(122)	(146)	(162)
60.	Nd	~6,36)	–	–	–	–	–	(111)	(129)	(147)	(171)

26   Chapter 2

TABLE 1 (Continued)

	I	II	III	IV	V	VI	VII	VIII	IX	X
61. -	-	-	-	-	-	-	-	-	-	-
62. Sm	~5.6	~11.4	-	-	-	-	-	(135)	(154)	(173)
63. En	5.64	11.2	-	-	-	-	-	-	(161)	(181)
64. Gd	6.7	-	-	-	-	-	-	-	-	(187)
65. Tb	6.7	-	-	-	-	-	-	-	-	-
66. Dy	~6.8[6])	-	-	-	-	-	-	-	-	-
70. Yb	6.22	~12	-	-	-	-	-	-	-	-
71. Cp	-	-	(19)	-	-	-	-	-	-	-
72. Hf	-	~14.8	(21)	(31)	-	-	-	-	-	-
73. Ta	-	-	(22)	(33)	(45)	-	-	-	-	-
74. W	7.94	-	(24)	(35)	(48)	(61)	-	-	-	-
75. Re	~8	(13)	(26)	(38)	(51)	(65)	(79)	-	-	-
76. Os	~8.7	(15)	(25)	(40)	(54)	(68)	(83)	(99)	-	-
77. Ir	~9.2	(16)	(27)	(39)	(57)	(72)	(88)	(104)	(121)	-
78. Pt	~8.9	18.5	(29)	(41)	(55)	(75)	(92)	(109)	(127)	(146)

The Single Electrode Potential

79. Au	9.23	20.0⁷⁾	(30)	(44)	(58)	(73)	(96)	(114)	(133)	(153)
80. Hg	10.44	18.8	~34⁷⁾	(46)	(61)	(77)	(94)	(120)	(139)	(159)
81. Tl	6.12	20.3	29.8	~50⁷⁾	(64)	(81)	(98)	(117)	(145)	(166)
82. Pb	7.42	15.0	31.9	42.1	69.4	(84)	(103)	(122)	(142)	(173)
83. Bi	(8.8)	(17)	26.6	(45)	55.7⁸⁾	~88	(107)	(127)	(148)	(169)
84. Po	(8.2)	(19)	(28)	(38)	(61)	(73)	(112)	(132)	(154)	(176)
85. -	(9.6)	(18)	(30)	(41)	(51)	(78)	(91)	(138)	(160)	(183)
86. Rn	10.75	(20)	(30)	(44)	(55)	(67)	(97)	(111)	(166)	(190)
87. -	-	(22)	(32)	(43)	(59)	(71)	(84)	(117)	(133)	(197)
88. Ra	5.27	10.1	(34)	(46)	(59)	(76)	(89)	(103)	(140)	(156)
89. Ae	-	-	-	(49)	(62)	(76)	(95)	(109)	(123)	(164)
90. Th	-	-	-	~29.4	(65)	(80)	(94)	(115)	(130)	(145)
91. Pa	-	-	-	-	-	(84)	(100)	(115)	(138)	(154)
92. U	~4	-	-	-	-	-	(104)	(121)	(137)	(162)

[a]Estimated values are in parentheses.
Source: Landolt-Börnstein, *Tabellen*, Vol. I, Part 1, Springer-Verlag, Berlin.

TABLE 2 Thermodynamic Properties of Solvation[a]

	$\Delta H°/\text{kJ mol}^{-1}$	$\Delta G°/\text{kJ mol}^{-1}$	$\Delta S°/\text{J mol}^{-1}\text{K}^{-1}$	$-\alpha_i°(\text{H}_2\text{O})/\text{kJ mol}^{-1}$
H^+	0.0	0.0	0.0	1090
Li^+	576.1	579.0	−10.	511.
Na^+	685.3	679.1	21.	411.
K^+	770.0	752.7	56.9	349.
Rb^+	795.0	774.0	69.0	316.
Cs^+	827.6	805.3	72	284.
Be^{2+}	−305			
Mg^{2+}	2594	274.1	49	1906.
Ca^{2+}	589.1	586.6	7.5	1593.
Sr^{2+}	736.8	732.6	14.2	1447.
Ba^{2+}	879.0	851.5	59	1318.
Ra^{2+}	920.			
Al^{3+}	−1387.	−1346.	−137	4616
Sc^{3+}	−641.4	−618.4	(−78)	3888
Y^{3+}	−343.5	−326	(−57)	3596
La^{3+}	−10.5			
Ga^{3+}	−1412.5	−1358	−184	4628
In^{3+}	−836.4	806.3	−100	4071
Tl^{3+}	−911.7			

The Single Electrode Potential

Ion			
Cu^+	496.6	(520)	(-78)	(569)
Ag^+	615.4	610.9	15.	479
Tl^+	764.8	746.8	61.	343
NH_4^+	774			
Cr^{2+}	332			
Mn^{2+}	336	348	-40	1831
Fe^{2+}	248	270	-72	1910
Co^{2+}	151.9	173.6	-73	2006
Ni^{2+}	87.0	112.1	-83	2068
Cu^{2+}	83.3	97.5	-42	2087
Zn^{2+}	137.2	152.3	-50	2028
Cd^{2+}	-375.7	378.7	-11	1801
Hg^{2+}		(354)		(1825)
Pb^{2+}	701.7	682.8	64	1497
Cr^{3+}	(-1130)			
Fe^{3+}	-1105	-1063	-140.5	4333
Ce^{3+}	-280			
Ce^{4+}	-2125			
F^-	-1602.9	-1524.2	-264	450
Cl^-	-1469.0	-1407.1	-207	317
Br^-	-1454.8	-1393.3	-192	303
I^-	-1397.0	-1346.8	-169	257

TABLE 2 (Continued)

	$\Delta H°/\text{kJ mol}^{-1}$	$\Delta G°/\text{kJ mol}^{-1}$	$\Delta S°/\text{J mol}^{-1}\text{ K}^{-1}$	$-\alpha_i^o(H_2O)/\text{kJ mol}^{-1}$
S^{2-}	−3554	(−3450)	−348	1270
OH^-	(−1552)			
SH^-	(−1427)			

[a] ΔH, ΔG, and ΔS are the standard enthalpy, Gibbs energy, and entropy change accompanying the solvation of a gaseous ion, with the simultaneous removal of z hydrogen ions from the solution into the gas phase (z is negative for anions). The standard states are 1 atm pressure for the gaseous state and 1 mol kg^{-1} in the solution. Estimated values are in parentheses.

Source: D. R. Rosseinsky, Chem. Rev., 65, 467 (1965).

TABLE 3 Enthalpies of Sublimation of Crystalline Elements at 298 K and 1 bar, A(c) → A(g)

Element	$\Delta H°/\text{kJ mol}^{-1}$	Element	$\Delta H°/\text{kJ mol}^{-1}$
Li	161.5 ± 1.7	Mo	658.1 ± 2
Be	324.3 ± 6.3	Ru	650.6 ± 6
B	571.1 ± 12	Rh	556 ± 4
C	716.68 ± 0.46	Pd	376.6 ± 2
Na	108.16 ± 0.63	Ag	284.1 ± 0.8
Mg	146.4 ± 1.3	Cd	111.80 ± 0.63
Al	392.3 ± 2.1	Ln	243 ± 4
Si	455.6 ± 4	Sn	302.1 ± 2
P (yellow)	332.2 ± 4.2	Sb	264.4 ± 2.5
S	277.0 ± 8.2	Tl	196.6 ± 2
K	89.62 ± 0.21	Cs	78.2 ± 0.4
Ca	178.2 ± 1.7	Ba	177.8 ± 4
Sc	377.8 ± 4	Ce	423 ± 12
Ti	469.9 ± 2.1	Sm	206.7 ± 2
V	514.2 ± 1.2	Er	317.1 ± 4
Cr	397 ± 4	Yb	152.09 ± 0.8
Mn	283.3 ± 4	Hf	619 ± 4
Fe	415.5 ± 1.2	Ta	782.0 ± 2.5
Co	428.4 ± 4	W	849.8 ± 4
Ni	430.1 ± 2.1	Re	774 ± 6
Cu	337.6 ± 1.2	Os	787 ± 6
Zn	130.75 ± 0.4	Ir	669 ± 4
Ga	273.6 ± 2.1	Pt	565.7 ± 1.2
Ge	374.5 ± 2.1	Au	368 ± 2
As	302.5 ± 12.3	Hg	61.46 ± 0.13
Se	227.2 ± 4.2	Tl	182.21 ± 0.41
Rb	82.0 ± 0.4	Pb	195.1 ± 1.2
Sr	163.6 ± 2.1	Bi	209.6 ± 2.1
Y	424.7 ± 2.1	U	527 ± 12
Zr	608.8 ± 4	Th	595.3 ± 2.1
Nb	721.3 ± 4	Pu	364 ± 17

Source: L. Brewer and G. M. Rosenblatt, Adv. High Temp. Chem., *2*, 1 (1969).

TABLE 4 Atomic Electron Affinities[a]

		A/kJ mol^{-1}			A/kJ mol^{-1}
1.	H	72.766	35.	Br	324.5 ± 0.3
2.	He	(−21.)	36.	Kr	(−39.)
3.	Li	59.8 ± 0.6	37.	Rb	46.88 ± 0.15
4.	Be	(−241.)	38.	Sr	(−167)
5.	B	23	42.	Mo	97 ± 19
6.	C	122. ± 1	47.	Ag	125.7 ± 0.7
7.	N	0 ± 20	49.	In	34 ± 15
8.	O	141 ± 5	50.	Sn	121 ± 10
9.	F	328 ± 5	51.	Sb	101 ± 5
10.	N	(−29)	52.	Te	190.16 ± 0.03
11.	N	52.9 ± 0.4	53.	I	295.3 ± 0.3
12.	Mg	(−230)	54.	Xe	(−40)
13.	Al	44	55.	Cs	45.5 ± 0.3
14.	Si	120 ± 3	56.	Ba	(−52)
15.	P	74 ± 5	73.	Ta	77 ± 29
16.	S	200.42 ± 0.05	74.	W	48 ± 29
17.	Cl	348.7 ± 0.3	75.	Re	14 ± 14
18.	Ar	(−34.)	78.	Pt	205.3 ± 0.2
19.	K	46.36 ± 0.04	79.	Au	227.75 ± 0.07
20.	Ca	(−156)	81.	Tl	48 ± 10
24.	Cr	64 ± 5	82.	Pb	101 ± 9
28.	Ni	111 ± 10	83.	Bi	101 ± 15
29.	Cu	123.1 ± 0.1	84.	Po	(170)
31.	Ga	36	85.	At	(270)
32.	Gl	116 ± 10	86.	Rn	(−41)
33.	As	77 ± 5	87.	Fr	(44.0)
34.	Se	194.91 ± 0.03			

[a]Calculated values are in parentheses.
Source: E. C. M. Chen and W. E. Wentworth, J. Chem. Educ., *52*, 486. (1975).

TABLE 5 Enthalpies of Dissociation of Diatomic Gaseous Elements at 298 K and 1 bar, $(1/2)A_2(g) \rightarrow A(g)$

Element	$\Delta H°/\text{kJ mol}^{-1}$
H_2	218.003 ± 0.004
N_2	472.67 ± 0.41
O_2	249.170 ± 0.100
F_2	79.1
Cl_2	124.290 ± 0.008
Br_2	111.838 ± 0.121
I_2	106.763 ± 0.041

Source: CODATA Bull. No. 2, Nov. 1970.

TABLE 6 Lattice Energies of Ionic Crystals[a]

	$\Delta H_1/\text{kJ mol}^{-1}$				
	F	Cl	Br	I	OH
Li	1029	849	849	762	979
Na	910	773	741	700	(845)
K	831	695	669	637	(770)
Rb	775	678	656	626	(736)
Cs	741	656	639	611	(703)
NH_4	799	(674)	(644)	(607)	
Cu		987	971	967	
Ag	954	900	887	883	
Tl		720	686	695	

	F	O	S
Be		4406	
Mg	2920	3820	3305
Ca	2632	3443	3042

(Continued)

TABLE 6 (Continued)

	F	O	S
Sr	2,498	3,268	2,908
Ba	2,360	3,100	2,766

	F				
MnF_2	2732	Li_2O	2887	Na_2S	2213
FeF_2	2066	Cu_2O	3272	Cu_2S	2874
NiF_2	2979	Ag_2O	2983	MnS	3360
CdF_2	2812	MnO	3862	ZnS	3569
PbF_2	2506	FeO	3933	CdS	3381
$SrCl_2$	2155	CoO	3953	HgS	3556
		NiO	4017	PbS	3084
		CdO	3807		
		SnO_2	11,632	CuSe	2912
		PbO_2	11,619	SrSe	2791
		Al_2O_3	15,100	BaSe	2644
		Cr_2O_3	19,531	Cu_2Se	2883
		ZnO	4037	MnSe	3280
				ZnSe	3531
				CdSe	3347
				HgSe	3548
				PbSe	3079

	SH	CN	H	O_2
Li			920	
Na	703	715	808	
K	649	653	690	(703)
Rb	628	628	678	(678)
Cs	598	595	653	(653)
NH_4	623	619		

[a]Values given in parentheses are theoretical. Errors on the experimental values are about ±20 kJ mol^{-1} for the 1:1 salts and about 5% for the rest.
Source: Landolt-Börnstein, *Tabellen*, Springer-Verlag, Berlin.

The Single Electrode Potential

$$_kE^R = (\alpha^S_{z+n} - \alpha^S_z)/ne \tag{2.49}$$

if both ions are present in solution. In this case there is clearly a simple relation between the electrode potential and the solvation energies of the ions.

The real Gibbs energies of solvation are tabulated in Table 2. Gibbs solvation energies of salts may be obtained from the Gibbs energy of solution of crystalline salts, that is, by studying the process

$$MA(c) \rightarrow M^+(aq) + A^-(aq) \tag{2.50}$$

together with the Gibbs energy accompanying the process

$$MA(c) \rightarrow M^+(g) + A^-(g) \tag{2.51}$$

Values of the related lattice energy (the enthalpy of this process at 0 K) are given in Table 6. The difference of the Gibbs energies for processes (2.50) and (2.51) gives the Gibbs energy of solvation of the salt MA, that is, that for the process

$$M^+(g) + A^-(g) \rightarrow M^+(aq) + A^-(aq) \tag{2.52}$$

If the left-hand side is considered as the reference state for electrochemical potentials, this Gibbs energy is equal to $\tilde{\mu}^S_{M^+} + \tilde{\mu}^S_{A^-}$ and because the salt as a whole is electrically neutral,

$$\tilde{\mu}^S_{M^+} + \tilde{\mu}^S_{A^-} = \alpha^S_{M^+} + \alpha^S_{A^-} = \mu^S_{M^+} + \mu^S_{A^-} \tag{2.53}$$

Consequently, individual real free energies of solvation may be obtained from the quantity given in (2.53) once one of them is known. As mentioned in Section IV, real Gibbs energies of solvation may be obtained from Volta potential measurements and the values tabulated in Table 2 are derived from the measurement made by Randles [5].

Finally, it is interesting to consider the distribution of potentials occurring at the three junctions of a cell such as (2.13) set out in (2.15). This cannot be done precisely because the Galvani potentials cannot be measured or calculated precisely. However, the corresponding distribution of Volta potentials,

$$E = (\psi^{M''} - \psi^{M'}) + (\psi^{M'} - \psi^S) + (\psi^S - \psi^M) \tag{2.54}$$

is in principle measurable and can be calculated from the real potentials since (2.54) can be written

$$E = (\alpha^{M''}_e - \alpha^{M'}_e) + (\alpha^{M'}_{M'^+} - \alpha^S_{M'^+}) + (\alpha^S_{M^+} - \alpha^M_{M^+}) \tag{2.55}$$

These terms will differ from the Galvani potential difference by the difference of the surface potentials of the clean free surfaces.

TABLE 7 Work Functions of Metals

Metal	Φ/eV	Metal	Φ/eV
Ag	4.30	Nd	3.1
Al	4.19	Ni	4.73
Au	5.32	Os	4.83
Ba	2.35	Pb	4.18
Be	5.08	Pd	5.00
Bi	4.36	Po	4.6
Ca	2.71	Pr	2.7
Cd	4.12	Pt	5.40
Ce	2.80	Rb	2.20
Co	4.70	Re	4.95
Cr	4.40	Ru	4.80
Cs	1.90	Sb	4.56
Cu	4.70	Sc	3.5
Fe	4.65	Sm	2.95
Ga	4.25	Sn	4.35
Hf	3.65	Sr	2.76
Hg	4.50	Ta	4.22
K	2.30	Te	4.70
In	4.08	Th	3.71
La	3.40	Ti	4.10
Li	3.10	Tl	4.02
Mg	3.66	U	3.50
Mn	3.90	V	4.44
Mo	4.30	W	4.55
Na	2.70	Zn	4.30
Nb	4.20	Zr	4.00

Source: Selected by S. Trasatti, J. Chem. Soc. Faraday Trans. I *68*, 229 (1972).

The first term on the right-hand side of (2.55) is the difference of two electronic work functions. From the values given in Table 7 it is evident that it may have values up to 3 or 4 V. Thus this term is of the order of magnitude of the emf of a galvanic cell. In the past it has been argued that this was the principal source of the emf. However, the other two terms may have a comparable magnitude. The real potential of ions in a metal may be obtained by considering the equilibrium of ions and electrons within the metal:

$$M^{z+} + ze^- \rightarrow M$$

for which

$$\mu_M^{M^{z+}} + z\tilde{\mu}_e^M = \mu_M^M \tag{2.56}$$

or

$$\alpha_M^{M^{z+}} + ze\psi^M + z\alpha_e^M - ze\psi^M = \alpha_M^{M^{z+}} + z\alpha_e^M = \mu_M^M \tag{2.57}$$

Consequently, the second term in (2.55) can be expressed as

$$\alpha_{M^+}^{M^1} - \alpha_M^S = \mu_M^{M^1} - z\alpha_e^{M^1} - \alpha_{M^+}^S \tag{2.58}$$

and the third term similarly. It may thus be concluded that each of the three phase boundaries in a cell such as (2.13) contributes significantly to the emf.

REFERENCES

1. E. Lange and K. Mischenko, Z. Phys. Chem., *149*, 1 (1930). See also R. Parsons, *Modern Aspects of Electrochemistry*, J. O'M. Bockris and B. E. Conway, eds., Vol. 1, Butterworth, London, 1954.
2. S. Trasatti, J. Chim. Phys., *74*, 60 (1977).
3. S. Trasatti, J. Electroanal. Chem., *52*, 313 (1974).
4. E. Kanevski, Zh. Fiz. Khim., *22*, 1397 (1948).
5. J. E. B. Randles, Trans. Faraday Soc., *52*, 1573 (1956).
6. P. McTigue and J. Farrell, J. Electroanal. Chem. *139*, 37 (1982).
7. J. E. B. Randles, Phys. Chem. Liquids, *7*, 63 (1977).
8. R. Gomer and G. Tryson, J. Chem. Phys., *66*, 4413 (1977).

3
Hydrogen*

PHILIP N. ROSS Lawrence Berkeley Laboratory, Berkeley, California

Monatomic hydrogen gas has the ground-state electronic configuration $s^1(^2S_{1/2})$. Molecular hydrogen gas has the ground-state electronic configuration $^1\Sigma_g^+$. In nature, two isotopic forms of hydrogen occur, protium, ^1H (99.985%), and deuterium, ^2H (0.015%). ^3H or tritium is an artificially produced isotope which finds use as a radiotracer in mechanistic studies. Because of the large percentage differences in mass between protium and deuterium, these isotopes show a larger variation in their physical and chemical properties than do the isotopes of any other element.

The proton has a nuclear spin $I = 1/2$, so that in the diatomic hydrogen (protium) molecule two resultant spin states occur ($J = 0, 1$) that give rise to two spin isomers, para ($J = 0$) and ortho ($J = 1$). Similar nuclear spin isomers exist for diatomic deuterium. The nuclear spin for the deuteron is $I = 1$, and the possible resultant spins are 0, 1, and 2. Resultant spins 0 and 2 belong to the ortho and spin 1 to the para modifications. The heteronuclear molecule, $^1H^2H$ or HD, has no nuclear spin isomers. The nuclear spin isomers have significantly different physical and chemical properties, and the equilibrium concentration ratio of the ortho and para forms is temperature dependent, with para favored at low temperatures. The conversion of ortho to para is an exothermic process, with a heat of conversion that is also temperature dependent (e.g., from -55.6 J mol^{-1} at 300 K to -1062.7 J mol^{-1} at 10 K) [1]. However, the variation in composition becomes significant only below 200 K, and at ambient or higher temperature the equilibrium mixture is 3 to 1 (ortho to para) for diatomic protium and 2 to 1 for diatomic deuterium. A recent, and very complete, compilation of the thermochemical properties of the diatomic hydrogen (protium) molecule was made by McCarty [1].

The thermodynamic standard state for the element hydrogen is the diatomic molecule in the gaseous state at 298.15 K containing an equilibrium mixture of the ortho and para isomers (normal hydrogen). By convention, the term *normal hydrogen* does not necessarily refer to isotopically pure hydrogen (protium), but is taken to mean hydrogen that contains deuterium at the levels of natural abundance or less. The relevant thermodynamic data on hydrogen are given in Table 1.

Reviewer: B. E. Conway, University of Ottawa, Ottawa, Canada.

TABLE 1 Thermodynamic Data on Hydrogen[a]

Formula	Description	State	$\Delta H°/\text{kJ mol}^{-1}$	$\Delta G°/\text{kJ mol}^{-1}$	$\Delta S°/\text{J mol}^{-1}\text{K}^{-1}$
H(^{1}H)	$^2S_{1/2}$	g	217.965	203.247	114.713
D(^{2}H)		g	221.673	206.506	123.349
H$^+$		g	1536.202		
	Std. state, $m = 1$	aq	0	0	0
D$^+$		aq		−1.2	
H$_2$	$^1\Sigma_g^+$	g	0	0	130.684
H$^-$	$^1S_0$	g	138.99		
H$^-$		aq		217.7	
D$_2$		g	0	0	144.960
HD		g	0.318	−1.464	143.801
H$_2^+$	$^1\sigma_g$	g	1494.65		

[a] 298.15 K and 1.01325 bar; National Bureau of Standards [2] values are in italic.

Hydrogen

I. OXIDATION STATES

The formal oxidation states of hydrogen are +1, 0, and −1. Acids and solvated photons are examples of the +1 oxidation state. The −1 oxidation state occurs in hydrogen compounds like metal hydrides, such as the alkali metal hydrides. The hydride ion has the electronic structure of helium ($^1S_0$) and thus resembles halide ions, which also have closed-shell configurations. The ionic radius of H^- in the alkali hydride is between that of F^- and Cl^- [3]. There are many hydrogen compounds referred to as metal hydrides in which the true oxidation state is uncertain but is clearly not −1. Examples include the technologically important "hydrides" of the groups IIIb to Vb transition metals, which form nonstoichiometric interstitial-type compounds with hydrogen where the oxidation state is probably zero (i.e., dissolved atomic hydrogen). Hydrogen forms covalent bonds with many elements, for which no oxidation state is normally given. The hydrogen ion (proton) in water is the predominant instance of the +1 state, and, by convention, is the reference state for establishing the energy levels of ions in solution. Studies on the detailed structure of the hydrated proton have been many and long standing, and an interesting survey of these studies is given by Conway [4].

II. H_2/H^+ COUPLE

By convention [5], the standard potential of the hydrogen electrode, for which the reaction is

$$2H^+(aq) + 2e^- \rightarrow H_2(g) \qquad E° = 0.000 \text{ V}$$

is zero at all temperatures. By the notation H^+(aq) it is understood that this represents the hydrated proton in aqueous solution without specifying the hydration sphere. When the proton activity or the hydrogen fugacity is not equal to unity, the electrode potential is given by the appropriate form of the Nernst equation,

$$E° = -(RT/2F) \log f_{H_2} + (RT/F) \log a_{H^+}$$

where f is the fugacity of hydrogen and a_{H^+} is the proton activity. The effect of nonideality of hydrogen gas may normally be ignored at ambient pressure, but becomes significant at high pressure.

The potential of the H_2/H^+ couple in pure water and in molal hydroxide solution can be calculated from the Nernst equation and the ionization constant of water at 298.15 K:

$$H_2O \rightleftharpoons H^+ + OH^- \qquad K = 1.008 \times 10^{-14}$$

$$2H^+ (10^{-7} \text{ M}) + 2e^- \rightleftharpoons H_2 \qquad E = -0.414 \text{ V}$$

$$2H_2O + 2e^- \rightleftharpoons H_2 + 2OH^- \qquad E°_B = -0.828 \text{ V}$$

Note again that this is the point at which the present IUPAC convention differs from that used in Latimer's book, so that the standard electrode potentials have the opposite sign. A couple more *negative* than 0.414 V should liberate hydrogen from water, and a couple more *negative* than 0.828 V should liberate hydrogen from 1 mol L^{-1} hydroxide solutions.

The realization of the equilibrium potential between hydrogen gas and protons in solution is usually accomplished by the use of a platinum black electrode (or high area forms of Pt such as a platinized Pt foil) immersed in electrolyte through which the hydrogen gas is bubbled. Thus, in practical work with a hydrogen reference electrode, the reference conditions are usually not exactly equal to the standard state conditions, and several corrections must be made to relate the potential measurement to the standard hydrogen electrode (SHE), for example, vapor pressure of water over the solution, the ambient or total pressure, and the proton activity. The correction for proton activity can be particularly difficult in many cases. It is often the case that when hydrogen reference electrodes are used, the potential scale is not corrected to the standard hydrogen electrode, but is left uncorrected. This has given rise to a convenient reference state defined as the reversible hydrogen electrode in the same electrolyte (RHE). The use of hydrogen gas is undesirable in many laboratory situations, and this has prompted the development of nonequilibrium hydrogen reference electrodes [6]. In these, a high-surface-area Pt electrode, usually a fuel cell type, is polarized at low current density (<1 mA cm^{-2}) such that it evolves hydrogen at a low rate. Because of the reversibility of the couple and the local saturation of the electrolyte by hydrogen, the potential at this electrode is displaced from the equilibrium potential (the RHE) by at most a few millivolts. Another type of hydrogen electrode that can be used without hydrogen gas bubbling is the palladium hydride-hydrogen electrode [7]. Palladium metal can be charged electrolytically with hydrogen to form the palladium hydride, and gives a steady potential of +0.050 V at 298.15 K with respect to a hydrogen electrode in the same solution. Rest potentials of the palladium hydride electrode versus the RHE at higher temperatures were measured by Dobson [8].

The equilibrium hydrogen electrode potential provided the basis for the definition of the conventional pH scale based on emf measurements with cells of the type

$$Cu(s) \mid Hg(\ell) \mid Hg_2Cl_2 \mid KCl(aq) \vdots \vdots \text{solnX} \mid H_2(g) \mid Pt(s) \mid Cu(s)$$
$$\text{(sat)} \quad \text{(sat)}$$

A review of the conventions and problems in the pH scale based on emf values is given by Chemla [9]. For practical purposes, however, the measurement of pH is now routinely done with the "glass" electrode [10], which is based on the surface potential of hydrogen ion on the glass membrane, not the H_2/H^+ equilibrium.

In principle, the attainment of equilibrium between hydrogen gas and hydrogen ion in solution does not require the use of a particular electrode material, and platinum is not the only material at which this equilibrium potential is observed. There are, however, two fundamental properties of the material required for reliable observation of the H_2/H^+ potential: (1) the corrosion potential for the material must be more positive than the equilibrium potential, and (2) the exchange current density for the hydrogen-hydrogen ion equilibration is at least 10^{-4} A cm^{-2}. The latter assures that the measured potential is not a mixed potential, representing anodic dissolution of the

TABLE 2 Exchange Current Densities for the
Hydrogen Electrode Reaction on Various
Metals in Acidic Electrolyte (pH 0)

\multicolumn{3}{c}{$\log i_0$ (A cm^{-2})}			
Type I[a]		Type II		Type III	
Ti	−6.9	Pt	−3.3	Zn	−10.5
V	−6.2	Ir	−3.3	Cd	−12.0
Zr	−6.7	Os	−4.0	Sn	−9.2
Nb	−7.3	Pd	−2.4	Sb	−8.7
Fe	−5.8	Rh	−2.5	Hg	−11.9
Co	−4.9	Ru	−3.3	Pb	−12.6
Ni	−5.2	Re	−5.1	Bi	−10.2

[a]Surfaces of Ti, V, Zr, and Nb may be
oxidized in water and remain so even with
hydrogen being evolved.
Source: Ref. 11.

material and cathodic evolution of hydrogen proceeding at the same rate. The former requirement assures that equilibrium is readily achieved and that mixed potentials from impurities or nonfaradaic (surface) processes are avoided. There are only a few of the elements that satisfy both these requirements, and at which the H_2/H^+ equilibrium potential has been observed. A classification according to Kita [11] of some common electrode materials as hydrogen electrode substrates is shown in Table 2. The materials termed type II are the low-overvoltage noble metals at which the equilibrium hydrogen potential is generally observed. Type I are the nonnoble transition metals with relatively low hydrogen overvoltage such that hydrogen evolution and anodic dissolution occur simultaneously (in acid solution) at rest potentials below the reversible potential for hydrogen. Type III are the corroding base metals with a high hydrogen overvoltage, whose rest potentials are also mixed potentials well below the reversible hydrogen potential, but with dissolution proceeding at a very slow rate because of the very slow kinetics of hydrogen evolution on the surfaces of these metals. If these base metals are placed in electrical contact with any of the type II metals (or even Cu or Ag), rapid anodic dissolution commences with vigorous hydrogen evolution from the noble metal.

III. H_2/H^- COUPLE

Hydride ion rapidly reduces water and is therefore unknown in aqueous solution. The potential for the H_2/H^- couple can, however, be estimated by thermodynamic calculation, as done by Latimer. Since Latimer's original calculation, there have been additional measurements of some thermodynamic quantities that improve the reliability of the estimate of the potential. In particular, the *JANF Tables* report new values [12] for the heats of vaporization and dissociation of the alkali hydrides from which lattice energies can be calculated from the cycle used by Rosseinsky [13]. The lattice energies of the alkali hydrides and halides provide a basis for interpolation of the data on anion heats of solution in order to estimate the heat of solution of hydride ion. As expected from the Goldschmidt [3] radii for H^- (0.156 nm) relative to the radii for F^- (0.133 nm) and Cl^- (0.181 nm), the lattice energies (Table 3) of the alkali hydrides fall between that for the two halide salts. It would be expected then that the heats of solution for the hydride would also fall between the heats of solution for the halide salts, and the value given in the table was interpolated from the F^- and Cl^- values using the dipole functional dependence $(r_i + r_{H_2O})^{-2}$ [14]. The entahlpy of hydration of H^- can be calculated from the heats of solution using the conventional enthalpies of hydration for Li, Na, and K [13], as shown in the last column in Table 3. The value arrived at here, -1495.8 kJ mol^{-1}, differs slightly from Latimer's estimate, -1504.6 kJ mol^{-1}. Together with the slightly different values for $H^+(g)$ and $H^-(g)$ given in Table 1, the present estimate for the enthalpy of formation of $H^-(aq)$ is 180.33 kJ mol^{-1}. Using the same procedure for the entropies of hydration, the interpolated value for the entropy of hydration of H^- is -235.14 J mol^{-1}, from which the standard entropy of $H^-(aq)$ can be calculated to be 5.9 J mol^{-1} K^{-1}. Again, this value is different from Latimer's estimate, but the present value is more consistent with the ionic radius of $H^-(aq)$. The final result is a Gibbs energy of formation of $H^-(aq)$ of 217.7 kK mol^{-1}:

$$(1/2)H_2 + e^- \rightarrow H^-(aq) \qquad E° = -2.25 \text{ V}$$

which is virtually identical to Latimer's original estimate for the Gibbs energy. Remarkably, the more negative enthalpy is canceled by less negative entropy

TABLE 3 Lattice Energies and Heats of Solution of Alkali Metal Hydrides, Fluorides, and Chlorides at 0 K

		F^-	H^-	Cl^-	$\Delta H_s(H^-)$
	r (nm)	(0.133)	(0.154)	(0.181)	(calc.)[a]
Li^+	U_L (kJ mol^{-1})	1025.5	908.8	835.5	-1501
	$\Delta H°_s$ (kJ mol^{-1})	4.6	-15.9	-36.0	
Na^+	U_L (kJ mol^{-1})	935.5	808.3	778.6	-1498
	$\Delta H°_s$ (kJ mol^{-1})	2.5	4.2	5.4	
K^+	U_L (kJ mol^{-1})	807.1	716.3	707.1	-1485
	$\Delta H°_s$ (kJ mol^{-1})	-17.2	0.4	18.4	
				Mean value =	-1495.8 ± 9

[a] Conventional enthalpy of hydration relative to $H^+(g) \rightarrow H^+(aq)$.

Hydrogen

term in these new estimates. There does not, therefore, appear to be any new data that suggest revisions of Latimer's original estimate for the H_2/H^-(aq) couple.

IV. D_2/D^+ COUPLE

A direct measurement of the D_2/D^+ potential versus the H_2/H^+ reference couple cannot be made because of the rapid isotopic exchange in solution. However, a number of attempts have been made to measure this potential using cells with common reference electrodes [15], but the results were inconsistent, with values ranging from -0.0034 to -0.034 V, and generally accurate corrections were not made for the substantial liquid junction potentials arising from the different mobilities of D^+ and H^+ [16]. The higher mass of the deuterium atom causes the zero point energy for D_2 to be 7.5404 kJ lower than that of H_2. The dissociation energy of D_2 (at 0 K) is therefore greater than that for H_2 by this difference in zero-point energies. The differences in the zero-point energies in liquid D_2O and H_2O, the vibrational and librational modes, and the larger moment of inertia of D_2O than H_2O also cause the ionic solution in D_2O to be distinctive from that in H_2O. It is to be expected that the solvation energy of D^+ in D_2O will be different from that for H^+ in H_2O. This point has been discussed at some length by Conway [4b]. Conway estimated the overall difference in solvation energy of D^+ in D_2O and H^+ in H_2O to be -5 kJ mol^{-1}, which would make the free energy of formation of D^+ about -1.2 kJ mol^{-1} and the equilibrium potential for the D_2/D^+ couple about -0.013 V.

When an aqueous solution containing a mixture of deuterium and hydrogen is electrolyzed, an enrichment in deuterium occurs due to the preferential evolution of H_2 over D_2. This preferential reduction of H^+, which has been exploited as a method for heavy water production, arises both from the difference in equilibrium potential and from kinetic factors. The relative degree of isotopic enrichment occurring upon electrolysis, usually termed the *separation factor*, is defined as

$$S_D = (c_H/c_D)_g / (c_H/c_D)_\ell$$

where c_H/c_D represents the atomic concentration ratio of hydrogen to deuterium and the subscripts g and ℓ refer to gas phase and solution, respectively. Values of S_D are typically 3 to 8, depending on the electrode material and the proton (deuteron) activity in the solution. Discussions of the mechanisms responsible for the observed separation factors are summarized in the review by Conway [4c].

V. SINGLE IONIC QUANTITIES

There has been and continues to be interest in establishing an absolute potential scale (see Chapter 2) in addition to the conventional potential scale. Conway [17] has reviewed the literature of extrathermodynamic calculations of absolute Gibbs energies, enthalpies and entropies of hydration indiviudal ions and derived "best" estimates of these values for H^+ in H_2O: $\Delta G^{o'} = -1065 \pm 17$ kJ mol^{-1}, $\Delta H^{o'} = -1104 \pm 17$ kJ mol^{-1}, and $\bar{S}^{o'} = -22.2 \pm 2$ J mol^{-1}

K^{-1}. Note that $\Delta G^{\circ\prime}$ is the standard value of μ_i defined in equation (2.2) and the $\Delta H^{\circ\prime}$ and $\overline{S}^{\circ\prime}$ are similarly defined as including only the bulk interaction and not the contribution of the surface dipolar layer. Their calculation therefore involves an extrathermodynamic assumption. Another useful physical property of the H^+ in H_2O tabulated by Conway is the absolute partial molar volume at infinite dilution, $\overline{V}^{\circ\prime}(H^+) = -4.7 \times 10^{-3}$ dm^3 mol^{-1}.

VI. DIAGRAMS FOR HYDROGEN

Acid Solution

$$\begin{array}{ccc} +1 & 0 & -1 \end{array}$$

$$H^+ \xrightarrow{0.00} H_2 \xrightarrow{2.25} H^-$$

Basic Solution

$$\begin{array}{ccc} +1 & 0 & -1 \end{array}$$

$$H_2O \xrightarrow{0.828} H_2 + OH^- \xrightarrow{2.25} H^-$$

REFERENCES

1. R. D. McCarty, in *Hydrogen: Its Technology and Implications*, K. E. Cox and K. D. Williamson, eds., Vol. 3, CRC Press, Cleveland, Ohio, 1975.
2. D. D. Wagman, W. H. Evans, V. B. Parker, R. H. Schumm, I. Halow, S. M. Bailey, K. L. Churney, and R. L. Nuttall, NBS Tables of Chemical Thermodynamic Properties, J. Phys. Chem. Ref. Data, *11*, Supplement 2 (1982).
3. V. Goldschmidt, Trans. Faraday Soc., *25*, 253 (1929).
4. B. E. Conway, *Ionic Hydration in Chemistry and Biophysics*, Elsevier, Amsterdam, 1981; (a) pp. 395-404; (b) pp. 551-568; (c) B. E. Conway, in *Modern Aspects of Electrochemistry*, J. O'M. Bockris and B. E. Conway, eds., Vol. 3, Butterworth, London, 1964, Chap. 2.
5. Comptes Rendux, 17th Conf. Union Int. Chim. Pure Appl. (1958); see also Chapter 1 for conventions.
6. J. Giner, J. Electrochem. Soc., *111*, 376 (1964).
7. M. Fleischmann and J. N. Hiddlestone, J. Sci. Instrum. Ser. 2, *1*, 667 (1968).
8. J. V. Dobson, J. Electroanal. Chem., *35*, 129 (1972).
9. M. Chemla, in *Encyclopedia of the Electrochemistry of the Elements*, A. J. Bard, ed., Vol. 9A, Marcel Dekker, New York, 1982, pp. 383-413.
10. R. G. Bates, *The Principles of pH: Theory and Practice*, Wiley, New York, 1973.
11. H. Kita, in *Encyclopedia of the Electrochemistry of the Elements*, A. J. Bard, ed., Vol. 9A, Marcel Dekker, New York, 1982, pp. 413-478.

12. D. R. Stull and H. Prophet, *JANAF Thermochemical Tables, Second Edition*, Nat. Bur. Stand. Publ. NBS-37, U.S. Government Printing Office, Washington, D.C., 1970.
13. D. R. Rosseinsky, Chem. Rev., *65*, 467 (1965).
14. F. G. Basolo and R. G. Pearson, *Mechanisms of Inorganic Reactions*, Wiley, New York, 1958, pp. 48-65.
15. I. Kolthoff, Recl. Trav. Chim., *49*, 401 (1930); E. Abel, E. Bratu, and O. Redlich, Z. Phys. Chem., *A170*, 153 (1934); *A173*, 353 (1935); G. N. Lewis and F. G. Doody, J. Am. Chem. Soc., *55*, 3504 (1933); E. Lange, Z. Elektrochem., *44*, 43 (1938).
16. B. E. Conway, Proc. R. Soc. Lond., *A247*, 400 (1958).
17. B. E. Conway, J. Solut. Chem., *7*, 721 (1978).

4
Oxygen*

JAMES P. HOARE General Motors Research Laboratories, Warren, Michigan

Since oxygen, having the electronic structure $1s^2 2s^2 2p^4$, forms compounds with all the elements except He, Ne, and Ar and because of its great importance in the electrochemistry of aqueous solutions, the free energies and electrode potentials of this element will be considered before the other elements.

In nature, three stable isotopic species occur—$^{16}O_2$ (99.759%), $^{17}O_2$ (0.0374%), and $^{18}O_2$ (0.2039%)—and concentrates of up to 97 atomic percent (a/o) $^{18}O_2$ are commercially available for tracer investigations of the mechanisms of oxygen reactions. The standard thermodynamic state of the element is the gaseous paramagnetic molecule O_2. Another allotropic form of oxygen is ozone, O_3. Although the O_2 molecule has two unpaired electrons (paramagnetic), true pairing of electrons to form a symmetrical O_4 species does not take place even in the solid state [1]. Because the concentrations of $^{17}O_2$ and $^{18}O_2$ are so low in natural oxygen, their presence can be ignored for all ordinary thermodynamic considerations. The thermodynamic data are given in Table 1.

I. OXIDATION STATES

In all true compounds of oxygen, the element has an oxidation number of -2 except in the oxygen fluorides, where the oxygen may be assigned an oxidation number of +2 because of the higher electronegativity of the F^- ion [2]. Several odd-electron compounds exist for which a simple oxidation number cannot be assigned. These include compounds of the dioxygenyl cation, O_2^+, such as O_2PtF_6 and O_2SbF_6; of the superoxide anion, O_2^-, such as NaO_2 and KO_2; and of the ozonide anion, O_3^-, such as KO_3 and NH_4O_3. All of these compounds are paramagnetic.

---

*Reviewers: V. S. Bagotsky, Institute of Electrochemistry, Moscow, USSR; J. O'M. Bockris, Texas A&M University, College Station, Texas; M. L. Hitchman, University of Salford, Salford, England; D. T. Sawyer, University of California, Riverside, California.

TABLE 1 Thermodynamic Data on Oxygen[a]

Formula	Description	State	$\Delta H°/\text{kJ mol}^{-1}$	$\Delta G°/\text{kJ mol}^{-1}$	$\Delta S°/\text{J mol}^{-1}\text{K}^{-1}$
O		g	249.170	231.747	160.946
O$^+$		g	1,569.293		
O^{2+}		g	4,967.49		
O^{3+}		g	10,275.40		
O^{4+}		g	17,751.0		
O^{5+}		g	28,747.4		
O^{6+}		g	42,078		
O^{7+}		g	113,411		
O^{8+}		g	197,497		
O$^-$		g	101.63	109.20	
O$^-$		aq		76.5	
O^{2-}		g	950		
O$_2(^3\Sigma_g^-)$		g	0	0	205.028
O$_2(^3\Sigma_g^-)$	Std. state, $m = 1$	aq	−11.7	16.3	110.9
O$_2(^1\Delta_g)$		g	?	?	
O$_2^+$		g	1,177.71		
O$_2^-$		aq		31.84	
O$_3$		g	142.7	163.2	238.82

Oxygen

O_3	aq	125.9	
O_3^-		?	?
OH	g	38.95	34.22
OH	aq	7.74	
OH^-	g	140.87	
OH^-	aq	−229.994	−157.293
OH^+	g	1,328.4	
H_2O	g	−238.915	−228.589
H_2O	ℓ	−285.830	−237.178
H_2O_2	g	−136.31	−105.60
H_2O_2	ℓ	−187.78	−120.41
H_2O_2 Undiss.; std. state, m = 1	aq	−191.17	−134.10
H_2O_2 In 0.1 H_2O	aq	−188.134	
H_2O_2 In 0.5 H_2O	aq	−189.108	
H_2O_2 In 1 H_2O	aq	−189.807	
H_2O_2 In 2 H_2O	aq	−190.456	
H_2O_2 In 3 H_2O	aq	−190.728	
H_2O_2 In 4 H_2O	aq	−190.874	

(with additional columns: 183.636, −10.75, 188.715, 69.91, 232.6, 109.6, 143.9 for respective rows O₃, OH⁻(aq), H₂O(g), H₂O(ℓ), H₂O₂(g), H₂O₂(ℓ), H₂O₂(aq std state))

Std. state, m = 1

TABLE 1 (Continued)

Formula	Description	State	$\Delta H°$/kJ mol^{-1}	$\Delta G°$/kJ mol^{-1}	$\Delta S°$/J mol^{-1} K^{-1}
H_2O_2	In 5 H_2O	aq	−190.949		
H_2O_2	In 10 H_2O	aq	−191.083		
H_2O_2	In 15 H_2O	aq	−191.129		
H_2O_2	In 20 H_2O	aq	−191.146		
H_2O_2	In 50 H_2O	aq	−191.154		
H_2O_2	In ∞ H_2O	aq	−191.17		
HO_2^-	Std. state, $m = 1$	aq	−160.33	−67.4	23.8
HO_2		g	21		

Oxygen

HO_2	aq		4.44
H_2O^+	g	980.3	
$H_2O_2^+$	g	923.4	
HO_2^+	g	1,134	
OD	g	36.86	32.47	189.623
D_2O	g	−249.199	−234.551	198.229
D_2O	ℓ	−294.600	−243.488	75.94
HDO	g	−245.300	−233.128	199.401
HDO	ℓ	−289.888	−241.906	79.29

[a]National Bureau of Standards [3] values are in italic.

II. OXYGEN-WATER COUPLE

The calculation at 25°C of the Gibbs energy of formation of liquid water from the best values for the heat, -285.830 kJ mol^{-1}, and the entropy, -163.18 J mol^{-1} K^{-1}, of formation of liquid water [3] give

$$H_2 + (1/2)O_2 \rightarrow H_2O \quad \Delta G° = 237.178 \text{ kJ mol}^{-1}$$

In acid solutions, the $E°$ value for the oxygen couple from this $\Delta G°$ is found to be

$$O_2 + 4H^+ + 4e^- \rightarrow 2H_2O \quad E° = 1.229 \text{ V} \quad (dE°/dT)_{isoth}$$
$$= -0.846 \text{ mV K}^{-1}*$$

From the Nernst equation, $E = (1.229 - 0.05916 \text{ pH})$ V, the oxygen-water couple in alkaline solutions (pH 14) is calculated to be

$$O_2 + 2H_2O + 4e^- \rightarrow 4OH^- \quad E°_B = 0.401 \text{ V} \quad (dE°/dT)_{isoth}$$
$$= -1.680 \text{ mV K}^{-1}$$

and in pure water (pH 7)

$$O_2 + 4H^+(10^{-7} M) + 4e^- \rightarrow 2H_2O \quad E° = 0.815 \text{ V}$$

With a value of the ionic product for water of 1.008×10^{-14} at 25°C [4], the free energy of the ionization of water is determined as 79.885 kJ mol^{-1}, giving a value of -157.293 kJ mol^{-1} for the $\Delta G°$ of OH$^-$ ion.

Accordingly, the oxidized form of any couple with a potential more noble than that of oxygen should, for a given pH, liberate oxygen from solution. As an example, the permanganate-manganous couple has an $E°$ of 1.51 V [5] and it is known that permanganate is unstable in acid solutions with respect to the oxidation of water to oxygen.

Any oxidizing agent in a couple more positive than 1.229 V in acid solutions of pH 0 or 0.401 V in alkaline solutions of pH 14 should liberate oxygen from an aqueous solution. Any reducing agent in a couple more negative than 0.0 V in acid solutions of pH 0 or -0.828 V in alkaline solutions of pH 14 should liberate hydrogen from an aqueous solution. As a result, one can define the region of stability of oxidizing and reducing agents in aqueous solutions as the area between the solid lines in Fig. 1. Depending on the electrode material (catalytic surface for the electrode reaction) present in the system, the existence of the characteristic H$_2$ and O$_2$ overvoltage for the given system permits one to expand more or less the region of redox stability beyond those lines in Fig. 1.

The oxygen-water couple is highly irreversible and it is virtually impossible to obtain equilibrium in systems involving this couple. Since the oxygen

---

*Isothermal temperature coefficients were obtained from de Béthune and Swendeman Loud [5].

Oxygen

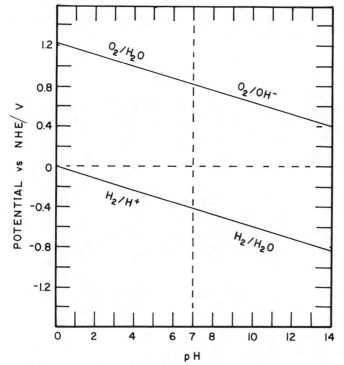

FIG. 1. Region of stability of oxidizing and reducing agents in aqueous solutions.

overvoltage is very high on most metal anodes [6,7] and is both strongly and nonlinearly dependent on the current density, it is difficult to predict the behavior of a given system from tables of standard electrode potentials alone.

For the electrochemical evolution of oxygen from aqueous electrolytes at metal anodes, the reaction is not homogeneous but occurs on the anode surface between adsorbed reactants and intermediates. Consequently, it is not possible to use the Gibbs energy tables to calculate the potentials of the intermediate steps of the oxygen electrode process without access to Gibbs energies of adsorption of the species of interest. Such data may be difficult to obtain.

Ideally, the oxygen electrode process takes place at an inert electrode since thermodynamics implies that the anode acts only as a source of or sink for electrons and the catalytic surface on which the electrode process proceeds. Since all metals, even the noble metals, interact with oxygen and, hence, are not inert, local cells are set up at the anode surface [8] and the reversible potential cannot be observed. At high anodic potentials, all metals—even the more electropositive metals such as iron and nickel—become passivated by the formation of an oxide layer on the anode surface. For the case of an electronically conducting oxide layer, the electrode process takes place on the surface of the oxide layer and the anode approaches more nearly the ideally inert electrode. The reversible oxygen potential of 1.229 V has been observed on strongly passivated Pt electrodes [7]. Because of this protective oxide film, water is oxidized instead of the metal at passivated anodes. For an account of the mechanisms of the oxygen electrode reactions, the reader may consult the literature (e.g., [6,7]).

III. OXYGEN ATOM-WATER COUPLE

Taking the Gibbs energy of atomic oxygen as 231.747 kJ mol^{-1} and that of liquid water as -237.178 kJ mol^{-1}, one calculates the standard potential for the reduction of atomic oxygen in acid solution as

$$O(g) + 2H^+ + 2e^- \rightarrow H_2O \quad E^\circ = 2.430 \text{ V} \quad (dE^\circ/dT)_{isoth} = -1.148 \text{ mV K}^{-1}$$

and in alkaline solution as

$$O(g) + H_2O + 2e^- \rightarrow 2OH^- \quad E^\circ_B = 1.602 \text{ V}$$

These very large positive E° values reflect the extremely strong oxidizing power of atomic oxygen.

IV. OXYGEN-PEROXIDE COUPLE

Although peroxide is never found in solution during the anodic oxidation of water to oxygen, it may be observed during the cathodic reduction of oxygen to water. Apparently, peroxide is an intermediate in cathodic but not in the anodic process because of the great difficulty encountered in the breaking of the O-O double bond. These facts are contributing factors to the irreversibility of the O_2 electrode. When oxygen is reduced at a metal cathode, oxygen molecules are adsorbed on the cathode surface with dissociative adsorption. If the rate of electron transfer is greater than the rate of O_2 dissociation to adsorbed O atoms, peroxide will be formed as a stable intermediate and O_2 is said to be reduced by a two-electron process. If, however, the rate of O_2 dissociation is greater than that of electron transfer, peroxide will not be an intermediate and O_2 is said to be reduced by a four-electron process. A good catalyst, then, for the reduction of O_2 must be not only a good electron transfer catalyst but also a good peroxide decomposing catalyst. At a Pt cathode, O_2 is reduced at low current densities virtually entirely by the four-electron path, but with higher current densities the two-electron path contributes a greater share to the total cathodic current. For a detailed account of O_2 reduction, the reader is referred to the literature [7].

Using the National Bureau of Standards (NBS) value of the Gibbs energy of formation of $H_2O_2(aq)$, -134.10 kJ mol^{-1}, one calculates for the reduction of oxygen in acid solution

$$O_2 + 2H^+ + 2e^- \rightarrow H_2O_2 \quad E^\circ = 0.695 \text{ V} \quad (dE^\circ/dT)_{isoth} = -1.033 \text{ mV K}^{-1}$$

In alkaline solutions, a value of -67.4 kJ mol^{-1} is given for the $\Delta G°$ for HO_2^-(aq). One obtains

$$O_2 + H_2O + 2e^- \rightarrow OH^- + HO_2^- \qquad E_B° = -0.0649 \text{ V}$$

Hydrogen peroxide is a weak acid with a pK of 11.7 and the second ionization constant is very small ($16 < pK < 18$).

With the newer Gibbs energy values obtained from the latest NBS tables [3], the standard potentials for the O_2/H_2O_2 couple in acid and alkaline solutions are changed considerably from the older values of 0.682 V and -0.076 V, respectively. Experimental support for the new $E°$'s values in acid solution may be found in the more recent literature. First, Kern [9] found a value of 0.695 V for the $E°$ of the O_2/H_2O_2 couple on Hg with a polarographic technique and then Tikhomirova et al. [10] measured 0.70 V on degassed bright Pt to which controlled amounts of oxygen gas were admitted.

For alkaline solutions the situation is still somewhat confusing. On carbon electrodes Yeager et al. [11] found that the O_2/HO_2^- electrode was highly reversible and the data exhibited the expected dependency on the concentrations of OH^- and HO_2^- ions (ca. 30 mV per logarithmic decade). However, the $E°$ value obtained from these data is -0.048 V for the O_2/HO_2^- electrode on carbon, assuming that the activity coefficients for the HO_2^- ion and the OH^- ion are equal. Polarographic measurements on a dropping mercury electrode carried out by Bagotskii and co-workers [12,13] resulted in an $E°$ value of -0.045 V. From a long extrapolation of rest potential data obtained on a bright Pt electrode in O_2-saturated 1 M KOH solutions containing hydrogen peroxide, Bowen et al. [14] arrived at a value of -0.066 V for $E°$ of the O_2/HO_2^- couple. It appears that experimental verification of the O_2/HO_2^- standard potential remains shrouded in doubt.

V. PEROXIDE-WATER COUPLE

Using the most recent NBS free-energy tables [3], one calculates for the standard potentials in the peroxide-water couple in acid solutions

$$H_2O_2 + 2H^+ + 2e^- \rightarrow 2H_2O \qquad E° = 1.763 \text{ V} \qquad (dE°/dT)_{isoth}$$
$$= -0.658 \text{ mV K}^{-1}$$

and in alkaline solutions

$$HO_2^- + H_2O + 2e^- \rightarrow 3OH^- \qquad E_B° = 0.867 \text{ V}$$

Such high positive potentials place peroxide in the group of the most powerful oxidizing agents known. Not only is peroxide unstable with respect to the oxidation of water but also to its own oxidation and reduction in both acid and alkaline solutions.

Strong oxidizing agents can oxidize peroxide to O_2, but the H_2O_2/H_2O couple has such a high potential that in many instances the reduced species of the oxidizing agent is oxidized back to its original state. The result of this behavior is the decomposition of peroxide to O_2 and H_2O. Consider, for example, the reaction of Br_2 with H_2O_2, where the $E°$ for the $Br_2(aq)/Br^-$ couple, 1.087 V, lies between that of the O_2/H_2O_2 and the H_2O_2/H_2O potentials. The sum of the peroxide oxidation reaction,

$$Br_2 + H_2O_2 \rightarrow 2H^+ + 2Br^- + O_2$$

and the oxidation of bromide ion reaction,

$$2Br^- + H_2O_2 + 2H^+ \rightarrow Br_2 + 2H_2O$$

yields the decomposition of peroxide,

$$2H_2O_2 \rightarrow O_2 + 2H_2O$$

A similar situation exists for the peroxide-stabilized oxygen electrode [6, p. 64], where a Pt electrode is placed in an O_2-saturated acid solution which contains H_2O_2 at concentrations greater than 10^{-3} M. The observed steady-state rest potential is 0.81 V and is accounted for by a local cell composed of the oxidation of H_2O_2 to O_2 through the O_2/H_2O_2 reaction and the reduction of H_2O_2 to H_2O through the H_2O_2/H_2O reaction. Accordingly, the overall process is the decomposition of peroxide and the electrode system behaves as though it were a reversible O_2/H_2O_2 electrode with a standard potential of 0.81 V [6, p. 134]. In both of the cases considered here, Br_2 and Pt may be considered as catalysts for the decomposition of H_2O_2. Many redox systems and metal surfaces serve as catalysts for peroxide decomposition.

To determine peroxide quantitatively, one must titrate with a redox system for which the oxidation of the reduced species by peroxide is strongly inhibited. Such a system is the MnO_4^-/Mn^{2+} couple. Peroxide is quantitatively oxidized by permanganate ion in alkaline solution.

VI. OZONE-WATER COUPLE

An interesting feature of the oxidizing action of ozone is the fact that only one of the oxygen atoms is reduced while the other two atoms form molecular oxygen. Using the appropriate free energy data for O_3 and H_2O from Table 1, the standard potential for the ozone-water couple in acid solutions is found to be

$$O_3 + 2H^+ + 2e^- \rightarrow O_2 + H_2O \qquad E° = 2.075 \text{ V}$$

$$(dE°/dT)_{isoth} = -0.483 \text{ mV K}^{-1}$$

and in alkaline solutions

$$O_3 + H_2O + 2e^- \rightarrow O_2 + 2OH^- \qquad E°_B = 1.246 \text{ V}$$

$$(dE°/dT)_{isoth} = -1.318 \text{ mV K}^{-1}$$

With these high potential values, ozone ranks next to fluorine in the list of powerful oxidizing agents. In solution, ozone is unstable with respect to the decomposition into oxygen,

$$O_3 + H_2O \rightarrow 2O_2 + 2H^+ + 2e^- \quad E° = -0.383 \text{ V}$$

The rest potential of an ozone electrode on bright Pt in alkaline solutions is more active than the theoretical value of 1.246 V by about 0.20 to 0.40 V, with a pH coefficient of $\partial E/\partial pH = 0.050$ V (15) in the pH range 7 to 11. This behavior is the result of local cell formation at the Pt electrode. Cathodic reduction of ozone,

$$O_3 + H_2O + 2e^- \rightarrow O_2 + 2OH^- \quad E°_B = 1.246 \text{ V}$$

and anodic evolution of O_2,

$$4OH^- \rightarrow O_2 + 2H_2O + 2e^- \quad E°_B = -0.401 \text{ V}$$

make up the partial reactions of the local cell.

Overvoltage curves for the reduction of ozone at a bright Pt cathode in alkaline solutions exhibit Tafel slopes ranging between 0.10 and 0.12 V [15]. This observation is considtent with a process limited by an electron transfer step, $O_3 + e^- \rightarrow O_3^-$. This step is followed by the reaction of an ozonide ion with water, $O_3^- + H_2O \rightarrow O_2 + OH + OH^-$, and of a second ozonide ion with the OH radical, $O_3^- + OH \rightarrow OH^- + O_3$. The sum of these steps gives the overall cathodic O_3/O_2 reaction given above. There are not enough thermodynamic data on the ozonide ion available in the literature to calculate the $E°$ of the O_3/O_3^- couple.

VII. IRRADIATED SOLUTIONS

If an aqueous solution is irradiated by a stream of particles such as an electron beam (3 to 5 MeV) in the pulsed radiolysis techniques, solvated electrons, e^-_{aq}, are generated in the solution [16]. These e^-_{aq} interact with species in solution to produce unusual ions and radicals. The primary oxidizing species in solution is the hydroxyl radical [17], which behaves as a weak acid:

$$OH \rightleftharpoons O^- + H^+ \quad pK = 11.8$$

In acid solutions, it exists as OH radicals, whereas in strong base, it exists as the oxygen radical anion, O^-.

With the technique of flash photolysis [18,19], a solution is exposed to the high-intensity flash of flash lamps which generate e^-_{aq} by the electron detachment from species in solution. The following sequence results in the formation of O^- in alkaline solutions:

$$OH^- \xrightarrow{h\nu} OH + e^-_{aq}$$

$$OH + OH^- \rightarrow O^- + H_2O$$

When an alkaline solution is aerated or saturated with oxygen, the ozonide ion, O_3^-, will be formed and the buildup and decay of its concentration can be followed spectrophotometrically by observing the optical absorption maximum at 430 nm,

$$O^- + O_2 \rightarrow O_3^-$$

Although the decay kinetics of the ozonide ion are very complicated [17], it is generally agreed that the first step is the loss of O_2,

$$O_3^- \rightarrow O^- + O_2$$

followed by the products of the interaction of the O^- radical with species in solution. For example, in pure alkaline solution,

$$O^- + H_2O \rightarrow OH^- + OH$$

in solutions containing peroxide,

$$OH + H_2O_2 \rightarrow H_2O + HO_2 \quad \text{(acid)}$$

or

$$O^- + HO_2^- \rightarrow OH^- + O_2^- \quad \text{(alkaline)}$$

and in solutions containing alcohols,

$$O^- + CH_3OH \rightarrow OH^- + CH_2OH$$

The hydroperoxyl radical, HO_2, can exist in two forms: as the radical in acid solutions and as the superoxide ion, O_2^-, in alkaline solutions,

$$HO_2 \rightarrow O_2^- + H^+ \quad pK = 4.8$$

Early literature values for the pK of HO_2 were quoted in the range 4.5 ± 0.15, but careful measurements of the adsorption spectra of O_2^- and HO_2 species in the pulsed radiolysis of sodium formate solutions by Behar et al. [20] yielded a value of 4.88 ± 0.1. In recent work in formate solutions of Bielski and co-workers, a value of 4.75 ± 0.08 [21] was reported for the pK of HO_2, and in later work [21a] a revaluated quantity of 4.69 ± 0.08 was recorded. Since there seems to be some difficulty in evaluating the pK for HO_2 unambiguously, we have chosen an intermediate value of 4.8, as Fee [21b] has done.

Hydroperoxyl radicals are generated in aerated acid solutions by pulsed radiolysis,

$$e_{aq}^- + H^+ \rightarrow H \xrightarrow{O_2} HO_2$$

Oxygen

and superoxide ions in aerated alkaline solutions,

$$e^-_{aq} + O_2 \rightarrow O_2^-$$

After pulsed radiolysis or flash photolysis, a number of chain reactions are present because of the presence of radicals in solution. Chain-breaking steps occur between radicals.

$$O_2^- + O_2^- + H_2O \rightarrow HO_2^- + OH^- + O_2$$

$$O_3^- + O_2^- + H_2O \rightarrow O_3 + HO_2^- + OH^-$$

$$OH + HO_2 \rightarrow H_2O_3 \rightarrow H_2O + O_2$$

$$HO_2 + HO_2 \rightarrow H_2O_2 + O_2$$

$$O_2^- + HO_2 \rightarrow HO_2^- + O_2$$

It has been determined [21,21a] that the rate of interaction of two O_2^- ions is very small and that of two HO_2 radicals is very large. Many of these reactions in the gas phase are important in studies of smog generation and structure of the stratospheric ozone layer [22].

VIII. SUPEROXIDE ION

When the superoxide ion is formed in solution either heterogeneously by cathodization at a metal electrode or homogeneously by radiolysis or photolysis of an oxygen-saturated alkaline solution, it may react with the solvent to form a peroxyl radical,

$$O_2 + e^- \rightarrow O_2^-$$

$$O_2^- + HA \rightarrow HO_2 + A^-$$

which may be reduced to a peroxyl ion by electron transfer, $HO_2 + e^- \rightarrow HO_2^-$ or react with another radical as a chain-breaking step, $HO_2 + HO_2 \rightarrow H_2O_2 + O_2$. To stabilize the O_2^- ion in solution, the second step must be greatly retarded either by using aprotic solvents or fused salt systems or by adding a surfactant to the aqueous solution. The porous adsorbed layer of surfactant on the electrode surface strongly inhibits the reaction between protons and the anion radicals by lowering the proton availability.

Divisek and Kastening [23] have studied the reduction of O_2 at a dropping mercury cathode polarographically in 0.1 M NaOH solution containing 0.05% triphenylphosphine oxide (TPPO). From a logarithmic plot of the data, a value of −0.284 V was determined for the O_2/O_2^- couple,

$$O_2 + e^- \rightarrow O_2^- \quad E° = -0.284 \text{ V}$$

In similar investigations, Chevalet et al. [24] used α-quinoline (0.017 M) in 1 M NaOH at a Hg cathode and recorded an $E°$ of -0.27 V, which was corrected [25] to -0.29 V.

From electron transfer studies between O_2^- ions generated in alkaline solutions by radiolysis and various quinone-semiquinone redox systems, Czapski and co-workers [25,26] report values of $E°$ for the O_2/O_2^- couple between -0.33 and -0.325 V. Fee and Valentine [22] have reviewed the chemistry of superoxide recently. It appears that the reduction of oxygen in many biological processes involves electron transfer from O_2^- ions to certain enzyme systems. Studies using superoxide dismutase [25] yield an $E°$ value of -0.32 V for the O_2/O_2^- couple. Because the radiolysis studies involve the thermodynamic calculations of equilibrium processes, we have chosen -0.33 V for the $E°$ of the O_2/O_2^- couple, from which a value for $\Delta G°$ of 31.84 kJ mol^{-1} is recorded in Table 1.

By adding up the $\Delta G°$ of the O_2/O_2^- couple ($E° = -0.33$ V), of the ionization of peroxyl radical ($HO_2 \to H^+ + O_2^-$, $pK = 4.8$), of the O_2/HO_2^- couple ($E_B^° = -0.0649$ V), and the ionization of water ($pK = 13.995$), the $\Delta G°$ of the HO_2/HO_2^- reaction gives

$$HO_2 + e^- \to HO_2^- \quad E° = -0.744 \text{ V}$$

From these data, the $\Delta G°$ for HO_2 radical in solution is calculated to be 4.44 kJ mol^{-1}.

The standard potentials of a number of other couples involving HO_2 radicals and O_2^- radicals may be calculated from the table of data.

$$HO_2 + H^+ + e^- \to H_2O_2 \quad E° = 1.44 \text{ V}$$

$$O_2^- + H_2O + e^- \to HO_2^- + OH^- \quad E_B^° = 0.20 \text{ V}$$

$$HO_2 + 3H^+ + 3e^- \to 2H_2O \quad E° = 1.65 \text{ V}$$

$$O_2^- + 2H_2O + 3e^- \to 4OH^- \quad E_B^° = 0.645 \text{ V}$$

$$O_2 + H^+ + e^- \to HO_2 \quad E° = -0.046 \text{ V}$$

$$HO_2 + O_2 + H^+ + e^- \to O_3 + H_2O \quad E° = 0.813 \text{ V}$$

In acid solution, the perhydroxyl radical is not only unstable with respect to the decomposition to ozone

$$2HO_2 \to H_2O + O_3 \quad \Delta G° = -65.12 \text{ kJ mol}^{-1}$$

but also with respect to the decomposition to oxygen,

$$2HO_2 \to H_2O_2 + O_2 \quad \Delta G° = -142.97 \text{ kJ mol}^{-1}$$

However, in alkaline solution, the O_2^- ion is much more stable:

Oxygen

$$2O_2^- + H_2O \rightarrow O_3 + 2OH^- \qquad \Delta G° = 22.09 \text{ kJ mol}^{-1}$$

$$2O_2^- + H_2O \rightarrow O_2 + HO_2^- + OH^- \qquad \Delta G° = -51.13 \text{ kJ mol}^{-1}$$

These results indicate that the decomposition of the HO_2 radical or O_2^- ion follows a path to peroxide and O_2 rather than ozone. It has been found experimentally that KO_2 treated with acid gives a virtually quantitative yield of O_2 and H_2O_2 with an occasional trace of ozone. The chemistry of the dioxygenyl cation, O_2^+, has been reviewed in the literature [27].

IX. OXYGEN RADICAL ANION

Considering the hypothetical cell composed of hydrogen gas and hydroxyl radical gas electrodes dipping in the same aqueous electrolyte, Hickling and Hill [28] arrived at a value for the standard potential of the OH/OH^- couple of 2.00 V. But using the latest Bureau of Standards data from Table 1, one obtains

$$OH + e^- \rightarrow OH^- \qquad E_B° = 1.985 \text{ V} \qquad (dE°/dT)_{isoth} = -2.689 \text{ mV K}^{-1}$$

From radiolysis studies, it has been suspected that this Hickling and Hill value for the OH/OH^- couple is too high because the Gibbs energy of hydration of the OH radical in solution has been neglected. From an estimate of the ΔG of hydration for the OH radical (-25.1 kJ mol^{-1}), Stein [29] arrived at a value of 1.25 V. Considering the results of permanganate-decomposition rate studies, Heckner and Lansberg [30] concluded that $E_B°$ should lie between 1.4 and 1.5 V. Recently, Berdinkov and Bazhin [31] report a value of 1.9 V from determinations of the heat and entropy of hydration of OH radicals (-25.9 kJ mol^{-1}). In view of the uncertainties involved in determining the potential of the OH/OH^- couple, a value of 1.55 V is chosen. Using the 1.55 V value, the $\Delta G°$ for $(OH)_{aq}$ is found to be -7.74 kJ mol^{-1}.

The standard potentials of other couples involving the hydroxyl radical determined with this average value are as follows:

$$OH + H^+ + e^- \rightarrow H_2O \qquad E° = 2.38 \text{ V} \qquad (dE°/dT)_{isoth} = -1.855 \text{ mV K}^{-1}$$

$$H_2O_2 + H^+ + e^- \rightarrow OH + H_2O \qquad E° = 1.14 \text{ V}$$

$$O_3 + H^+ + e^- \rightarrow O_2 + OH \qquad E° = 1.77 \text{ V}$$

$$HO_2^- + H_2O + e^- \rightarrow OH + 2OH^- \qquad E_B° = 0.184 \text{ V}$$

$$O_3 + H_2O + e^- \rightarrow O_2 + OH + OH^- \qquad E_B° = 0.943 \text{ V}$$

By adding the $\Delta G°$ for the ionization of OH radicals (pK = 11.8), the dissociation of O_2 (463.495 kJ mol^{-1}), the O_2/H_2O couple ($E°$ = 1.229 V), the OH/OH^- couple ($E°$ = 1.55 V), the ionization of water (pK = 13.995), and the hydration of O atoms (16.7 kJ mol^{-1})*, a $\Delta G°$ value of -155.24 kJ mol^{-1} is obtained for the O/O$^-$ couple, from which

$$O + e^- \rightarrow O^- \qquad E° + 1.61 \text{ V}$$

From these data the $\Delta G°$ of formation of O$^-$ ion in solution is 76.50 kJ mol^{-1}. Using these data, one may calculate the standard potential of the O$^-$/OH$^-$ couple,

$$O^- + H_2O + e^- \rightarrow 2OH^- \qquad E°_B = 1.59 \text{ V}$$

According to these results, as expected, the oxygen radical ion and the hydroxyl radical are strong oxidizing agents.

X. SINGLET OXYGEN

Dye-sensitized photooxygenation of organic compounds has been studied extensively and the accepted mechanism involves oxygen in the singlet state [32]. In the ground state, $^3\Sigma_g^-$ ($^3O_2$), there is one electron with parallel spin in each of the two molecular orbitals. The first excited state lies 92 kJ mol^{-1} (0.977 eV) above the $^3\Sigma_g^-$ state; it is a singlet state $^1\Delta_g$ ($^1O_2$) with two electrons with paired spins in one orbital, leaving the other empty. A very short lived second excited state, $^1\Sigma_g^+$, having an electron in each orbital with antiparallel spins, lies 154.8 kJ mol^{-1} (1.63 eV) above the $^3\Sigma_g^+$ state and contributes negligibly to the photo process.

According to the singlet oxygen mechanism, a dye molecule such as a fluorescence derivative or methylene blue is excited by photoillumination (visible light):

$$\text{dye} + h\nu \rightarrow \text{dye}*$$

The excited dye raisis the $^3O_2$ molecule to the singlet state,

$$\text{dye}* + {}^3O_2 \rightarrow \text{dye} + {}^1O_2$$

which oxygenates the given organic acceptor molecule, A,

$$^1O_2 + A \rightarrow AO_2$$

The advantage of this process is the ability to remove spin-conservation difficulties from the formation of singlet products from reactants converted to singlet multiplicity.

Singlet oxygen can be generated by passing a stream of oxygen through a strong electric field [33] and the $O_2(^1\Delta_g)$ was detected in a mass spectrograph. By the reaction of hydrogen peroxide and sodium hypochlorite in so-

---

*It is assumed that the free energy of hydration of oxygen atoms is similar to that for oxygen molecules. A value of 16.7 kJ mol^{-1} has been selected for $\Delta G°_{hydr}$ for O atoms.

Oxygen

lution, singlet oxygen is generated [34] accompanied by a red-orange chemiluminescence. In both of these cases, oxygenation of the same organic compounds, which were studied in the photo-oxidation process, were successfully carried out. Without the electric field, little or no reaction was obtained, but with it, virtually 100% oxygenation took place [35]. Under certain conditions, certain enzymes such as catalase and peroxidase may generate $^1O_2$, giving rise to luminescence [36]. Lifetimes of the $^1\Delta_g$ state range from a few microseconds to a few milliseconds, depending on the solvent used.

Although considerable effort has been expended in studying the chemistry of singlet oxygen, there is a dearth of knowledge concerning the thermodynamics and electrochemistry of this metastable form of O_2. With future investigations, it is hoped that this situation will be remedied.

XI. POTENTIAL DIAGRAMS FOR OXYGEN

Acid Solutions

$$O_2 \xrightarrow{-0.125} HO_2 \xrightarrow{1.51} H_2O_2 \xrightarrow{0.714} H_2O + OH \xrightarrow{2.813} 2H_2O$$

with $O_2 \xrightarrow{0.695} H_2O_2$, $H_2O_2 \xrightarrow{1.763} 2H_2O$, and $O_2 \xrightarrow{1.229} 2H_2O$

Alkaline Solutions

$$O_2 \xrightarrow{-0.33} O_2^- \xrightarrow{0.20} HO_2^- \xrightarrow{-0.251} OH + OH^- \xrightarrow{1.985} 2OH^-$$

with $O_2 \xrightarrow{-0.0649} HO_2^-$, $HO_2^- \xrightarrow{0.867} 2OH^-$, and $O_2 \xrightarrow{0.401} 2OH^-$

Acid Solutions

$$O_3 \xrightarrow{0.508} O_2 + OH \xrightarrow{1.985} O_2 + OH^-$$

with $O_3 \xrightarrow{1.246} O_2 + OH^-$

Alkaline Solutions

$$O_3 \xrightarrow{1.33} O_2 + OH \xrightarrow{2.813} O_2 + H_2O$$

with $O_3 \xrightarrow{2.075} O_2 + H_2O$

Acid Solutions

$$O_3 + H_2O \xrightarrow{-0.892} HO_2 + O_2 \xrightarrow{0.125} 2O_2$$

with $O_3 + H_2O \xrightarrow{-0.383} 2O_2$

REFERENCES

1. F. A. Cotton and G. Wilkinson, *Advanced Inorganic Chemistry*, 3rd ed., Wiley-Interscience, New York, 1972, p. 409.
2. A. G. Streng, Chem. Rev., 63, 607 (1963).
3. D. D. Wagman et al., Natl. Bur. Stand. Tech. Note 270-3, U.S. Government Printing Office, Washington, D.C., Jan. 1968.

4. B. E. Conway, *Electrochemical Data*, Elsevier, New York, 1952, p. 196.
5. A. J. de Béthune and N. A. Swendeman Loud, *Standard Aqueous Electrode Potentials and Temperature Coefficients at 25°C*, Hampel, Skokie, Ill., 1964.
6. J. P. Hoare, *The Electrochemistry of Oxygen*, Wiley-Interscience, New York, 1968, p. 81.
7. J. P. Hoare, in *Encyclopedia of the Electrochemistry of the Elements*, A. J. Bard, ed., Vol. 2, Marcel Dekker, New York, 1974, p. 191.
8. J. P. Hoare, J. Phys. Chem., 79, 2175 (1975).
9. D. M. H. Kern, J. Am. Chem. Soc., 76, 4208 (1954).
10. V. I. Tikhomirova, V. I. Luk'yanycheva, and V. S. Bagotskii, Elektrokhimiya, 1, 645 (1965).
11. E. Yeager, E. P. Krause, and K. V. Rao, Electrochim. Acta, 9, 1057 (1964).
12. V. S. Bagotskii and D. L. Motov, Dokl. Akad. Nauk SSSR, 71, 501 (1950).
13. I. E. Yablokova and V. S. Bagotskii, Dokl. Akad. Nauk SSSR, 85, 599 (1952).
14. R. J. Bowen, H. B. Urbach, and J. H. Harrison, Nature, 213, 592 (1967).
15. C. Fabyan, Monatsh. Chem., 106, 513 (1975).
16. S. Gordon et al., Discuss. Faraday Soc., 36, 193 (1963).
17. B. L. Gall and L. M. Dorfman, J. Am. Chem. Soc., 91, 2199 (1969).
18. E. Haydon and J. J. McGavey, J. Phys. Chem., 71, 1472 (1967).
19. V. R. Landi and L. J. Heidt, J. Phys. Chem., 73, 2361 (1969).
20. D. Behar, G. Czapski, J. Rabani, L. M. Dorfman, and H. A. Schwarz, J. Phys. Chem., 74, 3209 (1970).
21. B. H. J. Bielski and A. O. Allen, J. Phys. Chem., 81, 1048 (1977).
21a. B. H. J. Bielski, Photochem. Photobiol., 28, 645 (1978).
21b. J. A. Fee, Metal Ions Biol. Syst., 2, 209 (1980).
22. J. A. Fee and J. S. Valentine, in *Superoxide and Superoxide Dismutases*, A. M. Michelson, J. M. McCord, and I. Ridovich, eds., Academic Press, New York, 1977, p. 19.
23. J. Divisek and B. Kastening, J. Electroanal. Chem., 65, 603 (1975).
24. J. Chevalet, F. Rouelle, L. Gierst, and J. P. Lambert, J. Electroanal. Chem., 39, 201 (1972).
25. Y. A. Ilan, G. Czapski, and D. Meisel, Biochim. Biophys. Acta, 430, 209 (1976); Israel J. Chem., 12, 891 (1974).
26. G. Czapski, Annu. Rev. Phys. Chem., 22, 171 (1971).
27. R. J. Gillespie and J. Passmore, Acc. Chem. Res., 4, 413 (1971).
28. A. Hickling and S. Hill, Trans. Faraday Soc., 46, 557 (1950).
29. G. Stein, J. Chem. Phys., 42, 2986 (1965).
30. K.-H. Heckner and R. Lansberg, Z. Phys. Chem., 230, 63 (1965).
31. V. M. Berdnikov and N. M. Bazhin, Zh. Fiz. Khim., 44, 712 (1970).
32. C. S. Foote, Acc. Chem. Res., 1, 104 (1968).
33. S. N. Foner and R. L. Hudson, J. Chem. Phys., 25, 601 (1956).
34. A. U. Khan and M. Kasha, J. Chem. Phys., 39, 2105 (1963).
35. E. J. Corey and W. C. Taylor, J. Am. Chem. Soc., 86, 3881 (1964).
36. D. Pierre, Adv. Radiat. Res. Phys. Chem., 1, 217 (1973).

5
The Halogens

I. FLUORINE*

ANSELM T. KUHN† Institute of Dental Surgery, University of London, London, England
COLIN L. RICE University of Salford, Salford, England

Fluorine is the most active of the commonly found elements. The only stable oxidation state in water is -1. The oxides F_2O and F_2O_2 have both been prepared. They are powerful oxidizing agents and the potential for the reduction of the monoxide has been calculated by Latimer [1]. There appear to have been no further contributions to the literature since the publication of Latimer, and the more recent compilation of Milazzo and Caroli [2] confirms this. Interest has focused on the fluoride melts, and data also exist for mixed electrolytes. The thermodynamic data are given in Table 1.

A. Fluorine-Fluorine Couple

From the National Bureau of Standards (NBS) values of Gibbs energies, Latimer calculated

$$F_2 + 2e^- \rightarrow 2F^- \qquad E° = 2.87 \text{ V}$$

while Pourbaix [2] quotes the potential as 2.866 V. This is in approximate agreement with the value obtained also by Latimer [1], who calculated the reversible potential as 2.85 ± 0.04 V from the heat of formation of HF solution and the Gibbs energy of the resulting ions in hypothetical 1 M solution, giving a free energy of formation of HF(aq) of -274.8 kJ mol^{-1}. The temperature coefficient for this reaction has been given as -1.83 mV K^{-1} [3].

Latimer calculated the Gibbs energy of HF as -294.7 kJ mol^{-1} and of HF_2^- as -574.6 kJ mol^{-1}, but in his table of thermodynamic data quoted the latter as -575.4 kJ mol^{-1}. Pourbaix [4] cites a value of -575.0 kJ mol^{-1} for HF_2^-, averaging Latimer's two values, and quotes for

$$F_2 + 2H^+ + 2e^- \rightarrow 2HF(aq) \qquad E° = 3.053 \text{ V}$$

---

*Reviewer: J. Coetzee, University of Pittsburgh, Pittsburgh, Pennsylvania.
†Current affiliation: Harrow College of Higher Education, Harrow, England.

TABLE 1 Thermodynamic Data on Fluorine[a]

Formula	Description	State	$\Delta H°$/kJ mol^{-1}	$\Delta G°$/kJ mol^{-1}	$S°$/J mol^{-1} K^{-1}
F	$^2P_{3/2}$	g	78.91	61.83	158.64
F	$^2P_{1/2}$	g	139.79		
F	$^3P_2$	g	1766.23		
F^{2+}	$^4S_{3/2}$	g	5148.16		
F$^-$		g	−270.70	−267.18	145.47
F$^-$		aq	−332.63	−278.82	−13.81
F$_2$		g	0.0	0.0	202.71
F$_2$O		g	−21.76	−4.6	247.36
HF		g	−271.12	273.22	173.67
HF		aq	−332.63	−296.85	88.70
HF$_2^-$		aq	−649.9	−578.15	92.47
H$_6$F$_6$		g	−1782.38		

[a]National Bureau of Standards values.

The Halogens

The temperature coefficient is -0.60 mV K^{-1} [3]. The potential for the reaction

$$F_2 + H^+ + 2e^- \rightarrow HF_2^- \qquad E° = 2.979 \text{ V}$$

has also been derived from thermodynamic data.

B. Fluorine Oxide–Fluoride Couple

For the reduction of the oxide to fluoride ion,

$$F_2O + 2H^+ + 4e^- \rightarrow 2F^- + H_2O \qquad E° = 2.153 \text{ V}$$

is the best calculated value [4] and the temperature coefficient is -1.184 mV K^{-1} [3].

For the reduction to aqueous hydrogen fluoride,

$$F_2O + 4H^+ + 4e^- \rightarrow 2HF + H_2O \qquad E° = 2.246 \text{ V (Pourbaix)}$$

and finally for

$$F_2O + 3H^+ + 4e^- \rightarrow HF_2^- + H_2O \qquad E° = 2.209 \text{ V (Pourbaix)}$$

From all the foregoing potentials, it will be seen that these reversible potentials are entirely hypothetical, because at such potentials in aqueous media oxygen evolution will always occur (slow kinetics) and in consequence none of the values has been the subject of recent experimental investigation, which has been confined to nonaqueous media.

C. Potential Diagram for Fluorine

```
       0             -1
            2.866    F⁻(aq)
    ┌─────────────
    │ F₂(g) 2.979   HF₂⁻
    │
    └────3.053────  HF(aq)
```

REFERENCES

1. W. M. Latimer, J. Am. Chem. Soc., *48*, 2868 (1926).
2. G. Milazzo and S. Cavoli, *Tables of Standard Electrode Potentials*, Wiley, New York, 1978.
3. A. J. de Béthune and N. A. Swendeman Loud, "Table of Standard Aqueous Electrodes Potentials and Temperature Coefficients at 25°C," in *Encyclopedia of Electrochemistry*, C. A. Hampel, ed., Reinhold, New York, 1964.
4. M. Pourbaix, ed., *Atlas d'équilibres électrochimiques à 25°C*, Gauthier-Villars, Paris, 1963.

II. CHLORINE*

TORQUATO MUSSINI and PAOLO LONGHI   University of Milan, Milan, Italy

A. Oxidation States

The better known oxidation numbers of chlorine in its compounds are -1 for hydrochloric acid and its salts, the chlorides; $+1$ for the oxide Cl_2O, the corresponding hypochlorous acid, and its salts, the hypochlorites; $+3$ for chlorous acid and the chlorites; $+5$ for chloric acid and the chlorates; and $+7$ for the oxide Cl_2O_7, the corresponding perchloric acid $HClO_4$, and the perchlorates. The oxides ClO, Cl_2O_3, and Cl_2O_6 are also described but either as intermediate products in certain reactions or as unstable compounds whose chemistry is rather uncertain. The thermodynamic data are given in Table 2.

B. Chlorine-Chloride Couple

The best value for the standard potential $E°$ of the Cl_2/Cl^- couple is to be considered that redetermined critically in extensive work [1] based on electromotive force (emf) measurements of the cells

$$Pt \mid Ag \mid AgCl \mid HCl(aq) \mid Cl_2 \mid Pt + Ir$$

$$Pt \mid H_2 \mid HCl(aq) \mid Cl_2 \mid Pt + Ir$$

over the temperature range 298.15 to 353.15 K, where due account was also taken of the formation of the Cl_3^- ion, through a special procedure leading to the simultaneous determination of $E°$ and of the equilibrium constant K_{eq} of the reaction

(A) $Cl^-(aq) + Cl_2(g) \rightarrow Cl_3^-(aq)$

Thus for the relevant electrode reaction we can write

(I) $Cl_2(g) + 2e^- \rightarrow 2Cl^-(aq)$ $E°_{(I)} = 1.35828$ V

$$dE°_{(I)}/dT = -1.2444 \text{ mV K}^{-1}$$

at 298.15 K.

For interpolation purposes within the temperature range above, the following polynomial may be used:

$$E°_{(I)}(V) = 1.484867 + 3.958492 \times 10^{-4} \times T(K)$$
$$- 2.750639 \times 10^{-6} \times [T(K)]^2$$

However, since within the range of experiment the $E°$ versus T plot turns out to be very flat and quasi-linear, the equation above may be safely used over

---

*Reviewer: J. Coetzee, University of Pittsburgh, Pittsburgh, Pennsylvania.

The Halogens

a somewhat wider temperature range, say from 273 to 373 K, without introducing significant errors.

Owing to the foregoing equilibrium of formation of Cl_3^- ions, which may be also considered in the form

(B) $\quad Cl^-(aq) + Cl_2(aq) \rightarrow Cl_3^-(aq)$

the standard potential of the chlorine-chloride couple can be expressed in terms of the alternative schemes

(II) $Cl_2(aq) + 2e^- \rightarrow 2Cl^-(aq) \qquad E^\circ_{(II)} = 1.396$ V

$dE^\circ_{(II)}/dT = -0.716$ mV K^{-1}

or

(III) $Cl_3^-(aq) + 2e^- \rightarrow 3Cl^-(aq) \qquad E^\circ_{(III)} = 1.4152$ V

$dE^\circ_{(III)}/dT = -0.534$ mV K^{-1}

for which, for interpolation purposes, the following respective polynomials can be used:

$E^\circ_{(II)}(V) = 1.40922 - 4.30504 \times 10^{-4} \times T(K)$

$\qquad - 4.87922 \times 10^{-6} \times [T(K)]^2$

and

$E^\circ_{(III)}(V) = 1.359357 + 6.407284 \times 10^{-4} \times T(K)$

$\qquad - 1.969988 \times 10^{-6} \times [T(K)]^2$

As a result of the redetermination of standard potentials with related temperature coefficients above, the values of several basic thermodynamic data concerning species relevant to the chlorine domain have been updated, in particular the standard entropies of HCl and Cl$^-$ (see Table 2).

The standard potential of the Cl_2/Cl^- couple is higher than that of the O_2/H_2O couple at pH > 0. This means that aqueous solutions of chlorine are thermodynamically unstable. Fortunately, the kinetics of the reaction $H_2O + Cl_2 = 2HCl + O_2$ is rather slow both in homogeneous systems and in the presence of an (inert) metal electrode, so that the measurements of the Cl_2/Cl^- electrode potential may be safely accomplished provided that the following three conditions are fulfilled:

1. The "inert" electrode must be really inert; that is, it must not undergo corrosion in the presence of dissolved chlorine in HCl solutions, because corrosion would give rise to a mixed potential. At the same time, the inert electrode must ensure a sufficiently

TABLE 2 Thermodynamic Data on Chlorine[a]

Formula	Description	State	$\Delta H°$/kJ mol^{-1}	$\Delta G°$/kJ mol^{-1}	$S°$/J mol^{-1} K^{-1}
Cl		g	121.679	105.696	165.088
Cl$^+$		g	1383.23*	1360.323*	167.448*
Cl^{2+}		g	3686.82		
Cl$^-$		g	−233.89*	−240.116*	153.252*
Cl$^-$	Std. state, hyp., $m = 1$	aq	−166.8533	−131.0563	56.701
Cl$_2$		g	0	0	222.957
Cl$_2$	Std. state, hyp., $m = 1$	aq	−21.09	7.20	127.6
Cl$_3^-$	Std. state, hyp., $m = 1$	aq	−197.74	−120.08	138.9
ClO		g	101.84	98.11	226.52
ClO$^-$	Std. state, hyp., $m = 1$	aq	−107.1	−36.8	42
ClO$_2$		g	102.5	120.5	256.73
ClO$_2$	Std. state, hyp., $m = 1$	aq	74.9	117.6	173.2
ClO$_2^-$	Std. state, hyp., $m = 1$	aq	−66.5	17.2	101.3
ClO$_3$		g	155		

ClO_3^-	Std. state, hyp., $m = 1$	aq	−99.2	−3.3	162.3
ClO_4^-	Std. state, hyp., $m = 1$	aq	−129.33	−8.62	182.0
Cl_2O		g	80.3	97.9	266.10
Cl_2O_7		ℓ	238.1		
ClF		g	−54.48	−55.94	217.78
HCl		g	−92.307	−95.299	186.799
HCl	Std. state, hyp., $m = 1$	aq	−166.8533	−131.0563	56.701
HClO		g	−92.05*	−79.320*	236.59*
HClO	Undiss.; std. state, hyp., $m = 1$	aq	−120.9	−79.9	142
$HClO_2$	Undiss.; std. state, hyp., $m = 1$	aq	−51.9	5.9	188.3
$HClO_3$	Std. state, hyp., $m = 1$	aq	−99.2	−3.3	162.3
$HClO_4$		ℓ	−40.6		
$HClO_4$	Std. state, hyp., $m = 1$	aq	−129.33	−8.62	182.0
$HClO_4 \cdot H_2O$		c	−382.21		

aNational Bureau of Standards values are in italic. JANAF Thermochemical tables values are indicated by an asterisk.

high exchange current for the Cl_2/Cl^- couple and rapid attainment of the equilibrium potential, resulting in a close-to-zero bias potential. The electrode material that best satisfies these requirements turns out to be a 45 wt % Ir + Pt alloy chemically deposited onto a tantalum foil substrate [1].

2. The pH of the solution must be in the strongly acidic range, because at pH values not sufficiently low the disproportionation reaction of chlorine into chloride and hypochlorite would render the measured potential meaningless.

3. Finally, in measuring, and referring to, the Cl_2/Cl^- electrode potential one must take into consideration that the reaction of trichloride ion formation, (A) or (B), can substantially affect the Cl^- ion activity. The equilibrium constant for such a reaction is found [1] to be $K_A = 0.012$ with gaseous chlorine and $K_B = 0.21$ with aqueous chlorine. The best situation, in this respect, is found when working with a low partial pressure of Cl_2 (but not so low as to render the analytical determination of the concentration of chlorine in the gas mixture inaccurate), and with an appropriate concentration of Cl^- ions (HCl), any possible interference due to the behavior of the companion electrode in the cell—or to liquid junction potentials—being duly taken into account.

C. Chlorine Redox Couples Other Than the Chlorine-Chloride Couple

The standard potentials of these redox couples, including chlorine species with oxidation numbers of +1 or higher values, are hardly amenable to direct experimental determination. From the energy and entropy data in Table 2 we write the following (the "b" subscript is used here to denote basic solutions, and all chemical species in reactions are to be understood as "aqueous" unless otherwise noted):

$$ClO^- + H_2O + 2e^- \rightarrow Cl^- + 2OH^- \qquad E°_b = 0.890 \text{ V}$$

$$dE°_b/dT = -1.065 \text{ mV K}^{-1}$$

$$2HClO + 2H^+ + 2e^- \rightarrow Cl_2(g) + 2H_2O \qquad E° = 1.630 \text{ V}$$

$$dE°/dT = -0.268 \text{ mV K}^{-1}$$

$$2ClO^- + 2H_2O + 2e^- \rightarrow Cl_2(g) + 4OH^- \qquad E°_b = 0.421 \text{ V}$$

$$dE°_b/dT = -0.886 \text{ mV K}^{-1}$$

$$HClO_2 + 3H^+ + 4e^- \rightarrow Cl^- + 2H_2O \qquad E° = 1.584 \text{ V}$$

$$dE°/dT = -0.655 \text{ mV K}^{-1}$$

$$2HClO_2 + 6H^+ + 6e^- \rightarrow Cl_2(g) + 4H_2O \qquad E° = 1.659 \text{ V}$$

$$dE°/dT = -0.459 \text{ mV K}^{-1}$$

$$HClO_2 + 2H^+ + 2e^- \rightarrow HClO + H_2O \qquad E° = 1.674 \text{ V}$$

$$dE°/dT = -0.554 \text{ mV K}^{-1}$$

$ClO_2^- + H_2O + 2e^- \to ClO^- + 2OH^-$ $E_b^\circ = 0.681$ V

$dE_b^\circ/dT = -1.467$ mV K^{-1}

$ClO_2(g) + H^+ + e^- \to HClO_2$ $E^\circ = 1.188$ V

$dE^\circ/dT = -1.384$ mV K^{-1}

$ClO_2(g) + e^- \to ClO_2^-$ $E_b^\circ = 1.071$ V $dE_b^\circ/dT = -2.288$ mV K^{-1}

$ClO_2 + e^- \to ClO_2^-$ $E_b^\circ = 1.041$ V $dE_b^\circ/dT = -1.422$ mV K^{-1}

(It is interesting to compare the last E_b° value with such direct experimental values, 0.954 and 0.965 V, as determined through emf measurements of non-reversible cells by Naito [2] and by Troitskaya et al. [3], respectively.)

$ClO_3^- + 3H_2O + 6e^- \to Cl^- + 6OH^-$ $E_b^\circ = 0.622$ V

$dE_b^\circ/dT = -1.333$ mV K^{-1}

$2ClO_3^- + 12H^+ + 10e^- \to Cl_2(g) + 6H_2O$ $E^\circ = 1.468$ V

$dE^\circ/dT = -0.347$ mV K^{-1}

$ClO_3^- + 2H_2O + 4e^- \to ClO^- + 4OH^-$ $E_b^\circ = 0.488$ V

$dE_b^\circ/dT = -1.467$ mV K^{-1}

$ClO_3^- + 3H^+ + 2e^- \to HClO_2 + H_2O$ $E^\circ = 1.181$ V

$dE^\circ/dT = -0.180$ mV K^{-1}

$ClO_3^- + H_2O + 2e^- \to ClO_2^- + 2OH^-$ $E_b^\circ = 0.295$ V

$dE_b^\circ/dT = -1.467$ mV K^{-1}

$ClO_3^- + 2H^+ + e^- \to ClO_2(g) + H_2O$ $E^\circ = 1.175$ V

$dE^\circ/dT = 1.026$ mV K^{-1}

$ClO_3^- + H_2O + e^- \to ClO_2(g) + 2OH^-$ $E_b^\circ = -0.481$ V

$dE_b^\circ/dT = -0.646$ mV K^{-1}

$ClO_4^- + 8H^+ + 8e^- \to Cl^- + 4H_2O$ $E^\circ = 1.388$ V

$dE^\circ/dT = -0.477$ mV K^{-1}

$2ClO_4^- + 16H^+ + 14e^- \to Cl_2(g) + 8H_2O$ $E^\circ = 1.392$ V

$dE^\circ/dT = -0.367$ mV K^{-1}

$$ClO_4^- + 2H^+ + 2e^- \rightarrow ClO_3^- + H_2O \qquad E° = 1.201 \text{ V}$$

$$dE°/dT = -0.416 \text{ mV K}^{-1}$$

$$ClO_4^- + H_2O + 2e^- \rightarrow ClO_3^- + 2OH^- \qquad E°_b = 0.374 \text{ V}$$

$$dE°_b/dT = -1.252 \text{ mV K}^{-1}$$

D. Notes on the Chlorine Potentials

Hypochlorous acid and chlorous acid are unstable with regard to both disproportionation and oxygen evolution. Chloric acid has similar instabilities, but its reactions in dilute solution are very slow; even on boiling a solution of chlorate in excess hydrochloric acid, it takes some time for complete reduction. Hot solutions are reduced fairly rapidly by Fe^{2+}. Chlorate may also be reduced at a steel cathode upon electrolysis of chlorate solutions. It is noteworthy that there is a decrease in potential of each step in going from HClO to ClO_4^-. This fact is undoubtedly significant in the interpretation of the slowness of chlorate and perchlorate as oxidizing agents, as the lower potentials for the first step of the reduction act as a barrier to the net reaction. However, once the reaction starts, the production of heat may cause the reaction to proceed with explosive violence, and both the chlorate and the perchlorate reactions must be handled with extreme caution.

Some disproportionation reactions are of importance: for instance,

$$Cl_2(g) + 2OH^- \rightarrow Cl^- + ClO^- + H_2O \qquad \Delta G° = -90.4 \text{ kJ mol}^{-1}$$

which is practically complete in alkaline solution but is reversed to a measurable equilibrium in acid solution, and

$$ClO^- + 2HClO \rightarrow ClO_3^- + 2Cl^- + 2H^+ \qquad \Delta G° = -68.8 \text{ kJ mol}^{-1}$$

which is relevant to the industrial production of chlorates. The latter reaction requires the presence of both the ClO^- and the HClO species; the optimum operating (industrial) conditions are slightly acidic pH (6.2 to 7) as well as high temperatures (60 to 110°C). In practice, the industrial production of chlorates starts from the electrolysis of chlorides which are oxidized to chlorine, which in turn readily disproportionates to hypochlorous acid (in dissociation equilibrium with hypochlorite) and chloride ion as intermediate steps. The side reaction

$$12ClO^- + 6H_2O = 4ClO_3^- + 8Cl^- + 12H^+ + 3O_2(g) + 12e^-$$

whose mechanism is not yet completely clarified but which seems not to merely be the overlapping of the chemical disproportionation of hypochlorite and the electrochemical evolution of oxygen, results in a decrease of the current efficiencies for chlorate production. Another disproportionation reaction deserves mention:

$$2ClO_2(g) + 2OH^- \rightarrow ClO_2^- + ClO_3^- + H_2O \qquad \Delta G° = -149.7 \text{ kJ mol}^{-1}$$

which is known to proceed readily at pH > 10.5.

The Halogens

Chlorates are readily oxidized at platinum or lead dioxide anodes, at electrode potentials higher than 2.4 and 2.1 V, respectively [referred to the standard hydrogen electrode (SHE)], with good current efficiencies, the competing process being oxygen evolution. Little or no information is available about the relevant reaction mechanisms, for which a number of paths have been suggested [4]. The most favorable conditions are given by high current densities and low temperature.

E. Potential Diagrams for Chlorine

Acid Solution

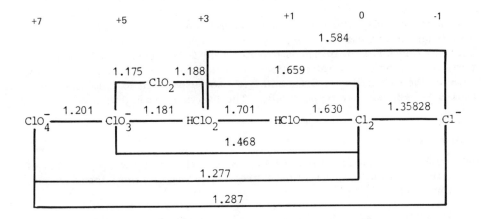

Basic Solution

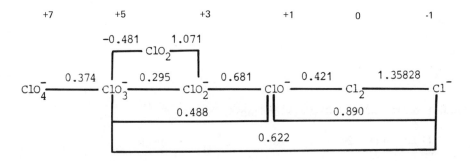

REFERENCES

1. T. Mussini and P. Longhi, J. Electrochem. Soc., *114*, 340 (1967); J. Chem. Eng. Data, *13*, 458 (1968); J. Electroanal. Chem., *20*, 411 (1969); A. J. Bard, ed., *Encyclopedia of the Electrochemistry of the Elements*, Vol. 1, Marcel Dekker, New York, 1973, p. 2.
2. T. Naito, Kogyo Kagaku Zasshi, *65*, 749 (1962).
3. N. V. Troitskaya, K. P. Mishchenko, and I. E. Flis, Zh. Fiz. Khim., *33*, 1614 (1959).
4. A. J. Bard, ed., *Encyclopedia of the Electrochemistry of the Elements*, Vol. 1, Marcel Dekker, New York, 1973, p. 50.

III. BROMINE*

TORQUATO MUSSINI and PAOLO LONGHI University of Milan, Milan, Italy

A. Oxidation States

The oxidation numbers for which the existence of bromine compounds can reliably be stated are −1 for hydrobromic acid and its salts, the bromides; +1 for hypobromous acid and its salts, the hypobromites; +5 for bromic acid and bromates; and +7 for perbromic acid and perbromates, which were synthesized only recently by Appelman [1]. This result is important not only because the perbromates are at last duly accommodated between the perchlorates and the periodates, as expected from homology, but also because the strongly established belief of nonsynthesizability and nonexistence of perbromates (for which many structural arguments were presented) is now cleared away. As for other oxidation states of bromine, bromous acid (+3) certainly is formed in intermediate steps in several reactions, and there is also evidence for the existence of the oxides Br_2O, BrO, and BrO_2 corresponding to the oxidation numbers +1, +2, and +4, respectively. The thermodynamic data are given in Table 3.

B. Bromine-Bromide Couple

Analogously to the standard potential of the chlorine-chloride couple, that of the bromine-bromide couple may be expressed in terms of the alternative electrode reactions

(I) $Br_2(\ell) + 2e^- \rightarrow 2Br^-(aq)$ $E°_{(I)} = 1.0652$ V

$dE°_{(I)}/dT = -0.629$ mV K^{-1}

or

(II) $Br_2(aq) + 2e^- \rightarrow 2Br^-(aq)$ $E°_{(II)} = 1.0874$ V

$dE°_{(II)}/dT = -0.5406$ mV K^{-1}

depending on the standard states chosen for the Br_2 species, and also

(III) $Br_3^-(aq) + 2e^- \rightarrow 3Br^-(aq)$ $E°_{(III)} = 1.0503$

$dE°_{(III)}/dT = -0.498$ mV K^{-1}

because of the well-known reaction of Br_2 with Br^- to form tribromide ions Br_3^-:

(A) $Br_2 + Br^- \rightarrow Br_3^-$

Even if at high concentrations of bromide, besides Br_3^-, the pentabromide ion Br_5^- may form

---

*Reviewer: J. Coetzee, University of Pittsburgh, Pittsburgh, Pennsylvania.

(B) $\quad 2Br_2 + Br^- \rightarrow Br_5^-$

this circumstance is of little or no importance for the determination of the standard potential of the bromine-bromide couple. The accurate $E°$ data above are drawn from two distinct sources: first, the electromotive force (emf) measurements at 298 K by Jones and Baeckström [2] on the cell

$$Pt \mid Ag \mid AgBr \mid KBr(aq) \mid Br_2 \mid Pt - Ir$$

in conjunction with the determination of the equilibrium constant $K_{(A)}$ of reaction (A) from solubility measurements, leading to $E°$ values in terms of the electrode reactions (I) and (II); and second, the extensive redetermination by Mussini and Faita [3] based on emf measurements from 273 to 323 K on the cell

$$Pt \mid Ag \mid AgBr \mid HBr(aq) \mid Br_2 \mid Pt - Ir$$

which were analyzed through a special scheme providing the simultaneous determination of $E°$ and $K_{(A)}$ and leading to $E°$ values in terms of electrode reactions (II) and (III). The agreement of results at 298 K is very good. The following respective polynomials may be used for interpolation purposes:

$$E°_{(II)}(V) = 0.797783 + 0.00248340 \times T(K) - 5.07121 \times 10^{-6} \times [T(K)]^2$$

$$E°_{(III)}(V) = 0.808872 + 0.00211746 \times T(K) - 4.38612 \times 10^{-6} \times [T(K)]^2$$

The equilibrium constant for reaction (A) with aqueous Br_2 is available over the same temperature range (273 to 323 K) as for $E°_{(II)}$ and $E°_{(III)}$, and its best value at 298 K is $K_{(A)} = 17.9$ [3,4]. The agreement with earlier data for 298 K is good [4], but this is not so with regard to the equilibrium constant $K_{(B)}$ for the pentabromide formation reaction (B) as given in the works of Liebhafsky [5] and Jones and Baeckström [2].

C. Bromine Redox Couples Other Than the Bromine-Bromide Couple

As for the standard potentials of these redox couples (that include bromine species with oxidation numbers of +1 or higher values), the lack of direct experimental data is even greater than that we can observe for the parallel domain of chlorine. However, from the free energy and entropy data in Table 3 we write the following (the "b" subscript is used here to indicate basic solutions, and all chemical species are to be understood as "aqueous" unless otherwise noted):

$HBrO + H^+ + 2e^- \rightarrow Br^- + H_2O \qquad E° = 1.341$ V

$dE°/dT = -0.622$ mV K^{-1}

$BrO^- + H_2O + 2e^- \rightarrow Br^- + 2OH^- \qquad E°_b = 0.766$ V

$dE°_b/dT = -0.933$ mV K^{-1}

TABLE 3 Thermodynamic Data on Bromine[a]

Formula	Description	State	$\Delta H°/\text{kJ mol}^{-1}$	$\Delta G°/\text{kJ mol}^{-1}$	$S°/\text{J mol}^{-1}\text{K}^{-1}$
Br		g	111.884	82.429	174.912
Br^+		g	1261.10		
Br^{2+}		g	3350.1		
Br^-		g	−233.9		
Br^-	Std. state, hyp., $m = 1$	aq	−121.55	−103.97	82.4
Br_2		ℓ	0	0	152.231
Br_2	Std. state, hyp., $m = 1$	aq	−0.92	4.18	135.23
Br_3^-	Std. state, hyp., $m = 1$	aq	−131.34	−105.77	208.0
Br_5^-	Std. state, hyp., $m = 1$	aq	−142.3	−103.8	316.7
BrO		g	125.77	108.24	237.44

BrO^-	Std. state, hyp., $m = 1$	aq	−94.1	−33.5	42
BrO_2		c	48.5		
BrO_3^-	Std. state, hyp., $m = 1$	aq	−83.7	1.7	163.2
BrO_4^-	Std. state, hyp., $m = 1$	aq	13.35	122.09	187.0
HBr		g	−36.40	−53.43	198.585
HBr	Std. state, hyp., $m = 1$	aq	−121.55	−103.97	82.4
HBrO	Undiss.; std. state, hyp., $m = 1$	aq	−113.0	−82.4	142
$HBrO_3$	Std. state, hyp., $m = 1$	aq	−83.7	1.7	163.2
$HBrO_4$	Std. state, hyp., $m = 1$	aq	13.35	122.09	187.0
BrF		g	93.85	109.16	228.87
			73.76*	58.58*	228.87*
BrCl		g	14.64	−0.96	239.99

[a]National Bureau of Standards values are in italic. JANAF Thermochemical tables values are indicated by an asterisk.

$2HBrO + 2H^+ + 2e^- \rightarrow Br_2(\ell) + 2H_2O \qquad E° = 1.604 \text{ V}$

$dE°/dT = -0.632 \text{ mV K}^{-1}$

$2BrO^- + 2H_2O + 2e^- \rightarrow Br_2(\ell) + 4OH^- \qquad E°_b = 0.455 \text{ V}$

$dE°_b/dT = -1.254 \text{ mV K}^{-1}$

$BrO_3^- + 3H_2O + 6e^- \rightarrow Br^- + 6OH^- \qquad E°_b = 0.584 \text{ V}$

$dE°_b/dT = -1.290 \text{ mV K}^{-1}$

$2BrO_3^- + 12H^+ + 10e^- \rightarrow Br_2(\ell) + 6H_2O \qquad E° = 1.478 \text{ V}$

$dE°/dT = -0.423 \text{ mV K}^{-1}$

$BrO_3^- + 5H^+ + 4e^- \rightarrow HBrO + 2H_2O \qquad E° = 1.447 \text{ V}$

$dE°/dT = -0.370 \text{ mV K}^{-1}$

$BrO_3^- + 2H_2O + 4e^- \rightarrow BrO^- + 4OH^- \qquad E°_b = 0.492$

$dE°_b/dT = -1.469 \text{ mV K}^{-1}$

$BrO_4^- + 2H^+ + 2e^- \rightarrow BrO_3^- + H_2O \qquad E° = 1.853 \text{ V}$

$dE°/dT = -0.435 \text{ mV K}^{-1}$

$BrO_4^- + H_2O + 2e^- \rightarrow BrO_3^- + 2OH^- \qquad E°_b = 1.025 \text{ V}$

$dE°_b/dT = -1.274 \text{ mV K}^{-1}$

D. Notes on the Bromine Potentials

Bromine is unstable in alkaline solutions with respect to the disproportionation reaction

(C) $\quad Br_2(\ell) + 2OH^- \rightarrow Br^- + BrO^- + H_2O \qquad \Delta G° = -60.1 \text{ kJ mol}^{-1}$

In turn, hypobromite is unstable with respect to the disproportionation reaction

(D) $\quad BrO^- + 2HBrO \rightarrow BrO_3^- + 2Br^- + 2H^+ \qquad \Delta G° = -7.9 \text{ kJ mol}^{-1}$

Similarly to the chlorate formation reaction (q.v.), the formation of bromate according to reaction (D) requires the presence of both the BrO⁻ and the HBrO species, so that the reaction is fast only within a specific narrow range of pH values. This pH range in the industrial production of bromate [starting from the anodic oxidation of bromides to bromine, associated with reaction (C)] is 9 to 10. Solid bromates are quite stable. Bromic acid is a faster oxidizing agent than chloric acid, and in strongly acid solutions it oxidizes water to oxygen.

Perbromates [1] can be synthesized electrolytically, chemically, or radiochemically, but they are best obtained through the chemical method by oxidation of bromates in alkaline solutions by molecular fluorine. Although thermodynamically unstable with respect to the oxidation of water, concentrated solutions of perbromic acid can be obtained.

E. Potential Diagrams for Bromine

Acid Solution

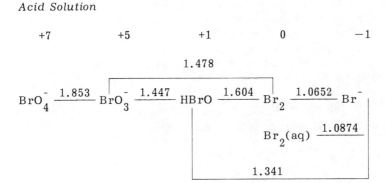

Basic Solution

$$\underset{+7}{BrO_4^-} \xrightarrow{1.025} \underset{+5}{BrO_3^-} \xrightarrow{0.492} \underset{+1}{BrO^-} \xrightarrow{0.455} \underset{0}{Br_2} \xrightarrow{1.0652} \underset{-1}{Br^-}$$

with 0.584 spanning BrO$_3^-$ to Br$_2$, and 0.766 spanning BrO$^-$ to Br$^-$.

The Pourbaix potential-pH diagram of bromine, updated and completed by the introduction of the electrode reactions involving the perbromates, was published recently [4].

REFERENCES

1. E. H. Appelman, J. Am. Chem. Soc., *90*, 1900 (1968); Inorg. Chem., *8*, 223 (1969).
2. G. Jones and S. Baeckström, J. Am. Chem. Soc., *56*, 1517, 1524 (1934).
3. T. Mussini and G. Faita, Ric. Sci., *36*, 175 (1966).
4. T. Mussini and G. Faita, in *Encyclopedia of the Electrochemistry of the Elements*, A. J. Bard, ed., Vol. 1, Marcel Dekker, New York, 1973, pp. 65-68.
5. H. A. Liebhafsky, J. Am. Chem. Soc., *56*, 1500 (1934).

IV. IODINE AND ASTATINE*

PIER GIORGIO DESIDERI, LUCIANO LEPRI, and DANIELA HEIMLER
University of Florence, Florence, Italy

A. Iodine

1. Oxidation States

Iodine is a solid element with a diatomic molecule. It has an oxidation number of -1 in hydriodic acid and its salts, $+1$ in hypoiodous acid and in complexes with the other halogens and pseudohalogens, and in $+3$ in iodine trichloride. It has the oxidation number of $+5$ in I_2O_5 and iodic acid and its salts and $+7$ in periodic acid and its derivatives. Also, the oxides I_2O_4 and I_4O_9 are known; for them, however, the real oxidation number of iodine has not been established.

2. Iodine-Iodide Couple

The standard potentials of this couple are

$$I_2(c) + 2e^- \to 2I^- \quad E° = 0.5355 \text{ V} \quad [1]$$

$$I_2(aq) + 2e^- \to 2I^- \quad E° = 0.621 \text{ V}$$

Although iodine is not a powerful oxidizing agent, its action is generally rapid. However, voltametric studies of the oxidation of iodide to iodine on platinum and graphite electrodes showed reversibility of the iodine-iodide couple only at low iodide concentrations ($c_{I^-} < 5 \times 10^{-4}$ M) [2]. In addition to the thermodynamic data of Table 4, a value of $\Delta H° = -56.27$ kJ mol^{-1} was found for iodide [3] by calorimetric measurements.

3. Triiodide-Iodide Couple

Iodine gives rise with iodide to the I_3^- species and in more concentrated iodine solutions also to the I_5^- species. From thermodynamic data the standard potential of the half-reaction

$$I_3^- + 2e^- \to 3I^-$$

is $E° = 0.536$ V. The value of the stability constant of I_3^- calculated from thermodynamic data ($K = 714$) is very close to that obtained from distribution, spectrophotometric, and conductometric methods [4].

4. Hypoiodite-Iodide Couple

The equilibrium constant of the reaction

$$I_2(aq) + H_2O \to HI)(aq) + I^- + H^+$$

is 4.6×10^{-13} [5]. On the basis of this value the Gibbs energy of HIO(aq) is -98.67 kJ mol^{-1}. Therefore, for the half-reaction

---

*Reviewer: J. Coetzee, University of Pittsburgh, Pittsburgh, Pennsylvania.

TABLE 4 Thermodynamic Data on Iodine[a]

Formula	$\Delta H°$/kJ mol^{-1}	$\Delta G°$/kJ mol^{-1}	$S°$/J mol^{-1} K^{-1}
$I(g)(^2P_{3/2})$	106.62	70.15	180.68
$I_2(c)$	0.0	0.0	116.73
$I_2(g)$	62.24	19.37	260.58
$I_2(aq)$	20.92	16.43	—
I^-	−55.94	−51.67	111.3
I_3^-	−51.88	−51.50	238.91
$HI(g)$	25.94	1.30	206.33
$HI(aq)$	3.05	3.14	122.76
$HIO(aq)$	—	−98.67	—
IO^-	—	−37.96	—
$I^+ \cdot H_2O$	—	−89.99	—
$ICl(c)$	−33.60	−13.56	102.51
$ICl(g)$	17.50	−5.57	247.37
$ICl(aq)$	—	−14.85	—
ICl_2^-	—	−158.70	—
$IF(g)$	51.12	24.27	236.20
$IBr(g)$	40.61	3.81	258.37
$IBr(aq)$	—	−3.81	—
IBr_2^-	—	−121.29	—
$ICN(aq)$	—	179.20	—
$ICl_3(c)$	−89.29	−22.65	169.01
$HIO_3(aq)$	—	−139.34	—
IO_3^-	—	−134.94	115.9
IO_4^-	−162.34	−53.14	—
$H_5IO_6(aq)$	−790.36	−537.14	—
$H_4IO_6^-$	−779.48	−518.35	—
$H_3IO_6^{2-}$	—	−480.11	—
$H_2IO_6^{3-}$	—	−410.47	—

[a] Data from Refs. 1, 8, 11, and 23 are calculated within this work.

$$HIO(aq) + H^+ + 2e^- \rightarrow I^- + H_2O \qquad E° = 0.985 \text{ V}$$

Hypoiodous acid is a very weak acid which behaves as an amphiprotic electrolyte:

$$HIO(aq) + H^+ \rightarrow I^+ \cdot H_2O \qquad K = 3.02 \times 10^{-2} \quad [6]$$

$$HIO(aq) \rightarrow H^+ + IO^- \qquad K = 2.29 \times 10^{-11} \quad [7]$$

The cation $I^+ \cdot H_2O$ exists in large amounts only in strongly acid solutions. From the equilibrium constants of the two above-mentioned reactions, the Gibbs energy of $I^+ \cdot H_2O$ is -89.989 kJ mol^{-1} and that of hypoiodite is -37.965 kJ mol^{-1}. Then, for the half-reactions

$$I^+ \cdot H_2O + 2e^- \rightarrow I^- + H_2O \qquad E° = 1.030 \text{ V}$$

$$IO^- + H_2O + 2e^- \rightarrow I^- + 2OH^- \qquad E° = 0.472 \text{ V}$$

Hypoiodous acid and its salts are unstable with respect to disproportionation:

$$5I^+ \cdot H_2O \rightarrow 2I_2(aq) + IO_3^- + 6H^+ + 2H_2O \qquad \Delta G° = -78 \text{ kJ mol}^{-1}$$

$$5HIO(aq) \rightarrow 2I_2(aq) + IO_3^- + 2H^+ + 2H_2O \qquad \Delta G° = -121 \text{ kJ mol}^{-1}$$

$$3IO^- \rightarrow 2I^- + IO_3^- \qquad \Delta G° = -134 \text{ kJ mol}^{-1}$$

5. Iodine Monochloride

Iodine, in the oxidation state of +1, gives rise in the presence of hydrochloric acid to ICl and ICl_2^-. From thermal data, the Gibbs energies of ICl(aq) and ICl_2^- are -14.853 kJ mol^{-1} and -158.699 kJ mol^{-1}, respectively. Applying these data to the reaction

$$ICl_2^- \rightarrow ICl(aq) + Cl^-$$

K is 6×10^{-3}, the same value found by Faull [9] from distribution measurements.

The standard potentials of the half-reactions are

$$2ICl(aq) + 2e^- \rightarrow I_2(c) + 2Cl^- \qquad E° = 1.20 \text{ V}$$

$$2ICl_2^- + 2e^- \rightarrow I_2(c) + 4Cl^- \qquad E° = 1.07 \text{ V}$$

From the equilibrium constant of the reaction

$$I_2(aq) + Cl^- \rightarrow I_2Cl^- \qquad K = 1.66 \quad [10]$$

the Gibbs energy of I_2Cl^- is -115.98 kJ mol^{-1}. Then for the half-reactions

$$2ICl(aq) + 2e^- \rightarrow I_2Cl^- + Cl^- \qquad E° = 1.12 \text{ V}$$

$$2ICl_2^- + 2e^- \rightarrow I_2Cl^- + 3Cl^- \qquad E° = 0.995 \text{ V}$$

From the voltametric curves for the oxidation of iodide in the presence of large amounts of hydrochloric acid the ICl_3^{2-} species has also been inferred and the formation constant of this complex has been calculated ($K = 0.346$) [2].

6. Iodine Trichloride

The Gibbs energy of iodine trichloride, $ICl_3(c)$, is -22.648 kJ mol^{-1} [12]. The standard potentials of its half-reactions are

$$ICl_3(c) + 2e^- \rightarrow ICl(c) + 2Cl^- \qquad E° = 1.312 \text{ V}$$

$$2ICl_3(c) + 6e^- \rightarrow I_2(c) + 6Cl^- \qquad E° = 1.281 \text{ V}$$

Solid ICl_3 decomposes to give only ICl and Cl_2 in the vapor. In concentrated hydrochloric acid solutions, the only important species present are ICl_3 and ICl_4^-; such species are, however, unstable and decompose to $ICl(aq)$ or ICl_2^- and chlorine.

7. Iodine Monobromide

From the equilibrium constants of the following reactions determined by Faull [9]:

$$IBr_2^- \rightarrow IBr(aq) + Br^- \qquad K = 2.7 \times 10^{-13}$$

$$2IBr(aq) \rightarrow I_2(aq) + Br_2(aq) \qquad K = 1.8 \times 10^{-5}$$

the Gibbs energies of IBr(aq) and IBr_2^- are -3.807 and -121.286 kJ mol^{-1}, respectively, and the potentials of the half-reactions are

$$2IBr(aq) + 2e^- \rightarrow I_2(c) + 2Br^- \qquad E° = 1.026 \text{ V}$$

$$2IBr_2^- + 2e^- \rightarrow I_2(c) = 4Br^- \qquad E° = 0.874 \text{ V}$$

In the presence of hydrochloric acid IBr gives rise to $IBrCl^-$:

$$IBr(aq) + Cl^- \rightarrow IBrCl^- \qquad K = 45.45 \text{ [9]}$$

The Gibbs energy of $IBrCl^-$ is -144.432 kJ mol^{-1}. Then

$$2IBrCl^- + 2e^- \rightarrow I_2(c) + 2Br^- + 2Cl^- \qquad E° = 0.928 \text{ V}$$

From the reaction

$$I_2Br^- \rightarrow I_2(aq) + Br^- \qquad K = 8.19 \times 10^{-2} \text{ [13]}$$

the Gibbs energy of I_2Br^- is -92.596 kJ mol^{-1}. Therefore,

$$2IBr(aq) + 2e^- \rightarrow I_2Br^- + Br^- \quad E^\circ = 0.973 \text{ V}$$

$$2IBr_2^- + 2e^- \rightarrow I_2Br^- + 3Br^- \quad E^\circ = 0.821 \text{ V}$$

8. Iodine Monocyanide

Yost and Stone [14] have applied their formation constants for the diiodocyanate

$$ICN(aq) + I^- \rightarrow I_2CN^- \quad K = 2.50$$

and the iodine dicyanide

$$ICN(aq) + CN^- \rightarrow I(CN)_2^- \quad K = 1.17$$

to Kovach's data [15] and obtained a corrected average value of 1.50 for the constant of the reaction

$$I_2(aq) + HCN(aq) \rightarrow ICN(aq) + I^- + H^+$$

From this value ΔG° of ICN(aq) is 179.2 kJ mol^{-1}, taking the Gibbs energy of HCN(aq) 112.13 kJ mol^{-1} [1]. The standard potentials of the couples are

$$2ICN(aq) + 2H^+ + 2e^- \rightarrow I_2(c) = 2HCN(aq) \quad E^\circ = 0.695 \text{ V}$$

$$2ICN(aq) + 2H^+ + 2e^- \rightarrow I_2(aq) + 2HCN(aq) \quad E^\circ = 0.609 \text{ V}$$

From emf measurements [16] the above-mentioned potentials are 0.711 V and 0.625 V, respectively.

9. Iodate

Iodate is reduced to hypoiodous acid, iodine, or iodide depending on the pH of the solution. The dissociation constant of HIO$_3$ is $K = 0.169$ [17]. The Gibbs energy of IO$_3^-$ from emf measurements is -134.938 kJ mol^{-1} [18] and therefore that of HIO$_3$ is -139.344 kJ mol^{-1}. The standard potentials of the half-reactions are

$$HIO_3(aq) + 5H^+ + 4e^- \rightarrow I^+ \cdot H_2O + 2H_2O \quad E^\circ = 1.098 \text{ V}$$

$$2IO_3^- + 12H^+ + 10e^- \rightarrow I_2(aq) + 6H_2O \quad E^\circ = 1.195 \text{ V}$$

$$IO_3^- + 3H_2O + 6e^- \rightarrow I^- + 6OH^- \quad E^\circ = 0.257 \text{ V}$$

For the iodate-iodine couple $E^\circ = 1.1942$ V has been obtained from emf measurements [19]. In acid media HIO$_3$ and IO$_3^-$ are strong and rapid oxidizing agents. They oxidize quantitatively iodide to iodine, and this reaction is important in the preparation of standard iodine solutions and in the analytical determination of iodide, iodate, and acid content. In concentrated hydrochloric acid solutions iodine is oxidized to ICl$_2^-$ by iodate. The standard potential of the half-reaction

The Halogens

$$HIO_3(aq) + 5H^+ + 2Cl^- + 4e^- \to ICl_2^- + 3H_2O$$

is $E° = 1.214$ V. Iodate behaves as chlorate and bromate do, but it is more stable than these last two anions toward its own oxidation and reduction, as the positive Gibbs energies of the following reactions demonstrate:

$$7IO_3^- + 7H^+ + 9H_2O \to I_2(aq) + 5H_5IO_6(aq) \qquad \Delta G° = 485 \text{ kJ mol}^{-1}$$

$$4IO_3^- + 3OH^- + 3H_2O \to I^- + 3H_3IO_6^{2-} \qquad \Delta G° = 251 \text{ kJ mol}^{-1}$$

Iodate is formed even by disproportionation of iodous acid according to the reaction

$$5HIO_2(aq) \to I_2(aq) + 3IO_3^- + 3H^+ + H_2O$$

10. Periodate-Iodate Couple

Of the solid acids of I(VII), H_5IO_6 and HIO_4, only the first is in equilibrium with water; its solubility is quite high. In aqueous solutions the species are H_5IO_6, $H_4IO_6^-$, IO_4^-, $H_3IO_6^{2-}$, $H_2IO_6^{3-}$, and $H_2I_2O_{10}^{4-}$ rather than $I_2O_9^{4-}$. The concentrations of the species are controlled by the following equilibria [20,21]:

$$H_5IO_6 \to H_4IO_6^- + H^+ \qquad K_1 = 5.1 \times 10^{-4}$$

$$H_4IO_6^- \to IO_4^- + 2H_2O \qquad K_D = 40$$

$$H_4IO_6^- \to H_3IO_6^{2-} + H^+ \qquad K_2 = 2 \times 10^{-7}$$

$$H_3IO_6^{2-} \to H_2IO_6^{3-} + H^+ \qquad K_3 = 6.3 \times 10^{-13}$$

$$2H_3IO_6^{2-} \to H_2I_2O_{10}^{4-} + 2H_2O \qquad K_d = 820 \qquad \Delta H° = -11.71 \text{ kJ mol}^{-1}$$

From the value of the Gibbs energy of IO_4^- reported by Brown [22] and from the experimental values of K_1, K_2, K_D, and K_3, the Gibbs energies of $H_3IO_6^{2-}$, $H_5IO_6(aq)$, $H_4IO_6^-$, and $H_2IO_6^{3-}$ have been calculated (see Table 4). The standard potentials of the half-reactions are

$$H_5IO_6(aq) + 2e^- + 2H^+ \to HIO_3(aq) + 3H_2O \qquad E° = 1.626 \text{ V}$$

$$H_5IO_6(aq) + 2e^- + H^+ \to IO_3^- + 3H_2O \qquad E° = 1.603 \text{ V}$$

$$H_4IO_6^- + 2e^- + 2H^+ \to IO_3^- + 3H_2O \qquad E° = 1.70 \text{ V}$$

$$H_3IO_6^{2-} + 2e^- \to IO_3^- + 3OH^- \qquad E° = 0.656 \text{ V}$$

$$H_2IO_6^{3-} + 2e^- + H_2O \to IO_3^- + 4OH^- \qquad E° = 0.603 \text{ V}$$

Periodic acid is a strong oxidizing agent but its action is slow. It oxidizes quantitatively manganous ion to permanganate in warm acid solutions and it is used for the selective oxidation of polyhydric organic compounds. The periodates are quite oxidizing even in alkaline solutions. Other than to iodate, in the presence of strong reducing agents the reduction goes on to iodide. Periodate can be obtained by electrochemical oxidation of iodate on lead dioxide anodes, on which the reaction is fast [2].

11. Potential Diagrams for Iodine

Acid Solution

$$H_5IO_6 \xrightarrow{1.60} \left[IO_3^- \xrightarrow{1.13} HIO(aq) \xrightarrow{1.44} I_2(c) \right] \xrightarrow{0.535} I^-$$

with $IO_3^- \xrightarrow{1.20} I_2(c)$ overall and $IO_3^- \xrightarrow{1.21} ICl_2^- \xrightarrow{1.07} I_2(c)$

Basic Solution

$$H_3IO_6^{2-} \xrightarrow{0.65} \left[IO_3^- \xrightarrow{0.15} IO^- \xrightarrow{0.42} I_2(c) \right] \xrightarrow{0.535} I^-$$

with $IO_3^- \xrightarrow{+0.26} I_2(c)$ and $IO^- \xrightarrow{0.48} I^-$

REFERENCES

1. W. M. Latimer, *Oxidation Potentials*, 2nd ed., Prentice-Hall, New York, 1952.
2. P. G. Desideri, L. Lepri, and D. Heimler, in *Encyclopedia of the Electrochemistry of the Elements*, A. J. Bard, ed., Vol. 1, Marcel Dekker, New York, 1973.
3. A. F. Vorob'ev, A. F. Broier, and S. M. Skuratov, Dokl. Akad. Nauk SSSR, *173*, 385 (1967).
4. A. I. Popov, in *Halogen Chemistry*, V. Gutman, ed., Vol. I, Academic Press, New York, 1967.
5. G. Horiguchi and H. Hagesawa, Bull. Inst. Phys. Chem. Res. (Tokyo), *22*, 661 (1943).
6. R. P. Bell and F. Gelles, J. Chem. Soc., 2734 (1951).
7. T. L. Allen and R. M. Keefer, J. Am. Chem. Soc., *77*, 2957 (1955).
8. J. C. McCoubrey, Trans. Faraday Soc., *51*, 743 (1955).
9. J. H. Faull, J. Am. Chem. Soc., *56*, 522 (1934).
10. D. L. Cason and H. N. Neumann, J. Am. Chem. Soc., *83*, 1822 (1961).
11. L. G. Cole and G. W. Eiverum, J. Chem. Phys., *20*, 1543 (1952).
12. R. H. Lamoreaux and W. F. Giauque, J. Phys. Chem., *73*, 755 (1969).
13. V. K. LaMer and M. H. Lewinsohn, J. Phys. Chem., *38*, 171 (1934).

14. D. M. Yost and W. E. Stone, J. Am. Chem. Soc., 55, 1889 (1933).
15. L. Kovach, Z. Phys. Chem., 80, 107 (1912).
16. D. F. Bowersox, E. A. Butler, and E. H. Swift, Anal. Chem., 28, 221 (1956).
17. C. A. Kraus and H. C. Parker, J. Am. Chem. Soc., 44, 2429 (1922).
18. W. O. Lundberg, C. S. Vestling, and J. E. Ahlberg, J. Am. Chem. Soc., 59, 264 (1937).
19. R. D. Spitz and H. A. Liefbafsky, J. Electrochem. Soc., 122, 363 (1975).
20. C. E. Crouthamel, A. M. Hayes, and D. S. Martin, J. Am. Chem. Soc., 73, 82 (1951).
21. G. J. Buist, W. C. P. Hipperson, and J. D. Lewis, J. Chem. Soc. A, 307 (1969).
22. M. G. Brown, Potential-pH diagrams for iodine compounds, Compt. Rend. 7th Meet. du CITCE (Lindau, 1955), Butterworth, London, 1957, pp. 244-256.
23. E. E. Mercer and T. Farrar, Can. J. Chem., 46, 2679 (1968).

B. Astatine

Astatine is a synthetic unstable element that can be worked with at concentrations of between 10^{-11} and 10^{-15} M by tracer scale experiments. Although the more stable isotope is ^{210}At (half-life 8.3 h), ^{211}At (half-life 7.2 h) is generally employed in the study of chemical properties. ^{211}At is prepared by the bombardment of bismuth with α particles with energy between 21 and 29 MeV.

Johnson and co-workers [1] pointed out that when astatine is dissolved in sulfuric acid and electrolyzed, no deposition at the cathode is observed until dichromate ion is added to the cell. The deposition potential in such a medium is 1.20 V and in nitric acid is 1.22 V.

The following oxidation states of astatine have been characterized: -1, 0, $+1$, $+3$, $+5$, and, more recently [2,3], $+7$. On the basis of the potential of the redox couples cited by Johnson and co-workers [1], Latimer [4] constructed the following potential schemes:

Acid Solution

$$\overset{+5}{HAtO_3} \xrightarrow{1.4} \overset{+1}{HAtO} \xrightarrow{0.7} \overset{0}{At_2} \xrightarrow{0.2} \overset{-1}{At^-}$$

Basic Solution

$$AtO_3^- \xrightarrow{0.5} AtO^- \xrightarrow{0.0} At_2 \xrightarrow{0.2} At^-$$

Appleman constructed the following potential scheme in 0.1 M perchloric acid solvent [5]:

$$AtO_3^- \xrightarrow{1.5} HatO \xrightarrow{1.0} At_2 \xrightarrow{0.3} At^-$$

Astatine gives rise to polyhalogen complexes such as AtI, AtBr, AtI_2^-, $AtIBr^-$, $AtICl^-$, $AtBr_2^-$, $AtCl_2^-$, and $AtCl_4^-$; for most of them the stability constants have been determined [6,7].

REFERENCES

1. G. L. Johnson, R. F. Leiniger, and E. Segrè, J. Chem. Phys., 17, 1 (1949).
2. E. Kroo, I. Dudova, and V. A. Khalkin, Magy. Kèm. Foly., 82, 1 (1976).
3. I. Dreyer, R. Dreyer, Yu. V. Norseev, and V. A. Chalkin, Radiochem. Radioanal. Lett., 33, 291 (1978).
4. W. M. Latimer, Oxidation Potentials, 2nd ed., Prentice-Hall, New York, 1952.
5. E. H. Appelmann, J. Am. Chem. Soc., 83, 805 (1961).
6. E. H. Appelmann, J. Phys. Chem., 65, 325 (1961).
7. L. Stein, in Halogen Chemistry, V. Gutman, ed., Vol. 1, Academic Press, New York, 1967, p. 191.

6
Sulfur, Selenium, Tellurium, and Polonium*

STEPHAN I. ZHDANOV Institute of the Ministry of Chemical Industry, Moscow, USSR

Oxygen, the first element in group VI, was considered in Chapter 4. The electronic structure of the atoms of sulfur, selenium, tellurium, and polonium is close to that of the inert gases following them in the periodic chart. These elements, like oxygen, tend to add two electrons to form a complete shell consisting of eight s and p electrons. With growing atomic weight and the respective increase in the distance between the atomic nuclei and the outer electrons, however, the stability of the compounds with the oxidation state -2 decreases. In contrast to oxygen, the valency of sulfur, selenium, tellurium, and polonium may differ from -2, since they can form bonds by giving away s and p electrons from the d orbitals.

On the whole, the group VI elements behave somewhat like nonmetals. The nonmetallic properties, however, become weaker as the atomic weight grows. Thus polonium and, to some extent, tellurium also display certain metallic properties. The group VI elements basically form covalent compounds. Thus one can speak about their oxidation states only formally.

The most stable compounds of the lighter group VI elements (except oxygen) are those where the oxidation states are +4 and +6; the oxidizing power of the latter grows with atomic weight. In the oxidation state +2 the elements form unstable compounds that decompose into the free element and a compound with the oxidation state +4 (except polonium). The elements cannot form compounds with odd oxidation numbers and completed electron octets, save for the compounds where two atoms of an element of this group share two electrons (i.e., Na_2S_2). Such compounds are, as a rule, less stable than those in which the oxidation state is even.

I. SULFUR

A. Oxidation States

In sulfides the oxidation state of sulfur is -2. By forming covalent chains, sulfur atoms may yield polysulfides S_x^{2-}. The suboxide SO with the oxidation state +2 is a highly reactive biradical, but certain sulfoxylates, the salts of H_2SO_2 corresponding to this oxide, were reported. The oxidation state of sulfur in thiosulfates $S_2O_3^{2-}$ and in pentathiomates $S_5O_6^{2-}$ also averages +2,

---

*Reviewer: J. David Stutts, Gannon University, Erie, Pennsylvania.

while in hydrosulfate $S_2O_4^{2-}$ it is equal to +3 and in sulfite SO_3^{2-}, to +4. Sulfurous acid in the free states does not exist. In dithionates $S_2O_6^{2-}$ the oxidation state of sulfur is +5 and in sulfates SO_4^{2-} it is equal to +6. Persulfate $S_2O_8^{2-}$ is a peroxy compound. In some compounds, such as trithionates $S_3O_6^{2-}$ and tetrathionates $S_4O_6^{2-}$, the average formal oxidation state of sulfur is a positive fractional number. It is convenient to divide all the oxygen-containing acids formed by sulfur into three groups of general formulas H_2SO_x (x = 2 to 5), $H_2S_2O_x$ (x = 2 to 8), and $H_2S_xO_6$ (x = 2 to 6). In compounds with halogens the oxidation states of sulfur are +1, +2, +4, +5, and +6.

The standard potentials of sulfur were for the most part calculated thermodynamically rather than found in direct experiments. Many redox systems of sulfur are irreversible. Therefore, the potential of an inert electrode immersed into the solution containing both the oxidized and the reduced forms is usually unsteady and the behavior of the electrode is inconsistent with the corresponding Nernst equation. The sulfate-sulfite system, for instance, is irreversible; the sulfite oxidation into sulfate occurs more readily than the inverse reaction. Such electrodes as sulfite-dithionate, sulfite-thiosulfate, thiosulfate-tetrathionate, and some others are also irreversible. However, the sulfite-dithionite electrode is a reversible system.

B. Sulfur-Sulfide Couple

The use of Gibbs energy values for $H_2S(g)$, $H_2S(aq)$, and HS^- accepted [1] by the National Bureau of Standards (NBS) (Table 1) leads to

$$S(s) + 2e^- + 2H^+ \rightarrow H_2S(aq) \qquad E° = 0.144 \text{ V}$$

$$S(s) + 2e^- + 2H^+ \rightarrow 2H^+ \rightarrow H_2S(g) \qquad E° = 0.174 \text{ V}$$

The accepted values of H_2S dissociation constants at 18°C in water [2] are equal to

$$H_2S(aq) \rightarrow H^+ + HS^- \qquad K_1 = 9.1 \times 10^{-8}$$

$$HS^- \rightarrow H^+ + S^{2-} \qquad K_2 = 1.2 \times 10^{-12}$$

For the reactions

$$S(s) + 2e^- \rightarrow S^{2-} \qquad E° = -0.447 \text{ V}$$

$$S(s) + 2e^- + H^+ \rightarrow HS^- \qquad E° = -0.062 \text{ V}$$

For the reaction

$$S(s) + 2e^- + H_2O \rightarrow HS^- + OH^-$$

the measured value is $E_B° = -0.52$ V [3]. According to the calculation, $E_B° = -0.476$ V.

TABLE 1 Thermodynamic Data on Sulfur[a]

Formula	Description	State	$\Delta H°$/kJ mol^{-1}	$\Delta G°$/kJ mol^{-1}	$S°$/J mol^{-1} K^{-1}
S	Rhombic	c	0.0	0.0	31.80
S	Monoclinic	c	0.377	0.188	32.43
S		g	278.81	238.28	167.711
S$^+$		g	1278.2	1232.8	163.510
S^{2-}		aq	33.05	86.31	−16.3
S$_2$		g	128.36	79.33	228.03
S$_2^{2-}$		aq	30.1	79.5	28.4
S$_3$		g	132.6		
S$_3^{2-}$		aq	25.9	73.6	65.5
S$_4$		g	136.8		
S$_4^{2-}$		aq	23.0	69.0	103.4
S$_5$		g	123.8		
S$_5^{2-}$		aq	21.3	65.7	140.6
S$_6$		g	102.5		
S$_7$		g	113.4		
S$_8$		g	102.30	49.66	431.0
SO		g	6.259	−19.836	221.84
SO$_2$		g	−296.830	−300.194	248.12

TABLE 1 (Continued)

Formula	Description	State	$\Delta H°/\text{kJ mol}^{-1}$	$\Delta G°/\text{kJ mol}^{-1}$	$S°/\text{J mol}^{-1} \text{K}^{-1}$
SO_2		aq	−322.980	−300.708	162.13
SO_3		g	−395.72	−371.08	256.72
SO_3^{2-}		aq	−634.5	−486.6	−26.7
SO_3^{2-}		aq	−909.27	−744.63	20.1
SO_4^{2-}		g	−109	−138.833	267.15
S_2O		aq	−652.3	(−518.8)	
$S_2O_3^{2-}$		aq	−753.5	−600.4	92
$S_2O_4^{2-}$		aq	(−973.2)	(−791)	
$S_2O_5^{2-}$		aq	−1198.3	(−966)	
$S_2O_6^{2-}$		aq	−1401.2		
$S_2O_7^{2-}$		aq	−1338.9	−1110.4	248.1
$S_2O_8^{2-}$		aq	−1199.6	(−958)	
$S_3O_6^{2-}$		aq	−1224.15	(−1022.2)	
$S_4O_6^{2-}$		aq	−1236.4	(−956.0)	
$S_5O_6^{2-}$		g	142.67	113.30	195.56
HS					

Sulfur, Selenium, Tellurium, and Polonium

HS^-	aq	−17.6	12.05	62.9
H_2S	g	−20.63	−33.56	205.69
H_2S	aq	−39.7	−27.87	129
H_2S_2	ℓ	−23.05		
H_2S_2	g	10.59		
H_2S_3	ℓ	−30.42		
H_2S_3	g	15.02		
H_2S_4	ℓ	−32.84		
H_2S_4	g	23.89		
H_2S_5	ℓ	−34.94		
H_2S_5	g	33.39		
H_2S_6	ℓ	−37.03		
HSO_3^-	aq	−629.22	−527.81	139.7
$HS_2O_4^-$	aq		−614.6	
HSO_4^-	aq	−887.34	−756.01	131.8
H_2SO_3	aq	−608.81	−537.90	232.2
H_2SO_4	ℓ	−813.989	−690.101	156.904
H_2SO_4	g	−743.9	−662.91	300.8
H_2SO_4	aq, Undiss.	−909.27	−744.63	20.1
$H_2S_2O_4$	aq		−616.7	
$H_2S_2O_6$	aq	−1198.3		

TABLE 1 (Continued)

Formula	Description	State	$\Delta H°/\text{kJ mol}^{-1}$	$\Delta G°/\text{kJ mol}^{-1}$	$S°/\text{J mol}^{-1} \text{K}^{-1}$
$H_2S_2O_7$		c	−1273.6	−1110.4	248.1
$H_2S_2O_8$		aq	−1338.9		
DS		g	144.8	115.9	201.42
D_2S		g	−24.3	−35.677	215.06
D_2SO_4		g	−751.4	−668.2	307.9
HDS		g	−22.6	−36.5	216.19
TS		g	144.9	116.2	204.81
T_2S		g	−25.8	−36.5	221.00
HTS		g	−23.4	−37.0	219.28
DTS		g	−23.8	−36.6	223.80
SF		g	76.132	47.882	228.03
SF_2		g	−137.36	−143.997	256.9
SF_4		g	−774.9	−741.4	291.9
SF_6		g	−1209	−1105.4	291.7
SF_6		aq	−1225.9	−1084.2	166.5
SOF_2		g	−713.51	−696.05	278.57
SO_2F_2		g	−858	−812	283.93
HSO_3F		ℓ	−778		
SCl		g	126.922	98.328	239.3

Sulfur, Selenium, Tellurium, and Polonium

Species	State			
SCl_2	l	-50.2		
SCl_2	g	-19.7		
SCl_4	l	-56.1		
S_2Cl_2	l	-59.4		
S_2Cl_2	g	-18.4	-31.8	331.4
S_3Cl_2	l	-51.9		
S_4Cl_2	l	-42.7		
SOCl	g	-105.345	-104.604	277.0
$SOCl_2$	g	-212.55	-198.3	310.03
SO_2Cl_2	g	-364.0	-320.1	311.83
SF_5Cl	g	-1048.1	-949.3	319.07
CS	g	234	184	210.5
CS_2	l	89.70	65.27	151.34
CS_2	g	117.36	67.15	237.73
COS	g	-142.09	-169.33	231.46
CNS^-	aq	76.44	92.68	144.3
HCNS	aq	76.44	92.68	144.3
HCNS	aq		97.53	
HNCS	g	127.6	113.0	247.7
N_4CNS	aq	-56.07	13.31	257.7
CH_3COSH*	g	-181.96	-154.01	313.21
CH_3NCS	g	131.0	144.4	289.91

(Thiocyanic acid ionized: CNS^-; Thiocyanic acid undiss.: HCNS; Isothiocyanic acid: HNCS; Thioacetic acid: CH_3COSH; Methyl isothiocyanate: CH_3NCS)

TABLE 1 (Continued)

Formula	Description	State	$\Delta H°/\text{kJ mol}^{-1}$	$\Delta G°/\text{kJ mol}^{-1}$	$S°/\text{J mol}^{-1}\text{K}^{-1}$
$(CH_3)_2SO_2$	Dimethyl sulfone	g	-371.1	-272.8	310.5
$(CH_3)_2SO$	Dimethyl fulfoxide	g	-150.46	-81.50	306.27
$(CH_3)_2S$	Dimethyl sulfide	g	-37.24	7.24	285.85
$(C_2H_5)_2S$*	Diethyl sulfide	g	-83.47	17.78	368.02
$(C_3H_7)_2S$*	Dipropyl sulfide	g	-125.35	33.22	448.36
$(C_4H_9)_2S$*	Dibutyl sulfide	g	-167.32	49.20	526.51
$CH_3SC_2H_5$*	Methylethylsulfide	g	-59.62	11.42	331.13
$CH_3SC_3H_7$*	Methylpropylsulfide	g	-81.76	18.40	371.71
$CH_3SC_4H_9$*	Methylbutylsulfide	g	-102.17	26.65	411.83
CH_3SSCH_3	Dimethyl disulfide	g	-23.60	15.23	336.64
$C_2H_5SSC_2H_5$*	Diethyl disulfide	g	-74.64	22.26	414.51
$C_3H_7SSC_3H_7$*	Dipropyldisulfide	g	-117.19	36.99	494.96
CH_3SH	Methanthiol	g	-22.34	-9.33	255.06
C_2H_5SH	Ethanthiol	g	-45.81	-4.39	296.10
C_3H_7SH*	Propanthiol	g	-67.86	2.18	336.39
C_4H_9SH*	Isobutanthiol	g	-88.07	11.09	375.22
C_6H_5SH*	Benzenethiol	g	111.55	147.61	336.85
C_4H_4S*	Thiophen	g	115.73	126.78	278.86
C_2H_4S	Thiacyclopropane	g	82.38	97.03	255.27
C_3H_6S*	Thiacyclobutane	g	61.13	107.49	285.22
C_4H_8S*	Thiacyclopentane	g	-33.81	46.02	309.36

[a]Values selected by the National Bureau of Standards [1] are in italic; the values from Ref. 9 are not specified; values on sulfur organic compounds, indicated by an asterisk, are from Ref. 10; the values from Ref. 11 are given in parentheses.

Sulfur, Selenium, Tellurium, and Polonium

Since sulfur dissolves in excessive sulfide to form polysulfides, it has proved impossible to measure the standard potential of this system experimentally. Measurements performed by Allen and Hickling [4] have shown that the potential of an inert electrode immersed into the solution 1 M NaOH + 1 M Na₂S + 1 M S does not depend on time, on agitation of the solution, or on the electrode material, but on changes with concentration of S, Na₂S, and Na₂S$_x$. The normal potential of this system differs considerably from the calculated value.

Under the action of iodine, Fe^{3+} ions and hot concentrated nitric acid hydrogen sulfide undergo rapid quantitative oxidation into sulfur. Stronger oxidants such as bromide and permanganate further oxidate sulfur to sulfates.

C. Polysulfides

The following values of standard potentials have been calculated with the use of the values of Gibbs energy of formation of polysulfide anions:

$$S_2^{-2} + Se^- \rightarrow 2S^{2-} \qquad E_B^\circ = -0.483 \text{ V}$$

$$S_2^{2-} + 2e^- + 2H^+ \rightarrow 2HS^- \qquad E^\circ = 0.287 \text{ V}$$

$$2S_3^{2-} + 2e^- \rightarrow 3S_2^{2-} \qquad E_B^\circ = -0.473 \text{ V}$$

$$S_3^{2-} + 4e^- + 3H^+ \rightarrow 3HS^- \qquad E^\circ = 0.097 \text{ V}$$

$$3S_4^{2-} + 2e^- \rightarrow 4S_3^{2-} \qquad E_B^\circ = -0.453 \text{ V}$$

$$S_4^{2-} + 6e^- + 4H^+ \rightarrow 4HS^- \qquad E^\circ = 0.036 \text{ V}$$

$$4S_5^{2-} + 2e^- \rightarrow 5S_4^{2-} \qquad E_B^\circ = -0.426 \text{ V}$$

$$S_5^{2-} + 8e^- + 10H^+ \rightarrow 5H_2S(g) \qquad E^\circ = 0.303 \text{ V}$$

$$S_5^{2-} + 8e^- + 5H^+ \rightarrow 5HS^- \qquad E^\circ = 0.007 \text{ V}$$

$$5S(s) + 2e^- \rightarrow S_5^{2-} \qquad E_B^\circ = -0.340 \text{ V}$$

D. Sulfur-Sulfite Couple

Modern physical methods have shown that aqueous solutions of SO_2 contain virtually no sulfurous acid, H_2SO_3. Therefore, the reaction of acid dissociation in such solutions ought to be written as

$$SO_2 \cdot xH_2O = H^+ + HSO_3^- + (x-1)H_2O$$

However, the difference in writing the formula does not affect the values of thermodynamic functions and potentials. Thus, to avoid uncertainty, below we use the more customary routine formula of SO_2 in acidic aqueous solutions.

$$H_2SO_3 \rightarrow H^+ + HSO_3^- \qquad K_1 = 1.7 \times 10^{-2}$$

$$HSO_3^- \rightarrow H^+ + SO_3^{2-} \qquad K_2 = 5 \times 10^{-6} \quad [2]$$

$$H_2SO_3 + 4e^- + 4H^+ \rightarrow S(s) + 3H_2O \qquad E° = 0.500 \text{ V}$$

$$SO_2(g) + 4e^- + 4H^+ \rightarrow S(s) + 2H_2O \qquad E° = 0.451 \text{ V}$$

$$SO_3^{2-} + 4e^- + 3H_2O \rightarrow S(s) + 6OH^- \qquad E°_B = -0.659 \text{ V}$$

The values of these potentials indicate that the aqueous acidic solution of SO_2 is a strong oxidizer; the oxidizing power of the sulfite ion is considerably lower. The SO_3^{2-} anions (alkaline solutions of SO_2) cannot be reduced cathodically, while the HSO_3^- anions (pH 6 to 3) may be reduced to dithionite $S_2O_4^{2-}$. In more acidic solution the reduction of SO_2 results in hydrosulfurous acid and elementary sulfur ($H_2S_2O_4 + S$), and in very acidic solutions sulfanes H_2S_x are formed [5]. The interaction of sulfites with reducing agents apparently obeys the same chemical mechanism as that of their electrochemical reduction.

E. Sulfate-Sulfite Couple

The Gibbs energies of formation of H_2SO_4(aq), HSO_4^-, and SO_4^{2-} have been determined with good accuracy (the values accepted by the NBS and included in reference tables). Hence the following equations may be written:

$$HSO_4^- \rightarrow H^+ + SO_4^{2-} \qquad K = 1.26 \times 10^{-2}$$

$$SO_4^{2-} + 2r^- + 4H^+ \rightarrow H_2SO_3 + H_2O \qquad E° = 0.158 \text{ V}$$

$$SO_4^{2-} + 2e^- + H_2O \rightarrow SO_3^{2-} + 2OH^- \qquad E°_B = -0.936 \text{ V}$$

Thus in 1 M solution sulfuric acid is a poor oxidant which is far weaker than sulfurous acid. However, concentrated sulfuric acid, especially when heated, is a stronger oxidant than sulfurous acid. This change in the properties of the sulfate-sulfite system is apparently due to two facts. On the one hand, the increase in temperature and the corresponding rapid increase in volatility of the resulting SO_2 accelerate the reaction of reduction of H_2SO_4. On the other hand, at higher temperatures and concentrations the thermodynamic properties of the sulfate-sulfite system components change. When passing from dilute sulfuric acid solutions to concentrated ones the ionic composition changes substantially. The concentration of SO_4^{2-} ions becomes negligible, the HSO_4^- ions dominate the solution, hydrogen ions solvated by sulfuric acid ($H_3SO_4^+$) appear, the activity of hydrogen ions grows sharply (much more rapidly than concentration), and the activity of water decreases. Quantitative data on the potential of the sulfate-sulfite system in concentrated sulfuric acid solutions are unavailable. A more detailed consideration of redox properties of the sulfate-sulfite couple must evidently allow also for the changes in the properties of second redox system, which interacts with the sulfate-sulfite system when one passes from standard conditions to the solutions in concentrated sulfuric acid. At present there is vast evidence that even weak bases form protonic complexes in very acidic media [6-8]. This, of course, must affect the redox properties of compounds. It is likely that at elevated temperatures sulfuric acid undergoes thermal decomposition into H_2O and SO_3, thus passing into an oleumlike state. Experiments on electrolysis of oleum on a smooth platinum cathode have shown that the rate of SO_2 formation is directly proportional to the concentration of SO_3 in oleum [5].

Electrochemical oxidation of sulfite on a platinum anode in neutral and alkaline solutions results in sulfate and dithionate $S_2O_6^{2-}$, whereas SO_2 oxidation in acidic solutions yields only sulfuric acid. Of many mechanisms put forward for the oxidation of sulfite, two seem the most plausible: the electrochemical mechanism involving slow loss of the first electron, and the mechanism postulating the participation of platinum oxide [5]. Similar ideas were developed by Taube et al. for homogeneous interaction of sulfite with oxygen-containing oxidants [12]. These workers suggest two mechanisms. According to the first one, good for the reaction of sulfite oxidation by nitrite, the reducing agent loses electrons and then adds oxygen supplied by the solvent molecules. During the reaction such intermediates as $HON(SO_3)_2^{2-}$ or $N(SO_3)_2^{3-}$ are formed. They may be isolated with high yields. The alternative mechanism involves direct transition of oxygen atoms. As proved by the labeled atoms technique (^{18}O), this mechanism is valid in the case of sulfite oxidation by oxygen-containing halogen anions (e.g., chlorate ClO_3^-).

F. Thiosulfate and Tetrathionate

For the sulfite-thiosulfate systems the following potentials were calculated with the use of reliable values of Gibbs energies:

$$2H_2SO_3 + 4e^- + 2H^+ \rightarrow S_2O_3^{2-} + 3H_2O \qquad E^\circ = 0.400 \text{ V}$$

$$2HSO_3^- + 4e^- + 4H^+ \rightarrow S_2O_3^{2-} + 3H_2O \qquad E^\circ = 0.453 \text{ V}$$

$$2SO_3^{2-} + 4e^- + 6H^+ \rightarrow S_2O_3^{2-} + 3H_2O \qquad E^\circ = 0.666 \text{ V}$$

$$2SO_3^{2-} + 4e^- + 3H_2O \rightarrow S_2O_3^{2-} + 6OH^- \qquad E^\circ_B = -0.576 \text{ V}$$

Many oxidizing agents convert thiosulfate into tetrathionate $S_4O_6^{2-}$. In the manner described above, the following potential values have been obtained for the sulfite-tetrathionate and tetrathionate-thiosulfate couples:

$$4H_2SO_3 + 6e^- + 4H^+ \rightarrow S_4O_6^{2-} + 6H_2O \qquad E^\circ = 0.507 \text{ V}$$

$$4SO_2(g) + 6e^- + 4H^+ \rightarrow S_4O_6^{2-} + 2H_2O \qquad E^\circ = 0.511 \text{ V}$$

$$4HSO_3^- + 6e^- + 8H^+ \rightarrow S_4O_6^{2-} + 6H_2O \qquad E^\circ = 0.577 \text{ V}$$

$$4SO_3^{2-} + 6e^- + 12H^+ \rightarrow S_4O_6^{2-} + 6H_2O \qquad E^\circ = 0.862 \text{ V}$$

$$S_4O_6^{2-} + 2e^- \to 2S_2O_3^{2-} \qquad E° = 0.080 \text{ V}$$

Comparison of potentials for the sulfate-tetrathionate and sulfate-sulfite couples indicates that any oxidant capable of oxidizing tetrathionate into sulfurous acid should inevitably oxidize the acid further into sulfate. Incidentally, the potential of the sulfite-tetrathionate system is very close to that of the iodine-iodide system. Since iodine does not oxidize tetrathionate, it is possible that at the first stage of $S_4O_6^{2-}$ oxidation the potential is higher than the overall potential of the process.

Connick and Awtrey [13], who studied the mechanism of thiosulfate oxidation by I_3^- ions, have concluded that iodine and I_3^- ions react rapidly with thiosulfate ion to yield $S_2O_3I^-$ ion as an intermediate. This ion also reacts rapidly with thiosulfate ion to form tetrathionate and iodide ions. If the initial ratio between thiosulfate and iodine is less than 2, a certain amount of $S_2O_3I^-$ remains in the solution after these two rapid reactions consume all the thiosulfate. The $S_2O_3I^-$ ions either react with each other to form tetrathionate ions and iodine or get oxidized by iodine to form sulfate and iodine ions. Sorum and Edwards [14] suggest that thiosulfate oxidation by persulfate is a free-radical chain reaction.

Thiosulfate may be oxidized to tetrathionate by means of electrolysis of alkali metal thiosulfates, the highest current efficiency being observed at low concentration of hydrogen ions (pH 5 to 7, platinized platinum, current density 0.2 A cm^{-2}). The current efficiency of electrolysis of thiosulfate ethylene glycol solutions into tetrathionate reaches almost 100%. The process always yields some sulfate, whose relative amount depends on the presence of certain compounds in the electrolyte. For instance, the sulfate/tetrathionate ratio is increased by molybdate. The mechanism of this reaction is not simple. It is likely that the added salts of metals of variable valency act as electron carriers. Thiosulfate oxidation potential is much higher than the calculated reversible potential; the overvoltage amounts to at least 1 V.

G. Sulfoxylic Acid

Salts of sulfoxylic acid H_2SO_2 have been reported. The acid itself, however, is unstable and decomposes according to the reaction

$$2H_2SO_2 \to S_2O_3^{2-} = 2H^+ + H_2O$$

Using the values of Gibbs energies of thiosulfate and water, one can easily show that the Gibbs energy of sulfoxylic acid is less negative than -378 kJ mol^{-1} and therefore

$$H_2SO_2 + 2e^- + 2H^+ \to S + 2H_2O \qquad E° > 0.5 \text{ V}$$

$$H_2SO_3 + 2e^- + 2H^+ \to H_2SO_2 + H_2O \qquad E° < 0.4 \text{ V}$$

H. Hydrosulfurous Acid

The constant of the first stage of its dissociation is high:

$$H_2S_2O_4 \to H^+ + HS_2O_4^- \qquad K_1 = 0.45$$

$$HS_2O_4^- \to H^+ + S_2O_4^{2-} \qquad K_2 = 3.16 \times 10^{-3}$$

The following values of potentials have been calculated with the use of selected values [1] of the energy of formation of $HS_2O_4^-$ and $S_2O_4^{2-}$

$$2H_2SO_3 + 2e^- + H^+ \to HS_2O_4^- + 2H_2O \qquad E° = -0.068 \text{ V}$$

$$2HSO_3^- + 2e^- + 3H^+ \to HS_2O_4^- + 2H_2O \qquad E° = 0.173 \text{ V}$$

$$2HSO_3^- + 2e^- + 2H^+ \to S_2O_4^{2-} + 2H_2O \qquad E° = 0.099 \text{ V}$$

$$2SO_3^{2-} + 2e^- + 4H^+ \to S_2O_4^{2-} + 2H_2O_4^{2-} + 2H_2O \qquad E° = 0.526 \text{ V}$$

$$2SO_3^{2-} + 2e^- + 2H_2O \to S_2O_4^{2-} + 4OH^- \qquad E° = -1.13 \text{ V}$$

Thus hydrosulfite in alkaline solutions is an active reducing agent. Being unstable, it decays according to an equation

$$2S_2O_4^{2-} + H_2O \to S_2O_3^{2-} + 2HSO_3^- \qquad \Delta G° = -136.4 \text{ kJ mol}^{-1}$$

The rate of decomposition is higher in acidic solutions.

Hydrosulfite may be obtained by means of cathodic reduction of HSO_3^- and SO_2 at pH ~ 3 to 6. The solution should not be too acidic to avoid the above-mentioned rapid decomposition. Nor should it be too alkaline, for in such a solution no reduction of SO_3^{2-} can be attained. The decomposition of hydrosulfite may be retarded by cooling and agitating the electrolyte as well as by adding Na_2SiO_3, NaF, or $HCOH$. The yield of hydrosulfite may be increased by passing an inert gas (CO_2 or N_2) through the electrolyte, by appropriate selection of the cathode material (a metal with high overvoltage of hydrogen), of $NaHSO_3$ concentration, current density, and so on. One can manage to prepare saturated solutions which yield a precipitate $Na_2S_2O_4 \cdot 2H_2O$. Studies on SO_2 reduction in dimethyl sulfoxide have shown that the electrode process involves the stage of reversible electron transfer, resulting in $SO_2^{\cdot-}$ radical anions and the stage of dimerization of these particles into dithionite anions $S_2O_4^{2-}$. Indeed, in hydrosulfite solutions the equilibrium

$$2SO_2^{\cdot-} = S_2O_4^{2-}$$

exists. Its equilibrium constant is $K \simeq 10^{-5}$ at 55°C [5].

I. Thionic Acids

The Gibbs energies of polythionate ions from $S_2O_6^{2-}$ to $S_5O_6^{2-}$ were calculated from the enthalpies of formation and approximate entropies. Using these values of Gibbs energy, one can find a number of important redox potentials.

The average oxidation state of sulfur in dithionic acid is +5, and the potentials of the following systems correspond to the transition from this oxidation state to +4 and +6 states:

$$S_2O_6^{2-} + 2e^- + 4H^+ \rightarrow 2H_2SO_3 \quad E° = 0.569 \text{ V}$$

$$S_2O_6^{2-} + 2e^- + 2H^+ \rightarrow 2HSO_3^- \quad E° = 0.464 \text{ V}$$

$$S_2O_6^{2-} + 2e^- \rightarrow 2SO_3^{2-} \quad E° = 0.037 \text{ V}$$

$$2SO_4^{2-} + 2e^- + 4H^+ \rightarrow S_2O_6^{2-} + 2H_2O \quad E° = -0.253 \text{ V}$$

Accordingly, for the process of $S_2O_6^{2-}$ decomposition we obtain

$$S_2O_6^{2-} + H_2O \rightarrow H_2SO_3 + SO_4^{2-} \quad E° = 0.41 \text{ V}$$

Although in alkaline solutions the Gibbs energy of this reaction is higher, experimental data indicate that its rate is greater in acidic solutions. Dithionate is rather insensitive to most reducing and oxidizing agents. The rate of its oxidation is determined by the rate of decomposition into sulfurous acid, since the oxidation of sulfurous acid is a rapid process.

Sulfite may be converted into dithionate by means of anodic oxidation. This reaction is largely similar to the above-described oxidation of thiosulfate. The yields are high on nickel, gold, and smooth platinum anodes. Preliminary anodic polarization and annealing of the latter increase the yield of $S_2O_6^{2-}$. The most favorable pH for this reaction is 7 to 9. The formation of dithionate is promoted also by addition of brucine, pyridine, ammonia, ammonium fluoride, and urea.

The potentials of oxidation of trithionate and pentathionate are equal to

$$3H_2SO_3 + 2e^- \rightarrow S_3O_6^{2-} + 3H_2O \quad E° = 0.290 \text{ V}$$

$$5H_2SO_3 + 10e^- + 8H^+ \rightarrow S_5O_6^{2-} + 9H_2O \quad E° = 0.416 \text{ V}$$

The mechanism of these reactions is undoubtedly complex. Any agent capable of oxidizing these thionates can provide for the further oxidation of sulfurous acid into sulfate.

Since the average oxidation state of sulfur is the same in thiosulfate and pentathionate, it is of interest to calculate the Gibbs energy of thiosulfate conversion into pentathionate:

$$5S_2O_3^{2-} + 6H^+ \rightarrow 2S_5O_6^{2-} + 3H_2O \quad \Delta G° = -29.6 \text{ kJ mol}^{-1}$$

The Gibbs energy of this reaction is apparently low. The fact that the reaction is slow undoubtedly arises from the complexity of its mechanism.

J. Persulfates

Thermodynamic calculations give the following values of standard potentials for the persulfate-sulfate system:

$$S_2O_8^{2-} + 2e^- \rightarrow 2SO_4^{2-} \qquad E° = 1.96 \text{ V}$$

$$S_2O_8^{2-} + 2e^- + 2H^+ \rightarrow 2HSO_4^- \qquad E° = 2.08 \text{ V}$$

Persulfate evidently belongs to the strongest oxidizers. Persulfuric acid hydrolyzes into peroxomonosulfuric acid (Caro's acid), H_2SO_5, whose Gibbs energy is unknown and further into hydrogen peroxide:

$$S_2O_8^{2-} + 2H_2O \rightarrow 2H^+ + 2SO_4^{2-} + H_2O_2 \qquad \Delta G° = -36 \text{ kJ mol}^{-1}$$

The Gibbs energy of this reaction is low and at high concentrations of sulfuric acid and hydrogen peroxide the reaction may take the reverse direction.

At room temperature the oxidation by persulfate is usually slow. It may be accelerated, however, by ions of silver, copper, and iron, which supposedly act as electron carriers. Oxidation of compounds such as mercaptans, isopropyl alcohol, and hydrogen peroxide by persulfate was proved to involve a radical mechanism with persulfate as the initial source of radicals [12]:

$$S_2O_8^{2-} \rightarrow 2SO_4^-$$

In industry persulfate is obtained by anodic oxidation of concentrated sulfuric acid or bisulfates of ammonium or potassium on smooth platinum. The process involves evolution of oxygen. Therefore, platinum, which has high overvoltage with respect to oxygen, is used as the anodic material. To inhibit the formation of oxygen, minor amounts of chlorides, fluorides, thiocyanates, and cyanides are added to the anolyte. The current yields are the highest when the sulfuric acid solutions of density 1.3 to 1.45 g cm^{-3} (510 to 800 g of H_2SO_4 per liter) are used and the anodic current density is 0.5 to 1 A cm^{-2}.

During the formation of peroxydisulfuric acid the anolyte accumulates Caro's acid. This is undesirable not only because the process results in higher energy expenditure but also because Caro's acid accelerates the decomposition of peroxydisulfuric acid. Thus the process is carried out at low temperatures (5 to 10°C) and at high volume density of current.

The electrolysis of ammonium bisulfate permits lower-volume current densities and higher temperatures, since in this case the side reactions of $(NH_4)_2S_2O_8$ decomposition and $(NH_4)_2SO_5$ formation are not so important.

Potassium persulfate resulting from the electrolysis of potassium sulfate forms a solid precipitate, while the impurities remain in the electrolyte. Therefore, electrolysis of potassium sulfate does not require particularly pure electrolytes. Experience shows that lead dioxide instead of platinum may well be used as the anodic material.

Numerous investigations have not yet shed enough light on the mechanism of oxidation of sulfates into persulfate. It is assumed that the first stage of the process is the discharge of SO_4^{2-} or HSO_4^- anions, resulting in radical particles. The state of the electrode surface is thought to be of importance. According to Veselovski et al., one of the stages of the process is that of electrochemical desorption:

$$Pt \cdot PtO[HSO_4]_{ads} + HSO_4^- - e^- \to Pt \cdot PtO + H_2S_2O_8$$

Smith and Hogland suggest the following mechanism:

$$M(X) + SO_4^{2-} \to M(X)SO_4^- + e^-$$

$$M(X)SO_4^- \to M(X)SO_4 + e^-$$

$$M(X)SO_4 + SO_4^{2-} \to M(X) + S_2O_8^{2-}$$

where M(X) is the surface region of the platinum electrode surface covered with sulfate [5].

K. Halogen Compounds of Sulfur

A rather large body of reliable thermodynamic data for various halogen-sulfur compounds is currently available. This permits calculation of potentials of redox reactions involving these compounds:

$$S_2Cl_2(g) + 2e^- \to 2S + 2Cl^- \qquad E^\circ = 1.19 \text{ V}$$

$$SF(g) + e^- \to S + F^- \qquad E^\circ = 3.36 \text{ V}$$

$$SF_3(g) + 2e^- \to S + 2F^- \qquad E^\circ = 2.12 \text{ V}$$

$$SF_4(g) + 4e^- \to S + 4F^- \qquad E^\circ = 0.97 \text{ V}$$

$$SF_6(g) + 6e^- \to S + 6F^- \qquad E^\circ = 0.96 \text{ V}$$

L. Sulfur Organic Compounds

Table 2 presents but a minor part of thermodynamic data for the organic compounds of sulfur. Many of these are mercaptans, sulfides, and disulfides. The values well fit the dependences of the thermodynamic parameter on the number of carbon atoms in the molecule for the compounds belonging to one class.

The use of the thermodynamic data reported makes it possible to calculate the potentials of certain reactions.

$$(CH_3)_2SO_2(g) + 2e^- + 2H^+ \to (CH_3)_2SO(g) + H_2O \qquad E^\circ = 0.238 \text{ V}$$

$$(CH_3)_2SO(g) + 2e^- + 2H^+ \to (CH_3)_2S(g) + H_2O \qquad E^\circ = 0.769 \text{ V}$$

TABLE 2 Temperature Coefficients of Standard Potentials

Reaction	$(dE°/dT)_{isoth}$ /mV K^{-1}	$(d^2E°/dT^2)_{isoth}$ /μV K^{-2}
$S_2O_8^{2-} + 2e^- \rightarrow 2SO_4^{2-}$	−1.26	—
$2SO_4^{2-} + 2e^- + 4H^+ \rightarrow S_2O_6^{2-} + 2H_2O$	+0.52	—
$SO_4^{2-} + 2e^- + H_2O \rightarrow SO_3^{2-} + 2OH^-$	−1.389	—
$SO_4^{2-} + 2e^- + 4H^+ \rightarrow H_2SO_3 + H_2O$	+0.81	—
$4H_2SO_3 + 6e^- + 4H^+ \rightarrow S_4O_6^{2-} + 6H_2O$	−1.31	—
$SO_4^{2-} + 6e^- + 8H^+ \rightarrow S(s) + 4H_2O$	−0.168	+1.278
$S_2O_6^{2-} + 2e^- + 4H^+ \rightarrow 2H_2SO_3$	+1.10	—
$2SO_3^{2-} + 2e^- + 2H_2O \rightarrow S_2O_4^{2-} + 4OH^-$	−0.71	—
$2H_2SO_3 + 4e^- + 2H^+ \rightarrow S_2O_3^{2-} + 3H_2O$	−1.26	—
$2SO_3^{2-} + 4e^- + 3H_2O \rightarrow S_2O_3^{2-} + 6OH^-$	−1.14	—
$H_2SO_3 + 4e^- + 4H^+ \rightarrow S(s) + 3H_2O$	−0.66	—
$S_4O_6^{2-} + 2e^- \rightarrow 2S_2O_3^{2-}$	−1.11	—
$S_2Cl_2 + 2e^- \rightarrow 2S(s) + 2Cl^-$	−0.64	−6.26
$S(s) + 2e^- + 2H^+ \rightarrow H_2S (aq)$	−0.209	—
$S + 2e^- \rightarrow S^{2-}$	−0.93	—

Source: Ref. 16.

These calculations indicate that dimethyl sulfone in acidic solutions must readily undergo reduction to form dimethyl sulfide at the second stage. In fact, however, high overvoltages are characteristic of both processes.

$$(CH_3)_2S_2(g) + 2e^- + 2H^+ \rightarrow 2CH_3SH(g) \qquad E° = 0.176 \text{ V}$$

$$(C_2H_5)_2S_2(g) + 2e^- + 2H^+ \rightarrow 2C_2H_5SH(g) \qquad E° = 0.161 \text{ V}$$

$$(C_3H_7)_2S_2(g) + 2e^- + 2H^+ \to 2C_3H_7SH(g) \qquad E° = 0.169 \text{ V}$$

$$C_4H_4S(g) + 4e^- + 4H^+ \to C_4H_8S(g) \qquad E° = 0.209 \text{ V}$$

These processes, too, involve high overvoltage. The hydration of thiophene is possible only with the help of electron carriers.

Clark's book [15] presents a review of potentials of organic compounds, including several sulfur-containing compounds obtained by redox titration. Among these are the systems containing derivatives of thiasine, thiourea, cystein, gluthathion, thioglycolic acid, and thiolactic acid. Since the coefficients of activity of these compounds are unknown, the reported potentials are not thermodynamic.

M. Stepwise Reduction of Sulfur Compounds

Given below are the diagrams of potentials of the systems where sulfur has various oxidation states. The diagrams may be employed in the solution of a number of problems. It is necessary, however, to bear in mind that as a rule, these systems are not quite reversible.

N. Potential Diagrams for Sulfur

Please see the diagrams for acid solutions and basic solutions which appear on page 111.

II. SELENIUM

A. Oxidation States

The oxidation states of selenium are -2 in selenides, $+4$ in selenites, and $+6$ in selenates. The oxygen compounds of selenium corresponding to the oxidation states +2, +3, and +5 are unknown, as are the selenium polyacids. Selenium forms polyselenides, which are less stable than polysulfides. The halogen compounds of selenium are similar to those of sulfur but possess greater stability. The thermodynamic data are given in Table 3.

B. Selenium-Selenide Couple

The dissociation constants of H_2Se in water and the potentials of the system are equal to

$$H_2Se(aq) \to H^+ + HSe^- \qquad K_1 = 1.88 \times 10^{-4}$$

$$HSe^- \to H^+ + Se^{2-} \qquad K_2 = 1 \times 10^{-14}$$

$$Se(c) + 2e^- + 2H^+ \to H_2Se(aq) \qquad E° = -0.115 \text{ V}$$

$$Se(c) + 2e^- + 2H^+ \to H_2Se(g) \qquad E° = -0.082 \text{ V}$$

$$Se(c) + 2e^- + H^+ \to HSe^- \qquad E° = -0.227 \text{ V}$$

$$Se(c) + 2e^- \to Se^{2-} \qquad E°_B = -0.670 \text{ V}$$

Sulfur, Selenium, Tellurium, and Polonium

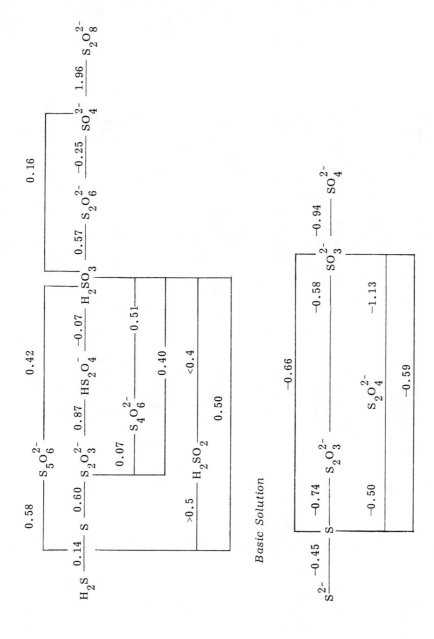

TABLE 3 Thermodynamic Data on Selenium[a]

Formula	Description	State	$\Delta H°$/kJ mol^{-1}	$\Delta G°$/kJ mol^{-1}	$S°$/J mol^{-1} K^{-1}
Se	Hexagonal black	c	0.0	0.0	42.42
Se	Monoclinic, red	c	6.7		
Se		g	227.07	187.07	176.58
Se	Glassy	Amorph.	5.4	2.7	51.5
Se$^+$		g	1174.629	1128.864	174.757
Se^{2-}		aq		129.3	
Se$_2$		g	146.0	96.2	252.7
Se$_6$		g	164.0		
SeO		g	53.35	26.82	233.92
SeO$_2$		c	−225.35	(−173.6)	
SeO$_2$		aq	−221.63		
SeO$_2$		g	−126.8	−132.043	264.8
SeO$_3$		c	−166.9		
SeO$_3^{2-}$		aq	−509.2	−369.9	13.3
SeO$_4^{2-}$		aq	−599.1	−441.4	54.1
Se$_2$O$_5$		c	−408.4		
HSe$^-$		aq	15.9	43.9	79.5
H$_2$Se		g	29.7	15.9	218.9

H$_2$Se	aq	19.2	22.2	*162.8*
HSeO$_3^-$	aq	−514.55	−411.54	*135.1*
HSeO$_4^-$	aq	−581.6	−452.3	*149.4*
H$_2$SeO$_3$	c	−524.46		
H$_2$SeO$_3$	aq	−517.48	−426.22	*207.9*
H$_2$SeO$_4$	c	−530.1		
H$_2$SeO$_4$	aq	−599.1	−441.4	*54.1*
D$_2$Se	g	32.493	20.184	*228.28*
HDSe	g	32.439	17.665	*229.37*
SeF$_6$	g	*−1117*	*−1017*	*313.76*
SeCl$_2$	g	−31.8		
SeCl$_4$	c	−183.3	(97.5)	
Se$_2$Cl$_2$	ℓ	−82.4	(−48.53)	
Se$_2$Cl$_2$	g	16.7		
SeOCl$_2$	g	−25		
SeBr$_2$	g	−20.9		
Se$_2$Br$_2$	g	29.3		

[a]Values selected by the National Bureau of Standards [1] are in italic; the values from Ref. 11 are given in parentheses.

C. Selenite-Selenium Couple

Thermodynamic calculations give the following potentials for this system:

$$H_2SeO_3 + 4e^- + 4H^+ \rightarrow Se(c) + 3H_2O \qquad E° = 0.739 \text{ V}$$

$$HSeO_3^- + 4e^- + 5H^+ \rightarrow Se(c) + 3H_2O \qquad E° = 0.777 \text{ V}$$

$$SeO_3^{2-} + 4e + 6H^+ \rightarrow Se(c) + 3H_2O \qquad E° = 0.885 \text{ V}$$

$$SeO_3^{2-} + 4e^- + 3H_2O \rightarrow Se(c) + 6OH^- \qquad E°_B = -0.357 \text{ V}$$

$$H_2SeO_3 \rightarrow H^+ + HSeO_3^- \qquad K_1 = 2.7 \times 10^{-3}$$

$$HSeO_3^- \rightarrow H^+ + SeO_3^{2-} \qquad K_2 = 2.63 \times 10^{-7}$$

Schott et al. [17] determined the potential of the H_2SeO_3/Se system by direct measurements of equilibrium constants in combination with the iodine-iodide system. The value they obtained is in agreement with the calculated potential. The same value was reported by Nevsky et al. [18], who titrated the hydrochloric solution of H_2SeO_3 by tin dichloride.

Selenious acid and selenites may react both as oxidizers and as reducers. for instance, they oxidize hydrogen sulfide, sulfurous acid, sulfites and hydrosulfites, and reduce chlorine, bichromate, permanganate, and hydrogen peroxide.

Hippel and Bloom [19] used electrolysis for precipitating metallic selenium from SeO_2-saturated hot 6 M H_2SO_4 solution. Their conclusion was that the high concentration of the acid promotes the metal precipitation by virtue of forming positive ions according to the reaction

$$H_2SeO_3 + H^+ \rightarrow H_3SeO_3^+$$

Nevsky et al. [18] report that in concentrated hydrochloric acid solutions, tetravalent selenium exists in the form of Se^{4+} cations. The reduction of H_2SeO_3 and $HSeO_3^-$ on a selenium cathode is considerably accelerated by electric lamp light [20].

D. Selenate-Selenite Couple

The values of standard potentials of the selenate-selenite system suggest that hexavalent selenium must be a rather strong oxidizer:

$$HSeO_4^- + 2e^- + 3H^+ \rightarrow H_2SeO_3 + H_2O \qquad E° = 1.094 \text{ V}$$

$$SeO_4^{2-} + 2e^- + 4H^+ \rightarrow H_2SeO_3 + H_2O \qquad E° = 1.151 \text{ V}$$

$$SeO_4^{2-} + 2e^- + 3H^+ \rightarrow HSeO_3^- + H_2O \qquad E° = 1.074 \text{ V}$$

$$SeO_4^{2-} + 2e^- + 2H^+ \rightarrow SeO_3^{2-} + H_2O \qquad E° = 0.895 \text{ V}$$

$$SeO_4^{2-} + 2e^- + H_2O \rightarrow SeO_3^{2-} + 2OH^- \qquad E°_B = 0.031 \text{ V}$$

$$HSeO_4^- \rightarrow H^+ + SeO_4^{2-} \qquad K = 8.9 \times 10^{-3}$$

However, the selenate-selenite system is irreversible. Therefore, selenate cannot be reduced even by such strong reducing agents such as hot concentrated HCl, Ti^{3+}, or SO_2 in hot concentrated sulfuric acid. Electrolytic reduction of selenate also proves impossible, incidating that the overvoltage is very high.

E. Halogen Compounds of Selenium

There is but a limited amount of thermodynamic data for selenium halides (Table 4). The standard potentials that may be calculated using these data are

$$Se_2Cl_2(\ell) + 2e^- \rightarrow 2Se + 2Cl^- \qquad E° = 1.108 \text{ V}$$

$$SeF_6 + 6e^- \rightarrow Se + 6F^- \qquad E° = 1.109 \text{ V}$$

F. Potential Diagrams for Selenium

$$-2 \qquad 0 \qquad +4 \qquad +6$$

Acidic Solution

$$H_2Se \xrightarrow{-0.11} Se \xrightarrow{0.74} H_2SeO_3 \xrightarrow{1.1} SeO_4^{2-}$$

Basic Solution

$$Se^{2-} \xrightarrow{-0.67} Se \xrightarrow{-0.36} SeO_3^{2-} \xrightarrow{0.03} SeO_4^{2-}$$

III. TELLURIUM

A. Oxidation States

Like selenium, in the case of tellurium only the compounds corresponding to the oxidation states −2, +4, and +6 are of importance. The monoxide TeO exists at temperatures above 1000°C as a gas; the dark brown oxide of composition TeO is in fact a mixture of black Te and white TeO_2. The thermodynamic data are given in Table 5.

B. Tellurium–Telluride Couple

Calculations give

TABLE 4 Temperature Coefficients of Standard Potentials

Reaction	$(dE°/dT)_{isoth}$ (mV K^{-1})
$SeO_4^{2-} + 2e + 4H^+ = H_2SeO_3 + H_2O$	+0.553
$SeO_4^{2-} + 2e + H_2O = SeO_3^{2-} + 2OH^-$	−1.187
$H_2SeO_3 + 4e + 4H^+ = Se + 3H_2O$	−0.520
$SeO_3^{2-} + 4e + 3H_2O = Se + 6OH^-$	−1.318
$Se + 2e + 2H^+ = H_2Se(aq)$	−0.028
$Se + 2e = Se^{2-}$	−0.89

Source: Ref. 16.

$$Te + 2e^- + 2H^+ \rightarrow H_2Te(aq) \qquad E° = -0.740 \text{ V}$$

$$Te + 2e^- + H^+ \rightarrow HTe^- \qquad E° = -0.817 \text{ V}$$

$$Te + 2e^- \rightarrow Te^{2-} \qquad E°_B = -1.143 \text{ V}$$

with

$$H_2Te \rightarrow H^+ + HTe^- \qquad K_1 = 2.3 \times 10^{-3}$$

$$HTe^- \rightarrow H^+ + Te^{2-} \qquad K_2 = 1 \times 10^{-11}$$

On the grounds of emf data, Kazarnovsky [21] obtained the following value of $E°$:

$$2Te + 2e^- \rightarrow Te_2^{2-} \qquad E° = -0.84 \text{ V}$$

which was used to calculate the Gibbs energy of the Te_2^{2-} ion (162.1 kJ mol^{-1}). This made it possible to estimate the energy of the reaction

$$Te^{2-} + Te \rightarrow Te_2^{2-} \qquad \Delta G° = -58.4 \text{ kJ mol}^{-1}$$

$$K = 1.8 \times 10^{-10}$$

Latimer [11] suggested that this value, as well as that of the enthalpy of formation of H_2Te, is somewhat too large. Current reference sources present a considerably lower value of the enthalpy of formation of $H_2Te(g)$, but the values for dissolved hydrogen telluride have not been considered.

TABLE 5 Thermodynamic Data on Tellurium[a]

Formula	State	$\Delta H°$/kJ mol^{-1}	$\Delta G°$/kJ mol^{-1}	$S°$/J mol^{-1} K^{-1}
Te	c	0.0	0.0	49.71
Te	g	*196.73*	*157.11*	*182.63*
Te	amorph.	11.3		
Te$^+$	g	1067.4	1022.0	180.740
Te^{2+}	g	2874		
Te^{4+}	aq		219.16*	
Te$_2$	g	*168.2*	118.0	*268.03*
Te^{2-}	aq		(220.5)	
Te$_2^{2-}$	aq		(162.1)	
HTe$^-$	aq		(157.7)	
H$_2$Te	g	99.6	(138.5)	
H$_2$Te	aq		(142.7)	
TeO	g	65.3	38.5	241.4
TeO$_2$	c	−322.6	−270.3	79.5
TeO$_2$	g	−51.9	−57.450	273.2
HTeO$_2^+$	aq		−261.54*	
TeO$_3^{2-}$	aq		−392.42*	
HTeO$_3^-$	aq		−436.56*	
H$_2$TeO$_3$	c	(−605.4)	(−484.1)	(199.6)
H$_2$TeO$_3$	aq		*−318.8*	
H$_4$TeO$_6^{2-}$	aq	*−1222.1*		
H$_5$TeO$_6^-$	aq	*−1261.5*		
H$_6$TeO$_6$	c	*−1298.7*	(−702.5)	
H$_6$TeO$_6$	aq	*−1284.5*		
TeO$_4^{2-}$	aq		−456.424*	
HTeO$_4^-$	aq		−515.753*	
H$_2$TeO$_4$	aq		−550.857*	
TeF$_6$	g	*−1318*	−1222.054	346
TeCl$_4$	c	−326.4	(−237.2)	(196)
TeBr$_4$	ℓ	−190.4		
TeCl$_6^{2-}$	aq		(−574.9)	

[a]Values selected by the National Bureau of Standards [1] are in italic; the values from Ref. 9 are not specified; the values from Ref. 11 are given in parentheses; and the values from Ref. 22 are indicated by an asterisk.

$$Te_2^{2-} + 2e^- + 2H^+ \rightarrow 2H_2Te \qquad E^\circ = -0.639 \text{ V}$$

$$Te_2^{2-} + 2e^- + 2H^+ \rightarrow 2HTe^- \qquad E^\circ = -0.794 \text{ V}$$

$$Te_2^{2-} + 2e^- \rightarrow 2Te^{2-} \qquad E_B^\circ = -1.445 \text{ V}$$

The high negative potentials of the tellurium-telluride and tellurium-ditelluride systems indicate that tellurium in oxidation state -2 is a strong reducing agent. Thus the potentials in these systems are difficult to measure because of oxidation processes. The H_2Te solutions are highly unstable in contact with air and rapidly yield elementary tellurium.

C. Tellurious Acid-Tellurium Couple

Schumann [23] has directly measured the potentials of the following systems:

$$TeO_2(s) + 4e^- + 4H^+ \rightarrow Te + 2H_2O \qquad E^\circ = 0.5213 \text{ V}$$

$$TeOOH^+ + 4e^- + 3H^+ \rightarrow Te + 2H_2O \qquad E^\circ = 0.559 \text{ V}$$

A calculation gives the value

$$TeO_2(s) + 4e^- + 4H^+ \rightarrow Te + 2H_2O \qquad E^\circ = 0.529 \text{ V}$$

Schumann has also shown that solid tellurium dioxide may be in an equilibrium with water and obtained the following values of the equilibrium constants:

$$TeO_2 + H^+ \rightarrow TeOOH^+ \qquad K = 8.9 \times 10^{-3}$$

$$TeO_2 + H_2O \rightarrow TeOOH^+ + OH^- \qquad K = 8.9 \times 10^{-17}$$

According to Kazarnovsky [21],

$$TeO_2(s) + H_2O \rightarrow H_2TeO_3(s) \qquad \Delta G^\circ = 23 \text{ kJ mol}^{-1}$$

These data were used to calculate the Gibbs energy of $H_2TeO_3(s)$ (-484.1 kJ mol^{-1}), the constant of $TeO(OH)_2$ basic dissociation,

$$TeO(OH)_2(s) \rightarrow TeOOH^+ + OH^- \qquad K = 1 \times 10^{-12}$$

and the potential

$$TeO_2(aq,s) + 4e^- + 4H^+ \rightarrow Te + 2H_2O \qquad E^\circ = 0.604 \text{ V}$$

Blanc [24] reports the dissociation constants of H_2TeO_3 as $K_1 = 3 \times 10^{-3}$ and $K_2 = 1.8 \times 10^{-8}$. The Gibbs energy of the TeO_3^{2-} ion may be calculated directly from the enthalpy of formation provided that its entropy is available.

Assuming that it is equal to -8 J mol^{-1} K^{-1} and using the above-mentioned value of K_2, we may calculate the Gibbs energy as -392.42 kJ mol^{-1} [11]:

$$TeO_3^{2-} + 4e^- + 3H_2O \rightarrow Te + 6OH^- \qquad E_B^\circ = -0.415 \text{ V}$$

Awad's measurements of anodic polarization curves with compact metallic tellurium in acidic solutions [25] make it possible to calculate E° for the reaction

$$H_2TeO_3 + 4e^- + 4H^+ \rightarrow Te + 3H_2O \qquad E^\circ = 0.589 \text{ V}$$

This value corresponds to solid H_2TeO_3. The calculation for dissolved H_2TeO_3 gives $E^\circ = 1.018$ V.

For the reactions

$$HTeO_3^- + 4e^- + 5H^+ \rightarrow Te + 3H_2O \qquad E^\circ = 0.713 \text{ V}$$

$$TeO_3^{2-} + 4e^- + 6H^+ \rightarrow Te + 3H_2O \qquad E^\circ = 0.827 \text{ V}$$

In very acidic solutions a simpler electrode reaction occurs [18,25,26]:

$$Te^{4+} + 4e^- \rightarrow Te \qquad E^\circ = 0.568 \text{ V}$$

It follows from the measurements of anodic polarization curves of tellurium in basic solutions performed by Komandenko and Totinian [27] that

$$Te(OH)_6^{2-} + 4e^- \rightarrow Te + 6OH^- \qquad E_B^\circ = -0.412 \text{ V}$$

A review of standard potentials of the tellurious acid–tellurium system shows that tellurium is a relatively noble element stable in the presence of water. Anodic dissolution of metallic tellurium results in tetravalent tellurium in both acidic and basic solutions. Nitric acid and concentrated sulfuric acid oxidize tellurium to a tetravalent state (Te^{4+}, $HTeO_2^+$, H_2TeO_3), and hydrogen peroxide oxidizes it to a hexavalent state.

D. Telluric Acid–Tellurious Acid Couple

On the basis of the available thermodynamic data and the constants of dissociation of tellurious and telluric acids [2]:

$$H_2TeO_4 \rightarrow H^+ + HTeO_4^- \qquad K_1 = 2.1 \times 10^{-8}$$

$$HTeO_4^- \rightarrow H^+ + TeO_4^{2-} \qquad K_2 = 6.5 \times 10^{-12} \text{ at } 18°C$$

the following standard potentials were calculated:

$$H_2TeO_4 + 2e^- + 6H^+ \rightarrow Te^{4+} + 4H_2O \qquad E^\circ = 0.926 \text{ V}$$

$$H_2TeO_4 + 2e^- + 3H^+ \rightarrow HTeO_2^+ + 2H_2O \quad E° = 0.959 \text{ V}$$

$$H_2TeO_4 + 2e^- + H^+ \rightarrow HTeO_3^- + H_2O \quad E° = 0.637 \text{ V}$$

$$HTeO_4^- + 2e^- + 2H^+ \rightarrow HTeO_3^- + H_2O \quad E° = 0.819 \text{ V}$$

$$HTeO_4^- + 2e^- + H^+ \rightarrow TeO_3^{2-} + H_2O \quad E° = 0.590 \text{ V}$$

$$TeO_4^{2-} + 2e^- + 2H^+ \rightarrow TeO_3^{2-} + H_2O \quad E° = 0.897 \text{ V}$$

$$H_2TeO_4 + 2e^- + 2H^+ \rightarrow TeO_2(s) + 2H_2O \quad E° = 1.004 \text{ V}$$

$$H_2TeO_4 + 2e^- + 2H^+ \rightarrow TeO_2aq(s) + 2H_2O \quad E° = 0.854 \text{ V}$$

$$HTeO_4^- + 2e^- + 3H^+ \rightarrow TeO_2(s) + 2H_2O \quad E° = 1.186 \text{ V}$$

$$HTeO_4^- + 2e^- + 3H^+ \rightarrow TeO_2aq(s) + 2H_2O \quad E° = 1.036 \text{ V}$$

$$TeO_4^{2-} + 2e^- + 4H^+ \rightarrow TeO_2(s) + 2H_2O \quad E° = 1.494 \text{ V}$$

$$TeO_4^{2-} + 2e^- + 4H^+ \rightarrow TeO_2aq(s) + 2H_2O \quad E° = 1.343 \text{ V}$$

$$TeO_3(s) + 2e^- + 2H^+ \rightarrow TeO_2(s) + H_2O \quad E° = 0.975 \text{ V}$$

$$TeO_4^{2-} + 2e^- + H_2O \rightarrow TeO_3^{2-} + 2OH^- \quad E°_B = 0.07 \text{ V}$$

These data are basically in agreement with the chemical properties of the components of the system.

E. Halogen Compounds of Tellurium

Reichinstein [28] reports that for the reaction

$$TeCl_6^- + 4e^- \rightarrow Te + 6Cl^- \quad E° = 0.55 \text{ V}$$

According to more recent measurements, in 2 to 3 M HCl, $E° = 0.630$ V [29], and in 4 to 10 M HCl, $E° = 0.625$ V [30].

The standard potentials (see Table 6) for the following reactions were calculated as

$$TeF_6 + 6e^- \rightarrow Te + 6F^- \quad E° = 0.755 \text{ V}$$

$$TeCl_4 + 4e^- \rightarrow Te + 4Cl^- \quad E° = 0.745 \text{ V}$$

TABLE 6 Temperature Coefficients of Standard Potentials

Reaction	$(dE°/dT)_{isoth}$ /mV K^{-1}
$H_6TeO_6 + 2e^- + 2H^+ \rightarrow TeO_2(s) + 4H_2O$	−0.13
$TeO_3^{2-} + 4e^- + 3H_2O \rightarrow Te + 6OH^-$	−1.23
$TeO_2(s) + 4e^- + 4H^+ \rightarrow Te + 2H_2O$	−0.370
$Te_2S + 2e^- \rightarrow 2Te + S^{2-}$	−0.94
$Te + 2e^- + 2H^+ \rightarrow H_2Te$	+0.280

Source: Ref. 16.

F. Potential Diagrams for Tellurium

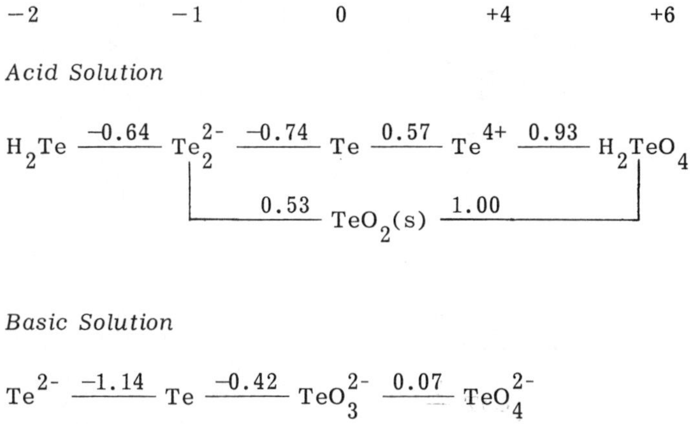

$$-2 \qquad -1 \qquad 0 \qquad +4 \qquad +6$$

Acid Solution

$$H_2Te \xrightarrow{-0.64} Te_2^{2-} \xrightarrow{-0.74} Te \xrightarrow{0.57} Te^{4+} \xrightarrow{0.93} H_2TeO_4$$
$$\xrightarrow{0.53} TeO_2(s) \xrightarrow{1.00}$$

Basic Solution

$$Te^{2-} \xrightarrow{-1.14} Te \xrightarrow{-0.42} TeO_3^{2-} \xrightarrow{0.07} TeO_4^{2-}$$

IV. POLONIUM

The standard state of the element is solid metallic polonium. The thermodynamic data are rather scarce and were obtained in most cases by approximate estimations (Table 7).

A. Oxidation States

As with selenium and tellurium, the oxidation states characteristic of polonium are −2, +2, +4, and +6. The most stable state is +4. Polonium forms a very unstable hydride H_2Po (oxidation state −2). In the oxidation state +2 polonium exists in aqueous solutions as a Po^{2+} ion and forms salts. In the dioxide PoO_2 the oxidation state is +4. The dioxide dissolves in acids to form salts. A hy-

TABLE 7 Thermodynamic Data on Polonium[a]

Formula	Description	State	$\Delta H°$/kJ mol^{-1}	$\Delta G°$/kJ mol^{-1}	$S°$/J mol^{-1} K^{-1}
Po	Cubic	c	0.0	0.0	62.8
Po		g	146.988	109.407	188.811
Po$^+$		g			186.954
Po^{2+}		g			175.431
Po^{2+}		aq		71	
Po^{4+}		aq		293	
PoO$_2$		g	145.507	98.257	284.01
PoO$_2$		c	−251	−912	71
H$_2$Po		g		(>192)	
PoO$_3$		c		(−138)	
PoO$_3^{2-}$		aq		(−423)	
Po(OH)$_4$		c	−544		
PoCl$_2$	Rhombic	c			130
PoCl$_4$		c			197
PoBr$_2$		c			155
PoBr$_4$	Cubic	c			230
PoCl$_6^{2-}$		aq		−577	
Po(OH)$_2^{4+}$		aq		−473	
PoS		c		218	

[a] Values selected by the National Bureau of Standards [1] are in italic; the values from Ref. 9 are not specified; the values from Ref. 11 are given in parentheses.

droxide, whose likely formula is PoO(OH)$_2$, may be obtained from solutions of the salts. Polonium dioxide under the action of strong oxidizers and via electrolysis may form unstable polonium trioxide PoO$_3$.

The chemical properties of polonium are dependent not only on the thermodynamic properties of its compounds but also on the radioactive decay of its nuclei with emanation of high-energy α particles. Owing to the effect of radiation bivalent polonium in aqueous solutions undergoes oxidation by ozone forming from water or by hydrogen peroxide.

B. Systems Containing Polonium

Most of the potentials were calculated from the potentials of precipitation of metallic polonium by cathodic reduction and from the potentials of precipitation of polonium dioxide by anodic oxidation. However, there is no evidence whatsoever that the electrode processes are reversible; and irreversibility must be inherent in systems where the electrode process involves formation or rupture of Po-O bonds. Therefore, the data thus obtained are only approximate. Only in very acidic media where polonium in the oxidation state from +2 to +4 apparently exists in the form of simple cations may one expect the processes to be more reversible. Such measurements, however, have not yet been performed.

The potential of the polonium-polonide system was determined by extrapolation of the corresponding data for sulfur, selenium, and tellurium.

$$Po + 2e^- + 2H^+ \rightarrow H_2Po(g) \qquad E° \approx 1.0 \text{ V}$$

$$Po + 2e^- \rightarrow Po^{2-} \qquad E°_B \approx 1.4 \text{ V}$$

$$Po^{2+} + 2e^- \rightarrow Po \qquad E° = 0.368 \text{ V}$$

$$PoO_2 + 2e^- + 4H^+ \rightarrow Po^{2+} + 2H_2O \qquad E° = 1.095 \text{ V}$$

$$PoO_3^{2-} + 4e^- + 3H_2O \rightarrow Po + 6OH^- \qquad E°_B = -0.494 \text{ V}$$

$$PoO_3^{2-} + 4e^- + 6H^+ \rightarrow Po + 3H_2O \qquad E° = 0.748 \text{ V}$$

$$PoO_3(s) + 2e^- + 2H^+ \rightarrow PoO_2(s) + H_2O \qquad E° = 1.509 \text{ V}$$

$$PoO_3(s) + 2e^- \rightarrow PoO_3^{2-} \qquad E° = 1.477 \text{ V}$$

C. Potential Diagrams for Polonium

Acid Solution

$$\text{H}_2\text{Po} \xrightarrow{ca. -1.0} \text{Po} \xrightarrow{0.37} \text{Po}^{2+} \xrightarrow{1.1} \text{PoO}_2 \xrightarrow{1.51} \text{PoO}_3$$

with overarching bracket 1.3 (Po^{2+} to PoO$_3$) and lower bracket 0.73 (Po to PoO$_2$)

Basic Solution

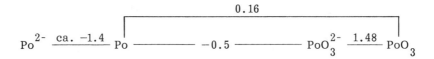

The diagrams of potentials of the group VI elements indicate that when passing from oxygen to polonium the oxidative power of the free elements decreases. The oxidative power of the acids corresponding to oxidation states +4 and +6 increases with atomic weight; the exceptions to this rule are the acids of selenium.

REFERENCES

1. D. D. Wagman, W. H. Evans, V. B. Parker, I. Halow, S. M. Bailey, and R. H. Schumm, *Selected Values of Chemical Thermodynamic Properties*, Natl. Bur. Stand. U.S. Government Office, Washington, D.C., Tech. Note 270-3, 1968.
2. D. Dobos, *Electrochemical Data*, Akadémiai Kidaó, Budapest, 1978.
3. C. Drucker, ed., *Abhandlungen der deutschen Bunsengesellschaft. X. Messungen elektromotorischer Kräfte galvanischer Ketten mit wässerigen Elektrolyten*, Verlag Chemie, Berlin, 1929.
4. P. L. Allen and A. Hickling, Trans. Faraday Soc., 53, 1626 (1957).
5. S. I. Zhdanov, in *Encyclopedia of the Electrochemistry of the Elements*, A. J. Bard, ed., Vol. 4, Marcel Dekker, New York, 1975, p. 335.
6. E. M. Arnett, in *Progress in Physical Organic Chemistry*, Vol. 1, S. G. Cohen, A. Streitwieser, Jr., and R. W. Taft, eds., Interscience, New York, 1963, p. 223.
7. J. E. Gordon, *The Organic Chemistry of Electrolyte Solutions*, Wiley-Interscience, New York, 1975.
8. H. H. Perkampus, in *Advances in Physical Organic Chemistry*, V. Gold, ed., Vol. 4, Academic Press, New York, 1966.
9. V. P. Glushko, ed., *Thermal Constants of Substances*, Moscow, Vol. II, 1966; Vol. IV, 1970.
10. D. R. Stull, E. F. Westrum, Jr., and G. C. Sinke, *The Chemical Thermodynamics of Organic Compounds*, Wiley, New York, 1969.
11. W. M. Latimer, *The Oxidation States of the Elements and their Potentials in Aqueous Solutions*, 2nd ed., Prentice-Hall, Englewood Cliffs, N.J., 1952.
12. T. A. Turnay, *Oxidation Mechanisms*, Butterworth, London, 1965.
13. R. E. Connick and A. D. Awtrey, J. Am. Chem. Soc., 73, 1842 (1951).
14. C. H. Sorum and J. O. Edwards, Am. Chem. Soc. Annu. Meet., Detroit, 1950.
15. W. M. Clark, *Oxidation-Reduction Potentials of Organic Systems*, Williams & Wilkins, Baltimore, 1960.
16. A. J. de Béthune and N. A. Swendeman Loud, in *Encyclopedia of Electrochemistry*, C. A. Hampel, ed., Reinhold, New York, 1964.
17. H. F. Schott, E. H. Swift, and D. M. Yost, J. Am. Chem. Soc., 50, 721 (1928).

18. O. B. Nevsky, A. D. Gerasimov, and N. N. D'yachkova, Elektrokhimiya, *4*, 624 (1968).
19. A. T. Hippel and M. C. Bloom, J. Chem. Phys., *18*, 1243 (1950).
20. Sh. D. Osman-Zade and A. T. Vagramian, Elektrokhimiya, *3*, 249 (1967).
21. I. Kazarnovsky, Z. Phys. Chem., *109*, 287 (1924).
22. M. Pourbaix, ed., *Atlas d'equilibres électrochimiques à 25°C*, Gauthier-Villars, Paris, 1963.
23. R. Schumann, J. Am. Chem. Soc., *47*, 356 (1925).
24. E. Blanc, J. Chim. Phys., *18*, 28 (1920).
25. S. A. Awad, Electrochim. Acta, *13*, 925 (1968).
26. F. H. Getman, Trans. Am. Electrochem. Soc., *64*, 206 (1933).
27. V. M. Komandenko and A. L. Rotinian, Zh. Prikl. Khim., *39*, 123 (1966).
28. D. Reichinstein, Z. Phys. Chem., *97*, 257 (1921).
29. V. S. Yakovleva and E. N. Andreev, Uch. Zap. Leningr. Gos. Pedagog. Inst., *160*, 181 (1959).
30. V. M. Komandenko, Zh. Prikl. Khim., *43*, 2480 (1970).

7
Nitrogen, Phosphorus, Arsenic, Antimony, and Bismuth

I. NITROGEN*

Joseph T. Maloy Seton Hall University, South Orange, New Jersey

The electrochemistry of nitrogen is enriched by the fact that this element is known to exist in at least 10 different oxidation states, one of which is non-integer. Those oxidation states are summarized in Table 1 as an index to the electrochemical and thermodynamic data reported in Tables 2 and 3. The electrochemical data have been abstracted from the original compilations by Latimer [1], Pourbaix and de Zoubov [2], and Plieth [3] and references reported therein. Thermodynamic data were obtained from the most recent National Bureau of Standards (NBS) Technical Note (270-3) [4].

All of these previous compilations use diagrams or graphical representations to simplify the complex thermodynamic relationships existing among the various oxidation states of the nitrogen compounds. The arrangement of the electrode reactions in Table 2 is by oxidation state. Similarly, the classification of all thermodynamic data in Table 3 according to oxidation state provides the information given in the Gibbs energy-oxidation state diagrams reported by Plieth [3] in a readily comprehended form, and these diagrams have not been reproduced.

N_2 is thermodynamically stable under nearly all the potential-pH conditions that are necessary for the stability of H_2O. Thus N_2 and H_2O can coexist readily in nature and N_2 is the most common nitrogen compound on the surface of the earth. Nitrogen fixation (conversion of this elemental nitrogen to the nonzero oxidation state) is predicted from this diagram either by reaction with H_2 or O_2, however:

$$N_2(g) + 3H_2(g) \rightarrow 2NH_3(aq)$$
$$2N_2(g) + 5O_2(g) + 2H_2O \rightarrow 4NO_3^- + 4H^+$$

Nitrogen is thermodynamically unstable under most of the potential-pH conditions required for O_2 or H_2 stability, and it should be thermodynamically feasible to decompose water electrolytically in the presence of N_2 to achieve fixation of the nitrogen at each electrode. In experimental practice this does not occur, presumably because of the kinetic irreversibility of both the reduction and the oxidation of elemental nitrogen.

Reviewer: Donald C. Thornton, Drexel University, Philadelphia, Pennsylvania.

TABLE 1 Oxidation States of Nitrogen

Oxidation state	Typical compounds
+5	HNO_3, NO_3^-
+4	NO_2, N_2O_4
+3	$HNO_2, NO_2^-, NO^+, NCl_3$
+2	NO
+1	$N_2O, H_2N_2O_2, N_2O_2^{2-}, NHCl_2$
0	N_2
−1/3	HN_3, N_3^-
−1	NH_3OH^+, NH_2OH, NH_2Cl
−2	$N_2H_5^+, N_2H_4$
−3	NH_4^+, NH_3

All of the half-reactions for the compounds of nitrogen exhibit this apparent electrochemical irreversibility. Several of the experimental sttempts to compare the thermodynamic values for $E°$ with the irreversible potentials achieved during electrolysis have been documented by Plieth [3]; few have been successful. Comparisons, where they exist, appear in Table 2. Users of the information contained in this compilation must be cautioned, therefore, that it is only potentiometrically (thermodynamically) valid. The reactivity of the various forms of nitrogen appears to be controlled primarily by kinetics. Thus predictions of thermodynamic reactivity among the nitrogen compounds are often rendered invalid by the charge transfer kinetics associated with the change in oxidation state.

In the case of the reduction reactions of NO_3^- and the oxidation reactions of NH_4^+ and NH_3, this problem of irreversibility is further complicated by pH relationships. Each of these electrode processes takes place at (thermodynamic) potentials that are quite close to the decomposition potentials for H_2O. Thus pH changes are possible in the vicinity of the electrode as H_2 or O_2 is evolved or as H^+ is consumed or produced during electrolysis. For example, the reduction of NO_3^- via the reaction

$$NO_3^- + 2H^+ + 2e^- \rightarrow NO_2^- + H_2O$$

proceeds quite slowly when the pH of the bulk solution is 3.0 because the pH in the vicinity of the electrode is quite high following the depletion of H^+ via the (slow) electrode process. Lowering the bulk pH increases the acidity in the vicinity of the electrode and increases the rate of NO_3^- reduction. The addition of a metal ion which can be precipitated as a hydroxide (e.g., Ni^{2+}, Cd^{2+}, Co^{2+}, Zn^{2+}) has two effects:

TABLE 2 Electrode Potentials of the Nitrogen Compounds

Oxidation states O → R	Electrode reaction	$E°/V$	Conditions	References
+5 +4	$2NO_3^- + 2H_2O + 2e^- \rightarrow N_2O_4(g) + 4OH^-$	-0.86^a		1
+5 +4	$2NO_3^- + 4H^+ + 2e^- \rightarrow N_2O_4(g) + 2H_2O$	0.803		1-3
+5 +4	$NO_3^- + 2H^+ + e^- \rightarrow NO_2(g) + H_2O$	0.775		2
+5 +3	$NO_3^- + 3H^+ + 2e^- \rightarrow HNO_2 + H_2O$	0.94		1
		0.980	2-12 M HNO_3	3
		0.950	pH 0-6	3
		0.864	pH 0-4	3
+5 +3	$NO_3^- + 2H^+ + 2e^- \rightarrow NO_2^- + H_2O$	0.835		2
+5 +3	$NO_3^- + H_2O + 2e^- \rightarrow NO_2^- + 2OH^-$	0.965	pH 4-7	3
+5,+2	$NO + NO_3^- + 2H^+ + e^- \rightarrow 2HNO_2$	0.01^a		1
+5 +2	$NO_3^- + 4H^+ + 3e^- \rightarrow NO(g) + 2H_2O$	0.517	HNO_3, 25°C	3
		0.957		1-2
		0.960	HNO_3, 25°C	3
+5 +1	$2NO_3^- + 10H^+ + 8e^- \rightarrow N_2O(g) + 5H_2O$	1.116		2

TABLE 2 (Continued)

Oxidation states O → R	Electrode reaction	$E°$/V	Conditions	References
+5 → 0	$2NO_3^- + 12H^+ + 10e^- \rightarrow N_2(g) + 6H_2O$	1.246		2
+5 → −3	$NO_3^- + 10H^+ + 8e^- \rightarrow NH_4^+ + 3H_2O$	0.875		2
+4 → +3	$NO_2(g) + H^+ + e^- \rightarrow HNO_2$	1.093		2
+4 → +3	$N_2O_4(g) + 2H^+ + 2e^- \rightarrow 2HNO_2$	1.07		1
+4 → +3	$NO_2(g) + e^- \rightarrow NO_2^-$	0.895		2
+4 → +3	$N_2O_4(g) + 2e^- \rightarrow 2NO_2^-$	0.867		1-2
+4 → +3	$N_2O_4(g) + 4H^+ + 2e^- \rightarrow 2NO^+ + 2H_2O$	0.77	50-80% H_2SO_4	3
+4 → +2	$NO_2 + 2H^+ + 2e^- \rightarrow NO(g) + H_2O$	1.045		2
+4 → +2	$N_2O_4 + 4H^+ + 4e^- \rightarrow 2NO(g) + 2H_2O$	1.039		1-2
+4 → +1	$2NO_2 + 6H^+ + 6e^- \rightarrow N_2O(g) + 3H_2O$	1.229		2
+4 → 0	$2NO_2(g) + 8H^+ + 8e^- \rightarrow N_2(g) + 4H_2O$	1.363		2
+4 → 0	$N_2O_4(g) + 8H^+ + 8e^- \rightarrow N_2(g) + 4H_2O$	1.357		2
+4 → −3	$NO_2(g) + 8H^+ + 7e^- \rightarrow NH_4^+ + 2H_2O$	0.897		2
+4 → −3	$N_2O_4(g) + 16H^+ + 14e^- \rightarrow 2NH_4^+ + 4H_2O$	0.89		2

+3	+2	$NO^+ + e^- \rightarrow NO(g)$	1.45	0.1-4 M H_2SO_4	3
+3	+2	$NO_2^- + 2H^+ + e^- \rightarrow NO(g) + H_2O$	1.202		2
+3	+2	$HNO_2 + H^+ + e^- \rightarrow NO(g) + H_2O$	0.996		1-2
+3	+2		0.983	0.1-4 M H_2SO_4; 50-80% H_2SO_4	
+3	+2	$NO_2^- + H_2O + e^- \rightarrow NO(g) + 2OH^-$	0.46^a		1
+3	+1	$2NO_2^- + 6H^+ + 4e^- \rightarrow N_2O(g) + 3H_2O$	1.396		2
+3	+1	$2HNO_2 + 4H^+ + 4e^- \rightarrow N_2O(g) + H_2O$	1.297		2
+3	+1	$2HNO_2 + 4H^+ + 2e^- \rightarrow H_2N_2O_2 + 2H_2O$	0.86		1
+3	+1	$2NO_2^- + 3H_2O + 4e^- \rightarrow N_2O(g) + 6OH^-$	0.15^a		1
+3	+1	$2NO_2^- + 2H_2O + 4e^- \rightarrow N_2O_2^{2-} + 4OH^-$	-0.14^a		1
+3	0	$2NO_2^- + 8H^+ + 6e^- \rightarrow N_2(g) + 4H_2O$	1.520		2
+3	0	$2HNO_2 + 6H^+ + 6e^- \rightarrow N_2(g) + 4H_2O$	1.454		2
+3	−3	$NO_2^- + 8H^+ + 6e^- \rightarrow NH_4^+ + 2H_2O$	0.897		2
+3	−3	$HNO_2 + 6H^+ + 6e^- \rightarrow NH_4^+ + 2H_2O$	0.864		2
+3	−3	$NO_2^- + 7H^+ + 6e^- \rightarrow NH_3(aq) + H_2O$	0.806		2
+3	−3	$NO_2^- + 7H^+ + 6e^- \rightarrow NH_3(g) + 2H_2O$	0.789		2

TABLE 2 (Continued)

Oxidation states		Electrode reaction	$E°/V$	Conditions	References
O	R				
+3	−3	$HNO_2 + 6H^+ + 6e^- \rightarrow NH_3(g) + 2H_2O$	0.755		2
+3	−3	$NCl_3 + 4H^+ + 6e^- \rightarrow 3Cl^- + NH_4^+$	−1.37		3
+2	+1	$2NO + 2H^+ + 2e^- \rightarrow N_2O(g) + H_2O$	1.59		1
+2	+1	$2NO + 2H^+ + 2e^- \rightarrow H_2N_2O_2$	0.71		1
+2	+1	$2NO + 2e^- \rightarrow N_2O_2^{2-}$	0.18		1
+2	0	$2NO(g) + 4H^+ + 4e^- \rightarrow N_2(g) + 2H_2O$	1.678		2
+2	−3	$NO(g) + 6H^+ + 5e^- \rightarrow NH_4^+ + H_2O$	0.836		2
+2	−3	$NO(g) + 5H^+ + 5e^- \rightarrow NH_3(aq)$	0.727		2
+1	0	$H_2N_2O_2 + 2H^+ + 2e^- \rightarrow N_2(g) + 2H_2O$	2.65		1
+1	0	$N_2O(g) + 2H^+ + 2e^- \rightarrow N_2(g) + H_2O$	1.77		1
+1	0	$N_2O_2^{2-} + 2H_2O + 2e^- \rightarrow N_2(g) + 4)H^-$	1.52[a]		1
+1	−1	$H_2N_2O_2 + 6H^+ + 4e^- \rightarrow 2NH_3OH^+$	0.387		3
+1	−1	$N_2O + 6H^+ + H_2O + 4e^- \rightarrow 2NH_3OH^+$	−0.05		1
+1	−1	$N_2O_2^{2-} + 7H_2O + 4e^- \rightarrow 2NH_2OH + 6OH^-$	−0.76[a]		1

+1	−1	$N_2O + 5H_2O + 4e^- \rightarrow 2NH_2OH + 4OH^-$	−1.05^a	1
+1	−3	$N_2O(g) + 10H^+ + 8e^- \rightarrow 2NH_4^+ + H_2O$	0.647	2
+1	−3	$N_2O(g) + H_2O + 8e^- \rightarrow 2NH_3(aq)$	0.510	2
+1	−3	$NHCl_2 + 3H^+ + 4e^- \rightarrow 2Cl^- + NH_4^+$	−1.39	3
0	−1/3	$3N_2(g) + 2H^+ + 2e^- \rightarrow 2HN_3(aq)$	−3.10	1
0	−1/3	$3N_2(g) + 2H^+ + 2e^- \rightarrow 2HN_3(g)$	−3.40	3
0	−1/3	$3N_2(g) + 2e^- \rightarrow 2N_3^-$	−3.40	1
0	−1	$N_2(g) + 2H_2O + 4H^+ + 2e^- \rightarrow 2NH_3OH^+$	−1.87	1
0	−1	$N_2(g) + 2H_2O + 2e^- \rightarrow 2NH_2OH + 2OH^-$	−3.04^a	1
0	−2	$N_2(g) + 5H^+ + 4e^- \rightarrow N_2H_5^+$	−0.23	1
0	−2	$N_2(g) + 4H_2O + 4e^- \rightarrow N_2H_4 + 4OH^-$	−1.16^a	1
0	−3	$N_2(g) + 8H^+ + 6e^- \rightarrow 2NH_4^+$	0.275	2
0	−3	$N_2(g) + 6H^+ + 6e^- \rightarrow 2NH_3(g)$	−0.057	2
0	−3	$N_2(g) + 6H^+ + 6e^- \rightarrow 2NH_3(aq)$	−0.092	2
−1/3	0, −3	$HN_3(aq) + 3H^+ + 2e^- \rightarrow N_2(g) + NH_4^+$	1.96	1
−1/3	−3	$HN_3(aq) + 11H^+ + 8e^- \rightarrow 3NH_4^+$	0.695	3

TABLE 2 (Continued)

Oxidation states O → R	Electrode reaction	$E°$/V	Conditions	References
−1 −2	$2NH_3OH^+ + H^+ + 2e^- \rightarrow N_2H_5^+ + 2H_2O$	1.41		1
−1 −2	$2NH_2OH + 2e^- \rightarrow N_2H_4 + 2OH^-$	0.73[a]		1
−1 −3	$NH_3OH^+ + 2H^+ + 2e^- \rightarrow NH_4^+ + H_2O$	1.35		1
−1 −3	$NH_2OH + 2H_2O + 2e^- \rightarrow NH_3(aq) + 2OH^-$	0.42[a]		1
−1 −3	$NH_2Cl + H_2O + 2e^- \rightarrow Cl^- + NH_3 + OH^-$	−0.81[a]		3
−1 −3	$NH_2Cl + NH_4^+ + 2e^- \rightarrow Cl^- + 2NH_3$	−1.4	liq. NH_3	3
−1 −3	$NH_2Cl + 2H^+ + 2e^- \rightarrow Cl^- + NH_4^+$	−1.48		3
−2 −3	$N_2H_5^+ + 3H^+ + 2e^- \rightarrow 2NH_4^+$	1.275		1
−2 −3	$N_2H_4 + 4H_2O + 2e^- \rightarrow 2NH_3(aq) + 2OH^-$	0.1[a]		1
−3	$2NH_4^+ + 2e^- \rightarrow 2NH_3(aq) + H_2$	−0.55	Pt, NH_4Cl, 25°C	1

[a] For alkaline conditions, $E°$ is reported vs. the hydrogen electrode with $a_{OH^-} = 1.0$.

TABLE 3 Thermodynamic Data on Nitrogen[a]

Oxidation state	Chemical formula	Material[b] state	ΔH_f°/kJ mol^{-1}	ΔG_f°/kJ mol^{-1}	S°/J mol^{-1} K^{-1}
+5	HNO_3	g	−135.1	−74.76	266.3
	HNO_3	ℓ	−174.1	−80.89	155.6
	HNO_3	aq	−207.4	−111.3	146.
	NO_3^-	aq	−207.4	−111.3	146
	N_2O_5	c	−43.1	114.	178.
+4	NO_2	g	33.2	51.30	240.0
	N_2O_4	g	9.16	97.82	304.2
	N_2O_4	ℓ	−19.5	97.44	209.
+3	HNO_2	aq	−119.	−55.6	153.
	NO_2^-	aq	−105.	−37.	140
	$NOCl$	g	51.71	66.07	261.6
	$NOBr$	g	82.17	82.42	273.5
	N_2O_3	g	83.72	139.4	312.2
+2	NO	g	90.25	86.57	210.65

TABLE 3 (Continued)

Oxidation state	Chemical formula	Material[b] state	ΔH_f°/kJ mol^{-1}	ΔG_f°/kJ mol^{-1}	S°/J mol^{-1} K^{-1}
	N^{2+}	g	4744.49	—	—
+1	N_2O	g	82.05	104.2	219.7
	$H_2N_2O_2$	aq	-64.4	36.	217
	$N_2O_2^{2-}$	aq	-17.	139.	28.
	$NH_2O_2^-$	aq	-39.	76.1	142
	$H_2NO_2^-$	aq	-51.9	—	—
	N^+	g	1881.08	—	—
0	N_2	g	0.0	0.0	191.50
	N	g	472.704	340.9	153.2
-1/3	HN_3	g	294.	328.	238.9
	HN_3	ℓ	264.	327.	141.
	HN_3	aq	260.1	322.	146.
-1/3, -3	NH_4N_3	c	115.	274.	113.
-1	$NH_2OH_2^+$	aq	-137.	-56.65	155.

NH_2OH	aq	−98.3	23.4	167.
NH_2OH	c	−114.	—	—
NH	g	330.	—	—
$N_2H_5^+$	aq	−7.5	82.4	150.
N_2H_4	aq	34.3	128.	140.
N_2H_4	g	95.40	159.3	238.4
N_2H_4	ℓ	50.63	149.2	121.2
$N_2H_4 \cdot H_2O$	ℓ	−242.5	—	—
$N_2H_4 \cdot H_2^{2+}$	aq	−17.	94.14	79.
NH_4^+	g	84.	—	—
NH_4^+	aq	−132.5	−79.37	113.
NH_3	g	−46.11	−16.5	192.3
NH_3	aq	−80.29	−26.6	111.
NH_4OH	aq	−366.12	−263.8	181.

aAll data reported at 298.15 K (25°C). Data in italic are taken from Ref. 1. Most data are taken from the most recent National Bureau of Standards source [4].
bData for aqueous solutions (aq) are reported for the standard state ($m = 1$); undissociated.

A. Potential Diagrams for Nitrogen

Acid Solution

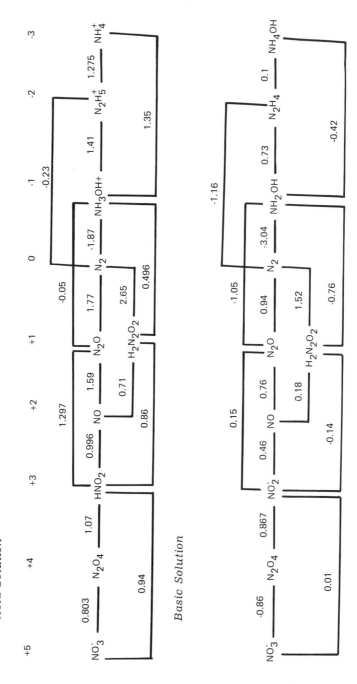

Basic Solution

1. The metal hydroxide is deposited on the electrode.
2. The rate of the NO_3^- reduction reaction is increased because of the concomitant lowering of the pH in the vicinity of the electrode.

This example, which forms the basis of metal hydroxide cathodic deposition processes in the battery industry [5], illustrates the importance of considering pH relationships in the electrochemistry of all the compounds of nitrogen.

A. Potential Diagrams for Nitrogen

Acid Solution

(See Art on page 138.)

Basic Solution

(See Art on page 138.)

REFERENCES

1. W. M. Latimer, *The Oxidation States of the Elements and Their Potentials in Aqueous Solution*, Prentice-Hall, Englewood Cliffs, N.J., 1952.
2. M. Pourbaix and N. de Zoubov, in *Atlas d'équilibres électrochimiques à 25°C*, M. Pourbaix, ed., Gauthier-Villars, Paris, 1963.
3. W. J. Plieth, in *Encyclopedia of the Electrochemistry of the Elements*, A. J. Bard, ed., Marcel Dekker, New York, 1973-1980.
4. D. D. Wagman, W. H. Evans, V. B. Parker, I. Hallow, S. M. Baily, and R. H. Schumm, *Selected Values of Chemical Thermodynamic Properties*, Natl. Bur. Stand. Tech. Note 270-3, U.S. Government Printing Office, Washington, D.C., 1968.
5. T. Palanisamy, Y. K. Kao, D. H. Fritts, and J. T. Maloy, J. Electrochem. Soc., *127*, 2535 (1980).

II. PHOSPHORUS*

K. S. V. SANTHANAM Tata Institute of Fundamental Research, Bombay, India, and University of Texas, Austin, Texas

Phosphorus exists in three main allotropic forms: white, red, and black. Each of the allotropic forms is polymorphic and there are in all about 11 modifications, some amorphous and others somewhat of indefinite identity and structure. The important compounds of the various oxidation states of phosphorus range from −3 to +5. Several other oxidation states of the elements, such as −0.5, +1, +6, and +7, are also known. Table 4 gives the oxidation states of a representative phosphorus compounds and their stabilities.

White phosphorus is taken as the standard state of the element. In recent years several oxidation-reduction reactions involving phosphorus or its compounds have been considered [1] and their Gibbs energies from equilibrium

---

*Reviewer: Donald Thornton, Drexel University, Philadelphia, Pennsylvania.

TABLE 4 Oxidation States of Phosphorus Compounds and Their Stabilities

Compound	Oxidation state	Stability
PH_3	−3	Readily oxidized by air
P_2H_4	−2	Decomposes on standing to polymeric amorphous solid
PH	−1	Unstable gas
H_3PO_2	1	
PCl_2	2	
H_3PO_3	3	Less stable than $P(OH)_2$; slowly hydrolyzed by moist air
$H_2P_2O_6$	4	
$POCl_3$	5	Hygroscopic; readily exchanges with Cl^-
POF_3	5	Stable in a limited temperature range; hydrolyzed by moist air
PF_5	5	Behaves as an acceptor

data have now become available through the persistent efforts of several authors discussed in the following pages.

A. Phosphorus-Phosphoric Acid Couple

In view of the importance of this couple in biological phosphorylation reactions, several attempts have been made to measure the equilibrium constant and its free energy of formation. The early work on this couple was due to Pitzer [2], who measured the equilibrium constants for the various stages of ionization:

$$H_3PO_4 \rightarrow H^+ + H_2PO_4^- \qquad K_1 = 7.5 \times 10^{-3} \tag{7.4}$$

$$H_2PO_4^- \rightarrow H^+ + HPO_4^- \qquad K_2 = 6.2 \times 10^{-8} \tag{7.5}$$

$$HPO_4^- \rightarrow H^+ + PO_4^- \qquad K_3 = 10^{-12} \tag{7.6}$$

These values have been largely confirmed in recent reports [3,4] except for $K_2 = 6.16 \times 10^{-8}$ ($pK_2 = 7.21$) and $K_3 = 2.1 \times 10^{-13}$ ($pK_3 = 12.67$). The pK_3 value has been arrived at indirectly by using esterification of the acid with pK_a increment of 0.66 for each ethoxy group replacing a hydroxy group. The incremental value was arrived at on the basis of a study of orthophosphate mono and diesters [3]. Among other reports, Grzybowsky [5] determined that $pK_2 = 7.2$ for $H_2PO_4^-$; Konopik and Leberl [6] determined a $pK_3 = 11.82$, while Vanderzee and Quist [7] gave a $pK_3 = 12.38$. The value determined by the latter authors appears to agree with a value of $pK_3 = 12.375$ reported by Bjerrum and Unmack [8]. Table 5 summarizes the pK_a values of orthorphosphoric acid. Based on the agreement in the values, $pK_1 = 2.12$, $pK_2 = 7.20$, and $pK_3 = 13.325$ appear to be reasonable values for H_3PO_4.

The pK_a values of substituted phosphorous acids depend on the nature of the substituent present in the acid. Presence of the $-OC_2H_5$ group decreases both the pK_1 and pK_2 of H_3PO_4; the same decreasing effect is observed with other acids, such as H_5PO_5 and HPO_3, suggesting the electron-withdrawing effect of the substituent [3]. The magnitude of such a decrease depends on the number of $-OC_2H_5$ groups present on the molecule. The enthalpy and Gibbs energy changes of the hydrolysis of phosphoric esters have also been measured [3] by the calorimetric method. Table 6 summarizes the values obtained with different esters.

An accurate $\Delta H_f^°(H_3PO_4)$ value in an aqueous medium has been obtained using a rotating bomb calorimeter [9]. White phosphorus combustion was carried out in a polyethylene tube in an atmosphere of oxygen and the products were hydrolyzed to orthophosphoric acid. The $\Delta H_f^°(H_3PO_4)$ value was determined as (-1294.3 ± 1.6) kJ mol^{-1}. This value differs by 26 kJ mol^{-1} from previous determinations [10-12]. The difference has been rising due to the uncertainty in $\Delta H_f^°[P_4H_{10} (c, hex)]$, whose value is not required in the bomb calorimetric approach adopted by Head and Lewis [9].

Kolthoff [13] measured the ionization of phosphorous acid as a function of ionic strength I and temperature. The extrapolated values of K for $I = 0$ at 291 K are as follows:

$$H_3PO_3 \rightarrow H^+ + H_2PO_3^- \qquad K_1 = 1.6 \times 10^{-2} \qquad (7.7)$$

$$H_2PO_3^- \rightarrow H^+ + HPO_3^- \qquad K_2 = 2.0 \times 10^{-7} \qquad (7.8)$$

Subsequent reports [14-19] have confirmed these values. Using Kolthoff's data, the $\Delta G_f^°(H_2PO_3^-) = -845.82$ kJ and $\Delta G_f^°(HPO_3^{2-}) = -810$ kJ mol^{-1}. These values lead to the following potentials for $H_3PO_3-H_3PO_4$.

TABLE 5 pK_a Values of Orthophosphoric Acid at 298 K

Method	pK_1	pK_2	pK_3	References
Heat of ionization	2.120	7.210	12.00	2
Group replacement			12.670	3
Indicator method			12.375	8
Electromotive force				
Cells with liquid junction	2.144		12.38	58,7
Cells without liquid junction		7.200		5
		7.197		8

TABLE 6 Enthalpies and Gibbs Energies of Addition and Elimination Reactions of Phosphoric Acid and Its Esters

Reactions	$\Delta H°$/kJ mol^{-1}	$\Delta G°$/kJ mol^{-1}	log K
$H_2O + H_3PO_4 \rightarrow P(OH)_5$	45.98 ± 8.36	71.06	−12.2 ± 2.2
$H_2O + EtOPO_3H_2 \rightarrow EtOP(OH)_4$	37.62	66.88	−11.6 ± 2.4
$H_2O + (EtO)_2PO_2H \rightarrow (EtO)_2P(OH)_3$	33.44	66.88	−11.6 ± 2.0
$H_2O + (EtO)_3PO \rightarrow (EtO)_3P(OH)_2$	29.66	58.52	−10.3 ± 2.7
$H_2O + H_2PO_4^- \rightarrow P(OH)_4O^-$		108.68	−18.6 ± 2.4
$H_2O + EtOPO_3H^- \rightarrow (EtO)P(OH)_3O^-$		104.5	−18.0 ± 2.6
$H_2O + (EtO)_2PO_2^- \rightarrow (EtO)_2P(OH)_2O^-$		104.5	−18.0 ± 2.2
$H_2O + HPO_4^{2-} \rightarrow P(OH)_3(O)_2^{2-}$		142.12	−24.7 ± 2.6
$H_2O + EtOPO_3^{2-} \rightarrow (EtO)P(OH)_2(O)_2^{2-}$		137.94	−24.0 ± 2.8
$H_2O + PO_4^{3-} \rightarrow P(OH)_2(O)_3^{3-}$		171.38	−30.0 ± 2.8

$HO^- + (EtO)_3PO \rightarrow (EtO)_3P(OH)O^-$	16.72	-2.8 ± 2.8
$HO^- + (EtO)_2PO_2^- \rightarrow (EtO)_2P(OH)(O)_2^{2-}$	91.76	-16.0 ± 2.4
$HO^- + (EtO)PO_3^{2-} \rightarrow (EtO)P(OH)(O)_3^{3-}$	154.66	-27.3 ± 3.0
$H_3PO_4 \rightarrow HPO_3 + H_2O$	133.76	-23.1 ± 1.5
$H_2PO_4^- \rightarrow PO_3^- + H_2O$	112.86	-19.6 ± 1.8
$HPO_4^{2-} \rightarrow PO_3^- + OH^-$	150.48	-26.4 ± 1.8
$EtOPO_3H_2 \rightarrow HPO_3 + EtOH$	121.22	-21.0 ± 1.8
$EtOPO_3H^- \rightarrow PO_3^- + EtOH$	104.50	-18.1 ± 2.0
$EtOPO_3^{2-} \rightarrow PO_3^- + EtO^-$	154.66	-27.0 ± 2.0
$(EtO)_2PO_3^- \rightarrow EtOPO_3 + EtO^-$	200.64	-35.2 ± 2.0
$(EtO)_2PO_3H \rightarrow EtOPO_3 + EtOH$	17.04	-20.5 ± 2.0

Source: Data taken from Ref. 3.

$$H_3PO_4 + 2H^+ + 2e^- \rightarrow H_3PO_3 + H_2O \qquad E^\circ = -0.276 \text{ V} \qquad (7.9)$$

$$PO_4^{3-} + 2H_2O + 2e^- \rightarrow HPO_3^{2-} + 3OH^- \qquad E^\circ = -1.12 \text{ V} \qquad (7.10)$$

The estimated standard entropies of H_3PO_3 and $H_2PO_3^-$ are 167.2 and 79.42 J mol^{-1} K^{-1} respectively.

Using ΔH_f° for $H_3PO_4(aq)$ as -1293.45 kJ mol^{-1}, the heats of formation of phosphoric acid solutions have been evaluated at 298.16 K. Table 7 gives the values of enthalpies of formation of H_3PO_4 solutions. Taking the enthalpy of fusion of H_3PO_4 as 12.96 kJ mol^{-1}, the increment $H_3PO_4(\ell,aq)$ is estimated as 26.50 kJ per mole of H_3PO_4. From a plot of relative apparent molal enthalpy of H_3PO_4 solutions it has been concluded [20] that the $\Delta H_f^\circ[H_3PO_4(aq)]$ value is in error by about 3.34 kJ mol^{-1}.

A precise determination of ΔG_f° of meta-H_3PO_4 has been difficult due to easy polymerization of the meta acid. Yost and Russell [21] reported the formation of acids $n(HPO_3)$, where $n = 2, 3, 4, 6$, and possibly 1. In a subsequent investigation Davis and Monk [22] confirmed the existence of tri-, tetra-, and hexametaphosphoric acids with the following ionization constants:

TABLE 7 Enthalpies of Formation of Phosphoric Acid Solutions at 298.16 K

nH_2O	$-\Delta H_f^\circ$/kJ mol^{-1}	nH_2O	$-\Delta H_f^\circ$/kJ mol^{-1}
1	1,277.7	100	1,292.2
2	1,281.7	200	1,292.6
3	1,284.1	300	1,292.8
4	1,285.72	400	1,292.8
5	1,286.77	500	1,292.9
6	1,287.6	700	1,293.0
8	1,288.6	1,000	1,293.1
10	1,289.3	2,000	1,293.2
12	1,289.8	3,000	1,293.2
15	1,290.3	4,000	1,293.3
20	1,290.8	5,000	1,293.3
25	1,291.1	7,000	1,293.3
30	1,291.3	10,000	1,293.3
40	1,291.6	20,000	1,293.3
50	1,291.8	50,000	1,293.4
75	1,292.1	100,000	1,293.4

Source: Data taken from Ref. 20.

$$HP_3O_9^{2-} \rightarrow H^+ + P_3O_9^{3-} \qquad K = 8.3 \times 10^{-8} \qquad (7.11)$$

$$HP_4O_{12}^{2-} \rightarrow H^+ + P_4O_{12}^{3-} \qquad K = 1.8 \times 10^{-3} \qquad (7.12)$$

$$Ca_2P_4O_{12} \rightarrow Ca^{2+} + CaP_4O_{12}^{2-} \qquad K = 2.2 \times 10^{-5} \qquad (7.13)$$

$$CaP_4O_{12}^{2-} \rightarrow Ca^{2+} + P_4O_{12}^{4-} \qquad K = 1.3 \times 10^{-5} \qquad (7.14)$$

For the trimetaphosphate

$$NaP_3O_{10}^{4-} \rightarrow Na^+ + P_3O_{10}^{5-} \qquad K = 3.0 \times 10^{-3} \qquad (7.15)$$

The dissociation constant of pyrophosphoric acid at various stages of ionization have also been measured or derived from conductivity measurements [23, 24], electrometric, or complex ion stability titrations [25-32]. The equilibria may be represented as

$$H_4P_2O_7 \rightarrow H^+ + H_3P_2O_7^- \qquad K_1 = 10^{-1} \qquad (7.16)$$

$$H_3P_2O_7^- \rightarrow H^+ + H_2P_2O_7^{2-} \qquad K_2 = 1.5 \times 10^{-2} \qquad (7.17)$$

$$H_2P_2O_7^{2-} \rightarrow H^+ + HP_2O_7^{3-} \qquad K_3 = 2.7 \times 10^{-7} \qquad (7.18)$$

$$HP_2O_7^{3-} \rightarrow H^+ + P_2O_7^{4-} \qquad K_4 = 2.4 \times 10^{-10} \qquad (7.19)$$

The thermodynamic quantities for other polyphosphoric acids are given in Table 8. A few general conclusions can be drawn from the ionization data. The ionization constant of ortho- and polyphosphoric acids is about 10^{-2} whenever terminal or middle PO_4 tetrahedron is bound to a strong hydrogen. When the PO_4 tetrahedron is also bonded to a weak hydrogen, the ionization constant is less by 10^{-8}. A similar trend has been observed in other thermodynamic functions based on the data in Table 8.

The ionization constant corresponding to the ionic reaction

$$H_2O + H_2P_2O_7^{2-} \rightarrow 2H_2PO_4^- \qquad K = 2.2 \times 10^2 \qquad (7.20)$$

has been reported [36] at 298 K. The temperature dependence of the equilibrium was also examined by the authors. From a plot of ln K versus T^{-1}, $\Delta H° = -19.64$ kJ mol^{-1} and $\Delta G° = -13.3$ kJ mol^{-1} have been determined for the hydration reaction. These low values for hydrolysis of $H_2P_2O_7^{2-}$ are in accord with low values for "high-energy phosphate" bonds [34-36].

The formation of pyrophosphoric acid from orthophosphoric acid has aslo been investigated [37]. The ionization

$$2H_3PO_4 \rightarrow H_4P_2O_7 + H_2O \qquad (7.21)$$

TABLE 8 Thermodynamic Quantities for Ionization of Phosphoric Acids at 298 K

H_aL^{n-}	pK_a	$\Delta G°$/kJ mol^{-1}	$\Delta H°$/kJ mol^{-1}	$\Delta S°^a$/J mol^{-1} K^{-1}
HPO_4^{2-}	12.36	70.43	10.80	−200.0
$H_2PO_4^{-}$	7.20	41.04	2.50	−130.0
H_3PO_4	2.12	12.08	−7.94	−67.18
$HP_2O_7^{3-}$	9.41	53.60	1.67	−175.5
$H_2P_2O_7^{2-}$	6.76	38.50	0.42	−128.0
$H_3P_2O_7^{-}$	2.64	15.04	−12.54	−92.5
$H_4P_2O_7$	−0.44	−2.50	−12.54	−33.69
$HP_3O_{10}^{4-}$	9.24	52.66	0.42	−175.0
$H_2P_3O_{10}^{3-}$	6.50	37.03	−6.27	−145.0
$H_3P_3O_{10}^{2-}$	2.30	13.12	−10.4	−79.0
$H_4P_3O_{10}^{-}$	1.20	6.85	−10.4	−57.8
$H_5P_3O_{10}$	−0.51	−2.92	−10.4	−25.0

aCalculated from $\Delta G° = \Delta H° - T\Delta S°$.
Source: Data taken from Ref. 34.

was examined by using specific electrical conductivity. The existence of equilibrium [18] results in decreasing electrical conductivity in the fusion of orthophosphoric acid and this decrease is proportional to the pyrophosphate concentration. The attainment of the equilibrium above takes about an hour at 353 K. The thermodynamic values for this equilibrium are $\Delta H° = 13.79$ kJ mol^{-1}, $\Delta G° = 7.92$ kJ mol^{-1}, and $\Delta S° = 16.76$ J mol^{-1} K^{-1} at 311 K.

Table 8 gives $\Delta S°$, which represents a sum of the differences between the entropies of the acid and conjugate base in the gaseous phase and in solution. On the basis of Pitzer's data [2] the gaseous contribution to $S°$ is only 12.5 to 25.0 J mol^{-1} K^{-1}. This difference has been interpreted as due to interaction of the two ions with the water structure and hydration numbers that accompany the addition of an electron to the nonionized species. The first of these factors would result in a positive entropy change since more charged conjugate base ion would disrupt the quasi-crystalline water structure. Since experimental results indicate a large negative $\Delta S°$, a second factor must be dominating, as this factor is due to dipole-dipole interactions (z^2/r, where z is the charge number of the ion and r is the crystal radius). Irani and Taulli [38] approximated the entropy variations as

$$\Delta S° = -11.78 - 7.6 \, \Delta z^2 \qquad (7.22)$$

$$\Delta S° = -6.2 - 5.2 \, \Delta z^2 \qquad (7.23)$$

$$\Delta S° = -3.0 - 4.2 \, \Delta z^2 \qquad (7.24)$$

for ortho-, pyro-, and tripolyphosphates.

B. Triphosphoric Acid: $H_5P_3O_{10}$

The pK_a values of $H_5P_3O_{10}$ were obtained [30,31,33,39,40] from electrometric titration curves. The values are pK_1 = very large, pK_2 = very large, pK_3 = 2.30, pK_4 = 6.50, and pK_5 = 9.24.

The proton dissociation of $H_2P_3O_{10}$ is compared with ribonucleotides and related compounds in Table 9. The $\Delta G°$ and $\Delta H°$ values given in the table suggests that the magnitude of $\Delta H°$ is a function of the proton donor atom and the type of bonding between it and the parent molecule (i.e., O; primary, secondary, tertiary, N, etc.). However, it does not appear to depend on the pK_a or the nature of the parent molecule. The entropies of dissociation show a trend. The cratic and unitary portions of $\Delta S°$ (cratic dealing with particle-solvent mixing, unitary solvent-particle interaction) are revealed by the data given in Table 9. The cratic portion of $\Delta S°$ = 12.5 to 25.0 J mol^{-1} K^{-1} [41]. The unitary portion of $\Delta S°$ becomes more negative as the negative charge increases, resulting in increased solvent ordering capacity of the aqueous species.

Watters et al. [42] considered the properties of tetraphosphoric acid as a base and a complexing agent. From electrometric titration they arrived at the pK_a values in Table 10; for comparison the pK values of other polyacids are also given in the table. The pK_{N-1} values increase and pK_N values decrease with increase in chain length. However, $H_5P_3O_{10}$ is an exception to this rule due to stable configuration of $HP_3O_{10}^{4-}$ ion.

C. Hypophosphoric Acid

The reaction

$$H_4P_2O_6 + H_2O \rightarrow H_3PO_3 + H_3PO_4 \qquad \Delta G° = -126.4 \text{ kJ mol}^{-1} \qquad (7.25)$$

is slow in cold dilute solutions and proceeds faster in acidic concentrated solutions [43]. Electrometric titrations yielded pK_1 = 2.20, pK_2 = 2.10, pK_3 = 6.77, and pK_4 = 9.48 at 293 K. On the basis of the data above, $\Delta G_f°(H_4P_2O_6)$ = -1638.5 kJ mol^{-1} has been estimated for $H_4P_2O_6$. The reactions

$$2H_3PO_4 + 2H^+ + 2e^- \rightarrow H_4P_2O_6 + 2H_2O \qquad E° = -0.93 \text{ V} \qquad (7.26)$$

$$H_4P_2O_6 + 2H^+ + 2e^- \rightarrow 2H_3PO_3 \qquad E° = 0.38 \text{ V} \qquad (7.27)$$

give a picture of the redox behavior of phosphoric-phosphorous couples. The redox potentials (Table 11) give a picture of the ease of oxidizability of the phosphorous acid. Thus a fairly powerful 1e oxidizing agent would be required

TABLE 9 Thermodynamic Values for Proton Dissociation from Ribonucleotides and Related Compounds[a]

Compound[b]	N_1H (pK = 4)			N_2H (pK = 9)			P–O–H (pK = 7)			P–O–H (pK = 9)		
	$\Delta G°$	$\Delta H°$	$\Delta S°$	$\Delta G°$	$\Delta H°$	$\Delta S°$	$\Delta G°$	$\Delta H°$	$\Delta S°$	$\Delta G°$	$\Delta H°$	$\Delta S°$
Adenine	23.8	20.5	−11.07	56.0	13.8	−141.6						
AMP	21.3	17.5	−12.57				36.7	−7.52	−148.38			
ADP	23.8	17.1	−22.62				39.7	−5.43	−148.38			
ATP	22.5	15.5	−23.88				39.7	−5.01	−148.38			
RP							38.0	−11.3	−165.30			
$H_2PO_4^-$							40.96	3.76	−124.80			
$H_2P_2O_7^-$							38.4	−1.25	−133.05	53.5	−7.1	−203.3
$H_2P_3P_{10}^{3-}$							35.5	−5.43	−137.3	50.5	−15.8	−222.7

[a] Units of $\Delta G°$ and $\Delta H°$ are kJ mol^{-1}; units of $\Delta S°$, J mol^{-1} K^{-1}.
[b] AMP, adenosine monophosphate; ADP, adenosine diphosphate; ATP, adenosine triphosphate; RP, ribose phosphate.

TABLE 10 pK Values of Polyphosphoric Acids

	$H_6P_4O_{13}$	$H_5P_3O_{10}$	$H_4P_2O_7$
		$I = 0.0$	
pK_N	9.11	9.24	9.42
pK_{N-1}	7.38	6.50	6.76
pK_{N-2}		2.30	2.64
		$I = 1.0\ (CH_3)_4NNO_3$	
pK_N	8.34	8.81	8.93
pK_{N-1}	6.63	5.83	6.13
pK_{N-2}	2.23	2.11	1.81
pK_{N-3}	1.36	1.06	0.82

to oxidize phosphorous acid to $H_2PO_3(H_4P_2O_6)$. The reaction will continue further due to ease of the next step with H_2PO_3 oxidized to phosphoric acid.

D. Hypophosphorous Acid

The ionization constant of hypophosphorous acid varies with ionic strength of the medium, as expected, and the following values were obtained: $I = 0$, $pK_1 = 1.34$; $I = 2.26$ mol L^{-1}, $pK_1 = 1.45$; and $I = 5.6$ mol L^{-1}, $pK_1 = 1.71$ [44]. The pK_1 value of another report [45] was 1.23 ($I = 0$). Furthermore, a recalculation of the pK_1 from the conductivity data reported in Ref. 8 gave a value of 1.09 [46]. On the basis of the various reports, the pK_1 of hypophosphoric acid appears to lie between 1.09 and 1.34. Taking $pK_1 = 1.23$, $\Delta G_f^\circ(H_2PO_2^-)$ can be calculated from the reaction

$$H_3PO_2 \to H^+ + H_2PO_2^- \qquad K = 0.058 \qquad (7.28)$$

as -519.57 kJ mol^{-1} and the potentials of the various couples are

$$H_3PO_3 + 2H^+ + 2e^- \to H_3PO_2 + H_2O \qquad E^\circ = -0.499\ \text{V} \qquad (7.29)$$

$$HPO_3^{2-} + 2H_2O + 2e^- \to H_2PO_2^- + 3OH^- \qquad E^\circ = -1.57\ \text{V} \qquad (7.30)$$

$$H_3PO_2 + H^+ + e^- \to 2H_2O + P \qquad E^\circ = -0.508\ \text{V} \qquad (7.31)$$

$$H_2PO_2^- + e^- \to 2OH^- + P \qquad E^\circ = -2.05\ \text{V} \qquad (7.32)$$

TABLE 11 Standard Potentials of Compounds of Phosphorus in Aqueous Solutions

Half-reaction	Standard potential, $E°/V$
P^{2-}/P^{3-}	
$P_2H_4 + 2H^+ + 2e^- \rightarrow 2PH_3$	0.006
P_4^{2-}/P_2^{4-}	
$P_4H_2 + 6H^+ + 6e^- \rightarrow 2P_2H_4$	−0.014
P_4^{2-}/P^{3-}	
$P_4H_2 + 10H^+ + 10e^- \rightarrow 4PH_3$	−0.006
P^0/P_2^{4-}	
$2P(r) + 4H^+ + 4e^- \rightarrow P_2H_4$	−0.169
$2P(w) + 4H^+ + 4e^- \rightarrow P_2H_4$	−0.100
P^0/P^{3-}	
$P(w) + 3H_2O + 3e^- \rightarrow PH_3 + 3OH^-$	−0.89
$P(r) + 3H^+ + 3e^- \rightarrow PH_3$	−0.111
$P(w) + 3H^+ + 3e^- \rightarrow PH_3$	−0.063
P^0/P_4^{2-}	
$4P(w) + 2H_2O + 2e^- \rightarrow P_4H_2 + 2OH^-$	−1.18
$4P(r) + 2H^+ + 2e^- \rightarrow P_4H_2$	−0.633
$4P(w) + 2H^+ + 2e^- \rightarrow P_4H_2$	−0.347
P^+/P^{3-}	
$H_3PO_2 + 4H^+ + 4e^- \rightarrow PH_3 + 2H_2O$	−0.174
$H_2PO_2^- + 5H^+ + 4e^- \rightarrow PH_3 + 2H_2O$	−0.145
P^+/P_4^{2-}	
$4H_3PO_2 + 6H^+ + 6e^- \rightarrow P_4H_2 + 8H_2O$	−0.455

TABLE 11 (Continued)

Half-reaction	Standard potential, $E°/V$
P^+/P^0	
$H_2PO_2^- + e^- \rightarrow P(w) + 2OH^-$	-2.05
$H_3PO_2 + H^+ + e^- \rightarrow 2H_2O + P(w)$	-0.508
$H_2PO_2^- + 2H^+ + e^- \rightarrow 2H_2O + P(w)$	-0.391
$H_3PO_2 + H^+ + e^- \rightarrow 2H_2O + P(r)$	-0.365
$H_2PO_2^- + 2H^+ + e^- \rightarrow 2H_2O + P(r)$	-0.248
P^{3+}/P^{3-}	
$H_3PO_3 + 6H^+ + 6e^- \rightarrow PH_3 + 3H_2O$	-0.282
$H_2PO_3^- + 7H^+ + 6e^- \rightarrow PH_3 + 3H_2O$	-0.265
$HPO_3^{2-} + 8H^+ + 6e^- \rightarrow PH_3 + 3H_2O$	-0.205
P^{3+}/P_4^{2-}	
$4H_3PO_3 + 14H^+ + 14e^- \rightarrow P_4H_2 + 12H_2O$	-0.480
$4H_2PO_3^- + 18H^+ + 14e^- \rightarrow P_4H_2 + 12H_2O$	-0.450
$4HPO_3^{2-} + 22H^+ + 14e^- \rightarrow P_4H_2 + 12H_2O$	-0.346
P^{3+}/P^0	
$H_3PO_3 + 3H^+ + 3e^- \rightarrow 3H_2O + P(w)$	-0.502
$H_2PO_3^- + 4H^+ + 3e^- \rightarrow 3H_2O + P(w)$	-0.467
$H_3PO_3 + 3H^+ + 3e^- \rightarrow 3H_2O + P(r)$	-0.454
$H_2PO_3^- + 4H^+ + 3e^- \rightarrow 3H_2O + P(r)$	-0.419
$HPO_3^{2-} + 5H^+ + 3e^- \rightarrow 3H_2O + P(r)$	-0.346
$HPO_3^{2-} + 5H^+ + 3e^- \rightarrow 3H_2O + P(w)$	-0.298
P^{3+}/P^+	
$HPO_3^{2-} + 2H_2O + 2e^- \rightarrow H_2PO_2^- + 3OH^-$	-1.57

TABLE 11 (Continued)

Half-reaction	Standard potential, $E°/V$
$HPO_3^{2-} + 3H^+ + 2e^- \rightarrow H_2PO_2^- + H_2O$	−0.504
$H_3PO_3 + 2H^+ + 2e^- \rightarrow H_3PO_2 + H_2O$	−0.499
$H_2PO_3^- + 3H^+ + 2e^- \rightarrow H_3PO_2 + H_2O$	−0.446
$HPO_3^{2-} + 3H^+ + 2e^- \rightarrow H_2PO_2^- + H_2O$	−0.323

P^{4+}/P^{3+}

Half-reaction	$E°/V$
$H_2P_2O_6^{2-} + 2e^- \rightarrow 2HPO_3^{2-}$	−0.061
$HP_2O_6^{3-} + H^+ + 2e^- \rightarrow 2HPO_3^{2-}$	0.275
$H_4P_2O_6 + 2e^- \rightarrow 2H_2PO_3^-$	0.275
$H_3P_2O_6^- + H^+ + 2e^- \rightarrow 2H_2PO_3^-$	0.340
$H_4P_2O_6 + 2H^+ + 2e^- \rightarrow 2H_3PO_3$	0.380
$H_4P_2O_6^{2-} + 2H^+ + 2e^- \rightarrow 2H_2PO_3^-$	0.423
$P_2O_6^{4-} + 2H^+ + 2e^- \rightarrow 2HPO_3^{2-}$	0.570

P^{5+}/P^{3-}

Half-reaction	$E°/V$
$H_3PO_4 + 8H^+ + 8e^- \rightarrow PH_3 + 4H_2O$	−0.281
$H_2PO_4^- + 9H^+ + 8e^- \rightarrow PH_3 + 4H_2O$	−0.265
$HPO_4^{2-} + 10H^+ + 8e^- \rightarrow PH_3 + 4H_2O$	−0.212
$PO_4^{3-} + 11H^+ + 8e^- \rightarrow PH_3 + 4H_2O$	−0.123

P^{5+}/P^{2-}

Half-reaction	$E°/V$
$4H_3PO_4 + 22H^+ + 22e^- \rightarrow P_4H_2 + 16H_2O$	−0.406
$4H_2PO_4^- + 26H^+ + 22e^- \rightarrow P_4H_2 + 16H_2O$	−0.383
$4H_2PO_2^- + 10H^+ + 6e^- \rightarrow P_4H_2 + 8H_2O$	−0.376
$4HPO_4^{2-} + 30H^+ + 22e^- \rightarrow P_4H_2 + 16H_2O$	−0.305
$4PO_4^{3-} + 34H^+ + 22e^- \rightarrow P_4H_2 + 16H_2$	−0.176

TABLE 11 (Continued)

Half-reaction	Standard potential, $E°/V$
P^{5+}/P^0	
$H_3PO_4 + 5H^+ + 5e^- \rightarrow 4H_2O + P(w)$	−0.411
$H_2PO_4^- + 6H^+ + 5e^- \rightarrow 4H_2O + P(w)$	−0.386V
$H_3PO_4 + 5H^+ + 5e^- \rightarrow 4H_2O + P(r)$	−0.383
$H_2PO_4^- + 6H^+ + 5e^- \rightarrow 4H_2O + P(r)$	−0.358
$HPO_4^{2-} + 7H^+ + 5e^- \rightarrow 4H_2O + P(w)$	−0.316
$HPO_4^{2-} + 7H^+ + 5e^- \rightarrow 4H_2O + P(r)$	−0.288
$PO_4^{3-} + 8H^+ + 5e^- \rightarrow 4H_2O + P(w)$	−0.156
$PO_4^{3-} + 8H^+ + 5e^- \rightarrow 4H_2O + P(r)$	−0.128
P^{5+}/P^{3+}	
$PO_4^{3-} + 2H_2O + 2e^- \rightarrow HPO_3^{2-} + 3OH^-$	−1.12
$H_3PO_4 + H^+ + 2e^- \rightarrow H_2PO_3^- + H_2O$	−0.329
$H_3PO_4 + 2H^+ + 2e^- \rightarrow H_3PO_3 + H_2O$	−0.276
$H_2PO_4^- + 2H^+ + 2e^- \rightarrow H_2PO_3^- + H_2O$	−0.260
$HPO_4^{2-} + 2H^+ + 2e^- \rightarrow HPO_3^{2-} + H_2O$	−0.234
$PO_4^{3-} + 3H^+ + 2e^- \rightarrow HPO_3^{2-} + H_2O$	−0.121
P^{5+}/P^{4+}	
$2HPO_4^{2-} + 2H^+ + 2e^- \rightarrow P_2O_6^{4-} + 2H_2O$	−1.039
$2H_2PO_4^- + 2H^+ + 2e^- \rightarrow H_2P_2O_6^{2-} + 2H_2O$	−0.955
$2H_3PO_4 + 2H^+ + 2e^- \rightarrow H_4P_2O_6 + 2H_2O$	−0.933
$2H_2PO_4^- + 3H^+ + 2e^- \rightarrow H_3P_2O_6^- + 2H_2O$	−0.872
$2H_2PO_4^- + 4H^+ + 2e^- \rightarrow H_4P_2O_6 + 2H_2O$	−0.807
$2HPO_4^{2-} + 3H^+ + 2e^- \rightarrow HP_2O_6^{3-} + 2H_2O$	−0.744

TABLE 11 (Continued)

Half-reaction	Standard potential, $E°/V$
$2HPO_4^{2-} + 4H^+ + 2e^- \rightarrow H_2P_2O_6^{2-} + 2H_2O$	−0.551
$2PO_4^{3-} + 4H^+ + 2e^- \rightarrow P_2O_6^{4-} + 2H_2O$	−0.328

Source: Ref. 1, except for the first reaction under P^0/P_4^{2-}, which is from Ref. 54.

On the basis of kinetic data, Hayward and Yost [47] proposed the existence of two forms of the acid:

$$\begin{array}{cc} O & O \\ \| & \| \\ H-P-H \rightleftharpoons H-P \\ | & | \\ OH & OH \end{array} \quad \text{(active form)} \tag{7.33}$$

Griffith and McKeown [48] determined the activation energy for the conversion as 76.07 kJ mol^{-1}.

E. Phosphine and Other Hydrogen Phosphides

A direct determination of the enthalpies of formation of PH$_3$ and its dimer gave $\Delta H_f° = 5.43 \pm 1.67$ kJ mol^{-1} for PH$_3$ and $\Delta H_f° = 20.90 \pm 4.18$ kJ mol^{-1} for P$_2$H$_4$ [49]. The $\Delta H_f°$(PH$_3$) value

$$2P(s,w) + 3H_2(g) \rightarrow 2PH_3(g) \tag{7.34}$$

has been calculated from equilibrium constant measurements at 298.1 K as −13.12 kJ mol^{-1} [50,51]. The data indicate a spontaneous reaction between white P and H$_2$ to form phosphine [52]. It is possible to reason why the formation of tetraphosphide is not favored in the reduction of phosphorus on the basis of the following potential arguments:

$$P(s) + 3H_2O + 3e^- \rightarrow PH_3(g) + 3OH^- \qquad E° = -0.89 \text{ V} \tag{7.35}$$

$$P(s) + 3H^+ + 3e^- \rightarrow PH_3(g) \qquad E° = -0.063 \text{ V} \tag{7.36}$$

and similarly for tetraphosphide,

$$4P(s) + 2H_2O + 2e^- \rightarrow P_4H_2 + 2OH^- \qquad E° = -1.18 \text{ V} \tag{7.37}$$

$$4P(s) + 2H^+ + 2e^- \rightarrow P_4H_2 \qquad E° = -0.347 \text{ V} \tag{7.38}$$

TABLE 12 Thermodynamic Data on Phosphorus at 298.15 K

Formula	Description	State	$\Delta H°$/kJ mol^{-1}	$\Delta G°$/kJ mol^{-1}	$S°$/J mol^{-1} K^{-1}
P		g	314.25	278.84	162.93
P	α, white	c	0.0	0.0	41.05
P	Red	c	−17.56	−12.12	22.78
P	Black	c	−39.29		
P$^+$		g	1378.9		
P^{2+}		g	3279.9		
P$_2$		g	146.0	144.21	217.8
P$_4$		g	58.85	24.32	279.6
PO$_4^{3-}$	Std. state, $m = 1$	aq	−1276.1	−1017.8	−221.5
P$_2$O$_7^{4-}$	St. state, $m = 1$	aq	−2268.9	−1917.3	−117.0
PH$_3$		g	5.43	13.38	209.9
P$_2$H$_4$		ℓ	−5.02		
P$_4$H$_2$		c	28.84	66.88	167.2

TABLE 11 (Continued)

Formula	Description	State	$\Delta H°$/kJ mol^{-1}	$\Delta G°$/kJ mol^{-1}	$S°$/J mol^{-1} K^{-1}
HPO_3		c	−947.6		
HPO_3^-		aq	−976.0		
$H_2PO_2^-$		aq	−613.2		
HPO_3^{2-}		aq	−968.1		
HPO_4^{2-}	Std. state, $m=1$	aq	−1290.9	−1088.2	−33.4
$H_2PO_3^-$		aq	−968.5		
$H_2PO_4^-$	Std. state, $m=1$	aq	−1294.9	−1087.5	90.2
H_3PO_2		c	−604.01		
		ℓ	−594.8		
H_3PO_4		c	−1277.8	−1118.1	110.4
		ℓ	−1265.7		
	Ionized state, $m=1$	aq	−1276.1	−1017.8	−221.5
H_5PO_3		c	−970.5		

H$_3$PO$_3$	c	−963.5	
H$_4$P$_2$O$_6$	aq	−963.9	
	aq	−917.93	
PF$_3$	g	−1638.5	272.8
PF$_5$	g	−1594.2	
PCl$_3$	ℓ	−319.3	
	c	−443.1	
PCl$_5$	g	−374.5	364.1
POCl$_3$	ℓ	−596.5	222.2
	g	−557.9	325.0
PBr$_3$	ℓ	−184.3	239.9
	g	−139.2	347.6
PBr$_5$	c	−269.6	
POBr$_3$	c	−458.1	
PH$_4$Br	c	−127.5	109.9
PI$_3$	c	−45.5	
PI$_2$	c	−23.82	

TABLE 11 (Continued)

Formula	Description	State	$\Delta H°$/kJ mol^{-1}	$\Delta G°$/kJ mol^{-1}	$S°$/J mol^{-1} K^{-1}
P_2I_4		c	−82.59	−82.7	184.7
PH_4I		c	−69.80	0.84	122.8
PN		g	136.70	114.5	210.9
PC		c	451.98		
P_4S_2		c	−104.0		
P_4S_3		c	−154.2		
P_4S_4		c	−170.5		
P_2S_5		c	−206.9		
P_4S_7		c	−257.4		
P_4S_9		c	−292.1		
P_4S_{10}		c	−308.9		

Source: Ref. 55.

F. Phosphorus-Halogen Compounds

The entropy and free energy of $PF_3(g)$ were determined by considering the reaction

$$3CaF_2 + 2PCl_3 \rightarrow 2PF_3 + 3CaCl_2 \tag{7.39}$$

as $\Delta S°(PF_3) = 229.20$ J mol^{-1} K^{-1} and $\Delta G° = 1013.06$ kJ mol^{-1} [56]. $\Delta H_f°(PF_3)$ determined by this method (-944.8 kJ mol^{-1}) agreed with a previous value of $\Delta H_f° = -927.75$ kJ mol^{-1} [55]. From the Gibbs energy of $PCl_3(g)$ and the Gibbs energy of vaporization, $\Delta G_f°[PCl_3(\ell)] = -286.7$ kJ mol^{-1} has been calculated [54]. The enthalpy of formation of PCl_3 has also been obtained as -310.9 kJ mol^{-1} [49,53]. In cold water the chloride hydrolyzes to phosphorous acid:

$$PCl_3 + 3H_2O \rightarrow H_3PO_3 + 3Cl^- + 3H^+ \tag{7.40}$$

The Gibbs energies of number of other halogen compounds have been estimated and the data are given in Table 12. An interesting phosphorous-halogen compound is PI_2 since it is one of the few compounds having the +2 oxidation state; $\Delta H_f° = -23.82$ kJ mol^{-1}. The solid PI_2 is easily hydrolyzed,

$$2PI_2(c) + 5H_2O \rightarrow H_3PO_3 + H_3PO_2 + 4H^+ + 4I^- \tag{7.41}$$

with $\Delta G° = -326.04$ kJ mol^{-1}.

G. Phosphorus-Sulfur Compounds

The enthalpies of formation of several phosphorus-sulfur compounds, such as P_4S_2, P_4S_3, P_4S_4, P_2S_5, P_4S_7, P_4S_9, and P_4S_{10}, have been measured [57]. Their values are given in Table 12.

H. Phosphorus-Nitrogen Compounds

The enthalpy of formation of the phosphorus-nitrogen compound PN has been determined [59] and the value is given in Table 12.

I. Phosphorus-Carbon Compounds

The enthalpy of formation of the phosphorus-carbon compound PC has been measured and the value is given in Table 12.

J. Potential Diagrams for Phosphorus

Acid Solution

(See diagram on page 160.)

Basic Solution

(See diagram on page 160.)

A consideration of the diagram leads to a discussion of a number of decomposition reactions in both acid and basic solutions. The disproportionation of phosphorus into +1 and −3 states occurs more readily in alkaline solu-

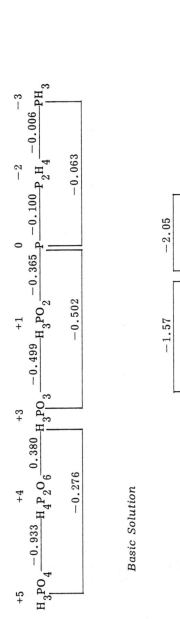

tions since the free-energy change for the reaction in alkaline solution is larger. Phosphorus also liberates H_2 in alkaline solution because of its enormous reducing power. Several other points may be noted: (1) hypophosphorous acid is unstable with respect to its decomposition into phosphorous acid and phosphine; (2) hypophosphorous acid will reduce water to H_2; (3) the decomposition of phosphorous acid into phosphine and phosphoric acid has a small positive free energy; and (4) phosphorus is unstable with respect to its hydrolysis into the +5 and −3 oxidation states.

REFERENCES

1. A. P. Tomilov and N. E. Chomutov, in *Encyclopedia of the Electrochemistry of the Elements*, A. J. Bard, ed., Vol. 3, Marcel Dekker, New York, 1975, Chap. 1.
2. K. S. Pitzer, J. Am. Chem. Soc., *59*, 2365 (1937).
3. J. P. Guthrie, J. Am. Chem. Soc., *99*, 3991 (1977).
4. R. C. Weast, ed. *Handbook of Chemistry and Physics*, 37th ed., Chemical Rubber Publishing Co., Cleveland, Ohio, 1956.
5. A. K. Grzybowsky, J. Phys. Chem., *62*, 555 (1958).
6. N. Konopik and O. Leberl, Monatsh. Chem., *80*, 655 (1949).
7. C. E. Vanderzee and A. S. Quist, J. Phys. Chem., *65*, 118 (1961).
8. N. Bjerrum and A. Unmack, Kl. Dan. Vidensk. Selsk. Math.-Fys. Medd., *9*, 5 (1929).
9. A. J. Head and G. B. Lewis, J. Chem. Thermodyn., *2*, 701 (1970).
10. E. P. Egan and B. B. Luff, Tennessee Valley Authority Report (Heats of formation of phosphorus oxides). June-Nov., 1963.
11. W. S. Holmes, Trans. Faraday Soc., *58*, 1916 (1962).
12. R. J. Irving and H. McKerrell, Trans. Faraday Soc., *63*, 2582 (1967).
13. I. M. Kolthoff, Recl. Trav. Chim., *46*, 350 (1927).
14. W. V. Bhagwat and N. R. Dhar, J. Indian Chem. Soc., *6*, 807 (1929).
15. C. Morton, Q. J. Pharm. Pharmacol., *3*, 438 (1930).
16. P. Nylem, Z. Anorg. Chem., *230*, 385 (1937).
17. R. O. Griffith, A. McKeown, and R. P. Taylor, Trans. Faraday Soc., *36*, 752 (1940).
18. K. Takhasi and N. Yui, Bull. Inst. Phys. Chem. Res. Tokyo, *20*, 521 (1941).
19. V. Frei, J. Podlahova, and J. Podlaha, Coll. Czech. Chem. Commun., *29*, 2587 (1964).
20. P. E. Egan, Jr., and B. B. Luff, J. Phys. Chem., *65*, 523 (1961).
21. D. Yost and H. Russell, *Systematic Inorganic Chemistry*, Prentice-Hall, Englewood Cliffs, N.J., 1946.
22. C. W. Davis and C. B. Monk, J. Chem. Soc., 413-427 (1949).
23. J. W. Mellor, *Comprehensive Inorganic Chemistry*, Vol. 8, Longmans, Green, London, p. 957, 1928.
24. C. B. Monk, J. Chem. Soc., 423 (1949).
25. J. Beukenkamp, W. Rieman, and S. Linderbaum, Anal. Chem., *26*, 505 (1954).
26. I. M. Kolthoff and W. Bosch, Recl. Trav. Chim., *47*, 826 (1928).
27. J. Muss, Z. Phys. Chem., A159, 268 (1932).
28. J. D. McGilvery and J. P. Crowther, Can. J. Chem., *32*, 174 (1954).
29. J. I. Watters and A. Aaron, J. Am. Chem. Soc., *75*, 611 (1953).
30. R. R. Irani, J. Phys. Chem., *65*, 1463 (1961).

31. S. M. Lamberts and J. I. Watters, J. Am. Chem. Soc., 79, 4262 (1957).
32. J. E. Ricci, J. Am. Chem. Soc., 70, 109 (1948).
33. G. E. Shaffer, Jr., Determination of ionization constants of polybasic acids, Thesis, Temple University, 1960; Diss. Abstr. 21, 1394 (1960).
34. R. R. Irani and T. A. Taulli, J. Inorg. Nucl. Chem., 28, 1011 (1966).
35. R. P. Mitra, H. C. Malhotra, and D. V. S. Jain, Trans. Faraday Soc., 62, 167 (1966).
36. C. D. Schmulbach, J. R. VanWazer, and R. R. Irani, J. Am. Chem. Soc., 81, 6347 (1959).
37. R. A. Munson, J. Phys. Chem., 69, 1761 (1965).
38. R. R. Irani and T. A. Taulli, J. Inorg, Nucl. Chem., 28, 1018 (1966).
39. R. G. Bates, J. Res. Natl. Bur. Stand., 47, 127 (1951).
40. R. G. Bates and S. F. Acree, J. Res. Natl. Bur. Stand., 30, 129 (1947).
41. J. J. Christensen and R. M. Izatt, J. Phys. Chem., 66, 1030 (1962).
42. J. I. Watters, P. E. Sturrock, and R. E. Simonaitis, Inorg. Chem., 2, 765 (1963).
43. G. Schwarzenbach and J. Zurc, J. Monatsch., 81, 202 (1950).
44. G. E. Schaffer, Jr., Ph.D. dissertation, Temple University 1960.
45. R. A. Paris and J. C. Merlin, Silicium Schwefel Phosphate Colloq. Sek. Anorg. Chem., Munich, 1954, p. 237.
46. T. D. Farr, Tennessee Valley Authority Chem. Eng., Rep. 8.
47. P. Hayward and D. Yost, J. Am. Chem. Soc., 71, 915 (1949).
48. R. O. Griffith and A. McKeown, Trans. Faraday Soc., 36, 752 (1940).
49. S. R. Gunn and L. G. Green, J. Phys. Chem., 65, 779 (1961).
50. D. P. Stevenson and D. M. Yost, J. Chem. Phys., 9, 403 (1941).
51. D. M. Yost and T. F. Anderson, J. Chem. Phys., 2, 624 (1934).
52. G. G. Devyatykh and A. S. Yushin, Zh. Fiz. Khim., 38, 957 (1964).
53. T. Charnley and H. A. Skinner, J. Chem. Soc., 450 (1953).
54. W. Latimer, *Oxidation Potentials*, 2nd ed., Prentice-Hall, Englewood Cliffs, N.J., 1952, p. 113.
55. R. L. Potter and V. N. Distefano, J. Phys. Chem., 65, 849 (1961).
56. H. C. Duus and D. P. Mykytiuk, J. Chem. Eng. Data, 9, 585 (1964).
57. H. Vincet and V. F. Christiane, Bull. Soc. Chim. Fr., 499 (1973).
58. C. B. Monk and M. F. Amira, J. Chem. Soc., Faraday Trans., 1170 (1978).
59. D. D. Wagman, W. H. Evans, V. B. Parker, I. Halow, S. M. Bailey, and R. H. Schumm, Natl. Bur. Stand. Tech. Note 270-3, U.S. Government Printing Office, Washington, D.C., 1968.

III. ARSENIC*

K. S. V. SANTHANAM AND N. S. SUNDARESAN[†] Tata Institute of Fundamental Research, Bombay, India

Arsenic exists in three allotropic forms: yellow (γ), black (β), and metallic gray (α) forms. At room temperature the gray form is the most stable form. The important oxidation states of arsenic are +5, +3, and −3, as in arsenic

---

Reviewer: Donald Thornton, Drexel University, Philadelphia, Pennsylvania.
†*Current affiliation*: Sophia College, Bombay, India.

acid, arsenious acid, and arsine, respectively. The other oxidation states, such as -1, -0.5, $+1$, $+2$, $+4$, and $+7$, are also known.

Gray arsenic is taken as the standard state of the element.

A. Arsine

The widely accepted value of the potential in acid solutions for the reaction

$$As + 3H^+ + 3e^- \rightarrow AsH_3(g) \tag{7.42}$$

of $E° = -0.60$ V, which was arrived at earlier [1] using $\Delta G_f° = 175.5$ kJ mol^{-1}, was questioned by Davis [2] and was reexamined by several authors [2-6]. This value was arrived at using an erroneous value of $\Delta H_f° = 171.38$ kJ mol^{-1} [1]. An estimate for the standard entropy of arsine of $S° = 217.57$ kJ mol^{-1} K^{-1} was used in this calculation. A reexamination of the reaction

$$SbH_3 + As \rightarrow AsH_3 + Sb \tag{7.43}$$

through a method of explosion in mixtures led to $\Delta H_f°(AsH_3) = 62.04$ kJ mol^{-1} [7]. This value was consistent with previous reports of 62.34 kJ mol^{-1} [8] and 62.76 kJ mol^{-1} [4]. With the standard entropy of $S°(AsH_3) = 222.4 \pm 0.4$ J mol^{-1} K^{-1} [5] and $\Delta S_f° = -8.37$ J mol^{-1} K^{-1}, the $\Delta G_f°(AsH_3) = 65.23$ kJ mol^{-1} has been estimated by Davis [2] using $\Delta H_f° = 62.76$ kJ mol^{-1}.

The National Bureau of Standards (NBS) reports [16] values of $\Delta H_f° = 66.4$ kJ mol^{-1}, $\Delta G_f° = 68.9$ kJ mol^{-1}, and $S° = 222.6$ J mol^{-1} K^{-1} for AsH$_3$, which are again closer to the values above. From $\Delta G_f° = 65.23$ kJ mol^{-1} [2], the standard aqueous electrode potential for AsH$_3$/As has been estimated as $E° = -0.225$ V. The pH dependence of the reversible potential for reaction (7.42) is

$$E = -0.225 - 0.0591 \text{ pH} - 0.01971 \log p(AsH_3) \tag{7.44}$$

From this it is seen that AsH$_3$ is a powerful reducing agent.

Arsine has been generated in alkaline solutions at arsenic cathode [1]. Also, at a mercury cathode, arsenious acid is reduced to arsine [1].

The enthalpy of formation of biarsine (As$_2$H$_4$), $\Delta H_f°(As_2H_4) = 117.87$ kJ mol^{-1}, was recently obtained, whose value is nearly twice the value of $\Delta H_f°(AsH_3)$ [7].

B. Arsenious Acid-Arsenic Couple

From the emf of the cell

$$Cu(s) | Pt(s) | H_2(g) | HClO_4 | As_2O_3(s) | As(s) | Cu(s) \tag{7.45}$$
$$\quad\quad\quad\quad\quad\quad\quad (m)$$

$$p = 1 \text{ atm}$$

Latimer [1] reported $E° = 0.234$ V for the following reaction:

$$As_2O_3 + 6H^+ + 6e^- \rightarrow 3H_2O + 2As \tag{7.46}$$

As$_2$O$_3$ dissolves slowly in acidic solutions to form HAsO$_2$. The Gibbs energy of the reaction

$$HAsO_2(aq) + 3H^+ + 3e^- \rightarrow 2H_2O + As \qquad (7.47)$$

has been calculated [8] as $\Delta G° = -71.65$ kJ mol^{-1}. The following Gibbs energies of formation have been reported [1]: As$_4$O$_6$(s) = -1151.1 kJ mol^{-1}, HAsO$_2$(aq) = -402.7 kJ mol^{-1}, and H$_3$AsO$_3$(aq) = -639.9 kJ mol^{-1}. The dissociation of arsenious acid was measured by Hughes [8]; the dissociation reaction is

$$H_3AsO_3 \rightarrow H^+ + H_2AsO_3^- \qquad (7.48)$$

with $K_1 = 6 \times 10^{-10}$ [8]. On the basis of this value, the Gibbs energy of formation of AsO$_2^-$ is -350.2 kJ mol^{-1} and the potential of the couple AsO$_2^-$/As,

$$AsO_2^- + 2H_2O + 3e^- \rightarrow 4OH^- + As \qquad (7.49)$$

is $E° = -0.68$ V. The second and third dissociation constants of H$_3$AsO$_3$ are $pK_2 = 12.13$ ($K_2 = 7.41 \times 10^{-13}$) and $pK_3 = 13.4$ ($K_3 = 5.01 \times 10^{-14}$) [11]. Recently, the second and third dissociation constants have been determined and the values are $pK_2 = 13.54$ and $pK_3 = 13.99$ [10] at $I = 1.0$ mol L^{-1}, whereas the previous value refers to $I = 0.0$. The Gibbs energy of AsO$^+$ was estimated from the dissociation reaction

$$HAsO_2 \rightarrow AsO^+ + OH^- \qquad (7.50)$$

as -163.8 kJ mol^{-1} [1].

C. Arsenic Acid–Arsenious Acid Couple

From the equilibrium constant value of $K = 1.6 \times 10^{-1}$ for the oxidation of arsenious acid by I$_3^-$,

$$H_3AsO_3 + I_3^- + H_2O \rightarrow H_3AsO_4 + 2H^+ + 3I^- \qquad (7.51)$$

the $E°$ value of H$_3$AsO$_4$/H$_3$AsO$_3$ has been evaluated by Latimer [1]. The energy of reaction (7.51) is estimated from its dissociation constant and its value is 4.55 kJ mol^{-1}. The Gibbs energy of formation of arsenic acid, H$_3$AsO$_4$(aq), is -768.2 kJ mol^{-1}. Washburn [9] determined by the conductivity method, the first dissociation constant of H$_3$AsO$_4$,

$$H_3AsO_4(aq) \rightarrow H^+ + H_2AsO_4 \qquad (7.52)$$

as $K_1 = 4.8 \times 10^{-3}$. This value was reexamined by Britton and Jackson [11a] and Agafonova and Agafonov [11b], who reported values of $K_1 = 5.60 \times 10^{-3}$ [11a] at 18°C and $K_1 = 5.98 \times 10^{-3}$ [11b] at 25°C. The second dissociation constant,

$$H_2AsO_4^- \rightarrow H^+ + HAsO_4^{2-} \tag{7.53}$$

was also reported by the authors above as $K_2 = 1.70 \times 10^{-7}$ [11a] at 18°C and $K_2 = 1.04 \times 10^{-7}$ [11b] at 25°C. The temperature dependence of the dissociation constant of H_3AsO_4 was also measured [11b] and as was expressed

$$\log K_1 = -2.014 - 5 \times 10^{-5}(t + 40)^2$$

$$\log K_2 = -6.971 - 5 \times 10^{-5}(t - 39.4)^2$$

The third dissociation constant,

$$HAsO_4^{2-} \rightarrow H^+ + AsO_4^{3-} \tag{7.54}$$

was reported by Konopik and Leberl [12] as $K_3 = 3.98 \times 10^{-12}$ at 20°C using a colorimetric method, which is consistent with $K_3 = 2.95 \times 10^{-12}$ [11a]. Thus the most recent values of Agafonova and Agafonov and Konopik and Leberl are reliable and are consistent with those of Britton and Jackson.

D. Arsenic-Sulfur Compounds

As_2S_3 can be precipitated by passing H_2S into arsenious acid,

$$3H_2S + 2HAsO_2(aq) \rightarrow 4H_2O + As_2S_3 \tag{7.55}$$

and the energy of this reaction is -190.2 kJ mol^{-1}. Estimates have been made [1] of the Gibbs energies of formation of other sulfides— -41.8 kJ mol^{-1} for

$$As_2S_3 + S^{2-} \rightarrow 2AsS_2^- \tag{7.56}$$

and -3.34 kJ mol^{-1} for

$$AsS_2^- + 3e^- \rightarrow As + 2S^{2-} \tag{7.57}$$

E. Arsenic-Halogen Compounds

The Gibbs energy of the hydrolysis reaction

$$AsCl_3(\ell) + 2H_2O \rightarrow HAsO_2 + 3H^+ + 3Cl^- \tag{7.58}$$

is estimated as -26.8 kJ mol^{-1} [1]. Similar reactions have been observed with other halides [17].

The standard potentials of arsenic couples are listed in Table 13. The data given in the table show that the redox potentials of the couples depend on the pH value of the medium in arsenic systems and the equations expressing this relationship are given by Pourbaix [13].

Pourbaix's results expressed as a diagram indicate that arsenic can be deposited at the cathodes over the entire pH range without evolving hydrogen from its salts, which are highly soluble in water.

TABLE 13 Standard Potentials of Compounds of Arsenic in Aqueous Solutions

Half-reaction	Standard potential, $E°$ (V)	References
As^0/As^{3-}		
$As(s) + 3H^+ + 3e^- \rightarrow AsH_3(g)$	-0.225	2
$As(s) + 3H_2O + 3e^- \rightarrow AsH_3(g) + 3OH^-$	-1.37	13-15
As^{3+}/As^0		
$AsS_2^- + 3e^- \rightarrow As + 2S^{2-}$	-0.75	13-15
$AsO_2^- + 2H_2O + 3e^- \rightarrow As + 4OH^-$	-0.68	13-15
$As_2O_3 + 6H^+ + 6e^- \rightarrow 2As + 3H_2O$	0.234	14,15
$H_3AsO_3 + 3H^+ + 3e^- \rightarrow As + 3H_2O$	0.24	13-15
$HAsO_2 + 3H^+ + 3e^- \rightarrow As + 2H_2O$	0.248	13
$AsO^+ + 2H^+ + 3e^- \rightarrow As + H_2O$	0.254	13
$AsO_2^- + 4H^+ + 3e^- \rightarrow As + 2H_2O$	0.429	13
As^{5+}/As^{3+}		
$AsO_4^{3-} + 2H_2O + 2e^- \rightarrow AsO_2^- + 4OH^-$	-0.67	13-15
$AsS_4^{3-} + 2e^- \rightarrow AsS_2^- + 2S^{2-}$	-0.60	13-15

Reaction	$E°$ (V)	Ref.
$H_3AsO_4 + 3H^+ + 2e^- \rightarrow AsO^+ + 3H_2O$	0.550	13-15
$H_3AsO_4 + 2H^+ + 2e^- \rightarrow H_3AsO_3 + H_2O$	0.559	13-15
$H_3AsO_4 + 2H^+ + 2e^- \rightarrow HAsO_2 + 2H_2O$	0.560	13-15
$2H_3AsO_4 + 4H^+ + 4e^- \rightarrow As_3O_3 + 5H_2O$	0.580	13
$HAsO_4^{2-} + 3H^+ + 2e^- \rightarrow AsO_2^- + 2H_2O$	0.609	13
$H_2AsO_4^- + 3H^+ + 2e^- \rightarrow HAsO_2 + 2H_2O$	0.666	13,14
$2AsO_4^{3-} + 10H^+ + 4e^- \rightarrow As_2O_3 + 5H_2O$	0.687	13
$As_2O_5 + 4H^+ + 4e^- \rightarrow As_2O_3 + 2H_2O$	0.721	13
$HAsO_4^{2-} + 4H^+ + 2e^- \rightarrow HAsO_2 + 2H_2O$	0.881	13
$2HAsO_4^{2-} + 8H^+ + 4e^- \rightarrow As_2O_3 + 5H_2O$	0.901	13
$AsO_4^{3-} + 4H^+ + 2e^- \rightarrow AsO_2^- + 2H_2O$	0.977	13
$2AsO_4^{3-} + 10H^+ + 4e^- \rightarrow As_2O_3 + 5H_2O$	1.270	13
As^{5+}/As^0		
$As_2O_5 + 10H^+ + 10e^- \rightarrow 2As + 5H_2O$	0.429	13
$AsO_4^{3-} + 8H^+ + 5e^- \rightarrow As + 4H_2O$	0.648	13

TABLE 14 Thermodynamic Data on Arsenic at 298.15 K[a]

Formula	Description	State	$\Delta H°$/ kJ mol^{-1}	$\Delta G°$/kJ mol^{-1}	$S°$/J mol^{-1} K^{-1}
As	$^4S_{3/2}$	g	302.5	261.0	174.1
As	α—gray metal	c	0	0	35.1
As	β	amorph	4.1		
As	γ—yellow	c	14.7		
As$^+$	$^3P_0$	g	1254.5		
As^{2+}	$^3P_{1/2}$	g	3208.7		
As$_2$		g	222.1	171.9	239.3
As$_4$		g	143.9	92.4	313.8
AsO		g	69.9		
AsCl$_3$		ℓ	−305.0	−256.7	207.5
AsCl$_3$		g	−258.5	−245.8	326.7
AsBr$_3$		c	−197.5	−160.2	161.0
As$_2$H$_4$			117.8		

Nitrogen, Phosphorus, Arsenic, Antimony, and Bismuth

HAsO$_2$	Std. state, $m = 1$	aq	−456.0	−402.7	126.6
	Nonionized, std. state, $m = 1$		(−456.5)	(−402.7)	(125.9)
HAsO$_4^{2-}$	Std. state, $m = 1$	aq	−898.8	−707.1	3.76
	Nonionized, std. state, $m = 1$		(−906.3)	(−714.7)	(−1.67)
H$_2$AsO$_3^-$	Std. state, $m = 1$	aq	−712.8	−587.5	
	Nonionized, std. state, $m = 1$		(714.7)	(−587.2)	(110.4)
H$_2$AsO$_4^-$	Std. state, $m = 1$	aq	−904.5	−748.5	117.0
	Nonionized, std. state, $m = 1$		(−909.6)	(−753.3)	(117.1)
H$_3$AsO$_3$	Std. state, $m = 1$	aq	−741.1	−639.9	196.6
	Nonionized, std. state, $m = 1$		(−742.2)	(−639.9)	(195.0)
AsF$_3$		g	−919.7	−898.4	289.00
AsF$_3$		ℓ	−955.3	−908.9	181.2
AsI$_3$		c	−58.2	−59.4	213.0
As$_2$S$_2$		g	−17.6		
As$_2$S$_2$		c	−142.5		
As$_2$S$_3$		c	−168.8	−168.4	163.4

TABLE 14 (Continued)

Formula	Description	State	$\Delta H°$ / kJ mol^{-1}	$\Delta G°$ / kJ mol^{-1}	$S°$ / J mol^{-1} K^{-1}
AsN		g	196.2	167.9	225.5
NH$_4$H$_2$AsO$_4$		c	−1058.2	−823.3	172.0
NH$_4$H$_2$AsO$_4$	Std. state, $m = 1$	aq	−1037.2		
			−1042.0	−832.7	230.5
AsO$^+$		aq		−163.8	
AsO$_2^-$	Std. state, $m = 1$	aq	−429.0	−350.2	41.2
AsO$_4^{3-}$		aq	−887.1	−647.5	−162.7
As$_2$O$_5$		aq	−950.4		
As$_2$O$_5$		c	−923.8	−781.4	105.4
As$_2$O$_5$·4H$_2$O		c	−2102.5		
As$_4$O$_6$	Octahedral	c	−1312.5	−1151.1	214.2
As$_4$O$_6$	Monoclinic	c	−1308.5	1153.0	234.3
AsH$_3$		g	66.4	68.9	222.6

[a] Recent National Bureau of Standards values are in parentheses.

The thermodynamic quantities of arsenic compounds are listed in Table 14. The values in parentheses are the values reported recently from NBS. The differences seen between the values reported in Refs. 1 and 16 are small and may have been caused by experimental difficulties. Extreme divergences were observed in the ΔH_f^o and ΔG_f^o values of As_2, AsO, AsH_3, and AsN. The values reported in Ref. 16 are more reliable for the compounds above.

F. Potential Diagrams for Arsenic

Acid Solution

$$\overset{+5}{H_3AsO_4} \xrightarrow{0.560} \overset{+3}{HAsO_2} \xrightarrow{0.240} \overset{0}{As} \xrightarrow{-0.225} \overset{-3}{AsH_3}$$

Basic Solution

$$AsO_4^{3-} \xrightarrow{-0.67} AsO_2^{-} \xrightarrow{-0.68} As \xrightarrow{1.37} AsH_3$$

Arsenic will not disproportionate to give the +3 and −3 states (compare phosphorus, which will disproportionate to the +3 and −3 states).

REFERENCES

1. W. Latimer, *Oxidation Potentials*, 2nd ed., Prentice-Hall, Englewood Cliffs, N.J., 1952, p. 113.
2. H. J. Davis, J. Electrochem. Soc., *124*, 722 (1977).
3. S. M. Ariya, Z. Obshch. Khim., *26*, 1813 (1956).
4. F. E. Sealfeld and H. J. Svec, U.S. AEC Rep. IS, 386 (1961).
5. R. H. Sherman and W. F. Giauque, J. Am. Chem. Soc., *77*, 2154, (1955).
6. K. K. Kelley and E. G. King, U.S. Bur. Mines Bull., 592 (1961).
7. W. S. Hughes, J. Chem. Soc., 491 (1928).
8. S. R. Gunn, Inorg. Chem., *11*, 796 (1972).
9. E. W. Washburn, J. Am. Chem. Soc., *30*, 31 (1908).
10. A. A. Ivakin, S. V. Vorob'eva, and E. M. Gertman, Zh. Neorg. Khim., *24*(1), 36 (1979); Chem. Abstr., *90*, 128337r (1979).
11a. H. T. S. Britton and P. Jackson, J. Chem. Soc., 1048 (1954).
11b. A. L. Agafonova and I. L. Agafonov, Zh. Fiz. Khim., *27*, 1137 (1953); Chem. Abstr., *48*, 5611e (1954).
12. N. Konopik and O. Leberl, Monatsh. Chem., *80*, 655 (1949); Chem. Abstr., *44*, 9218d (1950).
13. M. Pourbaix, ed., *Atlas of Electrochemical Equilibria in Aqueous Solutions*, Pergamon Press, Oxford, 1966.
14. A. P. Tomilov and N. E. Chomutov, in *Encyclopedia of the Electrochemistry of the Elements*, A. J. Bard, ed., Vol. 2, Marcel Dekker, New York, 1974, p. 21.
15. G. Milazzo, *Electrochemistry*, Elsevier, Amsterdam, 1963.

16. D. D. Wagman, W. H. Evans, V. B. Parker, I. Halow, S. H. Bailey, and R. H. Schuman, Natl. Bur. Stand. Tech. Note 270-3, U.S. Government Printing Office, Washington, D.C., 1968.
17. Natl. Bur. Stand. Tech. Note 270-1, U.S. Government Printing Office, Washington, D.C., 1965.

IV. ANTIMONY*

VELLO PAST   Tartu State University, Tartu, USSR

A. Oxidation States

The principal oxidation states of antimony in its compounds are −III, III, and V. Antimony forms hydride SbH_3 [Sb(−III)] and oxides Sb_4O_6 [Sb(III)], Sb_2O_5 [Sb(v)], and Sb_2O_4 [which is probably to be considered as Sb(III) and Sb(V) oxide]. The most important binary compounds of Sb(III) with sulfur and halides are Sb_2S_3, SbF_3, $SbCl_3$, $SbBr_3$, and SbI_3 and those of Sb(V) are Sb_2S_5, SbF_5, and $SbCl_5$. Chemically is Sb analogous to P and As but differs from them because of its notable metallic properties.

B. Sb—Sb(−III) Couple

Stibine is thermodynamically unstable. It may be produced by electrolysis of acid and alkaline solutions on an antimony electrode or by a reductive action on Sb salt solutions or on metallic Sb in the presence of water.

The thermodynamic properties of stibine have been found on the basis of calorimetric measurements or from vibrational spectra. The data given in Table 15 are recommended by the National Bureau of Standards (NBS) [1].

The standard Gibbs energy of formation of stibine is 147.74 kJ mol^{-1}. From this value for the antimony-stibine couple in acid solution the following can be computed:

$$Sb + 3H^+ + 3e^- \rightarrow SbH_3(g) \qquad E° = -0.510 \text{ V} \tag{7.59}$$

The value of $E° = 0.51$ V has been calculated by Latimer [2] from the enthalpy of formation and the entropy of SbH_3. The experimental data for $E°$ are not available because SbH_3 is unstable in the presence of aqueous solutions. The thermodynamic data of partially or completely deuterated and tritiated stibines are given in Ref. 3.

C. Sb(III)/Sb Couple

The accurate value of the standard potential of antimony against Sb^{3+} ions is not known since single Sb(III) ions exist in very small concentration in aqueous solutions. Solubility determinations of Sb_4O_6 in $HClO_4$ indicate [4,5] that dissolved trivalent antimony is mainly in the form SbO^+ (in the pH range 0 to 1). In basic solutions (pH > 11), trivalent antimony has been shown to exist as $Sb(OH)_4^-$ or simple as SbO_2^- ion. For a wide interval of pH from 2 to 10.4 the solubility of Sb_4O_6 does not depend on the value of pH, indicating a nonionic

---

*Reviewer: W. Plieth, Free University of Berlin, Berlin, Germany.

TABLE 15 Thermodynamic Data of Antimony[a]

Formula	Description	State	ΔH_f°/kJ mol^{-1}	ΔG_f°/kJ mol^{-1}	S°/J mol^{-1} K^{-1}
Sb	Metal	c	0	0	45.69
Sb	Explosive	amorph	10.63		
Sb		g	262.3	222.2	180.16
Sb$^+$		g	1102.3		
Sb^{2+}		g	2703.3		
Sb$_2$		g	235.6	187.0	254.81
Sb$_4$		g	205.0	141.4	351.5
SbO		g	199.45		
SbO$^+$		aq	−193.7	−175.64	22.33
SbO$_2^-$		aq	−417.6	−339.74	55.2
Sb$_4$O$_6$	Orthorhombic	c	−1403.15	−1249.34	282.0
Sb$_4$O$_6$	Cubic	c	−1423.2	−1264.4	265.3
Sb$_2$O$_4$		c	−907.5	−795.8	127.2
Sb$_2$O$_5$		c	−971.9	−829.3	125.1

TABLE 15 (Continued)

Formula	Description	State	ΔH_f°/kJ mol^{-1}	ΔG_f°/kJ mol^{-1}	S°/J mol^{-1} K^{-1}
Sb_2O_5		aq	−956.9		
Sb_6O_{13}		c	−2805.8		
SbH_3		g	145.10	147.74	232.67
$HSbO_2$	Undiss.	aq	−487.9	−407.5	46.4
$Sb(OH)_3$	Undiss.	aq	−773.6	−644.8	116.3
$Sb(OH)_4^-$		aq	−989.60	−813.20	190.4
H_3SbO_4		aq	−907.1		
$HSb(OH)_6$		aq	−1478.6		
SbF_3		c	−915.5	−836	105.4
SbF_3		aq	−910.9		
H_3SbF_6		aq	−1876.1		
$SbOF$	Undiss.	aq		−487.4	

Compound	State		
Sb(OH)$_2$F	Undiss. aq	−724.7	
SbCl$_3$	c	−382.27	−323.72	184.1
SbCl$_3$	g	−313.8	−301.2	337.69
SbCl$_5$	ℓ	−440.2	−350.2	301.2
SbCl$_5$	g	−394.34	−334.34	401.83
SbOCl	c	−374.0		
Sb$_4$O$_5$Cl$_2$	c	−1451.4		
SbBr$_3$	c	−259.4	−239.3	207.1
SbBr$_3$	g	−194.6	−223.0	372.75
SbI$_3$	c	−100.4	−94	(212)
Sb$_2$S$_3$	Black c	−174.9	−173.6	182.0
Sb$_2$S$_3$	Orange amorph	−147.3		
Sb$_2$S$_4^{2-}$	aq	−219.2	−98.7	−52.3
SbS$_3^{2-}$	aq		(−134)	

[a]National Bureau of Standards data are given in italics, approximate data in parentheses.

solute. This solute has been considered as meta-antimonious acid $HSbO_2$ or $SbO(OH)$.

The antimony trioxide-antimony electrode standard potential has been determined [6,7] from electromotive force (emf) measurements in the cells

$$(Pt)H_2 | HClO_4(aq) | Sb_4O_6 | Sb \qquad (7.60)$$

and

$$Hg | HgO | NaOH(aq) | Sb_4O_6 | Sb \qquad (7.61)$$

where the concentration of $HClO_4$ or $NaOH$ in the solution has been varied from 0.3 to 2.5 M. These measurements at 25°C gave for the acidic solutions

$$Sb_4O_6 + 12H^+ + 12e^- \rightarrow 4Sb + 6H_2O \qquad E° = 0.1504 \text{ V} \qquad (7.62)$$

and for the alkaline solutions

$$Sb_4O_6 + 6H_2O + 12e^- \rightarrow 4Sb + 12OH^- \qquad E° = -0.6831 \text{ V} \qquad (7.63)$$

The corresponding Gibbs energies of formation of Sb_4O_6 are -1249.34 and -1255.33 kJ mol^{-1}. In between these values is the value $\Delta G°(Sb_4O_6) = -1253.11$ kJ mol^{-1} accepted by NBS [1].

The previous values for the Sb_4O_6/Sb electrode standard potential in acidic solution 0.152 V [4] and 0.1512 V [8], are not too different from the value $E° = 0.1504$ V given above.

The isothermal temperature coefficient of the Sb_4O_6/Sb electrode potential in acidic solution $dE°/dT = -0.41$ mV K^{-1} at 25°C [6]. The accuracy of the temperature coefficient measurement is not high; therefore, calculation of the thermodynamic parameters from this coefficient gives values of higher inaccuracy than that of values determined by other methods.

There are two known crystallographic modifications of antimony trioxide. The cubic form, stable up to 570°C, is composed of discrete Sb_4O_6 molecular clusters. The high-temperature form is an orthorhombic polymer structure. The two forms of antimony trioxide have different thermodynamic characteristics, as shown in Table 15. From the Gibbs energies of both forms, the Gibbs energy for transformation can be found:

$$Sb_4O_6(\text{orthorhombic}) \rightarrow Sb_4O_6(\text{cubic}) \qquad \Delta G° = -15.06 \text{ kJ mol}^{-1} \qquad (7.64)$$

The oxide used in the epxeriments [4,6-8] contains a mixture of both crystals, with the orthorhombic form greatly predominating. Using cubic crystals of Sb_4O_6, Roberts and Fenwick [8] obtained the value $E°(Sb_4O_6/Sb) = 0.1445$ V in acidic solution. They have also found the average difference in the potential of Sb_4O_6 (orthorhombic)/Sb and of Sb_4O_6(cubic)/Sb electrodes in the same solution to be 0.013 V.

The results of potentiometric study [6] have shown that

$$SbO^+ + 2H^+ + 3e^- \rightarrow Sb + H_2O \qquad E° = 0.2040 \text{ V} \qquad (7.65)$$

This would lead to the Gibbs energy of formation $\Delta G_f^\circ(SbO^+) = -178.20$ kJ mol^{-1}.

The value $\Delta G_f^\circ(SbO^+) = -175.64$ kJ mol^{-1} in Table 15 has been calculated from solubility measurements. Handbooks usually give the value $E^\circ(SbO^+/Sb) = 0.212$ V obtained in 1924 [4].

The potential measurements of the Sb_4O_6/Sb electrode in alkaline solutions give [7]

$$SbO_2^- + 2H_2O + 3e^- \rightarrow Sb + 4OH^- \qquad E^\circ = -0.6389 \text{ V} \qquad (7.66)$$

The corresponding Gibbs energy $\Delta G_f^\circ(SbO_2^-) = -339.74$ kJ mol^{-1} is shown in Table 15. The previous values for $E^\circ(SbO_2^-/Sb)$ are -0.66 V [2] and -0.675 V [9]. The isothermal temperature coefficient of the electrode potential at 25°C is -0.37 mV K^{-1} for reaction (7.65) and 0.22 mV K^{-1} for reaction (7.66) [6,7].

D. Sb(V)/Sb(III) Couple

The equilibrium between Sb(V) and Sb(III) has been studied electrochemically in KOH solutions [9]. In alkaline solutions the antimonate ion was shown to be $Sb(OH)_6^-$. The calculation based on the experimental data of the equilibrium potentials demonstrated for the antimonate-antimonite couple [5]

$$Sb(OH)_6^- + 2H^+ + 2e^- \rightarrow Sb(OH)_4^- + 2H_2O \qquad E^\circ = 0.363 \text{ V} \qquad (7.67)$$

From Gibbs energy ΔG_f° values given in Table 15 we can compute

$$2Sb_2O_5 + 8H^+ + 8e^- \rightarrow Sb_4O_6 + 4H_2O \qquad E^\circ = 0.699 \text{ V} \qquad (7.68)$$

$$Sb_2O_5 + 6H^+ + 4e^- \rightarrow 2SbO^+ + 3H_2O \qquad E^\circ = 0.605 \text{ V} \qquad (7.69)$$

The experimental values of E° for these reactions are not known.

E. Antimony Tetroxide

The most reliable value of ΔG_f° for Sb_2O_4, -795.8 kJ mol^{-1} [1], differs from the value -694.2 kJ mol^{-1} used by previous authors [2,5]. Because of that the calculated E° values for reactions in which Sb_2O_4 takes part are very different from the corresponding values of E° accepted by Refs 2 and 5.

Using the values of ΔG_f° given in Table 15, the standard electrode potentials of following reactions are computed:

$$2Sb_2O_4 + 4H^+ + 4e^- \rightarrow Sb_4O_6 + 2H_2O \qquad E^\circ = 0.342 \text{ V} \qquad (7.70)$$

$$Sb_2O_5 + 2H^+ + 2e^- \rightarrow Sb_2O_4 + H_2O \qquad E^\circ = 1.055 \text{ V} \qquad (7.71)$$

There are no experimental data for these electrode potentials.

F. Antimony Chlorides

Sb(III) forms at high concentrations of HCl different complex chlorides (e.g., $SbCl_4^-$, $SbCl_5^{2-}$, and $SbCl_6^{3-}$). According to Vagramian and co-workers [10], the antimony electrode in concentrated HCl solutions (3.5 to 7 M) is in equilibrium with $SbCl_4^-$ ions:

$$SbCl_4^- + 3e^- \rightarrow Sb + 4Cl^- \qquad E^\circ = 0.17 \text{ V} \qquad (7.72)$$

At lower HCl concentrations $SbCl_4^-$ hydrolyzes and gives basic chlorides ($SbOHCl_3^-$ and others).

The results of spectroscopic measurements indicate that Sb(V) exists in concentrated HCl solutions as $SbCl_6^-$. The tendency of Sb(V) chloro complexes to hydrolyze by dilution is very noticeable.

The formal equilibrium potential of the couple Sb(V)/Sb(III) in concentrated HCl solutions has been measured in Ref. 11. The dependence of the formal potential on the acid concentration and the time has been explained by the hydrolysis of $SbCl_6^-$.

G. Antimony Sulfides

Using the Gibbs energy of formation of the ion $Sb_2S_4^{2-}$, $\Delta G_f^\circ = -98.7$ kJ mol^{-1}, it can be stated that

$$Sb_2S_4^{2-} + 6e^- \rightarrow 2Sb + 4S^{2-} \qquad E^\circ = -0.763 \text{ V} \qquad (7.73)$$

The value $\Delta G_f^\circ(SbS_2^-)$ roughly estimated from the equilibrium data by Latimer [2] led to an E° value of -0.85 V for reaction (7.73).

Sorochan and Fridman [12] determined the following on the basis of the potentiometric titration curves:

$$SbS_3^{3-} + 3e^- \rightarrow Sb + 3S^{2-} \qquad E^\circ = -1.45 \text{ V} \qquad (7.74)$$

$$SbSO^- + H_2O + 3e^- \rightarrow Sb + 2OH^- + S^{2-} \qquad E^\circ = -0.350 \text{ V} \qquad (7.75)$$

The E° values of reactions (7.66) and (7.73) to (7.75) are in agreement with the fact that the reduction of the ions SbSO$^-$ occurs easily in the alkaline solution. The reduction of SbO_2^- is not easy, whereas the SbS_2^- ions and in particular the SbS_3^{3-} ions are reduced with difficulty [12].

There are no Gibbs energy data on the antimonic sulfide Sb_2S_5. Rough estimation has shown [2] that

$$SbS_4^{3-} + 2e^- \rightarrow SbS_2^- + 2S^{2-} \qquad E^\circ = -0.6 \text{ V} \qquad (7.76)$$

H. Potential Diagrams for Antimony

Strong Acid Solution

$$\overset{+5}{Sb_2O_5}\xrightarrow{0.605}\overset{+3}{SbO^+}\xrightarrow{+0.204}\overset{0}{Sb}\hspace{2cm}-III$$

Acid and Neutral Solution

$$Sb_2O_5\xrightarrow{1.055}Sb_2O_4\xrightarrow{0.342}Sb_4O_6\xrightarrow{0.150}Sb\xrightarrow{-0.510}SbH_3$$
$$\underset{0.699}{\underline{\hspace{4cm}}}$$

Basic Solution

$$Sb(OH)_6^-\xrightarrow{-0.465}Sb(OH)_4^-\xrightarrow{-0.639}Sb\xrightarrow{-1.338}SbH_3$$

REFERENCES

1. *Selected Values of Chemical Thermodynamic Properties*, Natl. Bur. Stand. Tech. Note 270-3, U.S. Government Printing Office, Washington, D.C., 1968, Table 20, p. 99.
2. W. M. Latimer, *The Oxidation States of the Elements and Their Potentials in Aqueous Solutions*, 2nd ed., Prentice-Hall, Englewood Cliffs, N.J., 1952, p. 116.
3. P. N. Daykin and S. Sundaram, Z. Phys. Chem. (NF), *32*, 222 (1962).
4. R. Schuhmann, J. Am. Chem. Soc., *46*, 52 (1924).
5. A. L. Pitman, M. Pourbaix, and N. de Zoubov, J. Electrochem. Soc., *104*, 594 (1957).
6. V. P. Vasiliev and V. I. Shorokhova, Electrokhimiya, *8*, 185 (1972).
7. V. P. Vasiliev, V. I. Shorokhova, and S. V. Kovonova, Electrokhimiya, *9*, 1006 (1973).
8. J. Roberts and F. Fenwick, J. Am. Chem. Soc., *50*, 2125 (1928).
9. G. Grube and F. Schweigardt, Z. Elektrochem., *29*, 257 (1923).
10. I. G. Narakidze, V. A. Kazakov, and A. T. Vagramian, Electrokhimiya, *4*, 1464 (1968).
11. R. A. Brown and E. H. Swift, J. Am. Chem. Soc., *71*, 2719 (1949).
12. R. I. Sorochan and Ya. D. Fridman, Tr. Otd. Gorn. Dela Metall. Akad. Nauk Kirg. SSR, *1*, 69-III (1958).

V. BISMUTH*

B. LOVREČEK University of Zagreb, Zagreb, Yugoslavia
I. MEKJAVIĆ University of Split, Split, Yugoslavia
M. METIKOŠ-HUKOVIĆ University of Zagreb, Zagreb, Yugoslavia

A. Oxidation States

Bismuthine BiH_3 (−3 state) can be prepared [1], but due to its instability, the yield is low. According to Weeks and Druce [3], Latimer [2] mentions the preparation of solid Bi_2H_2 (−1), but in more recent literature there are no additional data regarding this compound. The species BiX (+1) (where X = Cl, Br, and I) occurs in equilibrium with the mixture $Bi + BiX_3$ in the vapor phase [4]. However, the most stable oxidation state is +3, with the most important representative being Bi_2O_3. Bismuth compounds of the +5 state, which are strong oxidizing agents, are known, while the Bi^{5+}/Bi^{3+} couple is able to oxidize water to oxygen.

B. Oxidation State −3

Bismuthine must be stored at low temperature (liquid nitrogen), and measurements at room temperature must be taken within a few minutes. For the following reaction the standard potential is calculated [5] and results in

$$Bi(s) + 3H^+ + 3e^- = BiH_3(g) \qquad E° = -0.97 \text{ V}$$

C. Oxidation State +3

Bi_2O_3 dissolves in strong acid, giving a solution that contains Bi^{3+} ions. Using a perchlorate medium, Olin [6] established the existence of a whole series of complexes (depending on solution pH) up to the appearance of $Bi(OH)_3$ precipitation. At lower pH, in fact, a simple $BiOH^{2+}$ complex exists which occurs according to the reaction

$$Bi^{3+}(aq) + H_2O(\ell) = BiOH^{2+} + H^+ \qquad \log \beta_1 = -1.58 \pm 0.05$$

At somewhat higher pH there is formed a more complicated polynuclear complex ion $Bi_6(OH)_{12}^{6+}$, according to the reaction

$$6Bi^{3+}(aq) + 12H_2O(\ell) = Bi_6(OH)_{12}^{6+} + 12H^+ \qquad \log \beta_{12,6} = 0.33 \pm 0.1$$

At 0.01 M total bismuth this ion predominates at pH 1.5. With the lowering of acidity a hydrolysis of $Bi_6(OH)_{12}^{6+}$ ion occurs together with the appearance of polynuclear aggregates, according to the following reactions:

$$1.5 Bi_6(OH)_{12}^{6+} + 2H_2O(\ell) = Bi_9(OH)_{20}^{7+} + 2H^+ \qquad \log K = -3.5 \pm 0.1$$

$$Bi_9(OH)_{20}^{7+} + H_2O(\ell) = Bi_9(OH)_{21}^{6+} + H^+ \qquad \log K = -3.2 \pm 0.1$$

Reviewer: E. D. Moorhead, University of Kentucky, Lexington, Kentucky.

$$Bi_9(OH)_{21}^{6+} + H_2O(\ell) = Bi_9(OH)_{22}^{5+} + H^+ \qquad \log K = -2.6 \pm 0.1$$

These higher species predominate at pH > 2.5 at 0.01 M total dissolved bismuth. At pH 5, however, precipitation occurs. Numerous early attempts to measure the standard potential of the Bi^{3+}/Bi electrode were unsuccessful due to the above-mentioned hydrolysis of the Bi^{3+} ion, or due to the formation of nonhydroxy complexes with a great number of anions. In 1969 Vasiljev and Glavina [7] measured the potential of the bismuth electrode in a solution of perchloric acid of different concentrations (0.9 to 4.0 M) at 15, 25, 40, and 50°C. The standard electrode potential for zero ionic strength was calculated by use of an equation based on Debye-Hückel theory using the least-squares method. At 25°C they obtained for the reaction

$$Bi^{3+}(aq) + 3e^- = Bi(s) \qquad E° = 0.3172 \pm 0.0006 \text{ V}$$

The temperature coefficient $(dE/dT)_{25°C}$ was 0.13 ± 0.015 mV K^{-1}.

Using the listed equilibrium constants for the series of bismuth hydroxy complexes, the standard Gibbs energies of formation (see Table 16) are calculated, and with the help of these values the standard potentials are further evaluated for some bismuth couples involving complexes. The data are as follows*:

$$BiOH^{2+} + 3e^- = Bi(s) + OH^- \qquad E° = 0.072 \text{ V}$$

$$Bi_6(OH)_{12}^{6+} + 18e^- = 6Bi(s) + 12OH^- \qquad E° = -0.234 \text{ V}$$

$$Bi_9(OH)_{22}^{5+} + 27e^- = 9Bi(s) + 22OH^- \qquad E° = -0.336 \text{ V}$$

Depending on pH (up to 5), however, there exist different proportions of bismuth hydroxy complexes in the solution, and these proportions are governed by total dissolved bismuth.

It should be mentioned that on the basis of data from earlier literature, Latimer [8] pointed out the existence of the ionic species $Bi(OH)_2^+$ (i.e, BiO^+). However, Olin [6], who could not establish its presence, postulated that this concept is probably due to the fact that the hexameric ion $Bi_6(OH)_{12}^{6+}$ is so stable in relation to $Bi(OH)_2^+$ (i.e., BiO^+) that equilibrium concentration of this ion is too small to be detected. When chloride ions are added to perchlorate solution (above a certain concentration) [9], the separation of a crystallized precipitate of $Bi(OH)_2Cl$ or $BiOCl$ occurs. The rate of separation depends on the concentration of both bismuth and chloride ions. The solubility product ($K_s = [Bi^{3+}][OH^-]^2[Cl^-]$) has a low value and according to Ahrland and Grenthe [10] it equals $\log K_s = -34.93$.

Bi_2O_3 is not very soluble in 1 M NaOH. From direct measurement of the bismuth electrode potential in 1 M NaOH solution saturated with Bi_2O_3, the value $E = -0.413$ V was obtained by Baur and Lattmann [11]. However,

---

*Without going into the kinetic probability of high multielectron reactions being potential determining.

TABLE 16 Thermodynamic Data on Bismuth, $t = 25°C$[a]

Formula	Description	State	$\Delta H°/\text{kJ mol}^{-1}$	$\Delta G°/\text{kJ mol}^{-1}$	$S°/\text{J mol}^{-1}\text{ K}^{-1}$
Bi	Metal	c	0	0	56.74
Bi		g	207.1	168.2	186.90
Bi_2		g	219.7	172.4	273.6
Bi_4		g	241.0		
BiH_3		g	277.8		
BiCl		g	28.5	0.2	262.8
BiBr		g			301 (935 K)
BiI		g			307 (913 K)
Bi_2O_3		c	−573.9	−493.7	151.5
Bi^{3+}		g	5005.7		
Bi^{3+}	Std. state, $m = 1$	aq	80.60	91.82	−176.75
$BiOH^{2+}$	Std. state, $m = 1$	aq		−136.3	
$Bi_6(OH)_{12}^{6+}$	Std. state, $m = 1$	aq		−2293.3	
$Bi_9(OH)_{20}^{7+}$	Std. state, $m = 1$	aq		−3894.4	

$Bi_9(OH)_{21}^{6+}$	Std. state, $m = 1$	aq		−4113.3	
$Bi_9(OH)_{22}^{5+}$	Std. state, $m = 1$	aq		−4335.6	
$Bi(OH)_3$		c	−711.3		
$BiCl^{2+}$	Std. state, $m = 1$	aq		−52.7	
$BiCl^+$	Std. state, $m = 1$	aq		−190.3	
$BiCl_3$		g	−264.8	−254.7	357.3
		c	−379.1	−314.6	174.5
$BiCl_4^-$	Std. state, $m = 1$	aq		−332.0	
$BiCl_5^{2-}$	Std. state, $m = 1$	aq		−467.5	
$BiCl_6^{3-}$	Std. state, $m = 1$	aq		−602.2	
	Std. state, $m = 1$	aq		−732.6	
$BiOCl$		c	−369.35	−319.32	103.05
$BiBr^{2+}$	Std. state, $m = 1$	aq		−24.8	
$BiBr_2^+$	Std. state, $m = 1$	aq	−141.1		
$BiBr_3$		c	−276.1	−247.7	190.4
	Std. state, $m = 1$	aq		−255.5	

TABLE 16 (Continued)

Formula	Description	State	$\Delta H°$/kJ mol^{-1}	$\Delta G°$/kJ mol^{-1}	$S°$/J mol^{-1} K^{-1}
$BiBr_4^-$	Std. state, $m = 1$	aq		−367.3	
$BiBr_5^{2-}$	Std. state, $m = 1$	aq		−480.1	
$BiBr_6^{3-}$	Std. state, $m = 1$	aq		−584.7	
$BiOBr$		c		−297.1	
BiI_3		c	−150	−148.7	224.7
BiI_4^-	Std. state, $m = 1$	aq		−198.4	
BiI_5^{2-}	Std. state, $m = 1$	aq		−260.4	
BiI_6^{3-}	Std. state, $m = 1$	aq		−323.2	
Bi_2S_3		c	−143.1	−140.6	200.4
Bi_2Se_3		c	−149.8		
Bi_2Te_3		c	−77.4	−77.0	261

[a]For halogen complexes, see the text. National Bureau of Standards values are in italic.

using the Gibbs energies of formation, the following value for $E°$ for this couple was calculated:

$$Bi_2O_3(s) + 3H_2O(\ell) + 6e^- = 2Bi(s) + 6OH^- \qquad E° = -0.452 \text{ V}$$

or

$$Bi_2O_3(s) + 6H^+ + 6e^- = 2Bi(s) + 3H_2O(\ell) \qquad E° = 0.376 \text{ V}$$

In the presence of chloride, bromide, and iodide, bismuth forms different complexes. Many authors using various methods have obtained values of equilibrium constants for such complexes in perchlorate medium at $c_{H+} = 1$ M (with different halide ion concentrations) which agreed well. In Table 17 the data are summarized, according to Ahrland and Grenthe [10], and were obtained for ionic strength $I = 2$ M, 1 M [H$^+$] (and 20°C) for the reaction

$$Bi^{3+}(aq) + nX^- = BiX_n^{3-n}$$

These data have been used for calculating the standard Gibbs energies of formation of the halide bismuth complexes (see Table 16), which then makes possible the evaluation of standard potentials. It should be mentioned that the $\Delta G°$ for the reactions according to the equation above are calculated for 20°C, whereas the other data needed for evaluation of the Gibbs energy of formation (i.e., $E°$) refer to 25°C. That unfortunately introduces a certain error, but it can be assumed that this is practically negligible. Values of the standard potentials for several chloride complexes are as follows:

$$BiCl^{2+} + 3e^- = Bi(s) + Cl^- \qquad E° = 0.271 \text{ V}$$

$$BiCl_2^+ + 3e^- = Bi(s) + 2Cl^- \qquad E° = 0.249 \text{ V}$$

$$BiCl_4^- + 3e^- = Bi(s) + 4Cl^- \qquad E° = 0.199 \text{ V}$$

$$BiCl_6^{3-} + 3e^- = Bi(s) + 6Cl^- \qquad E° = 0.190 \text{ V}$$

TABLE 17

	$\log \beta_n$					
X^-	$n = 1$	2	3	4	5	6
Cl$^-$	2.36	3.5	5.35	6.10	6.72	6.56
Br$^-$	2.26	4.45	6.30	7.70	9.28	9.38
I$^-$	—	—	—	14.95	16.8	18.8

Source: Ref. 10.

It should be mentioned that in the range 10^{-3} to 10 M [Cl$^-$], various proportions of individual complexes exist [12]. Thus, for 1 M [H$^+$] at 10^{-3} M [Cl$^-$], we have 86 mol % Bi^{3+} and 14 mol % BiCl^{2+}; and with 10 M [Cl$^-$], 8 mol % BiCl$_5^{2-}$ and 92 mol % BiCl$_6^{3-}$.

Analogous equilibria involving bromide complexes have closely similar values for $E°$:

$$BiBr^{2+} + 3e^- = Bi(s) + Br^- \qquad E° = 0.273 \text{ V}$$

$$BiBr_2^+ + 3e^- = Bi(s) + 2Br^- \qquad E° = 0.231 \text{ V}$$

$$BiBr_4^- + 3e^- = Bi(s) + 4Br^- \qquad E° = 0.168 \text{ V}$$

$$BiBr_6^{3-} + 3e^- = Bi(s) + 6Br^- \qquad E° = 0.135 \text{ V}$$

Values for iodide complexes are more negative:

$$BiI_4^- + 3e^- = Bi(s) + 4I^- \qquad E° = 0.027 \text{ V}$$

$$BiI_6^{3-} + 3e^- = Bi(s) + 6I^- \qquad E° = -0.047 \text{ V}$$

Vasiljev and Grechina [13] measured the potential of the bismuthoxychloride electrode at different concentrations of perchloric acid (1.0 to 3.0 M) and at temperatures of 15, 25, 35, and 50°C. The standard potential given by these authors, referred to zero ionic strength and a temperature 25°C, is

$$BiOCl(s) + 2H^+ + 3e^- = Bi(s) + Cl^- + H_2O(\ell) \qquad E° = 0.1697 \pm 0.0004 \text{ V}$$

while the temperature coefficient $(dE/dT)_{25°C}$ is -0.40 ± 0.06 mV K^{-1}. For the analogous reaction with bismuthoxybromide the free energy of formation yields the following value for $E°$:

$$BiOBr(s) + 2H^+ + 3e^- = Bi(s) + Br^- + H_2O(\ell) \qquad E° = 0.152 \text{ V}$$

D. Oxidation State +5

Ford-Smith and Habeeb [14] determined the approximative value fo the electrode potential for this oxidation state on the basis of a potentiometric titration of Bi^{5+} with I$^-$ in 0.5 M [H$^+$] at ionic strength $I = 2.0$ M:

$$Bi^{5+}(aq) + 2e^- = Bi^{3+}(aq) \qquad E = 2.0 \pm 0.2 \text{ V}$$

(These species are probably far more complex than formally shown here.)

E. Potential Diagram for Bismuth

$$\overset{+5}{Bi^{5+}} \xrightarrow{ca.\ 2*} \overset{+3}{Bi^{3+}} \xrightarrow{0.317} \overset{0}{Bi^{0}} \xrightarrow{-0.97} \overset{-3}{BiH_3}$$

REFERENCES

1. J. D. Smith, in *Comprehensive Inorganic Chemistry*, Vol. 2, A. F. Trotman-Dickenson, ed., Pergamon Press, Oxford, 1973, p. 585.
2. W. M. Latimer, *The Oxidation States of the Elements and Their Potentials in Aqueous Solutions*, 2nd ed., Prentice-Hall, Englewood Cliffs, N.J., 1952, p. 121.
3. E. J. Weeks and J. G. F. Druce, J. Chem. Soc., *127*, 1799 (1925).
4. D. Cubicciotti, J. Phys. Chem., *64*, 791 (1969); J. D. Smith, in *Comprehensive Inorganic Chemistry*, Vol. 2, A. F. Trotman-Dickenson, ed., Pergamon Press, Oxford, 1973, p. 601.
5. J. D. Smith, in *Comprehensive Inorganic Chemistry*, Vol. 2, A. F. Trotman-Dickenson, ed., Pergamon Press, Oxford, 1973, p. 582.
6. A. Olin, Acta Chem. Scand., *11*, 1445 (1957); *13*, 1791 (1959); Sven. Kem. Tidskr., *73*, 482 (1961).
7. V. P. Vasiljev and S. R. Glavina, Elektrokhimiya, *5*, 413 (1969).
8. W. M. Latimer, *The Oxidation States of the Elements and Their Potentials in Aqueous Solutions*, 2nd ed., Prentice-Hall, Englewood Cliffs, N.J., 1952, p. 122.
9. A. A. Moussa and H. M. Sammour, J. Chem. Soc., 2151 (1960).
10. S. Ahrland and I. Grenthe, Acta Chem. Scand., *11*, 1111 (1957); *15*, 932 (1961).
11. E. Baur and W. Lattmann, Z. Elektrochem., *40*, 582 (1934).
12. A. M. Bond, Electrochim. Acta, *17*, 769 (1972).
13. V. P. Vasiljev and N. K. Grechina, Elektrokhimiya, *5*, 426 (1969).
14. M. H. Ford-Smith and J. J. Habeeb, Chem. Commun., 1445 (1969).

*Conditions in the text.

# 8
Carbon, Silicon, Germanium, Tin, and Lead*

ZBIGNIEW GALUS University of Warsaw, Warsaw, Poland

The elements of this group show greater diversity of properties than do elements in groups distant from the center of the periodic table. For instance, carbon is nonmetal and lead is a typical metal, while silicon and germanium exhibit semiconducting properties.

Compounds with hydrogen are stable and common for carbon, but the number of such compounds for silicon and germanium becomes limited, and for lead, only one unstable compound of that type is known. The chemistry of the divalent state is more important for heavier elements and the basicity of oxides increases from carbon to lead.

I. CARBON

A. Oxidation States

Carbon forms a great number of compounds with different oxidation states. The +4 state is characteristic for carbon in CO_2 and carbonates, while −4 formally may be ascribed to carbon in various hydrocarbons such as CH_4. Intermediate oxidation numbers may be formally calculated for such stable compounds as CH_3OH, HCOH, and HCOOH.

The characteristic feature of carbon is significant stability in both the +4 and −4 valence state compounds. Only the potentials of those compounds of carbon that are traditionally grouped with inorganic substances are considered here. As a rule the electrode reactions of these compounds are highly irreversible and therefore all available potentials are calculated on the basis of thermodynamic data. In the case of organic carbon compounds there are systems that exhibit reversible behavior, and their potentials are measured with good accuracy. One such system is the quinone-hydroquinone couple, sometimes used as a pH indicator.

All calculated potentials of the carbon-containing redox couples are given by Latimer [1] and Pourbaix [2]. Some of the Latimer data were recalculated later by de Béthune and Swendeman Loud [3]. Thermodynamic data used in these sometimes modified calculations are given in Table 1. The potentials of oxygen and hydrogen compounds of carbon are discussed next.

---
*Reviewer: Barry Miller, Bell Laboratories, Murray Hill, New Jersey.

Chapter 8

TABLE 1 Thermodynamic Data on Carbon[a]

Formula	Description	State	$\Delta H°$/kJ mol^{-1}	$\Delta G°$/kJ mol^{-1}	$S°$/J mol^{-1} K^{-1}
C	$^3P_0$	g	718.384	672.975	157.992
C	Diamond	c	1.895	2.832	2.37
C	Graphite	c	0.0	0.0	5.6940
C$^+$	$^2P_{1/2}$	g	1,806.04		
C^{2+}	$^1S_0$	g	4,164.27		
C^{3+}		g	8,788.91		
C^{4+}		g	15,016.46		
C$_2$		g	982.0		
C$_3$		g		727.38	238.0
CH		g	594.5		
CH$_3$		g	134.		
CH$_4$		g	-74.847	-50.794	186.2
C$_2$H$_2$		g	226.75	209.	200.82
C$_2$H$_2$		aq	212.		

Carbon, Silicon, Germanium, Tin, and Lead

$C_2H_2F_2$	g		−313.5	265.2
C_2H_4	g	52.283	86.124	219.4
$C_2H_4F_2$	g		433.27	282.34
C_2H_6	g		−32.89	229.5
nC_6H_{14}	l		−4.35	296.0
C_6H_6	g	82.927	129.66	269.2
C_6H_6	l	49.028	124.50	172.8
C_6H_{12}	l		26.6	204.3
CO	g	−110.523	−137.268	197.91
CO_2	g	−393.513	−394.383	213.64
CO_2	aq	−412.9	−386.2	121.
H_2CO_3	aq	−698.7	−623.42	191.
HCO_3^-	aq	−691.1	−587.06	95.0
CO_3^{2-}	aq	−676.26	−527.90	−53.1
HCOOH	g	−362.6	−335.7	251.
$(HCOOH)_2$	g	−785.3	−685.3	348.
HCOOH	l	−409.	−346.	128.9

TABLE 1 (Continued)

Formula	Description	State	ΔH°/kJ mol^{-1}	ΔG°/kJ mol^{-1}	S°/J mol^{-1} K^{-1}
HCOOH		aq	−410.0	−356.0	163.6
HCOO$^-$		aq	−410.	−351.00	91.6
CH$_3$COOH		g	−487.0	−373.6	282.
CH$_3$COOH		ℓ	−488.452	−392.	160.
CH$_3$COOH		aq	−488.871	−399.6	
CH$_3$COO$^-$		aq	−826.8	−369.40	
(COOH)$_2$		c	−818.26	−697.9	120.
(COOH)$_2$		aq	−818.8	−697.0	
HC$_2$O$_4^-$		aq	−818.8	−690.86	
C$_2$O$_4^{2-}$		aq	−116.	−674.04	45.6
HCHO		g		−110.	218.6
HCHO		aq	−201.2	−129.7	
CH$_3$OH		g	−238.64	−161.9	238.
CH$_3$OH		ℓ		−166.3	127.

CH$_3$OH	aq	−246.	−174.5	132.3
C$_2$H$_5$OH	g	−235.3	−168.6	282.
C$_2$H$_5$OH	ℓ	−277.63	−174.8	161.
C$_2$H$_5$OH	aq	−288.1	−177.	
(CH$_3$)$_2$O	g	−185.	−114.	266.6
(CH$_3$)$_2$O	aq	−218.		
CO(NH$_2$)$_2$	aq		−202.6	175.6
CN	g	307.9	397.8	202.5
C$_2$N$_2$	g	131.	296.3	242.1
HCN	g	105.	120.	201.8
HCN	ℓ	105.	121.	112.8
HCN	aq	151.	112.	129.
CN$^-$	aq	−147.	166.	118.
HCNO	aq	−140.	−121.	182.
CNO$^-$	aq		−98.7	130.
CF	g		218.46	212.92
CF$_4$	g	−679.9	−635.1	262.

TABLE 1 (Continued)

Formula	Description	State	$\Delta H°$/kJ mol^{-1}	$\Delta G°$/kJ mol^{-1}	$S°$/J mol^{-1} K^{-1}
COF		g	−171	−245.145	247
COF$_2$		g	627.2 ± 4.2		244.8 ± 8.4
CCl$_4$		g	−106.7	−64.22	309.4
CCl$_4$		ℓ	−139.5	−68.74	214.4
COCl$_2$		g	−233.0	−210.4	289.2
CBr$_4$		g	50.	−56.530	357.995
Cl$_4$		g		260.90	391.6
CS$_2$		g	115.3	65.06	237.8
CS$_2$		ℓ	97.9	63.6	151.0
COS		g	−137.2	−169.2	234.5
(CNS)$_2$				325.9	
CNS$^-$		aq	72.0	88.7	(151.)

[a]National Bureau of Standards values are in italic; estimated values are in parentheses.
Source: Refs. 1, 2, and 11-14.

B. Potentials of Carbon Compounds with Oxygen and Hydrogen

Potentials of several redox systems in which the oxidized form is CO_2 have been calculated. Latimer [1] gives $E° = -0.49$ V for the reaction

$$2CO_2(g) + 2H^+ + 2e^- \rightarrow H_2C_2O_4 \qquad E° = -0.475 \text{ V is accepted}$$

For another reaction of CO_2, with a transfer of two electrons,

$$CO_2(g) + 2H^+ + 2e^- \rightarrow HCOOH(aq)$$

he reported $E° = -0.196$ V, while de Béthune and Swendeman Loud [3] found $E° = -0.199$ V which should be accepted. For this potential the latter authors calculated an isothermal coefficient (for a definition, see Ref. 3) of -0.036 mV K^{-1}. For the reaction

$$CO_2 + 2H^+ + 2e^- \rightarrow CO + H_2O$$

Pourbaix [2] reports $E° = -0.106$ V (slightly corrected).

The thermodynamic potential for electroreduction of CO_2 to C or CH_4 occurs at positive potentials:

$$CO_2 + 4H^+ + 4e^- \rightarrow C + 2H_2O \qquad E° = 0.206 \text{ V}$$

$$CO_2 + 8H^+ + 8e^- \rightarrow CH_4 + 2H_2O \qquad E° = 0.169 \text{ V}$$

H_2CO_3 and carbonates may be reduced to various substances. The potentials calculated for H_2CO_3 are written below together with the appropriate equations:

$$2H_2CO_3 + 2H^+ + 2e^- \rightarrow H_2C_2O_4 + 2H_2O \qquad E° = -0.391 \text{ V}$$

$$H_2CO_3 + 2H^+ + 2e^- \rightarrow HCOOH + H_2O \qquad E° = -0.166 \text{ V}$$

$$H_2CO_3 + 4H^+ + 4e^- \rightarrow HCHO + 2H_2O \qquad E° = -0.050 \text{ V}$$

$$H_2CO_3 + 6H^+ + 6e^- \rightarrow CH_3OH + 2H_2O \qquad E° = 0.044 \text{ V}$$

$$H_2CO_3 + 4H^+ + 4e^- \rightarrow C + 3H_2O \qquad E° = 0.228 \text{ V}$$

The corresponding reactions with participation of carbonates are

$$2CO_3^{2-} + 4H^+ + 2e^- \rightarrow C_2O_4^{2-} + 2H_2O \qquad E° = 0.478 \text{ V}$$

$$CO_3^{2-} + 3H^+ + 2e^- \rightarrow HCOO^- + H_2O \qquad E° = 0.311 \text{ V}$$

$$CO_3^{2-} + 6H^+ + 4e^- \rightarrow HCHO + 2H_2O \qquad E° = 0.197 \text{ V}$$

$$CO_3^{2-} + 8H^+ + 6e^- \rightarrow CH_3OH + 2H_2O \qquad E^\circ = 0.209 \text{ V}$$

$$CO_3^{2-} + 6H^+ + 4e^- \rightarrow C + 3H_2O \qquad E^\circ = 0.475 \text{ V}$$

Potentials of other carbon oxygen acids or their anions may be also calculated using thermodynamic data:

$$HCOOH + 2H^+ + 2e^- \rightarrow HCHO + H_2O \qquad E^\circ = 0.034 \text{ V}$$

$$2CO_2 + 2H^+ + 2e^- \rightarrow H_2C_2O_4 \qquad E^\circ = -0.481$$

$$H_2C_2O_4 + 2H^+ + 2e^- \rightarrow 2HCOOH \qquad E^\circ = 0.254 \text{ V}$$

$$HCOOH + 4H^+ + 4e^- \rightarrow CH_3OH + H_2O \qquad E^\circ = 0.100 \text{ V}$$

$$HCOOH + 2H^+ + 2e^- \rightarrow C + 2H_2O \qquad E^\circ = 0.523 \text{ V}$$

and for anions

$$C_2O_4^{2-} + 2H^+ + 2e^- \rightarrow 2HCOO^- \qquad E^\circ = 0.145 \text{ V}$$

$$HCOO^- + 3H^+ + 2e^- \rightarrow HCOH + H_2O \qquad E^\circ = 0.082 \text{ V}$$

$$HCOO^- + 5H^+ + 4e^- \rightarrow CH_3OH + H_2O \qquad E^\circ = 0.157$$

To consider the stepwise electroreduction of $CO_2(g)$ to $CH_4(g)$ as Latimer [1] did, the potentials of the following systems should also be considered:

$$HCHO(aq) + 2H^+ + 2e^- \rightarrow CH_3OH(aq) \qquad E^\circ = 0.232 \text{ V [2]}$$

$$CH_3OH(aq) + 2H^+ + 2e^- \rightarrow CH_4(g) + H_2O \qquad E^\circ = 0.588 \text{ V}$$

The potential of the latter reaction was calculated by Latimer [1] to be 0.586 V and a practically identical result, 0.588 V, was obtained by de Béthune and Swendeman Loud [3]. These authors have calculated the isothermal coefficient of the potential of this reaction to be -0.035 mV K^{-1}.

The calculated standard potentials of the following carbon couples are also available:

$$CO + 6H^+ + 6e^- \rightarrow CH_4 + H_2O \qquad E^\circ = 0.260 \text{ V}$$

$$CO + 2H^+ + 2e^- \rightarrow C + H_2O \qquad E^\circ = 0.517 \text{ V}$$

$$C + 4H^+ + 4e^- \rightarrow CH_4 \qquad E^\circ = 0.132 \text{ V}$$

For the last reaction an identical potential was reported by Latimer [1] and de Béthune and Swendeman Loud [3]. The latter authors give -0.209 mV K^{-1} as the isothermal coefficient of the potential of this reaction.

The schematic representation of potentials of carious couples that may be formed when CO_2 is reduced to CH_4 for acidic solutions may be represented in the following way:

$$CH_4 \xrightarrow{0.59} CH_3OH(aq) \xrightarrow{0.232} HCHO(aq) \xrightarrow{0.034} HCOOH(aq) \xrightarrow{-0.20} CO_2(g)$$

A similar diagram for alkaline solutions reported by Latimer [1] is characterized by, as expected, more negative potentials:

$$CH_4(g) \xrightarrow{-0.2} CH_3OH(aq) \xrightarrow{-0.59} HCHO(aq) \xrightarrow{-1.07} HCOO^- \xrightarrow{-1.01} CO_3^{2-}(aq)$$

The majority of reactions presented proceed with a large overvoltage.

The kinetics of CO_2 electroreduction to HCOOH at mercury electrodes was studied by Paik et al. [4]. For the reaction

$$CO_2 + H_2O + 2e^- \rightarrow HCOO^- + OH^-$$

Ryn et al. [5] calculated the following potentials at several temperatures:

$t/°C$	2	10	25	40	50	60
$E°/V$	−0.699	−0.708	−0.723	−0.739	−0.750	−0.760

These reactions are also quite slow.

In theory CO_2 as well as H_2CO_3 or carbonates may be further reduced to carbon or methane. The thermodynamic potentials of these reactions are quite near 0 V. However, due to large overpotentials, these reactions were not observed in aqueous solutions. Some of these reactions may proceed in molten salts. Reversible potentials would also suggest a possibility of oxidizing carbon to CO_2, H_2CO_3, or carbonates, but such reactions do not proceed in water under normal conditions.

Formaldehyde may be electrolytically reduced to methyl alcohol, but further reduction of this substance does not occur under normal conditions. However, methanol may be oxidized to carbonic acid or to carbonates, but this reaction also proceeds in a very irreversible way. The –COOH group is not reduced at the mercury electrode in aqueous solutions, but Latimer [1] reports that HCOOH is reduced to HCHO and CH_3OH. Formaldehyde may be oxidized to H_2CO_3 or carbonates. According to Latimer [1], this reaction may occur in alkaline solutions. The electrode reaction of the system CO_2/CO is also irreversible.

There is now a large literature on the irreversible adsorption of CO_2, CO, HCHO, and CH_3OH on platinum electrodes. This adsorption is combined with the chemical transformation of adsorbed molecules in the overall electrochemical reaction. These studies, as well as the oxidation of CO to CO_2 or carbonates and reduction of CO_2, were reviewed by Randin [6].

According to Latimer [1], H_2CO_3 and CO_2 may be oxidized anodically to peroxycarbonates:

$$C_2O_6^{2-} + 4H^+ + 2e^- \rightarrow 2H_2CO_3 \qquad E° > 1.7 \text{ V}$$

The stepwise change from CH_4 to CO_2 is represented by the following diagram valid for acidic solutions:

$$\overset{+4}{CO_2} \xrightarrow{-0.106} \overset{+2}{CO} \xrightarrow{0.517} \overset{0}{C} \xrightarrow{0.132} \overset{-4}{CH_4}$$

C. Potentials of Carbon Compounds with Nitrogen and Sulfur

Carbon may form with nitrogen and sulfur $(CN)_2$ and $(SCN)_2$ compounds with properties similar to halogens. As pseudohalogens $(CN)_2$ and $(SCN)_2$ may be reduced in simple reactions

$$(CN)_2(g) + 2H^+ + 2e^- \rightarrow 2HCN(aq)$$

De Béthune and Swendeman Loud [3] gave an $E°$ value of 0.373 V for this reaction. These authors reported -0.596 mV K^{-1} as the isothermal temperature coefficient. A similar $E°$ value was reported earlier by Latimer [1] and Deltombe and Pourbaix [7], $E° = 0.375$ V is suggested. For

$$(CN)_2(g) + 2e^- \rightarrow 2CN^-$$

$E°$ is -0.176 V. According to Gauguin [8], these two systems should be reversible, but he reported later [9] that the potentials of these systems do not agree with calculated values.

In the case of $(SCN)_2$ the potential of its simple reduction to SCN^- ions is more positive:

$$(SCN)_2 + 2e^- \rightarrow 2SCN^-(aq) \qquad E° = 0.77 \text{ V}$$

Such a potential was reported by Gauguin [9] and others [1,3]. Gauguin carried out a study of this reaction in aqueous solutions and found that it is irreversible, due to fast disproportionation of $(SCN)_2$ according to the reaction

$$3(SCN)_2 + 4H_2O \rightarrow 5SCN^- + SO_4^{2-} + HCN + 7H^+$$

Gauguin [9] gave the potential of the reaction

$$SO_4^{2-} + CN^- + 8H^+ + 6e^- \rightarrow SCN^- + 4H_2O$$

as $E° = 0.515$ V. This reaction proceeds irreversibly [10] and the potential was not dependent on CN^- ion concentration.

Several potentials are reported for reactions in which HOCN or OCN⁻ ion is an oxidant:

$$2HOCN(aq) + 2H^+ + 2e^- \rightarrow (CN)_2(g) + 2H_2O$$

De Béthune and Swendeman Loud [3] and Deltombe and Pourbaix [7] reported $E° = -0.330$ V for this reaction. The isothermal temperature coefficient of the potential of this reaction is -0.588 mV K^{-1} [3].

In the case of further reduction of HOCN to HCN or CN⁻ ions, the following reactions were considered:

$$HOCN(aq) + 2H^+ + 2e^- \rightarrow HCN(aq) + H_2O \qquad E° = 0.02 \text{ V}$$

$$OCN^- + 2H^+ + 2e^- \rightarrow CN^- + H_2O \qquad E° = -0.14 \text{ V}$$

$$HOCN(aq) + 2H^+ + 2e^- \rightarrow HCN(g) + H_2O \qquad E° = -0.02 \text{ V}$$

$$OCN^- + 3H^+ + 2e^- \rightarrow HCN(g) + H_2O \qquad E° = 0.09 \text{ V}$$

All these potentials are reported by Deltombe and Pourbaix [7]. Gauguin [8] found that the CNO⁻/CN⁻ system is irreversible.

For the reaction

$$CNO^- + H_2O + 2e^- \rightarrow CN^- + 2OH^-$$

$E° = -0.97$ V was reported by Pourbaix [2] and later by de Béthune and Swendeman Loud [3]; -1.211 mV K^{-1} was reported [3] as the isothermal coefficient.

Considering $(CN)_2$ and $(SCN)_2$ as pseudohalogens, the following order for simple reactions of halogens and pseudohalogens may be presented: F_2/F^-, 2.87 V; Cl_2/Cl^-, 1.358 V; Br_2/Br^-, 1.087 V; $(SCN)_2/SCN^-$, 0.77 V; I_2/I^-, 0.535 V; and $(CN)_2/CN^-$, -0.176 V. It follows from this comparison that $(CN)_2$ has the lowest oxidation power in this series, while the oxidizing properties of $(SCN)_2$ are significant.

Although numerous potentials for carbon were given in this chapter, almost all of these values were calculated from thermodynamic data, while experimental measurements usually led to disagreement, owing to the slowness of electrode reactions of carbon compounds. Data used in potential calculations are given in Table 1.

REFERENCES

1. W. M. Latimer, *Oxidation Potentials*, Prentice-Hall, Englewood Cliffs, N.J., 1952.
2. M. Pourbaix, ed., *Atlas of Electrochemical Equilibria in Aqueous Solutions*, Pergamon Press, Oxford, 1966.

3. A. J. de Béthune and N. A. Swendeman Loud, *Standard Aqueous Electrode Potentials and Temperature Coefficients at 25°C*, Hampel, Skokie, Ill., 1964.
4. W. Paik, T. N. Andersen, and H. Eyring, Electrochim. Acta, *14*, 1217 (1969).
5. J. Ryn, T. N. Andersen, and H. Eyring, J. Phys. Chem., *76*, 3278 (1972).
6. J.-P. Randin, in *Encyclopedia of The Electrochemistry of the Elements*, A. J. Bard, ed., Vol. 7, Marcel Dekker, New York, 1976, p. 2.
7. E. Deltombe and M. Pourbaix, Comité Int. Thermodyn. Cinétique Electrochim., Compt. Rend., 6th Meet., Poitiers, 1954, 1955.
8. R. Gauguin, Bull. Soc. Chim. Fr., *15*, 1048 (1948).
9. R. Gauguin, Ann. Chim., *4*, 832 (1949).
10. R. Gauguin, Comité Int. Electrochim. Thermodyn. Kinet., 2nd Meet., Tamburini, 1950, Milan, 1951, p. 269.
11. O. Kubaschewski and E. L. Evans, *Metallurgical Thermochemistry*, Pergamon Press, Oxford, 1958.
12. *Selected Values of Chemical Thermodynamic Properties*, Nat. Bur. Stand. Circ. 500, U.S. Government Printing Office, Washington, D.C., 1952; NBS Technical Notes 270.3 and 270.4.
13. I. Barin and O. Knacke, *Thermochemical Properties of Inorganic Substances*, Springer-Verlag, Berlin, 1973.
14. M. Kh. Karapetyants, *Introduction in the Theory of Chemical Processes* (in Russian), Vysshaya Shkola, Moscow, 1975.

II. SILICON

A. Oxidation States

Silicon is a nonmetallic element [1]. The most important state is +4. In this oxidation state simple silicon compounds usually have a tetrahedral arrangement. In some compounds Si(IV) forms octahedral inner-sphere complexes such as SiF_6^{2-}.

The chemistry of this element in the +4 oxidation state is quite complex. There are a large number of different acids, especially of the polyacid type. These complex forms exist in equilibrium, which obviously may be shifted by a change of pH.

In contrast to other elements of fourth group, two-valent cations in aqueous solutions are not known, and no potentials for this oxidation state under such conditions are reported. Compounds of Si^{2+} are unstable at normal temperature.

SiO exists in the gas phase. It may also be formed in the course of decomposition of SiO_2 at high temperatures and in vapor over $Si + SiO_2$, also at high temperatures. The existence of $SiCl_2$ and SiS has been reported.

Silicon forms compounds with hydrogen, but not nearly as many compounds as in the case of carbon. However, several compounds of the type Si_nH_{2n+2} are known. Potentials are reported for SiH_4. Since the free energies of conversion from simple to more complex forms are known, the potentials of these forms with respect to simpler ones may, in principle, be estimated. The thermodynamic data are given in Table 2.

TABLE 2 Thermodynamic Data on Silicon[a]

Formula	Description	State	$\Delta H°$/kJ mol^{-1}	$\Delta G°$/kJ mol^{-1}	$S°$/J mol^{-1} K^{-1}
Si		g	368.4	323.9	167.9
Si		c	0.000	0.000	18.70
Si	Amorph.	c	4.2		
Si$^+$		g	1,160.8		
Si^{2+}		g	2,736.5		
Si^{3+}		g	5,971.0		
Si^{4+}		g	10,332.0		
SiH		g	334.4	197.9	
SiH$_4$		g	55.21		204.1
Si$_2$H$_6$			(126.)		(274.)
Si(CH$_3$)$_4$		g		-176.05	360.
SiO		g	-111.80	-137.11	206.10
SiO				-386.6 ± 22.6	
SiO$_2$		g		-327.5	227.6
SiO$_2$	Quartz	c	-859.4	-825.1 ± 3.3	41.84

TABLE 2 (Continued)

Formula	Description	State	$\Delta H°$/kJ mol^{-1}	$\Delta G°$/kJ mol^{-1}	$S°$/J mol^{-1} K^{-1}
SiO_2	Cristobalite	c	−857.7	−821.7 ± 3.3	42.63
SiO_2	Tridymite	c	−856.9	−802.9	43.35
SiO_2	Vitreous	glass	−847.3	−817.1 ± 3.3	46.9
H_2SiO_3		c	−1,132.6	−1,023.0	
$H_2Si_2O_5$		c	−1,976.		
$HSiO_3^-$		aq		−955.46	
SiO_3^{2-}		aq		−887.	
$Si(OH)_4$		aq		−1,276. ± 4.6	
$Si(OH)_3O^-$		aq		−1,223.4 ± 5.4	
$Si(OH)_2O_2^{2-}$		aq		−1,152.7 ± 6.3	
$Si_4O_8(OH)_4^{4-}$		aq		−3,969.8 ± 18.4	
$Si_4O_6(OH)_6^{2-}$		aq		−4,079.8 ± 18.4	
$Si_2O_3(OH)_4^{2-}$		aq		−2,211.2 ± 9.2	
$SiOF_2$		g		−950.7	271.2
SiF_2		g	−615.0	−691.5	256.5
SiF_4		g	−1,614.94	−1,574.8	282.14

SiF_6^{2-}	aq	−2,336.8	−2,138	−50.
H_2SiF_6	aq	−2,331.3		
$SiCl_2$	g		(−171.2)	(279.)
$SiCl_2H_2$	g		(−293.)	(287.)
$SiCl_4$	g	−609.6	−569.9	331.4
$SiCl_4$	ℓ	−604.1	−572.8	239.3
$SiBr_2$	g	−42.7	−133.1	303.3
$SiBr_3H$	g		(−303.)	(348.)
$SiBr_4$	ℓ	−397.9		
SiI_2	g		(23.9)	(317.)
SiI_4	c	−132.2		
Si_3N_4	c		−642.7	101.2
SiS	g		(51.8)	(223.5)
SiS_2	c	−212.1	−232.1	66.9
$SiSe$	g	202.9	132.8	235.1
SiC	c	−73.2	−78.2	16.6

[a]National Bureau of Standards values are in italic.
Source: Refs. 2, 3, 5, and 8-11.

B. Silicon Potentials

The electrode reactions of silicon species are slow under normal conditions, which may follow from the fact that the bonds of silicon with other atoms are quite strong.

The potentials of silicon species are calculated on the basis of existing thermodynamic data. Some were calculated by Latimer [2] and Pourbaix [3] and later by Carasso and Faktor [4]. For the reaction

$$SiO_2 + 4H^+ + 4e^- \rightarrow Si + 2H_2O$$

several potentials are reported, depending on the form of SiO_2. Their values are given in Table 3. Temperature coefficients of the potential of this reaction (for quartz) were also reported [5]; -0.374 mV K^{-1} was given for the isothermal coefficient.

It follows from the data in Table 3 that silicon is quite a strong reductant. It can also disproportionate with water to form SiO_2, or silicate acid, and hydrogen, or SiH_4. $\Delta G°$ of the reaction

$$2Si + 2H_2O \rightarrow SiH_4(g) + SiO_2$$

is -369.5 kJ mol^{-1}.

However, probably due to the above-mentioned high irreversibility of silicon reactions, it is slow at room temperature. SiO_2 on the surface of silicon may also inhibit this reaction [3].

One may add that free silicon cannot be obtained by electrolysis of aqueous solutions. The free element may be obtained from SiO_2, when strong reductants are used (e.g., carbon at high temperature).

Silicon forms a number of different acids in solutions. Carasso and Faktor [4] estimated the standard Gibbs energy of formation of orthosilicic acid as -1274.9 ± 4.6 kJ mol^{-1}, and for $H_3SiO_4^-$ and $H_2SiO_4^{2-}$, -1222.2 ± 5.4 kJ mol^{-1} and -1151.6 ± 6.3 kJ mol^{-1}, respectively. It seems, however, that the use of the values of pK as determined by Lagerström [6] is probably not quite correct in view of other data tabulated by Perrin [7]. In consequence, these $\Delta G°$ values may be slightly modified.

In solutions of silicic acid, polymeric species may be formed. An instance of complicated equilibria existing in such solutions show the reactions

$$2H_2SiO_4 + 2OH^- \rightarrow H_4Si_2O_7^{2-} + 3H_2O$$

$$4H_4SiO_4 + 4OH^- \rightarrow H_4Si_4O_{12}^{4-} + 8H_2O$$

TABLE 3 Standard Potentials of the SiO_2/Si System

SiO_2 form	$E°$/V
Quartz	-0.909
Cristobalite	-0.900
Vitreous	-0.888

Carasso and Faktor [4] give $\Delta G°$ values for some polynuclear species of that type. They calculated the electrode potentials for some of the systems with participation of different acids and anions of silicon. However, reactions with participation of polynuclear compounds are not considered, because of the lack of reliable thermodynamic data.

$$H_4SiO_4 + 4H^+ + 4e^- \rightarrow Si + 4H_2O \qquad E° = -0.848 \text{ V}$$

$$H_3SiO_4^- + 5H^+ + 4e^- \rightarrow Si + 4H_2O \qquad E° = -0.712 \text{ V}$$

The later value may be influenced to some extent by the choice of pK_1; $pK_1 = 9.2 \pm 0.2$ was assumed [4] as a value extrapolated to zero ionic strength (for orthosilicic acid). It should again be stressed that, because of the possibility of formation of polynuclear species, at least in some pH ranges, these data may be erroneous to some extent.

In alkaline solutions for the reaction

$$SiO_3^{2-} + 3H_2O + 4e^- \rightarrow Si + 6OH^- \qquad E°_B = -1.69 \text{ V}$$

Si is less reducing in fluoride:

$$SiF_6^{2-} + 4e^- \rightarrow Si + 6F^- \qquad E° = -1.37 \text{ V}$$

The isothermal coefficient of the potential of this reaction was reported as -0.65 mV K^{-1} [5].

Potentials have also been calculated for systems involving Si(IV) and Si(II). The potential of the reaction occurring between two solid oxides of silicon,

$$SiO_2(\text{glassy}) + 2H^+ + 2e^- \rightarrow SiO + H_2O$$

was calculated as -0.967 V [4]. Slightly different potentials could be calculated for reactions with participation of other forms of SiO_2 in accordance with thermodynamic data given in Table 2.

Potentials of reactions with silicic acid reacting instead of SiO_2 were found to be less negative [4]:

$$H_4SiO_4 + 2H^+ + 2e^- \rightarrow SiO + 3H_2O \qquad E° = -0.887 \text{ V}$$

$$H_3SiO_4^- + 3H^+ + 2e^- \rightarrow SiO + 3H_2O \qquad E° = -0.615 \text{ V}$$

$$H_2SiO_4^{2-} + 4H^+ + 2e^- \rightarrow SiO + 3H_2O \qquad E° = -0.248 \text{ V}$$

The potential of the SiO/Si system was calculated from existing thermodynamic data:

$$SiO + 2H^+ + 2e^- \rightarrow Si + H_2O \qquad E^\circ = -0.808 \text{ V}$$

Carasso and Faktor [4] have also calculated the potentials of systems with SiH_4 as a reduced form:

$$Si + 4H^+ + 4e^- \rightarrow SiH_4 \qquad E^\circ = -0.143 \text{ V}$$

$$SiO_2 + 8H^+ + 8e^- \rightarrow SiH_4 + 2H_2O \qquad E^\circ = -0.516 \text{ V}$$

$$H_4SiO_4 + 8H^+ + 8e^- \rightarrow SiH_4 + 4H_2O \qquad E^\circ = -0.495 \text{ V}$$

These potentials are quite different from those reported by Pourbaix [3]. The results of Carasso and Faktor [4] seem to be better, as they are based on more recent thermodynamic data (Table 2).

Some thermodynamic data concerning higher analogs of SiH_4 are available, but it seems that the precision of their determination is rather low. In addition, limited use of such systems makes calculation of these potentials in practice less valuable.

Oxide films on silicon cannot be reduced cathodically in aqueous solutions prior to hydrogen gas evolution [12].

REFERENCES

1. This section is based partly on F. A. Cotton and G. Wilkinson, *Advanced Inorganic Chemistry*, 2nd ed., Interscience, New York, 1966.
2. W. M. Latimer, *Oxidation Potentials*, 2nd ed., Prentice-Hall, Englewood Cliffs, N.J., 1952, p. 142.
3. M. Pourbaix, ed., *Atlas of Electrochemical Equilibriua in Aqueous Solutions*, Pergamon Press, Oxford, 1966, p. 458.
4. J. I. Carasso and M. M. Faktor, in *The Electrochemistry of Semiconductors*, P. J. Holmes, ed., Academic Press, London, 1962, Chap. 5.
5. A. J. de Béthune and N. A. Swendeman Loud, *Standard Aqueous Electrode POtentials and Temperature Coefficients at 25°C*, Hampel, Skokie, Ill., 1964.
6. G. Lagerström, Acta Chem. Scand., 13, 722 (1959).
7. D. D. Perrin, Pure Appl. Chem., 20, 133 (1969).
8. O. Kubaschewski and E. L. Evans, *Metallurgical Thermochemistry*, Pergamon Press, Oxford, 1958.
9. I. Barin and O. Knacke, *Thermochemical Properties of Inorganic Substances*, Springer-Verlag, Berlin, 1973.
10. M. Kh. Karapetyants, *Introduction in the Theory of Chemical Processes* (in Russian), Vysshaya Shkola, Moscow, 1975.
11. *Selected Values of Chemical Thermodynamic Properties*, Nat. Bur. Stand. Circ. 500, U.S. Government Printing Office, Washington, D.C., 1952.
12. D. R. Turner, in *The Electrochemistry of Semiconductors*, P. J. Holmes, ed., Academic Press, London, 1962, p. 155.

III. GERMANIUM

A. Oxidation States

Germanium usually occurs in one structure—that of diamond. It is unstable in the presence of water. In aqueous solutions germanium should dissolve with formation of GeO_2. In practice, germanium is not attacked by water and solutions of HCl and NaOH over long periods.

The chemistry of germanium is to some extent similar to that of silicon [1]. As in the case of silicon, the most stable compounds are those of Ge(IV); however, Ge(II) compounds are more stable than those of Si(II). Both GeO and $Ge(OH)_2$ are known and the latter compound may be prepared from aqueous solutions. Halogen compounds of Ge(II) such as $GeCl_2$ may also be prepared, although they easily decompose to Ge and $GeCl_4$.

Both GeO and GeO_2 have amphoteric character, although the basic properties of GeO_2 are very low. Thermodynamic parameters of these substances are given in Table 4. Carasso and Faktor [2] estimate that the standard free energy of formation of a glass form of GeO_2 should not differ significantly from that of a hexagonal form. This estimation is based on the solubility data of GeO_2.

Both metagermanate $Me_2^{(I)}GeO_3$ and orthogermanate $Me_2^{(I)}GeO_4$ are known. In aqueous solutions ions of Ge(IV) may have different structure. Mostly $[GeO(OH)_3]^-$ and $[GeO_2(OH)_2]^{2-}$ ions are present.

Germanium also forms compounds with hydrogen. Hydrides of the type Ge_nH_{2n+2} are known with n ranging from 1 to 9. All these compounds are unstable and undergo reactions with oxygen to yield GeO_2 and H_2O. This reaction proceeds more slowly than in the corresponding case of silanes.

B. Germanium Potentials

The electrode reaction considered by Lovrecek and Bockris [3]

$$GeO_2 + 4H^+ + 4e^- \rightarrow Ge + 2H_2O$$

is slow and its potential cannot be measured precisely experimentally. The standard potentials calculated from thermodynamic data for this reaction are -0.019 V and -0.058 V for the hexagonal and the tetragonal forms of GeO_2, respectively. The temperature coefficient of the potential of this reaction was reported [4] as -0.335 mV K^{-1} for the isothermal coefficient.

Potentials of the Ge(IV)/Ge and Ge(IV)/Ge(II) couples were calculated by Reid [5]. The following values were reported: $E°(Ge^{4+}/Ge) = 0.124$ V and $E°(Ge^{4+}/Ge^{2+}) = 0.00$ V.

Potentials of reactions involving H_2GeO_3 and its anions were calculated by Lovrecek and Bockris [3] and later corrected by Carasso and Faktor [2]:

$$H_2GeO_3 + 4H^+ + 4e^- \rightarrow Ge + 3H_2O \qquad E° = 0.012 \text{ V}$$

$$HGeO_3^- + 5H^+ + 4e^- \rightarrow Ge + 3H_2O \qquad E° = 0.141 \text{ V}$$

$$GeO_3^{2-} + 6H^+ + 4e^- \rightarrow Ge + 3H_2O \qquad E° = 0.328 \text{ V}$$

TABLE 4 Thermodynamic Data on Germanium[a]

Formula	Description	State	$\Delta H°$/kJ mol^{-1}	$\Delta G°$/kJ mol^{-1}	$S°$/J mol^{-1} K^{-1}
Ge		g	328.19	209.8	167.803
Ge		c	0.000	0.000	42.42
Ge$^+$		g	1,118.47		
Ge^{2+}		g	2,663.12		
Ge^{2+}		aq		47.7 ± 5.4	
Ge^{3+}		g	5,970.6		
Ge^{4+}		g	10,386.4		
GeO		g	−95.4	−118.0	219.91
GeO		c	−305	−276	(50)
GeO	Anhydr. black	c	−248.9 ± 14.6	−220.1 ± 14.6	48.1
GeO	Hydr. brown	c		−218.9 ± 4.2	
GeO	Hydr. yellow	c		−187.9 ± 4.2	
GeO$_2$	Amorph.	c	−536.8	−477.8 ± 2.5	
GeO$_2$	Hex. soluble	c	−537.6	−481.6 ± 2.5	(55.2 ± 0.4)

Tetrahedral

GeO$_2$	c		−496.6 ± 2.5	
H$_2$GeO$_2$	aq	−833.9		
H$_2$GeO$_3$	aq	−849	−707.1 ± 2.5	(180)
HGeO$_3^-$	aq	−828	−657.3 ± 2.5	(88)
GeO$_3^{2-}$	aq		−584.9 ± 2.5	
HGeO$_2^-$	aq		−385.765	
GeF$_2$	g		(−483.8)	(271.1)
GeF$_4$	g		−1,149.9	310.4
GeCl	g		(59.8)	(245.)
GeCl$_4$	g		(−466.0)	(348.)
GeCl$_4$	ℓ	−569	−498	251
GeBr	g		(101.)	(257)
GeBr$_2$	g		(−108.9)	(318.)
GeBr$_4$	g	−298.7	−323.0	396.2
GeI$_2$	c	−113	−109	(151)
GeI$_4$	c	−176	−171	(268)

TABLE 4 (Continued)

Formula	Description	State	$\Delta H°$/kJ mol^{-1}	$\Delta G°$/kJ mol^{-1}	$S°$/J mol^{-1} K^{-1}
GeI$_4$		g		−86.98	428.8
GeH$_4$		g		113.	217.5
Ge$_2$H$_6$		g		(209.)	(297.)
GeS		g		(48.)	(235.5)
GeS		c	−76.1	(−71.)	(66.0)
GeS$_2$		c		−152.5	87.4
GeSe		c		(−39.9)	78.2
GeSe$_2$		c		(−61.9)	112.5
GeTe		c		(−33.5)	(83.7)
GeP		c	−27.2	−45.410	61.1
Ge$_3$N$_4$		c	−61.9		

[a]National Bureau of Standards values are in italic; estimated values are in parentheses.
Source: Refs. 2, 13, 14, and 15.

The reaction occurring in alkaline solution is also considered

$$HGeO_3^- + 2H_2O + 4e^- \rightarrow Ge + 5OH^- \qquad E_B^\circ = -0.89 \text{ V}$$

De Béthune and Swendeman Loud [4] report the value of -1.03 and the temperature coefficient as -1.29 mV K^{-1} (isothermal).

In solutions of Ge(IV), other more complicated forms may exist in equilibrium, such as

$$\text{monogermanate} \rightleftharpoons \text{pentagermanate} \rightleftharpoons \text{heptagermanate}$$

The equilibria are displaced to the right by the increase of Ge(OV) concentration and to the left by an increase of pH above 9. These equilibria are not well known; however, no strict thermodynamic data exist.

Halogen compounds of germanium undergo hydrolysis in contact with water; however, in very acid solution mixed complexes of the type $[Ge(OH)_nCl_{6-n}]^{2-}$ are formed. No potentials are presented for such species.

The standard Gibbs energy change for the reaction

$$GeO_2(\text{hex}) + 2H^+ + 2e^- \rightarrow GeO(\text{brown,hydrous}) + H_2O$$

was evaluated by Jolly and Latimer [7] as -22.6 ± 2.1 kJ mol^{-1}. The standard potential of this reaction is -0.132 V for hexagonal GeO$_2$ and for the tetragonal form of GeO$_2$, -0.210 V.

The potential for the system GeO$_2$/GeO(brown) was determined experimentally by Jolly and Latimer [7] to be -0.118 ± 0.010 V, which agrees well with the value calculated for the hexagonal form of GeO$_2$.

The standard Gibbs energy of the transformation

$$GeO(\text{yellow,hydrous}) \rightarrow GeO(\text{brown,hydrous})$$

is -31.0 kJ mol^{-1}, and consequently, the potential of reaction

$$GeO_2 + 2H^+ + 2e^- \rightarrow GeO(\text{yellow,hydrous}) + H_2O$$

calculated from thermodynamic data is -0.293 and -0.370 V for the hexagonal and tetragonal forms of GeO$_2$, respectively. The participation of H$_2$GeO$_3$ or its anions in these reactions instead of GeO$_2$ does not change the potentials significantly. For the reaction

$$H_2GeO_3 + 2H^+ + 2e^- \rightarrow GeO + 2H_2O$$

E° is -0.72 and -0.232 V for the brown and yellow forms of GeO, respectively.

Potentials of reactions similar type to that given above, but with HGeO$_3^-$ or GeO$_3^{2-}$ instead of H$_2$GeO$_3$ were calculated by Lovrecek and Bockris [3] and reconsidered by Carasso and Faktor [2]. In such a case the potentials move to positive values. For the reaction

$$HGeO_3^- + 3H^+ + 2e^- \to GeO(brown) + 2H_2O$$

and taking a mean value of pK_1 of H_2GeO_3 as 9.0, $E° = 0.186$ V. On the basis of experiments by Jolly [8] on the solubility of brown GeO, the standard Gibbs energy of the reaction

$$GeO(brown) + 2H^+ \rightleftharpoons Ge^{2+}(aq) + H_2O$$

was calculated by Carasso and Faktor [2] to be $\Delta G° = 28.4 \pm 1.2$ kJ mol^{-1}.

This value, together with other thermodynamic parameters, enables the standard Gibbs energy of the dissolved hydrated Ge(II) to be calculated as $+47.6 \pm 5.4$ kJ mol^{-1}. With this value the potential of the reaction

$$H_2GeO_3 + 4H^+ + 2e^- \to Ge^{2+} + 3H_2O$$

was found to be -0.223 V [2]. The potentials of similar reactions with $HGeO_3^-$ and GeO_3^{2-} ions are considerably more positive [2].

Taking the pK_1 of germanic acid as 9.0 at 25°C [9], the potential for the couple $HGeO_3^-/Ge^{2+}$ is equal $+0.042$ V. However, it should be noted that the value of the dissociation constant may be influenced to some extent by polymerization equilibria.

The potential of Ge(II) with respect to free germanium was also calculated [2]:

$$Ge^{2+} + 2e^- \to Ge \qquad E° = 0.247 \text{ V}$$

It should be remembered, however, that hydrated Ge^{2+} is not stable. More stable are complexes of Ge^{2+} with ligands such as halides. The potentials of the reaction

$$GeO + 2H^+ + 2e^- \to Ge + H_2O$$

are 0.095 and 0.255 V for the brown and yellow forms of GeO, respectively. Extended experimental measurements of the potentials of germanium couples were carried out by Lovrecek and Bockris [3], with some by other workers, but usually, mixed potentials were measured.

The potential of only one hydride of germanium GeH_4 was calculated from existing thermodynamic data

$$Ge + 4H^+ + 4e^- \to GeH_4 \qquad E° = -0.42 \text{ V}$$

It should be remembered that GeH_4 is not stable in aqueous solutions.

Some of the potentials given above do not agree with those reported by Pourbaix [10]. Also other potentials were corrected when more appropriate thermodynamic data were available. The experimental studies of Lovrecek and Bockris [3] show the agreement of measured results with calculated data in the case of the system GeO/Ge. The agreement should be also observed in case of the system Ge^{2+}/Ge. However, in this case under experimental conditions, Ge(II) in contact with water forms GeO.

It was also found by Lovrecek and Bockris [3] that n-type germanium electrodes have a mean potential value about 5 mV more negative than that of p-type ones. This is not expected on the basis of the thermodynamic considerations, but it may be expalined by taking into account the lower exchange current for n-type germanium [3].

Potential-pH diagrams for the germanium-water system published by Pourbaix [10] were later reproduced by Vijh [11]. Similar, but not so extensive diagrams have also been given by other workers [2,3]. Turner [12] reports that oxide films and etching residues on germanium are completely reduced at a potential about −0.55 V.

REFERENCES

1. This section is based partly on F. A. Cotton and G. Wilkinson, *Advanced Inorganic Chemistry*, 2nd ed., Interscience, New York, 1966.
2. J. I. Carasso and M. M. Faktor, in *The Electrochemistry of Semiconductors*, P. J. Holmes, ed., Academic Press, London, 1962, Chap. 5.
3. B. Lovrecek and J. O'M. Bockris, J. Phys. Chem., 63, 1368 (1959).
4. A. J. de Béthune and N. A. Swendeman Loud, *Standard Aqueous Electrode Potentials and Temperature Coefficients at 25°C*, Hampel, Skokie, Ill., 1964.
5. W. E. Reid, Jr., J. Phys. Chem., 69, 3168 (1965); 69, 2269 (1965).
6. W. M. Latimer, *Oxidation Potentials*, 2nd ed., Prentice-Hall, Englewood Cliffs, N.J., 1952, p. 145.
7. W. L. Jolly and W. M. Latimer, J. Am. Chem. Soc., 74, 5751 (1952).
8. W. L. Jolly, Thesis, University of Carolina, 1952; cited in Ref. 2.
9. D. D. Perrin, Pure Appl. Chem., 20, 133 (1969).
10. M. Pourbaix, ed., *Atlas of Electrochemical Equilibria in Aqueous Solutions*, Pergamon Press, Oxford, 1966, p. 464.
11. A. K. Vijh, in *Encyclopedia of the Electrochemistry of the Elements*, A. J. Bard, ed., Vol. 5, Marcel Dekker, New York, 1956.
12. D. R. Turner, in *The Electrochemistry of Semiconductors*, P. J. Holmes, ed., Academic Press, London, 1962, p. 155.
13. *Selected Values of Chemical Thermodynamic Properties*, Natl. Bur. Stand. Circ. 500, U.S. Government Printing Office, Washington, D.C., 1952.
14. M. Kh. Karapetyants, *Introduction in the Theory of Chemical Processes* (in Russian), Vysshaya Shkola, Moscow, 1975.
15. O. Kubaschewski and E. L. Evans, *Metallurgical Thermochemistry*, Pergamon Press, Oxford, 1958.

IV. TIN

A. Oxidation States

Tin and its compounds show properties to some extent similar to those of germanium. Metallic tin exists in two different forms. From Table 5 it can be seen that the thermodynamic parameters of both forms are quite similar.

As in the case of germanium in aqueous solutions two oxidation states exist; however, in the case of tin, the stability of the +2 valence state is con-

TABLE 5 Thermodynamic Data on Tin[a]

Formula	Description	State	ΔH°/kJ mol^{-1}	ΔG°/kJ mol^{-1}	S°/J mol^{-1} K^{-1}
Sn		g	301.	268.	168.385
Sn	Gray	c	2.5	4.6	44.8
Sn	White	c	0.0	0.0	51.5
Sn$^+$		g	1,015.0		
Sn^{2+}		g	2,432.6		
Sn^{2+}		aq		−27.24	−22.7
Sn^{3+}		g	5,396.8		
Sn^{4+}		g	9,334.		
Sn^{4+}		aq		2.72	
Sn^{5+}		g	17,129.		
SnO		g	20.857	−48.313	232.00
SnO		c	−286.	−257.	56.6
SnO$_2$		c	−580.7	−519.9	52.3
SnO$_2$	Hydr. white	c		−477.5	

Carbon, Silicon, Germanium, Tin, and Lead

$Sn(OH)^+$	Std. state hyp., $m = 1$	aq		−255.6	
$HSnO_2^-$	Std. state hyp., $m = 1$	aq		−410.0	
$Sn(OH)_2$		c	−578.6	−492.0	96.6
$Sn(OH)_4$		c	−1,131.8	−951.9	(121.)
SnO_3^{2-}		aq		−574.965	
$Sn(OH)_6^{2-}$		aq		−1,299.	
$H_2Sn(OH)_6$		c		−1,433.3	
SnS		g	105.069	32.869	242.153
SnS		c		−108.2	77.0
SnS_2		c	−153.5	−179.623	87.4
$SnSe$	β phase	c	−94.5	−123.813	98.11
$SnSe_2$		c	−124.7	−159.862	118.0
$SnTe$		c	−60.7	−90.107	98.7
SnF_6^{2-}		aq	−1,986.1	−1,757.	(0.)
$SnCl_2$		c	−349.8	−302.1	(122.6)

TABLE 5 (Continued)

Formula	Description	State	$\Delta H°$/kJ mol^{-1}	$\Delta G°$/kJ mol^{-1}	$S°$/J mol^{-1} K^{-1}
SnCl$^+$		aq		−167.9	106.7
SnCl$_4$		ℓ	−545.2	−474.0	258.6
Sn$_3$(PO$_4$)$_2$		c		−2,540.6	
SnBr$_2$		c	−266.1	−248.9	(146.0)
SnBr$^+$		aq		−138.2	95.27
SnBr$_4$		ℓ	−406.3		
SnBr$_4$		g	−314.6	−437.425	411.83
SnBr$_4$		c	−377.4	−456.22	264.4
SnI$_2$		c	−143.9	−143.9	(168.6)
SnSO$_4$		c	−1,014.6	−1,055.94	138.57
Sn(SO$_4$)$_2$		c	−1,646.0	−1,451.0	(155.2)
Sn(SO$_4$)O$_2$		c	−1,629.2	−1,673.91	149.8
SnH$_4$		g		414.0	

[a]National Bureau of Standards values are in italic; estimated values are in parentheses.
Source: Refs. 4, 7, and 14–16.

siderably higher. Oxidation of Sn(II) to Sn(IV) proceeds easily in aqueous solutions by dissolved oxygen. Sn(IV) is stable in aqueous solutions. Depending on pH, it may exist in different forms, exhibiting amphoteric character. SnH_4 of low stability is also known.

B. Sn(II)/Sn System

In general, electrode reactions occurring in the Sn(II)/Sn system are fast [1]. In this connection, both experimentally determined and calculated potentials have been reported. Both pure tin and saturated tin amalgam have been used as a reduced form of the Sn(II)/Sn electrode.

Haring and White [2] have measured the potential difference of the tin and saturated tin amalgam electrodes using the cell

$$Sn(Hg) | SnCl_2 + HCl | Sn$$

The mean value of a number of experiments shows that the tin electrode is 1.0 mV more negative than the tin saturated amalgam. This value is in good agreement with an earlier result [3]. Direct determination of the potential of the Sn(II)/Sn system is not easily carried out because of various side reactions that can occur.

Latimer [4], utilizing the International Critical Tables, gives $E° = -0.136$ V for the reaction

$$Sn^{2+} + 2e^- \to Sn$$

However, the value reported later by El Wakkad et al. [5] is slightly different. Using the cell

$$H_2 | HClO_4(m_1) \| HClO_4(m_1)Sn(ClO_4)_2(m_2) | Sn$$

with m_2 much lower than m_1 to make the liquid junction potential negligible, they obtained in the presence of air a potential similar to that reported above. In the absence of air the potential found was -0.1375 ± 0.0005 V. This result is to be preferred over that determined in the presence of air. The isothermal temperature coefficient of the potential of this reaction is reported as -0.282 mV K^{-1} [6].

Latimer [4] estimates the potential of the reaction

$$HSnO_2^- + H_2O + 2e^- \to Sn + 3OH^-$$

to be -0.91 V, with $\Delta G°(HSnO_2^-)$ value of -410.0 kJ mol^{-1}.

Pourbaix [7], for the reaction

$$HSnO_2^- + 3H^+ + 2e^- \to Sn + 2H_2O$$

reports the potential 0.333 V. The same $\Delta G°$ value of -410.0 kJ mol^{-1} for $HSnO_2^-$ was assumed in both works. The potentials of other Sn(II) ions or compounds are known with less precision.

In the presence of phosphates at pH higher than 12.5, where in only a small amount of HPO_4^{2-} ions are present in the solution, the electrode reaction of tin may be written

$$Sn_3(PO_4)_2 + 6e^- \rightarrow 3Sn + 2PO_4^{3-}$$

The Gibbs energy of formation of $Sn_3(PO_4)_2$ at 30°C is -2543.0 kJ mol^{-1}. The standard potential of this reaction measured by Awad and Kassab [8] is -0.865 V.

Potentials of other Sn(II) compounds are reported by Pourbaix [7]:

$$SnO(black) + 2H^+ + 2e^- \rightarrow Sn + H_2O \qquad E° = -0.104 \text{ V}$$

$$Sn(OH)_2 + 2H^+ + 2e^- \rightarrow Sn + 2H_2O \qquad E° = -0.091 \text{ V}$$

The potential of the reaction

$$SnS + 2e^- \rightarrow Sn + S^{2-}$$

was also calculated. Latimer [4] reported a value of -0.94 V (taking the solubility product 1×10^{-20}), but, from other thermodynamic data, a value of -1.00 V may be given. The authors [6] gave the isothermal temperature coefficient of the potential of this reaction as -1.01 mV K^{-1}.

Hoar [9] reported the potential for the reaction

$$Sn(OH)Cl \cdot H_2O(s) + 2e^- \rightarrow Sn + OH^- + Cl^- + H_2O \qquad E° = -0.631 \text{ V}$$

As mentioned above, as a rule electrode reactions occurring in the $Sn^{2+}/Sn(Hg)$ system in various solutions are fast. In connection with that, formal potentials found by using different electroanalytical methods may be used as an approximate guide in the work. The tabulation of such potentials was given earlier [1].

C. Sn(IV)/Sn(II) System

The electrode reaction occurring in this system is highly irreversible in non-complexing media [1]. Thus the determination of its standard potential may not be very precise. Nonetheless, Huey and Tartar [10] have studied the potentials of the reaction

$$Sn^{4+} + 2e^- \rightarrow Sn^{2+}$$

in HCl solutions. In 0.09678 mol kg^{-1} HCl the normal potential was found to be 0.0702 V and in 2.023 mol kg^{-1} HCl, 0.1325 V. The potential determined by extrapolation to zero acid concentration is 0.154 V. However in view of the irreversibility of the reaction, it is proper to suggest it as 0.15 V [4].

Prytz [11] attempted to measure the potential of the Sn^{4+}/Sn^{2+} system in perchlorate solutions, although his results are not easily interpreted. The electrode reaction rate of the system studied is considerably lower in perchlorates than in chloride solutions [1].

The potentials of the Sn(IV)/Sn(II) couple in other solutions are reported by Pourbaix [7]:

$$SnO_2 + 2H^+ + 2e^- \rightarrow SnO + H_2O \qquad E° = +0.088 \text{ V}$$

This potential is for black stannous oxide and white stannic oxide. Potentials of this reaction with participation of other oxide forms of Sn(IV) and Sn(II) are usually less positive. White stannic oxide is reduced to Sn^{2+} gives

$$SnO_2 + 4H^+ + 2e^- \rightarrow Sn^{2+} + 2H_2O \qquad E° = +0.125 \text{ V}$$

and

$$SnO_2 + H^+ + 2e^- \rightarrow HSnO_2^- \qquad E° = -0.350 \text{ V}$$

In the case of the participation in redox reaction of SnO_3^{2-} the potentials of the following reactions were calculated:

$$SnO_3^{2-} + 6H^+ + 2e^- \rightarrow Sn^{2+} + 3H_2O \qquad E° = 0.849 \text{ V}$$

$$SnO_3^{2-} + 3H^+ + 2e^- \rightarrow HSnO_2^- + H_2O \qquad E° = 0.375 \text{ V}$$

$$SnO_3^{2-} + 4H^+ + 2e^- \rightarrow SnO(\text{black}) + 2H_2O \qquad E° = 0.810 \text{ V}$$

The thermodynamic data used in these calculations are given in Table 5.

On the basis of the fact that stannous sulfide is oxidized by polysulfide, Latimer [4] suggests that the potential for the reaction

$$SnS_3^{2-} + 2e^- \rightarrow SnS + 2S^{2-}$$

is more negative than -0.6 V. From thermal data Latimer [4] estimated the equilibrium constant of the reaction

$$Sn^{4+} + 6F^- \rightarrow SnF_6^{2-} \qquad K = 10^{18}$$

The potential of the redox couple

$$SnF_6^{2-} + 4e^- \rightarrow Sn + 6F^- \qquad E° = -0.200 \text{ V}$$

50 mV more negative potential was reported [4,6] together with the isothermal coefficient of -0.693 mV K^{-1}.

Pourbaix [7] reports the potential for the reaction

$$Sn + 4H^+ + 4e^- \rightarrow SnH_4 \qquad E° = -1.074 \text{ V}$$

Though from other $\Delta G°$ much less negative potential is found, Deltombe and de Zoubov [12] measured $E° = -1.07 \pm 0.05$ V for this reaction. All potentials given here which were taken from Pourbaix's book [7] refer to unit activity of substances participating in the reaction.

It follows from the potentials reported that tin should react to evolve hydrogen in both strongly acidic and alkaline solutions. In practice this reaction in acids is very slow, in the absence of other oxidizing agents, because of the high overvoltage of hydrogen evolution on pure tin [13].

As Pourbaix points out [7], the film of SnO_2 on the metal in the absence of complex-forming and insoluble compound-forming substances will protect tin against corrosion in solutions of different pH, with the exception of strongly acidic and alkaline solutions.

REFERENCES

1. Z. Galus, in *Encyclopedia of the Electrochemistry of the Elements*, A. J. Bard, ed., vol. 4, Marcel Dekker, New York, 1975, p. 223.
2. M. M. Haring and J. C. White, Trans. Am. Electrochem. Soc., *73*, 211 (1938).
3. W. J. van Heteren, Z. Anorg. Chem., *42*, 129 (1904).
4. W. M. Latimer, *Oxidation Potentials*, Prentice-Hall, Englewood Cliffs, N.J., 1952.
5. S. E. S. El Wakkad, T. M. Salem, and J. A. El Sayed, J. Chem. Soc., 3770 (1957).
6. A. J. de Béthune and N. A. Swendeman Loud, *Standard Aqueous Electrode Potentials and Temperature Coefficients at 25°C*, Hampel, Skokie, Ill., 1964.
7. M. Pourbaix, ed., *Atlas of Electrochemical Equilibria in Aqueous Solutions*, Pergamon Press, Oxford, 1966.
8. S. A. Awad and A. Kassab, J. Electroanal. Chem., *20*, 203 (1969).
9. T. P. Hoar, Trans. Faraday Soc., *33*, 1152 (1937).
10. C. S. Huey and H. V. Tartar, J. Am. Chem. Soc., *56*, 2585 (1934).
11. M. Prytz, Z. Anorg. Chem., *219*, 89 (1934).
12. N. de Zoubov and E. Deltombe, Int. Comité Electrochem. Thermodyn. Kinet., 7th Meet., Lindau, 1955, Butterworth, London, 1957, p. 240.
13. A. G. Pecherskaya and V. V. Stender, Zh. Fiz. Khim., *24*, 856 (1950).
14. *Selected Values of Chemical Thermodynamic Properties*, Natl. Bur. Stand. Circ. 500, U.S. Government Printing Office, Washington, D.C., 1952.
15. I. Barin and O. Knacke, *Thermochemical Properties of Inorganic Substances*, Springer-Verlag, Berlin, 1973.
16. I. Barin, O. Knacke, and O. Kubaschewski, *Thermochemical Properties of Inorganic Substances*, Springer-Verlag, Berlin, 1977.

V. LEAD

A. Oxidation States

The most important oxidation state is +2. Plumbous ion is stable both in acid and in strongly alkaline solutions as Pb^{2+} and $HPbO_2^-$ ions, respectively. This shows the amphoteric character of this oxidation state. Pb^{2+} forms a number of weakly soluble salts or complexes with different anions.

The chemistry of Pb(IV) compounds is limited. PbO_2 is the best known example. It also has amphoteric properties, but it is quite inert to the action of H^+ and OH^- ions. Pb(IV) may be prepared in a soluble form by anodic oxidation of Pb(II) in concentrated (10 M) NaOH. PbO_3^{2-} ions form under such conditions.

The oxide Pb_3O_4 may be considered as plumbous plumbate. This is indicated by its reaction with strong hydroxide solutions:

$$Pb_3O_4 + 4OH^- \rightleftharpoons PbO_3^{2-} + 2HPbO_2^- + H_2O$$

Lead also forms an unstable hydride. The thermodynamic data are given in Table 6.

B. Pb(II)/Pb and Pb(II)/Pb(Hg) Systems

Since the solubility of lead in mercury is high and mercury electrodes were frequently used to study the behavior and potentials of the Pb(II)/Pb(Hg) system, an intercomparison of potentials of both electrodes is instructive.

Carmody [1], who used the cell

$$Pb(stick) \mid PbCl_2 \mid Pb(Hg)$$

measured the electromotive force (emf) to be equal 5.8 mV, the solid Pb electrode being more negative. Pb(Hg) was the saturated Pb amalgam not rich in the solid phase. This value is quite close to that reported by Gerke [2].

C. Pb(II)/Pb System

Using the cell

$$Pg(Hg) \mid PbCl_2, AgCl \mid Ag$$

Carmody [1] measured its emf at different ionic strengths to 0.1 mV. From the method of Randall [3], by extrapolation to infinite dilution, an emf of 0.3426 V was found. Taking the potential of the silver chloride electrode (0.2221 V) and the lead-lead amalgam potential given above, he obtained for the reaction

$$Pb^{2+} + 2e^- \rightarrow Pb$$

a potential of -0.1263 V.

A practically identical result, -0.1264 V, was reported by Lingane [4]. However, Haring et al. [5] found slightly different values. The cell

$$Pt, H_2(1 \text{ atm}) \mid HClO_4(x \; m) \mid HClO_4(x \; m) + Pb(ClO_4)_2(y \; m) \mid Pb(\text{amalg.,sat.})$$

was used in their experiments, which permitted direct comparison of the lead and hydrogen electrodes. By proper selection of x and y the liquid junction potential was of the order of 50 μV, far below the experimental error. The equation relating the emf of this cell to concentration is

TABLE 6 Thermodynamic Data on Lead[a]

Formula	Description	State	$\Delta H°$/kJ mol^{-1}	$\Delta G°$/kJ mol^{-1}	$S°$/J mol^{-1} K^{-1}
Pb		g	193.9	161.0	175.27
Pb	Metal	c	0.0	0.0	64.89
Pb$^+$		g	915.54		
Pb^{2+}		g	2,371.9		
Pb^{2+}	Std. state, hyp., $m = 1$	aq	1.63	−24.31	21.3
Pb^{3+}		g	5,473.1		
Pb^{4+}		g	9,556.2		
Pb^{4+}		aq		302.5	
Pb^{5+}		g	16,292		
Pb$_2$		g	326.		
PbO	Yellow	c	−217.9	−188.5	69.4
PbO	Red	c	−219.2	−189.3	67.8
PbO	Hydr. white, hex.	c		−183.72	

PbO	g	47.2		240.0
Pb_2O_3	c	−411.78		
Pb_3O_4	c	−616.2		211.0
PbO_2	c	−218.99	−276.65	76.6
$PbO_{1.57}$	c	−211.21		
$HPbO_2^-$	aq	−339.0		
	Std. state, hyp., $m = 1$			
$Pb(OH)_2$	c	−420.9	−514.6	88.
PbO_3^{2-}	aq	−272.7		
PbO_4^{4-}	aq	−282.1		
PbS	c	−98.78		91.2
PbS_2O_3	c	−560.6	−628.0	(148.1)
PbS_3O_6	c	−894.5	−1,184.1	(231.8)
$PbSO_4 \cdot PbO$	c	1,083.2	−1,1182.0	(203.8)
$PbSO_4$	c	−813.83		140.2
PbSe	c	−64.4	−75.3	(112.5)
PbSe*	c	−130.562	−100.0	102.5

TABLE 6 (Continued)

Formula	Description	State	$\Delta H°$/kJ mol^{-1}	$\Delta G°$/kJ mol^{-1}	$S°$/J mol^{-1} K^{-1}
PbSeO$_3$		c	−532.6	−570.919	128.4
PbSeO$_4$		c	−619.	−510	(155.)
PbTe		c	−73.2	−75.7	(115.5)
PbWO$_4$		c		−1,050.	168.
PbCO$_3$		c		−625.88	131.
PbO·PbCO$_3$		c	−920.5	−818.4	202.9
2PbO·PbCO$_3$		c	−1,142.	−1,012.	272
Pb$_3$(OH)$_2$(CO$_3$)$_2$		c		−1,711.7	
PbSiO$_3$		c	−1,082.8	−1,000.0	113.
Pb$_2$SiO$_4$		c	−1,308.3	−1,195.4	180.
PbF$_2$		g		−444.79	293.
PbF$_2$		c	−663.2	−619.6	121.
PbF$_4$		c	−930.1	−745.2	(148.5)
PbCl$_2$		c	−359.20	−313.97	136.4

Carbon, Silicon, Germanium, Tin, and Lead

PbCl$_2$	g		−182.0	316.0
PbCl$_4$	g	−314.	−428.442	384.5
PbBr$_2$	c	−277.02	−260.41	161.5
PbBr$_2$·PbO	c	−511.7		
PbBr$_4$	g	−184.1	−311.089	425.9
PbI$_2$	g	107.5	24.125	279.74
PbI$_4$	g	−1.7	−140.641	466.1
PbI$_3^-$	aq		−203.13	
PbI$_4^{2-}$	aq		−263.67	
PbH$_2$	g		−291.	
Pb(N$_3$)$_2$	c	436.4	565.1	(207.1)
Pb(NO$_3$)$_2$	c	−449.15	−252.3	(213.0)
Pb(OH)NO$_3$	c	−493.3	−303.7	(157.7)
Pb$_3$(PO$_4$)$_2$	c		−2,371.6	
PbHPO$_4$	c		−1,182.7	(133.5)
PbC$_2$O$_4$	c	−858.1	−754.4	(138.9)
Pb(CH$_3$)$_4$	g		269.8	

TABLE 6 (Continued)

Formula	Description	State	$\Delta H°$/kJ mol^{-1}	$\Delta G°$/kJ mol^{-1}	$S°$/J mol^{-1} K^{-1}
Pb(C$_2$H$_5$)$_4$		ℓ	*217.*		
PbCrO$_4$		c	−942.2	−851.9	(152.7)
PbMoO$_4$		c	−1,112.1	−969.4	(161.1)
Pb(HCO$_2$)$_2$		c	−837.2		(148.5)
Pb(CH$_3$COO)$_2$		c	−964.4		(167.)
Pb(KCS)$_2$		c	*115.1*		

[a]National Bureau of Standards values are in italics; estimated values are in parentheses.
Source: Refs. 8, 10, 21, and 37-40.

$$E = E° - (RT/2F) \ln[(\gamma_{H^+} m_{H^+})^2 / \gamma_{Pb^{2+}} m_{Pb^{2+}}]$$

On the basis of a large number of careful experiments and recalculation to the solid lead electrode, the value of -0.1251 V $\pm$ 0.5 mV was obtained. This value is recommended. The isothermal temperature coefficient of potential of this electrode is -0.451 mV K^{-1} [6].

The divergence of $E°$ values for this electrode found by various workers is due to the choice of activity coefficients rather than to the error of the experiment. For instance, the result of Carmody [1] recalculated with the use of Brønsted coefficients [7] leads to $E° = -0.1234$ V. The suggested value is less than 1 mV more positive than the value given by Latimer [8].

One should mention that Vasilev and Glavin [9] reported an even lower value for this electrode potential, -0.1237 V. This value seems to be too low, although why is not clear since only the summary of the work has been published.

Chartier [10] calculated the potential of the reaction

$$HPbO_2^- + H_2O + 2e^- \rightarrow Pb + 3OH^-$$

to be -0.502 V. A slightly more negative value equal -0.540 V is suggested.

Delahay et al. [11] calculated $E° = 0.242$ V for the reaction

$$Pb(OH)_2 + 2H^+ + 2e^- \rightarrow Pb + 2H_2O$$

$E° = 0.277$ V is suggested. If, in this reaction, PbO(red) is considered instead of Pb(OH)$_2$, the standard potential is 0.248 V.

For the reaction

$$PbO(red) + H_2O + 2e^- \rightarrow Pb + 2OH^-$$

de Béthune and Swendeman Loud [6] give $E° = -0.580$ V and -1.163 mV K^{-1} as the isothermal temperature coefficient, which are accepted.

D. Pb Halide/Pb Systems

The emfs of the cells were measured

$$Pb(Hg) | PbCl_2, HCl(1\ M) + PbCl_2, AgCl | Ag$$

$$Pb(Hg) | PbCl_2, HCl(1\ M), PbCl_2 + Hg_2Cl_2 | Hg$$

with the reactions

$$PbCl_2 + 2Ag \rightarrow 2AgCl + Pb \qquad (8.1)$$

$$PbCl_2 + 2Hg \rightarrow Hg_2Cl_2 + Pb \qquad (8.2)$$

Gerke [2], using solid lead electrodes, found standard emfs of 0.4900 and 0.5356 V for chloride solutions for reactions (8.1) and (8.2) respectively.

Taking into account the standard potentials of silver-silver chloride and calomel electrodes from both cells, one arrives at a standard potential of -0.268 V for the $PbCl_2/Pb$ electrode. Such a value is reported by Latimer [8]. An identical value was later found by Altynov and Ptitsyn [12]. Thermal and isothermal temperature coefficients of this potential are $+0.396$ and -0.475 mV K^{-1}, respectively [6]. Cells in which saturated lead amalgam was used instead of a solid lead exhibited potentials less negative by about 6 mV, in agreement with the above.

Similarly, in the case of lead-iodide, Gerke [2] found the following:

$$PbI_2 \rightarrow Pb + I_2 \qquad E = 0.8993 \text{ V}$$

$$PbI_2 + 2Ag \rightarrow Pb + 2AgI \qquad E = 0.2135 \text{ V}$$

He also reported temperature coefficients for the emfs of these cells.

Taking into account standard potentials of $I_2/2I^-$ and AgI/Ag systems one obtaines for the reaction

$$PbI_2 + 2e^- \rightarrow Pb + 2I^-$$

a standard potential of -0.365 V. The isothermal temperature coefficient of this potential is -0.124 mV K^{-1} [6]. This value of $E°$ was reported by Latimer [8] and a similar one by Cann and Taylor [13].

The standard potential of the reaction

$$PbBr_2 + 2e^- \rightarrow Pb + 2Br^-$$

was reported by Cann and Sumner [14] to be -0.280 V. The thermodynamic parameters of $PbBr_2$ are also reported by these authors. De Béthune and Swendeman Loud [6] report -0.341 mV K^{-1} as the isothermal coefficient of the potential of this reaction.

Ivett and De Vries [15] have measured -0.334, -0.344, and -0.355 V for the Pb, PbF_2, F^- electrode at 15, 25, and 35°C, respectively. In the case of a two-phase lead amalgam electrode, these potentials were 5.8 mV less negative. The value reported for 25°C is in good agreement with the solubility product of 3.7×10^{-8} reported for PbF_2.

E. Pb(II)-Phosphate/Pb System

Awad and Elhady [16] have studied the potential of the reaction

$$PbHPO_4 + 2e^- \rightarrow Pb + HPO_4^{2-}$$

Both by experiment and theoretical calculation (taking the standard Gibbs energy of formation of $PbHPO_4$ as -1182.69 kJ mol^{-1}), this potential was found to be -0.465 V (experimental value at 30°C).

The measurements of the potential of the lead electrode in aqueous phosphate solutions in function of pH have shown that such a reaction occurs up to pH 8.7. At higher pH values lead functions as a $Pb_3(PO_4)_2/Pb$ electrode according to the reaction

$$Pb_3(PO_4)_2 + 6e^- \rightarrow 3Pb + 2PO_4^{3-}$$

The measured standard potential of this reaction is -0.54 V at 30°C.

Taking the solubility product of $Pb_3(PO_4)_2$ as 3×10^{-44} [17], which gives $\Delta G°$ of $Pb_3(PO_4)_2$ as -2373.9 kJ mol^{-1}, the calculated standard potential of the latter reaction is -0.557 V, which agrees rather well with the experimental value.

F. $PbSO_4/Pb$ System

The $PbSO_4/Pb$ system is important as one of the electrodes of the lead storage battery. This system was studied by Shrawder and Cowperthwaite [18]. Using two-phase lead amalgam as an electrode in the cell

$$Pb(Hg)\,|\,PbSO_4(s),H_2SO_4(m)\,|\,H_2$$

$E°$ values were measured for different concentrations of H_2SO_4. Measured $E°$ values were plotted against $m^{1/2}$. A new method was devised for extrapolating to infinite dilution, taking into account the incomplete dissociation of bisulfate ions, by means of which $E°$ was obtained for several temperatures. The following values were found:

$t/°C$	0	12.5	25.0	37.5	50
$E°/V$	-0.3281	-0.3392	-0.3505	-0.3619	-0.3738

In agreement with Gerke [2] these potentials should be 5.8 mV more negative for the Pb^{2+}/Pb electrode, or, at 25°C, -0.3563 V. Such a value was earlier reported by Latimer [8] and later by other workers [6].

Harned and Hamer [19] proposed a general equation for the potential of this electrode in the temperature range 0 to 60°C; -1.015 mV K^{-1} is reported as the isothermal coefficient [6]. If the saturated lead amalgam is used instead of metallic lead, $E° = -0.3505$ V. The isothermal temperature coefficient of this potential is -0.914 mV K^{-1} [6].

G. Other Systems

Assuming the solubility product of PbS as 7×10^{-29}, Latimer [8] calculated $E° = -0.98$ V for the reaction

$$PbS + 2e^- \rightarrow Pb + S^{2-}$$

However, de Béthune and Swendeman Loud [6] give $E° = -0.93$ V and -0.90 mV K^{-1} as the isothermal coefficient of this potential. $E° = -0.954$ V is suggested, as it is based on more accurate data of the thermodynamic parameters.

Latimer [8] calculated $E° = -0.506$ V for

$$PbCO_3 + 2e^- \to Pb + CO_3^{2-}$$

assuming the solubility product of $PbCO_3$ to be 1.5×10^{-13}. The same value was reported by Delahay et al. [11]. However, de Béthune and Swendeman Loud [6] report $E° = -0.509$ V for this reaction. The latter value is recommended. The isothermal coefficient is equal to -1.29 mV K^{-1} [6].

Wolf and Bonilla [20] report the potentials of the lead electrode in various salts of sodium and potassium (0.1 M in cations) such as $NaCH_3COO$, $NaNO_3$, NaF, Na_3PO_4, Na_2SiO_3, and others. Because some of these potentials have rather strange values, they are not reported. Pourbaix [21] gives $E° = -1.507$ V for the reaction

$$Pb + 2H^+ + 2e^- \to PbH_2$$

H. Pb(IV)/Pb(II) System

The most extensively studied system of this type is the $PbO_2/PbSO_4$ electrode, with the reaction

$$PbO_2 + SO_4^{2-} + 4H^+ + 2e^- \to PbSO_4 + 2H_2O$$

Earlier results of Harned and Hamer [19] reported by Latimer [8] and Hamer [22] now appear to be too low.

This reaction was found to behave reversibly over the temperature range 5 to 55°C and up to a sulfuric acid concentration of 8 mol kg^{-1} [23]. The potential of this reaction depends on the sort of PbO_2 used. The standard potential of the electrode at 25°C with α-PbO_2 was reported by Ruetschi et al. [24] to be 1.698 V at a hydrogen ion activity equal to unity. A similar result was given by Bone et al. [25]. The measured emf of the cell

$$Pt(H_2) | H_2SO_4(m) | PbSO_4 | \alpha\text{-}PbO_2(Pt)$$

is given by the equation

$$E = E° + (RT/2F) \ln(a_{H_2SO_4}/a^2_{H_2O})$$

and thus

$$E° = E - (RT/2F) \ln(a_{H_2SO_4}/a^2_{H_2O})$$

which should be constant.

In testing this relation the activity coefficients of H_2SO_4 and activities of water at various molalities of the acid have been taken from Stokes [26]. The authors found a mean value a bit more positive than 1.697 V. A lower value of $E°$ found from experiments carried out in 6.095 mol kg^{-1} H_2SO_4 was attributed to the gradual conversion of the α to the β modification of PbO_2 under such conditions.

In the case of β-PbO_2, careful studies by Covington et al. [27] found +1.6902 and +1.6899 V for $E°$, depending on the choice of the adjustable parameter in the Debye-Hückel equation. The result also depends on the value of the equilibrium constant of the HSO_4^- ion. The authors estimate that an uncertainty of at least ±0.30 mV in the $E°$ value is possible according to the choice of these parameters.

In conclusion the authors give $E° = 1.6901$ V with the uncertainty of ±0.15 mV. Since for this system $E° = 1.690$ V was reported by Ruetschi et al. [24], we may suggest the latter value as a standard potential. It follows that this electrode is thermodynamically unstable in sulfuric acid solutions, since this value is about 0.47 V more positive than that of the oxygen electrode. The spontaneous oxidation of water to oxygen is very slow, however, because of the high oxygen overvoltage on PbO_2 [28].

Temperature coefficients of emf and the enthalpy of the cell reaction

$$PbO_2 + H_2SO_4 + H_2 \rightarrow PbSO_4 + 2H_2O$$

at 25°C were reported by Beck et al. [23]. Taking this reaction scheme, and considering that this reaction proceeds in a reversible way, one may write for emf of the cell

$$E = E° + (RT/2F) \ln(4m^3 \gamma_\pm^3 / a_w^2)$$

from which $E°$ can be calculated. The temperature coefficient of emf and enthalpy of the cell reaction at 25°C are reported as a function of H_2SO_4 concentration. For instance, for 0.1 mol kg^{-1} H_2SO_4 these values are -0.432 mV K^{-1} and 326.71 kJ mol^{-1}, respectively.

Using the values given above for the $PbO_2/PbSO_4$ electrode, as well as the standard potentials of $PbSO_4/Pb$ and $Pb(II)/Pb$ couples, one may calculate the potentials of the reaction

$$PbO_2 + 4H^+ + 2e^- \rightarrow Pb^{2+} + 2H_2O$$

as 1.468 V and 1.460 V for α-PbO_2 and β-PbO_2, respectively. Latimer [8] estimated the standard Gibbs energy of the reaction

$$PbO_2 + 4H^+ \rightarrow Pb^{4+} + 2H_2O$$

as $\Delta G° = 45.35$ kJ mol^{-1}. This in turn leads to the estimation of the potential of

$$Pb^{4+} + 2e^- \rightarrow Pb^{2+}$$

as about 1.69 V. Berka et al. [29] measured formal potentials of this system in 1 to 5.8 M $HClO_4$ and found values from 1.65 to 1.70 V.

Latimer [8] reports $E° = 0.248$ V for the reaction

$$PbO_2 + H_2O + 2e^- \rightarrow PbO(red) + 2OH^-$$

A practically identical result, 0.247 V, was reported by de Béthune and Swendeman Loud [6]. They give -1.194 mV K^{-1} as the isothermal coefficient of this potential $E° = 0.249$ V is suggested.

It should be pointed out that the potential of this reaction cannot be measured directly in alkaline solution becuase in such a case intermediate oxides may be formed (Pb_3O_4 or Pb_2O_3). Chartier [10] calculated several potentials of the Pb(IV)/Pb(II) couple.

$$PbO_3^{2-} + 2H_2O + 2e^- \rightarrow HPbO_2^- + 3OH^- \qquad E° = 0.330 \text{ V}$$

This potential is 30 mV more positive than the value reported earlier by Glasstone [30].

$$\beta\text{-}PbO_2 + H_2O + 2e^- \rightarrow HPbO_2^- + OH^- \qquad E° = 0.208 \text{ V}$$

$$PbO_4^{4-} + 3H_2O + 2e^- \rightarrow HPbO_2^- + 5OH^- \qquad E° = 0.683 \text{ V}$$

Earlier calculations of some of these potentials are summarized by Latimer [8].

The potentials of other systems in alkaline solutions may be also given.

$$PbO_3^{2-} + 3H_2O + 4e^- \rightarrow Pb + 6OH^- \qquad E = -0.127 \text{ V}$$

$$PbO_{1.33} + 0.33H_2O + 0.33OH^- + 0.67e^- \rightarrow HPbO_2^- \qquad E° = 0.024 \text{ V}$$

$$PbO_{1.57} + 0.57H_2O + 1.14e^- \rightarrow HPbO_2^- + 0.14OH^- \qquad E° = 0.133 \text{ V}$$

The thermodynamic data used in these calculations are given in Table 6.

Standard potentials of several reactions occurring in alkaline media were reported by Delahay et al. [11]. In their calculations slightly different standard free energies of some lead compounds were taken from those used by Chartier [10].

Fleischmann and Liler [31] reported the equilibrium potentials of β-PbO_2 and lead amalgam electrodes in acetate solutions at pH 2.76 to 5.59 (at 25°C). These potentials may be of some value in practical work and the reader should consult this paper.

The entropy change for the cell reaction

$$PbO_2 + H_2 + H_2SO_4(x\ M) = PbSO_4 + 2H_2O$$

calculated [32] from the third law of thermodynamics was found to be in excellent agreement with dE/dT determined by Beck et al. [23].

The following reaction potentials were also calculated:

$$PbO_3^{2-} + SO_4^{2-} + 6H^+ + 2e^- \rightarrow PbSO_4 + 3H_2O \qquad E° = 2.34 \text{ V}$$

$$PbO_2 + CO_3^{2-} + 4H^+ + 2e^- \rightarrow PbCO_3 + 2H_2O \qquad E° = 1.83 \text{ V}$$

Formal and reversible half-wave potentials of lead in various solutions have been reported by Sharpe [33]. These parameters may be of some practical value.

I. Lead Storage Battery

The reactions occurring at both electrodes may be written:

Negative electrode: $PbSO_4 + H^+ + 2e^- \rightarrow Pb + HSO_4^-$

Positive electrode: $PbSO_4 + 2H_2O \rightarrow PbO_2 + 3H^+ + HSO_4^- + 2e^-$

For the total reaction

$$PbO_2 + Pb + 2H_2SO_4(\text{pure}) \rightarrow 2PbSO_4 + 2H_2O(\text{pure})$$

Duisman and Giauque [32] give $\Delta G° = -502.44$ kJ mol^{-1} and $\Delta H° = -506.45$ kJ mol^{-1} at 298.15 K.

Harned and Hamer [19] considered the emf of the lead storage battery as a function of the H_2SO_4 concentration and the temperature from 0 to 60°C. Their equation is of the form

$$E = E° + at + bt^2$$

where $E°$ is the emf at 0°C, t is the Celsius temperature, and a and b are constants different for each H_2SO_4 concentration. They are given in Table 7.

On the basis of work of Vosburgh and Craig [34] and Binal and Craig [35], Duisman and Giauque [32] give the equation that indirectly relates the emf of the lead storage battery with the concentration of H_2SO_4 at 298.15 K. The temperature coefficient of the emf of this battery is $dE/dT = 0.203$ mV K^{-1}. Papers dealing with the meachanism of reactions occurring in lead storage battery electrodes have been discussed in reviews [33,36].

VI. POTENTIAL DIAGRAMS FOR THE GROUP

We present only potentials of the systems in acidic solutions. In view of the different forms of compounds constituting redox couples, such a presentation for alkaline solutions would be difficult.

TABLE 7 Constants for the Equation $E = E° + at + bt^2$

$m_{H_2SO_4}$ (mol kg^{-1})	$E°$ (V)	$10^6 a$ (V K^{-1})	$10^8 b$ (V K^{-2})
0.05	1.7687	−310	134
0.10	1.8021	−265	129
0.20	1.8349	−181	128
0.50	1.8791	−45	126
1.0	1.9174	56.1	108
2.0	1.9664	159	103
3.0	2.0087	178	97
4.0	2.0479	177	91
5.0	2.0850	167	87
6.0	2.1191	162	85
7.0	2.1507	153	80

```
       +4                  +2              0       −2     −4

CO₂ ──−0.106── CO ────0.517──── C ───0.132─── CH₄

SiO₂ ──−0.967── SiO ───−0.808─── Si ──−0.143── SiH₄
    (glassy)

GeO₂ ──−0.370── GeO ────0.255──── Ge ──−0.29── GeH₄
  (tetragonal)   (yellow,hydrous)

Ge⁴⁺ ──0.00── Ge²⁺ ──0.247

SnO₂ ──−0.088── SnO ───−0.104─── Sn ──−1.071── SnH₄
    (white)        (black)

Sn⁴⁺ ──0.15── Sn²⁺ ──−0.137

Pb⁴⁺ ──1.69── Pb²⁺ ──−0.1251── Pb ──−1.507── PbH₂
```

REFERENCES

1. W. R. Carmody, J. Am. Chem. Soc., *51*, 2905 (1929).
2. R. H. Gerke, J. Am. Chem. Soc., *44*, 1684 (1922).
3. M. Randall, Trans. Faraday Soc., *23*, 498 (1927).
4. J. J. Lingane, J. Am. Chem. Soc., *60*, 724 (1938).
5. M. M. Haring, M. R. Hatfield, and P. P. Zapponi, Trans. Electrochem. Soc., *75*, 473 (1939).
6. A. J. de Béthune and N. A. Swendeman Loud, *Standard Aqueous Electrode Potentials and Temperature Coefficients at 25°C*, Hampel, Skokie, Ill., 1964.

7. J. N. Brønsted, Z. Phys. Chem., *56*, 645 (1906).
8. W. M. Latimer, *Oxidation Potentials*, Prentice-Hall, Englewood Cliffs, N.J., 1952.
9. V. P. Vasilev and S. R. Glavin, Elektrokhimya, *7*, 1395 (1971).
10. P. Chartier, Bull. Soc. Chim. Fr., 2706 (1967).
11. P. Delahay, M. Pourbaix, and P. Van Rysselberghe, J. Electrochem. Soc., *98*, 57 (1951).
12. V. I. Altynov and B. V. Ptitsyn, Zh. Neorgan, Khim., *9*, 2407 (1964).
13. J. Y. Cann and A. C. Taylor, J. Am. Chem. Soc., *59*, 1987 (1937).
14. J. Y. Cann and R. A. Sumner, J. Phys. Chem., *36*, 2615 (1932).
15. R. W. Ivett and T. De Vries, J. Am. Chem. Soc., *63*, 2821 (1941).
16. S. A. Awad and Z. A. Elhady, J. Electroanal. Chem., *20*, 79 (1969).
17. M. Jowett and H. I. Price, Trans. Faraday Soc., *28*, 668 (1932).
18. J. Shrawder, Jr., and I. A. Cowperthwaite, J. Am. Chem. Soc., *56*, 2340 (1934).
19. H. S. Harned and W. J. Hamer, J. Am. Chem. Soc., *57*, 33 (1935).
20. E. F. Wolf and C. F. Bonilla, Trans. Electrochem. Soc., *79*, 307 (1941).
21. M. Pourbaix, ed., *Atlas of Electrochemical Equilibria in Aqueous Solutions*, Pergamon Press, Oxford, 1966.
22. W. J. Hamer, J. Am. Chem. Soc., *57*, 9 (1935).
23. W. W. Beck, K. P. Singh, and W. F. K. Wynne-Jones, Trans. Faraday Soc., *55*, 331 (1959).
24. P. Ruetschi, J. Sklarchuk, and R. T. Angstadt, Electrochim. Acta, *8*, 333 (1963).
25. S. J. Bone, K. P. Singh, and W. F. K. Wynne-Jones, Electrochim. Acta, *4*, 288 (1961).
26. R. H. Stokes, Trans. Faraday Soc., *44*, 298 (1948).
27. A. K. Covington, J. V. Dobson, and W. F. K. Wynne-Jones, Trans. Faraday Soc., *61*, 2050 (1965).
28. P. Ruetschi, R. T. Angstadt, and B. D. Cahan, J. Electrochem. Soc., *106*, 547 (1959).
29. A. Berka, V. Dvorak, I. Nemec, and J. Zýka, J. Electroanal. Chem., *4*, 150 (1962).
30. S. Glasstone, J. Phys. Chem., *121*, 1456 (1922).
31. M. Fleischmann and M. Liler, Trans. Faraday Soc., *54*, 1370 (1958).
32. J. A. Duisman and W. F. Giauque, J. Phys. Chem., *72*, 562 (1968).
33. T. F. Sharpe, in *Encyclopedia of the Electrochemistry of the Elements*, A. J. Bard, ed., Vol. 1, Marcel Dekker, New York, 1973, p. 235.
34. W. C. Vosburgh and D. N. Craig, J. Am. Chem. Soc., *51*, 2009 (1929).
35. G. W. Vinal and D. N. Craig, J. Res. Natl. Bur. Stand., *14*, 449 (1935).
36. J. Burbank, A. C. Simon, and E. Willihnganz, *Advances in Electrochemistry and Electrochemical Engineering*, C. W. Tobias, ed., Vol. 8, Wiley-Interscience, New York, 1971, p. 157.
37. O. Kubaschewski and E. L. Evans, *Metallurgical Thermochemistry*, Pergamon Press, Oxford, 1958.
38. *Selected Values of Chemical Thermodynamic Properties*, Natl. Bur. Stand. Circ. 500, U.S. Government Printing Office, Washington, D.C., 1952.
39. I. Barin and O. Knacke, *Thermochemical Properties of Inorganic Substances*, Springer-Verlag, Berlin, 1973.
40. M. Kh. Karapetyants, *Introduction in the Theory of Chemical Processes* (in Russian), Vysshaya Shkola, Moscow, 1975.

9
Gallium, Indium, and Thallium *

V. V. LOSEV Karpov Institute of Physical Chemistry, Moscow, USSR

The highest oxidation state of gallium, indium, and thallium is +3. These elements may also exist in lower oxidation states, but gallium and indium are less stable in these states than in the +3 state, whereas thallium is considerably more stable in the +1 state. Furthermore, the chemical properties of Tl(III) are not similar to those of Al(III), whereas those of Ga(III) and In(III) are very close to Al(III) in many respects. Hydroxides of Ga(III) and In(III) as well as Al(OH)$_3$ are amphotheric; their salts are partly hydrolyzed. Gallium, indium, and thallium differ from Al in lower enthalpies of formation of their oxides and in the ease of their reduction to the metallic state.

I. GALLIUM

A. Oxidation States

The principal oxidation state of gallium is +3, which is represented by the solvated ion Ga^{3+} in acid solution. Depending on the pH value, Ga(III) may also exist in the form of the following ions: Ga(OH)$^{2+}$, GaO$^+$, GaO$_2^-$, H$_2$GaO$_3^-$, Ga(OH)$_4^-$, HGaO$_3^{2-}$, and GaO$_3^{3-}$. Polynuclear cations [Ga$_{2m}$(OH)$_{5m}$]$^{5+}$$_{aq}$ exist in alkaline as well as acid solutions [1]. In basic solutions gallium may exist in the +1 state due to the low rate of its oxidation to Ga(III). The oxide Ga$_2$O is formed by the reduction of Ga$_2$O$_3$ by the metal at high temperatures. The thermodynamic data are given in Table 1.

B. Ga/Ga(III) Couple

The best value for the potential of the Ga/Ga(III) couple

$$\text{Ga}^{3+} + 3\text{e}^- \rightarrow \text{Ga} \qquad E° = -0.529 \text{ V (at 301 K)}$$

has been obtained from measurements of the electromotive force (emf) of a cell

---

Reviewer: Edward D. Moorhead, University of Kentucky, Lexington, Kentucky.

TABLE 1 Thermodynamic Data on Gallium

Formula	State	$\Delta H°$/kJ mol^{-1}	$\Delta G°$/kJ mol^{-1}	$S°$/J mol^{-1} K^{-1}
Ga	c	0	0	40.9
Ga	g	277	239	169
Ga$^+$	g	862		
Ga^{2+}	g	2847		
Ga^{2+}	aq		−88	
Ga^{3+}	g	5815		
Ga^{3+}	aq	−212	−159	−330
GaO	g	279	253	231
GaO$_3^{3-}$	aq		−619	
Ga$_2$O	c	−356		
Ga$_2$O	g	−88		
Ga$_2$O$_3$	c	−1089	−998.3	85.0
GaH	g	220	194	195.3
GaOH	g	−115		
Ga(OH)$^{2+}$	aq		−380	
HGaO$_3^{2-}$	aq		−686	
Ga(OH)$_2^+$	aq		−597.4	
H$_2$GaO$_3^-$	aq		−745	
Ga(OH)$_3$	c	−964.4	−831.3	100
GaF	g	−252		
GaF^{2+}	aq	−536.8	−464.8	−226
GaF$_2^+$	aq	−1163	−1085	84
GaCl	g	−79.9	−106	240
GaCl$_3$	c	−524.6	−454.8	142
GaBr	g	−49.8	−90.0	252

TABLE 1 (Continued)

Formula	State	$\Delta H°$/kJ mol^{-1}	$\Delta G°$/kJ mol^{-1}	$S°$/J mol^{-1} K^{-1}
GaBr$_3$	c	−387	−360	180
GaI	g	29		
GaI$_3$	c	−239		
GaH	c	−110		

with solid gallium in acid chloride solutions [1,2]. The temperature coefficient is [1]

$$(dE°/dT)_{isoth} \to 0.871 \text{ mV K}^{-1} \text{ (at 298 K)}$$

$E°$ values for the Ga/Ga(III) couple for the above-mentioned gallium ions according to the Pourbaix diagram, as well as the dependencies of the potentials of the corresponding couples upon the ClO_4^-, Cl^-, and Br^- concentrations, are given in Ref. 1.

In alkaline solution the potential of the couple Ga/H$_2$GaO$_3^-$ determined from thermodynamic data [3] is

$$H_2GaO_3^- + H_2O + 3e^- \to Ga + 4OH^- \qquad E° = -1.22 \text{ V}$$

C. Ga(II)/Ga(III) Couple

Gallium dichloride is formed by treating a concentrated aqueous solution of GaCl$_3$ with the metal. Upon dilution this solution evolves hydrogen. These chemical properties are consistent with the following approximate potentials [1]:

$$Ga^{3+} + e^- \to Ga^{2+} \qquad E° \approx -0.67 \text{ V}$$

$$Ga^{2+} + 2e^- \to Ga \qquad E° \approx -0.45 \text{ V}$$

D. Ga$_2$O

Using the enthalpy of formation of Ga$_2$O, −343 kJ mol^{-1} [4] and estimating $S° = 92$ J mol^{-1} K^{-1}, Latimer calculated the Gibbs energy of Ga$_2$O, $\Delta G° = -315$ kJ mol^{-1}, and potentials of the reactions of Ga$_2$O [3]:

$$Ga_2O + 2H^+ + 2e^- \to 2Ga + H_2O \qquad E° = -0.4 \text{ V}$$

$$Ga_2O_3 + 4H^+ + 4e^- \to Ga_2O + 2H_2O \qquad E° = -0.5 \text{ V}$$

These data show that Ga$_2$O is unstable with respect to its decomposition into Ga$_2$O$_3$ and the metal.

II. INDIUM

A. Oxidation States

The only stable ion in aqueous solutions in In^{3+}. Some halides, sulfides, and tellurides of In(I) and In(II) may be prepared. The thermodynamic data are given in Table 2.

B. In/In(III) Couple

The exact determination of $E°$ of the couple In/In^{3+} is complicated because its equilibrium is distorted in acid solutions (0.1 M H^+ and higher) due to oxidation of In^+ ions (unstable intermediates of the overall electrode process) by hydrogen ions [5]. The best results have been obtained by Covington et al. [6] for dilute chloride solutions:

$$In^{3+} + 3e^- \rightarrow In \qquad E° = -0.3382 \text{ V}$$

TABLE 2 Thermodynamic Data on Indium

Formula	State	$\Delta H°$/kJ mol^{-1}	$\Delta G°$/kJ mol^{-1}	$S°$/J mol^{-1} K^{-1}
In	c			57.82
In	g	243.3	208.7	173.7
In^+	g	807.8		
In^+	aq		−12	
In^{2+}	g	2634.6		
In^{2+}	aq		−50.6	
In^{3+}	g	5345.3		
In^{3+}	aq	−105	−97.9	−150
InO	g	387	364	236
In_2O_3	c	−925.73	830.73	104
InH	g	215	190.3	207.5
InOH	g	−79		
$In(OH)^{2+}$	aq	−370	313	88
$In(OH)^+$	aq	−619	−525.1	25
InF	g	−203.4		
InCl	c	−186		

TABLE 2 (Continued)

Formula	State	$\Delta H°$/kJ mol^{-1}	$\Delta G°$/kJ mol^{-1}	$S°$/J mol^{-1} K^{-1}
InCl$_3$	c	−537		
InCl$_3$	g	−374		
InBr	c	−175	−169	113
InBr	g	−56.9	−94.31	259.4
InBr$_3$	c	−428.9		
InBr$_3$	g	−282		
InI	c	−116	−120	130
InI	g	7.5	−38	267.2
InI$_3$	c	−238		
InI$_3$	g	−120		
InS	c	−138	−132	67
InS	g	377		
In$_2$S	g	63	130	318
In$_2$S$_3$	c	−427	−412.6	163.6
InSe	c	−120		
In$_2$Se$_3$	c	−343		
InTe	c	−96		
In$_2$Te$_3$	c	−197		
InH	c	−17.6		
InP	c	−88.7	−77.0	59.8
InAs	c	−58.6	−53.6	75.7
InSb	c	−31	−26	86.2
InSb	g	344		
InSb$_2$	g	314		
In(CNS)$_3$	aq		165	

$dE°/dT = 40$ mV K^{-1} (at 298 K)

Hence the Gibbs energy of In^{3+} is -7.9 kJ mol^{-1}. Indium dissolves fairly readily in acid solutions with the evolution of hydrogen. In concentrated alkaline solutions In(III) exists in the form of hydroxy complexes In(OH)$_m^{(3-m)+}$ (indates).

C. Compounds of In(I) and In(II)

In$^+$ ions are formed both when metallic indium comes into contact with a solution of In(III) salt and during anodic polarization of indium and its amalgam. They are easily oxidized by hydrogen ions, while in the presence of metallic surfaces they are subject to heterogeneous disproportionation which takes place as a combination of the coupled charge transfer steps In$^+$ + e$^-$ → In and In$^+$ → In^{3+} + 2e$^-$ [5]. These complications, as well as the results of experiments with indicator electrodes [7-11], raise doubts [5] about the conclusion of some authors on the existence in the system In/In^{3+} of a noticeable concentration of In^{2+} ions [3,12]. The most reliable data on $E°$ for the couple In/In$^+$ were obtained by Visco [7] in acid perchlorate solution:

$$\text{In}^+ + \text{e}^- \rightarrow \text{In} \qquad E° = -0.126 \text{ V}$$

Hence the Gibbs energy of In$^+$ is -12 kJ mol^{-1}. In iodide solutions In$^+$ ions form an insoluble reddish-brown deposit of InI. The couple In/InI behaves in NaI solutions as a reversible electrode of the second kind [5,13].

The results of direct potentiometric determination of the potential of the In$^+$/In^{3+} couple [14,15] are questionable [5]. The calculation of this potential by means of the above-mentioned potentials of the In/In^{3+} and In/In$^+$ couples gives

$$\text{In}^{3+} + 2\text{e}^- \rightarrow \text{In}^+ \qquad E° = -0.444 \text{ V}$$

Chlorides and bromides of In(I) and In(II) may be prepared by the interaction of chlorine and bromine with metallic indium at high temperatures. They are decomposed by water with formation of metallic indium and corresponding salts of In(III). Using the heat of formation of InCl and taking the entropy of InCl as 97.1 J mol^{-1} K^{-1} (compared with 96.2 J mol^{-1} K^{-1} for AgCl), Latimer calculated the potential of the In/InCl couple [3]:

$$\text{InCl} + \text{e}^- \rightarrow \text{In} + \text{Cl}^- \qquad E° = -0.34 \text{ V}$$

The sulfides of In(I) and In(II) are formed by melting of the elements. Although the data on the formation and properties of In$_2$O and InO are contradictory [16-19], it has been proved that In$_2$O is formed when In$_2$O$_3$ is heated in vacuum, as well as from In(III) oxalate at high temperatures [19,20].

III. THALLIUM

Thallium forms compounds of +1 and +3 oxidation states. Tl(I) compounds are most stable, whereas Tl(III) compounds are strong oxidants and are easily reduced to the +I oxidation state. TlOH is a strong base and similar to alkaline metal hydroxides. The thermodynamic data are given in Table 3.

TABLE 3 Thermodynamic Data on Thallium

Formula	State	$\Delta H°$/kJ mol^{-1}	$\Delta G°$/kJ mol^{-1}	$S°$/J mol^{-1} K^{-1}
Tl	c	0	0	64.18
Tl	g	182.2	147.4	180.85
Tl$^+$	g	777.734		
Tl$^+$	aq	5.36	−32.4	126
Tl^{2+}	g	2754.94		
Tl^{3+}	c	5639.2		
Tl^{3+}	aq	197	215	−192
Tl$_2$O	c	−179	−147	126
Tl$_2$O$_3$	c		−312	
TlOH	c	−239		
TlOH	aq	−224.6	−189.7	115
Tl(OH)$_3$	c		−507.1	
TlF	c	−325		
TlCl	c	−204.1	−184.9	111.3
TlCl	g	−67.8		
TlCl	aq	−161.80	−163.63	182.0
TlCl$_3$	c	−315		
TlCl$_3$	aq	−351		
TlClO$_3$	aq	−93.7	−36	286
TlBr	c	−173	−167	120
TlBr	g	−38		
TlBr$_3$	aq	−250		
TlBrO$_3$	c	−136.4	−53.14	169
TlBrO$_3$	aq	−78.2	−30.5	287

TABLE 3 (Continued)

Formula	State	$\Delta H°$/kJ mol^{-1}	$\Delta G°$/kJ mol^{-1}	$S°$/J mol^{-1} K^{-1}
TlI	c	−123.8	−125.39	127.6
TlI	g	7.1		
TlIO$_3$	c	−267	−191.9	177
TlIO$_3$	aq	−216	−160	244
Tl$_2$S	c	97.1	−93.7	151
Tl$_2$SO$_4$	c	−931.8	−830.48	231
Tl$_2$SO$_4$	aq	−898.56	−809.39	271
Tl$_2$Se	c	−59	−59	172
Tl$_2$Te	c	−92		
TlN$_3$	c	233.5	294.5	146.8
TlNO$_3$	c	−243.9	−152.46	160.7
TlNO$_3$	aq	−202.0	−143.72	272
Tl$_2$CO$_3$	c	−700.0	−614.6	155
TlCNS	c	28	39	163
TlCNS	aq	81.80	60.29	270

A. Tl/Tl$^+$ Couple

Precise measurements of the electromotive force of a cell with a Tl/Tl$^+$ electrode has led to the value [3]

$$Tl^+ + e^- \rightarrow Tl \qquad E° = -0.3363 \text{ V}$$

The corresponding Gibbs energy of Tl$^+$ is −32.45 kJ mol^{-1} and the temperature coefficient of $E°$ [21]

$$(dE°/dT)_{\text{isoth}} = -1.327 \text{ mV K}^{-1}$$

The potentials of some Tl/Tl$^+$ couples calculated from thermodynamical data are given in Table 4.

B. Tl(I)/Tl(III) Couple

From the results of potentiometric measurements in perchloric acid solutions with different ionic strength, Latimer estimated the potential

$$Tl^{3+} + 2e^- \rightarrow Tl^+ \qquad E° = 1.25 \text{ V}$$

and the Gibbs energy of thallic ion Tl^{3+}, 209 kJ mol^{-1} [3]. This ion is a strong oxidant and is readily hydrolyzed; hence the potential of the couple Tl(I)/Ti(III) is very dependent on pH. From the solubility product of $Tl(OH)_3$, Latimer calculated the Gibbs energy of $Tl(OH)_3(c)$, -515 kJ mol^{-1}, and the potential of the Tl(I)/Tl(III) couple in alkaline solution [3]:

$$Tl(OH)_3(c) + 3e^- \rightarrow Tl(OH)(c) + 2OH^- \qquad E° = -0.05 \text{ V}$$

The temperature coefficient of $E°$ is [21]

$$(dE°/dT)_{isoth} = -0.940 \text{ mV K}^{-1}$$

In the presence of metallic thallium the Tl^{3+} ion is unstable with respect to the reaction $Tl^{3+} + 2Tl = 2Tl^+$ and the potential of the metallic thallium is determined by the highly reversible reaction $Tl^+ + e^- = Tl$. Hence the potential of the Tl/Tl^{3+} couple cannot be measured directly. Its value can be calculated from the potentials of the couples Tl/Tl^+ and Tl^+/Tl^{3+}:

TABLE 4 Potentials of Some Tl/Tl^+ Couples

Couple	$E°$ / V	$(dE°/dT)_{isoth}$ (mV K^{-1})	Reference
$TlOH + e^- \rightarrow Tl + OH^-$	-0.343	-0.868	21
$TlCl + e^- \rightarrow Tl + Cl^-$	-0.5568	-0.560	21
$TlBr + e^- \rightarrow Tl + Br^-$	-0.658	-0.413	21
$TlI + e^- \rightarrow Tl + I^-$	-0.752	-0.150	21
$Tl_2S + 2e^- \rightarrow 2Tl + S^{2-}$	-0.90	-0.94	21
$TlSCN + e^- \rightarrow Tl + SCN^-$	-0.56		22
$TlIO_3 + e^- \rightarrow Tl + IO_3^-$	-0.666		23
$Tl_2CrO_4 + e^- \rightarrow 2Tl + CrO_4^{2-}$	-1.056		23
$TlN_3 + e^- \rightarrow Tl + N_3^-$	-0.554		23

$$Tl^{3+} + 3e^- \to Tl \qquad E° = 0.72 \text{ V}$$

The potential of the couple Tl(I)/Tl(III) is very dependent on the presence of complexing ions [21]. In hydrochloric acid solution the potential of this couple is [3]

$$TlCl_3 + 2e^- \to TlCl + 2Cl^- \qquad E° = 0.890 \text{ V}$$

IV. POTENTIAL DIAGRAMS FOR THE GROUP IN ACID SOLUTION

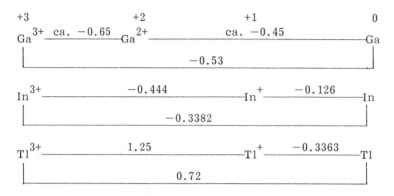

The principal point of interest is the increase in the oxidizing power of the +3 ion with increasing atomic weight.

REFERENCES

1. A. J. Bard, ed., *Encyclopedia of the Electrochemistry of the Elements,* Vol. 8, Marcel Dekker, New York, 1974, p. 207 (Gallium).
2. W. Saltman and N. Nachtrieb, J. Electrochem. Soc., *100,* 126 (1953).
3. W. M. Latimer, *The Oxidation States of the Elements and Their Potentials in Aqueous Solutions,* Prentice-Hall, Englewood Cliffs, N.J., 1952.
4. W. Klemm and I. Schnick, Z. Anaorg. Chem., *143,* 184 (1925).
5. A. J. Bard, ed., *Encyclopedia of the Electrochemistry of the Elements* Vol. 6, (Indium), Marcel Dekker, New York, 1976, p. 1.
6. A. K. Covington, M. A. Hakeem, and W. F. K. Wynne-Jones, J. Chem. Soc., 4394, (1963).
7. R. E. Visco, J. Phys. Chem., *69,* 202 (1965).
8. A. P. Pchelnikov and V. V. Losev, Zashch. Met., *1,* 482 (1965).
9. B. Miller and R. E. Visco, J. Electrochem. Soc., *115,* 251 (1968).
10. L. Kiss and J. Farkas, Ann. Univ. Sci. Budapest, Sect. Chim., *12,* 123 (1971).
11. A. I. Molodov, G. N. Markosyan, and V. V. Losev, Elektrokhimiya, *9,* 1368 (1973).
12. L. G. Hepler, Z. Z. Hugus, and W. M. Latimer, J. Am. Chem. Soc., *75,* 5652 (1953).

13. V. V. Losev, Electrochim. Acta, *15*, 1095 (1970).
14. W. Kangro and F. Weingärtner, Z. Elektrochem., *58*, 505 (1954).
15. G. Biedermann and T. Wallin, Acta Chem. Scand., *14*, 594 (1960).
16. A. Thiel and A. Luckman, Z. Anorg. Chem., *172*, 362 (1928).
17. M. F. Stubbs, A. Schufle, and A. Tompson, J. Am. Chem. Soc., *74*, 6201 (1952).
18. F. Gastinger, Naturwissenschaften, *4*, 195 (1955).
19. Yu. V. Rumjantsev and N. A. Khvorostukhina, *Fizikôkhimicheskie Osnovy pirometallurgii indija*, Izdatelstvo "Nauka," Moscow, 1965.
20. S. A. Shchukarev, G. A. Semenov, and I. A. Ratkovskii, Zh. Prikl. Khim., *35*, 1454 (1962).
21. A. J. Bard, ed., *Encyclopedia of the Electrochemistry of the Elements*, Vol. 4, Marcel Dekker, New York, 1975, p. 179.
22. R. P. Bell and J. H. B. George, Trans. Faraday Soc., *49*, 619 (1953).
23. G. Milazzo and S. Cavoli, *Tables of Standard Electrode Potentials*, Wiley, New York, 1978.

10
Zinc, Cadmium, and Mercury

I. ZINC*

RALPH J. BRODD † Inco Research and Development Center, Inc., Suffern, New York
JOHN WERTH Englehard Industries, Menlo Park, New Jersey

Zinc is an active metal above hydrogen in the electromotive force (emf) scale. It normally liberates hydrogen when immersed in aqueous solution. It also has a high hydrogen overpotential depending somewhat on pH. This kinetically hinders the hydrogen evolution reaction and renders zinc fairly stable in aqueous solutions. This fact makes it a useful material for negative electrodes in batteries, corrosion protection of steel, and so on. Also because of the high hydrogen overpotential, the zinc ion can be reduced to metal, permitting quantitative determination of zinc by cathodic deposition. The analytical estimation is more favorable at high pH when the zincate or complex cyanide is the predominant solution species.

Zinc has a $3d^{10}4s^2$ outer electronic orbit configuration. Thus zinc normally loses the two $4s$ electrons to form the divalent ion. There is little evidence that zinc forms a univalent ion in solution, although it is postulated as an intermediate for many reaction schemes. Evidence in radiation studies indicates a transitory stability of univalent zinc of about $t_{1/2} = 10^{-6}$ s in aqueous media. The zinc ion is expected to form sp^3 hybridized orbitals with sigma covalent bond formation to give tetrahedral coordinated ions in solution. The primary coordination number is 4, although others have been shown to exist. In aqueous solution zinc seldom exists as the simple ion and most probably exists as the tetra-coordinate aqueous ion $Zn(H_2O)_4^{2+}$. Complex ion formation generally follows the pattern of successive addition of the complexing ion, probably displacing water to form the zinc complex. Oxidation states higher that +2 are unknown except for the gaseous phase.

The value of the standard potential calculated from the National Bureau of Standards (NBS) data (Table 1) is significantly lower than that reported for many investigators [1]. Based on the value of the works summarized previously [2], we select -0.7626 V as the potential of the zinc-zinc ion couple in the standards state. This gives the Gibbs energy of the zinc ion as -147.160 kJ mol^{-1}:

---

*Reviewer: J. Koryta, J. Heyrovsky Institute of Physical Chemistry and Electrochemistry, Prague, Czechoslovakia.
†Current affiliation: Amoco Research Center, Naperville, Illinois.

TABLE 1 Thermodynamic Data on Zinc[a]

Formula	Description	State	$\Delta H°$/kJ mol^{-1}	$\Delta G°$/kJ mol^{-1}	$S°$/J mol^{-1} K^{-1}
Zn		g	130.73	95.178	160.88
Zn		c	0.0	0.0	41.6
Zn$^+$		g	1043.29		
Zn^{2+}		g	2743.7		
Zn^{2+}		aq	−152.84	−147.16	107.53
Zn^{3+}		g	6620.8		
ZnO		c	−348.28	−318.32	43.64
Zn(NH$_3$)$_4^{2+}$		aq	577.	−307.	301.
Zn(OH)$^+$		aq		−342.	
Zn(OH)$_4^{2-}$		aq		877.4	
Zn(OH)$_2$	γ	c		−553.88	
Zn(OH)$_2$	β	c	−641.91	−553.59	81.2
Zn(OH)$_2$	ε	c	−643.25	−555.13	81.6
Zn(OH)$_2$	Precipitate	c	−642.2		
Zn(OH)$_2$		aq	−613.88	−461.62	−133.

Zinc, Cadmium, and Mercury

Zn(OH)$_2$	Undiss.	aq		−521.1	
ZnCO$_3$		c	−812.78	−731.57	82.4
ZnCO$_3\cdot$H$_2$O		c		−970.7	
Zn(BO$_2$)$_2$		c		−1563.1	
ZnAl$_2$O$_4$		c	2101.		
Zn(CNS)$_2$	Undiss.	aq		33.	
ZnC$_2$O$_4$		aq	−979.1	−820.9	−66.5
ZnC$_2$O$_4\cdot$2H$_2$O		c	−1565.	−1346.0	195
Zn(CH$_3$)$_2$		ℓ	23.		
Zn(C$_2$H$_5$)$_2$		ℓ	10.5		
Zn(CHO$_2$)$_2$		c	−986.6		
Zn(CH$_3$CO$_2$)$_2$		c	−1078.6		
Zn(CH$_3$CO$_2$)$_2$		aq	−1125.9	−885.84	61.1
Zn(CHO$_2$)$_2\cdot$2H$_2$O		c	−1584.9		
Zn(CN)$_2$		c	95.8	121.	(95.8)
Zn(CN)$_4^{2-}$		aq	257.7	403.3	(226.)
ZnSiO$_3$		c	−1260.	−1150.	(89.5)

TABLE 1 (Continued)

Formula	Description	State	$\Delta H°$/kJ mol^{-1}	$\Delta G°$/kJ mol^{-1}	$S°$/J mol^{-1} K^{-1}
ZnSiO$_3$		c	−1260.	−1150.	131.
Zn$_2$SiO$_4$		c	−1636.7	−1523.2	131.
ZnF$_2$		c	−764.4	−713.4	73.68
ZnCl		g	4.	20.9	244.
ZnCl$_2$		c	−415.1	−369.43	111.5
ZnBr$_2$		c	−328.7	−312.1	139.
ZnI		g	63.	12.5	264.
ZnI$_2$		c	208.0	209.	161.
ZnS	Sphalerite	c	−206.0	−201.3	57.5
ZnS	Wurtzite	c	−192.6	−185.	(57.7)

Zinc, Cadmium, and Mercury

ZnS	Precipitate	c	−185	−181	
Zn(ClO$_4$)$_2$		aq	−412.5	−164.3	252
Zn(IO$_3$)$_2$		c		−438.80	
ZnSO$_4$		c	−982.8	−871.5	111
ZnSO$_4$·H$_2$O		c	−1304.5	−1132.1	139
ZnSO$_4$·6H$_2$O		c	−2777.5	−2324.8	364
ZnSO$_4$·7H$_2$O		c	−3077.8	−2563.1	389
ZnSe		c	−163	−163	84
Zn$_3$(PO$_4$)$_2$		c	−2892.		
Zn(NO$_3$)$_2$		c	−483.7	−298.82	194

[a]National Bureau of Standards values are in italic; estimated values are in parentheses.

$$Zn^{2+} + 2e^- \rightarrow Zn \qquad E^\circ = -0.7626 \text{ V}$$

In alkali the soluble species is the zincate ion $Zn(OH)_4^{2-}$, as would be predicted on the basis of the electronic configuration [3-5]. Based on the equilibrium constants of Gubeli and Ste-Maire [6], the Gibbs energy of zincate ion is -209.7 kJ mol^{-1}, which gives

$$Zn(OH)_4^{2-} + 2e^- \rightarrow Zn + 4OH^- \qquad E^\circ = -1.285 \text{ V}$$

Ruben [7] has reported the value of the mercury oxide zinc hydroxide couple in alkali as

$$Zn + HgO + H_2O \rightarrow Zn(OH)_2 + Hg(\ell) \qquad E = -1.344 \text{ V}$$

This gives

$$Zn(OH)_2 + 2e^- \rightarrow Zn + 2OH^- \qquad E^\circ = -1.246 \text{ V}$$

$$Zn(OH)_2 \rightarrow Zn^{2+} + 2OH^- \qquad K = 4.5 \times 10^{-17}$$

and a solubility for $Zn(OH)_2$ of about 9.8×10^{-6} mol L^{-1}. In concentrated alkali the solubility is much greater and amounts to about 1.42 mol L^{-1} in 6 M KOH [8].

Several forms of $Zn(OH)_2$ are known with stability roughly in the order $\gamma < \beta < \varepsilon$, the stable form being $\varepsilon-Zn(OH)_2$. The many forms of $Zn(OH)_2$, each with its own free energy, as well as the tendency of its solutions to supersaturate and to form colloidal complexes, have led to conflicting values for solubility. Also, nonionized or undissociated $Zn(OH)_2$ exists in alkali. To complicate the situation further, in concentrated alkali ZnO rather than $Zn(OH)_2$ is reported to be the stable solid species [9]. Thus we write

$$ZnO + H_2O + 2e^- \rightarrow Zn + 2OH^- \qquad E^\circ = -1.248 \text{ V}$$

In practice it is almost impossible to distinguish between the ZnO or $Zn(OH)_2$ reactions based on measured potentials in alkali solutions.

Luks et al. [10] have determined the ammonia complex formation constants from which we calculate -307.1 kJ mol^{-1} for the Gibbs energy of $Zn(NH_3)_4^{2+}$. Bjerrum [11] chose $K = 3.4 \times 10^{-10}$ for the dissociation $Zn(NH_3)_4^{2+} = Zn^{2+} + 4NH_3(aq)$. Both give

$$Zn(NH_3)_4^{2+} + 2e^- \rightarrow Zn + 4NH_3 \qquad E^\circ = -1.04 \text{ V}$$

Izatt et al. [12] report equilibrium constants for cyanide complexes. Using their value for formation constant of $Zn(CN)_4^{2-}$ of 1.74×10^{-27}, we write

$$Zn(CN)_4^{2-} + 2e^- \rightarrow Zn + 4CN^- \qquad E^\circ = -1.34 \text{ V}$$

From the Gibbs energy of zinc carbonate

$$ZnCO_3 + 2e^- \rightarrow Zn + CO_3^{2-} \qquad E^\circ = -1.06 \text{ V}$$

and the solubility product

$$ZnCO_3 \rightarrow Zn^{2+} + CO_3^{2-} \qquad K = 2 \times 10^{-10}$$

According to Sillen and Martell [13], $K = 1.9 \times 10^{-11}$, which would make E° about -1.08 V. Based on Gibbs energy, the solubility products of zinc sulfides are:

$$ZnS(\text{sphalerite}) \rightarrow Zn^{2+} + S^{2-} \qquad K = 2.2 \times 10^{-27}$$

$$ZnS(\text{wurtzite}) \rightarrow Zn^{2+} + S^{2-} \qquad K = 1.6 \times 10^{-23}$$

$$ZnS(\text{precipitated}) \rightarrow Zn^{2+} + S^{2-} \qquad K = 8.7 \times 10^{-23}$$

The precipitated material has slightly higher solubility and depends somewhat on the composition of the media. Using wurtzite as the stable form, we calculate

$$ZnS + 2e^- \rightarrow Zn + S^{2-} \qquad E^\circ = -1.44 \text{ V}$$

Rowlands and Monk [14] reported for the equilibrium

$$Zn^{2+} + C_2O_4^{2-} \rightarrow Zn\,C_2O_4 \qquad K_1 = 7.1 \times 10^4$$

$$ZnC_2O_4 + C_2O_4^{2-} \rightarrow Zn(C_2O_4)_2^{2-} \qquad K_2 = 5.0 \times 10^2$$

From these values we calculate

$$Zn(C_2O_4)_2^{2-} + 2e^- \rightarrow Zn + 2C_2O_4^{2-} \qquad E^\circ = -0.99 \text{ V}$$

For tartrate solution we calculate

$$Zn(C_4O_6H_4)_4^{6-} + 2e^- \rightarrow Zn + 4\,(C_4O_6H_4)^{2-} \qquad E^\circ = -1.15 \text{ V}$$

based on the equilibrium constant determined by Verdier et al. [15].

$$Zn^{2+} + 4(C_4O_6H_4)^{2-} \rightarrow Zn(C_4O_6H_4)_4^{6-} \qquad K = 6.17 \times 10^6$$

Zinc ions also form complexes with halides and sulfates. Brodd and Leger [2] selected the values in Table 2 for the formation constants.

TABLE 2 Overall Formation Constants for Various Zinc Complexes

Complexing agent	Log formation constant			
	β_1	β_2	β_3	β_4
F^-	0.92			
Cl^-	0.72	0.49	−0.19	0.18
Br^-	0.22	−0.10	−0.74	−1.00
I^-	−0.47	−2.00	−0.74	−1.25
SO_4^{2-}	2.08			

A. Potential Diagrams for Zinc

Acid Solution

$$Zn^{2+} \xrightarrow{-0.7626} Zn$$
(+2) (0)

Basic Solution

$$Zn(OH)_4^{2-} \xrightarrow{-1.285} Zn$$

$$Zn(OH)_2 \xrightarrow{-1.246} Zn$$

Acknowledgment

The authors gratefully acknowledge the help of S. Cieszewski in the preparation of the tables and checking of references.

REFERENCES

1. D. D. Wagman, W. H. Evans, V. B. Parker, I. Hallow, S. M. Bailey, and R. H. Schumm, Natl. Bur. Stand. Tech. Note 270-3, 1968.
2. R. J. Brodd and V. Z. Leger, in *Encyclopedia of the Electrochemistry of the Elements*, A. J. Bard, ed., Marcel Dekker, New York, 1976, Chap. 1.
3. T. P. Dirkse, J. Electrochem. Soc., *101*, 328 (1954).
4. G. H. Newman and G. E. Blomgren, J. Chem. Phys., *43*, 2744 (1965).
5. J. S. Fordyce and R. L. Baum, J. Chem. Phys., *43*, 843 (1965).
6. A. O. Gubeli and J. Ste-Maire, Can. J. Chem., *45*, 827 (1967).
7. S. Ruben, 17th CITCE Meet., Tokyo, 1966.
8. C. T. Baker and I. Trachtenberg, J. Electrochem. Soc., *114*, 1045 (1967).

9. R. J. Bennett, W. G. Darland, R. J. Brodd, and R. A. Powers, in *Power Sources*, D. H. Collins, ed., Oriel Press, Newcastle upon Tyne, England, 1969, p. 461.
10. C. Luca, V. Margearu, and G. Popa, J. Electroanal. Chem., *12*, 45 (1966).
11. J. Bjerrum, Chem. Rev., *46*, 381 (1950).
12. R. M. Izatt, J. J. Christensen, J. W. Hansen, and G. D. Watt, Inorg. Chem., *4*, 718 (1965).
13. I. G. Sillen and A. E. Martell, Chem. Soc. Spec. Publ. 25, Chemical Society, London, 1972.
14. D. L. G. Rowlands and C. B. Monk, Trans Faraday Soc., *62*, 945 (1966).
15. E. Verdier, R. Bennes, and B. Balette, J. Electroanal. Chem., *31*, 463 (1971).

II. CADMIUM*

YUTAKA OKINAKA[†]   Bell Telephone Laboratories, Murray Hill, New Jersey

A. Oxidation States

The only stable oxidation state of cadmium in aqueous solutions is +2. The existence of cadmium in the +1 state as an intermediate has been discussed recently by Kisova et al. [1].

B. Cd(II)/Cd(0) Couples[‡]

The $E°$ values for the Cd^{2+}/Cd couple were determined potentiometrically using cells with either a cadmium amalgam (11 wt %, saturated, two phase) [2,3] or a solid cadmium (polycrystalline, plated) [4]. Values of the electromotive force (emf) measured with the amalgam were corrected for the potential of the Cd/Cd amalgam couple determined independently [5]. The $E°$ values obtained in this manner range from -0.4020 to -0.4026 V. Measurement with solid cadmium yielded -0.4035 V. The Gibbs energy values listed in the most recent National Bureau of Standards (NBS) compilation [6] (Table 3) give -0.4020 V. Thus, in agreement with Berecki-Biedermann et al. [7], we shall write

$$Cd^{2+} + 2e^- \to Cd \qquad E° = -0.4025 \text{ V}$$

Potentiometrically determined $E°$ values for the Cd^{2+}/Cd amalgam (11 wt %, saturated, two phase) range from -0.3514 to -0.3521 V [3,4,7-9]. From the Gibbs energy values we obtain

$$Cd^{2+} + Hg + 2e^- \to Cd(Hg) \qquad E° = -0.3515 \text{ V}$$

---

*Reviewers: J. Koryta, J. Heyrovsky Institute of Physical Chemistry and Electrochemistry, Prague, Czechoslovakia; J. H. Sluyters, University of Utrecht, Utrecht, The Netherlands.
[†]*Current affiliation*: Bell Communications Research, Inc., Murray Hill, New Jersey.
[‡]Formal potentials are not included. They are listed in Refs. 8 and 14.

TABLE 3 Thermodynamic Data on Cadmium[a]

Formula	Description	State	$\Delta H°$/kJ mol^{-1}	$\Delta G°$/kJ mol^{-1}	$S°$/J mol^{-1} K^{-1}
Cd	Metal γ	c	0	0	51.76
Cd	Metal α	c	−0.59	−0.59	51.76
Cd	In Hg; two-phase amalgam		−21.246	−9.740	13.159
Cd^{2+}		aq	−75.90	−77.612	−73.2
CdO		c	−258.2	−228.4	54.8
CdO$_2^{2-}$		aq		−284.4	
HCdO$_2^-$		aq		−363.5	
CdOH$^+$		aq		−261.1	
Cd(OH)$_2$	Precipitated	c	−560.7	−473.6	96.
Cd(OH)$_2$	Undiss.	aq		−442.6	
Cd(OH)$_3^-$		aq		−600.7	
Cd(OH)$_4^{2-}$		aq		−758.4	
CdF$_2$		c	−700.4	−647.7	77.4
CdCl$^+$		aq	−240.6	−224.39	43.5

Zinc, Cadmium, and Mercury

$CdCl_2$		c	−391.50	−343.93	115.27
$CdCl_2$	Undiss.	aq	−405.0	−359.29	121.8
$CdCl_3^-$		aq	−561.1	−487.0	202.9
$CdBr^+$		aq	−200.8	−193.94	39.7
$CdBr_2$		c	−316.18	−296.31	137.2
$CdBr_2$	Undiss.	aq		−302.1	
$CdBr_2 \cdot 4H_2O$		c	−1492.56	−1247.837	316.3
$CdBr_3^-$		aq		−407.5	
CdI^+		aq	−141.0	−141.4	43.1
CdI_2		c	−203.3	−201.38	161.1
CdI_2	Undiss.	aq		−201.3	
CdI_3^-		aq		−259.4	
CdI_4^{2-}		aq	−341.8	−315.9	326.
CdS		c	−161.9	−156.5	64.9
$CdSO_4$		c	−993.28	−822.72	123.039
$CdSO_4 \cdot H_2O$		c	−1239.55	−1068.73	154.030
$CdSO_4 \cdot (8/3)H_2O$		c	−1729.4	−1465.141	229.630

TABLE 3 (Continued)

Formula	Description	State	$\Delta H°$/kJ mol^{-1}	$\Delta G°$/kJ mol^{-1}	$S°$/J mol^{-1} K^{-1}
CdSe		c	−136.0	(−136.4)	96.
CdTe		c	−92.5	−92.0	*100.*
Cd(NO$_3$)$_2$		c	−456.31	−255.	(197.9)
Cd(NH$_3$)$^{2+}$		aq		−118.8	
Cd(NH$_3$)$_2^{2+}$		aq	−266.1	−158.9	144.8
Cd(NH$_3$)$_4^{2+}$		aq	−450.2	−226.1	336.4
CdCO$_3$		c	−750.6	−669.4	92.5
Cd(CN)$_2$		c	162.3	207.9	(104.2)
Cd(CN)$_4^{2-}$		aq	428.0	507.6	322.

[a]National Bureau of Standards values are in italic; estimated values are in parentheses.

The electrode potentials of the couples involving $Cd(OH)_2$, CdO, $Cd(OH)_3^-$, and $Cd(OH)_4^{2-}$ are of interest in connection with rechargeable alkaline batteries. From the solubility product of $Cd(OH)_2$, $K_{sp} = 10^{-14.35}$ [10], the $E°$ value for the $Cd(OH)_2(s)/Cd$ couple is calculated to be -0.827 V. The Gibbs energy values give

$$Cd(OH)_2(s) + 2e^- \rightarrow Cd + 2OH^- \qquad E° = -0.824 \text{ V}$$

These values are significantly different from -0.809 V calculated from the old $\Delta G°$ values [11] and quoted in previous publications [8,9,12-14].

For the $CdO(s)/Cd$ couple in alkaline solutions, the $\Delta G°$ values yield

$$CdO(s) + H_2O + 2e^- \rightarrow Cd + 2OH^- \qquad E° = -0.783 \text{ V}$$

The $E°$ values calculated from Gibbs energies for the couples involving $Cd(OH)_3^-$ and $Cd(OH)_4^{2-}$ are

$$Cd(OH)_3^- + 2e^- \rightarrow Cd + 3OH^- \qquad E° = -0.668 \text{ V}$$

$$Cd(OH)_4^{2-} + 2e^- \rightarrow Cd + 4OH^- \qquad E° = -0.670 \text{ V}$$

For the $CdCO_3/Cd$ couple, relevant $\Delta G°$ values yield

$$CdCO_3(s) + 2e^- \rightarrow Cd + CO_3^{2-} \qquad E° = -0.734 \text{ V}$$

which is not significantly different from -0.74 V calculated from the old $\Delta G°$ values and listed in previous compilations [8,9,12,14].

Six cadmium ammine complexes, $Cd(NH_3)_n^{2+}$ with $n = 1$ to 6, are known [10]. For $Cd(NH_3)_4^{2+}$ the $\Delta G°$ values yield

$$Cd(NH_3)_4^{2+} + 2e^- \rightarrow Cd + 4NH_3 \qquad E° = -0.622 \text{ V}$$

as compared to -0.61 V and -0.613 V appearing in previous listings [8,9,12,14].

Cadmium forms cyanide complexes $Cd(CN)_n^{2-n}$ with $n = 1$ to 4. For $Cd(CN)_4^{2-}$ the $\Delta G°$ values yield

$$Cd(CN)_4^{2-} + 2e^- \rightarrow Cd + 4CN^- \qquad E° = -0.943 \text{ V}$$

The stability constant of this complex ion, $K = 10^{17.92}$ [10], gives -0.933 V. These values differ greatly from the values quoted previously, -1.09 V [12] and -1.028 V [8,14,15].

The stability constant of $CdCl_4^{2-}$, $K = 10^{1.7}$ [10], yields

$$CdCl_4^{2-} + 2e^- \rightarrow Cd + 4Cl^- \qquad E° = -0.453 \text{ V}$$

For $CdBr_4^{2-}$ with $K = 10^{2.9}$ [10] we obtain

$$CdBr_4^{2-} + 2e^- \to Cd + 4Br^- \qquad E° = -0.488 \text{ V}$$

The corresponding iodide complex, CdI_4^{2-}, is more stable with $K = 10^{6.0}$ [10], which gives

$$CdI_4^{2-} + 2e^- \to Cd + 4I^- \qquad E° = -0.580 \text{ V}$$

while the Gibbs energy values yield -0.568 V. The $E°$ values for lower halide complexes have been calculated from relevant Gibbs energy data, and they are included in Table 4.

Cadmium chalcogenides are of interest as materials for semiconductor photoelectrodes. For the CdS(s)/Cd couple the $\Delta G°$ values yield

$$CdS(s) + 2e^- \to Cd + S^{2-} \qquad E° = -1.255 \text{ V}$$

as compared to -1.24 V [12] and -1.175 V [8,9,14,15] in the previous compilations.

Available thermodynamic data for CdSe are in disparity, and the NBS table gives no values for this compound. On the basis of four determinations by independent investigators, Goldfinger and Jeunehomme [16] proposed the $\Delta H°$ and $S°$ values quoted in Table 3. From these values and relevant entropy data listed in the NBS tables [6], we obtain $\Delta G° = -136.4$ kJ mol^{-1} for CdSe, which yields

$$CdSe(s) + 2e^- \to Cd + Se^{2-} \qquad E° = -1.38 \text{ V}$$

The NBS values of $\Delta G°$ for CdTe and $\Delta G° = 220$ kJ mol^{-1} for Te^{2-} [12] yield

$$CdTe(s) + 2e^- \to Cd + Te^{2-} \qquad E° = -1.62 \text{ V}$$

while the solubility product of CdTe, $K_{sp} = 10^{-41.5}$ [10] gives -1.63 V.

TABLE 4 Standard Potentials and Isothermal Temperature Coefficients at 25°C

Electrode reaction	$E°$/V	$(dE°/dT)$/mV K^{-1} [a]
$Cd^{2+} + 2e^- \to Cd$	-0.4025	-0.029
		$-0.053*$
$Cd^{2+} + Hg + 2e^- \to Cd(Hg)$	-0.3515	-0.229
		$-0.250*$

TABLE 4 (Continued)

Electrode reaction	$E°/V$	$(dE°/dT)/mV\ K^{-1}$ a
		−0.252*
		−0.292*
$Cd(OH)_2(s) + 2e^- \to Cd + 2OH^-$	−0.824	−1.019
$CdO(s) + H_2O + 2e^- \to Cd + 2OH^-$	−0.783	−1.166
$Cd(OH)_3^- + 2e^- \to Cd + 3OH^-$	−0.668	—
$Cd(OH)_4^{2-} + 2e^- \to Cd + 4OH^-$	−0.670	—
$CdCO_3(s) + 2e^- \to Cd + CO_3^{2-}$	−0.734	−1.183
$Cd(NH_3)_4^{2+} + 2e^- \to Cd + 4NH_3$	−0.622	0.155
$Cd(CN)_4^{2-} + 2e^- \to Cd + 4CH^-$	−0.943	−0.127
$CdCl^+ + 2e^- \to Cd + Cl^-$	−0.483	−0.341
$CdCl_2 + 2e^- \to Cd + 2Cl^-$	−0.502	−0.454
$CdCl_3^- + 2e^- \to Cd + 3Cl^-$	−0.483	−0.582
$CdCl_4^{2-} + 2e^- \to Cd + 4Cl^-$	−0.453	—
$CdBr^+ + 2e^- \to Cd + Br^-$	−0.466	−0.187
$CdBr_2 + 2e^- \to Cd + 2Br^-$	−0.488	—
$CdBr_3^- + 2e^- \to Cd + 3Br^-$	−0.495	—
$CdBr_4^{2-} + 2e^- \to Cd + 4Br^-$	−0.488	—
$CdI^+ + 2e^- \to Cd + I^-$	−0.467	−0.055
$CdI_2 + 2e^- \to Cd + 2I^-$	−0.508	—
$CdI_3^- + 2e^- \to Cd + 3I^-$	−0.542	—
$CdI_4^{2-} + 2e^- \to Cd + 4I^-$	−0.568	0.207
$CdS(s) + 2e^- \to Cd + S^{2-}$	−0.255	−0.820
$CdSe(s) + 2e^- \to Cd + Se^{2-}$	−1.38	—
$CdTe(s) + 2e^- \to Cd + Te^{2-}$	−1.62	—

aAn asterisk indicates direct experimental values [15].

C. Temperature Coefficients

The original tabulation of temperature coefficients by de Béthune and co-workers [9,15] includes values calculated from entropy and heat capacity data taken from Latimer [12] or the old NBS table [11]. Because many such data have since been revised, the temperature coefficients have been recalculated using values given in the more recent NBS table [6]. They are listed in Table 4 together with some values directly determined by experiments and quoted by de Béthune and co-workers [9,15].

D. Potential Diagrams for Cadmium

Acid Solution

$$Cd^{2+} \xrightarrow{-0.4025} Cd$$

Basic Solution

$$Cd(OH)_2(s) \xrightarrow{-0.824} Cd$$

REFERENCES

1. L. Kisova, M. Golendzinowski, and J. Lipkowski, J. Electroanal. Chem., 95, 29 (1979).
2. H. S. Harned and M. E. Fitzgerald, J. Am. Chem. Soc., 58, 2624 (1936).
3. W. B. Treumann and L. M. Ferris, J. Am. Chem. Soc., 80, 5048 (1958).
4. J. L. Gurnett and M. H. Zirin, J. Inorg. Nucl. Chem., 28, 902 (1966).
5. W. G. Parks and V. K. La Mer, J. Am. Chem. Soc., 56, 90 (1934).
6. D. D. Wagman, W. H. Evans, V. B. Parker, R. H. Schumm, I. Halow, S. M. Bailey, K. L. Churney, and R. L. Nuttall, *NBS Tables of Chemical Thermodynamic Properties*, J. Phys. Chem. Ref. Data, 11, Supplement 2 (1982).
7. C. Berecki-Biedermann, G. Biedermann, and L. G. Sillén, Report to Analytical Section, IUPAC, July 1953. (Cited in Ref. 13.)
8. G. Milazzo and S. Caroli, *Tables of Standard Electrode Potentials*, Wiley, New York, 1978.
9. A. J. de Béthune, T. S. Licht, and N. Swendeman, J. Electrochem. Soc., 106, 616 (1959).
10. R. M. Smith, and A. E. Martell, *Critical Stability Constants*, Vol. 4: *Inorganic Complexes*, Plenum Press, New York, 1976.
11. F. D. Rossini, D. D. Wagman, W. H. Evans, S. Levine, and I. Jaffe, *Selected Values of Chemical Thermodynamic Properties*, Natl. Bur. Stand. Circ. 500, U.S. Government Printing Office, Washington, D.C., 1952.
12. W. M. Latimer, *Oxidation Potentials*, 2nd ed., Prentice-Hall, Englewood Cliffs, N.J., 1952.

13. J. Bjerrum, G. Schwarzenbach, and L. C. Sillén, *Stability Constants*, Spec. Publ. 6, Chemical Society, London, 1957.
14. R. J. Latham and N. A. Hampson, in *Encyclopedia of the Electrochemistry of the Elements*, A. J. Bard, ed., Vol. 1, Marcel Dekker, New York, 1973, Chap. I-4.
15. A. J., de Béthune and N. A. Swendeman Loud, in *Encyclopedia of Electrochemistry*, C. A. Hampel, ed., Reinhold, New York, 1964, pp. 414-426.
16. P. Goldfinger and M. Jeunehomme, Trans. Faraday Soc., *59*, 2851 (1963).

III. MERCURY*

JAN BALEJ† Institute of Inorganic Chemistry, Czechoslovak Academy of Sciences, Prague, Czechoslovakia

A. Oxidation States

Mercury forms compounds of the double mercurous ion, Hg_2^{2+}, +1 oxidation state, and the mercuric ion, Hg^{2+}, +2 state. The binuclear ion, $^+Hg\text{-}Hg^+$, can readily be obtained by reduction of mercuric salts, and can also easily be oxidized into Hg^{2+}. Even at very low concentrations, the aqeuous mercurous ion appears to be undissociated. For this dissociation

$$Hg_2^{2+}(aq) \rightarrow 2Hg^+(aq) \tag{10.1}$$

Moser and Voigt [1] estimated the limiting value of the dissociation constant $K_{(1)} < 10^{-7}$. The existence of $Hg^+(aq)$ remains doubtful [2,3]. Other ions of the form Hg_3^{2+} and Hg_4^{2+} are also known, but only in such nonaqueous solvents as $SO_2 + AsF_5$ or $AlCl_3 + NaCl$ melt [4,5].

Many mercury compounds may readily be precipitated from aqueous solutions. Soluble compounds are mostly rather weak electrolytes. The mercurous ion forms few complexes, due, in part, to a low tendency to form coordination bonds, but mainly due to the fact that with most ligands Hg^{2+} will form even more stable complexes. Addition of complexing agents to Hg(I) compounds in aqueous systems often leads to disproportionation to Hg^{2+} complexes and metallic mercury. Therefore, no mercurous hydroxide, sulfide, or cyanide can be precipitated by adding the appropriate anion to aqueous Hg_2^{2+}.

B. Chemistry of the Element

In nature, mercury occurs only in the form of its compounds. The only important ore is cinnabar, HgS, which by roasting yields oxide that decomposes at about 500°C into mercury and oxygen.

Metallic mercury is inert to nonoxidizing acids. At 300 to 350°C, it reacts with air oxygen to the oxide HgO, which from about 400°C and above decomposes rapidly into the elements. Mercury reacts directly with halogens,

---

*Reviewer: J. Koryta, J. Heyrovský Institute of Physical Chemistry and Electrochemistry, Czechoslovak Academy of Sciences, Prague, Czechoslovakia.
†Current address: Teichstrasse 9, D-5170 Jülich, Federal Republic of Germany.

sulfur, selenium, phosphorus, and similar nonmetals. It combines with many other metals to amalgams [6,7], some of which form intermetallic compounds of defined compoisition, especially in the solid state (e.g., $NaHg_4$, $NaHg_2$, etc.). Some of the transition metals do not form amalgams directly with mercury (e.g., iron and platinum metals); they may, however, be amalgamated by other easily soluble amalgams (e.g., by liquid sodium amalgams). Sodium amalgams and amalgamated zinc have been used as reducing agents for chemical syntheses as well as for analytical purposes [8].

C. Mercury-Mercurous Ion Couple

From thorough recalculations by Vanderzee and Swanson [9] of results of a number of electrochemical investigations, the "best value" of the standard potential of the electrode reaction

$$Hg_2^{2+}(aq) + 2e^- \rightarrow 2Hg(\ell) \qquad E^\circ_{(2)} = 0.7960 \text{ V} \qquad (10.2)$$

has been adopted, which is in full agreement with the authors own calorimetric determination of ΔG_f° for $Hg_2^{2+}(aq)$. The same value was included by Hepler and Olofson [10] in their excellent critical compilation of thermodynamic properties of mercury and its compounds,* as well as by Marcus [15] in his recent critical compilation and evaluation of solubility data in the Hg(I) chloride-water system. The isothermal temperature coefficient $dE^\circ_{(2)}/dT = -0.23$ mV K^{-1}.

The couple is capable of attaining equilibrium rapidly [16,17] and when used as a reference electrode, mostly of the second kind, it is readily reversible when all equilibria in the investigated solution have been attained.

D. Mercury-Mercurous Halide Couples

From the value of ΔG_f° for $Hg_2F_2(c)$ from Table 5 and with the use of $\Delta G_f^\circ = -278.82$ kJ mol^{-1} for F^-(aq) (see Ref. 11), we find for the electrode reaction

$$Hg_2F_2(c) + 2e^- \rightarrow 2Hg(\ell) + 2F^-(aq) \qquad E^\circ_{(3)} = 0.6562_5 \text{ V} \qquad (10.3)$$

corresponding to the value $\Delta G^\circ_{(3)} = -126.64$ kJ mol^{-1} and to the solubility product of Hg_2F_2: $K_{sp} = 1.89 \times 10^{-5}$.

---

*The compilation by Hepler and Olofson [10] served as a source of most thermochemical data given in Table 5 for its critical selection and recalculation of hitherto presented data, including those reported in NBS Technical Notes 270-3, 270-4, and 270-7 [11,12] and the JANAF tables [13]. Their results were checked and full agreement was found in most cases. Most of their critically selected data may therefore be considered as the "best values." In the review [10] more detailed information about the procedure of data selection is also given. The critical selection by Hepler and Olofson [10] has been prefered in view of the consistency of electrochemically measured thermodynamic quantities, especially standard electrode potentials and solubility products, with other independently determined thermodynamic quantities of given compounds and their reactions, which is fulfilled in the NBS Technical Notes and JANAF tables, but not in the recent *Tables of Standard Electrode Potentials* by Milazzo and Caroli [14].

TABLE 5 Thermodynamic Data on Mercury[a]

Formula	Description	State	ΔH_f°/kJ mol^{-1}	ΔG_f°/kJ mol^{-1}	S°/J mol^{-1} K^{-1}
Hg	Metal	ℓ	0	0	76.02
Hg		g	61.317	31.853	174.85
Hg		aq	13.93	37.2	−2.1
Hg$^+$		g	1074.53		
Hg^{2+}		g	2890.43		
Hg^{2+}		aq	170.16	164.703	−36.23
Hg^{3+}		g	6192		
Hg$_2$		g	109	68.2	287.9
Hg$_2^{2+}$		aq	166.82	153.607	65.77
HgO	Red, Orthorh.	c	−90.83	−58.555	70.29
HgO	Yellow, Orthorh.	c	−90.83	−58.450	69.87
HgO	Red, Hexag.	c	−90.33	−58.325	71.25
HgO		g			241.8
HgH		g	239.3	216.0	219.5

TABLE 5 (Continued)

Formula	Description	State	ΔH_f°/kJ mol^{-1}	ΔG_f°/kJ mol^{-1}	S°/J mol^{-1} K^{-1}
Hg(OH)$^+$		aq	−84.5	(−52.01)	69.0
HHgO$_2^-$		aq		(−190.0)	
Hg(OH)$_2$	Undiss.	aq	−359.8	(−274.5)	126.4
HgF		g	4.2	−10.0	225.85
			~2.9	~−18.4	248.28
HgF$^+$		aq	(−159.0)	(−123.0)	(−8)
Hg$_2$F$_2$		c	−485	−431	167*
HgCl		g	84.1	62.8	259.8
HgCl$^+$		aq	(−19.7)	(−5.0)	(71)
HgCl$_2$		c	−225.9	−180.3	146.0
HgCl$_2$		g	−143.1	−141.8	294.68
HgCl$_2$	Undiss.	aq	(−217.1)	(−172.8)	(151)
HgCl$_3^-$		aq	(−389.5)	(−308.8)	(205)
HgCl$_4^{2-}$		aq	(−554.8)	(−446.4)	(289)

Hg_2Cl_2		c	−265.45	−210.374	191.86
$HgBr$		g	~105	~67	271.5
$HgBr^+$		aq	5.4	9.2	75
$HgBr_2$		c	−170.7	−153.1	172
$HgBr_2$		g	86.6	113.8	320.12
$HgBr_2$	Undiss.	aq	(−161.5)	(−142.7)	(167)
$HgBr_3^-$		aq	(−294.1)	(−259.0)	(255)
$HgBr_4^{2-}$		aq	(−431.8)	(−370.7)	(305)
Hg_2Br_2		c	−206.94	−181.084	217.6
HgI		g	132.38	88.45	281.42
HgI^+		aq	(42.3)	(40.2)	(75)
HgI_2	Red	c	−105.4	−101.7	180
HgI_2	Yellow	c	−102.9		
HgI_2	Undiss.	g	−17.2	−59.8	336.02
HgI_2		aq	(−80.3)	(−74.9)	(172)
HgI_3^-		aq	(−153.6)	(−148.1)	(297)
HgI_4^{2-}		aq	(−236.0)	(−211.3)	(356)

TABLE 5 (Continued)

Formula	Description	State	ΔH_f°/kJ mol^{-1}	ΔG_f°/kJ mol^{-1}	S°/J mol^{-1} K^{-1}
Hg_2I_2		c	−121.34	−111.002	233.5
$Hg(IO_3)_2$		c		−167.2	
$Hg_2(IO_3)_2$		c		−179.9	
HgS	Red	c	−54.0	−46.4	82.4
HgS	Black	c	−50.2	−44.4	88.7
$HgSO_4$		c	−707.5	−594	142*
$HgSO_4$	Undiss	aq		(−587.9)	
Hg_2SO_4		c	−743.58	−626.34	200.66
HgSe		c	−45.2	−38.1	94.1
$HgSeO_3$		c		(−284.1)	
Hg_2SeO_3		c		−297.5	
HgTe		c	−33.9	−28.0	106.7
Hg_2CrO_4		c		−623.8	
$Hg(N_3)_2$	Undiss.	aq		774.0	
$Hg_2(N_3)_2$		c	592.0	746.4	208.8

Zinc, Cadmium, and Mercury

$Hg(NO_3)_2 \cdot 0.5H_2O$		c	−392.46	
$Hg_2(NO_3)_2 \cdot 2H_2O$		c	−868.18	
$Hg(NO_3)_2$	Undiss.	aq		−50.2
$Hg(NO_3)^+$		aq		51.5
$Hg(NO_2)_4^{2-}$		aq		−46.9
$(Hg_2N)_2O$		c	318.8	
$Hg(NH_3)^{2+}$		aq		87.9
$Hg(NH_3)_2^{2+}$		aq	(−94.6)	11.7	172
$Hg(NH_3)_3^{2+}$		aq	(−188.3)	−20.5	259
$Hg(NH_3)_4^{2+}$		aq	(−283.7)	−51.5	335
Hg_2HPO_4		c		−966.57	
$Hg_2P_2O_7^{2-}$		aq		−1820	
$Hg_2OH(P_2O_7)^{3-}$		aq		−2012	
$Hg_2(OH)_2(P_2O_7)^{4-}$		aq		−2197	
$Hg_2(P_2O_7)_2^{6-}$		aq		−3694	
Hg_2CO_3		c	−553.5	−468.2	180
$Hg(COO)_2$		c	−678.2	

TABLE 5 (Continued)

Formula	Description	State	ΔH_f°/kJ mol^{-1}	ΔG_f°/kJ mol^{-1}	S°/J mol^{-1} K^{-1}
Hg$_2$(COO)$_2$		c	−565.6	−593.8	162.0
Hg$_2$(COO)$_4^{2-}$		aq		−1234.3	
Hg$_2$(OH)(COO)$_2^-$		aq		−752.3	
Hg(CH$_3$)		g	167		
Hg(CH$_3$)$_2$		ℓ	59.8	140.2	209
Hg(CH$_3$)$_2$		g	94.39	146.0	305
Hg(CH$_3$)(C$_2$H$_5$)		ℓ	46.4		
Hg(C$_2$H$_5$)$_2$		ℓ	30.1		
Hg(C$_2$H$_5$)$_2$		g	75.3		
Hg$_2$(HCOO)$_2$		c	−751.41	−592.8	173.4
Hg(CH$_3$COO)$_2$		c	−816.7		
Hg$_2$(CH$_3$COO)$_2$		c	−838.4	−640.14	311.8
Hg(CH$_3$)Cl		c	−116.3		
Hg(CH$_3$)Cl		g	−52.3		
Hg(C$_2$H$_5$)Cl		c	−139.3		

Zinc, Cadmium, and Mercury

Hg(C₂H₅)Cl	g	−62.8		
HgCl₂·CH₃OH	c	(−474.9)	(−347.7)	243
HgCl₂·2CH₃OH	c	−720	−514.6	335
Hg(CH₃)Br	c	−85.8		
Hg(CH₃)Br	g	−18.4		
Hg(C₂H₅)Br	c	−106.7		
Hg(C₂H₅)Br	g	−30.1		
Hg(CH₃)I	c	−42.7		
Hg(CH₃)I	g	21.8		
Hg(C₂H₅)I	c	−65.7		
Hg(C₂H₅)I	g	13.8		
HgCN₂	c	276		
Hg(CN)₂	c	263.6		
Hg(CN)₂	g	381		
Hg(CN)₂	aq, Undiss	(277.4)	(312.5)	161.1
Hg(CN)⁺	aq	(224.7)	(238.5)	66.1
Hg(CN)₃⁻	aq	(396.2)	(463.6)	(219.7)

TABLE 5 (Continued)

Formula	Description	State	ΔH_f°/kJ mol^{-1}	ΔG_f°/kJ mol^{-1}	S°/J mol^{-1} K^{-1}
Hg(CN)Cl	Undiss.	aq		67.4	
Hg(CN)$_2$Cl$^-$		aq		(182.8)	
Hg(CN)$_3$Cl^{2-}		aq		335	
Hg(CN)Br	Undiss.	aq		84.1	
Hg(CN)$_3$Br^{2-}		aq		356	
Hg(CN)I	Undiss.	aq		121.3	
Hg(SCN)$_2$	Undiss.	aq	(195.4)	253.1	151
Hg(SCN)$^+$		aq		206.3	
Hg(SCN)$_3^-$		aq		329.7	
Hg(SCN)$_4^{2-}$		aq	(325.5)	411.7	452
Hg(SCN)OH	Undiss.	aq		461.9	
Hg(SCN)Cl	Undiss.	aq		(35.6)	
Hg(SCN)Br	Undiss.	aq		(51.5)	

Zinc, Cadmium, and Mercury

Compound	State		
Hg(SCN)(CN)$_3^{2-}$	aq		(553.5)
Hg$_2$(SCN)$_2$	c		226.4
Hg(ONC)$_2$ (fulminate)	c	268	
Hg(CH$_3$NH$_2$)$^{2+}$	aq		136.0
Hg(CH$_3$NH$_2$)$_2^{2+}$	aq	−55.6	104.2	265.3
HgCl(CH$_3$NH$_2$)$^+$	aq	−148.5	−34.7	168.6
Hg(NH$_2$CH$_2$COO)$_2$	aq Undiss.	(−860.2)	(−574.9)	(264)
Hg(NH$_2$CH$_2$COO)$^+$	aq		(−208.8)	
HgCl(NH$_2$CH$_2$COO)	aq Undiss.	(−545.2)	(−376.1)	(192)
Hg(CN)$_2$(tu) (tu = thiourea)	aq	206.3		
Hg(CN)$_2$(tu)$_2$	aq	105.0		
Hg(en)$_2^{2+}$ (en = enthylenediamine)	aq	−68.2		
HgCl(en)$^+$	aq	(−141.8)		

aValues in italic were taken from Ref. 11, 12, or 13. Parentheses indicate that the value is based on data of Ref. 11, adjusted to take into account the new values of $\Delta H_f^°$ and $S°$ for Hg^{2+}(aq). An asterisk indicates an estimated value.
Note: The most recent NBS tables [59] present rather different values of thermodynamic properties for some substances; however, for most reactions treated in this chapter the differences are not too significant.

For the electrode reaction

$$Hg_2Cl_2(c) + 2e^- \rightarrow 2Hg(\ell) + 2Cl^-(aq) \qquad E°_{(4)} = 0.26816 \text{ V} \qquad (10.4)$$

has been chosen as the best value of the standard electrode potential [9,10, 18]. Its isothermal temperature coefficeint is $dE°_{(4)}/dT = -0.298$ mV K^{-1}. Both these values are consistent with the values of $\Delta G°_f$ and $S°$ for $Hg_2Cl_2(c)$ in Table 5, with the use of $\Delta G°_f = -131.061$ kJ mol^{-1} and $S° = 56.48$ J mol^{-1} K^{-1} for Cl$^-$(aq) from Ref. 11, respectively.

From the recommended values of $E°_{(2)}$ and $E°_{(4)}$ the solubility product of $Hg_2Cl_2(c)$ in water, $K_{sp} = 1.42_7 \times 10^{-18}$, can be obtained, which is in an excellent agreement with the value $K_{sp} = 1.43_3 \times 10^{-18}$ recommended by Marcus [15] in his recent critical compilation of solubility data for this compound. This value is consistent with other reported thermodynamic quantities given in Table 5, but is slightly higher than that presented by Hepler and Olofson [10]. It also agrees very well with the directly measured value, $K_{sp} = 1.49 \times 10^{-18}$ [19].

With respect to the use of the calomel electrode as the most common reference electrode, it must be taken into account that the value of its potential is rather dependent on several factors contributing to the irreversibility of the system [dissolved oxygen; the mode of preparation of the calomel and its excess in the electrode; the wedge effect of the solution; calomel-mercury interaction; the aging effect (also called "first-day and second-day calomel electrode"); the geometry of the electrode design, in particular its area/solution volume ratio; etc.] [20,21]. Due to a certain temperature hysteresis, for exact measurements it is recommended that new cells be prepared and used at each temperature without thermal cyclization [21].

For the electrode reaction

$$Hg_2Br_2(c) + 2e^- \rightarrow 2Hg(\ell) + 2Br^-(aq) \qquad (10.5)$$

the best value of the standard electrode potential $E°_{(5)} = 0.13920$ V has been chosen as an average of the results of Gupta et al. [22] and Leuschke and Schwabe [23], using the new value of $E°_{(2)}$ in the recalculation. The value of $E°_{(5)}$ is consistent with $\Delta G°_f$ of $Hg_2Br_2(c)$ in Table 5, using $\Delta G°_f = -103.97$ kJ mol^{-1} for Br$^-$(aq) from Ref. 11. On the basis of entropy data given in Table 5 and using $S° = 82.42$ J mol^{-1} K^{-1} for Br$^-$(aq) (see Ref. 11), the isothermal temperature coefficeint $dE°_{(5)}/dT = -0.162$ mV K^{-1} can be obtained. Similarly, the solubility of $Hg_2Br_2(c)$ in water, accompanied by complete dissociation,

$$Hg_2Br_2(c) \rightarrow Hg_2^{2+}(aq) + 2Br^-(aq) \qquad (10.6)$$

is characterized by the solubility product $K_{sp} = 6.23 \times 10^{-23}$ calculated on the basis of appropriate thermodynamic data in Table 5. The independently found value [24] $K_{sp} = 6.43 \times 10^{-23}$ is in satisfactory agreement with that calculated here.

From thermodynamic data given in Table 5 and from $\Delta G°_f = -51.589$ kJ mol^{-1} for I$^-$(aq) (see Ref. 11), the standard electrode potential for the electrode reaction

$$Hg_2I_2(c) + 2e \to 2Hg(\ell) + 2I^-(aq) \qquad E°_{(7)} = -0.0405_4 \text{ V} \qquad (10.7)$$

may be calculated. This value is in very good agreement with that found experimentally by Bates and Vosburgh [25] ($E°_{(7)} = -0.0405$ V). From the thermodynamic data given in Table 5, from $\Delta G°_f$ for I^-(aq) given above, and from $S° = 111.3$ J mol^{-1} K^{-1} for the same species (see Ref. 11), a value for the temperature coefficient of $dE°_{(7)}/dT = 0.055$ mV K^{-1}, and a value for the solubility product of Hg_2I_2 (in water) $K_{sp} = 5.23 \times 10^{-29}$ may be calculated. The latter value is somewhat different from that obtained on the basis of data taken from NBS Technical Notes 270-3 and 270-4 only ($K_{sp} = 5.34 \times 10^{-29}$), owing to the difference in $\Delta G°_f$ for Hg_2^{2+}(aq) (see Table 5). (Hepler and Olofson [10] reported $K_{sp} = 5.19 \times 10^{-29}$, obtained by the same method of calculation.) For the thermodynamic properties of HgI(g) the revised values listed in NBS Technical Note 270-7 [12] were adopted in Table 5; values accepted by Hepler and Olofson [10] from the same source are somewhat erroneous.

E. Additional Mercury-Mercurous Salt Couples

From $\Delta G°_f$ for Hg_2CO_3(c) (see Table 5) and from $\Delta G°_f = -527.90$ kJ mol^{-1} for CO_3^{2-}(aq) [11] it follows for the electrode reaction

$$Hg_2CO_3(c) + 2e^- \to 2Hg(\ell) + CO_3^{2-}(aq) \qquad (10.8)$$

that $\Delta G°_{(8)} = -59.70$ kJ mol^{-1} and the corresponding standard electrode potential $E°_{(8)} = 0.3094$ V at 25°C. This value is quite different from the old one given by Latimer [26] but is in good agreement with the result of Saegusa [27]. With the use of the new thermodynamic data, the solubility product of Hg_2CO_3(c) in water is $K_{sp} = 3.54 \times 10^{-17}$.

In spite of a number of recent measurements for the electrode reaction

$$Hg_2SO_4(c) + 2e^- \to 2Hg(\ell) + SO_4^{2-}(aq) \qquad (10.9)$$

no excellent agreement for $E°_{(9)}$ has been obtained (for a review, see Ref. 10). On the basis of all these results, $E°_{(9)} = 0.613$ V may be accepted as the best value of the standard electrode potential. In combination with $\Delta G°_f = -744.63$ kJ mol^{-1} for SO_4^{2-}(aq) from Ref. 11, it leads to the value of $\Delta G°_f = -626.34$ kJ mol^{-1} for Hg_2SO_4(c) (see Table 5) as well as to the solubility product $K_{sp} = 6.5 \times 10^{-7}$ of this compound. On comparing reliable heat capacity measurements [28,29] leading to the $S°$ value for Hg_2SO_4(c) with published $dE°_{(9)}/dT$ results [30], it was concluded [28] that the $dE°/dT$ values were in some error. Therefore, the third-law entropy $S°$ for Hg_2SO_4(c) [28,29] listed in Table 5 was used for calculation of $\Delta H°_f$ of this compound. With the use of the mentioned $S°$ value for Hg_2SO_4(c) and of $S° = 20.08$ J mol^{-1} K^{-1} for SO_4^{2-}(aq) (see Ref. 11), the isothermal temperature coefficient $dE°_{(9)}/dT = -0.825$ mV K^{-1} can be calculated. This value is, however, in quite good agreement with the value of this quantity, $dE°_{(9)}/dT = -0.802$ mV K^{-1}, derived from the original tabulated data of $E°$ at various temperatures by Harned and Hamer [30], after their recalculation and conversion to absolute volts.

For the electrode reaction

$$Hg_2(IO_3)_2(c) + 2e^- \to 2Hg(\ell) + 2IO_3^-(aq) \qquad (10.10)$$

two earlier determined values of standard electrode potential exist [31,32]. After recalculating to absolute volts and with the use of the new value of $E^°_{(2)}$, one obtains the standard electrode potential $E^°_{(10)} = 0.3947$ V. This value leads to the corresponding solubility product $K_{sp} = 2.7 \times 10^{-14}$ for $Hg(IO_3)_2(c)$, which agrees quite well with the higher of the two values given by Sillén and Martell [33] (1.95×10^{-14}). The lower one, accepted [33] from Brodsky's calculation [34] (1.3×10^{-18}) on the basis of measurements of Spencer [35], must be denoted as erroneous. The recalculation of Spencer's original data (given in normal and not in standard electrode potentials) for reactions (10.2) and (10.10) leads to $K'_{sp} = 3.7 \times 10^{-14}$, which corresponds better to the new standard value given above.

On the basis of $E^°_{(10)}$ and $\Delta G^°_f = -128.03$ kJ mol^{-1} for IO_3^-(aq) [11], the value of $\Delta G^°_f = -179.9$ kJ mol^{-1} for $Hg_2(IO_3)_2$(c) has been obtained (see Table 5). Its comparison with analogous value from NBS tables [11] is unfortunately impossible, as it is not given there, perhaps due to the uncertainty mentioned in the value of K_{sp} on the basis of earlier calculations.

For the electrode reaction

$$Hg_2HPO_4(c) + 2e^- \rightarrow 2Hg(\ell) + HPO_4^{2-} (aq) \tag{10.11}$$

from two measurements of the standard electrode potential [36,37], the more exact experimentally found values of Larson [37] have been used as a basis of complete recalculation of $E°$, taking into account some uncertainty in the first dissociation constant of orthophosphoric acid (it moves, according to newer data [33,38], in the range from 7.11×10^{-3} to 7.59×10^{-3}). After this recalculation, $E^°_{(11)} = 0.635_8$ V as the best value of the standard electrode potential has been obtained, in good agreement with the original value given by Larson [37] ($E^°_{(11)} = 0.6359$ V), even though some of his partial results of calculations are somewhat erroneous. This value, in combination with $E^°_{(2)}$, leads to the theoretical solubility product of Hg_2HPO_4(c) in water, $K_{sp} = 3.84 \times 10^{-6}$. This value is completely different from both values given by Sillén and Martell [33]. [The first one, log $K_{sp} = -12.4$, with reference to lower exact measuremens [36], seems, however, compared with the new calculated value, to be given in natural rather than in common (Briggs) logarithms; the second value, log $K_{sp} = -14.52$, calculated by Brodsky [34], is based on very old and low exact measurements of Immerwahr [39], and may be rejected.] The value $K_{sp} = 3.84 \times 10^{-6}$ also differs from the value reported by Hills and Ives [40] ($K_{s\bar{p}} = 4 \times 10^{-3}$, with reference to the same sources [34,36]). The latter value seems to fit more closely the solubility rather than the solubility product of the compound mentioned, although the exactness of its estimation is not clear.

On the basis of $E^°_{(11)}$ and $\Delta G^°_f = -1089.26$ kJ mol^{-1} for HPO_4^-(aq) from Ref. 11, the value $\Delta G^°_f = -966.57$ kJ mol^{-1} for Hg_2HPO_4(c) has been obtained, as listed in Table 5. Its comparison with the analogous value from NBS tables [11] is, however, impossible, as it is not given there, perhaps due to reasons similar to those in the case of reaction (10.10).

For the electrode reaction

$$Hg_2(HCOO)_2(c) + 2e^- \rightarrow 2Hg(\ell) + 2HCOO^-(aq) \tag{10.12}$$

the standard electrode potential $E^°_{(12)} = 0.5664$ V has been redetermined [41] (the older paper of Chauchard and Gauthier [42] reported $E^°_{(12)} = 0.567$ V).

This value, in combination with $\Delta G_f^{\circ} = -351.0$ kJ mol^{-1} for HCOO$^-$(aq) from Ref. 11, leads to $\Delta G_f^{\circ} = -592.8$ kJ mol^{-1} for Hg$_2$(HCOO)$_2$(c), as listed in Table 5, and in combination with $E_{(2)}^{\circ}$ to the solubility product $K_{sp} = 1.73 \times 10^{-8}$ mol^2 kg^{-2}. Original calculations [41] of other thermodynamic properties of this compound were recalculated with the use of an experimentally determined average value of $dE_{(12)}^{\circ}/dT = 0.167$ mV K^{-1} (see Ref. 41) and resulting values, somewhat different from those obtained originally [41], are listed in Table 5.

For the mercury-mercurous acetate couple

$$Hg_2(CH_3COO)_2(c) + 2e^- \rightarrow 2Hg(\ell) + 2CH_3COO^-(aq) \quad (10.13)$$

the best value of the standard electrode potential $E_{(13)}^{\circ} = 0.5113$ V has been chosen as an average value of recent recalculated results [43-45]. This value, together with $\Delta G_f^{\circ} = -369.40$ kJ mol^{-1} for CH$_3$COO$^-$(aq) [11], leads to $\Delta G_f^{\circ} = -640.14$ kJ mol^{-1} for Hg$_2$(CH$_3$COO)$_2$(c) (see Table 5), and to the solubility product of this compound, $K_{sp} = 2.35 \times 10^{-10}$. The average temperature coefficient of the reaction [12] determined on the basis of mentioned measurements [43-45], $dE_{(13)}^{\circ}/dT = 0.607$ mV K^{-1}, leads to $S^{\circ} = 311.8$ J mol^{-1} K^{-1} and $\Delta H_f^{\circ} = -838.4$ kJ mol^{-1} for Hg$_2$(CH$_3$COO)$_2$(c) (see Table 5), values that are in good agreement with values estimated by Hepler and Olofson [10] for both quantities (ca. 310 J mol^{-1} K^{-1} and -841 kJ mol^{-1}, respectively).

For the mercury-mercurous propionate couple

$$Hg_2(C_2H_5COO)_2(c) + 2e^- \rightarrow 2Hg(\ell) + 2(C_2H_5COO)^-(aq) \quad (10.14)$$

no sufficient agreement has been found for the standard electrode potential $E_{(14)}^{\circ}$. Chauchard and Gauthier [42] give 0.499 V, Basu and Aditya [46] give 0.5031 V, and recently Lebedj and Bondarev [45] give 0.51293 V for this quantity at 25°C. A similar situation exists in the case of the isothermal temperature coefficient $dE_{(14)}^{\circ}/dT$, for which the following values have been found: -0.856 (Ref. 46) and -0.686 mV K^{-1} [45]. The differences between individual values of both quantities are too high to make it possible to choose one as the best. According to Ref. 45, these data should be more reliable than those given in Ref. 46; however, neither the numerical values of individual measured data nor the detailed manner of their treatment are given, so that there is no possibility of checking them. Taking an approximate value $E_{(14)}^{\circ} = 0.51$ V for the standard electrode potential, one can obtain, with the use of $E_{(2)}^{\circ}$, a theoretical approximate value of $K_{sp} = 2.1 \times 10^{-10}$ for the solubility product of Hg$_2$(C$_2$H$_5$COO)$_2$(c). This value differs from that given originally by Chauchard and Gauthier [42] ($K_{sp} = 1.3 \times 10^{-11}$), which, however, is based on an erroneous calculation (their corrected value should be $K_{sp} = 7.4 \times 10^{-11}$).

For the mercury-mercurous oxalate couple

$$Hg_2(COO)_2(c) + 2e^- \rightarrow 2Hg(\ell) + (COO)_2^{2-}(aq) \quad (10.15)$$

from two measured values of the standard electrode potential [47,48], the latter, $E_{(15)}^{\circ} = 0.4158$ V, has been adopted as the best value, since it agrees better with the experimentally determined difference of standard electrode potentials of the mercurous oxalate-silver oxalate couple [48]. With the use of $\Delta G_f^\circ = -674.04$ kJ mol^{-1} for (COO)$_2^{2-}$(aq) according to Ref. 11, the value of $\Delta G_f^\circ = -593.8$ kJ mol^{-1} for Hg$_2$(COO)$_2$(c) has been calculated (see Table 5), which is somewhat different from that given in Ref. 11 (-593.3 kJ mol^{-1}). From the isothermal temperature coefficient $dE^\circ_{(15)}/dT = -0.49$ mV K^{-1} at 25°C, derived from measurements [48], and the value $S^\circ = 45.6$ J mol^{-1} K^{-1} for (COO)$_2^{2-}$(aq) [11], it follows that $S^\circ = 162$ J mol^{-1} K^{-1} and $\Delta H_f^\circ = -565.6$ kJ mol^{-1} for Hg$_2$(COO)$_2$(c), as listed in Table 5. The solubility product of this compound, $K_{sp} = 1.4 \times 10^{-13}$, calculated on the basis of the new data given, is more exact than that calculated formerly by Brodsky [34] and taken over by Latimer [26].

For the mercury-mercurous benzoate couple

$$Hg_2(C_6H_5COO)_2(c) + 2e^- \rightarrow 2Hg(\ell) + 2C_6H_5COO^-(aq) \qquad (10.16)$$

a very good consistency of individual data for $E^\circ_{(16)}$ exists between two measurements [49,50] at temperatures in the range 20 to 40°C, from which $E^\circ_{(16)} = 0.4263$ V at 25°C, and $dE^\circ_{(16)}/dT = -0.774$ mV K^{-1} at the same temperature [50]. The recent measurement [45] of this couple for the temperature range 10 to 50°C gives another series of results differing quite remarkably from the former ones [49,50], as the new standard electrode potential $E^\circ_{(16)} = 0.42994$ V at 25°C, and $dE^\circ_{(16)}/dT = -0.710$ mV K^{-1} have been found, respectively [45]. As, unfortunately, no numerical values of individual measured data are given in Ref. 45, it is impossible to choose a more reliable value of $E^\circ_{(16)}$ from both given here. It must only be noted that in the original paper [45], the coefficients C in the equation for the temperature dependence of E° for the electrode reactions (10.14) and (10.16) must have the opposite sign ("minus"). Further, from the temperature coefficient $dE^\circ_{(16)}/dT = -0.774$ mV K^{-1} [50], it follows that $\Delta S^\circ_{(16)} = -149.4$ J mol^{-1} K^{-1}, and hence, using $E^\circ_{(16)} = 0.4263$ V, $\Delta H^\circ_{(16)} = -126.8$ kJ mol^{-1} at 25°C, whereas the originally reported values of these quantities [49] are incorrect. Combining both given values of $E^\circ_{(16)}$ with $E^\circ_{(2)}$, the solubility product of mercurous benzoate, $K_{sp} = (3.2 - 4.2) \times 10^{-13}$ at 25°C, may be obtained.

For the mercury-mercurous picrate couple

$$Hg_2(C_6H_2(NO_2)_3O)_2(c) + 2e^- \rightarrow 2Hg(\ell) + 2C_6H_2(NO_2)_3O^-(aq) \qquad (10.17)$$

the standard electrode potential $E^\circ_{(17)} = 0.4924$ V has been obtained with the use of a sodium-responsive glass electrode [51]. Combining $E^\circ_{(17)}$ with $E^\circ_{(2)}$ one obtains $K_{sp} = 5.44 \times 10^{-11}$ for the solubility product of mercurous picrate in water, in quite good agreement with the value $K_{sp} = 4.9 \times 10^{-11}$ determined spectrophotometrically by the same authors [51].

F. Mercurous-Mercuric Couple

In accordance with the thorough revision by Vanderzee and Swanson [9] of a number of electrochemical investigations, for the electrode reaction

$$2Hg^{2+}(aq) + 2e^- \rightarrow Hg_2^{2+}(aq) \qquad (10.18)$$

$E^\circ_{(18)} = 0.9110$ V has been adopted as the best value of the standard electrode potential.

Zinc, Cadmium, and Mercury

G. Mercury-Mercuric Ion Couple

In accordance with the thorough revision by Vanderzee and Swanson [9], for the electrode reaction

$$Hg^{2+}(aq) + 2e^- \rightarrow 2Hg(\ell) \tag{10.19}$$

$E^°_{(19)} = 0.8535$ V has been adopted as the best value of the standard electrode potential.

Combination of the values of $E^°_{(19)}$ and $E^°_{(2)}$ leads to the equilibrium constant of the disproportionation reaction

$$Hg^{2+}(aq) + Hg(\ell) \rightarrow Hg_2^{2+}(aq) \qquad K_{(20)} = 87.9 \tag{10.20}$$

The experimentally found value for this equilibrium, $K_{(20)} = 90$ [42], is in good accordance with the calculated value. This equilibrium is readily reversible and if a mercuric compound with a given anion is much less soluble than the corresponding mercurous compound, the latter becomes unstable with respect to decomposition into mercury and the mercuric compound.

Since the Hg^{2+}/Hg_2^{2+} couple is a somewhat stronger oxidizing agent than the Hg^{2+}/Hg couple, a reducing agent will first reduce mercuric ion to mercurous ion, but as the values of standard potential for both couples are relatively close, an excess of reducing agent will take the mercurous ion down to mercury. It also follows from these potentials that fairly powerful oxidizing agents, such as bromine water or hot nitric acid, are required to oxidize mercurous ion to mercuric one.

H. Mercury-Mercuric Oxide Couple

For the electrode reaction

$$HgO(c,red) + 2H^+(aq) + 2e^- \rightarrow Hg(\ell) + H_2O(\ell) \tag{10.21}$$

it follows from tabulated data [11] that the best value of the standard electrode potential, $E^°_{(21)} = 0.9256$ V, may be adopted, as was concluded in a critical review [10]. The same selection was accepted for other thermodynamic values of HgO(c,red,orthorhombic) listed in Table 5.

For the same couple in basic solution,

$$HgO(c,red) + H_2O(\ell) + 2e^- \rightarrow Hg(\ell) + 2OH^-(aq) \tag{10.22}$$

it follows that $E^°_{(22)} = 0.0977$ V. This value, in combination with $E^°_{(19)}$, leads to the solubility product $K_{sp} = 2.8 \times 10^{-26}$ for HgO(c,red) in water, which is in reasonable agreement with values obtained from independent non-emf solubility measurements (for a review, see Ref. 10).

I. Mercury-Mercuric Sulfide Couple

With the use of $\Delta G^°_f$ values for both forms of HgS from Table 5, and from $\Delta G^°_f = -27.87$ kJ mol^{-1} for H$_2$S(aq,undiss.) from Ref. 11, the following values of standard electrode potentials of electrode reactions can be obtained:

$$\text{HgS(c,red)} + 2\text{H}^+(\text{aq}) + 2e^- \to \text{Hg}(\ell) + \text{H}_2\text{S}(\text{aq}) \qquad E^\circ_{(23)} = -0.096 \text{ V}$$
(10.23)

$$\text{HgS(c,black)} + 2\text{H}^+(\text{aq}) + 2e^- \to \text{Hg}(\ell) + \text{H}_2\text{S}(\text{aq}) \qquad E^\circ_{(24)} = -0.085 \text{ V}$$
(10.24)

Although at "ordinary" temperatures the red form is thermodynamically more stable than the black form (see Table 5), it is the black form that is usually precipitated from aqueous solutions. Therefore, in most tables of standard electrode potentials [14,26,53,54] only values for the black form of HgS have been listed.

Corresponding calculation leads to the solubility product $K_{sp} = 2 \times 10^{-52}$ for HgS(c,black) in water [10]. There is, however, evidence that the second dissociation constant of $\text{H}_2\text{S}(\text{aq})$ is not 1×10^{-13} (which is consistent with ΔG°_f for $\text{S}^{2-}(\text{aq})$ according to Ref. 11), but 10^{-17} or even smaller [55]. Because the determinations of ionization constants of $\text{H}_2\text{S}(\text{aq})$ in both series seem to be carried out reasonably, there is, at present, no satisfactory way to resolve this problem. Besides, there is convincing evidence for the existence of complex species $\text{HgS}_2^{2-}(\text{aq})$, $\text{Hg(HS)}_2(\text{aq})$, $\text{HgS(HS)}_2^{2-}(\text{aq})$, and so on [56,57]. Because of combined uncertainties in the Gibbs energies of HgS(c,red) and HgS(c,black) and in the second ionization constant of $\text{H}_2\text{S}(\text{aq})$, no other reactions with related equilibrium constants for complex sulfide species have been considered here.

J. Additional Mercury–Mercuric Ion Couple

For the electrode reaction

$$\text{Hg(IO}_3)_2(\text{c}) + 2e^- \to \text{Hg}(\ell) + 2\text{IO}_3^-(\text{aq}) \tag{10.25}$$

the standard electrode potential $E^\circ_{(25)} = 0.460_4$ V may be adopted in accordance with the recent determination by Nash [58], after having accepted the new value of $E^\circ_{(10)} = 0.3947$ V. This leads to the solubility product $K_{sp} = 5 \times 10^{-14}$ for $\text{Hg(IO}_3)_2(\text{c})$ in water at 25°C. As in the preceding cases, one obtains $\Delta G^\circ_f = -167.2$ kJ mol^{-1} for $\text{Hg(IO}_3)_2(\text{c})$, as listed in Table 5.

For the standard electrode potential of reaction (10.25) the value $E^\circ_{(25)} = 0.394$ V was given by Latimer [26] with reference to the paper of Takacs [32]; in the paper quoted [32], however, only reaction (10.10) has been investigated, not reaction (10.25). Hence the corresponding value of $E^\circ_{(10)}$ found by Takacs has erroneously been applied by Latimer [26] to reaction [10.25].

K. Potential Diagrams for Mercury

Acid Solution

```
      +2              +1             0
       ┌────────── 0.8535 ──────────┐
       │                            │
      Hg²⁺  ── 0.9110 ── Hg₂²⁺ ── 0.7960 ── Hg
```

Basic Solution

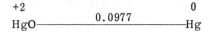

REFERENCES

1. H. C. Moser and A. F. Voigt, J. Am. Chem. Soc., **79**, 1837 (1957).
2. S. Fujita, H. Horii, and S. Taniguchi, J. Phys. Chem., **77**, 1868 (1973).
3. S. Fujita, H. Horii, T. Mori, and S. Taniguchi, J. Phys. Chem., **79**, 960 (1975).
4. B. D. Cutforth, C. G. Davies, P. A. W. Dean, R. J. Gillespie, P. R. Ireland, and P. K. Ummat, Inorg. Chem., **12**, 1343 (1973).
5. I. D. Brown, B. D. Cutforth, G. J. Davies, R. J. Gillespie, and P. R. Ireland, Can. J. Chem., **52**, 791 (1974).
6. M. T. Kozlovskii, A. I. Zebreva, and V. P. Gladyshev, *Amalgamy i ikh primenenie*, Izdatelstvo "Nauka," Alma-Ata, 1971,
7. L. F. Kozin, R. Sh. Nigmetova, and M. B. Dergacheva, *Termodinamika binarnykh amalgamnykh sistem*, Izdatelstvo "Nauka," Alma-Ata, 1977.
8. V. A. Smirnov, *Vosstanovlenie amalgamami*, Izdatelstvo "Khimiya," Leningrad, 1970.
9. C. E. Vanderzee and J. A. Swanson, J. Chem. Thermodyn., **6**, 827 (1974).
10. L. G. Hepler and G. Olofson, Chem. Rev., **75**, 585 (1975).
11. D. D. Wagman, W. H. Evans, V. B. Parker, I. Halow, S. M. Bailey, and R. H. Schumm, *Selected Values of Chemical Thermodynamic Properties*, Nat. Bur. Stand. Tech. Notes 270-3 and 270-4, U.S. Government Printing Office, Washington, D.C., 1968 and 1969, resp.
12. R. H. Schumm, D. D. Wagman, S. M. Bailey, W. H. Evans, and V. B. Parker, *Selected Values of Chemical Thermodynamic Properties*, Nat. Bur. Stand. Tech. Note 270-7, U.S. Government Printing Office, Washington, D.C., 1973.
13. D. R. Stull and H. Prophet, *JANAF Thermochemical Tables*, 2nd ed., NSRDS-NBS 37, U.S. Government Printing Office, Washington, D.C., 1971.
14. G. Milazzo and S. Caroli, *Tables of Standard Electrode Potentials*, Wiley, New York, 1978.
15. Y. Marcus, J. Phys. Chem. Ref. Data, **9**, 1307 (1980).
16. N. Tanaka, R. Tamamushi, Electrochim. Acta, **9**, 963 (1964).
17. P. Bindra, A. P. Brown, M. Fleischmann, and D. Pletcher, J. Electroanal. Chem., **58**, 31, 39 (1975).
18. J. C. Ahluwahlia and J. W. Cobble, J. Am. Chem. Soc., **86**, 5381 (1964).
19. W. J. Galloway, Thesis, University of New Zeeland (Canterbury), 1961 (cited in Ref. 10).
20. G. J. Hills and D. J. G. Ives, in *Reference Electrodes*, D. J. G. Ives and G. J. J. Janz, eds., Academic Press, New York, 1961, p. 127.
21. A. K. Covington, J. V. Dobson, and W. F. K. Wynne-Jones, Electrochim. Acta, **12**, 513, 525 (1967).
22. S. R. Gupta, G. J. Hills, and D. J. G. Ives, Trans. Faraday Soc., **59**, 1886 (1963).

23. W. Leuschke and K. Schwabe, J. Electroanal. Chem., *25*, 219 (1970).
24. A. J. Read, Thesis, University of New Zeeland (Canterbury), 1960 (cited in Ref. 10).
25. R. G. Bates and W. C. Vosburgh, J. Am. Chem. Soc., *59*, 1188 (1937).
26. W. M. Latimer, *Oxidation Potentials*, 2nd ed., Prentice-Hall, Englewood Cliffs, N.J., 1952, p. 175.
27. F. Saegusa, Sci. Rep. Tohoku Univ., Ser. 1., *34*, 1, 55 (1950).
28. T. E. Brackett, E. W. Hornung, and T. E. Hopkins, J. Am. Chem. Soc., *82*, 4155 (1960).
29. M. N. Papadopoulos and W. F. Giauque, J. Phys. Chem., *66*, 2049 (1962).
30. H. S. Harned and W. J. Hamer, J. Am. Chem. Soc., *57*, 27 (1935).
31. M. M. Haring and P. P. Zapponi, Trans. Electrochem. Soc., *80*, 203 (1941).
32. I. Takacs, Magy. Chem. Foly., *49*, 33, 100 (1943).
33. L. G. Sillén and A. E. Martell, *Stability Constants*, Spec. Publ. 17, The Chemical Society, London, 1964; *Stability Constants Supplement 1*, Spec. Publ. 25, London, Chemical Society, 1971.
34. A. N. Brodsky, Z. Elektrochem., *35*, 833 (1929).
35. J. F. Spencer, Z. Phys. Chem., *80*, 701 (1912).
36. T. De Vries and D. Cohen, J. Am. Chem. Soc., *71*, 1114 (1949).
37. W. D. Larson, J. Phys. Chem., *54*, 310 (1950).
38. D. D. Perrin, *Dissociation Constants of Inorganic Acids and Bases in Aqueous Solutions*, Butterworth, London, 1969, p. 189.
39. G. Immerwahr, Z. Elektrochem., *7*, 477 (1901).
40. G. J. Hills and D. J. G. Ives, in *Reference Electrodes*, D. J. G. Ives and G. J. J. Janz, eds., Academic Press, New York, 1961, p. 169.
41. N. I. Ostannii, T. A. Zharkova, B V. Erofeev, and G. D. Kazakova, Zh. Fiz. Khim., *48*, 358 (1974); Russ. J. Phys. Chem. (Engl. trans.), *48*, 209 (1974).
42. J. Chauchard and J. Gauthier, Compt. Rend. Acad. Sci. Paris, *261*, 1528 (1965).
43. A. K. Covington, P. K. Talukdar, and H. R. Thirsk, Trans. Faraday Soc., *60*, 412 (1964).
44. B. K. Choudhary and B. Prasad, Indian J. Chem., *11*, 931 (1973).
45. V. J. Lebedj and N. K. Bondarev, Elektrokhimiya, *17*, 398 (1981).
46. A. K. Basu and S. Aditya, Electrochim. Acta, *23*, 1341 (1978).
47. W. D. Larson and W. J. Tomsicek, J. Am. Chem. Soc., *63*, 3329 (1941).
48. D. T. Ferrell, Jr., I. Blackburn, and W. C. Vosburgh, J. Am. Chem. Soc., *70*, 3812 (1948).
49. J. Bertram and S. J. Bone, Trans. Faraday Soc., *63*, 415 (1967).
50. T. P. Russell and J. F. Reardon, J. Chem. Eng. Data, *22*, 370 (1977).
51. A. K. Covington and K. V. Srinivasan, J. Chem. Thermodyn., *3*, 795 (1971).
52. R. H. McKeown, Thesis, University of New Zeeland (Canterbury), 1962 (cited in Ref. 10).
53. A. J. de Béthune and N. A. Swendeman-Loud, *Standard Aqueous Electrode Potentials and Temperature Coefficients at 25°C*, Hampel, Skokie, Ill., 1964.

54. G. Charlot, A. Collumea, and M. J. C. Marchon, *Selected Constants: Oxidation-Reduction Potentials of Inorganic Substances in Aqueous Solution*, Butterworth, London, 1971, p. 23.
55. W. Giggenbach, Inorg. Chem., *10*, 1333 (1971).
56. G. Schwarzenbach and M. Widmer, Helv. Chim. Acta, *46*, 2613 (1963).
57. H. L. Barnes, S. B. Romberger, and M. Stemprok, Econ. Geol. Bull. Soc. Econ. Geol., *62*, 957 (1967) (cited in Ref. 10).
58. Ch. P. Nash, J. Electrochem. Soc., *125*, 875 (1978).
59. D. D. Wagman, W. H. Evans, V. B. Parker, R. H. Schumm, I. Halow, S. M. Bailey, K. L. Churney, and R. L. Nuttall, *NBS Tables of Chemical Thermodynamic Properties*, J. Phys. Chem. Ref. Data, *11*, Supplement 2 (1982).

11
Copper, Silver, and Gold

I. COPPER*

UGO BERTOCCI AND DONALD D. WAGMAN[†] National Bureau of Standards, Gaithersburg, Maryland

Copper forms compounds in two oxidation states, +1 and +2. The +3 oxidation state is also known, but trivalent copper tends to be unstable in aqueous solutions.

A. Thermodynamic Data

Thermodynamic data for copper and its compounds at 298.15 K are collected in Table 1. The values have been taken mainly from a book by King et al. [1], a paper by Hepler and co-workers [2], and National Bureau of Standards (NBS) Technical Note 270-4 [3]. CODATA Bulletin 28 [4] has also been consulted. These data are also in agreement with those reported in a more recent work [5]. However, the ΔG_f^o of $Cu(NH_3)_2^+$(aq) has been calculated from the stability constant given by Couturier and Petitfaux [6] and that of $Cu(CN)_2^-$(aq) from the work of Kappenstein and Hugel [7].

For gaseous cuprous hydride and for the cuprous halides, the source is a book by Huber and Herzberg [8]. The ΔH_f^o for crystalline CuH is taken from a paper by Warf [9].

The values for the copper oxides are somewhat uncertain, particularly for CuO. Recent, accurate thermodynamic measurements [10] give ΔH_f^o for CuO(c) in granular form as -161.5 kJ mol^{-1} (-38.6 kcal mol^{-1}). This is the value reported in Table 1. However, values more positive by 1 to 2 kcal mol^{-1} have been obtained for fine powders, and still more positive values are deduced from measurements of aqueous equilibria [11].

B. Equilibrium Potentials

Standard potentials calculated from the data in Table 1 are listed in Table 2. The thermodynamic data necessary for the calculation of the E^o values not in Table 1 have been taken either from Ref. 4, 12, or 13.

Reviewers: L. Hepler, University of Lethbridge, Alberta, Canada; Y. Okinaka, Bell Telephone Laboratories, Murray Hill, New Jersey.
[†]Retired.

TABLE 1 Thermodynamic Data on Copper[a]

Substance	State	ΔH_f°/kJ mol^{-1}	ΔG_f°/kJ mol^{-1}	S°/J mol^{-1} K^{-1}
Cu	g	338	298.3	166.52
Cu	c	0	0	33.2
Cu$^+$	g	1090.7	1052.5	160.8
Cu$^+$	aq	72.1	50.3	41
Cu^{2+}	g	3057.5	3015	173.2
Cu^{2+}	aq	65.78	65.7	−97.2
CuH	g	288.7	259.4	196.6
CuH	c	33.1		
Cu$_2$O	c	−171.0	−148.1	92.52
CuO	c	−161.7	−134	42.7
Cu(OH)$_2$	c	−443.7	−359.5	87.2
HCuO$_2^-$	aq		−258.9	
CuO$_2^{2-}$	aq		−183.9	
CuF	g	−13.0	−40.2	226.7

Copper, Silver, and Gold

CuF$_2$	g	−281.6	−287.4	255.5
CuF$_2$	c	−545.5	−496.5	71
CuF$_2 \cdot$2H$_2$O	c	−1158.5	−993.9	151
CuF$^+$	aq	−261.5	−220.8	−67
CuCl	g	76.7	70.0	237.4
CuCl	c	−137.0	−119.8	87.6
CuCl$_2$	c	−218.3	−173.9	108.2
CuCl$_2 \cdot$2H$_2$O	c	−818.3	661.2	195.7
CuCl$_2^-$	aq		−240.5	(201)
KCuCl$_3$	c	−667	−603	(281)
K$_2$CuCl$_4$	c	−1104	−1006	355.9
K$_2$CuCl$_4 \cdot$2H$_2$O	c	−1697	−1483	248.6
CuBr	g	116	74.6	
CuBr	c	−105	−101	96.20
CuBr$_2$	c	−140	−123	129.0
CuBr$_2 \cdot$4H$_2$O	c	−1322		

TABLE 1 (Continued)

Substance	State	ΔH_f°/kJ mol^{-1}	ΔG_f°/kJ mol^{-1}	S°/J mol^{-1} K^{-1}
CuI	g	126.5	77.5	256.0
CuI	c	−67.9	−69.6	96.4
Cu(IO$_3$)$_2$	c	−388	−226	222
Cu(IO$_3$)$_2$·H$_2$O	c	−690	−469	257
CuSO$_4$	c	−772.2	−663	109
CuSO$_4$·H$_2$O	c	−1084	−916.4	145.3
CuSO$_4$·3H$_2$O	c	−1684	−1399	222.5
CuSO$_4$·5H$_2$O	c	−2280	−1880	301.7
Cu$_2$S	c, α	−81.3	−87.6	118.8
CuS	c	−52.8	−53.2	66.96
Cu$_2$SO$_4$	c	−748		
Cu(NO$_3$)$_2$	c	−302	−109	193

CuN_3	c	280.3	366	102
$Cu(NH_3)_2^+$	aq		−63	
$Cu(NH_3)_2^{2+}$	aq	−39	15.6	12
$Cu(NH_3)_2^{2+}$	aq	−142	−30.5	111.5
$Cu(NH_3)_3^{2+}$	aq	−246	−73.2	200
$Cu(NH_3)_4^{2+}$	aq	−349	−111.5	274
$CuCN$	c	95.1	108.5	90.13
$Cu(CN)_2^-$	aq	274	302	205
$Cu(CN)_3^{2-}$	aq	334	401	243
$Cu(CN)_4^{3-}$	aq	437.8	565	209
$CuSCN$	c		63	
$Cu(SCN)^+$	aq	129.5	145	48.2
$Cu(OH)_2CO_3$	Malachite	−1055	−901	186
$2CuCO_3 \cdot Cu(OH)_2$	Azurite	−1630	−1437	

[a]Estimated values are in parentheses.

TABLE 2 Standard Potentials of Copper

Electrode reaction	$E°$/V
1. $Cu^+(aq) + e^- \rightarrow Cu(c)$	0.520
2. $Cu^{2+}(aq) + 2e^- \rightarrow Cu(c)$	0.340
3. $Cu^{2+}(aq) + e^- \rightarrow Cu^+(aq)$	0.159
4. $Cu_2O(c) + H_2O + 2e^- \rightarrow 2Cu(c) + 2OH^-(aq)$	−0.365
5. $2CuO(c) + H_2O + 2e^- \rightarrow Cu_2O(c) + 2OH^-(aq)$	−0.22
6. $CuO(c) + H_2O + 2e^- \rightarrow Cu + 2OH^-(aq)$	−0.29
7. $CuCl(c) + e^- \rightarrow Cu(c) + Cl^-(aq)$	0.121
8. $Cu^{2+}(aq) + Cl^-(aq) + e^- \rightarrow CuCl(c)$	0.559
9. $CuBr(c) + e^- \rightarrow Cu(c) + Br^-(aq)$	0.033
10. $Cu^{2+}(aq) + Br^-(aq) + e^- \rightarrow CuBr(c)$	0.654
11. $CuI(c) + e^- \rightarrow Cu(c) + I^+(aq)$	−0.182
12. $Cu^{2+}(aq) + I^-(aq) + e^- \rightarrow CuI(c)$	0.861
13. $Cu(NH_3)_2^+(aq) + e^- \rightarrow Cu(c) + 2NH_3(aq)$	−0.100
14. $Cu(NH_3)_4^{2+}(aq) + e^- \rightarrow Cu(NH_3)_2^+(aq) + 2NH_3(aq)$	0.10
15. $Cu_2S(c,\alpha) + 2e^- \rightarrow 2Cu + S^{2-}(aq)$	0.898
16. $2CuS(c) + 2e^- \rightarrow Cu_2S(c,\alpha) + S^{2-}(aq)$	−0.542
17. $Cu(CN)_2^-(aq) + e^- \rightarrow Cu(c) + 2CN^-(aq)$	−0.44
18. $Cu^{2+}(aq) + 2CN^-(aq) + e^- \rightarrow Cu(CN)_2^-(aq)$	1.12
19. $CuCNS(c) + e^- \rightarrow Cu(c) + CNS^-(aq)$	−0.310
20. $CuCNS^+(aq) + e^- \rightarrow CuCNS(c)$	0.854

Experimental determinations of the equilibrium potential for reaction 2 in Table 2 have given values ranging from 0.342 to 0.347 V for the standard potential. Some of these differences have been attributed to the effect of crystal orientation of the metallic copper. For a more detailed discussion of this point, see Ref. 14. The range of values mentioned for reaction (2) is indicative of the uncertainty of the most reliable values collected in Table 2.

Reflecting the value of $\Delta G_f°$ for CuO(c) given in Table 1, the $E°$ values of reactions 5 and 6 in Table 2 are quite different from earlier values. It is unclear, however, whether the $E°$ values calculated from thermochemical data can be experimentally observed, since the solid CuO in contact with aqueous solutions probably differs sufficiently from the material employed in dry oxidation and reduction experiments as to give rise to quite different electrode potentials.

The $E°$ values of reactions involving ammonia and cyanide complexes are believed to be uncertain by some decivolts. Since Cu^{2+} is complexed by SCN^- [15], reaction 20 in Table 2 is given in a form different from that originally published by Latimer.

C. Potential Diagram for Copper

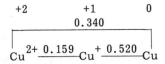

REFERENCES

1. E. G. King, A. D. Mah, and L. B. Pankratz, *Thermodynamic Properties of Copper and Its Inorganic Compounds*, International Copper Research Association, 1973.
2. L. M. Gedansky, E. M. Woolley, and L. G. Hepler, J. Chem. Thermodyn., 2, 561 (1970).
3. D. D. Wagman, W. H. Evans, V. B. Parker, I. Halow, S. M. Bailey, and R. H. Schumm, Natl. Bur. Stand. Tech. Note 270-4, U.S. Government Printing Office, Washington, D.C., 1969.
4. CODATA Bull. 28, International Council of Scientific Unions, 1977.
5. P. Duby, *The Thermodynamic Properties of Aqueous Inorganic Copper Systems*, International Copper Research Association, 1977.
6. Y. Couturier and C. Petitfaux, Bull. Soc. Chim. Fr., 439, (1973).
7. C. Kappenstein and R. Hugel, J. Inorg. Nucl. Chem., 36, 1821 (1974).
8. K. P. Huber and G. Herzberg, *Molecular Spectra and Molecular Structure*, Vol. IV, Van Nostrand Reinhold, New York, 1978.
9. J. C. Warf, J. Inorg. Nucl. Chem., 28, 1031 (1966).
10. L. Nuñez, G. Pilcher, and H. A. Skinner, J. Chem. Thermodyn., 1, 31 (1969).
11. A. O. Gübeli, G. Hébert, P. A. Coté, and R. Taillon, Helv. Chim. Acta, 53, 186 (1970).
12. D. D. Wagman, W. H. Evans, V. B. Parker, I. Halow, S. M. Bailey, and R. H. Schumm, Natl. Bur. Stand. Tech. Note 270-3, U.S. Government Printing Office, Washington, D. C., 1968.
13. V. B. Parker, D. D. Wagman, and D. Garvin, *Selected Thermochemical Data Compatible with the CODATA Recommendations*, NBSIR 75-968, U.S. Government Printing Office, Washington, D.C., 1976.
14. U. Bertocci and D. R. Turner, in *Encyclopedia of the Electrochemistry of the Elements*, A. J. Bard, ed., Vol. 2, Marcel Dekker, New York, 1974, Chap. 6.
15. L. Kullberg, Acta Chem. Scand., A28, 829 (1974).

II. SILVER*

G. V. ZHUTAEVA AND N. A. SHUMILOVA Academy of Sciences of the USSR, Moscow, USSR

A. Oxidation States

Like mercury, silver shows low solubility in water, 2.99×10^{-7} M [1]. The silver ion Ag^+ forms numerous slightly soluble compounds and complex ions. For most of these the equilibrium constants have been determined and thus it is possible to calculate their free energies of formation. The silver ions Ag^{2+} and Ag^{3+} are known to exist. They are unstable in aqueous solutions, however, since they act as strong oxidants and decompose water by the reaction

$$4Ag^{2+} + 2H_2O \rightarrow 4Ag^+ + O_2 + 4H^+$$

according to Noyes et al. [2]. Extremely low stability of Ag^{2+} ions in aqueous solutions is indicated [3] by the equilibrium constant of the system

$$2Ag^+(aq) \rightarrow Ag + Ag^{2+}(aq)$$

which is

$$[Ag^{2+}]/[Ag^+]^2 = 10^{-20} \text{ at } 25°C$$

In solution the Ag^{2+} ion oxidizes Mn^{2+} to MnO_4^-, Tl^+ to Tl^{3+}, vanadyl to vanadate, and so on [4]. As a strong oxidizing agent, Ag^{2+} is used for analytical purposes in acid solutions. Some methods of volumetric determination of vanadium, chromium, manganese, and cerium have been described in the literature [3]. The data on the chemistry of bi- and trivalent silver are given fairly completely in the works of Ray and Sen [3] and McMillan [5].

B. Ag/Ag+ Systems

Literature data for $\Delta H°Ag(g)$ differ somewhat (280.5 versus 289.3). Table 3 gives the value taken from National Bureau of Standards Technical Note 270-4 [6] and "Thermic Constants of Substances" by Glushko [7]. For $Ag^+(aq)$ the value of $\Delta G° = 77.16$ kJ mol^{-1} (accurate to 0.01) has commonly been used for more than 30 years. The potential of the corresponding half-reaction is 0.7991 V.

Silver fluoride stands apart from other silver halides in that it is moderately soluble in water and forms stable hydrates, such as $AgF \cdot 4H_2O(c)$, which can be isolated from saturated aqueous solutions.

Silver chloride is important in many aspects, in particular because of its low solubility in water and use in reversible Ag/AgCl electrodes. Since numerous determinations of the potential of the corresponding half-reaction show fair agreement, $E°$ was taken to be equal to 0.2223 V. Roberts [8] was the first to determine $E°$ to an accuracy of four decimal places. A review of many determinations of $E°$ was published by Bates and Macaskill [9]. The system Ag/AgCl was studied in a wide temperature range in Refs. 10 and 11.

---

*Reviewers: A. Hamelin and N. Gizard, Laboratoire d'Electrochimie Interfaciale du CNRS, Meudon, France.

TABLE 3 Thermodynamic Data on Silver[a]

Formula	Description	State	$\Delta H°$/kJ mol^{-1}	$\Delta G°$/kJ mol^{-1}	$S°$/J mol^{-1} K^{-1}
Ag		g	284.69	245.80	172.97
Ag	Metal	c	0	0	42.57
Ag	Metal	aq	—	37.26	—
Ag$_2$		g	410.19	358.95	257.15
Ag$^+$		g	1019.67	979.21	167.23
Ag$^+$		aq	105.63	77.16	72.71
Ag^{2+}		g	3093.04	—	—
Ag^{2+}		aq	268.74	269.16	-87.91
Ag^{3+}		g	6448.95	—	—
AgO$^+$		aq	—	225.63	—
AgO$^-$		aq	—	-22.98	—
Ag$_2$O		c	-31.06	-11.22	121.39
Ag$_2$O$_2$		c	-24.28	27.63	117.21
Ag(OH)$_2^-$		aq	—	-260.37	—

TABLE 3 (Continued)

Formula	Description	State	$\Delta H°$/kJ mol^{-1}	$\Delta G°$/kJ mol^{-1}	$S°$/J mol^{-1} K^{-1}
AgOH		aq	−124.47	−80.21	61.95
AgO		c	−12.14	14.65	57.35
Ag$_2$O$_3$		c	33.91	121.39	100.46
AgO$_2$		c	−30.06	−10.99	184.02
AgH		g	277.52	2486.11	204.82
AgF		g	7.12	−20.25	235.76
AgF		c	−204.70	−187.98	83.72
AgF		aq	−227.17	−201.81	59.02
AgF·H$_2$O		c	−503.99	−426.13	(114.70)
AgF·2H$_2$O		c	−801.20	−671.43	−174.97
AgF·4H$_2$O		c	−1388.91	−1147.80	267.90
AgF$_2$		c	−365.44	−165.35	263.72
AgHF$_2$		aq	−550.04	—	—
AgCl		g	92.57	65.13	245.93
AgCl		c	−127.13	−109.86	96.28

Copper, Silver, and Gold

AgCl	aq	−61.61	−54.16	129.35
AgCl$_2^-$	aq	−245.30	−215.58	231.49
AgCl$_3^{2-}$	aq	—	−345.97	—
AgCl$_4^{3-}$	aq	—	−478.46	—
AgClO$_2$	c	8.79	75.77	134.62
AgClO$_2$	aq	38.93	94.18	174.14
AgClO$_3$	c	−25.53	71.20	149.86
AgClO$_3$	aq	6.28	73.67	235.25
AgClO$_4$	c	−31.14	88.32	163.25
AgClO$_4$	aq	−23.78	68.52	254.93
Ag(ClO$_4$)$_2$(4MHClO$_4$)	aq	10.05	—	—
AgBr	c	−100.42	−96.95	107.16
AgBr	aq	−15.99	−26.87	155.30
AgBr$_2^-$	aq	—	−172.46	—
AgBr$_3^{2-}$	aq	—	−284.65	—
AgBr$_4^{3-}$	aq	—	−387.67	—
AgBrO$_2$	c	—	—	140.65

TABLE 3 (Continued)

Formula	Description	State	$\Delta H°$/kJ mol^{-1}	$\Delta G°$/kJ mol^{-1}	$S°$/J mol^{-1} K^{-1}
AgBrO$_3$		c	−27.21	54.42	152.79
AgBrO$_3$		aq	21.77	78.70	236.09
AgCl$_3$Br^{3-}		aq	—	−465.90	—
AgClBr$_3^{3-}$		aq	—	−420.27	—
AgJ		c	−61.87	−66.22	115.53
AgJ		aq	50.40	25.53	184.18
AgJ$_2^-$		aq	—	−87.07	—
AgJ$_3^{2-}$		aq	−182.09	−154.04	253.25
AgJ$_4^{3-}$		aq	—	−209.72	—
AgJO$_2$		c	—	—	143.16
AgJO$_3$		c	−171.21	−93.77	149.44
AgJO$_3$		aq	−115.95	−51.07	191.30
Ag$_2$H$_3$JO$_6$		c	—	—	248.82
3AgJ·HJ·7H$_2$O		c	−2252.49	—	—

Copper, Silver, and Gold

Ag_2S	Orthorhombic	c	-32.61	-40.69	144.08
Ag_2S	β	c	-29.43	-39.47	150.70
$AgHS$		c	—	-156.97	—
Ag_2SO_3		c	-491.02	-411.48	158.23
Ag_2SO_3		aq	-424.46	-332.37	116.37
$AgSO_3^-$		aq	—	-440.37	—
$Ag(SO_3)_2^{3-}$		aq	-1223.15	-946.50	-10.16
Ag_2SO_4		c	-716.22	-618.77	200.51
$AgSO_4$		aq	-698.43	-590.64	165.77
$AgSO_4^-$		aq	-797.85	-675.20	133.95
$AgS_2O_3^-$		aq	-592.32	-490.18	136.87
$Ag(S_2O_3)_2^{3-}$		aq	-1286.36	-1033.65	98.92
$Ag_2S_2O_6$		aq	-987.48	—	—
$Ag_2S_2O_6 \cdot 2H_2O$		c	-1604.49	—	—
Ag_2Se		c	-37.67	-44.37	150.78
Ag_2SeO_3		c	-365.44	-304.32	230.23
$AgSeO_3$		aq	-298.04	-215.58	159.07

TABLE 3 (Continued)

Formula	Description	State	$\Delta H°$/kJ mol^{-1}	$\Delta G°$/kJ mol^{-1}	$S°$/J mol^{-1} K^{-1}
Ag_2SeO_4		c	−420.69	−334.46	248.65
$AgSeO_4$		aq	−388.04	−287.16	199.67
$AgTe$		c	−37.26	−43.12	154.88
AgN_3		c	308.93	376.32	104.23
AgN_3		aq	380.88	425.30	180.84
Ag_3N		c	199.25	—	—
$AgNO_2$		c	−45.08	19.09	128.26
$AgNO_2$		aq	0.84	39.77	213.07
$AgNO_3$		c	−124.45	−33.49	140.98
$AgNO_3$		aq	−101.85	−34.24	219.35
$Ag_2N_2O_2$		c	85.81	123.07	(175.81)
$Ag_2N_2O_2$		aq	194.23	—	—
$Ag_2(NO_2)_2^-$		aq	—	−10.05	—
$Ag(NH_3)^+$		aq	—	31.65	—

Ag(NH$_3$)$_2^+$	aq	−111.35	−17.25	245.30
Ag(NH$_3$)$_2$NO$_3$	c	−335.81	—	—
Ag(NH$_3$)$_2$NO$_3$	aq	−318.81	−128.64	391.81
Ag(NH$_3$)$_3$NO$_2$	c	−441.62	—	—
AgCl·NH$_3$	c	−223.95	−131.86	139.81
Ag$_2$CO$_3$	c	−506.09	−437.02	167.44
Ag$_2$C$_2$O$_4$	c	−673.53	−584.37	209.30
Ag(CH$_3$CO$_2$)	c	−399.55	−307.88	147.59
Ag$_2$MoO$_4$	c	−838.87	755.57	242.79
Ag$_2$WO$_4$	c	−965.71	−830.92	114.70
Ag$_2$CrO$_4$	c	−728.36	−640.46	217.25
Ag$_2$Cr$_2$O$_7$	c	—	−1166.97	—
Ag$_4$Fe(CN)$_6$	c	—	−788.64	
AgCN	c	146.09	156.97	107.25
AgCN	aq	256.18	249.49	167.02
Ag(CN)$_2^-$	aq	270.42	305.58	192.56

TABLE 3 (Continued)

Formula	Description	State	$\Delta H°$/kJ mol^{-1}	$\Delta G°$/kJ mol^{-1}	$S°$/J mol^{-1} K^{-1}
Ag$_2$(CN)$_2$		c	215.58	—	—
Ag(CN)$_3^{2-}$		aq	417.34	466.61	312.52
AgCN$_2$		c	235.25	—	—
AgCNO		c	−95.44	−58.19	121.39
AgONC		c	180.0	—	—
AgCNS		c	87.91	101.43	131.02
Ag(CNS)$_2^-$		aq	—	215.16	—
Ag(CNS)$_3^{2-}$		aq	—	300.97	—
Ag(CNS)$_4^{3-}$		aq	—	392.23	—

AgP$_2$	c	−46.05	—	(87.91)
AgP$_3$	c	−69.49	—	(105.49)
Ag$_3$PO$_4$	c	−988.31	−879.06	258.28
Ag$_3$PO$_4$	aq	−872.49	−781.60	—
Ag$_4$P$_2$O$_7$	c	−1896.26	−1738.03	—
Ag$_3$AsO$_4$	c	−634.60	−542.92	—
Ag$_3$AsO$_4$	aq	−571.68	−417.34	55.67
AgMnO$_4$	c	(−46.05)	(−408.18)	—
Ag$_2$SiO$_3$	c	—	—	177.49
Ag$_2$C$_2$	c	350.79	—	—

[a]National Bureau of Standards values are in italic; estimated values are in parentheses.
Source: Refs. 6 and 7.

The solubility data for AgCl show that AgCl(aq) is a weak electrolyte and that under certain conditions the Ag^+ and Cl^- ions undergo association to form various complex ions (Ag_2Cl^+, $AgCl_2^-$, $AgCl_3^{2-}$, etc.). Jonte and Martin [12] showed that the dependence of the solubility S of AgCl on the NaCl solution concentration has an unusual shape. At first S drops, passing through a minimum ($\sim 5 - 6 \times 10^{-7}$ M) in the range 10^{-2} to 10^{-3} M Cl^-, and then rises again. The above-mentioned authors interpreted these data in terms of the equilibrium of Ag^+, AgCl(c), AgCl(aq), and $AgCl_2^-$ particles. Jonte and Martin [12] and Ramette [13] computed the stability constants of chlorine-containing complexes of silver ions from data on their solubility. As regards the thermodynamic parameters of these complex ions, literature data show poor agreement. Association of Ag^+ ions with halogen anions is also observed in the case of bromides and iodides [14,15]. Lyalikov and Piskunova [16] determined the equilibrium constants of complex silver bromides, K_2AgBr_3 and $(NH_4)_2AgBr_3$. The chloride-bromide silver complexes $AgClBr_3^{3-}$ and $AgBrCl_3^{3-}$ were examined by Chateau and Hervier [17], who determined their equlibrium constants.

The thermodynamic data of the system Ag/AgCl in DCl in heavy water was studied by Gary et al. [18]. The standard electromotive force (emf) of the system at 25°C by 9.78 mV (on the molality scale) or by 4.31 mV (on the mole fraction scale) is less than the standard emf of the corresponding cell with a hydrogen electrode and HCl solution in ordinary water. Later, Lietzke and Nall [19] repeated the investigation and obtained for $E°$ the value 0.2131 V. Lietzke and Lemmonds [20] measured the standard potential of an Ag/AgBr electrode in DBr solutions and found $E° = 0.06014$ V. $E°$ of Ag/AgBr and Ag/AgCl electrodes in HBr and HJ solutions in ordinary water was measured at temperatures up to 200°C in Refs. 21 and 22.

Table 4 lists the standard potentials of silver. Table 5 gives the values of the solubility products of some silver compounds.

The temperature coefficients for $E°$ of some silver electrodes are listed here (with t = Celsius temperature). The Ag/AgCl standard potential is

$$E°/V = 0.23695 - 4.8564 \times 10^{-4}t - 3.4205 \times 10^{-6}t^2 - 5.869 \times 10^{-9}t^3$$

at 0 to 95°C in Ref. 10, and

$$E°/V = 0.23735 - 5.3783 \times 10^{-4}t - 2.3728 \times 10^{-6}t^2$$

at 25 to 275°C in Ref. 11. The extrapolation gave the potential at 300°C as $E° = 0.138$ V.

The temperature dependence of the Ag/AgBr electrode standard potential as given in Ref. 28 is at 0 to 50°C:

$$E°/V = 0.07109 - 4.87 \times 10^{-4}(t - 25) - 3.08 \times 10^{-6}(t - 25)^2$$

and in Ref. 21 is at 25 to 200°C:

$$E°/V = 0.08289 - 4.0647 \times 10^{-4}t - 2.3986 \times 10^{-6}t^2$$

TABLE 4 Standard and Formal Potentials for Ag/Ag Compounds

Reactions	Standard potential, $E°$/V	References
$Ag_2S(c) + 2e^- \rightarrow 2Ag(c) + S^{2-}(aq)$	−0.691	23, 24
$Ag(CN)_3^{2-}(aq) + e^- \rightarrow 2Ag(c) + 3CN^-(aq)$	−0.5	4
$Ag(CN)_2^{1-}(aq) + e^- \rightarrow Ag(c) + 2CN^-(aq)$	−0.31	23, 25
$AgI + e^- \rightarrow Ag(c) + I^-(aq)$	−0.1522	22, 26
$AgCN + e^- \rightarrow Ag(c) + CN^-(aq)$	−0.017	23, 25
$Ag(S_2O_8)_2^{3-}(aq) + e^- \rightarrow Ag(c) + 2S_2O_8^{2-}(aq)$	−0.01	27
$Ag(S_2O_3)_2^{3-}(aq) + e^- \rightarrow Aq(c) + 2S_2O_3^{2-}(aq)$	−0.017	23, 25
$AgBr(c) + e^- \rightarrow Ag(c) + Br^-(aq)$	0.0711	26, 28
$AgSCN(c) + e^- \rightarrow Ag(c) + SCN^-(aq)$	0.0895	29
$Ag(SCN)_4^{3-}(aq) + e^- \rightarrow Ag(c) + 4SCN^-(aq)$	0.214	30
$Ag(SCN)_3^{2-}(aq) + e^- \rightarrow Ag(c) + 3SCN^-(aq)$	0.231	30
$Ag(SCN)_2^-(aq) + e^- \rightarrow Ag(c) + 2SCN^-(aq)$	0.304	30
$Ag_4[Fe(CN)_6](c) + 4e^- \rightarrow 4Ag(c) + Fe(CN)_6^{4-}(aq)$	0.1478	30
$AgCl(c) + e^- \rightarrow Ag(c) + Cl^-(aq)$	0.2223	8–10

TABLE 4 (Continued)

Reactions	Standard potential, $E°$/V	References
$AgN_3(c) + e^- \rightarrow Ag(c) + N_3^-(aq)$	0.2933	32
$Ag_3[Co(CN)_6](c) + 3e^- \rightarrow 3Ag(c) + Co(CN)_6^{3-}(aq)$	0.298	33
$Ag_3PO_4(c) + 3e^- \rightarrow 3Ag(c) + PO_4^{3-}(aq)$	0.4525	34
$AgIO_3(c) + e^- \rightarrow Ag(c) + IO_3^-(aq)$	0.354	23, 25, 27
$Ag_2SeO_3(c) + 2e^- \rightarrow 2Ag(c) + SeO_3^{2-}$	0.363	4
$Ag(NH_3)_2^+(aq) + e^- \rightarrow Ag(c) + 2NH_3(aq)$	0.373	23, 25
$AgCNO(c) + e^- \rightarrow Ag(c) + CNO^-(aq)$	0.41	23
$Ag(SO_3)_2^{3-}(aq) + e^- \rightarrow Ag(c) + 2SO_3^{2-}(aq)$	0.43	23
$Ag_2CrO_4(c) + 2e^- \rightarrow 2Ag(c) + CrO_4^{2-}(aq)$	0.4491	34
$Ag_2C_2O_4(c) + 2e^- \rightarrow 2Ag(c) + C_2O_4^{2-}(aq)$	0.4647	35
$Ag_2WO_4(c) + 2e^- \rightarrow 2Ag(c) + WO_4^{2-}(aq)$	0.466	36

$Ag_2CO_3(c) + 2e^- \rightarrow 2Ag(c) + CO_3^{2-}(aq)$	0.47	4,23,25
$Ag_2MoO_4 + 2e^- \rightarrow 2Ag(c) + MoO_4^{2-}(aq)$	0.486	25
$AgBrO_3(c) + e^- \rightarrow Ag(c) + BrO_3^-(aq)$	0.546	4,23
$Ag_2SeO_4(c) + 2e^- \rightarrow 2Ag(c) + SeO_4^{2-}(aq)$	0.55	4
$AgNO_2(c) + e^- \rightarrow Ag(c) + NO_2^-(aq)$	0.564	23,25
$AgCH_3COO(c) + e^- \rightarrow Ag(c) + CH_3COO^-(aq)$	0.643	23,25
$Ag_2SO_4(c) + 2e^- \rightarrow 2Ag(c) + SO_4^{2-}(aq)$	0.654	25
$Ag(IO_3)_2^-(c) + e^- \rightarrow Ag(c) + IO_3^-(aq)$	0.725	36
$AgF(c) + e^- \rightarrow Ag(c) + F^-(aq)$	0.779	36
$AgClO_4(c) + e^- \rightarrow Ag(c) + ClO_4^-(aq)$	0.787	36
$Ag^+(aq) + e^- \rightarrow Ag(c)$	0.7991	23,25,27
$Ag_3AsO_4(c) + 3e^- \rightarrow 3Ag(c) + AsO_4^{3-}(aq)$	0.4010	34

TABLE 5 Solubility Products, K_{sp}

Substance	K_{sp}	Substance	K_{sp}
Ag_2O	2.0×10^{-8}	Ag_2S	1×10^{-50}
$AgCl$	1.77×10^{-10}	Ag_2SO_3	2×10^{-14}
$AgBr$	5.0×10^{-13}	Ag_2SO_4	1.2×10^{-5}
AgI	8.7×10^{-17}	Ag_2CO_3	8×10^{-12}
$AgBrO_3$	5.4×10^{-5}	Ag_3AsO_4	6.2×10^{-21}
$AgIO_3$	3.1×10^{-8}	$Ag_2C_2O_4$	1.1×10^{-11}
$AgCN$	1×10^{-15}	Ag_2CrO_4	2.7×10^{-12}
$AgCN_2$	1×10^{-10}	Ag_2SeO_3	2×10^{-15}
$AgN(CN)_2$	1×10^{-9}	Ag_2SeO_4	4×10^{-9}
$AgCNS$	1.0×10^{-12}	Ag_2MoO_4	2.8×10^{-12}
$AgCNO$	2.3×10^{-4}	Ag_2WO_4	4.8×10^{-13}
AgN_3	2.9×10^{-9}	$AgC_2H_3O_2$	2.3×10^{-3}
$AgNO_2$	1.2×10^{-4}	$Ag_3Fe(CN)_6$	1×10^{-20}
$Ag_2(N_2O_2)$	1×10^{-19}	$Ag_4Fe(CN)_6$	8.6×10^{-45}
Ag_3PO_4	2.3×10^{-18}	$Ag_3Co(CN)_6$	3.9×10^{-26}

The temperature dependence of the Ag/AgI electrode standard potential is

$$E°/V = -0.15242 - 3.19 \times 10^{-4}(t-25) - 2.84 \times 10^{-6}(t-25)^2$$

at 0 to 50°C in Ref. 28, and

$$E°/V = -0.14590 - 1.696 \times 10^{-4}t - 2.934 \times 10^{-6}t^2$$

at 20 to 200°C in Ref. 22. Vanderzee and Smith [29] showed that $E°$ of the Ag/AgSCN electrode varied with temperature at 5 to 45°C according to

$$E°/V = 0.08949 - 1.00 \times 10^{-4}(t-25) - 1.61 \times 10^{-6}(t-25)^2$$

C. Compounds of Ag^{2+} and Ag^{3+}

Luther and Pokorny [37] were the first to discover the existence of bi- and trivalent silver. They determined the standard potentials of the systems Ag_2O/Ag, AgO/Ag_2O, and Ag_2O_3/AgO, which were equal to 1.17, 1.41, and 1.57 V, respectively. The generally accepted standard potential values [38] are very close to these. The potential of the system Ag^{2+}/Ag^+ in acid solutions, first determined by Noyes and Kossiakoff [39] (the formal potential in 1 M HNO_3 is 1.914 V), and that in alkaline solutions are sufficiently high. But these potentials decrease considerably on complexing. The studies on silver complexes are reviewed in the book by Ray and Sen [3].

Electrolytic and chemical oxidation of Ag^+ in alkaline solution both yield a solid compound of composition AgO. At first it was considered to be silver peroxide, then an oxide of bivalent silver. Later, on the basis of investigation of its magnetic properties and crystalline structure [40], the compound was assumed to be a mixed oxide $Ag_2O \cdot Ag_2O_3$. This assumption is confirmed by the kinetic study of the isotope exchange in $HClO_4$ carried out by Gordon and Wahl [41], who showed that the exchange reaction

$$2Ag^{2+} \rightarrow Ag^+ + Ag^{3+}$$

occurs.

AgO is an electronic semiconductor. The electrical conductivity of a specimen subjected to the pressure 12,000 kg cm^{-2} at 20°C is 7×10^{-2} Ω^{-1} cm^{-1}, the temperature coefficient in the range -40 to 20°C is positive [42]. In acid solution AgO dissolves at room temperature: $AgO + 2H^+ \rightarrow Ag^{2+} + H_2O$, but Ag^{2+} immediately reacts with water, evolving O_2, and is reduced to Ag^+.

The standard free energy [43] of this reaction $\Delta G^{\circ}_{298} = -226$ kJ mol^{-1}. In water AgO is stable up to 100°C. Theoretically, spontaneous decomposition according to the equation

$$4AgO + 2H_2O \rightarrow 4Ag^+ + 4OH^- + O_2$$

is thermodynamically impossible. In weakly alkaline solutions AgO dissolves without decomposition. The following reactions are assumed to occur:

$$AgO + H_2O \rightarrow Ag(OH)_2$$

$$AgO + OH^- + H_2O \rightarrow Ag(OH)_3^-$$

$$AgO + 2OH^- + H_2O \rightarrow Ag(OH)_4^{2-}$$

The Gibbs energies of formation ΔG° have been determined for $Ag(OH)_2^-$, $Ag(OH)_3^-$, and $Ag(OH)_4^-$. They are -238.87, -357.40, and -475.95 kJ mol^{-1}, respectively [5]. So far the oxide Ag_2O_3 has not been isolated in the free state. Table 6 lists the standard potentials of the system silver-water.

TABLE 6 Standard Potentials for Silver-Water System

Half-reaction	Standard potential, $E°/V$	References
$Ag_2O + H_2O + 2e^- \rightarrow 2Ag + 2OH^-$	0.342	44
$Ag + e^- \rightarrow AgO^- + H_2O$	0.351	38
$2AgO + H_2O + 2e^- \rightarrow Ag_2O + 2OH^-$	0.604	45
$Ag_2O_3 + H_2O + 2e^- \rightarrow 2AgO + 2OH^-$	0.739	25
$Ag^+ + e^- \rightarrow Ag$	0.7991	23,25,27,38
$Ag_2O_3 + 2H^+ + 4e^- \rightarrow 2AgO^- + H_2O$	0.960	38
$Ag_2O + 2H^+ + 2e^- \rightarrow 2Ag + H_2O$	$1.173^a; 1.505^b$	37,38
$AgO^+ + 2e^- \rightarrow AgO^-$	1.288	38
$Ag_2O_3 + 6H^+ + 2e^- \rightarrow 2Ag^{2+} + 3H_2O$	1.360	38
$2AgO + 2H^+ + 2e^- \rightarrow Ag_2O + H_2O$	$1.398^a; 1.066^b$	27,38
$Ag_2O_3 + 2H^+ + 2e^- \rightarrow 2AgO + H_2O$	1.569	37,38
$Ag_2O_3 + 6H^+ + 4e^- \rightarrow 2Ag^+ + 3H_2O$	1.670	27,38
$Ag_2O_3 + H_2O + 2e^- \rightarrow Ag_2O_2 + 2OH^-$	1.711	46
$Ag_2O_3 + 3H_2O + 6e^- \rightarrow 2Ag^+ + 6OH^-$	1.757	46
$AgO + 2H^+ + e^- \rightarrow Ag^+ + H_2O$	1.772	38
$Ag^{2+} + e^- \rightarrow Ag^+$	1.980	25,38
$AgO^+ + 2H^+ + 2e^- \rightarrow Ag^+ + H_2O$	1.998	27,38
$AgO^+ + 2H^+ + e^- \rightarrow Ag^{2+} + H_2O$	2.016	38
$AgO^- + 2H^+ + e^- \rightarrow Ag + H_2O$	2.220	27,38

[a]The nonhydrated oxides as in the Pourbaix diagram.
[b]The hydrated oxides as in the Pourbaix diagram.

D. Potential Diagrams for Silver

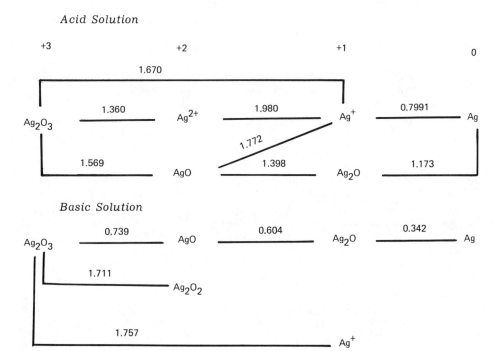

REFERENCES

1. J. Dobrowolski and J. Oglaza, Nucleonika, 8, 79 (1963).
2. A. A. Noyes, J. L. Hoard, and K. S. Pitzer, J. Am. Chem. Soc., 57, 1221, 1229, 1238 (1935).
3. P. Ray and D. Sen, *Chemistry of Bi- and Tripositive Silver*, National Institute of Sciences of India, New Delhi, 1960.
4. L. M. Gedansky and L. G. Hepler, Engelhard Ind. Tech. Bull., 9, 117 (1969).
5. J. A. McMillan, Chem. Rev., 62, 65 (1962).
6. *Selected Values of Chemical Thermodynamic Properties*, Natl. Bur. Stand. Tech. Note 270-4, U.S. Government Printing Office, Washington, D.C., 1969.
7. V. P. Glushko, *Thermic Constants of Substances*, Vol. 6, Part I, Academy of Sciences of the USSR, Moscow, 1972, p. 153.
8. E. J. Roberts, J. Am. Chem. Soc., 52, 3877 (1930).
9. R. G. Bates and J. B. Macaskill, Pure Appl. Chem., 50, 11-12, 1701 (1978).
10. R. G. Bates and V. E. Bower, J. Res. Natl. Bur. Stand., 53, 283 (1954).
11. R. S. Greeley, W. T. Smith, R. W. Stoughton, and M. H. Lietzke, J. Phys. Chem., 64, 652, 1445 (1960).
12. J. H. Jonte and D. S. Martin, J. Am. Chem. Soc., 74, 2052 (1952).
13. R. Ramette, J. Chem. Educ., 37, 348 (1960).

14. L. G. Sillén and A. E. Martell, *Stability Constants of Metal-Ion Complexes*, Spec. Publ. 17, Chemical Society, London, 1964.
15. E. Berne and I. Leden, Z. Naturforsch., *8A*, 719 (1953).
16. Yu. S. Lyalikov and V. N. Piskunova, Zh. Fiz. Khim., *28*, 127 (1954).
17. H. Chateau and B. Hervier, J. Chim. Phys., *54*, 356 (1957).
18. R. Bary, R. G. Bates, and R. A. Robinson, J. Phys. Chem., *68*, 1186 (1964).
19. M. H. Lietzke and D. S. Nall, J. Inorg. Nucl. Chem., *38*, 1541 (1976).
20. M. H. Lietzke and T. J. Lemmonds, J. Inorg. Nucl. Chem., *36*, 2299 (1974).
21. M. B. Towns, R. S. Greeley, and M. H. Lietzke, J. Phys. Chem., *64*, 1861 (1960).
22. G. Kortüm and W. Häussermann, Ber. Bunsenges. Phys. Chem., *69*, 594 (1965).
23. W. M. Latimer, *The Oxidation States of the Elements and Their Potentials in Aqueous Solutions*, 2nd ed., Prentice-Hall, Englewood Cliffs, N.J., 1952.
24. M. Golding, J. Chem. Soc., 1838 (1959).
25. A. de Béthune and N. A. Swendeman Loud, *The Encyclopedia of Electrochemistry*, C. A. Hampel, ed., Reinhold, New York, 1964, p. 414; A. de Béthune and N. A. Swendeman Loud, *Standard Aqueous Electrode Potentials and Temperature Coefficients at 25°C*, Hampel, Skokie, Ill., 1964.
26. G. Janz and H. Taniguchi, Chem. Rev., *53*, 397 (1953).
27. *Spravochnik Khimika* (A Chemist's Handbook), Vol. 3, Izdatelstvo "Khimiya," Moscow, 1965.
28. H. B. Hetzer, R. A. Robinson, and R. G. Bates, J. Phys. Chem., *66*, 1423 (1962); *68*, 1929 (1964).
29. C. F. Vanderzee and W. E. Smith, J. Am. Chem. Soc., *78*, 721 (1956).
30. G. C. B. Gave and D. N. Hume, J. Am. Chem. Soc., *75*, 2893 (1953).
31. P. A. Rock and R. E. Powell, Inorg. Chem., *3*, 1593 (1964).
32. S. Suzuki, J. Chem. Soc. Jpn., *73*, 153 (1952).
33. P. A. Rock, Inorg. Chem., *4*, 1667 (1965).
34. U. N. Dash and M. C. Padhi, Thermochim. Acta, *48*(1-2), 241 (1981).
35. P. B. Mathur and S. M. A. Naqvi, Indian J. Chem., *6*, 311 (1968).
36. G. Millazzo and S. Cardi, *Tables of Standard Electrode Potentials*, Wiley, New York, 1978.
37. R. Luther and F. Pokorny, Z. Inorg. Allg. Chem., *57*, 290 (1908).
38. M. Pourbaix, ed., *Atlas d'équilibres électrochimiques à 25°C*, Gauthier-Villars, Paris, 1963, p. 393.
39. A. A. Noyes and A. Kossiakoff, J. Am. Chem. Soc., *57*, 1238 (1935).
40. J. A. McMillan, J. Inorg. Nucl. Chem., *13*, 28 (1960).
41. B. M. Gordon and A. C. Wahl, J. Am. Chem. Soc., *80*, 273 (1958).
42. A. B. Neiding and I. A. Kazarnovskii, Dokl. Akad. Nauk SSSR, *78*, 713 (1951).
43. T. P. Dirkse and B. Wiers, J. Electrochem. Soc., *106*, 284 (1959).
44. W. J. Hamer and D. N. Craig, J. Electrochem. Soc., *104*, 206 (1957).
45. J. F. Bonk and A. B. Carrett, J. Electrochem. Soc., *106*, 612 (1959).
46. B. Stehlik, Chem. Zvesti, *17*, 6 (1963).

III. GOLD*

GERHARD M. SCHMID University of Florida, Gainesville, Florida

A. Oxidation States

Gold occurs in the 0, +1, +2, and +3 oxidation states. The free aurous ion (Au^+) is not stable in water; the free auric ion (Au^{3+}) is a strong oxidant. With the exception of the insoluble AuS, all +2 compounds of gold are unstable in water. However, gold forms very stable complexes in aqueous solution in both the +1 and the +3 oxidation state. The thermodynamic data are given in Table 7.

B. Au(I)/Au Couple

Because of the instability of the Au^+ ion, standard potential measurements for the reaction

$$Au^+ + e^- \rightarrow Au \tag{11.1}$$

cannot be made directly. The potential was estimated as $E° = 1.83$ V, using stability data for the chloride and bromide complexes and Edwards' equation [1]. Diluting Au(I) solutions in pure acetonitrile with 0.1 M aqueous perchloric acid to 0.4 to 3.8% residual acetonitrile yields $E°' = 1.695 \pm 0.01$ V in 0.1 M $HClO_4$ [2].

For

$$AuCl_2^- + e^- \rightarrow Au + 2Cl^- \tag{11.2}$$

Lingane [3] finds $E° = 1.154$ V by direct potentiometry in a cell without liquid junction, and this value is corroborated by Pouradier et al. [4]. The standard potential for

$$AuBr_2^- + e^- \rightarrow Au + 2Br^- \tag{11.3}$$

is within 2 to 3 mV of 0.960 V as found from direct potentiometric measurements in a cell without liquid junction and from potentiometric titrations [5-7]. At ionic strengths between 0.16 and 1 mol L^{-1}, the potential for

$$AuI_2^- + e^- \rightarrow Au + 2I^- \tag{11.4}$$

is 0.578 V, independent of ionic strength [8]. Together with the experimentally determined equilibrium constant for $AuI + I^- \rightarrow AuI_2^-$, $pK = 0.82 \pm 0.14$ (K dimensionless), one obtains $E° = 0.530$ V [8] for

$$AuI(s) + e^- \rightarrow Au + I^- \tag{11.5}$$

The standard potential for the reaction

$$Au(SCN)_2^- + e^- \rightarrow Au + 2SCN^- \tag{11.6}$$

---

*Reviewer: Y. Okinaka, Bell Telephone Laboratories, Murray Hill, New Jersey.

TABLE 7 Thermodynamic Data on Gold[a]

Formula	State	$\Delta H°$/kJ mol^{-1}	$\Delta G°$/kJ mol^{-1}	$S°$/J mol^{-1} K^{-1}
Au	g	366	326	180.39
Au	c	0	0	47.40
Au$^+$	g	1262.4		
Au$^+$	aq		176	
Au^{2+}	g	3.247×10^3		
Au^{3+}	aq		440	
AuO$_3^{3-}$	aq		−51.9	
HAuO$_3^{2-}$	aq		−142	
H$_2$AuO$_3^-$	aq		−218	
Au(OH)$_3$	c	−424.7	−317.0	190
Au(OH)$_3$	aq		−283.5	
AuCl	c	−35		
AuCl$_2^-$	aq		−151.0	
AuCl$_3$	c	−118		

Copper, Silver, and Gold

AuCl$_3$·2H$_2$O	c	−715.0	
AuCl$_4^-$	aq	−322	−234.6	267
AuBr	c	−14.0	
AuBr$_2^-$	aq	−128	−115.0	220
AuBr$_3$	c	−53.26	
AuBr$_3$	aq	−39.3	
AuBr$_4^-$	aq	−192	−167	336
HAuBr$_4$·5H$_2$O	c	−1668	
AuI	c	0	−0.5	
AuI$_2^-$	aq		−47.6	
AuI$_4^-$	aq		−45	
Au(CN)$_2^-$	aq	242	286	1.7×10^2
Au(SCN)$_2^-$	aq		252	
Au(SCN)$_4^-$	aq		561.5	

[a]National Bureau of Standards values are in italic.

is given by Pouradier and Gadet as $E° = 0.662 \pm 0.005$ V [9]. These authors also studied the reaction

$$Au(S_2O_3)_2^{3-} + e^- \rightarrow Au + 2S_2O_3^{2-} \tag{11.7}$$

and determined $E°' = 0.153$ V at 25°C in solutions containing thiosulfate and potassium chloride as supporting electrolyte [10]. The complexes that Au(I) forms with thiourea have been studied in solutions of pH 3 containing Au(I)-thiourea complex and perchloric acid. The average potential for

$$Au[CS(NH_2)_2]_2^+ + e^- \rightarrow Au + 2CS(NH_2)_2 \tag{11.8}$$

is $E°' = 0.380 \pm 0.010$ V [11]. From potential measurements made on the system

$$Au(CN)_2^- + e^- \rightarrow Au + 2CN^- \tag{11.9}$$

by Beltowska-Brzezinska et al. [12] in 1×10^{-3} to 1×10^{-1} M KAu(CN)$_2$ and 5×10^{-2} to 5×10^{-1} M KCN at a constant ionic strength of 1.25 mol L^{-1} (with K$_3$PO$_4$), one can deduce that $E°' = -0.595 \pm 0.002$ V. Potentiometric titrations in 10 M ammonium nitrate yield $E°' = 0.563 \pm 0.006$ V [13] for

$$Au(NH_3)_2^+ + e^- \rightarrow Au + 2NH_3 \tag{11.10}$$

C. Au(III)/Au Couple

Measurements of the potential of the Au/Au(III) oxide electrode have been made by Gerke and Rourke [14] in 0.05 to 1.03 mol kg^{-1} sulfuric acid and by Buehrer and Roseveare [15] in 0.01 to 1.02 mol kg^{-1} sulfuric acid. The average of their results is $E°' = 1.362 \pm 0.002$ V. This value, a formal potential, seems to be the accepted standard potential for this system and the basis for the $\Delta G° = 317.0$ kJ mol^{-1} listed for crystalline Au(III) hydroxide [16].

Again using Edwards' equation, the potential of

$$Au^{3+} + 3e^- \rightarrow Au \tag{11.11}$$

can be calculated using stability and potential data for the chloride complex, $E° = 1.52$ V [1].

Lingane obtained $E° = 1.002$ V for

$$AuCl_4^- + 3e^- \rightarrow Au + 4Cl^- \tag{11.12}$$

from spectrophotometric concentration measurements as the $E°$ value for Au/Au(I) chloride [2,3]. For the corresponding bromide complex,

$$AuBr_4^- + 3e^- \rightarrow Au + 4Br^- \tag{11.13}$$

Evans and Lingane [5] find $E° = 0.854 \pm 0.002$ V, in close agreement with Peshchevitskii et al. [6] and Pouradier and Gadet [7]. For

$$AuI_4^- + 3e^- \to Au + 4I^- \tag{11.14}$$

the estimated standard potential is $E° = 0.56$ V [17].

The application of Luther's rule to the corresponding Au/Au(I) and Au(I)/Au(III) complexes gives $E° = 0.636$ V [9] for

$$Au(SCN)_4^- + 3e^- \to Au + 4SCN^- \tag{11.15}$$

Similarly, one obtains for

$$Au(NH_3)_4^{3+} + 3e^- \to Au + 4NH_3 \tag{11.16}$$

$E°' = 0.325 \pm 0.003$ V [13].

D. Au(III)/Au(I) Couple

The standard potential for

$$Au^{3+} + 2e^- \to Au^+ \tag{11.17}$$

$E° = 1.36$ V can be calculated from the potentials for the Au/Au(I) and Au/Au(III) couples, (11.1) and (11.11).

Similarly, the potential for

$$AuCl_4^- + 2e^- \to AuCl_2^- + 2Cl^- \tag{11.18}$$

consistent with the values for (11.2) and (11.12), is $E° = 0.926$ V [3]. From the experimentally determined equilibrium constant for

$$AuBr_4^- + 2Au + 2Br^- \to 3AuBr_2^-$$

and the standard potential for (11.3), Evans and Lingane calculate $E° = 0.802 \pm 0.002$ V [5] for

$$AuBr_4^- + 2e^- \to AuBr_2^- + 2Br^- \tag{11.19}$$

in close agreement with $E° = 0.801$ V calculated from (11.3) and (11.13). For the corresponding iodide reaction one calculates $E° = 0.55$ V from Erenburg and Peshchevitskii's data [8,17]. The standard potential for

$$Au(SCN)_4^- + 2e^- \to Au(SCN)_2^- + 2SCN^- \tag{11.20}$$

is $E° = 0.623$ V [9], and for the corresponding ammonia complex one obtains $E°' = 0.206 \pm 0.003$ V [13].

The standard potentials of the gold couples treated here are summarized in Table 8.

TABLE 8 Summary of Standard Potentials of Gold Couples in Aqueous Solutions

Half-reaction	Standard potential, E°/V	References
Au(I)/Au(0)		
$Au^+ + e^- \rightarrow Au$	1.83	1
$AuCl_2^- + e^- \rightarrow Au + 2Cl^-$	1.154	3,4
$AuBr_2^- + e^- \rightarrow Au + 2Br^-$	0.960	5,6,7
$AuI_2^- + e^- \rightarrow Au + 2I^-$	0.578	8
$AuI(s) + e^- \rightarrow Au + I^-$	0.530	8
$Au(SCN)_2^- + e^- \rightarrow Au + 2SCN^-$	0.662	9
Au(III)/Au(0)		
$Au^{3+} + 3e^- \rightarrow Au$	1.52	1
$AuCl_4^- + 3e^- \rightarrow Au + 4Cl^-$	1.002	3
$AuBr_4^- + 3e^- \rightarrow Au + 4Br^-$	0.854	5,6,7
$AuI_4^- + 3e^- \rightarrow Au + 4I^-$	0.56	17
$Au(SCN)_4^- + 3e^- \rightarrow Au + 4SCN^-$	0.636	9
Au(III)/Au(I)		
$Au^{3+} + 2e^- \rightarrow Au^+$	1.36	
$AuCl_4^- + 2e^- \rightarrow AuCl_2^- + 2Cl^-$	0.926	2,10
$AuBr_4^- + 2e^- \rightarrow AuBr_2^- + 2Br^-$	0.802	5
$AuI_4^- + 2e^- \rightarrow AuI_2^- + 2I^-$	0.55	8,17
$Au(SCN)_4^- + 2e^- \rightarrow Au(SCN)_2^- + 2SCN^-$	0.623	9

E. Other Gold Couples

Thermodynamic data for Au(II) compounds, additional Au(III) compounds, and Au(IV) compounds may be found in the literature [18,19]. For calculated standard potentials the reader is referred to Ref. 20.

F. Potential Diagrams for Gold

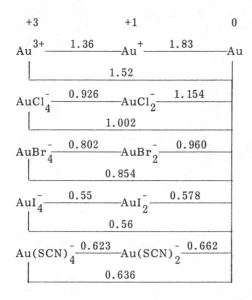

REFERENCES

1. L. H. Skibsted and J. Bjerrum, J. Indian Chem. Soc., 54, 102 (1977); Acta Chem. Scand., A31, 155 (1977).
2. P. R. Johnson, J. M. Pratt, and R. I. Tilley, J. Chem. Soc. Chem. Commun., 606, (1978).
3. J. J. Lingane, J. Electroanal. Chem., 4, 332 (1962).
4. J. Pouradier, M. C. Gadet, and H. Chateau, J. Chim. Phys., 62, 203 (1965).
5. D. H. Evans and J. J. Lingane, J. Electroanal. Chem., 6, 1 (1963).
6. B. I. Peshchevitskii, V. P. Kazakov, and A. M. Erenburg, Zh. Neorg. Khim., 8, 853 (1963).
7. J. Pouradier and M. C. Gadet, J. Chim. Phys., 62, 1181 (1965).
8. A. M. Erenburg and B. I. Peshchevitskii, Zh. Neorg. Khim., 14, 932 (1969).
9. J. Pouradier and M. C. Gadet, J. Chim. Phys., 63, 1467 (1966).
10. J. Pouradier and M. C. Gadet, J. Chim. Phys., 66, 109 (1969).
11. V. P. Kazabov, A. I. Lapshin, and B. I. Peshchevitskii, Zh. Neorg. Khim., 9, 1299 (1964).
12. M. Beltowska-Brzezinska, E. Dutkiewicz, and W. Lawicki, J. Electroanal. Chem., 99, 341 (1979).
13. L. H. Skibsted and J. Bjerrum, Acta Chem. Scand., A28, 764 (1974).
14. R. H. Gerke and M. D. Rourke, J. Am. Chem. Soc., 49, 1855 (1927).
15. T. F. Buehrer and W. E. Roseveare, J. Am. Chem. Soc., 49, 1989 (1927).

16. Natl. Bur. Stand. Tech. Note 270-4, U.S. Government Printing Office, Washington, D.C., May 1969.
17. A. M. Erenburg and B. I. Peshchevitskii, Zh. Neorg. Khim., *14*, 2714 (1969).
18. T. Marcovic, P. Zivkovic, N. Latilagic, and V. Sargic, Werkst. Korros., *17*, 1039 (1966).
19. M. Pourbaix, ed., *Atlas of Electrochemical Equilibria in Aqueous Solutions*, Pergamon Press, Oxford, 1966.
20. G. M. Schmid and M. E. Curley-Fiorino, in *Encyclopedia of the Electrochemistry of the Elements*, A. J. Bard, ed., Vol. 4, Marcel Dekker, New York, 1975, Chap. 3.

12
Nickel, Palladium, and Platinum

I. NICKEL*

ALEJANDRO J. ARVIA AND DIONISIO POSADAS   Universidad Nacional de La Plata, La Plata, Argentina

A. Oxidation States

Nickel, together with iron and cobalt, completes one group of the transition series in which 10 electrons have been added to the argon electronic core according to the electron distribution $A4s^23d^8$. The relatively small energy of the electrons present in the s and d levels leads to nickel compounds with oxidation numbers ranging from 0 to +4. In complex ion formation nickel exhibits a strong tendency to share enough additional electrons to complete the 3d, 4s, and 4p orbitals. The dsp^2 arrangement corresponds to the square-planar configuration, while the d^2sp^3 arrangement corresponds to the octahedrical geometry. The zero oxidation state is represented by the compounds $Ni(CO)_4$ and $K_4[Ni(CN)_4]$, where the eight additional electrons are donated by the complexing species. The +2 state is by are the most important. The chemistry of nickel in aqueous solutions is principally that of Ni^{2+} and its related compounds. NiO prepared at temperatures above 200°C corresponds to NaCl crystal structure. Only a few complexes of low stability are reported for the +3 state. Although Ni_2O_3 may be prepared, it probably corresponds to a mixture of +2 and +4 oxides. The +3 state involves species of strong oxidizing power.

The Ni^{2+} ion in solution can either be oxidized to Ni^{3+} in the form of chelates (oximes) or reduced to Ni^+ or to Ni^0 in the presence of suitable complexing agents. The +4 state represented by NiO_2 ($NiO_2 \cdot H_2O$) is related to very powerful oxidizing agents. The +6 and +8 states, in the form of oxides, have also been described [1-3].

B. Standard Electrode Potentials

Most of the standard potentials of electrode reactions involving different nickel species are computed from thermal data (Table 1). Standard potentials either calculated or measured are shown in Table 2. They are arranged according to the change in oxidation state per unit charge in the electroreduction half-reaction. Occasionally, the temperature coefficient of the standard electrode potential, either thermal or isothermal or both, is included in Table 2.

---

*Reviewer: Karl Kadish, University of Houston, Houston, Texas.

TABLE 1 Thermodynamic Data on Nickel[a]

Formula	Description	State	$\Delta H°$/kJ mol^{-1}	$\Delta G°$/kJ mol^{-1}	$\Delta S°$/kJ mol^{-1} K^{-1}
Ni		g	425.14	379.8	182.39
Ni		c	0.0	0.0	30.1
Ni$^+$		g	1167.8	—	—
Ni^{2+}		g	2932	—	—
Ni^{2+}	Std. state, hyp, $m = 1$	aq	−64.0	−46.4	−159
NiO		g	248	217	240
NiO		c	−244	−216	38.6
NiH		g	390	—	—
NiH		c	−11	—	—
NiH$_2$		c	−26	—	—
Ni(OH)$_2$		c	−538.1	−453.1	79
Ni(OH)$_3$		c	−678.2	−541.8	(81.6)
NiF$_2$		c	−667.3	−622.6	(83.3)
NiF$_2 \cdot$4H$_2$O		c	—	−1589	—
NiCl		g	63	—	—

Nickel, Palladium, and Platinum

NiCl$_2$		c	−316	−272	107
NiCl$_2 \cdot$2H$_2$O		c	−923.8	—	—
NiCl$_2 \cdot$4H$_2$O		c	−1526	—	—
NiCl$_2 \cdot$6H$_2$O		c	−2116	−1718	315
NiBr$_2$		c	−227	−213	(137)
NiBr$_2 \cdot$3H$_2$O		c	−1162	—	—
NiI$_2$		c	−85.8	−89.1	(158)
Ni(IO$_3$)$_2$		c	−520.9	−363	(228)
Ni(IO$_3$)$_3$		c	−1108	—	—
Ni(IO$_3$)$_2$		c	−1104	—	—
Ni(IO$_3$)$_2 \cdot$4H$_2$O		c	−1681	—	—
NiS	α	c	−73.2	−74.1	—
NiS	γ	c	−77.8	−114	—
Ni$_3$S$_2$		c	−182	—	—
NiSO$_4$		c	−891.2	−773.9	77.8
NiSO$_4$		aq	−971.5	−778.3	−142
NiSO$_4 \cdot$6H$_2$O	Green	c	−2698.6	—	—

TABLE 1 (Continued)

Formula	Description	State	$\Delta H°/\text{kJ mol}^{-1}$	$\Delta G°/\text{kJ mol}^{-1}$	$\Delta S°/\text{kJ mol}^{-1} \text{K}^{-1}$
$NiSO_4 \cdot 6H_2O$	Blue	c	−2688	−2222	306
$NiSO_4 \cdot 7H_2O$		c	−2983	—	—
$Ni_2S_2O_6 \cdot 6H_2O$		c	−2962	—	—
$NiSe$		c	−42	—	—
$NiTe$		c	−38	—	—
$Ni(N_3)_2 \cdot H_2O$	Nickel azide hydrate	c	190	—	—
$Ni(NO_3)_2$		c	−427.6	−236	(192)
$Ni(NO_3)_2 \cdot 6H_2O$		c	−2223	—	—
$[Ni(NH_3)_4]^{2+}$	Std. state, hyp, $m = 1$	aq	—	−192	—
$[Ni(NH_3)_6]^{2+}$	Std. state, hyp, $m = 1$	aq	—	−250	—

Ni$_2$P	c	-184	—	—
Ni$_3$P	c	-222	—	—
Ni$_5$P$_2$	c	-439	—	—
NiSb	c	-65.3	—	—
Ni$_5$Sb$_2$	c	-152	—	—
NiCO$_3$	c	-664	-613.8	(92)
Ni(CO)$_4$	g	—	—	405
Ni(CO)$_4$	liq	—	—	310
Ni(CN)$_2$	c	113	(155)	(94.1)
[Ni(CN)$_4$]$^{2+}$	aq	364	489.9	(138)

[a]National Bureau of Standards values are in italic; estimated values are in parentheses.
Source: Data from Refs. 4-6.

TABLE 2 Standard Electrode Potentials

Oxidation number change	Half-reaction	$E°/V$	Remarks	References
2 → 0	$Ni^{2+} + 2e^- \rightarrow Ni$	-0.257 ± 0.008		8
		-0.2496	25°C, 1 mol kh^{-1} Ni^{2+}	10
		-0.2480	1 mol kg^{-1} Ni^{2+} activity	10
		-0.231	0.5 M NiSO$_4$	11
		-0.232	Computed from isothermal data	12
		-0.246	Computed from exptl. measurements	13
		-0.250	Calcd.	4
		-0.228	Calcd. from exptl. thermal data	14
			$(dE°/dT)_T = 0.93$ mV K^{-1}	9
			$(dE°/dt)_i = 0.06$ mV K^{-1}	9
	$NiO + 2H^+ + 2e^- \rightarrow Ni + H_2O$	0.116	Calcd.	7
	$Ni(OH)_2 + 2H^+ + 2e^- \rightarrow Ni + 2H_2O$	0.110	Calcd.	7,8,9
	$Ni(OH)_2 + 2e^- \rightarrow Ni + 2OH^-$	-0.72	$(dE°/dt)_i = -1.04$ mV K^{-1}	8,9

	Reaction	E	Note	Ref.
	$HNiO_2^- + 3H^+ + 2e^- \rightarrow Ni + 2H_2O$	0.648	Calcd.	7
	$[Ni(NH_3)_6]^{2+} + 2e^- \rightarrow Ni + 6NH_3$	−0.476		9
	$NiCO_3 + 2e^- \rightarrow Ni + CO_3^{2-}$	−0.45	Calcd. solubility $(dE^\circ/dt)_i^+ = -1.271$ mV K^{-1}	8, 9
	$NiS(\alpha) + 2e^- \rightarrow 2Ni + S^{2-}$	−0.814	Calcd. solubility	9, 33
	$NiS(\beta) + 2e^- \rightarrow 2Ni + S^{2-}$	−0.960	Calcd. solubility	33
	$NiS(\gamma) + 2e^- \rightarrow 2Ni + S^{2-}$	−1.07	Calcd. solubility	9, 33
	$Ni(S_2O_3) + 2e^- \rightarrow Ni + S_2O_3^{2-}$	−0.372	Calcd. solubility	15
	$[Ni(CN)_4]^{2-} + e^- \rightarrow [Ni(CN)_3]^{2-} + CN^-$	−0.401		32
2 → 1				
2.67 → 2	$Ni_3O_4 + 2H_2O + 2e^- \rightarrow 3HNiO_2^- + H^+$	−0.718	Calcd.	7
	$Ni_3O_4 + 2H^+ + 2e^- \rightarrow 3NiO + H_2O$	0.876	Calcd.	7
	$Ni_3O_4 + 2H^+ + 2H_2O + 2e^- \rightarrow 3Ni(OH)_2$	0.897	Calcd.	7
	$Ni_3O_4 + 8H^+ + 2e^- \rightarrow 3Ni^{2+} + 4H_2O$	1.977	Calcd.	7
3 → 2	$Ni(OH)_3 + e^- \rightarrow Ni(OH)_2 + OH^-$	0.48	Calcd.	16
	$Ni_2O_3 + 2H^+ + 2e^- \rightarrow 2NiO + H_2O$	1.020	Calcd.	7
	$Ni_2O_3 + H_2O + 2H^+ + 2e^- \rightarrow 2Ni(OH)_2$	1.032	Calcd.	7
	$Ni_3O_3 + 6H^+ + 2e^- \rightarrow 2Ni^{2+} 3H_2O$	1.753	Calcd.	7

TABLE 2 (Continued)

Oxidation number change	Half-reaction	$E°/V$	Remarks	References
$3 \to 2.67$	$Ni(OH)_3 + 3H^+ + e^- \to Ni^{2+} + 3H_2O$	2.08	Calcd.	16
	$3Ni_2O_3 + 2H^+ + 2e^- \to 2Ni_3O_4 + H_2O$	1.305	Calcd.	7
$4 \to 2$	$NiO_2 + 2H_2O + 2e^- \to Ni(OH)_2 + 2OH^-$	−0.490	Calcd.	9
	$2NiO_2 + 2H^+ + 2e^- \to Ni_2O_3 + H_2O$	1.434	Calcd.	7
	$NiO_2 + 4H^+ + 2e^- \to Ni^{2+} + 2H_2O$	1.593	Calcd.	7
		1.678	Calcd.	9
$4 \to 3$	$Ni(OH)_4 + e^- \to Ni(OH)_3 + OH^-$	0.6	Calcd.	16
$6 \to 2$	$NiO_4^{2-} + 8H^+ + 4e^- \to Ni^{2+} + 4H_2)$	>1.6	Calcd.	4

The potential of the Ni/Ni^{2+} couple cannot be given accurately because a truly reversible equilibrium appears difficult to attain. Thus the following values were reported: -0.227 V [17], -0.231 V [11], and -0.248 V [10]. However, calculations [14] from the Gibbs energy of the reaction

$$Ni^{2+}(aq) + 2e^- \rightarrow Ni^0$$

give the value -0.228 V [14], in satisfactory agreement with that reported by Carr and Bonilla [12]. The Gibbs energy data given by the National Bureau fo Standards corresponds to -0.241 V. However, this is based on the entropy of Ni^{2+}(aq) equal to -159 J K^{-1}. This value is relatively low in comparison to Fe^{2+} and Cu^{2+}, although there may be a considerable divergence because of the multiplicity of the ground states. Therefore, on the basis of the Gibbs energy change, the formation of Ni^{2+}(aq) is taken as -46.4 kJ mol^{-1}, as indicated by Milasso and Caroli [8] for the above-mentioned reaction, $E° = -0.257$ V. The dispersion of standard potential values derived from electromotive force (emf) measurements [10,11,13,17] is probably due either to the interference of oxygen or to the onset of mixed potentials instead of reversible potentials.

The standard potential values of other equilibria referred to in Table 2 involve the equilibrium constants of different Ni^{2+}(aq) species (Table 3). Then, from the Gibbs energy of nickel hydroxide, the $E°$ value of the reaction

$$Ni(OH)_2 + 2e^- \rightarrow Ni + 2OH^- \qquad E° = -0.720 \text{ V}$$

yields

$$Ni(OH)_2 \rightarrow Ni^{2+} + 2OH^- \qquad K_b = 1.6 \times 10^{-16}$$

This K_b value is in reasonably good agreement with the reported data [18-20]. However, in view of the uncertainty about the Gibbs energy of Ni^{2+}(aq), the experimental value K_b is preferred to that derived from thermal data. The formal first hydrolysis constant of Ni^{2+}(aq) is given by $pK_a' = 1692/T + 4.09$ [21].

The standard potential of the reaction

$$[Ni(NH_3)_6]^{2+} + 2e^- \rightarrow Ni + 6NH_3(aq) \qquad E° = -0.49 \text{ V}$$

was derived from the equilibrium data given by Bjerrum et al. [22] for the reactions:

$$[Ni(NH_3)_4]^{2+} \rightarrow Ni^{2+} + 4NH_3(aq) \qquad K = 1 \times 10^{-8}$$

$$[Ni(NH_3)_6]^{2+} \rightarrow Ni^{2+} + 6NH_3(aq) \qquad K = 1.8 \times 10^{-9}$$

However, an $E°$ value equal to -0.476 V has also been reported [9]. Both $E°$ values have the same degree of uncertainty.

Nickelous ion forms the [Ni(CN)$_4$]$^{2-}$ complex with cyanide ions. Using a very approximate value for the entropy of the complex ion, Latimer calculated [4]

TABLE 3 Equilibrium Constants of Reactions Involving Ni^{2+} in Aqueous Solutions

Reaction	Equilibrium constant	Conditions	References
$Ni(OH)_2 \rightarrow Ni^{2+} + 2OH^-$	$K = 1.6 \times 10^{-16}$	Calcd.	4
	6.5×10^{-18}	Exptl. measurement	18
	1.6×10^{-14}	Equil. measurement	19
	6.2×10^{-16}	Exptl. measurement	20
$Ni^{2+} + H_2O \rightarrow NiOH^+ + H^+$	$K = 2.3 \times 10^{-11}$	$I = 0$ (corr.)	22, 27
	1.2×10^{-9}	$I = 0$ (corr.)	22, 26
	5.0×10^{-9}	$NiSO_4$ var.	22, 23
	3.2×10^{-10}	$NiSO_4$	25
	3.2×10^{-7}	$NiCl_2$	22, 23
	5.1×10^{-10}	$NiCl_2$	25
	2.5×10^{-9}	$NiCl_2$ dil.	22, 28
	4.0×10^{-10}	$30°C, I = 0.1$ (KCl)	22, 24
	5.95×10^{-10}	$Ni(NO_3)_2$	25
	5.02×10^{-10}	$Ni(ClO_4)_2$	25

Reaction	K	Notes	Ref
$Ni(OH)_2 \rightarrow H^+ + HNiO_2^-$	1.7×10^{-10}	$0.25 < I < 1$ (NaClO$_4$)	21
	$K = 6 \times 10^{-19}$	Expt. measurement	18
$NiOH^+ + H_2O \rightarrow Ni(OH)_2 + H^+$	$K = 2.7 \times 10^{-5}$ to 8.5×10^{-3}	Calcd.	25
$Ni^{2+} + H_2O \rightarrow NiO + 2H^+$	—	$\log(Ni^{2+}) = 12.18 - 2pH$ (hydrated oxide)	7
	—	$\log(Ni^{2+}) = 12.41 - 2pH$ (unhydrated oxide)	7
$NiO + H_2O \rightarrow HNiO_2^- + H^+$	—	$\log(HNiO_2^-) = -18.22 + pH$ (hydrated oxide)	7
$[Ni(NH_3)_4]^{2+} \rightarrow Ni^{2+} + 4NH_3(aq)$	$K = 1 \times 10^{-8}$	—	4
$[Ni(NH_3)_6]^{2+} \rightarrow Ni^{2+} + 6NH_3(aq)$	$K = 1.8 \times 10^{-9}$	—	4
$[Ni(CN)_4]^{2+} \rightarrow Ni^{2+} + 4CN^-$	$K = 10^{-22}$	—	29
$NiCO_3 \rightarrow Ni^{2+} + CO_3^{2-}$	$K = 1.36 \times 10^{-7}$	—	30
$NiS(\alpha) \rightarrow Ni^{2+} + S^{2-}$	$K = 3 \times 10^{-21}$	—	31
$NiS(\beta) \rightarrow Ni^{2+} + S^{2-}$	$K = 1 \times 10^{-26}$	—	31
$NiS(\gamma) \rightarrow Ni^{2+} + S^{2-}$	$K = 2 \times 10^{-28}$	—	31

$$[Ni(CN)_4]^{2-} \to Ni^{2+} + 4CN^- \qquad K = 10^{-22}$$

This value has been checked by Hume and Kolthoff [29]. Powerful reducing agents reduce $[Ni(CN)_4]^{2-}$ to the +1 complex ion $[Ni(CN)_4]^{3-}$. The potential of the reaction

$$[Ni(CN)_4]^{2-} + e^- \to [Ni(CN)_3]^{2-} + CN^-$$

as determined by polarography [32], gives $E° = -0.401$ V.

The solubility product of nickel carbonate reported by Ageno and Balla [30] is

$$NiCO_3 \to Ni^{2+} + CO_3^{2-} \qquad K_s = 1.36 \times 10^{-7}$$

From this K_s value, Latimer computed the standard Gibbs energy change of $NiCO_3$ as -615.0 kJ mol^{-1} and

$$NiCO_3 + 2e^- \to Ni + CO_3^{2-} \qquad E° = -0.45 \text{ V}$$

Theil and Gessner [31] have reported three forms of nickel sulfide: the α, β, and γ forms. Their solubility products are shown in Table 3. The α form, mixed with some β and γ, is precipitated in cold alkaline solution. The conversion of α-nickel sulfide to the other forms is favored by acids. Based on the solubility products, the $E°$ values are given for the following reactions:

$$NiS(\alpha) + 2e^- \to S^{2-} + Ni \qquad E° = -0.83 \text{ V}$$

$$NiS(\gamma) + 2e^- \to S^{2-} + Ni \qquad E° = -1.04 \text{ V}$$

The values calculated from the solubility data given by Ringbom [33] are slightly different, as can be seen in Table 2.

1. Electrode Potentials and Galvanic Cells Involving Oxo- and Hydroxonickel Species

The oxidation of nickelous hydroxide in alkaline solution appears to form solid solutions of Ni^{2+} and Ni^{3+} compounds with a variable water content. With fairly high concentrations of hydroxide, the probable formation of pure $NiO_2(s)$ was assumed [34]. This oxide mixture is the oxidizing agent used in the Edison storage battery, and the electrode is often considered as the $Ni(OH)_2/NiOOH$ couple [35]. For the Ni^{2+}/NiO_2 couple in acid, Latimer computed

$$NiO_2 + 4H^+ + 2e^- \to Ni^{2+} + 2H_2O \qquad E° = +1.68 \text{ V}$$

This potential is sufficient to cause the rapid oxidation of water, so the oxide is unstable and acid solutions.

TABLE 4 Electrode Potentials (vs. NHE) Involving Hydroxonickel Species

Half-cell	E/V	Remarks	Reference
Ni, NiO(OH)|0.5 M NiCl$_2$	$E_{initial}$ = 1.17	Diminishes with time; pH 5	46
Ni, NiO(OH)|Ni(ClO$_4$)$_2$	1.19	pH 5	46
Ni, NiO(OH)|NiBr$_2$	$E_{initial}$ = 1.02	Diminishes with time; pH 5	46
Ni, NiO(OH)|NiSO$_4$	1.19	pH 5	46
Ni, NiO(OH)|M NaOH	0.56	pH 13.8	7
Ni, NiO(OH)|2.8 M KOH	$E_{initial}$ = 0.60	Diminishes with time	46
Ni, NiO(OH)|0.05 M Na$_2$B$_4$O$_7$	0.85	pH 9.2	7
Ni, NiO(OH)|0.5 M Na$_2$CO$_3$	0.68	pH 12	7
Pt|Ni$_3$O$_2$(OH)$_4$, NiO(OH)|0.1 M NaH	0.15	N$_2$ atm	47
Pt|Ni$_3$O$_2$(OH)$_4$, NiO(OH)|0.5 M Na$_2$CO$_3$	0.24	—	47
Ni|Ni(OH)$_2$|0.1 M NaOH	−0.60	N$_2$ atm	47
Ni|Ni(OH)$_2$|25% KOH	−0.33	Ref.: NCE	48
Ni|Ni(OH)$_2$|0.5 M Na$_2$CO$_3$	−0.51	—	47
Pt|Ni(OH)$_2$·Ni$_3$O$_2$(OH)$_4$|0.1 M nAOH	−0.35	N$_2$ atm	16
Pt|Ni(OH)$_2$·Ni$_3$O$_2$(OH)$_4$|0.5 M Na$_2$CO$_3$	−0.26	pH 11.5	16

TABLE 5 Open-Circuit Potentials of Galvanic Cells Incorporating Nickel Species

Galvanic cell	E/V	Conditions	References
NiO(OH)|25% KOH|H_2	1.305 (10°C)	E_{calcd} from Ni(OH)$_3$ → Ni(OH)$_2$ at 10°C, 1.308 V	41, 42
NiO(OH)|25% KOH|H_2	1.266 (65°C)		41, 42
NiO(OH)|25% KOH + 20 g L^{-1} Zn(OH)$_2$|Zn(Hg)	1.75 (10–20°C) 10%	Zn–90% Hg amalgam	41
NiO(OH)|0.5 M NiSO$_4$|Zn	1.630	—	49
NiO(OH)|satd. (NH$_4$)$_2$Ni(SO$_4$)$_2$|Zn	1.643	—	49
NiO(OH)|0.5 M NiSO$_4$|Pb	1.139	—	49
NiO(OH)|satd. (NH$_4$)$_2$Ni(SO$_4$)$_2$|Pb	1.138	—	49
NiO(OH)|0.5 M NiSO$_4$|Ni	1.086	—	49
NiO(OH)|satd. (NH$_4$)$_2$Ni(SO$_4$)$_2$|Ni	1.075	—	49

NiO(OH)|0.5 M NiSO$_4$|Cu	0.551	—	49
NiO(OH)|satd. (NH$_4$)$_2$Ni(SO$_4$)$_2$|Cu	0.560	—	49
NiO(OH), Ni(OH)$_2$|KOH|HgO, Hg	$E \approx \log c_{KOH}$ (up to 1 M)	$0.01\,M \leqslant c_{KOH} \leqslant 8\,M$	50
Ni|NiO$_{1.10}$|M KOH|HgO|Hg	0.423 ± 0.005	—	39, 40
Ni|NiO$_{1.20}$|M KOH|HgO|Hg	0.427	—	39, 40
Ni|NiO$_{1.25}$|M KOH|HgO|Hg	0.423	—	39, 40
Ni|NiO$_{1.25}$|0.01 M KOH|HgO|Hg	0.429	—	39, 40
Ni|NiO$_{1.25}$|0.1 M KOH|HgO|Hg	0.425	—	39, 40
Ni|NiO$_{1.25}$|1.01 M KOH|HgO|Hg	0.419	—	39, 40
Ni|NiO$_{1.25}$|7.3 M KOH|HgO|Hg	0.390	—	39, 40
Ni|NiO$_{1.25}$|14.6 M KOH|HgO|Hg	0.297	—	39, 40

TABLE 6 Electrode Potentials of Nickel Sulfide Electrodes

Half-electrode	E/V	Reference
Ni_3S_2|0.5 M $NiSO_4$||SCE	−0.235	51
Ni_2S_2|200 $NiSO_4 \cdot 7H_2O$ + 4ONa_2 + 2OH_3BO_3 + 3NaCl||NHE	0.07	52
Ni_3S_2|100 g L^{-1} H_2SO_4||SCE	0.202	53
NiS|satd. NaCl||SCE	−0.330 (20°C)	54

When nickelous hydroxide is fused with potassium nitrate and potassium hydroxide, the nickelate compound, K_2NiO_4, is formed. It is also unstable in acid solutions because of its high oxidizing power:

$$NiO_4^{2-} + 8H^+ + 4e^- \rightarrow 4H_2O + Ni^{2+} \qquad E^\circ > 1.8 \text{ V}$$

Claims are also made for the formation by the same process of some pernickelate, K_2NiO_5, but the evidence is not conclusive.

The interpretation of the experimental data related to the electrode potential involving either oxygen or hydroxo-containing nickel species are open to question because nickel oxide electrodes may involve nonstoichiometric oxidation states, and consequently equilibrium potentials are difficult to achieve [35-37]. Hickling and Spice [38] measured the potential of the cell

$$Pt\,|\,NiO(Ni)\,|\,NaOH(xM)\,|\,NHE$$

and obtained a potential equal to 1.39 V for 1 M NaOH, after correction for the hydrogen electrode in the basic solution. This figure is close to the standard potential of the β-NiO(OH)/NiO$_2$ couple [7]. But as NiO$_2$ has never been isolated, under steady-state conditions the oxidation state of the metal should be related to Ni^{3+}. Experimental electrode potentials of hydroxonickel electrodes are assembled in Table 4.

Table 5 refers to the electromotive force of galvanic cells with the participation of nickel-oxygen compounds. Data reported by Bourgault and Conway [39] and Conway and Gileadi [40] are considered to be the most reliable. The real reversible potential of the Ni^{2+}/Ni^{3+} couple equal to 0.424 V (vs. HgO electrode) has been determined at the intersection point of the anodic and cathodic potential/log(time) decay curves. From the open-circuit potential of the positive plate of the alkali accumulator, Zedner [41,42], Foerster [43, 44] and Foerster and Krueger [45] obtained a value between 0.47 and 0.49 V (vs. HgO electrode).

2. *Electrode Potential of Nickel Sulfide Electrodes*

The potential of nickel sulfide electrodes in various electrolytes measured against different reference electrodes [52-54] are shown in Table 6. The same objections already reported for the other electrodes also apply to the experimental data of nickel sulfide electrodes. These potentials should be considered as rest potential values.

C. Potential Diagrams for Nickel

Acid Solution

$$\begin{array}{cccc} +6 & +4 & +2 & 0 \end{array}$$

$$NiO_4^{2-} \xrightarrow{>1.8} NiO_2 \xrightarrow{1.593} Ni^{2+} \xrightarrow{-0.257} Ni$$

$$\underline{\qquad >1.6 \qquad}$$

Basic Solution

$$NiO_2^{2-} \xrightarrow{>0.4} NiO_2 \xrightarrow{0.490} Ni(OH)_2 \xrightarrow{-0.72} Ni$$

REFERENCES

1. F. A. Cotton and G. Wilkinson, *Advanced Inorganic Chemistry*, Wiley, New York, 1966.
2. G. C. Demitras, C. R. Russ, J. F. Salmon, J. H. Weber, and G. S. Weis, *Inorganic Chemistry*, Prentice-Hall, Englewood Cliffs, N.J., 1972.
3. *Gmelins Handbuch der Anorganische Chemie*, 8th ed., *Nickel*, Part B-2, Verlag Chemie, Weinheim, West Germany, 1966.
4. W. M. Latimer, *Oxidation Potentials*, 2nd ed., Prentice-Hall, Englewood Cliffs, N.J., 1952, p. 200.
5. *Selected Values of Chemical Thermodynamic Properties*, Natl. Bur. Stand. 500, 1952.
6. O. Kubaschevski and E. L. Evans, *Metallurgical Thermochemistry*, Pergamon Press, Oxford, 1958.
7. M. Pourbaix, ed., *Atlas of Electrochemical Equilibria in Aqueous Solutions*, Pergamon-CEBELCOR, Brussels, 1966.
8. G. Milazzo and S. Caroli, *Tables of Standard Electrode Potentials*, Wiley, New York, 1978.
9. A. J. de Béthune and N. A. Swendeman Loud, *Standard Aqueous Electrode Potentials and Temperature Coefficients*, Hampel, Skokie, Ill., 1964.
10. F. Murate, Bull. Chem. Soc. Jpn., *3*, 57 (1928).
11. M. M. Haring and E. G. van den Bosche, J. Phys. Chem., *33*, 161 (1929).
12. D. S. Carr and C. F. Bonilla, J. Electrochem. Soc., *99*, 475 (1952).
13. M. A. López-López, C. R. Acad. Sci., Paris, *256*, 2594 (1963); *255*, 3170 (1962).
14. J. W. Larson, P. Cerutti, H. K. Garber, and L. O. Kepler, J. Phys. Chem., *72*, 2902 (1968).
15. T. O. Denney and C. B. Monk, Trans. Faraday Soc., *47*, 992 (1951).
16. A. Prokopcikas, Liet. TSR Mokslu Akad. Dar. Ser. B, 31 (1962); Chem. Abstr., *58*, 1915 (1963).
17. I. Colombier, C. R. Acad. Sci., Paris, *199*, 273 (1934).
18. K. H. Gayer and A. B. Garret, J. Am. Chem. Soc., *71*, 2953 (1949).
19. H. J. Wijs, Recl. Trav. Chim. Pays-Bas, *44*, 663 (1925).
20. R. Näsänen, Ann. Acad. Fenn., *A-59*, 3 (1943).
21. J. A. Bolzán, E. A. Jáuregui, and A. J. Arvía, Electrochim. Acta, *8*, 841 (1963).
22. J. Bjerrum, G. Schwarzenbach, and L. G., Sillén, *Stability Constants*, Vol. 2: *Inorganic Ligands*, Chemical Society, London, 1958, p. 13.
23. H. G. Denham, J. Chem. Soc., *93*, 41 (1908).
24. S. Chaberek, R. C. Courtney, and A. E. Martell, J. Am. Chem. Soc., *74*, 5557 (1952).
25. Z. Ksandr and M. Hejtmanek, Sb. Celostatni Pracovni Konf. Anal. Chem., 1st, Prague, 1952, p. 42; Chem. Abstr., *50*, 3150g (1956).
26. F. Cuta, Z. Ksandr, and M. Hejtmanek, Chem. Listy, *50*, 1064 (1956).
27. K. H. Gayer and N. Woonter, J. Am. Chem. Soc., *74*, 1437 (1952).
28. C. Kullgren, Z. Phys. Chem., *85*, 466 (1913).
29. D. H. Hume and I. M. Kolthoff, J. Am. Chem. Soc., *72*, 4423 (1950).
30. F. Ageno and E. Valla, Atti Accad. Naz. Lincei, Rend., Cl. Sci. Fis. Mat. Nat., [5], *20*, 706 (1911).
31. A. Thiel and H. Gessner, Z. Anorg. Allg. Chem., *86*, 49 (1914).
32. V. Gaglioti, G. Sartori, and P. Silvestroni, Ric. Sci., *17*, 624 (1947).

33. A. Ringbom, *Solubility of Sulfides*, Report to Analytical Section, IUPAC, July 1953.
34. O. R. Howell, J. Chem. Soc., *123*, 1772 (1923).
35. W. Fischer, Z. Phys. Chem NF, *103*, 77 (1976).
36. I. A. Dibrov and T. V. Grigorieva, Elektrokhimiya, *13*, 979 (1977).
37. I. A. Dibrov, Elektrokhimiya, *14*, 114 (1978).
38. A. Hickling and J. E. Spice, Trans. Faraday Soc., *43*, 762 (1947).
39. P. L Bourgault and B. E. Conway, Can. J. Chem., *38*, 1557 (1960).
40. B. E. Conway and E. Gileadi, Can. J. Chem., *40*, 1933 (1962).
41. J. Zedner, Z. Elektrochem., *11*, 809 (1905); *12*, 463 (1906).
42. J. Zedner, Z. Elektrochem., *13*, 752 (1907).
43. F. Foerster, Z. Elektrochem., *13*, 414 (1907).
44. F. Foerster, Z. Elektrochem., *14*, 17 (1908).
45. F. Foerster and F. Krueger, Z. Elektrochem., *33*, 406 (1927).
46. K. Georgi, Z. Elektrochem., *38*, 681 (1932); *38*, 714 (1932).
47. S. E. S. El Wakkad and S. E. Emara, J. Chem. Soc., 3504 (1953).
48. M. deKay Thompson and H. K. Richardson, Trans. Am. Electrochem. Soc., *7*, 95 (1905).
49. W. Pfanhauser, Z. Elektrochem., *7*, 678 (1901).
50. F. Kornfiel, Proc. Ann. Battery Res. Dev. Conf., 12th, Fort Monmouth, N.J., 1958, p. 18.
51. A. I. Zurin, Tr. Leningr. Politekhn. Inst., *6*, 38 (1953).
52. A. A. Bulakh and O. A. Khan, Zh. Prilk. Khim., *27*, 166 (1954).
53. B. Z. Ustinskii and D. M. Chizhikov, Zh. Prikl. Khim., *22*, 1249 (1949).
54. N. S. Fortunatov and V. I. Mikhailovskaya, Ukr. Khim. Zh., *16*, 667 (1951).

II. PALLADIUM AND PLATINUM*

FRANCISCO COLOM   Institute of Physical Chemistry, CSIC, Madrid, Spain

A. Palladium

1. Oxidation States

In solid compounds, palladium takes up mainly the II and IV oxidation states (PdO, PdO_2) together with a minor occurrence of I and III species (Pd_2O, Pd_2O_3). In aqueous solutions, only the oxidation states II and IV. which appear in the form of palladium complexes with coordination numbers 4 and 6, respectively, are known.

The element shows a great capacity for hydrogen absorption, a property that depends greatly on the physical condition of the metal. This phenomenon takes place with changes in the crystalline structure of the metal with formation mainly of α and β solid solutions and the Pd_2H (or Pd_4H_2) hydride. The thermodynamic data are given in Table 7.

2. Palladium-Palladous Couples

The half-cell potential of the Pd^{2+}/Pd couple was measured by Izatt et al. [3] in $HClO_4$ solutions at zero ionic strength, giving the value

---

*Reviewer: Karl Kadish, University of Houston, Houston, Texas.

TABLE 7 Thermodynamic Data on Palladium[a]

Formula	Description	State	$\Delta H°$/kJ mol^{-1}	$\Delta G°$/kJ mol^{-1}	$S°$/J mol^{-1} K^{-1}
Pd	Std. state	c	0	0	37.71
Pd		g	378.2	339.7	-167.05
Pd$^+$		g	1188.746		
Pd^{2+}		g	3069.4		
Pd^{2+}	Std. state, m = 1	aq	149.0	176.5	-184
Pd^{3+}		g	6251		
PdO		c	-85.4		
PdO		g	348.9	325.9	218
Pd$_2$H		c	-19.7		
Pd(OH)$_2$	Precipitated	c	-395.0		
Pd(OH)$_4$	Precipitated	c	-715.9		
PdCl$^+$	Std. state, m = 1	aq	-38	22.6	-117
PdCl$_2$		c	-198.7		
PdCl$_2$	Undiss.; std. state, m = 1	aq		-128.8	

Nickel, Palladium, and Platinum

Species	Notes	State			
$PdCl_3^-$	Std. state, $m = 1$	aq		−276.1	
$PdCl_4^{2-}$	Std. state, $m = 1$; in 1 M HCl	aq	−550.2	−417.1	167
$PdCl_6^{2-}$	Std. state, $m = 1$; in 1 M HCl	aq	−598	−430.0	272
H_2PdCl_6		aq	−551.9		
$PdBr_2$		c	−104.2		
$PdBr_3^-$	Std. state, $m = 1$	aq		−204.2	
$PdBr_4^{2-}$	Std. state, $m = 1$	aq	−384.9	−318.0	247
$PdBr_6^{2-}$	Std. state, $m = 1$	aq		−335.1	
PdI_2		c	−63.2	−11.5	180
PdI_4^{2-}	Std. state, $m = 1$	aq		−159.0	
PdI_6^{2-}	Std. state, $m = 1$	aq		−170.3	
PdS		c	−75	−67	46
PdS_2		c	−81.2	−74.5	79
Pd_4S		c	−67	−67	180
$Pd(NO_2)_4^{2-}$	Std. state, $m = 1$	aq		−68	
$Pd(NH_3)_4^{2+}$	Std. state, $m = 1$	aq		−75	
$PdCl_2 \cdot 2NH_3$		c	−460.		

TABLE 7 (Continued)

Formula	Description	State	ΔH°/kJ mol^{-1}	ΔG°/kJ mol^{-1}	S°/J mol^{-1} K^{-1}
PdCl$_2$·4NH$_3$		c	−686		
Pd(N$_3$)Cl$_2^{2-}$	Std. state, $m = 1$	aq		557	
PdI$_2$·2NH$_3$		c	−297		
PdI$_2$·4NH$_3$		c	−498.7		
PdCl$_3$(C$_2$H$_4$)$^-$	Std. state, $m =1$	aq	−370.3	−210.3	176
Pd(CN)$^+$	Std. state, $m =1$	aq		289	
Pd(CN)$_2$		c	205.4		
Pd(CN)$_4^{2-}$	Std. state, $m =1$	aq		628	
Pd(CNS)$_2$		c		243.3	
Pd(CNS)$_2$	Undiss.; std. state, $m = 1$	aq		314.3	
Pd(CNS)$_4^{2-}$	Std. state, $m =1$	aq		410.5	
PdI$_2$(CNS)$^-$	Std. state, $m =1$	aq		23.9	

[a]National Bureau of Standards values are in italic.
Source: Data from Refs. 1 and 2.

$$Pd^{2+}(aq) + 2e^- \rightarrow Pd(s) \qquad E° = 0.915 \pm 0.005 \text{ V}$$

This potential is lower than the formal one reported by Templeton and co-workers [4,5], $E°' = 0.987$ V for the same solution at ionic strength $I = 4$ mol L^{-1}, which when adjusted to zero ionic strength yields $E° = 0.945$ V. A formal potential of $E° = 0.979 \pm 0.005$ V was determined by the former authors [6] in 3.94 mol kg^{-1} HClO$_4$ solutions. Polarographic measurements [7] confirmed the Izatt et al. value.

From thermodynamic data for the hydroxide, Pourbaix et al. [8] calculated

$$Pd(OH)_2(s) + 2H^+ + 2e^- \rightarrow Pd(s) + 2H_2O \qquad E° = 0.897 \text{ V}$$

which coincides with that derived by Goldberg and Hepler [5] from pH, spectrophotometric, and solubility measurements [3]. However, the $\Delta G° = -301.25$ kJ mol^{-1} deduced from these measurements and calculations leads to the cited potential, which has not been considered reliable by the National Bureau of Standards (NBS) and is not listed in Table 7.

Divergent values have been obtained for the potential of the half-reaction

$$PdO(s) + 2H^+ + 2e^- \rightarrow Pd(s) + H_2O$$

according to the different data from which it has been deduced. Thus the standard potential estimated from thermochemical [8-12] data and electromotive force (emf) measurements [13] coincides with the value $E° = 0.79$ V determined experimentally by Hoar [14], but it differs from that given by Pourbaix et al. [8] ($E° = 0.917$ V) based on thermodynamic calculations.

On the basis of their potential measurements, several authors [4,15] have reported $E°' = 0.62$ V for the half-reaction

$$PdCl_4^{2-}(aq) + 2e^- \rightarrow Pd(s) + 4Cl^-(aq)$$

in 1 M HCl solutions and $E°' = 0.60$ V for 1 M KCl. These values were confirmed by Kravtsov and Zelenskii [16] for 1 M HCl, $E°' = 0.59$ V, and by the study of the equilibrium constants for the stepwise dissociation

$$Pd^{2+}(aq) + 4Cl^-(sq) \rightarrow PdCl_4^{2-}(aq) \qquad \log \beta_4 = 11 \; [5,17]$$

The polarographic determinations [7] gave a slightly higher value, $E° = 0.64 \pm 0.01$ V. For the corresponding bromide complex, potential determinations [15] and estimations from stability constants [5,17] with $\log \beta_4 = 15$ yield

$$PdBr_4^{2-}(aq) + 2e^- \rightarrow Pd(s) + 4Br^-(aq) \qquad E°' = 0.49 \text{ V}$$

A similar agreement [5] is obtained for the half-reaction

$$PdI_4^{2-}(aq) + 2e^- \rightarrow Pd(s) + 4I^-(aq) \qquad E°' = 0.18 \text{ V}$$

with the stability constant log β_4 = 25 for the equilibrium

$$Pd^{2+}(aq) + 4I^-(aq) \rightarrow PdI_4^{2-}(aq)$$

The temperature dependence of the standard potential of the system $PdCl_4^{2-}/Pd$ was determined by Nikolaeva et al. [18] through emf measurements in the range 50 to 150°C and ionic strength I = 1 mol L^{-1}. The results fit the equation $E^{o'}/V = 0.619 - 8.8 \times 10^{-4}[T/K - 323]$. A lower-temperature coefficient was found for the $PdBr_4^{2-}/Pd$ couple:

$$E^{o'}/V = 0.52 - 6.3 \times 10^{-4}[T/K - 323]$$

Direct emf measurements in 0.5 M KNO_2 solutions [15] lead to $E^{o'}$ = 0.34 V for the half-reaction

$$Pd(NO_2)_4^{2-}(aq) + 2e^- \rightarrow Pd(s) + 4NO_2^-$$

For the amine complex, the following potential has been calculated:

$$Pd(NH_3)_4^{2+}(aq) + 2e^- \rightarrow Pd(s) + 4NH_3 \qquad E^{o'} = 0.0 \text{ V}$$

on the basis of stability constants (logβ_4 = 30.5) measured for the equilibrium $Pd^{2+}(aq) + 4NH_3(aq) = Pd(NH_3)_4^{2+}(aq)$ in ammonia solutions at I = 1 mol L^{-1} [5]. From polarographic determinations [7], the value E° = 0.82 ± 0.01 V has been deduced for the half-reaction $Pd(SO_4)_2^{2-}(aq) + 2e^- \rightarrow Pd(s) + 2SO_4^{2-}$, where the stability constant is logβ_2 = 3.16 ± 0.15.

3. Palladium in the IV and VI States

The potentials for the half-reactions

$$PdO_2(s) + 2H^+(aq) + 2e^- \rightarrow PdO(s) + H_2O \qquad E^\circ = 1.263 \text{ V}$$

$$PdO_2(s) + 4H^+(aq) + 2e^- \rightarrow Pd^{2+}(aq) + 2H_2O \qquad E^\circ = 1.194 \text{ V}$$

$$PdO_2(s) + 2H^+(aq) + 2e^- \rightarrow Pd(OH)_2(s) \qquad E^\circ = 1.283 \text{ V}$$

have been calculated by Pourbaix et al. [8] from thermodynamic data, whereas Goldberg and Hepler [5] have estimated from thermochemical results

$$Pd(OH)_4(s) + 2H^+(aq) + 2e^- \rightarrow Pd(OH)_2(s) + 2H_2O \qquad E^\circ = 1.258 \text{ V}$$

$$Pd(OH)_4(s) + 4H^+ + 4e^- \rightarrow Pd(s) + 4H_2O \qquad E^\circ = 1.228 \text{ V}$$

These values are doubtful because of the lack of dependable values for the Gibbs energy of formation of the palladium oxides and hydroxides. Thus the standard potential of the PdO$_2$/Pd couple measured potentiometrically by Hoar [14] in acid solutions is notably higher, $E° = 1470 ± 10$ mV.

The earlier potentiometric value found by Wellman [19] for the half-reaction

$$PdCl_6^{2-}(aq) + 2e^- \rightarrow PdCl_4^{2-}(aq) + 2Cl^-(aq) \qquad E° = 1470 ± 10 \text{ mV.}$$

was confirmed later by Grinberg and Shamsiev [20] through direct measurements of the system in 1 M HCl solutions. The authors also reported the following data:

$$PdBr_6^{2-}(aq) + 2e^- \rightarrow PdBr_4^{2-}(aq) + 2Br^-(aq) \qquad E°' = 0.99 \text{ V}$$

$$PdI_6^{2-}(aq) + 2e^- \rightarrow PdI_4^{2-}(aq) + 2I^-(aq) \qquad E°' = 0.48 \text{ V}$$

for 1 M KBr and 1 M KI solutions, respectively.

From data on anodic formation of PdO$_3$ in alkaline solutions Pourbaix et al. [8] have deduced

$$PdO_3(s) + 2H^+(aq) + 2e^- \rightarrow PdO_2(s) + H_2O \qquad E° = 2.030 \text{ V}$$

Babaeva and Khananova [21] have reported the following potentials for ethylenediamine complexes from emf determinations at 25°C in $10^{-5}-10^{-6}$ M nitrate solutions:

$$PdCl_2(en)_2^{2+}(aq) + 2e^- \rightarrow Pd(en)_2^{2+}(aq) + 2Cl^-(aq) \qquad E°' = 1.13 \text{ V}$$

$$PdBr(en)_2^{2+}(aq) + 2e^- \rightarrow Pd(en)_2^{2+}(aq) + 2Br^-(aq) \qquad E°' = 0.69 \text{ V}$$

$$PdI_2(en)_2^{2+}(aq) + 2e^- \rightarrow Pd(en)_2^{2+}(aq) + 2I^-(aq) \qquad E°' = 0.63 \text{ V}$$

4. Potential Diagram for Palladium

$$Pd^{6+} \xrightarrow{2.03} Pd^{4+} \xrightarrow{1.263} Pd^{2+} \xrightarrow{0.915} Pd(0)$$

B. Platinum

1. Oxidation States

The chemistry of platinum involves a great variety of species in which the element is mainly present in the II and IV oxidation states, and although some Pt(III) and Pt(I) compounds have been reported, they are poorly defined. As is characteristic of the transition elements, platinum shows a great capacity to form complex species with coordination numbers 4 and 6 for the II and IV oxidation states, respectively. The chemistry of platinum in aqueous solutions deals almost exclusively with these complexes.

The metal is most stable in aqueous media at almost any pH level in the absence of complexing agents or chemicals. Only in very acid solutions at high potentials is the hydrated oxide PtO$_2 \cdot n$H$_2$O or unstable Pt(OH)$_2$ formed. The thermodynamic data are given in Table 8.

TABLE 8 Thermodynamic Data on Platinum[a]

Formula	Description	State	$\Delta H°$/kJ mol^{-1}	$\Delta G°$/kJ mol^{-1}	$S°$/J mol^{-1} K^{-1}
Pt	Std. state	c			41.63
Pt		g	565.3	520.5	192.406
Pt$^+$		g	1402.1		
Pt^{2+}		g	3199.1		
Pt^{2+}	Std. state, $m = 1$	aq		254.8	
PtO$_2$		g	171.5	167.8	
Pt$_3$O$_4$		c	−163		
Pt(OH)$_2$		c	−351.9		
PtF$_6$	Cubic	c			235.6
PtF$_6$		g			348.34
PtCl		c	−56.5		
PtCl$_2$		c	−123.4		
PtCl$_2$	Std. State, $m = 1$	aq		−73.2	
PtBr$_3^-$	Undiss.; std. state, $m = 1$	aq		−144.8	

Nickel, Palladium, and Platinum

Formula		State			
PtCl$_3$		c	−182.0		
PtCl$_3^-$	Std. state, $m = 1$	aq		−221.7	
PtCl$_4$		c	−231.8		
PtCl$_4$		aq	−314.2		
PtCl$_4 \cdot$5H$_2$O		c	−1752.7		
PtCl$_4^{2-}$	Std. state, $m = 1$	aq	−499.2	−361.4	155
PtCl$_6^{2-}$	Std. state, $m = 1$	aq	−668.2	−482.7	219.7
HPtCl$_5 \cdot$2H$_2$O		c	−1012.5		
H$_2$PtCl$_6$	Std. state, $m = 1$	aq	−668.2	−482.7	219.7
H$_2$PtCl$_6 \cdot$6H$_2$O		c	−2371.1		
[PtCl$_2$(OH)$_2$]$^{2-}$	Std. state, $m = 1$	aq		−470.6	
[PtCl$_2$(OH)H$_2$O]$^-$	Std. state, $m = 1$	aq		−517.8	
[PtCl$_2$(H$_2$O)$_2$]	Undiss.; std. state, $m = 1$	aq		−547.6	
[PtCl$_3$(OH)]$^{2-}$	Std. state, $m = 1$	aq		−418.3	
[PtCl$_3$(H$_2$O)]$^-$	Std. state, $m = 1$	aq		−458.5	
PtBr		c	−38.5		

348   Chapter 12

TABLE 8 (Continued)

Formula	Description	State	$\Delta H°$/kJ mol^{-1}	$\Delta G°$/kJ mol^{-1}	$S°$/J mol^{-1} K^{-1}
PtBr$_2$		c	−82.0		
PtBr$_3$		c	−120.9		
PtBr$_4$		c	−156.5		
PtBr$_4$		aq	−197.9		
PtBr$_4^{2-}$		aq	−368.2	−262.7	121
PtBr$_6^{2-}$		aq	−470.7	−332.2	121
H$_2$PtBr$_6$		aq	−470.7		
H$_2$PtBr$_6$·9H$_2$O		c	−3054.7		
PtI$_4$		c	−72.8		
PtI$_6^{2-}$		aq	−213.4	−108.8	167
Pt(NH$_3$)$_2$(OH)$_2$	Cis	aq	−213	−334.1	
PtS		c	−81.6	−76.1	55.06
PtS$_2$		c	−108.8	−99.6	74.68
Pt(NH$_3$)$_4^{2+}$	Std. state, $m = 1$	aq	−361.9	−52.9	42.

Compound	Description	State			Ref.
$Pt[(NH_3)_2(OH)H_2O]^+$	Cis; std. state, $m = 1$	aq		−375.9	
$Pt[(NH_3)_2(H_2O)_2]^{2+}$	Cis, std. state, $m = 1$	aq		−407.3	
$Pt[(NH_3)_4(OH)_2]$		aq	−824.7		
$Pr[(NH_3)_4(NO_3)_2]$		c	−836.8		197
$Pt[(NH_3)Cl_3]^-$	Std. state, $m = 1$	aq	−464.0	−303.2	
$Pt[(NH_3)Cl_5]^-$	Std. state, $m = 1$	aq	−467.4	−432.5	
$[Pt(NH_3)_2Cl_2]$	Cis	c	−467.4		
$[Pt(NH_3)_2Cl_2]$	Trans	c	−480.3		
$Pt(NH_3)_2Cl_2$	Cis, undiss.; std. state, $m = 1$	aq		−228.7	
$Pt(NH_3)_2Cl_2$	Trans, undiss.; std. state, $m = 1$	aq		−222.8	
$NH_4[Pt(NH_3)Cl_3]$		c	−625.1		
$Pt(NH_3)_2Cl_4$	Cis, undiss.; std. state, $m = 1$	aq		−359.6	
$Pt(NH_3)_2Cl_4$	Trans, undiss.; std. state, $m = 1$	aq		−353.8	
$(NH_4)_2PtCl_4$		c	−803.3		

TABLE 8 (Continued)

Formula	Description	State	$\Delta H°$/kJ mol^{-1}	$\Delta G°$/kJ mol^{-1}	$S°$/J mol^{-1} K^{-1}
$(NH_4)_2PtCl_4$	Std. state, $m = 1$	aq	−769.9		
$[Pt(NH_3)Cl]^+$	Std. state, $m = 1$	aq	−396.2	−142.9	113
$[Pt(NH_3)_3Cl]Cl$		c	−603.8		
$[Pt(NH_3)_3Cl_3]^+$	Std. state, $m = 1$	aq	−725.5	−282.6	
$[Pt(NH_3)_4]Cl_2$		c	−696.2		
$Pt(NH_3)_4Cl_2$		aq	−1016.7		
$Pt(NH_3)_4Cl_2 \cdot H_2O$		c			
$[Pt(NH_3)_4Cl_2]^{2+}$	Std. state, $m = 1$	aq	−814.6	−200.6	
$PtCl_2 \cdot 5NH_3$		c			
$Pt(NH_3)_4(PtCl_4)$	Rose	c			
$Pt(NH_3)_4(PtCl_4)$	Green	c			
$[Pt(NH_3)_3Cl][Pt(NH_3)Cl_3]$		c	−959.8		
$[Pt(NH_3)_4][Pt(NH_3)Cl_3]_2$		c	−1412.5		
$[Pt(NH_3)_3Cl]_2[PtCl_4]$		c	−1429.3		

Nickel, Palladium, and Platinum

				180
[Pt(NH₃)Cl₂(H₂O)]	Cis, undiss.; std. state, m = 1	aq	−574.0	−392.7
[Pt(NH₃)Cl₂(H₂O)]	Trans, undiss.; std. state, m = 1	aq		−398.2
[Pt(NH₃)Cl(H₂O)₂]⁺	Std. state, m = 1	aq		−473.5
[Pt(NH₃)₂Cl(H₂O)]⁺	Cis, std. state, m = 1	aq		−320.7
[Pt(NH₃)₂Cl(H₂O)]	Trans, std. state, m = 1	aq		−311.3
[Pt(NH₃)₃Cl]ClO₄		c	−577.4	
[Pt(NH₃)₄Br₂]²⁺	Std. state, m = 1	aq		−149.1
[Pt(NH₃)₂I₂]	Cis, undiss.; std. state, m = 1	aq		−91.1
[Pt(NH₃)₂I₂]	Trans, undiss.; std. state, m = 1	aq		−88.2
[Pt(NH₃)₄]I₂		c	−534.7	
[Pt(NH₃)₄I₂]²⁺	Std. state, m = 1	aq		−73.0
[Pt(NH₃)₄]SO₄		c	−1269.4	
Pt(CN)₄²⁻	Std. state, m = 1	aq		−710.5

TABLE 8 (Continued)

Formula	Description	State	$\Delta H°$/kJ mol^{-1}	$\Delta G°$/kJ mol^{-1}	$S°$/J mol^{-1} K^{-1}
[Pt(CH$_3$NH$_2$)$_2$Cl$_2$]	Cis, undiss.; std. state, $m = 1$	aq		*−140.4*	
[Pt(CH$_3$NH$_2$)$_2$Cl$_2$]	Trans, undiss.; std. state, $m = 1$	aq		*−141.6*	
[Pt(CH$_3$NH$_2$)$_2$I$_2$]	Cis, undiss.; std. state, $m = 1$	aq		*−4.8*	
[Pt(CH$_3$NH$_2$)$_2$I$_2$]	Trans, undiss.; std. state, $m = 1$	aq		*−2.3*	

[a]National Bureau of Standards values are in italic.
Source: Data from Refs. 1 and 2.

2. Platinum-Platinous Couples

Pourbaix et al. [8] have calculated from thermodynamic data the potentials of the following half-reactions:

$$Pt^{2+}(aq) + 2e^- \rightarrow Pt(s) \qquad E^\circ = 1.188 \text{ V}$$

$$PtO(s) + 2H^+ + 2e^- \rightarrow Pt(s) + H_2O \qquad E^\circ = 0.980 \text{ V}$$

The value for the Pt^{2+}/Pt couple is in agreement with that derived by Ginstrup [22] from stability constants for the $PtCl_4^{2-}$ and $PtBr_4^{2-}$ ions in 0.5 M $HClO_4$ solutions at 25°C, $E^{\circ\prime} = 1.18 \pm 0.03$ V, and at 60°C, $E^{\circ\prime} = 1.22 \pm 0.05$ V.

The potential for the systems

$$PtCl_4^{2-}(aq) + 2e^- \rightarrow Pt(s) + 4Cl^-(aq) \qquad E^\circ = 0.758 \text{ V}$$

$$PtBr_4^{2-}(aq) + 2e^- \rightarrow Pt(s) + 4Br^-(aq) \qquad E^\circ = 0.698 \text{ V}$$

at 25°C and zero ionic strength has been deduced by the same authors [22, 23] from careful potentiometric measurements against Ag/AgCl electrodes at several temperatures in 3 M $HClO_4$ and HCl or HBr solutions. At 60°C, the standard potentials are $E^\circ = 0.767$ and 0.697 V, respectively. These values are in good agreement with those obtained previously under other experimental conditions [5].

Direct emf determinations yielded for the half-reaction,

$$PtI_4^{2-}(aq) + 2e^- \rightarrow Pt(s) + 4I^-(aq) \qquad E^{\circ\prime} = 0.40 \text{ V}$$

in 0.5 M NaI solutions at 18°C [25].

3. Platinous-Platinic Couples

The platinic ion Pt^{4+} does not exist in aqueous solution because its salts are rapidly hydrolyzed, but the formal potential of the half-reaction

$$Pt^{4+}(aq) + 4e^- \rightarrow Pt(s) \qquad E^{\circ\prime} = 1.150 \pm 0.070 \text{ V}$$

has been derived from the stability constant of $PtBr_6^{2-}$ in 0.5 M $HClO_4$ at 25°C [22].

The equilibrium reactions between the higher platinum oxides have been calculated from thermodynamic data by Pourbaix et al. [8], who report

$$PtO_2(s) + 2H^+(aq) + 2e^- \rightarrow PtO(s) + H_2O \qquad E^\circ = 1.045 \text{ V}$$

$$PtO_2(s) + 4H^+(aq) + 2e^- \rightarrow Pt^{2+}(aq) + 2H_2O \qquad E^\circ = 0.837 \text{ V}$$

$$PtO_3(s) + 2H^+(aq) + 2e^- \rightarrow PtO_2(s) + H_2O \qquad E^\circ = 2.00 \text{ V}$$

In the haloplatinic acid solutions, the $PtCl_6^{2-}/Pt$ and $PtBr_6^{2-}/Pt$ couples were studied through direct emf measurements against Ag/AgCl electrodes at different temperatures in 3 M HCl or HBr media [22,23]. Extrapolation to zero ionic strength resulted in

$$PtCl_6^{2-}(aq) + 4e^- \rightarrow Pt(s) + 6Cl^-(aq) \qquad E^\circ = 0.744 \text{ V}$$

$$PtBr_6^{2-}(aq) + 4e^- \rightarrow Pt(s) + 6Br^-(aq) \qquad E^\circ = 0.657 \text{ V}$$

which confirm the values earlier published [5].
Combination of the PtI_4^{2-}/Pt and PtI_4^{2-}/PtI_6^{2-} couple potentials leads to

$$PtI_6^{2-}(aq) + 4e^- \rightarrow Pt(s) + 6I^-(aq) \qquad E^{\circ\prime} = 0.40 \text{ V in } 1 \text{ } M \text{ } N_2I \text{ [25]}$$

The reversible potential of the Pt halo-complex systems has been studied several times [5,22-28] by direct emf measurements, and the more reliable values are given below. The potential appears to decrease with the increasing atomic number of the halogen atom and points out the greater instability of the Pt(IV) complex with respect to the Pt(II) complex.

At 25°C and zero ionic strength,

$$PtCl_6^{2-}(aq) + 2e^- \rightarrow PtCl_4^{2-}(aq) + 2Cl^-(aq) \qquad E^\circ = 0.726 \text{ V [22]}$$

$$PtBr_6^{2-}(aq) + 2e^- \rightarrow PtBr_4^{2-}(aq) + 2Br^-(aq) \qquad E^\circ = 0.613 \text{ V [22]}$$

At 60°C, the redox potential decreases to 0.718 and 0.608 V, respectively. A decrease with temperature was found by Tanaka and co-workers [28] for the Pt chloro system in 1 M HCl solutions, so that the formal values were 0.757, 0.747, and 0.730 V at 90, 70, and 35°C, respectively.

In 1 M NaI + 3 M H_2SO_4 solutions,

$$PtI_6^{2-}(aq) + 2e^- \rightarrow PtI_4^{2-}(aq) + 2I^-(aq) \qquad E^{\circ\prime} = 0.329 \pm 0.003 \text{ V [26]}$$

which is slightly lower than that reported earlier [25], $E^{\circ\prime} = 0.398$ V, at different ionic strength or measured by thin-layer voltammetry [24].

Kukushkin and Dkhara [29] carried out the potential determination of

$$Pt(SCN)_6^{2-} + 2e^- \rightarrow Pt(SCN)_4^{2-}(aq) + 2SCN^-(aq) \text{ (in 1 } M \text{ NaSCN)}$$

$$E^{\circ\prime} = 0.468 \text{ V}$$

The same authors also studied the ammine-halo complexes, which are very sensitive to light. They report the following half-reactions in aqueous 1 M NaCl solutions:

$$PtNH_3Cl_5^-(aq) + 2e^- \rightarrow PtNH_3Cl_3^-(aq) + 2Cl^-(aq) \qquad E^{\circ\prime} = 0.731 \text{ V [30]}$$

$Pt(NH_3)_2Cl_4(aq) + 2e^- \rightarrow Pt(NH_3)_2Cl_2(aq) + 2Cl^-(aq)$ $E^{o'} = 0.669$ V [31]

$Pt(NH_3)_3Cl_3^+(aq) + 2e^- \rightarrow Pt(NH_3)_3Cl^+(aq) + 2Cl^-(aq)$ $E^{o'} = 0.651$ V [30]

$Pt(NH_3)_4Cl_2^{2+}(aq) + 2e^- \rightarrow Pt(NH_3)_4^{2+}(aq) + 2Cl^-(aq)$ $E^{o'} = 0.619$ V [32]

$Pt(NH_3)_6^{4+}(aq) + 2e^- \rightarrow Pt(NH_3)_4^{2+}(aq) + 2NH_3(aq)$ $E^{o'} = 0.401$ V [29]

in 1 M NH$_4$Cl.

Also in 1 M NaCl solutions:

$Pt(NH_3)_2Br_2Cl_2 + 2e^- \rightarrow Pt(NH_3)_2Br_2(aq) + 2Cl^-(aq)$ $E^{o'} = 0.650$ V [31]

$Pt(NH_3)_2(NO_2Cl)Cl_2(aq) + 2e^- \rightarrow Pt(NH_3)_2(NO_2Cl)(aq) + 2Cl^-(aq)$
$E^{o'} = 0.689$ V [31]

$Pt(NH_3)_2(NO_2)_2Cl_2(aq) + 2e^- \rightarrow Pt(NH_3)_2(NO_2)_2(aq) + 2Cl^-(aq)$
$E^{o'} = 0.707$ V [31]

$Pt(NH_3)_2(SCN)_2Cl_2(aq) + 2e^- \rightarrow Pt(NH_3)_2(SCN)_2(aq) + 2Cl^-(aq)$
$E^{o'} = 0.593$ V [31]

$Pt(NH_3)_2(NO_2Cl_2)Cl(aq) + 2e^- \rightarrow Pt(NH_3)_2(NO_2Cl)(aq) + 2Cl^-(aq)$
$E^{o'} = 0.678$ V [32]

$cis\text{-}Pt(NH_3NO_2)_2Cl_2(aq) + 2e^- \rightarrow cis\text{-}Pt(NH_3NO_2)_2(aq) + 2Cl^-(aq)$
$E^{o'} = 0.759$ V [34]

$trans\text{-}Pt(NH_3NO_2)_2Cl_2(aq) + 2e^- \rightarrow trans\text{-}Pt(NH_3NO_2)_2(aq) + 2Cl^-(aq)$
$E^{o'} = 0.763$ V [34]

$trans\text{-}Pt(NH_3)_2(NO_2Cl)_2(aq) + 2e^- \rightarrow trans\text{-}Pt(NH_3)_2(NO_2Cl) + NO_2^- + Cl^-$
$E^{o'} = 0.722$ V [34]

$PtNH_3NO_2NH_3ClNO_2Cl(aq) + 2e^- \rightarrow PtNH_3NO_2NH_3Cl(aq) + NO_2^-(aq) + Cl^-(aq)$
$E^{o'} = 0.778$ V [34]

In 0.1 M potassium halide and hydroxide, respectively, the following values were found [33]:

$Pt(NH_3)_4Br_2^{2+}(aq) + 2e^- \rightarrow Pt(NH_3)_4^{2+} + 2Br^-(aq)$ $E^{o'} = 0.559$ V

$$Pt(NH_3)_4I_2^{2+}(aq) + 2e^- \rightarrow Pt(NH_3)_4^{2+} + 2I^-(aq) \quad E^{o\prime} = 0.405 \text{ V}$$

$$Pten(NH_3)_2Cl_2^{2+}(aq) + 2e^- \rightarrow Pten(NH_3)_2^{2+}(aq) + 2Cl^-(aq)$$

$$E^{o\prime} = 0.578 \text{ V}$$

where en = ethylenediamine.

Chernyaev et al. [34] measured the emf of the Pt halo-nitro-ammine systems in aqueous solutions at 0.25×10^{-3} M concentration:

$$cis\text{-}Pt(NH_3NO_2)_2Br_2(aq) + 2e^- \rightarrow cis\text{-}Pt(NH_3NO_2)_2(aq) + 2Br^-(aq)$$

$$E^{o\prime} = 0.782 \text{ V}$$

$$trans\text{-}Pt(NH_3NO_2)_2Br_2(aq) + 2e^- \rightarrow trans\text{-}Pt(NH_3NO_2)(aq) + 2Br^-(aq)$$

$$E^{o\prime} = 0.786 \text{ V}$$

$$PtNH_3NO_2NH_3BrNO_2Br(aq) + 2e^- \rightarrow PtNH_3NO_2NH_3Br(aq) + NO_2^-(aq) + Br^-(aq)$$

$$E^{o\prime} = 0.807 \text{ V}$$

Zheligovskaya et al. [35] studied by potentiometry the halo-nitro-tri-ammine complexes in 0.1 M KCl solutions at 25°C:

$$Pt(NH_3)_3NO_2Cl_2^+(aq) + 2e^- \rightarrow Pt(NH_3)_3NO_2^+(aq) + 2Cl^-(aq) \quad E^{o\prime} = 0.646 \text{ V}$$

$$PtNH_3CH_3NH_2NO_2NH_3Cl_2^+(aq) + 2e^- \rightarrow PtNH_3CH_3NH_2NO_2NH_3^+(aq) + 2Cl^-(aq)$$

$$E^{o\prime} = 0.654 \text{ V}$$

$$PtNH_3CH_3NH_2NO_2CH_3NH_2Cl_2^+(aq) + 2e^- \rightarrow PtNH_3CH_3NH_2NO_2CH_3NH_2^+(aq) +$$

$$2Cl^-(aq) \quad E^{o\prime} = 0.653 \text{ V}$$

$$Pt(CH_3NH_2)_3NO_2Cl_2^+(aq) + 2e^- \rightarrow Pt(CH_3NH_2)_3NO_2^+(aq) + 2Cl^-(aq)$$

$$E^{o\prime} = 0.664 \text{ V}$$

$$PtenNH_3NO_2Cl_2^+(aq) + 2e^- \rightarrow PtenNH_3NO_2^+(aq) + 2Cl^-(aq) \quad E^{o\prime} = 0.627 \text{ V}$$

$$PtenCH_3NH_2NO_2Cl_2^+(aq) + 2e^- \rightarrow PtenCH_3NH_2NO_2^+(aq) + 2Cl^-(aq)$$

$$E^{o\prime} = 0.642 \text{ V}$$

$$Pten(CH_3)_2NHNO_2Cl_2^+(aq) + 2e^- \rightarrow Pten(CH_3)_2NHNO_2^+(aq) + 2Cl^-(aq)$$

$$E^{o\prime} = 0.688 \text{ V}$$

$$Pt(NH_3)_2CH_3NH_2NO_2Cl_2^+(aq) + 2e^- \rightarrow Pt(NH_3)_2CH_3NH_2NO_2^+(aq) + 2Cl^-(aq)$$

$$E^{o'} = 0.650 \text{ V}$$

$$Pt(CH_3NH_2)_2NO_2NH_3Cl_2^+(aq) + 2e^- \rightarrow Pt(CH_3NH_2)_2NO_2NH_3^+(aq) + 2Cl^-(aq)$$

$$E^{o'} = 0.651 \text{ V}$$

In 0.01 M KBr, the following were obtained:

$$Pt(NH_3)_3NO_2Br_2^+(aq) + 2e^- \rightarrow Pt(NH_3)_3NO_2^+(aq) + 2Br^-(aq)$$

$$E^{o'} = 0.592 \text{ V}$$

$$Pt(CH_3NH_2)_3NO_2Br_2^+(aq) + 2e^- \rightarrow Pt(CH_3NH_2)_3NO_2^+(aq) + 2Br^-(aq)$$

$$E^{o'} = 0.608 \text{ V}$$

$$PtenNH_3NO_2Br_2^+(aq) + 2e^- \rightarrow PtenNH_3NO_2^+(aq) + 2Br^-(aq) \qquad E^{o'} = 0.581 \text{ V}$$

$$PtenCH_3NH_2NO_2Br_2^+(aq) + 2e^- \rightarrow PtenCH_3NH_2NO_2^+(aq) + 2Br^-(aq)$$

$$E^{o'} = 0.585 \text{ V}$$

The reversible potentials of Pt chloro-alkyl-ammine complexes were studied by Grinberg et al. [36] in different solutions:

$$Pt(CH_3NH_2)_2(NH_3)_2Cl_2^{2+}(aq) + 2e^- \rightarrow Pt(CH_3NH_2)_2(NH_3)_2^{2+}(aq) + 2Cl^-(aq)$$

$$E^{o'} = 0.615 \text{ V in 1 } M \text{ NaCl [29]}$$

$$Pt(CH_3NH_2)_4Cl_2^{2+}(aq) + 2e^- \rightarrow Pt(CH_3NH_2)_4^{2+}(aq) + 2Cl^-(aq) \text{ in 1 } M \text{ NaCl}$$

$$E^{o'} = 0.617 \text{ V}$$

$$Pt(Et-NH_2)_4Cl_2^{2+}(aq) + 2e^- \rightarrow Pt(Et-NH_2)_4^{2+}(aq) + 2Cl^-(aq) \text{ in 1 } M \text{ NaCl}$$

$$E = 0.681 \qquad \text{and in 1 } M \text{ Na}_2SO_4 \qquad E^{o'} = 0.678 \text{ V}$$

$$Pt(Et-NH_2)_2(NH_3)_2Cl_2^{2+}(aq) + 2e^- \rightarrow Pt(Et-NH_2)_2(NH_3)_2^{2+}(aq) + 2Cl^-(aq)$$

in 1 M NaCl $\qquad E^{o'} = 0.625 \text{ V [29]}$

$$Pt(Pr-NH_2)_2(NH_3)_2Cl_2^{2+}(aq) + 2e^- \rightarrow Pt(Pr-NH_2)_2(NH_3)_2^{2+}(aq) + 2Cl^-(aq)$$

$$E^{o'} = 0.654 \text{ V in 1 } M \text{ NaCl [29]}$$

$$Pt(Pr-NH_2)_4Cl_2^{2+}(aq) + 2e^- \rightarrow Pt(Pr-NH_2)_4^{2+}(aq) + 2Cl^-(aq) \qquad E^{o\prime} = 0.712 \text{ V}$$

in 1 M Na_2SO_4 and also $Pt(Bu-NH_2)_4Cl_2^{2+}(aq) + 2e^- \rightarrow Pt(Bu-NH_2)_4^{2+}$

$(aq) + 2Cl^-(aq) \qquad E^{o\prime} = 0.741$ V [36]

In 1 M NaCl solutions [29]:

$$Pt(Bu-NH_2)_2(NH_3)_2Cl_2^{2+}(aq) + 2e^- \rightarrow Pt(Bu-NH_2)_2(NH_3)_2^{2+}(aq) + 2Cl^-(aq)$$
$$E^{o\prime} = 0.661 \text{ V}$$

$$Pt((CH_3)_2NH)_4Cl_2^{2+}(aq) + 2e^- \rightarrow Pt((CH_3)_2NH)_4^{2+}(aq) + 2Cl^-(aq)$$
$$E^{o\prime} = 0.736 \text{ V}$$

$$Pt((CH_3)_2NH)_2(NH_3)_2Cl_2^{2+}(aq) + 2e^- \rightarrow Pt((CH_3)_2NH)_2(NH_3)_2^{2+}(aq) + 2Cl^-(aq)$$
$$E^{o\prime} = 0.683 \text{ V}$$

$$Pt((CH_3-CH_2)_2NH)_2(NH_3)_2Cl_2^{2+}(aq) + 2e^- \rightarrow Pt((CH_3-CH_2)_2NH)_2(NH_3)_2^{2+}(aq) +$$
$$2Cl^-(aq) \qquad E^{o\prime} = 0.701 \text{ V}$$

Also in the same solution, the monoammine chloroplatinate complexes give [37]

$$PtCH_3NH_2Cl_5^-(aq) + 2e^- \rightarrow PtCH_3NH_2Cl_3^-(aq) + 2Cl^-(aq) \qquad E^{o\prime} = 0.728 \text{ V}$$

$$PtC_6H_5CH_2-NH_2Cl_5^-(aq) + 2e^- \rightarrow PtC_6H_5CH_2-NH_2Cl_3^-(aq) + 2Cl^-(aq)$$
$$E^{o\prime} = 0.720 \text{ V}$$

$$PtC_2H_5NH_2Cl_5^-(aq) + 2e^- \rightarrow PtC_2H_5NH_2Cl_3^-(aq) + 2Cl^-(aq) \qquad E^{o\prime} = 0.737 \text{ V}$$

$$PtC_5H_{10}NHCl_5^-(aq) + 2e^- \rightarrow PtC_5H_{10}NHCl_3^-(aq) + 2Cl^-(aq) \qquad E^{o\prime} = 0.755 \text{ V}$$

$Pt(DMSO)Cl_5^-(aq) + 2e^- \rightarrow Pt(DMSO)Cl_3^-(aq) + 2Cl^-(aq)$ in 0.1 M NaCl
$$E^{o\prime} = 0.872 \text{ V} \text{ [38]}$$

$Pt(DESO)Cl_5^-(aq) + 2e^- \rightarrow Pt(DESO)Cl_3^-(aq) + 2Cl^-(aq)$ in 0.1 M NaCl
$$E^{o\prime} = 0.877 \text{ V}$$

where DMSO = dimethyl sulfoxide and DESO = diethyl sulfoxide [38].

In the case of systems with alcohol-ammine ligands [29] in 1 M NaCl:

$$Pt(OH\text{-}(CH_2)_2\text{-}NH_2)_4Cl_2^{2+} + 2e^- \rightarrow Pt(OH\text{-}(CH_2)_2\text{-}NH_2)_4^{2+}(aq) + 2Cl^-$$

$E^{o\prime} = 0.701$ V

$$Pt(OH\text{-}(CH_2)_2\text{-}NH_2)_2(NH_3)_2Cl_2^{2+}(aq) + 2e^- \rightarrow Pt(OH\text{-}(CH_2)_2\text{-}NH_2)_2(NH_3)_2^{2+}(aq)$$

$E^{o\prime} = 0.641$ V

$$Pt((OH\text{-}(CH_2)_2)_2NH)_2(NH_3)_2Cl_2^{2+}(aq) + 2e^- \rightarrow Pt((OH\text{-}(CH_2)_2)_2NH)_2(NH_3)_2^{2+}$$

$(aq) + 2Cl^-(aq) \qquad E^{o\prime} = 0.593$ V

$$Pt(OH\text{-}(CH_2)_3\text{-}NH_2)_4Cl_2^{2+}(aq) + 2e^- \rightarrow Pt(OH\text{-}(CH_2)_3\text{-}NH_2)_4^{2+}(aq) + 2Cl^-$$

$E^{o\prime} = 0.647$ V

$$Pt(OH\text{-}(CH_2)_3\text{-}NH_2)_2(NH_3)_2Cl_2^{2+}(aq) + 2e^- \rightarrow Pt(OH\text{-}(CH_2)_3\text{-}NH_2)_2(NH_3)_2^{2+}(aq) +$$

$2Cl^- \qquad E^{o\prime} = 0.636$ V

In the field of the Pt tetraamino complexes, the following half-reactions have been measured [39] in 1 M KCl media:

$$Pt(NH_3CH_3NH_2)_2Cl_2^{2+}(aq) + 2e^- \rightarrow Pt(NH_3CH_3NH_2)_2^{2+} + 2Cl^-(aq)$$

$E^{o\prime} = 0.625$ V

$$Pt(NH_3C_2H_5NH_2)_2Cl_2^{2+}(aq) + 2e^- \rightarrow Pt(NH_3C_2H_5NH_2)_2^{2+}(aq) + 2Cl^-(aq)$$

$E^{o\prime} = 0.634$ V

$$Pt(NH_3n\text{-}C_3H_7NH_2)_2Cl_2^{2+}(aq) + 2e^- \rightarrow Pt(NH_3n\text{-}C_3H_7NH_2)_2^{2+}(aq) + 2Cl^-(aq)$$

$E^{o\prime} = 0.643$ V

$$Pt(CH_3NH_2C_2H_5NH_2)_2Cl_2^{2+}(aq) + 2e^- \rightarrow Pt(CH_3NH_2C_2H_5NH_2)_2^{2+}(aq) + 2Cl^-(aq)$$

$E^{o\prime} = 0.649$ V

$$Pt(CH_3NH_2)_2(C_2H_5NH_2)_2Cl_2^{2+}(aq) + 2e^- \rightarrow Pt(CH_3NH_2)_2(C_2H_5NH_2)_2^{2+}(aq) +$$

$2Cl^-(aq) \qquad E^{o\prime} = 0.647$ V

$$Pt(CH_3NH_2)_2(n-C_4H_9NH_2)_2Cl_2^{2+}(aq) + 2e^- \rightarrow Pt(CH_3NH_2)_2(n-C_4H_9NH_2)_2^{2+}(aq) +$$
$$2Cl^-(aq) \qquad E^{o\prime} = 0.639 \text{ V}$$

$$Pt(CH_3NH_2)_2((CH_3)_2NH)_2Cl_2^{2+}(aq) + 2e^- \rightarrow Pt(CH_3NH_2)_2((CH_3)_2NH)_2^{2+}(aq) +$$
$$2Cl^-(aq) \qquad E^{o\prime} = 0.698 \text{ V}$$

$$Pt(NH_3 i\text{-}C_3H_7NH_2)_2Cl_2^{2+}(aq) + 2e^- \rightarrow Pt(NH_3 i\text{-}C_3H_7NH_2)_2^{2+}(aq) + 2Cl^-(aq)$$
$$E^{o\prime} = 0.701 \text{ V}$$

$$Pt(NH_3)_2(i\text{-}C_3H_7NH_2)_2Cl_2^{2+}(aq) + 2e^- \rightarrow Pt(NH_3)_2(i\text{-}C_3H_7NH_2)_2^{2+}(aq) + 2Cl^-(aq)$$
$$E^{o\prime} = 0.698 \text{ V}$$

The redox potentials of some Pt complexes with aryl ligands have been measured, so the pyridine-chloro complexes in 1 M NaCl yield

$$PtpyCl_5^-(aq) + 2e^- \rightarrow PtpyCl_3^-(aq) + 2Cl^-(aq) \qquad E^{o\prime} = 0.763 \text{ V [30]}$$

$$Pt(py)_4Cl_2^{2+}(aq) + 2e^- \rightarrow Pt(py)_4^{2+}(aq) + 2Cl^-(aq) \qquad E^{o\prime} = 0.932 \text{ V and}$$
$$0.982 \text{ V in 1 } M \text{ H}_2SO_4 \text{ [40]}$$

$$Ptpy(NH_3)_2Cl_3^+(aq) + 2e^- \rightarrow Ptpy(NH_3)_2Cl^+(aq) + 2Cl^-(aq)$$
$$E^{o\prime} = 0.718 \text{ V [30]}$$

$$Ptpy(NH_3)_2NO_2Cl_2^+(aq) + 2e^- \rightarrow Ptpy(NH_3)_2NO_2^+(aq) + 2Cl^-(aq)$$
$$E^{o\prime} = 0.746 \text{ V [30]}$$

$$Ptpy(NH_3)_3Cl_2^{2+}(aq) + 2e^- \rightarrow Ptpy(NH_3)_3^{2+}(aq) + 2Cl^-(aq)$$
$$E^{o\prime} = 0.706 \text{ V [32]}$$

$$cis\text{-}Pt(py)_2(NH_3)_2Cl_2^{2+}(aq) + 2e^- \rightarrow cis\text{-}Pt(py)_2(NH_3)_2^{2+}(aq) + 2Cl^-(aq)$$
$$E^{o\prime} = 0.776 \text{ V [32]}$$

$$trans\text{-}Pt(py)_2(NH_3)_2Cl_2^{2+}(aq) + 2e^- \rightarrow trans\text{-}Pt(py)_2(NH_3)_2^{2+}(aq) + 2Cl^-(aq)$$
$$E^{o\prime} = 0.795 \text{ V [32]}$$

$$\text{Pt(py)}_3\text{NH}_3\text{Cl}_2^{2+}(aq) + 2e^- \rightarrow \text{Pt(py)}_3\text{NH}_3^{2+}(aq) + 2\text{Cl}^-(aq) \qquad E^{o\prime} = 0.868 \text{ V}$$

$$\text{Ptpy(CH}_3\text{NH}_2)_3\text{Cl}_2^{2+}(aq) + 2e^- \rightarrow \text{Ptpy(CH}_3\text{NH}_2)_3^{2+}(aq) + 2\text{Cl}^-(aq)$$
$$E^{o\prime} = 0.723 \text{ V [32]}$$

$$trans\text{-Pt(py)}_2(\text{CH}_3\text{NH}_2)_2\text{Cl}_2^{2+}(aq) + 2e^- \rightarrow trans\text{-Pt(py)}_2(\text{CH}_3\text{NH}_2)_2^{2+}(aq) +$$
$$2\text{Cl}^-(aq) \qquad E^{o\prime} = 0.825 \text{ V [32]}$$

$$\text{Pt(py)}_3(\text{CH}_3\text{NH}_2)\text{Cl}_2^{2+}(aq) + 2e^- \rightarrow \text{Pt(py)}_3(\text{CH}_3\text{NH}_2)^{2+}(aq) + 2\text{Cl}^-(aq)$$
$$E^{o\prime} = 0.880 \text{ V [32]}$$

$$\text{Pt}(\alpha\text{-pic})_2(\text{NH}_3)_2\text{Cl}_2^{2+}(aq) + 2e^- \rightarrow \text{Pt}(\alpha\text{-pic})_2(\text{NH}_3)_2^{2+}(aq) + 2\text{Cl}^-(aq)$$
$$E^{o\prime} = 0.801 \text{ V}$$

in 1 M NaCl and 0.774 V in 1 M HCl [29], where pic = picoline

$$\text{Pt}(\beta\text{-pic})_4\text{Cl}_2^{2+}(aq) + 2e^- \rightarrow \text{Pt}(\beta\text{-pic})_4^{2+}(aq) + 2\text{Cl}^-(aq) \qquad E^{o\prime} = 0.930 \text{ V}$$

in 1 M NaCl and 0.877 V and 1.042 V in 1 M HCl and H_2SO_4, respectively [40]

$$\text{Pt}(\beta\text{-pic})_2(\text{NH}_3)_2\text{Cl}_2^{2+}(aq) + 2e^- \rightarrow \text{Pt}(\beta\text{-pic})_2(\text{NH}_3)_2^{2+}(aq) + 2\text{Cl}^-(aq)$$
$$E^{o\prime} = 0.726 \text{ V in 1 } M \text{ NaCl [29]}$$

$$\text{Pt}(\gamma\text{-pic})_4\text{Cl}_2^{2+}(aq) + 2e^- \rightarrow \text{Pt}(\gamma\text{-pic})_4^{2+}(aq) + 2\text{Cl}^-(aq) \qquad E^{o\prime} = 0.896,$$
0.823, and 1.021 V in 1 M NaCl, 1 M HCl, and H_2SO_4, respectively [40]

$$\text{Pt}(\gamma\text{-pic})_2(\text{NH}_3)_2\text{Cl}_2^{2+}(aq) + 2e^- \rightarrow \text{Pt}(\gamma\text{-pic})_2(\text{NH}_3)_2^{2+}(aq) + 2\text{Cl}^-(aq)$$
$$E^{o\prime} = 0.713 \text{ V in 1 } M \text{ NaCl [29]}$$

$$\text{Pt}(\alpha\text{-}\gamma\text{-pic})_2(\text{NH}_3)_2\text{Cl}_2^{2+}(aq) + 2e^- \rightarrow \text{Pt}(\alpha\text{-}\gamma\text{-pic})_2(\text{NH}_3)_2^{2+}(aq) + 2\text{Cl}^-(aq)$$
$$E^{o\prime} = 0.618 \text{ V in 1 } M \text{ NaCl, where } \alpha\text{-}\gamma\text{-pic} = o\text{-}p\text{-dimethyl pyridine}$$

$$\text{Pt}(\alpha\text{-}\alpha'\text{-py})_2(\text{NH}_3)_2\text{Cl}_2^{2+}(aq) + 2e^- \rightarrow \text{Pt}(\alpha\text{-}\alpha'\text{-py})_2(\text{NH}_3)_2^{2+}(aq) + 2\text{Cl}^-(aq)$$
$$E^{o\prime} = 0.685 \text{ V in 1 } M \text{ NaCl solutions [29]}$$

The same authors [29] have studied the following systems in 1 M NaCl:

$$\text{Pt(quin)}_2(\text{NH}_3)_2\text{Cl}_2^{2+}(\text{aq}) + 2e^- \rightarrow \text{Pt(quin)}_2(\text{NH}_3)_2^{2+}(\text{aq}) + 2\text{Cl}^-(\text{aq})$$

$$E^{\circ\prime} = 0.746 \text{ V}$$

$$\text{Pt(phen)}_2(\text{NH}_3)_2\text{Cl}_2^{2+}(\text{aq}) + 2e^- \rightarrow \text{Pt(phen)}_2(\text{NH}_3)_2^{2+}(\text{aq}) + 2\text{Cl}^-(\text{aq})$$

$$E^{\circ\prime} = 0.718 \text{ V}$$

$$\text{Pt(an)}_2(\text{NH}_3)_2\text{Cl}_2^{2+}(\text{aq}) + 2e^- \rightarrow \text{Pt(an)}_2(\text{NH}_3)_2^{2+}(\text{aq}) + 2\text{Cl}(\text{aq})$$

$$E^{\circ\prime} = 0.569 \text{ V}$$

$$\text{Pt(tol-am)}_4\text{Cl}_2^{2+}(\text{aq}) + 2e^- \rightarrow \text{Pt(tol-am)}_4^{2+}(\text{aq}) + 2\text{Cl}^-(\text{aq}) \qquad E^{\circ\prime} = 0.571 \text{ V}$$

(in 1 M HCl)

$$\text{Pt(tol-am)}_2(\text{NH}_3)_2\text{Cl}_2^{2+}(\text{aq}) + 2e^- \rightarrow \text{Pt(tol-am)}_2(\text{NH}_3)_2^{2+}(\text{aq}) + 2\text{Cl}^-(\text{aq})$$

$$E^{\circ\prime} = 0.648 \text{ V},$$

where quin = quinoline, phen = phenanthroline, an = aniline, and (tol-am) = benzylamine.

The inclusion of bidentate instead of monodentate ligands into the Pt complex lowers the oxidizing potential of the Pt(IV) halo-compound systems: for example, in chloroethylenediamine complexes in 1 M NaCl solutions:

$$\text{PtenCl}_4(\text{aq}) + 2e^- \rightarrow \text{PtenCl}_2(\text{aq}) + 2\text{Cl}^-(\text{aq}) \qquad E^{\circ\prime} = 0.654 \text{ V [31]}$$

$$\text{Pt(en)}_2\text{Cl}_2^{2+}(\text{aq}) + 2e^- \rightarrow \text{Pt(en)}_2^{2+}(\text{aq}) + 2\text{Cl}^-(\text{aq}) \qquad E^{\circ\prime} = 0.579 \text{ V [36]}$$

$$\text{PtenBr}_2\text{Cl}_2(\text{aq}) + 2e^- \rightarrow \text{PtenBr}_2(\text{aq}) + 2\text{Cl}^-(\text{aq}) \qquad E^{\circ\prime} = 0.647 \text{ V [31]}$$

$$\text{Pt(en)}_2\text{Br}_2^{2+}(\text{aq}) + 2e^- \rightarrow \text{Pt(en)}_2^{2+}(\text{aq}) + 2\text{Br}^-(\text{aq}) \text{ in 1 } M \text{ NaBr}$$

$$E^{\circ\prime} = 0.537 \text{ V [31]}$$

$$\text{PtenI}_2\text{Cl}_2(\text{aq}) + 2e^- \rightarrow \text{PtenI}_2(\text{aq}) + 2\text{Cl}^-(\text{aq}) \qquad E^{\circ\prime} = 0.564 \text{ V [31]}$$

$$\text{Pt(en)}_2\text{I}_2^{2+}(\text{aq}) + 2e^- \rightarrow \text{Pt(en)}_2^{2+}(\text{aq}) + 2\text{I}^-(\text{aq}) \text{ in 1 } M \text{ NaI}$$

$$E^{\circ\prime} = 0.398 \text{ V [31]}$$

The effect of different types of ammine ligands in the Pt ethylenediamine complex on the standard redox potential of the system was studied by Kukushkin et al. [41] in 1 M NaCl solutions:

$$\text{Pt en (NH}_3)_2\text{Cl}_2^{2+}(\text{aq}) + 2e^- \rightarrow \text{Pt en (NH}_3)_2^{2+}(\text{aq}) + 2\text{Cl}^-(\text{aq})$$

$$E^{\circ\prime} = 0.597 \text{ V [29]}$$

$$\text{Pten(py)}_2\text{Cl}_2^{2+}(aq) + 2e^- \rightarrow \text{Pten(py)}_2^{2+}(aq) + 2\text{Cl}^-(aq) \qquad E^{\circ\prime} = 0.740 \text{ V [32]}$$

$$\text{Pten(NH}_2\text{-C}_2\text{H}_4\text{-NHC}_6\text{H}_5)\text{Cl}_2^{2+}(aq) + 2e^- \rightarrow \text{Pten(NH}_2\text{-C}_2\text{H}_4\text{-NHC}_6\text{H}_5)^{2+}(aq) +$$

$$2\text{Cl}^-(aq) \qquad E^{\circ\prime} = 0.622 \pm 0.003 \text{ V and } 0.595 \pm 0.003 \text{ V in } 1\,M \text{ HCl}$$

$$\text{Pten(NH}_2\text{-C}_2\text{H}_4\text{-SC}_6\text{H}_5)\text{Cl}_2^{2+}(aq) + 2e^- \rightarrow \text{Pten(NH}_2\text{-C}_2\text{H}_4\text{-SC}_6\text{H}_5)^{2+}(aq) +$$

$$2\text{Cl}^-(aq) \qquad E^{\circ\prime} = 0.715 \pm 0.003 \text{ V and } 0.689 \pm 0.003 \text{ V in } 1\,M \text{ HCl}$$

The influence of the nitro and cyano ligands on these Pt complexes is shown in the following half-reactions in $1\,M$ NaCl media:

$$\text{Pten(NO}_2\text{Cl)Cl}_2(aq) + 2e^- \rightarrow \text{Pten(NO}_2\text{Cl)}(aq) + 2\text{Cl}^-(aq) \qquad E^{\circ\prime} = 0.667 \text{ V [31]}$$

$$\text{Pten(NO}_2)_2\text{Cl}_2(aq) + 2e^- \rightarrow \text{Pten(NO}_2)_2(aq) + 2\text{Cl}^-(aq) \qquad E^{\circ\prime} = 0.690 \text{ V [31]}$$

$$\text{Pten(SCN)}_2\text{Cl}_2(aq) + 2e^- \rightarrow \text{Pten(SCN)}_2(aq) + 2\text{Cl}^-(aq) \qquad E^{\circ\prime} = 0.640 \text{ V [31]}$$

In $1\,M$ KBr solutions:

$$\text{Pten(CN)}_2\text{Br}_2(aq) + 2e^- \rightarrow \text{Pten(CN)}_2(aq) + 2\text{Br}^-(aq)$$

$$E^{\circ\prime} = 0.645 \pm 0.001 \text{ V [42]}$$

$$\text{Pten(NO}_2)_2\text{Br}_2(aq) + 2e^- \rightarrow \text{Pten(NO}_2)_2(aq) + 2\text{Br}^-(aq) \qquad E^{\circ\prime} = 0.622 \text{ V [42]}$$

4. Potential Diagram for Platinum

$$\begin{array}{cccc} +6 & +4 & +2 & 0 \\ \text{PtO}_3 \xrightarrow{2.0} & \text{PtO}_2 \xrightarrow{1.045} & \text{PtO} \xrightarrow{0.98} & \text{Pt} \\ & & \text{Pt}^{2+} \xrightarrow{1.188} & \end{array}$$

REFERENCES

1. G. T. Furukawa, M. L. Reilly, and J. S. Gallagher, J. Phys. Chem. Ref. Data, 3, 163 (1974).
2. D. D. Wagman, W. H. Evans, V. B. Parker, R. H. Schumm, I. Halow, S. M. Bailey, K. L. Churney, and R. L. Nuttall, *NBS Tables of Chemical Thermodynamic Properties*, J. Phys. Chem. Ref. Data, Vol. 11, Supplement 2 (1982).
3. R. M. Izatt, D. Eatough, and J. J. Christensen, J. Chem. Soc. A, 1301 (1967).
4. D. H. Templeton, G. W. Watt, and C. S. Garner, J. Am. Chem. Soc., 65, 1608 (1943).

5. R. N. Goldberg and L. H. Hepler, Chem. Rev., *68*, 229 (1968).
6. R. M. Izatt, D. Eatough, J. J. Christensen, J. Delbert, and C. Morgan, J. Chem. Soc. A, 2514 (1970).
7. E. Jackson and D. A. Pantony, J. Appl. Electrochem., *1*, 283 (1971).
8. M. Pourbaix, ed., *Atlas d'équilibres électrochimiques à 25°C*, Gauthier-Villars, Paris, 1963, p. 359.
9. W. E. Bell, R. E. Inyard, and M. Tagami, J. Phys. Chem., *70*, 3735 (1966).
10. N. G. Schmal and E. Minzl, Z. Phys. Chem. (Frankfurt), *47*, 142 (1965).
11. J. S. Warner, J. Electrochem. Soc., *114*, 68 (1967).
12. G. Bayer and H. G. Wiedemann, Thermochim. Acta, *11*, 79 (1975).
13. H. Kleykamp, Z. Phys. Chem. (Frankfurt), *71*, 142 (1970).
14. J. P. Hoare, J. Electrochem. Soc., *111*, 611 (1964).
15. A. B. Fasman, G. G. Kutyukov, and D. V. Sokolskii, Zh. Neorg. Khim., *10*, 1338 (1965).
16. V. I. Kravtsov and M. I. Zelenskii, Elektrokhimiya, *2*, 1138 (1966).
17. T. Ryhl, Acta Chem. Scand., *26*, 2961 (1972).
18. N. M. Nikolaeva, L. D. Tsvelodub, and A. M. Ehrenburg, Izv. Sib. Otd. Nauk SSSR, Ser. Khim., 44 (1978).
19. H. B. Wellman, J. Am. Chem. Soc., *52*, 985 (1930).
20. A. A. Grinberg and A. S. Shamsiev, Zh. Obshch. Khim., *12*, 55 (1942).
21. A. V. Babaeva and E. Y. Khananova, Zh. Neorg. Khim., *10*, 2579 (1965).
22. D. Ginstrup, Acta Chem. Scand., *26*, 1527 (1972).
23. D. Ginstrup and I. Leden, Acta Chem. Scand., *22*, 1163 (1968).
24. A. T. Hubbard and F. C. Anson, Anal. Chem., *38*, 58 (1966).
25. A. A. Grinberg, V. B. Ptitsyn, and V. N. Lavrentiev, Zh. Fiz. Khim., *10*, 661 (1937); A. A. Grinberg and M. I. Gelfman, Dokl. Akad. Nauk SSSR, *133*, 1081 (1960).
26. V. I. Kravtsov and B. V. Simakov, Elektrokhimiya, *2*, 646 (1966).
27. V. I. Kravtsov and B. V. Simakov, Elektrokhimiya, *2*, 406 (1966).
28. H. Yamamoto, S. Tanaka, T. Nagai and T. Takei, Denki Kagaku, *32*, 43 (1964).
29. Y. N. Kukushkin and S. Ch. Dkhara, Zh. Neorg. Khim., *15*, 1585 (1970).
30. Y. N. Kukushkin and S. Ch. Dkhara, Zh. Neorg. Khim., *14*, 853 (1969).
31. Y. N. Kukushkin and S. Ch. Dkhara, Zh. Neorg. Khim., *14*, 2816 (1969).
32. Y. N. Kukushkin and S. Ch. Dkhara, Zh. Neorg. Khim., *14*, 1012 (1969).
33. M. Felin, L. V. Popov, N. N. Zheligovskaya, and V. I. Spitsyn, Izv. Akad. Nauk SSSR Ser. Khim., 930 (1972).
34. I. I. Chernyaev, G. S. Muraveiskaya, and L. S. Korablina, Zh. Neorg. Khim., *11*, 1339 (1966).
35. N. N. Zheligovskaya, A. Khidzhazi, and V. I. Spitsyn, Izv. Akad. Nauk SSSR, Ser. Khim., 160 (1973).
36. A. A. Grinberg, S. Ch. Dkhara, and M. J. Gelfman, Zh. Neorg. Khim., *13*, 2199 (1968).
37. Y. N. Kukushkin, E D. Ageeva, and V. N. Spevak, Zh. Neorg. Khim., *19*, 1126 (1974).

38. Y. N. Kukushkin and V. N. Spevak, Zh. Neorg. Khim., *17*, 1686 (1972).
39. N. N. Zheligovskaya, N. N. Kamalov, and V. I. Spitsyn, Izv. Akad. Nauk SSSR, Ser. Khim., 2396 (1975).
40. Y. N. Kukushkin, S. Ch. Dkhara, and M. J. Gelfman, Zh. Neorg. Khim., *13*, 2530 (1968).
41. Y. N. Kukushkin, E. A. Andronov, and G. N. Krasova, Zh. Obshch. Khim., *46*, 1321 (1976).
42. N. N. Zheligovskaya and N. N. Kamalov, Izv. Akad. Nauk SSSR, Ser. Khim., 2396 (1975).

13
Cobalt, Rhodium, and Iridium

I. COBALT*

NOBUFUMI MAKI Shizuoka University, Hamamatsu City, Japan
NOBUYUKI TANAKA Tohoku University, Sendai, Miyagi, Japan

Like the neighboring elements of the first transition series, nearly all the simple compounds of cobalt are of Co(II) oxidation state, since the Co(III) salts are practically confined to oxides, sulfides, sulfates, fluorides, and acetates; in fact, each of the latter three is considered to be a sort of complex. The simple salts, however, would actually be a sort of complex if they were examined in more detail: the crystal structure of cobaltous chloride, $CoCl_2 \cdot 6H_2O$ (the pink hexahydrate), can be said to have the structure of an octahedral complex, $[Co^{II}Cl_2(OH_2)_4] \cdot 2H_2O \cdot$(dichlorotetraaquocobalt(II) dihydrate), with two molecules of water placed close to the Cl^- ions. The $[Co^{III}(OH_2)_6]^{3+}$ ion is such a powerful oxidizing agent that it decomposes water even at low temperatures to give the $[Co^{II}(OH_2)_6]^{2+}$ ion, evolving oxygen. Such an unstable $[Co^{III}(OH_2)_6]^{3+}$ ion is produced from the stable $[Co^{II}(OH_2)_6]^{2+}$ ion only by the action of the strongest oxidizing agent. This is in marked contrast to the Co(III) complexes with ligands possessing strong ligand fields, where the Co(III) state is usually the stable and predominant form. More than 600 kinds of mixed liganded Co(III) complexes with various ligands have already been prepared and identified. Thermodynamic data on cobalt are given in Table 1.

A. Standard Electrode Potentials of Cobalt Systems

Konrád and Vlček [1] reported the standard electrode potential of −0.255 V for the $[Co\ en_3]^{3+}/[Co\ en_3]^{2+}$ couple in 1 M KCl + 3 M $NaClO_4$ at 25°C from the analysis of the polarographic behavior of the $[Co\ en_3]^{2+} \to [Co\ en_3]^{3+}$ anodic reaction. They concluded that the cathodic reduction of $[Co\ en_3]^{3+}$ proceeds reversibly to Co(II) even under conditions where the $[Co\ en_3]^{2+}$ ion is thermodynamically completely unstable. In the observed mechanism of the electrode process the necessity of having the same stoichiometric composition for both forms of the reduced and oxidized states was emphasized. These forms are converted rapidly into each other, the conversion being ac-

---

Reviewer: M. Gross, Louis Pasteur University, Strasbourg, France.

TABLE 1 Thermodynamic Data on Cobalt

Formula[a]	Description	State	$\Delta H°$/kJ mol^{-1}	$\Delta G°$/kJ mol^{-1}	$\Delta S°$/J mol^{-1} K^{-1}
Co		g	439		179.41
Co	Metal	c	0.0	0.0	28.5
Co$^+$		g	1201.94		
Co^{2+}		g	2850.1		
Co^{3+}		g	6082.7		
Co^{2+}		aq	−2916		
Co^{3+}		aq	−5996		
CoO		c	−211.3		43.9
Co(OH)$_2$		c	−453.5		82.0
Co$_3$O$_4$		c	−745		149.8
CoOOH		c	−369.4		
[Co^{III}(acac)$_3$]		g	−1289.5		
[Co^{III}(acac)$_3$]		c	−1364.4		
[Co^{III}(acac)$_3$]		aq	−7811.5		

Species	State			
$[Co^{III}(NH_3)_6]Cl_3$	c	−1175.7	−621.7	−1857.3
$[Co^{III}(NH_3)_6]Br_3$	c	−1050.6	−549.4	−1681.5
$[Co^{III}(NH_3)_6]I_3$	c	−868.2	−402.9	−1561.1
$[Co^{III}(NH_3)_6](NO_3)_3$	c	−1329.7	−572.4	−2539.3
$[Co^{III}(NH_3)_5OH_2]Cl_3$	c	−1336.0	−826.8	−1707.9
$[Co^{III}(NH_3)_5Cl]Cl_2$	c	−1064.8	−630.9	−1454.8
$[Co^{III}(NH_3)_5Cl]Br_2$	c	−977.8	−573.2	−1360.6
$[Co^{III}(NH_3)_5Cl]I_2$	c	−855.6	−465.7	−1307.5
$[Co^{III}(NH_3)_5Cl](NO_3)_2$	c	−1161.5	−579.9	−1951.0
$[Co^{III}(NH_3)_5Cl](C_2O_4)$	c	−1538.9	−1041.0	−1669.8
$[Co^{III}(NH_3)_5Br]Cl_2$	c	−987.4		
$[Co^{III}(NH_3)_5I]Cl_2$	c	−930.1		
$[Co^{II}(NH_3)_6]^{2+}$	aq	−635.1		
$[Co^{II}(NH_3)_6]Cl_2$	g	−1697.0		
$[Co^{II}edta]^{2-}$	aq	−17.2	−89.5	242.7
$[Co^{II}(dcta)]^{2-}$	aq	−11.72	−107.1	325.5

TABLE 1 (Continued)

Formula[a]	Description	State	$\Delta H°$/kJ mol^{-1}	$\Delta G°$/kJ mol^{-1}	$\Delta S°$/J mol^{-1} K^{-1}
[Co^{II}(py)(NCS)$_2$]		aq	−69.5	−27.6	−142.7
[Co^{II}(pc)Cl]		aq	28.5	11.80	54
[Co^{III}(NH$_3$)$_6$](N$_3$)$_3$		c	128.4		
[Co^{III}(NH$_3$)$_5$N$_3$](N$_3$)$_2$		c	210.9		
[Co^{III}(NH$_3$)$_4$(N$_3$)$_2$]N$_3$	Cis	c	380.3		
[Co^{III}(NH$_3$)$_4$(N$_3$)$_2$]N$_3$	Trans	c	378.2		
[Co^{III}(NH$_3$)$_3$(N$_3$)$_3$]		c	400.4		

[a] pc, phthalocyanine; py, pyridine; dcta, trans-1,2-diaminocyclohexane-N,N'-tetraacetate ion; acac, acetylacetonate ion, CH$_3$COCH$_2$COCH$_3$.
Source: Data from Refs. 38–40.

companied by considerable change in the distance Co-N (from about 0.2 nm with Co^{III} to about 0.25 nm with Co^{II}) and by a change in the electronic structure on the basis of the following scheme:

$$[Co\ en_3]^{3+} \rightarrow [Co\ en_3]^{2+} \rightarrow Co\ en_2^{2+} \rightarrow Co\ en^{2+} \rightarrow Co^{2+}(aq)$$

Laitinen and Grieb [2] established and Konrád and Vlček [1] confirmed, both by potentiometry, that $Co\ en_2^{2+}$ and $Co\ en^{2+}$ are electrochemically inactive and that the only oxidizable Co(II)en species in this potential region is the $[Co\ en_3]^{2+}$ ion, the behavior of which was the same as that of ammonia Co(II) complexes. Bartelt and Skilandat [3] have determined the kinetic parameters of the reversible $[Co\ en_3]^{3+}/[Co\ en_3]^{2+}$ system at the RDE by the method of following the variation of exchange current density near equilibrium. Evidence was found for a hydrolysis process which caused partial coverage of the platinum electrode surface, and hence a decrease in the exchange current density. By varying the concentration of the free ligand (en) in excess, it was possible to show that only ions of similar stoichiometry take part in the electron exchange reaction, whereas ions such as $[Co(OH)_4en]^{2-}$ and $[Co(OH)_2en_2]^0$ are electrochemically inactive. Fischerova et al. [4] derived the theoretical equation for the kinetics of a first-order chemical reaction of the primary reduction product, $[Co^{II}en_3]^{2+}$, in the presence of an excess of ethylenediamine and applied this to the $[Co\ en_3]^{3+}/[Co\ en_3]^{2+}$ couple. The homogeneous rate constant for subsequent chemical reaction of the $[Co\ en_3]^{2+}$ ion was estimated to be $1.85 \times 10^{-7}\ M\ s^{-1}$ at 20°C. Thus the $[Co\ en_3]^{3+}/[Co\ en_3]^{2+}$ redox couple is certainly a reversible electrode process in the presence of excess ethylenediamine, but this electrode system is not reversible polarographically. Kim and Rock [5] reported the standard electrode potential of -0.18 ± 0.002 V at 25°C for the $[Co\ en_3]^{3+}/[Co\ en_3]^{2+}$(aq) couple in cells without a liquid junction using cation-sensitive glass electrodes. The following two cells were used:

$$Au(s)\,|\,[Co\ en_3]Cl_3(\ell), [Co\ en_3]Cl_2(\ell), NaCl(\ell), en(\ell)\,|\,glass \qquad (13.1)$$

$$Hg(\ell)\,|\,Hg_2Cl_2(s), NaCl(\ell)\,|\,glass \qquad (13.2)$$

In the net cell reaction of the hypothetical double cell combined with the two cells above, all the properties of the particular glass electrode used cancel out, and in addition the double cell has no liquid junctions. They determined the temperature coefficient of $0.0107\ V\ K^{-1}$ for the standard potential of the $[Co\ en_3]^{3+}/[Co\ en_3]^{2+}$ electrode at 25°C. From the value of $dE°/dT$, together with other available thermodynamic data, the entropy, Gibbs energy, and the enthalpy changes between the oxidized and reduced forms of the $[Co\ en_3]^{3+}/[Co\ en_3]^{2+}$ couple were evaluated at 25°C as follows:

$$S° = 167\ J\ mol^{-1}\ K^{-1}$$

$$\Delta G° = 17.2\ J\ mol^{-1}$$

$$\Delta H° = 49.0\ J\ mol^{-1}$$

It is well known that the hexaaquocobalt(III) ion, $[Co(OH_2)_6]^{3+}$, is one of the most powerful oxidizing agents in aqueous media and that the standard electrode potential of the $[Co(OH_2)_6]^{3+}/[Co(OH_2)_6]^{2+}$ couple is about 1.8 V

[6,7]. Johnson and Sharpe [8] have adopted the following redox system with 4 M perchloric acid at 15°C as a working medium:

$$[Co(OH_2)_6]^{3+}(aq) + [Fe(OH_2)_6]^{2+}(aq) \rightarrow [Co(OH_2)_6]^{2+}(aq) + [Fe(OH_2)_6]^{3+}(aq)$$

The tetrahedral perchlorate anion involving Cl^{7+} as a nucleus has one of the lowest values of coordinating ability as an electron donor ligand of all the donor anions in aqueous media. In this medium the heat of the reduction was found to be $\Delta H = -110.0$ kJ mol^{-1}. The close similarity in size, mass, and charge of the pairs of ions involved indicates that the entropy change for the reaction should be very small, although there is an increase of 8 to 46 J mol^{-1} K^{-1} in the magnetic entropy, arising from the fact that only Co^{3+}(aq) in 4 M HClO$_4$ is a low-spin species. The coordination of ClO$_4^-$ ions must be considered, but it is only a substantial difference in the complexing of Fe^{3+} and Co^{3+}, or of Fe^{2+} and Co^{2+}, under identical conditions, that would appreciably affect $\Delta E°$, the difference between the standard potentials of the Fe^{3+}/Fe^{2+} and Co^{3+}/Co^{2+} systems. The Gibbs energy change derived from the changes in enthalpy and magnetic entropy for the redox reaction above leads to $\Delta E° = 1.18$ V at 15°C. Combining this value with that of +0.745 V for the formal redox potential of the Fe^{3+}/Fe^{2+} electrode in 0.5 M HClO$_4$ at 15°C, the formal electrode potential of the Co^{3+}/Co^{2+} couple is obtained as +1.93 V at 15°C. The true standard electrode potential of the [Co^{III}(OH$_2$)$_6$]$^{3+}$/[Co^{II}(OH$_2$)$_6$]$^{2+}$ couple would be 1.95 ± 0.1 V at 25°C. They suggested that an error of ±0.1 V in $E°$ would correspond to one of 8 to 12 kJ mol^{-1} in $\Delta H°$, or of 33 J mol^{-1} K^{-1} in $\Delta S°$, and that the formation of a more stable perchlorate complex by Co(III) than by Fe(III) would raise the value of $E°$.

Independently, Huchital et al. [9] and Warnqvist [10] have determined directly the standard potential of the [Co(OH$_2$)$_6$]$^{3+}$/[Co(OH$_2$)$_6$]$^{2+}$ couple to be 1.92 ± 0.02 V in 4 M HClO$_4$ at 25°C, a value lower than that (1.95 ± 0.1 V) measured indirectly under the same conditions [8]. They also calculated the equilibrium constant for the following redox reaction from the formal electrode potentials of the Ag(II)/Ag(I) and Co(III)/Co(II) couples [9]:

$$Co(III) + Ag(I) = Co(II) + Ag(II)$$

That is, the formal electrode potentials of these two couples are 2.00 V in 4 M HClO$_4$ and 1.84 V in 4 M HNO$_3$, respectively, at 25°C. The lower value for the latter suggests that extensive complex formation occurs between Co(III) and nitrate ions. From these values they calculated that $K = (4 \pm 2) \times 10^{-2}$, in satisfactory agreement with the equilibrium constant obtained from direct kinetic measurements. In spite of these data there is still some uncertainty and even controversy concerning the species present in acidic cobalt(III) perchlorate solutions, the problem being, whether the predominant aquo-cobalt(III) species in perchloric acid solutions is a dimeric and/or a polymeric complex [11]. Recently, Warnqvist [10] concluded from the potentiometric study of the [Co(OH$_2$)$_6$]$^{3+}$/[Co(OH$_2$)$_6$]$^{2+}$ couple that cobalt(III) is mainly monomeric [i.e., [Co(OH$_2$)$_6$]$^{3+}$(aq)] under experimental conditions where a 3 M NaClO$_4$ solution contained 0.50 M HClO$_4$ and 0.50×10^{-3} to 5.0×10^{-3} M Co(III) ion at 23 and 3°C, respectively. That is, if Co(III) exists in solution in the form of dimeric ions, the Nernst slope would be half of that expected for monomeric cobalt(III) at a constant cobalt(II) concentration. The slope obtained agreed with that predicted from the monomeric cobalt(III) species at a constant cobalt(II) concentration. The cell used for the electromotive force (emf) measurements was ref. electrode $|$ 0.50 M HClO$_4$,3 M NaClO$_4$(aq)$\|$0.50 M HClO$_4$,3 M NaClO$_4$(aq),

[Co(OH$_2$)$_6$]$^{3+}$,[Co(OH$_2$)$_6$]$^{2+}$,(Ag$^+$)$|$Au(s)

The reference electrode was a calomel electrode in which the KCl has been replaced with 4.0 M NaCl to avoid the precipitation of KClO$_4$. As a potential mediator, a few drops of AgClO$_4$ were added in order to obtain electrode equilibrium quickly. Table 2 shows the upper limit of the cobalt(III) dimerization quotient K_D and the formal electrode potentials of the [Co(OH$_2$)$_6$]$^{3+}$/[Co(OH$_2$)$_6$]$^{2+}$ couple in 3 M NaClO$_4$ solution containing 3 to 0.05 M HClO$_4$ on the hydrogen scale. In computing $E°$, the potentials of the 4 M NaCl calomel electrode were assumed to be 0.242 and 0.249 V, respectively, at 25 and 3°C.

Rotinjan et al. [12] have determined the standard potential of the [Co(OH$_2$)$_6$]$^{3+}$/[Co(OH$_2$)$_6$]$^{2+}$ couple on the basis of the calorimetric method for the following redox reaction:

[Co(OH$_2$)$_6$]$^{3+}$(aq) + [Fe(OH$_2$)$_6$]$^{2+}$(aq) → [Co(OH$_2$)$_6$]$^{2+}$(aq) + Fe(OH$_2$)$_6$]$^{3+}$(aq)

They evaluated the $E°$ value for the same couple by the method of extrapolating the cathodic and anodic Tafel curves to their intersection. The average value of the $E°$ obtained by both methods was 1.45 V in 5.6 M HClO$_4$ at 25°C. They also found from the data on potentials of cobalt oxides in an alkali medium that the standard potential of the Co^{3+}/Co^{2+}(aq) couple must be within the range 1.45 to 1.53 V. They assert that this value agrees with that of 1.45 V obtained in acidic media. In any case, uncertainties in liquid junction potentials are the major contributors to the error estimates. The net sign of the liquid junction potentials may in fact have been such that the actual $E°$ values are closer to the upper limit quoted [13].

The [Co^{II}(NH$_3$)$_6$]$^{2+}$ ion is not present in solution to an appreciable extent, even in the presence of excess ammonia ligands. Calculations for the [Co(NH$_3$)$_6$]$^{2+}$ ion reveal that even at NH$_3$(aq) concentrations as high as 6 M, the equilibrium mixture of the cobalt(II)-hydroxoammonia system comprises less than 50% of the [Co^{II}(NH$_3$)$_6$]$^{2+}$ ion in solution. The substitution-labile

TABLE 2 Upper-Limit Estimates of Hexaaquocobalt(III) Dimerization Quotient, K_D, and Standard Potentials for [Co(OH$_2$)$_6$]$^{3+}$/[Co(OH$_2$)$_6$]$^{2+}$ Couple in 3 M NaClO$_4$[a]

Temperature (°C)	h/mol L^{-1}	K_D (approx. upper limit), /M^{-1}	$E°$/V (vs. NHE)[b]
23	3	20	1.86
23	0.5	500	1.85
3	3	10	1.83
3	0.5	20	1.83
3	0.1	20	1.82
3	0.05	10	1.83

[a][Co(III)-dimer]/[Co(III)-monomer]^2 = K_D.
[b]Error estimated at ±(0.01 to 0.02) V.

Co(II) complexes can always exist in an equilibrium state with free ligands (i.e., ammonia and hydroxide ions) present in solution. The change of the electronic configuration from the low-spin state to high-spin state accompanies the redox electrode process, and a slow and irreversible process could always be expected for the $[Co(NH_3)_6]^{3+}/[Co(NH_3)_6]^{2+}$ system in aqueous media.

At high ammonia concentrations Co(II) reacts stoichiometrically with triiodide, I_3^-, to yield the $[Co^{III}(NH_3)_5I]^{2+}$ ion. This reaction is reversible and the equilibrium constanst for the formation of iodopentamminecobalt(III) ion is $(5.1 \pm 0.6) \times 10^5$, corresponding to the following reaction stoichiometry [14]:

$$2[Co(NH_3)_5OH_2]^{2+} + I_3^- \rightarrow 2[Co(NH_3)_5I]^{2+} + I^- + 2H_2O$$

At 25°C the second-order rate constant is 1.0×10^{-6} M s^{-1}; the enthalpy and entropy changes of this redox reaction were -96 kJ mol^{-1} and $+33$ J mol^{-1} K^{-1}, respectively, at 25°C.

In connection with this, the standard potential of the $[Co(NH_3)_5OH_2]^{3+}/[Co(NH_3)_5OH_2]^{2+}$ (actually $[Co(NH_3)_5OH]^{2+}/[Co(NH_3)_5OH]^+$) couple was $+0.33$ V in 1 M NH$_4$NO$_3$ at 25°C. This value appears to be reasonable, judging from the values of the other members of hydroxoammonia series and their ligand field strength. Bartelt et al. [15] investigated the $[Co(NH_3)_6]^{3+}/[Co(NH_3)_6]^{2+}$, [Co en$_3$]$^{3+}$/[Co en$_3$]$^{2+}$, [Co dien$_2$]$^{3+}$/[Co dien$_2$]$^{2+}$, and [Co phen$_3$]$^{3+}$/[Co phen$_3$]$^{2+}$ couples by means of polarization resistance measurements at a platinum rotating disk electrode. The two effects of hydrolysis and ion-pairing association decrease the exchange current density, j_0 (μA cm^{-2}) and operate on the $E°$ value of the redox couple. It is possible, however, to eliminate these effects by extrapolating to zero on the exchange current density-potential curves. Table 3 shows the results obtained at $25 \pm 0.2°$C. In the table only the [Co phen$_3$]$^{2+}$ species can exist stably without any disruption at the Co(II) state among the reduction products. The $E°$ value for this couple is fairly positive, suggesting that the electron-mediating effect of an unsaturated π-aromatic ring is favorable for the adiabatic electron-transfer process.

Hin-Fat and Higginson [16] have determined the formal redox potentials of tris(oxalato)-, diaquobis(oxalato)-, and tris(glycinato)-cobalt systems in 1 M KCl aqueous solutions, which had been summarized in Table 4. They attempted to estimate the proportions of the [Co ox$_3$]$^{3-}$ ion present in aqueous solutions, since a portion of them may exist as the aquobis(oxalato)cobaltate(III)

TABLE 3 Standard Electrode Potentials of Co(III)/Co(II) Couples $(25 \pm 0.2°C)^a$

Couple	Conditions	$E°$/V
$[Co(NH_3)_6]^{3+}/[Co(NH_3)_6]^{2+}$	7 M NH$_3$(aq), 1 M NH$_4$Cl	0.058 ± 0.002
[Co dien$_2$]$^{3+}$/[Co dien$_2$]$^{2+}$	0.1 M dien, 1 M KCl	-0.233 ± 0.002
[Co phen$_3$]$^{3+}$/[Co phen$_3$]$^{2+}$	8 mM phen, 1 M KCl	0.327 ± 0.02

[a]dien, diethylenetriamine; NH$_2$CH$_2$CH$_2$NH·CH$_2$CH$_2$NH$_2$; phen, 1,10-phenanthroline.
Source: Ref. 15.

TABLE 4 Formal Electrode Potentials of Co(III)/Co(II) Couples (in a 1 M KCl Aqueous Solution, 25°C)

Couple[a]	$E^{o'}/V$
$[Co\ ox_3]^{3-}/[Co\ ox_3]^{4-}$	0.57 ± 0.02
$[Co\ ox_2(OH_2)_2]^{-}/[Co\ ox_2(OH_2)_2]^{2-}$	0.78 ± 0.04
$trans(N)\text{-}[Co\ gly_3]^{0}/[Co\ gly_3]^{-}$	0.20
$[Co\ edta]^{-}/[Co\ edta]^{2-}$	0.38 ± 0.01

[a]ox, oxalate anion, $C_2O_4^{2-}$; gly, glycinate anion, $NH_2CH_2COO^-$; edta, ethylenediaminetetraacetate ion, $N(CH_2COO)_2CH_2CH_2N(CH_2COO)_2^{4-}$; these abbreviations denote the ligands from which all hydrogen ions capable of dissociating are detached.
Source: Ref. 16.

ion, in which the third oxalate group in coordinated to cobalt by one bond only. They concluded that a portion of the $[Co\ ox_3]^{3-}$ ion would actually be present in the form $[Co\ ox_3(OH_2)]^{3-}$ rather than $[Co(OH)ox_2(Hox)]^{3-}$ in a 1 M KCl aqueous medium containing no excess ox^{2-} ions; the $[Co\ ox_3(OH_2)]^{3-}$ ion involves an ox^{2-} ion as an unidentate ligand. The fraction of the $[Co\ ox_3]^{3-}$ ion present as the $[Coox_3(OH_2)]^{3-}$ ion at equilibrium was given by $K_a/K_b = 10^{-3}$.

In contrast to the Co(III)/Co(II) complex systems with ligands of π-bonding feature in which the degraded Co(II) complex releases its ligands rapidly to give an equilibrium mixture with free ligands present in solution, the outer-sphere redox reactions of Co(III)/Co(II) systems with π-bonding ligands do not involve bond cleavage during the electron-transfer process. Hence it is possible to study the kinetics of oxidation of a common reductant, $[Co\ terpy_2]^{2+}$, by a series of cobalt(III) oxidants, $[Co\ dip_3]^{3+}$, $[Co\ phen_3]^{3+}$, and $[Co\ tmp_3]^{3+}$, or of cobalt(II) reductants, $[Co\ dip_3]^{2+}$, $[Co\ phen_3]^{2+}$, and $[Co\ terpy_2]^{2+}$ (tmp = 3,5,6,8-tetramethyl-1,10-phenanthroline; dip = 2,2'-dipyridyl; phen = 1,10-phenanthroline; terpy = 2,2',2"-terpyridyl; py = pyridine). The effect of the Gibbs energy drive, the relative reactivity of aquo and hydroxo species, and electronic configuration, especially low- and high-spin cobalt(II), on the rates of these outer-sphere redox reactions has been systematically examined by Farina and Wilkins [17]. The standard electrode potentials of Co(III)/Co(II) couples obtained at 0°C have been summarized in Table 5.

It is well known that the cyanide couple of Co(II)/Co(I) is not reversible (due to the protonation and dimerization of $[Co^I(CN)_5]^{4-}$ species) [18]. The addition of cyanide ions to a Co(II)(aq) solution results in the formation of the pentacyanocobaltate(II) ion which may be oxidized into aquopentacyanocobaltate(III) ion. Nowadays it is established and accepted beyond doubt that the stable Co(II) species is the five-coordinate complex ion $[Co^{II}(CN)_5]^{3-}$ (yellow), in which the sixth ligand, OH_2, was detached (i.e., actually uncoordinated), although in solid salts isolated from aqueous media this usually dimerizes to form the $[Co_2(CN)_{10}]^{6-}$ ion (violet). The standard enthalpy of forming

TABLE 5 Standard Potentials of Co(III)/Co(II) Systems with π-Bonding Ligands at 0°C

Half-reaction	$E°$/V
$[\text{Co terpy}_2]^{3+} + e^- \rightarrow [\text{Coterpy}_2]^{2+}$	0.31
$[\text{Co dip}_3]^{3+} + e^- \rightarrow [\text{Codip}_3]^{2+}$	0.34
$[\text{Co phen}_3]^{3+} + e^- \rightarrow [\text{Cophen}_3]^{2+}$	0.40
$[\text{Co(OH}_2)_2\text{phen}_2]^{3+} + e^- \rightarrow [\text{Co(OH}_2)_2\text{phen}_2]^{2+}$	0.68
$[\text{Co(OH}_2)_4\text{dip}]^{3+} + e^- \rightarrow [\text{Co(OH}_2)_4\text{dip}]^{2+}$	0.84
$[\text{Co(OH}_2)_4\text{phen}]^{3+} + e^- \rightarrow [\text{Co(OH}_2)_4\text{phen}]^{2+}$	0.84
$[\text{Co(OH}_2)_6]^{3+} + e^- \rightarrow [\text{Co(OH}_2)_6]^{2+}$	1.92 ± 0.02 (25°C)

Source: Ref. 17.

$[\text{Co}^{II}(\text{CN})_5]^{3-}$ is 257 kJ mol^{-1}. The $[\text{Co}^{II}(\text{CN})_5]^{3-}$ ion is oxidized chemically or electrochemically to the Co(III) state [i.e., $[\text{Co}^{III}(\text{CN})_5\text{OH}_2]^{2-}$ (an electrochemically active species)]. Conversely, the $[\text{Co}^{III}(\text{CN})_5\text{OH}_2]^{2-}$ ion is reduced in two steps as follows:

$$[\text{Co}^{III}(\text{CN})_5\text{OH}_2]^{2-} \xrightarrow{e^-} [\text{Co}^{II}(\text{CN})_5]^{3-} \xrightarrow{e^-} [\text{Co}^{I}(\text{CN})_5]^{4-} \xrightarrow{H^+}$$

$$[(\text{NC})_5\text{Co}^{I}\text{-H-Co}^{I}(\text{CN})_5]^{7-}$$
electrochemically inactive
chemically active

Sharpe [19] suggests that the quoted value of $E°$ for the Co(III)/Co(II) couple in complex cyanides probably involves cyanide transfer as well as electron transfer. Certainly, the $[\text{Co}^{I}(\text{CN})_5]^{3-}$ ion is extremely reactive and undergoes a very wide range of addition and oxidative reaction; Hanzlík and Vlček [20] called this the redox addition reaction and Halpern [21] used the term "oxidative addition reaction." Sharpe's suggestions, however, should be excluded from the results of electrode redox reactions. Independent of whether excess cyanide ions are present, the $[\text{Co}^{III}(\text{CN})_5\text{OH}]^{3-}$, $[\text{Co}^{III}(\text{CN})_5\text{OH}_2]^{2-}$, and $[\text{Co}^{II}(\text{CN})_5]^{3-}$ ions may exist stably as electrochemically active species without any addition of CN^- to the sixth site of the pentacyano coordination sphere, whereas the $[\text{Co}^{III}(\text{CN})_6]^{3-}$ and $[\{\text{Co}^{I}(\text{CN})_5\}_2\text{H}]^{7-}$ ions are neither reducible nor oxidizable at the electrode (electrochemically inactive). Nevertheless, the latter Co(I) species is very sensitive to a trace of oxygen and chemically oxidized into the Co(III) state (chemically active). Thus the chemical and electrochemical redox reactions should strictly be distinguished from each other. Probably, the inner-sphere mechanism of electron-transfer processes operates on the electrochemical redox reaction, whereas the outer-sphere mechanism may function on the chemical redox reaction. Only under deprotonating conditions (i.e., in a strongly alkaline solution of 12 M KOH) can the sixth site of $[Co^I(CN)_5]^{4-}$ be retained vacant, through which electrochemical electron transfer proceeds reversibly. Hanzlík and Vlcek [20] showed on the basis of measurements using a Kalousek commutator that the $[Co^{II}(CN)_5]^{3-}/[Co^I(CN)_5]^{4-}$ couple is reversible electrochemically as well as chemically without releasing cyanide ions under such deprotonating conditions. The $E°$ value for this redox couple was approximately -0.96 V and hence $\Delta G° = 93$ kJ mol^{-1}. The logarithm of the equilibrium constant for protonation to the $[Co^I(CN)_5]^{4-}$ ion was estimated as being pK = ca. 18 to 19. The μ-hydrido dimer restores its electrochemical activity only after the removal of hydrogen.

Izatt et al. [22] have determined the value of $\Delta H°$ for the reaction

$$Co^{2+}(aq) + 5CN^-(aq) = [Co^{II}(CN)_5]^{3-}(aq) \qquad \Delta H° = -257 \pm 2 \text{ kJ mol}^{-1}$$

They used a wide range of initial molar concentration of Co^{2+}(aq) in the presence of NaOH and extrapolated the concentration of the latter to zero, since the OH$^-$ ion retards decomposition of $[Co^{II}(CN)_5]^{3-}$.

The $\Delta H°$ value for the following protonation reactions was evaluated [23-25]:

$$2[Co^{II}(CN)_5]^{3-} + H_3O^+ \rightarrow [Co^{III}(CN)_5OH_2]^{2-} + [Co^{III}(CN)_5H]^{3-}$$

$$\Delta H° = -133.9 \pm 2.1 \text{ kJ mol}^{-1}$$

$$2Co^{2+}(aq) + 10CN^-(aq) + H_3O^+ \rightarrow [Co^{III}(CN)_5OH_2]^{2-} + [Co^{III}(CN)_5H]^{3-}$$

$$\Delta H° = -391.2 \pm 2.1 \text{ kJ mol}^{-1}$$

$$[Co^{III}(CN)_5H]^{3-}(aq) + H_3O^+(aq) \rightarrow [Co^{III}(CN)_5OH_2]^{2-}(aq) + H_2(g)$$

$$\Delta H° = -105.6 \text{ kJ mol}^{-1}$$

$$2[Co^{II}(CN)_5]^{3-}(aq) + H_2(g) \rightarrow 2[Co^{III}(CN)_5H]^{3-}(aq)$$

$$\Delta H° = -46.9 \text{ kJ mol}^{-1}$$

$$2[Co^{II}(CN)_5]^{3-}(aq) + 2H_3O^+(aq) \rightarrow H_2(g) + 2[Co^{III}(CN)_5OH_2]^{2-}(aq)$$

$$\Delta H° = -244.3 \text{ kJ mol}^{-1}$$

The equilibrium constant for the reaction [26]

$$2[Co^{II}(CN)_5]^{3-}(aq) + H_2O(\ell) \rightarrow 2[Co^{III}(CN)_5H]^{3-}(aq)$$

has been measured over the temperature range 0 to 35°C in 0.86 and 0.065 M KCN; from the dependence of the equilibrium constant K on the temperature, $\Delta H° = -47.0$ kJ mol^{-1}. At 25°C in a 0.5 M KCl aqueous solution the K value was calculated to be 1.5×10^5. The calculated value of K is in agreement with the observed equilibrium constant [26].

The enthalpy of formation of $K_3[Co(CN)_6]$(s) is unknown, but the standard entropy of $[Co^{III}(CN)_6]^{3-}$(aq) is estimated as 234 J mol^{-1} K^{-1} [27]. The overall formation constant of $[Co^{III}(CN)_6]^{3-}$ is a quantity of great interest, but unfortunately it is remarkably difficult to get even an estimate of its magnitude.

Hieber and Hübel [28] have determined the standard electrode potential of -0.4 V for the $[Co(CO)_4]_2/2[Co(CO)_4]^-$ couple at 20°C in alkaline aqueous solutions. The reversible redox reaction was as follows"

$$[Co(CO)_4]_2 + 2e^- \rightarrow 2[Co(CO)_4]^- \qquad E° = -0.4 \text{ V}$$

The results were compared with those of iron tetracarbonyl systems in both alkaline and acidic aqueous solutions;

$$[Fe(CO)_4]_3 + 6e^- \rightarrow 3[Fe(CO)_4]^{2-} \qquad E° = -0.74 \text{ V (20°C)}$$
in alkaline aq. soln.

$$[Fe(CO)_4]_3 + 3H^+ + 6e^- \rightarrow 3[Fe(CO)_4H]^- \qquad E° = -0.35 \text{ V (20°C)}$$
in acidic aq. soln.

The $E°$ value for the cobalt tetracarbonyl system is more positive by 340 mV than that of the corresponding iron tetracarbonyl system, suggesting that the dimerized cobalt tetracarbonyl with an electronic configuration of a rare gas type tends to be more readily ionized to yield the carbonyl complex monoanion in aqueous media. In acidic aqueous solutions the $E°$ value was difficult to determine due to the decomposition of the tetracarbonyl anion according to the following reaction:

$$2[Co[CO)_4]^- + 2H_3O^+ \rightarrow [Co(CO)_4]_2 + H_2 + 2H_2O$$

Koepp et al. [29] have determined the standard electrode potential of the Co(I)/Co(0) couples for bis(cyclopentadienyl)cobalt(II) (i.e., cobaltocene) and triethylenetetraminedisalicylidinecobalt(II) complexes in water. The $E°$ values of $[Co^I(C_5H_5)_2]^-/[Co^0(C_5H_5)_2]^{2-}$ and of $[Co^I\text{triensal}]^-/[Co^0\text{triensal}]^{2-}$ couples were -0.918 ± 0.01 V and -0.403 ± 0.01 V, respectively, in aqueous media at 25°C, where the abbreviation of triensal denotes a hexadentate ligand of triethylenetetraminedisalicylidine.

Only a few investigations have been made on the stability constants of Co(III) complexes [30]. This is because Co^{3+}(aq) forms mostly substitution-inert complexes; the formation of such stable complexes discourages considerably determination of the stability constant. The $E°$ value of complex couples that dissociate reversibly can be calculated from their stability constants if the $E°$ value of the Co^{3+}(aq)/Co^{2+}(aq) couple is known. That is, the $E°$ value can be obtained from the stability constants extrapolated to infinite dilution. The stability constants for Co(III) complexes with a hexadentate ligand of ethylenediaminetetraacetate (edta^{4-}), propylenediaminetetraacetate (pdta^{4-}), trimethy-

TABLE 6 Formal Electrode Potentials of Co(III)/Co(II) Couples for Complexes with EDTA-Like Ligands (25°C)

Redox couple[a]	$E^{o\prime}/V$
$[\text{Co edta}]^-/[\text{Co edta}]^{2-}$	0.376
$[\text{Co pdta}]^-/[\text{Co pdta}]^{2-}$	0.36
$[\text{Co trdta}]^-/[\text{Co trdta}]^{2-}$	0.29
$[\text{Co cydta}]^-/[\text{Co cydta}]^{2-}$	0.36

[a] edta, ethylenediaminetetraacetate; pdta, propylenediaminetetraacetate; trdta, trimethylenediaminetetraacetate; cydta, 1,2-cyclohexanediaminetetraacetate; the hexadentate ligands from which all hydrogen ions capable of dissociating are detached.

lenediaminetetraacetate (trdta^{4-}), and 1,2-cyclohexanediaminetetraacetate (cydta^{4-}), which were all substitution inert, have been determined by several investigators [31,32], and hence the standard potentials for the Co(III)/Co(II) systems with EDTA-like ligands can be evaluated. Tanaka and co-workers [32] have determined the E^o values and thermodynamic parameters of Co(III)/Co(II) systems for the EDTA-like complexes and confirmed that the E^o values calculated from the stability constants are in good agreement with those evaluated by the polarographic method. Table 6 shows the most reliable E^o values of the Co(III)/Co(II) couple for complexes of EDTA-like type in aqueous solutions at 25°C.

The formal redox potentials of the "μ-superoxo" species, $[\text{Co}_2(\text{NH}_3)_{10}(\mu\text{-O}_2)]^{5+}$, $[\text{Co}_2(\text{NH}_3)_8(\mu\text{-O}_2,\mu\text{-NH}_2)]^{4+}$, and $[\text{Co}_2(\text{CN})_{10}(\mu\text{-O}_2)]^{4-}$, have been determined in aqueous solutions as follows [33]:

$$[\text{Co}_2(\text{NH}_3)_{10}(\mu\text{-O}_2)]^{5+} \quad E^{o\prime} = 0.9 \pm (0.95 \text{ V})$$

$$[\text{Co}_2(\text{NH}_3)_8(\mu\text{-O}_2,\mu\text{-NH}_2)]^{4+} \quad E^{o\prime} = 0.75 \pm (0.60 \text{ V})$$

$$[\text{Co}_2(\text{CN})_{10}(\mu\text{-O}_2)]^{4-} \quad E^{o\prime} = 0.90 \pm (0.06 \text{ V})$$

The first values cited above are the voltammetric values (CV), and values in parentheses are obtained by potentiometry corrected for complex decomposition. The oxidation process of the redox couple for $[\text{Co}_2(\text{NH}_3)_{10}(\mu\text{-O}_2)]^{5+}$ was obscured by a breakdown of aquation. The electron-transfer reaction of cobalt-bound dioxygen is due to that of the O_2/O_2^{2-} or O_2/O_2^- couple. That is, the redox reaction is not the metal-based reaction but the ligand-based one.

B. Anodic Behavior of Cobalt Electrodes and The Cobalt Oxidation Potential Co(s)/Co^{2+}(aq)

Compared with a number of studies on nickel and iron electrodes, very few studies have been made on cobalt. The anodic formation of particular cobalt oxides on the cobalt electrode has been reported by several investigators.

The cobalt electrode in alkaline solutions shows two stages of anodic oxidation: the transitions $Co/Co(OH)_2$ and $Co(OH)_2/Co(OH)_3$ under potentiostatic conditions. Cowling and Riddiford [34] have examined the cobalt electrode in 1 M and 0.1 M NaOH solutions under potentiostatic and galvanostatic conditions. The anodic oxidation of a clean cobalt surface takes place in two stages; the first and the second oxidation steps were attributed to the transitions $Co \rightarrow Co(OH)_2$ and $Co(OH)_2 \rightarrow CoOOH$ [or $Co(OH)_3$]. The third process of forming the unstable oxide CoO_2 was not found between the rest potential and the point of oxygen evolution on either the potentiostatic or galvanostatic anodic polarization curves. The rest potential [i.e., steady-state, open-circuit potential of cobalt(s) in 1 M NaOH] was about -0.63 V. The equilibrium potentials of Co/CoO and $Co/Co(OH)_2$ are -0.662 and -0.733 V at pH 14, respectively, while the second step of anodic oxidation began at $+2.00$ V in a 1 M NaOH solution and at 0.250 V in a 0.1 M NaOH solution, respectively. The current-time curves show that in a 1 M NaOH solution the passivating layer grows in three dimensions (on the simple model, hemispherical growth being assumed) and that progressive nucleation occurs.

Boldin [35] has examined the anodic and cathodic polarization curves at constant current densities with a platinum electrode immersed in 10 M KOH saturated with $Co(OH)_2$. The oxide films formed at 600 $\mu A\ cm^{-2}$ and 13 $\mu A\ cm^{-2}$ underwent reduction at -0.04 and -0.1 V (vs. Hg/HgO electrode), respectively. The steady-state potential of oxide film formed at 600 $\mu A\ cm^{-2}$ varies with the KOH concentration in accord with the following equation:

$$E = E° + (RT/2F) \log[(a_{H_2O})^2/(a_{KOH})^3]$$

The overall reaction was inferred to be

$$2Co(OH)_2 + HgO + 2KOH = 2KCoO_2 + Hg + 3H_2O$$

For the half-reaction,

$$Co(OH)_2 + KOH + OH^- = KCoO_2 + 2H_2O + e^-$$

the standard electrode potential of the $Co(OH)_2/CoO_2^-$ couple is -0.22 V. The oxide formed at the constant current density of 13 $\mu A\ cm^{-2}$ was deduced to be the intermediate between $KCoO_2$ and $CoOOH$.

Behl and Toni [36] have studied the anodic oxidation of cobalt in 0.2 to 8.0 M KOH solutions by cyclic voltammetry at the stationary and rotating cobalt disks and at rotating cobalt ring-disk electrodes. The anodic oxidation of cobalt occurred only in the solid phase and resulted in the formation of a $Co(OH)_2$ film on the electrode. $Co(OH)_2$ was subsequently oxidized to Co_3O_4 and $CoOOH$ at more positive potentials. $Co(OH)_2$ also dissolves at all KOH concentrations (0.2 to 8.0 M) to produce Co^{2+} species in solution. Both Co_3O_4 and $CoOOH$ are either insoluble or have negligible solubility, so that only Co^{2+} species are predominantly present in KOH solutions. It was shown that although the oxidation of $Co(OH)_2$ to $CoOOH$ occurred readily, the reduction of $CoOOH$ to $Co(OH)_2$ did not occur at any appreciable rate. Further, the charge used in the oxidation of cobalt was not completely recovered, the charge used in the oxidation of cobalt were not completely recovered, so that at the end of the experiment the electrode was always covered with brown film (CoOOH) or blue film [Co(OH)$_2$ and/or CoO], depending on the potential to which the electrode was scanned.

Göhr [37] has thoroughly investigated the identification of anodically formed oxide layers at the cobalt electrode; free enthalpies of formation of CoO, Co(OH)$_2$, Co$_3$O$_4$, CoOOH, and CoO$_2$ derived from calorimetric and electrochemical potentiostatic measurements and from equilibrium data were employed to account for the experimental results made so far, and limits of error were given. Formal potentials of formation and oxidation of oxide layers (with respect to normal hydrogen electrode) were estimated for the first time.

C. Potential Diagrams for Cobalt

Acid Solution

$$\underset{+4}{CoO_2} \xrightarrow{1.416} \underset{+3}{Co^{3+}} \xrightarrow{1.92} \underset{+2}{Co^{2+}} \xrightarrow{-0.277} \underset{0}{Co}$$

Basic Solution

$$CoO_2 \xrightarrow{0.7} Co(OH)_3 \xrightarrow{0.17} Co(OH)_2 \xrightarrow{-0.733} Co$$

REFERENCES

1. D. Konrad and A. A. Vlček, Coll. Czech. Chem. Commun., 28, 595, 808 (1963).
2. H. A. Laitinen and M. W. Grieb, J. Am. Chem. Soc., 77, 5201 (1955).
3. H. Bartelt and H. Skilandat, J. Electroanal. Chem., 23, 407 (1969).
4. E. Fischerova, O. Dracka, and M. Meloun, Coll. Czech. Chem. Commun., 33, 473 (1968).
5. J. J. Kim and P. A. Rock, Inorg. Chem., 8, 563 (1969).
6. G. Milazzo and S. Caroli, *Tables of Standard Electrode Potentials*, Wiley, New York, 1978, pp. 336-337.
7. H. L. M. Van Gaal and J. G. M. Van der Linden, Coord. Chem. Rev., 47, 41 (1982); P. A. Rock, Inorg. Chem., 7, 837 (1968).
8. D. A. Johnson and A. G. Sharpe, J. Chem. Soc., 3490 (1964).
9. D. H. Huchital, N. Sutin, and B. Warnqvist, Inorg. Chem., 6, 838 (1967).
10. B. Warnqvist, J. Phys. Chem., 68, 174 (1964); Inorg. Chem., 9, 682 (1970).
11. C. F. Wells, Discuss. Faraday Soc., 46, 197 (1968).
12. A. L. Rotinjan, L. M. Borisowa, and R. W. Boldin, Electrochim. Acta, 19, 43 (1974).
13. G. Davies and B. Warnqvist, Coord. Chem. Rev., 5, 349 (1970).
14. R. G. Yalman, Inorg. Chem., 1, 16 (1962).
15. H. Bartelt, Electrochim. Acta, 16, 307, 629 (1971); H. Bartelt and M. Prügel, Electrochim. Acta, 16, 1815 (1971).

16. L. Hin-Fat and W. C. E. Higginson, J. Chem. Soc. A, 298 (1967).
17. R. Farina and R. G. Wilkins, Inorg. Chem., 7, 514 (1968).
18. N. Maki, Inorg. Chem., 13, 2180 (1974).
19. A. G. Sharpe, *The Chemistry of Cyano Complexes of Transition Metals*, Academic Press, New York, 1976, pp. 159-207.
20. J. Hanzlík and A. A. Vlček, Inorg. Chim. Acta, 8, 247 (1974); Chem. Commun., 47 (1969).
21. J. Halpern, Acc. Chem. Res., 3, 386 (1970).
22. R. M. Izatt, G. D. Watt, C. H. Bartholomew, and J. J. Christensen, Inorg. Chem., 7, 2236 (1968).
23. J. M. Pratt and R. J. P. Williams, J. Chem. Soc. A, 1291 (1967).
24. B. DeVries, J. Catal., 1, 489 (1962).
25. W. P. Griffith and G. Wilkinson, J. Chem. Soc., 2757 (1959).
26. M. G. Burnett, P. J. Connolly, and C. Kemball, J. Chem. Soc. A, 800 (1969).
27. L. G. Hepler, J. R. Sweet, and R. A. Jesser, J. Am. Chem. Soc., 82, 304 (1960).
28. W. Hieber and W. Hübel, Z. Elektrochem., 57, 331 (1953).
29. H. M. Koepp, H. Wendt, and H. Strehlow, Z. Elektrochem., 64, 483 (1960).
30. D. D. Perrin, ed., *Stability Constants of Metal Ion Complexes*, I: *Organic Ligands*, IUPAC Chemical Data Series 22, Pergamon Press, Oxford, 1979; J. Bjerrum, G. Schwarzenbach, and L. G. Sillén, *Stability Constants of Metal Ion Complexes*, II: *Inorganic Ligands*, Spec. Publ. 25, Chemical Society, London, 1971.
31. R. G. Wilkins and R. E. Yelin, J. Am. Chem. Soc., 92, 1191 (1970).
32. N. Tanaka, Nippon Kagaku Zassi, 92, 919 (1971); A. Yamada and N. Tanaka, Sci. Rep. Tohoku Univ. Ser. I, 52(2), 73 (1969); N. Tanaka and H. Ogino, Bull. Chem. Soc. Jpn., 38, 1054 (1965).
33. G. McLendon and W. F. Mooney, Inorg. Chem., 19, 12 (1980).
34. R. D. Cowling and A. C. Riddiford, Electrochim. Acta, 14, 981 (1969).
35. R. V. Boldin, Elektrokhimiya, 3(10), 1259 (1967) (in Russ.): Chem. Abstr., 68, 1725 (1968).
36. W. K. Behl and J. E. Toni, J. Electroanal. Chem., 31, 63 (1971).
37. H. Göhr, Electrochim. Acta, 11, 827 (1966).
38. J. J. Christensen and R. M. Izatt, *Handbook of Metal Ligand Heats*, Marcel Dekker, Inc., New York, 1970.
39. P. George and D. S. McClure, *Progress in Inorganic Chemistry*, Vol. 1 (edited by F. A. Cotton), Wiley, New York, 1959, pp. 381-461.
40. T. M. Donovan, C. H. Shomate, and T. B. Joyner, J. Phys. Chem., 64, 378 (1960).

II. RHODIUM AND IRIDIUM*

FRANCISCO COLOM   Institute of Physical Chemistry, CSIC, Madrid, Spain

RHODIUM

A. Oxidation States

Rhodium takes up a variety of oxidation states, which range from −I to +VI, although the most stable and important of them seems to be Rh(III).

Reviewer: Karl Kadish, University of Houston, Houston, Texas.

Cobalt, Rhodium, and Iridium

The lower and higher values are detected only in carbonyl complexes and alkaline solutions, respectively. Rhodium metal, like all the transition elements, has a great tendency to form complex ions in aqueous solutions, generally with coordination number 6. The best known and most reliable thermodynamic data on rhodium compounds are listed in Table 7.

B. Oxidation Potentials of Some Rhodium Compounds

Very few dependable electrochemical and thermodynamic data are available for the oxidation and reduction processes of rhodium compounds, mainly because of uncertainty about the chemical species involved. In contrast to Ru and Os in this metals group, no oxoanions or higher oxides have been identified for Ir and Rh in aqueous solutions [3]; therefore, all the oxidation processes so far proposed that imply the existence of those species should generally be questioned.

For the III oxidation state, Goldberg and Hepler [4] have calculated the potential of the half-reaction

$$Rh^{3+}(aq) + 3e^- \rightarrow Rh(s) \qquad E° = 0.7 \text{ V}$$

on the basis of thermodynamic data for Rh_2O_3 and the solubility product of the hydroxide. A similar value, $E° = 0.799$ V, was calculated by Pourbaix et al. [5] from thermodynamic data. Both values are in good agreement with that obtained by Amosse et al. [6]. $E° = 0.758 \pm 0.002$ V through direct measurement in $HClO_4$ solutions at pH 5 and extrapolated to zero ionic strength.

Only the oxides Rh_2O_3 and RhO_2 have been identified and isolated for this metal, but Pourbaix et al. [5] have calculated from thermodynamic data the following series of electrochemical reactions for rhodium oxides in acid media:

$$Rh_2O(s) + 2H^+ + 2e^- \rightarrow 2Rh(s) + H_2O \qquad E° = 0.796 \text{ V}$$

$$2RhO(s) + 2H^+ + 2e^- \rightarrow Rh_2O(s) + H_2O \qquad E° = 0.882 \text{ V}$$

$$Rh_2O_3(s) + 2H^+ + 2e^- \rightarrow 2RhO(s) + H_2O \qquad E° = 0.871 \text{ V}$$

$$2RhO_2(s) + 2H^+ + 2e^- \rightarrow Rh_2O_3(s) + H_2O \qquad E° = 1.730 \text{ V}$$

$$Rh_2O_3(s) + 4H^+ + 4e^- \rightarrow Rh_2O(s) + H_2O \qquad E° = 0.877 \text{ V}$$

Hoare [7] has measured the potential of a rhodium electrode covered by a film of adsorbed oxygen in aqueous solutions. According to the experimental conditions, the author detected two types of surface oxide that were assumed to be formed through the reactions

$$RhO(s) + 2H^+ + 2e^- \rightarrow Rh(s) + H_2O \qquad E° = 0.81 \text{ V}$$

$$Rh_2O_3(s) + 6H^+ + 6e^- \rightarrow 2Rh(s) + 3H_2O \qquad E° = 0.88 \text{ V}$$

In the latter case, the Gibbs energy of formation deduced from the standard potential is less negative than that for $Rh_2O_3(c)$, which, if these reactions are correct, suggests a lower stability for the surface oxide.

TABLE 7 Thermodynamic Data on Rhodium[a]

Formula	State	$\Delta H°$/kJ mol^{-1}	$\Delta G°$/kJ mol^{-1}	$S°$/J mol^{-1} K^{-1}
Rh	c	0.	0.	31.56
Rh	g	556.9	510.8	185.808
Rh$^+$	g	1290.60		
Rh^{2+}	g	3040.9		
Rh^{3+}	g	6042.		
RhO	g	385.		
RhO$^+$	g	1289.		
RhO$_2$	g	184.		
RhO$_2^+$	g	1115.		
Rh$_2$O$_3$	c	−343.		
RhCl$_2$	g	126.8		
RhCl$_3$	c	−299.2		
RhCl$_3$	g	66.9		
RhCl$_6^{3-}$	aq	−848.5		
RhCl$_3 \cdot 3(C_2H_5)_2$S	c	−695.		

[a]National Bureau of Standards values are in italic.
Source: Data from Refs. 1 and 2.

For the reduction of rhodium oxides to metal ions in acid solutions Pourbaix et al. [5] have calculated the following from thermodynamic data:

$$Rh_2O_3(s) + 6H^+(aq) + 4e^- \rightarrow 2Rh^+(aq) + 3H_2O \qquad E° = 0.975 \text{ V}$$

$$Rh_2O_3(s) + 6H^+(aq) + 2e^- \rightarrow 2Rh^{2+}(aq) + 3H_2O \qquad E° = 1.349 \text{ V}$$

$$RhO_2(s) + 4H^+(aq) + e^- \rightarrow Rh^{3+}(aq) + 2H_2O \qquad E° = 1.881 \text{ V}$$

These rhodium ions are solely stable in acid solutions in the presence of complexing agents or chemicals.

The potential of the half-reaction

$$RhCl_6^{3-}(aq) + 3e^- \rightarrow Rh(s) + 6Cl^- \qquad E° = 0.5 \text{ V}$$

was calculated from the stability constant determined by Cossi and Pantani [8] in HCl solutions at 25°C.

In an earlier paper, Willis [9] estimated the standard potential for the Rh(III) cyanide complex in aqueous solutions from polarographic results as

$$Rh(CN)_6^{3-}(aq) + e^- \rightarrow Rh(CN)_6^{4-}(aq) \qquad E° = 0.9 \text{ V}$$

C. Potential Diagrams for Rhodium

$$\text{Rh}(+3) \xrightarrow{\overset{+3}{} \quad 0.76 \quad \overset{0}{}} \text{Rh}$$

IRIDIUM

A. Oxidation States

Iridium chemistry deals with a wide range of species in which the metal is present in oxidation states that vary from 0 to +VI, although only the I, III, and IV compounds are commonly formed and studied. The I compounds consist mainly of square planar complexes with π-bonding ligands (CO, tertiary phosphines, olefins, etc.), while the III and IV states form complex ions in aqueous solutions with coordination number 6. The thermodynamic data are given in Table 8.

B. Iridium Oxides and Their Ions

Hoare [11] has studied the formation of surface oxides on iridium electrodes in 1 M H_2SO_4 solutions and has distinguished two types of films that the author assumes to be formed after the reactions

$$IrO(s) + 2H^+(aq) + 2e^- \rightarrow Ir(s) + H_2O \qquad E° = 0.870 \pm 0.020 \text{ V}$$

$$IrO_2(s) + 4H^+(aq) + 4e^- \rightarrow Ir(s) + 2H_2O \qquad E° = 0.935 \pm 0.005 \text{ V}$$

TABLE 8 Thermodynamic Data on Iridium[a]

Formula	State	ΔH°/kJ mol^{-1}	ΔG°/kJ mol^{-1}	S°/J mol^{-1} K^{-1}
Ir	c	0.	0.	35.48
Ir	g	665.3	617.9	193.578
Ir$^+$	g	1540.0		
IrO$_2$	c	−274.1		
IrO$_3$	g	7.9		
IrO$_3^+$	g	1163.		
IrF$_6$	c	−579.65	−461.66	247.7
IrF$_6$	g	−544	−460	357.8

IrCl	c	−81.6
IrCl$_3$	c	−245.6
IrCl$_6^{2-}$	aq	−573
IrCl$_6^{3-}$	aq	−736
IrS$_2$	c	−138
Ir$_2$S$_3$	c	−234
IrCl$_3\cdot$3(C$_2$H$_5$)$_2$S	c	−649
IrBr$_3$	c	−223.0 ± 0.3 *130.5*

[a]National Bureau of Standards values are in italic.
Source: Data from Refs. 1, 2, and 10.

On the other hand, Pourbaix et al. [12] have calculated the thermodynamic equilibrium relations

$$Ir_2O_3(s) + 6H^+(aq) + 6e^- \rightarrow 2Ir(s) + 3H_2O \qquad E^\circ = 0.926 \text{ V}$$

$$IrO_2(s) + 4H^+(aq) + 4e^- \rightarrow Ir(s) + 2H_2O \qquad E^\circ = 0.926 \text{ V}$$

$$2IrO_2(s) + 2H^+(aq) + 2e^- \rightarrow Ir_2O_3(s) + H_2O \qquad E^\circ = 0.926 \text{ V}$$

$$IrO_2(s) + 4H^+(aq) + e^- \rightarrow Ir^{3+}(aq) + 2H_2O \qquad E^\circ = 0.233 \text{ V}$$

$$Ir^{3+}(aq) + 3e^- \rightarrow Ir(s) \qquad E^\circ = 1.156 \text{ V}$$

C. Iridium Potentials in Hydrohalogen Solutions

From the large number of Ir(III) and Ir(IV) complexes known, only the halo compounds have been studied from an electrochemical point of view. The Ir(IV) halo complexes in aqueous solutions have octahedral geometry and are easily reduced to the III state in basic, neutral, or slightly acidic solutions with retention of their configuration; in acid media, they are quite stable and kinetically inert. In contrast, the Ir(III) halo complexes are thermodynamically stable but kinetically labile, tending to form aquo-halo complexes in aqueous solutions. In basic media, the aquo complexes are readily converted into hydroxyaquo complexes.

From the heat of solution of K_2IrCl_6 and thermodynamic data for $IrCl_6^{3-}$ and $IrCl_6^{2-}$, Goldberg and Hepler [4] have derived the following potentials:

$$IrCl_6^{2-}(aq) + 4e^- \rightarrow Ir(s) + 6Cl^-(aq) \qquad E^\circ = 0.86 \text{ V}$$

$$IrCl_6^{3-}(aq) + 3e^- \rightarrow Ir(s) + 6Cl^-(aq) \qquad E^\circ = 0.86 \text{ V}$$

George et al. [13] have determined the redox potential of the couple $IrCl_6^{2-}/IrCl_6^{3-}$ with Rh electrodes in 0.3 M HCl solutions, which at zero ionic strength resulted in

$$IrCl_6^{2-}(aq) + e^- \rightarrow IrCl_6^{3-}(aq) \qquad E^\circ = 0.867 \text{ V}$$

The variation of the formal potential at $I = 0.3$ mol L^{-1} with temperature between 15 and 40°C yields an isothermal temperature coefficient of 1.37 mV/K.

The more positive values found for this system in Refs. 14 and 15 are probably due to the higher ionic strength of the solutions employed. This experimental factor can also account for the higher formal potentials obtained potentiometrically by Kravtsov and co-workers at 10^{-3} M concentration of iridium ions in 1 M HCl and 1 M NaCl solutions [16] and in 3 M HCl [17], $E^{\circ\prime}$ = 0.941 ± 0.002 V, or by polarography (DME) in 0.2 M HCl [18], $E^{\circ\prime}$ = 1.05 V. The George et al. value was verified by potentiometry in 0.1 M NaClO$_4$ and confirmed by Jackson and Pantony through voltammetric measurements with rhodium and boron carbide electrodes [19].

The equilibrium potentials of the aquo-halo complexes were directly determined by Garner and co-workers at 25°C:

$IrCl_5(H_2O)^-(aq) + e^- \rightarrow IrCl_5(H_2O)^{2-}(aq)$ in 0.2 M HNO_3 $\quad E^{o\prime} = 1.0$ V [20]

trans-$IrCl_4(H_2O)_2(aq) + e^- \rightarrow IrCl_4(H_2O)_2^-(aq)$ in 0.4 M HNO_3 or 0.4 M $HClO_4$ + 0.2 M NaH_2PO_4 $\quad E^{o\prime} = 1.22 \pm 0.02$ V [21]

$IrCl_3(H_2O)_3^+(aq) + e^- \rightarrow IrCl_3(H_2O)_3(aq)$ in 0.4 M $HClO_4$ $\quad E^{o\prime} = 1.30 \pm 0.01$ V [21]

The results show a decrease in stability with the increase of aquo ligands.

For the Ir(IV)/Ir(III) bromo system, Jackson and Pantony [19] have obtained

$IrBr_6^{2-}(aq) + e^- \rightarrow IrBr_6^{3-}(aq) \quad E^\circ = 0.805 \pm 0.001$ V

by potentiometric and polarographic measurements with rhodium and boron carbide electrodes in several solutions from 0.1 to 1 M $NaClO_4$, $HClO_4$ or NaBr and extrapolation to zero ionic strength. This value is lower than those earlier reported by Kravtsov et al. [22], $E^{o\prime} = 0.883 \pm 0.003$ V, at 25°C in 1 M $HClO_4$ and other authors [15,23], $E^\circ = 0.99$ V, at zero ionic strength. However, it is consistent with those recorded for the Pt(IV)-Pt(II)-halogen system sequence, where redox potentials increase in the ligand series I < Br < Cl < F < H_2O, as the behavior of iridium toward halide ions is similar to that of platinum. The higher values in previous determinations have been attributed to the presence of appreciable concentration of the aquated Ir(III) bromo complex in the solution.

The potential of the half-reaction

$IrI_6^{2-}(aq) + e^- \rightarrow IrI_6^{3-}(aq) \quad E^\circ = 0.49$ V

was measured in 1 M KI by Ptitsyn [15], but the reality of the system has been placed in doubt [19].

D. Potential Diagrams for Iridium

$$\begin{array}{c}
\overset{+4}{IrO_2} \xrightarrow{0.223} \overset{+3}{Ir^{3+}} \xrightarrow{1.156} \overset{0}{Ir} \\
\underline{\qquad\qquad 0.926 \qquad\qquad} \\
IrCl_6^{2-} \xrightarrow{0.867} IrCl_6^{3-} \xrightarrow{0.86} \\
IrBr_6^{2-} \xrightarrow{0.805} IrBr_6^{3-} \\
IrI_6^{2-} \xrightarrow{0.49} IrI_6^{3-}
\end{array}$$

REFERENCES

1. G. T. Furukawa, M. L. Reilly, and L. S. Gallagher, J. Phys. Chem. Ref. Data, *3*, 163 (1974).
2. D. D. Wagman, W. H. Evans, V. B. Parker, R. H. Schumm, I. Halow, S. M. Bailey, K. L. Churney, and R. L. Nuttall, *NBS Tables of Chemical Thermodynamic Properties*, J. Phys. Chem. Ref. Data, Vol. II, Supplement 2 (1982).
3. F. A. Cotton and G. Wilkinson, *Advanced Inorganic Chemistry*, 2nd ed., Wiley-Interscience, New York, 1966, p. 1009.
4. R. N. Goldberg and L. H. Hepler, Chem. Rev., *68*, 229 (1968).
5. M. Pourbaix, ed., *Atlas d'équilibres Thermodinamiques à 25°C*, Gauthier-Villars, Paris, 1963, p. 350.
6. J. Amosse, M. Ruband, and M. J. Barbier, C. R. Acad. Sci. Paris, Ser. C, *273*, 1708 (1971).
7. J. P. Hoare, J. Electrochem. Soc., *111*, 232 (1964).
8. D. Cossi and F. Pantani, J. Inorg. Nucl. Chem., *8*, 385 (1958).
9. J. B. Willis, J. Am. Chem. Soc., *66*, 1067 (1944).
10. N. I. Kolbin and V. M. Samoilov, Zh. Neorg. Khim., *14*, 631 (1969).
11. J. P. Hoare, J. Electrochem. Soc., *111*, 988 (1964).
12. M. Pourbaix, ed., *Atlas d'équilibres Thermodinamiques à 25°C*, Gauthier-Villars, Paris, 1963, p. 373.
13. P. George, G. I. H. Hanania, and D. H. Irvine, J. Chem. Soc., 3048 (1957).
14. A. A. Grinberg and A. S. Shamsiev, Zh. Obshch. Khim., *12*, 55 (1942).
15. B. V. Ptitsyn, Zh. Obshch. Khim., *15*, 277 (1945).
16. V. I. Kravtsov and G. M. Petrova, Zh. Neorg. Khim., *9*, 1010 (1964).
17. V. I. Kravtsov and G. P. Tsayun, Elektrokhimiya, *6*, 1485 (1970).
18. J. A. Page, Talanta, *9*, 365 (1962).
19. E. Jackson and D. A. Pantony, J. Appl. Electrochem., *1*, 113 (1971).
20. J. C. Chang and C. S. Garner, Inorg. Chem., *4*, 209 (1965).
21. A. A. El-Awady, E. J. Bounsall, and C. S. Garner, Inorg. Chem., *6*, 79 (1967).
22. V. I. Kravtsov, V. N. Titova, and G. P. Tsayun, Elektrokhimiya, *6*, 573 (1970).
23. F. P. Dwyer, H. A. McKenzie, and R. S. Nyholm, J. Proc. R. Soc. N. S. Wales, *81*, 216 (1947).

14
Iron, Ruthenium, and Osmium

I. IRON*

KONRAD E. HEUSLER Technical University of Clausthal, Federal Republic of Germany
W. J. LORENZ Karlsruhe University, Karlsruhe, Federal Republic of Germany

A. Oxidation States

The common oxidation states of iron in aqueous solutions are Fe(II) and Fe(III). In alkaline solutions, the unstable oxidation states Fe(IV) and Fe(VI) are known to exist. Fe(V) salts and complexes containing Fe(I), Fe(O), Fe(−I), and Fe(−II) have been prepared, but these oxidation states are usually not stable in aqueous solutions.

The aquo complex Fe^{2+}, which is present in acid solutions, hydrolyzes by formation of $FeOH^+$ and precipitates as the hydroxide $Fe(OH)_2$ upon further increase of the pH value. In alkaline solutions, anion complexes such as FeO_2^{2-} are formed. Similarly, the aquo complex Fe^{3+} can be obtained in certain acid solutions. It hydrolyzes readily to $Fe(OH)^{2+}$, $Fe(OH)_2^+$, $Fe_2(OH)_2^{4+}$, and so on. Fe(III) hydroxide is precipitated in the neutral range of pH values. Its solubility increases again by formation of various ferrate(III) complexes in alkaline solutions. One should also consider the possibility that weak anion complexes can be formed in acid aqueous solutions (e.g., $FeSO_4$, $FeSO_4^+$, or $FeCl^{2+}$). Similarly, the ferrates may form weak cation complexes. The stability of these weak complexes depends on both the oxidation state and the composition of the electrolyte. These effects must be taken into account when deriving equilibrium potentials in a particular electrolyte from the standard potentials.

Several very stable complexes of iron are not readily hydrolyzed in aqueous solutions (e.g., $[Fe(CN)_6]^{4-}$ and several organic complexes). Large organic complexes of low charge number (e.g., ferrocene and ferricenium) are used to correlate standard potentials in different nonaqueous solvents to those in water on an extrathermodynamic basis [1]. Stability constants for iron complexes in aqueous solution can be found elsewhere [2].

B. Table of Thermodynamic Data

In Table 1 the data in italic are taken from the compilation of the National Bureau of Standards (NBS) [3]. The other data are reported by Barin et al.

---

*Reviewers: W. Reynolds, University of Minnesota, Minneapolis, Minnesota; L. Hepler, University of Lethbridge, Lethbridge, Alberta, Canada.

TABLE 1 Thermodynamic Data on Iron[a]

Formula	Description	State	$\Delta H°$/kJ mol^{-1}	$\Delta G°$/kJ mol^{-1}	$S°$/J mol^{-1} K^{-1}
Fe	α	c	0	0	27.29
		c	0	0	27.12
		g	416.3	370.7	180.381
Fe$^+$		g	1184.74		
Fe^{2+}		g	2752.2		
		aq	−89.1	−78.87	−137.7
		aq	−92.5	−91.2	−107.1
		aq	−92.7	−91.8	−107.
		aq		−91.2	−107.1
		aq		−88.9	
		aq	−92.4	−91.5	−106.
Fe^{3+}		g	5714.9		
		aq	−48.5	−4.6	−315.9
		aq	−50.2	−16.7	−280.3
		aq		−16.8	
Fe$_{0.942}$O	Wustite	c	−266.27	−245.14	57.49
		c	−266.27	−245.211	57.589
FeO		c	−272.0		

Iron, Ruthenium, and Osmium

FeO_2^{2-}	c	−272.04	−251.454	60.75	
	aq		−455.2		
	aq		−301.	−98.	
Fe_2O_3	c	−824.2	−742.2	87.40	Hematite
	c	−825.5	−743.608	87.4	
$Fe_2O_3 \cdot H_2O$	c	−1118.0	−984.03	118.8	
Fe_3O_4	c	−1118.4	−1015.5	146.4	Magnetite
	c	−1118.4	−1015.359	146.4	
$FeOH^+$	aq	−324.7	−277.4	−29.	
	aq		−274.3	−33.	
	aq		−268.0	−94.6	
$FeOH^{2+}$	aq	−290.8	−229.41	−142.	
$FeO(OH)$	c	−559.0			Goethite
$HFeO_2^-$	aq		−377.8		
	aq		−376.4	63.	
$Fe(OH)_2$	c	−569.0	−486.6	88.	Precipitated
	c	−574.	−492.017	87.9	
	aq		−441.0	29.1	
$Fe(OH)_2^+$	aq		−438.1		

TABLE 1 (Continued)

Formula	Description	State	$\Delta H°/\text{kJ mol}^{-1}$	$\Delta G°/\text{kJ mol}^{-1}$	$S°/\text{J mol}^{-1}\text{ K}^{-1}$
$Fe(OH)_3$	Precipitated	c	−823.0	−696.6	106.7
		c	−833.	−705.535	105.
	Undiss.	aq		−659.4	
		aq		−660.	75.4
$Fe(OH)_3^-$		aq		−615.0	
		aq		−605.1	29.9
$Fe(OH)_4^-$		aq		−842.2	24.5
$Fe(OH)_4^{2-}$		aq		−769.9	
$Fe_2(OH)_2^{2+}$		aq	−612.1	−467.27	−356.
FeF^{2+}		aq	−371.1	−322.6	−163
FeF_2		c			86.99
		c	−705.8	−663.557	86.99
		g	−389.5	−354.326	265.3
		aq	−754.4	−636.51	−165.3
FeF_2^+		aq	−696.2	−628.4	−63.
FeF_3		c	−1042.	−919.876	98.3
		g	−820.9	−767.047	304.2

FeCl	aq	−1046.4	−841.0	−357.3
FeCl^{2+}	g	251.		
FeCl$_2$	aq	180.3	143.9	−113.
	c	−341.79	−302.33	117.95
	c	−342.3	−303.43	120.1
	g	−148.5		
FeCl$_2$	aq	−432.4	−341.37	−24.7
FeCl$_2 \cdot$2H$_2$O	c	−953.1		
FeCl$_2 \cdot$4H$_2$O	c	−1549.3		
FeCl$_2^+$	aq	−399.49	−279.1	
FeCl$_3$	c	−399.40	−334.05	142.3
	c	−254.0	−333.974	142.335
	g	−253.0		344.104
	g	−550.2	−398.3	146.4
	aq Undiss.	−2223.8	−404.6	
FeCl$_3 \cdot$6H$_2$O	c	−439.		
(FeCl$_2$)$_3$	g	−654.8		
Fe$_2$Cl$_6$	g			

TABLE 1 (Continued)

Formula	Description	State	$\Delta H°/\text{kJ mol}^{-1}$	$\Delta G°/\text{kJ mol}^{-1}$	$S°/\text{J mol}^{-1}\text{ K}^{-1}$
FeOCl		g	−654.4	−598.755	537.02
		c	−377.0		
		c	−407.1	−359.234	80.8
$FeClO_4^{2+}$		aq	−17.		
$Fe(ClO_4)_2$		aq	−347.7	−96.11	226.4
$Fe(ClO_4)_2 \cdot 6H_2O$		c	−2068.6		
$FeBr^{2+}$		aq	−144.8	−112.1	−138.
$FeBr_2$		c	−249.8		
		c	−248.9	−237.375	−140.76
		g	−46.		
		g	−41.4	−15.029	337.2
		aq	−332.2	−286.81	27.2
$FeBr_3$		c	−268.2		
		c	−267.4	−242.919	173.6
		g	−123.8		
		aq	−413.4	−316.7	−68.6
$(FeBr_2)_2$		g	−264.		

Fe_2Br_6	g	−253.1	−153.063	515.9
$FeBrCl_2$	g	−394.6		
	c	−348.9		
FeJ^{2+}	aq	−504.6	−371.1	−120.5
FeJ_2	aq		−66.9	
	c	−113.0		
	c	−105.		167.
	g	60.7	−119.87	349.4
	g	88.	115.206	84.9
	aq	−199.6	−182.05	
FeJ_3	g	71.		
	aq	−214.2	−159.4	18.0
Fe_2J_4	g	29.		
	g	8.4	109.336	543.5
Fe_3J_6	g	16.7		
$Fe_{1.000}S$ Iron-rich pyrrhotite, α	c	−100.0	−100.4	60.29
	c	−95.4	−97.826	67.4
	c	−100.4	−100.738	60.29
FeS_2 Pyrite	c	−178.2	−166.9	52.93

TABLE 1 (Continued)

Formula	Description	State	$\Delta H°$/kJ mol^{-1}	$\Delta G°$/kJ mol^{-1}	$S°$/J mol^{-1} K^{-1}
	Marcasite	c	−177.4	−166.012	52.93
		c	−172.	−160.155	52.93
		c	−154.8		
Fe_7S_8	Sulfur-rich pyrrhotite	c	−736.4	−748.5	485.8
$FeSO_4$		c	−928.4	−820.9	107.5
		c	−928.8	−824.996	120.96
		aq	−998.3	−823.49	−117.6
$FeSO_4^+$		aq	−931.8	−772.8	−130.
$FeSO_4 \cdot H_2O$		c	−1243.69		
$FeSO_4 \cdot 4H_2O$		c	−1292.4		
$FeSO_4 \cdot 7H_2O$		c	−3014.57	−2510.27	409.2
$Fe(SO_4)_2^-$		aq		−1524.6	
$Fe_2(SO_4)_3$		c	−2581.5		
		c	−2582.99	−2263.065	307.5
		aq	−2825.0	−2243.0	−571.5
FeSe		c	−75.3		

Iron, Ruthenium, and Osmium

Formula	Description	State		
FeSe$_2$	Amorphous	c	−63.6	
FeSe$_2$		c		86.82
Fe$_{1.042}$Se		c		72.09
		c	−67.	69.20
Fe$_3$Se$_4$		c	−66.588	279.78
Fe$_7$Se$_8$		c		613.8
FeTe		c	−62.8	
FeTe$_2$	ε Phase	c		100.17
		c	−64.597	100.16
		c	−72.4	88.99
Fe$_{1.111}$Te		c	−23.22	80.08
FeN$_3^{2+}$		c	−23.305	−137.2
		aq	220.1	
		c	315.9	101.3
Fe$_2$N		c	−3.8	155.
Fe$_4$N		c	−10.5	
		c	3.8	155.6
		c	−10.9	155.6
		c	−11.09	
		c	32.221	
Fe(NO)$^{2+}$		aq	−9.2	32.054
				−213.
Fe(NO$_3$)$^{2+}$		aq		82.8
		aq		−121.8

TABLE 1 (Continued)

Formula	Description	State	$\Delta H°$/kJ mol^{-1}	$\Delta G°$/kJ mol^{-1}	$S°$/J mol^{-1} K^{-1}
Fe(NO$_3$)$_3$		aq	−670.7	−338.5	123.4
Fe(NO$_3$)$_3 \cdot$9H$_2$O		c	−3285.3		
Fe(NO)Cl$_2$		aq	−374.9		
FeCl$_2 \cdot$NH$_3$		c	−480.3		
FeCl$_2 \cdot$2NH$_3$		c	−605.0		
FeCl$_2 \cdot$6NH$_3$		c	−993.7		
FeCl$_3 \cdot$6NH$_3$		c	−894.1		
FeCl$_2 \cdot$10NH$_3$		c	−1292.4		
FeBr$_2 \cdot$NH$_3$		c	−381.1		
FeBr$_2 \cdot$2NH$_3$		c	−508.4		
FeBr$_2 \cdot$6NH$_3$		c	−915.5		
FeBr$_3 \cdot$6NH$_3$		c	−824.7		
FeI$_2 \cdot$2NH$_3$		c	−385.3		
FeI$_2 \cdot$6NH$_3$		c	−808.3		
FeSO$_4 \cdot$NH$_3$		c	−1060.2		

Iron, Ruthenium, and Osmium

FeSO$_4$·2NH$_3$	c	−1181.1		
FeSO$_4$·3NH$_3$	c	−1286.2		
FeSO$_4$·4NH$_3$	c	−1389.1		
FeSO$_4$·6NH$_3$	c	−1587.8		
FeP	c	−126.		
FeP$_2$	c	−192.		
Fe$_2$P	c	−163.		
Fe$_3$P	c	−163.		
FePO$_4$	c	−1297.5		
FePO$_4$·2H$_2$O	Strengite	c	−1888.2	−1657.7	111.25
FeAsS	c	−42.	−50.	121.
FeSb	c	−10.0		
FeSb$_2$	c	−15.1		
Fe$_3$C	α-Cementite	c	25.1	25.1	104.6
	c	22.6	26.104	101.3
FeCO$_3$	Siderite	c	−740.57	−666.72	92.9
	c	−740.6	−666.715	92.9

TABLE 1 (Continued)

Formula	Description	State	$\Delta H°$/kJ mol^{-1}	$\Delta G°$/kJ mol^{-1}	$S°$/J mol^{-1} K^{-1}
FeC$_2$O$_7$·2H$_2$O		c	−740.6	−666.715	92.9
		c	−1482.4		
Fe(CO)$_5$		ℓ	−774.0	−705.4	338.1
		ℓ	−764.0	−695.146	337.6
		g	−733.9	−697.27	445.2
Fe(C$_2$O$_4$)$_3$		aq	−2572.3	−2031.3	−495.0
Fe(C$_2$H$_3$O$_2$)$_3$		aq	−1506.7	−1112.9	−56.1
Fe(CO)$_4$Br$_2$		c	−807.1		
Fe(CO)$_4$I$_2$		c	−735.1		
Fe(CN)$_6^{3-}$		aq	561.9	729.3	270.3
Fe(CN)$_6^{4-}$		aq	455.6	694.92	95.0
Fe$_4$[Fe(CN)$_6$]$_3$		c	1184.		
Fe(CO)(CN)$_5^{3-}$		aq	192.9		
Fe$_2$(CO)(CN)$_5$		c	414.		
HFe(CN)$_6^{3-}$		aq	455.6	671.11	176.

$H_2Fe(CN)_6^{2-}$		aq	455.6	658.44	218.
$H_4Fe(CN)_6$		c	462.8		
$(NH_4)_4Fe(CN)_6 \cdot 6H_2O$		c	-1802.9		
$HFe(CO)(CN)_6$		aq	192.5		
$H_2Fe(CO)(CN)_5^-$		aq	193.7		
$H_3Fe(CO)(CN)_5 \cdot H_2O$		c	-67.		
$Fe(SCN)^{2+}$	Thiocyanate	aq	23.4	71.1	-130.
FeSi		c	-73.6	-73.6	46.0
		c	-78.7	-83.496	62.3
$FeSi_2$	β-Lebeanite	c	-81.2	-78.2	55.6
$FeSi_{2.33}$	α-Lebeanite	c	-59	-59.	69.5
Fe_3Si		c	-93.7	-94.6	103.8
Fe_5Si_3		c			209.6
$FeSiO_3$		c	-1205.		
		c	-1155.	-1075.408	87.4
Fe_2SiO_4	Fayalite	c	-1479.9	-1379.0	145.2
		c	-1479.9	-1379.024	145.2
$2FeI_2 \cdot PbI_2$		c	-425.1		

TABLE 1 (Continued)

Formula	Description	State	$\Delta H°/\text{kJ mol}^{-1}$	$\Delta G°/\text{kJ mol}^{-1}$	$S°/\text{J mol}^{-1} \text{K}^{-1}$
$Pb_2Fe(CN)_6 \cdot 3H_2O$		c		-168.2	
Fe_2Ti		c	-87.4	-84.245	74.5
$FeTi$		c	-40.6	-39.032	52.7
$FeTiO_3$		c	-1246.4	-1168.757	105.9
		c	-1235.459	-1158.050	105.9
$FeAl$		c	-50.2		
$FeAl_2$		c	-79.1		
$FeAl_3$		c	-79.1		
$FeAl_2O_4$		c	-1966.	-1849.	106.3
		c	-1984.97	-1869.367	106.3
		c	-1975.607	-1860.004	106.3
$TlFe(CN)_6^{3-}$		aq	464.8	644.8	292.9
$Tl_4Fe(CN)_6 \cdot 2H_2O$		c		26.4	
$ZnFe_2O_4$		c	-1169.4	-1063.6	151.67
$Zn_2Fe(CN)_6 \cdot 2H_2O$		c		-163.2	
$CdFe_2O_4$		c	-1069.8		

$Cd_2Fe(CN)_6 \cdot 7H_2O$		c		−1220.1	
$CuFeO_2$	Cupreous ferrite	c	−532.6	−479.9	88.6
$CuFe_2O_4$	Cupric ferrite	c	−965.21	−858.8	141.0
$FeWO_4$		c	−1190.01	−1114.277	131.8
Fe_3W_2		c	−31.4	−31.162	146.
FeB		c	−71.	−69.505	27.70
Fe_2B		c	−71.	−70.002	56.65
Fe_3Mo_2		c	−4.2	−6.402	146.
$FeCr_2O_4$		c	−1407.16	−1320.611	−146.
$Cu_{0.75}Fe_{2.25}O_4$		c			151.5
$Cu_2Fe(CN)_6$		c		736.4	
$Ag_4Fe(CN)_6 \cdot H_2O$		c		514.6	
$NiFe_2O_4$		c	−1081.1	−973.2	131.8
$Ni_{0.1}Zn_{0.9}Fe_2O_4$		c			154.0
$Ni_{0.2}Zn_{0.8}Fe_2O_4$		c			154.4
$Ni_{0.3}Zn_{0.7}Fe_2O_4$		c			154.4
$Ni_{0.4}Zn_{0.6}Fe_2O_4$		c			149.8

TABLE 1 (Continued)

Formula	Description	State	$\Delta H°$/kJ mol^{-1}	$\Delta G°$/kJ mol^{-1}	$S°$/J mol^{-1} K^{-1}
FeCoO$_4$		c			125.5
CoFe$_2$O$_4$		c			134.7
		c	−1087.8	−938.364	142.7

[a]National Bureau of Standards values are in italic.

C. Standard Electrode Potentials

1. Fe^{2+}/Fe Couple

The standard potential of the electrochemical reaction

$$Fe^{2+} + 2e^- \rightarrow Fe$$

is in the range $E° = -0.44 \pm 0.04$ V as estimated by Heusler [10]. This value is based on direct measurements of equilibrium cell voltages and on different ways of calculating from Gibbs energy data. There are several reasons for the relatively large error limit. Measurements with electrochemical cells are subject to systematic errors working in opposite directions. Corrosion cannot be avoided completely. It shifts the electrode potentials to positive values. On the other hand, overly negative electrode potentials were observed with finely divided or microcrystalline iron. In the experiments of Hampton [11] and of Randall and Frandsen [12] the two effects are believed to cancel. Errors due to corrosion decrease with temperature, mainly because the exchange current density of the iron electrode grows with temperature much faster than the exchange current density of the hydrogen electrode at iron. Extrapolation to 298 K of equilibrium cell voltages measured at temperatures between 373 and 473 K with iron in ferrous sulfate solutions (pH 1.5 at 298 K) vs. the mercury sulfate electrode by Zarechanskaya et al. [13] yields a formal potential of about $E°' = -0.56$ V in 1 M $FeSO_4$. Assuming an activity coefficient similar to those given by Robinson and Stokes [14] for the sulfates of manganese, nickel, copper, zinc, and cadmium, one obtains a standard potential of about $E° = -0.48$ V.

Calculation of $E°$ from the NBS thermodynamic data in Table 1 yields a rather positive value of $E° = -0.409$ V. If the standard potential is calculated from the solubility of iron(II) hydroxide and its Gibbs energy of formation as proposed by Lamb [15], one obtains a mean value of $E° = -0.43$ V [10]. The uncertainty of ± 0.04 V is due to the scatter of the literature values given for the relevant Gibbs energies.

Solubilities of magnetite were measured at elevated temperatures in solutions saturated with hydrogen at 298 K and $p(H_2) = 1$ atm by Sweeton and Baes [6]. On this basis the standard potential was calculated at $E° = -0.475 \pm 0.010$ V using a Gibbs energy of formation of magnetite believed to be the best value. However, it would have been necessary to consider the dependence of the solubility of magnetite on hydrogen pressure. The dependence of magnetite solubility on hydrogen pressure was taken into account by Tremaine and Le Blanc [8]. Their result is $E° = -0.461 \pm 0.003$ V based on $\Delta G°(Fe_3O_4) = -1015.5$ kJ mol^{-1} for the standard Gibbs energy of formation of magnetite at 298 K.

Larson et al. [5] calculated the standard potential from enthalpies of formation and solution of $FeSO_4 \cdot 7H_2O$, $FeCl_2$, and $FeBr_2$. Together with entropies of formation of Fe^{2+} containing estimated activity coefficients in addition to experimental data they arrived at $E° = -0.473$ V, a value also adopted by Tremaine and Le Blanc [8] as well as by Sadiq and Lindsay [9]. The total uncertainty was believed to be ± 0.010 V. It comprises unknown systematic errors due to the use of enthalpies of dilution and activity coefficients measured with nickel sulfate instead of iron sulfate.

In principle, it is also possible to derive the standard potential from equilibrium potentials obtained from kinetic measurements. Hurlen [16-18] extrapolated Tafel lines of iron dissolution and deposition measured in acidified solutions of iron(II) sulfate and chloride to their intersection. From such equilibrium potentials a standard potential $E° = -0.44$ V was calculated at 293 K for sulfate solutions and $E° = -0.47$ V for chloride solutions. These values contain systematic errors from the estimation of activity coefficients of the iron salts and from the neglect of junction potentials. The junction potential was believed to make the experimental values of equilibrium potentials too negative. It should also be mentioned that experimental errors for steady-state polarization curves are fairly large. This was avoided by Heusler [19] in measurements of transient polarization curves in 1 M iron(II) sulfate solution at pH 2.9 and 292.5 K. The experiments yielded $E° = -0.475$ V using the activity coefficient of nickel sulfate and again neglecting the junction potential toward the saturated potassium chloride solution of the calomel reference electrode.

Taking all the evidence together, the standard potentials cover a range between -0.41 V and -0.48 V. More recent experiments favor the range -0.43 V to -0.48 V.

2. Fe^{3+}/Fe^{2+} Couple

The standard potential $E° = +0.771$ V of the redox reaction

$$Fe^{3+} + e^- \rightarrow Fe^{2+}$$

is based on measurements by Schumb et al. [20] using the cell

$$Pt | (1-x) \, M \, Fe(ClO_4)_2 + x \, M \, Fe(ClO_4)_3 + y \, M \, HClO_4 \| y \, M \, HClO_4 | H_2, | Pt$$

where $0.5 \times 10^{-3} \leqslant x \leqslant 60 \times 10^{-3}$ and $0.025 \leqslant y \leqslant 0.3$. More recently, Whittemore and Langmuir [21] arrived at $E° = +0.770 \pm 0.002$ V from measurements between 278 and 308 K after correction for hydrolysis of the iron ions. For the cell reaction, the standard enthalpy was found to be $\Delta H° = -42.7 \pm 1.7$ kJ mol^{-1}, and the standard entropy $\Delta S° = 106 \pm 6$ J mol^{-1} K^{-1}. The standard potential of the reaction

$$Fe^{3+} + 3e^- \rightarrow Fe$$

can only be calculated as $E° = -0.037$ V using the standard potentials of the Fe^{2+}/Fe and Fe^{3+}/Fe^{2+} couples [22].

3. $[Fe(CN)_6]^{3-}/[Fe(CN)_6]^{4-}$ Couple

The most reliable value of the standard potential of the redox reaction

$$[Fe(CN)_6]^{3-} + e^- \rightarrow [Fe(CN)_6]^{4-}$$

appears to be $E° = +0.3610 \pm 0.0005$ V. It was obtained by electromotive force (emf) measurements in cells without a liquid junction using cation-sensitive glass electrodes [23]. Cells with a liquid junction yielded similar values (see [10]). In the presence of supporting electrolyte, the apparent standard

4. Redox Reactions of Organic Iron Complexes

Standard potentials of Fe(II)/Fe(III) complexes with organic ligands are summarized in Table 2. The data were obtained by polarography, coulometric titration, emf measurements, and photometry.

5. Other Equilibrium Potentials

For a large number of electrochemical reactions in the systems iron-water and iron-water-sulfur, Biernat and Robins [31] published tables of calculated standard potentials at temperatures of $298 \text{ K} \leqslant T \leqslant 573 \text{ K}$. The calculations are based on the NBS thermodynamic data [3] and on the entropy correspondence principle of Criss and Cobble [32]. In many cases the standard potentials involving specific complexes cannot be determined experimentally because several complexes are present simultaneously.

Standard potentials published between 1945 and 1973 were comprehensively but uncritically compiled by Milazzo et al. [33]. Some of these potentials cannot be measured experimentally or are only rough estimates from thermodynamic data.

The standard potential of the reaction

$$HFeO_2^- + H_2O + 2e^- \rightarrow Fe + 3OH^-$$

was estimated at $E° = -0.747$ V by Pourbaix [22]. The NBS thermodynamic data yield $E° = -0.740$ V. Losev and Kabanov [34] determined the equilibrium potential $E = -0.12$ vs. the hydrogen electrode in the same solution saturated with hydrogen at atmospheric pressure from the intersection of the Tafel lines for deposition and dissolution of iron in a solution containing 10 M NaOH and 1.3×10^{-3} M Fe(II) at 353 K. Assuming that $\log a_{OH^-} = 1.9 \pm 0.4$, one finds for the standard potential about $E°_{353K} = -0.93 \pm 0.07$ V. Further uncertainties arise from the possibility that iron complexes other than $HFeO_2^-$ are present under the experimental conditions.

The measurements of Grube and Gmelin [35] at 353 K in 40% NaOH with $\log a_{OH^-} \approx 2.5 \pm 0.5$ yield approximately $E° = -0.54$ V, but potentials are certainly too positive because corrosion was neglected. They also contain further errors due to diffusion polarization and the assumption of a temperature-independent potential of the calomel electrode in 0.1 M KCl of $E = 0.286$ V vs. the standard hydrogen electrode.

The measurement of the equilibrium potential of the redox couple

$$FeO_2^- + H_2O + e^- \rightarrow HFeO_2^- + OH^-$$

also contains the latter errors. Under the conditions described above, Grube and Gmelin [35] found $E = -0.685$ V in the presence of 6×10^{-3} M Fe(III) and 6×10^{-3} M Fe(II). For the reaction

$$HFeO_4^- + 2H_2O + 3e^- \rightarrow FeO_2^- + 4OH^-$$

they estimate an equilibrium potential $E = 0.55$ V at equal concentrations of Fe(VI) and Fe(III). For the reaction

TABLE 2 Standard Potentials of Fe(II)/Fe(III) Complexes with Organic Ligands

Reaction[a]	$E°$/V	$\Delta H°$/kJ mol^{-1}	$S°$/J mol^{-1} K^{-1}	References
Fe(C$_5$H$_5$)$_2$ + e$^-$ → Fe(C$_5$H$_5$)$_2$	0.400 ± 0.007			24
Fe(phen)$_3^{3+}$ + e$^-$ → Fe(phen)$_3^{2+}$	1.13 ± 0.01	−137.8 ± 2.1	−87.0 ± 8.4	24-27
Fe(tmphen)$_3^{3+}$ + e$^-$ → Fe(tmphen)$_3^{2+}$	0.895 ± 0.005			24
Fe(bipy)$_3^{3+}$ + e$^-$ → Fe(bipy)$_3^{2+}$	1.11 ± 0.01	−136.8 ± 2.1	−99.6 ± 8.4	25, 26
Fe(dimbipy)$_3^{3+}$ + e$^-$ → Fe(dimbipy)$_3^{2+}$	0.94	−112.1	−73.2	28
Fe(pyal)$_3$ + e$^-$ → Fe(pyal)$_3^-$	0.348 ± 0.007	−86.6 ± 0.2	−178.2 ± 6.6	29
Fe(pydial)$_2^-$ + e$^-$ → Fe(pydial)$_2^{2-}$	0.201 ± 0.0005	−88 ± 5	−230 ± 10	30
Fe(phensu)$_3$ + e$^-$ → Fe(phensu)$_3^-$	1.225 ± 0.010			24
Fe(tadsal)$_3^+$ + e$^-$ → Fe(tadsal)$_3$	−0.258 ± 0.003			24
Fe(pyasal)$_3^+$ + e$^-$ → Fe(pyasal)$_3$	−0.050 ± 0.010			24
Fe(etdipyr)$_3^+$ + e$^-$ → Fe(etdipyr)$_3$	−0.200 ± 0.010			24

[a]phen, 1,10-phenanthroline; tmphen, 3,4,7,8-tetramethyl-1,10-phenanthroline; bipy, 2,2'-bipyridil; dimbipy, 4,4'-dimethyl-2,2'-bipyridil; pyal, pyridine-2-aldoxime; pydial, 2,6-pyridinedialdoxime; phensu, 1,10-phenanthroline-6-sulfonic acid; tadsal, ethylenetetraminedisalicylidine; pyasal, 2-methylpyridineaminosalicylidine; etdipyr, ethylenetetraminedipyrrolidine.

$$[Fe(CN)_6]^{4-} + 2e^- \rightarrow Fe + 6CN^-$$

a value of $E° = -1.16$ V was reported by Stephenson and Morrow [36].

Conditional standard potentials of the iron(II/III)-oxalate electrode reactions in solutions containing a relatively high concentration of sodium oxalate in the supporting electrolyte (>0.2 M),

$$Fe(C_2O_4)_3^{3-} + e^- \rightarrow Fe(C_2O_4)_3^{4-} \qquad E^{o'} = 0.005 \text{ V}$$

and in solutions containing a lower excess of sodium oxalate (<0.1 M),

$$Fe(C_2O_4)_3^{3-} + e^- \rightarrow Fe(C_2O_4)_2^{2-} + C_2O_4^{2-} \qquad E^{o'} = 0.035 \text{ V}$$

were summarized by Vetter [37].

D. Potential Diagrams for Iron

Acid Solution (pH 0)

```
          +3              +2            0
     Fe³⁺  0.771   Fe²⁺  -0.44   Fe
       |_____ -0.04 _____|
  [Fe(CN)₆]³⁻  0.361  [Fe(CN)₆]⁴⁻  -1.16
```

Alkaline Solution (pH 14)

```
    +6              +3              +2              0
  FeO₄²⁻  ca. 0.55  FeO₂⁻  ca. -0.69  HFeO₂⁻  ca. -0.8  Fe
```

REFERENCES

1. G. Gritzner and J. Kuta, Pure Appl. Chem., 54, 1527 (1982).
2. L. G. Sillén, *Stability Constants of Metal-Ion Complexes*, 2nd ed., Spec. Publ. 17, Chemical Society, London, 1964; Supplement No. 1, Chemical Society, London, 1971.
3. D. D. Wagman, W. H. Evans, J. Halow, V. B. Parker, S. M. Bailey, and B. H. Schumm, Natl. Bur. Stand. Tech. Note 270-4, U.S. Government Printing Office, Washington, D.C., 1969, Table 41.
4. I. Barin, O. Knacke, and O. Kubaschewski, *Thermochemical Properties of Inorganic Substances*, 1973, and *Supplement*, 1977, Springer-Verlag, Berlin/Verlag Stahleisen, Düsseldorf.
5. J. W. Larson, P. Cerutti, H. K. Gerber, and L. G. Hepler, J. Phys. Chem., 72, 2902 (1968).
6. F. H. Sweeton and C. F. Baes, J. Chem. Thermodyn., 2, 479 (1970).
7. P. R. Tremaine, R. von Massow, and G. R. Shierman, Thermochim. Acta, 19, 287 (1977).

8. P. R. Tremaine and J. C. Le Blanc, J. Solut. Chem., 9, 415 (1980).
9. M. Sadiq and W. L. Lindsay, Arabian J. Sci. Eng., 6, 95 (1981).
10. K. E. Heusler, in *Encyclopedia of the Electrochemistry of the Elements*, Vol. 9A, A. J. Bard, ed., Marcel Dekker, New York, 1982, p. 229.
11. W. H. Hampton, J. Phys. Chem., 30, 980 (1926).
12. M. Randall and M. Frandsen, J. Am. Chem. Soc., 54, 47 (1932).
13. V. V. Zarechanskaya, M. A. Zhamagortzyants, and A. T. Vagramyan, Elektrokhimiya 8, 180 (1972).
14. R. A. Robinson and R. H. Stokes, *Electrolyte Solutions*, Butterworth, London, 1959.
15. A. B. Lamb, J. Am. Chem. Soc., 32, 1214 (1910).
16. T. Hurlen, Tek. Ukeblad, 105, 101, 119 (1958).
17. T. P. Hoar and T. Hurlen, *CITCE VIII*, Butterworth, London, 1958, p. 445.
18. T. Hurlen, Acta Chem. Scand., 14, 1533 (1960).
19. K. E. Heusler, *Habilitationsschrift*, Stuttgart, 1966; cited in Ref. 10.
20. W. C. Schumb, M. S. Sherril, and S. B. Sweetser, J. Am. Chem. Soc., 59, 2360 (1937).
21. D. O. Whittemore and D. Langmuir, J. Chem. Eng. Data, 17, 288 (1972).
22. M. Pourbaix, ed., *Atlas d'équilibres électrochimiques à 25°C*, Gauthier-Villars, Paris, 1963.
23. R. C. Murray and P. A. Rock, Electrochim. Acta, 13, 969 (1968).
24. H. M. Koepp, H. Wendt, and H. Strehlow, Z. Elektrochem., Ber. Bunsenges. Phys. Chem., 64, 483 (1960).
25. F. P. Dwyer, N. A. Gibson, and E. C. Gyarfas, J. Proc. R. Soc. N.S. Wales, 84, 80 (1951).
26. P. George, G. I. H. Hanania, and D. H. Irvine, J. Chem. Soc., 2548 (1959).
27. D. N. Hume and I. M. Kolthoff, J. Am. Chem. Soc., 65, 1895 (1943).
28. D. H. Irvine, J. Chem. Soc., 2977 (1959).
29. G. I. H. Hanania, D. H. Irvine, and F. R. Shurayh, J. Phys. Chem., 72, 1355 (1968).
30. G. I. H. Hanania, M. S. Michallides, and D. H. Irvine, J. Phys. Chem., 81, 1382 (1977).
31. R. H. Biernat and R. G. Robins, Electrochim. Acta, 17, 1261 (1972).
32. C. M. Criss and J. W. Cobble, J. Am. Chem. Soc., 86, 5385 (1964).
33. G. Milazzo, S. Caroli, and V. K. Sharma, *Tables of Standard Electrode Potentials*, Wiley, New York, 1978.
34. V. Losev and B. N. Kabanov, Zh. Fiz. Khim., 28, 824, 914 (1954).
35. G. Grube and H. Gmelin, Z. Elektrochem., 26, 466 (1920).
36. C. C. Stephenson and J. C. Morrow, J. Am. Chem. Soc., 78, 275 (1956).
37. K. J. Vetter, *Electrochemical Kinetics*, Academic Press, New York, 1967, p. 495.

II. RUTHENIUM AND OSMIUM*

FRANCISCO COLOM   Institute of Physical Chemistry, CSIC, Madrid, Spain

A. Ruthenium

1. Oxidation States

The range of oxidation states for the ruthenium metal extends from 0 (metal carbonyls) to VIII, and because of the capacity of its ions to form polynuclear complexes, apparent fractional valencies are also known. However, the more usual states are Ru(IV) and (III). In aqueous solutions, the Ru(II) to Ru(IV) compounds tend to form complex ions with coordination number 6 and, to a lesser extent, 4.

In noncomplexing aqueous solutions and the presence of oxygen, Ru(IV) is the more stable oxidation state except at high pH, where the more stable form is Ru(VI). Under these conditions, all the ruthenium ions tend to assume the IV and VI valence states by spontaneous oxidation or reduction reactions. In basic solutions, the more stable oxidation state is Ru(VI), but the structure of the ion depends on the solution pH. Thus in very alkaline solution Ru(VI) exists in the form of orange ruthenate ion RuO_4^{2-}, which at a pH value near 12 undergoes a dismutation to solid black RuO_2 and green RuO_4^- ions. At this pH, the RuO_4^- ions are unstable and oxidize the H_2O molecule, liberating oxygen and becoming RuO_4^{2-}; at pH 7 that dismutation may produce RuO_4 or unstable H_2RuO_5. In very basic solutions, the volatile yellow RuO_4 dissolves to form $HRuO_5^-$, which rapidly decomposes to RuO_4^- and RuO_4^{2-}. In acid media, Ru(IV), Ru(III), and Ru(II) are the more stable oxidation states. The thermodynamic data are given in Table 3.

2. Ruthenium Potentials

From the Gibbs energy of formation of the hydrous dioxide and RuO_4, Goldberg and Hepler [4] have deduced the potentials

$$RuO_2(s,hyd) + 4H^+(aq) + 4e^- \rightarrow Ru(s) + 2H_2O \qquad E^\circ = 0.68 \text{ V}$$

$$RuO_4(aq) + 8H^+(aq) + 8e^- \rightarrow Ru(s) + 4H_2O \qquad E^\circ = 1.04 \text{ V}$$

$$RuO_4(aq) + 4H^+(aq) + 4e^- \rightarrow RuO_2(s,hyd) + 2H_2O \qquad E^\circ = 1.387 \text{ V}$$

The Ru^{3+}(aq) ion is now well established as $Ru(H_2O)_6^{3+}$ and the $Ru(H_2O)_6^{2+}$ ion can be stabilized in tetrafluoroborate or p-toluenesulfonate anion solutions. Thus Buckley and Mercer [5] have measured the potential of the reaction

$$Ru(H_2O)_6^{3+}(aq) + e^- \rightarrow Ru(H_2O)_6^{2+}(aq) \qquad E^\circ = 0.249 \text{ V}$$

against a glass electrode in p-toluenesulfonic acid media at different concentrations and extrapolating to zero ionic strength at 25°C. A slightly lower

---

Reviewer: M. Gross, Louis Pasteur University, Strasbourg, France.

TABLE 3 Thermodynamic Data on Ruthenium[a]

Formula	Description	State	$\Delta H°$/kJ mol^{-1}	$\Delta G°$/kJ mol^{-1}	$S°$/J mol^{-1} K^{-1}
Ru		c	0.	0.	28.61
Ru		g	642.7	595.8	186.507
Ru$^+$		g	1350.30		
Ru^{2+}		g	2974.0		
Ru^{3+}		g	5728.		
RuO$_2$		c	−305.0		
RuO$_2$	Hydrated	amorph		−214.6	
RuO$_3$		g	−78.2		
RuO$_4$		c	−239.3	−152.2	146.4
RuO$_4$		ℓ	−228.4	−152.2	183.3
RuO$_4$		g	−184.1	−139.7	290.1

RuO_4	Std. state; $m = 1$	aq	−239.7	−147.2	130.
RuO_4^-	Std. state; $m = 1$	aq		−245.5	
RuO_4^{2-}	Std. state; $m = 1$	aq		−303.7	
$Ru(OH)_2^{2+}$	Std. state; $m = 1$	aq		−213.	
RuF_5		c	−892.9		
RuF_5		g	−791.		
$RuCl_3$	Black	c	−205.		
$RuCl_3$		g	−1.3		
$RuCl_4$		g	−51.9		
$RuCl_5(OH)^{2-}$	Std. state; $m = 1$	aq		−705.7	
$RuBr_3$		c	−138.		
RuI_3		c	−65.7		
RuS_2		c	−197.		

[a]National Bureau of Standards values are in italic.
Source: Data from Refs. 1, 2, and 3.

value, $E^{o'} = 0.217$ V, was obtained in a 1.0 M CF_3-SO_3H solution for this system from kinetic data [6]. Under the latter experimental conditions the potential for the equilibrium

$$Ru(H_2O)_5Cl^{2+}(aq) + e^- \rightarrow Ru(H_2O)_5Cl^+(aq) \qquad E^{o'} = 0.086 \text{ V}$$

was also deduced.

Data relevant to the equilibrium

$$Ru(CN)_6^{3-}(aq) + e^- \rightarrow Ru(CN)_6^{4-}(aq) \qquad E^\circ = 0.86 \pm 0.05 \text{ V}$$

have been reported from potentiometric and voltammetric studies with stationary and rotating disk electrodes in different solutions of KCl [7].

Research on the couple Ru(IV)/Ru(III) has been rather scarce, due mainly to difficulties in identifying the chemical species involved. Earlier potentiometric measurements in HCl and HBr solutions [8,9] gave quite divergent values from those in $HClO_4$ [10]. These results should be revised in the light of new knowledge on the coordination number of Ru ions and the formation of polymeric ruthenium complexes in hydrohalic acid solutions [11,12].

For the reduction of Ru(VIII), Ru(VII), and Ru(VI) to Ru(IV) in aqueous solution, Pourbaix et al. [13] have calculated from thermodynamic data

$$H_2RuO_5(aq) + 4H^+(aq) + 4e^- \rightarrow RuO_2(s,hyd) + 3H_2O \qquad E^\circ = 1.40 \text{ V}$$

$$RuO_4^-(aq) + 4H^+(aq) + 3e^- \rightarrow RuO_2(s,hyd) + 2H_2O \qquad E^\circ = 1.533 \text{ V}$$

$$RuO_4^{2-}(aq) + 4H^+(aq) + 2e^- \rightarrow RuO_2(s,hyd) + 2H_2O \qquad E^\circ = 2.005 \text{ V}$$

The calculated potentials of the RuO_4/Ru(IV) couple are supported by earlier potentiometric measurements of the system, which rendered the approximate values 1.40, 1.43, and 1.51 V in 1, 6, and 9 M $HClO_4$ solutions, respectively [14]. Nevertheless, recent measurements with Donnan membranes [15] indicate a charge of +4 for ions of Ru(IV) in aqueous perchloric solutions, which suggests the formation of polymeric complexes.

Eichner [16] has calculated the potential-pH diagram of ruthenium metal in solutions at high pH values. The results point out that the VII and VIII oxidation states are unstable throughout this region. These states can be obtained at the anode only if the reactions of the OH^-/O_2 system proceed very slowly. Thus the change from Pt to vitreous carbon electrodes in the electrolysis of solutions at pH 14 increases the overpotential of that reaction from 0.6 V to 1.0 V, which allows the oxidation step of RuO_4^- to RuO_4 and the reduction of this product to be reached.

Latimer [17] estimated from thermodynamic data the potential of the reaction

$$RuO_4(aq) + e^- \rightarrow RuO_4^-(aq) \qquad E^\circ = 0.99 \text{ V}$$

which is in good agreement with the polarographic results obtained by Silverman and Levy [18] and was confirmed by equilibrium measurements [19,20] and more recently by cyclic voltammetry [21] in basic solutions (NaOH + NaClO$_4$) at an ionic strength of 1 M.

The reduction of per-ruthenate to ruthenate,

$$RuO_4^-(aq) + e^- \rightarrow RuO_4^{2-}(aq) \qquad E° = 0.593 \text{ V}$$

has been studied through reversible cell measurements [22], polarography [18], and equilibrium constants [20] with very consistent results. The ammine complexes of Ru in aqueous solutions form a large group of stable Ru(III)/Ru(II) systems which have been widely investigated.

Equilibrium and potentiometric measurements in HClO$_4$ medium at infinite dilution at 25°C lead to

$$Ru(NH_3)_6^{3+}(aq) + e^- \rightarrow Ru(NH_3)_6^{2+}(aq) \qquad E° = 0.10 \pm 0.01 \text{ V [23]}$$

The formal potential of this system as determined by equilibrium measurements in CF$_3$-SO$_3$Na [24] and cyclic voltammetry [25] in 0.1 M NaBF$_4$ is $E°' = 0.066 \pm 0.004$ and 0.051 V, respectively. Also,

$$Ru(en)_3^{3+} + e^- \rightarrow Ru(en)_3^{2+} \qquad E° = 0.210 \pm 0.005 \text{ V [23]}$$

The study of these systems by cyclic voltammetry and polarography in aqueous 0.2 M CF$_3$–COONa [25], 0.1 M HCl [26,27], 0.30 M CH$_3$–SO$_3$Na adjusted to pH 2.0 $\pm$ 0.1 at 25°C [28], and 0.1 M p-toluenesulfonic acid/0.1 M p-toluene potassium sulfonate [29] solutions gave the following formal potentials:

$$Ru(NH_3)_5(H_2O)^{3+}(aq) + e^- \rightarrow Ru(NH_3)_5(H_2O)^{2+}(aq) \qquad E°' = 0.066 \text{ V}$$

[25,28,29]

or 0.084 V [24] and the standard potential $E° = 0.16 \pm 0.01$ V [23]

$$cis\text{-}Ru(NH_3)_4(H_2O)_2^{3+}(aq) + e^- \rightarrow cis\text{-}Ru(NH_3)_4(H_2O)_2^{2+}(aq)$$

$$E°' = 0.100 \text{ V [25,28]}$$

$$trans\text{-}Ru(NH_3)_4(H_2O)_2^{3+}(aq) + e^- \rightarrow trans\text{-}Ru(NH_3)_4(H_2O)_2^{2+}(aq)$$

$$E°' = 0.020 \text{ V [25]} \quad \text{and} \quad E°' = 0.040 \text{ V [28]}$$

$$Ru(NH_3)_5(OH)^{2+}(aq) + e^- \rightarrow Ru(NH_3)_5(OH)^+(aq) \text{ in } 0.1 \text{ } M \text{ NaOH}$$

$$E°' = -0.420 \text{ V [25,27]}$$

$Ru(NH_3)_5Cl^{2+}(aq) + e^- \rightarrow Ru(NH_3)_5Cl^+(aq)$ $E^{o\prime} = -0.04$ V [25,28,29]

and also in 0.2 M $HClO_4$ [26]

$cis\text{-}Ru(NH_3)_4Cl_2^+(aq) + e^- \rightarrow cis\text{-}Ru(NH_3)_4Cl_2(aq)$ $E^{o\prime} = -0.100$ V [25]

and -0.082 V [28]

$trans\text{-}Ru(NH_3)_4Cl_2^+(aq) + e^- \rightarrow trans\text{-}Ru(NH_3)_4Cl_2(aq)$ $E^{o\prime} = -0.180$ [25]

and -0.166 V [28]

$cis\text{-}Ru(NH_3)_4(H_2O)Cl^{2+}(aq) + e^- \rightarrow cis\text{-}Ru(NH_3)_4(H_2O)Cl^+(aq)$

$E^{o\prime} = 0.03$ V [28]

$trans\text{-}Ru(NH_3)_4(H_2O)Cl^{2+}(aq) + e^- \rightarrow trans\text{-}Ru(NH_3)_4(H_2O)Cl^+(aq)$

$E^{o\prime} = -0.050$ V [28]

$Ru(NH_3)_5Br^{2+}(aq) + e^- \rightarrow Ru(NH_3)_5Br^+(aq)$ in 0.1 M $NaClO_4$

$E^{o\prime} = -0.034$ V [25]

$Ru(NH_3)_5(SCN)^{2+}(aq) + e^- \rightarrow Ru(NH_3)_5(SCN)^+(aq)$ $E^{o\prime} = 0.133$ V [25]

$Ru(NH_3)_5(SO_2)^{3+}(aq) + e^- \rightarrow Ru(NH_3)_5(SO_2)^{2+}(aq)$ $E^{o\prime} = 0.250$ V [28]

$Ru(NH_3)_5N_2^{3+}(aq) + e^- \rightarrow Ru(NH_3)_5N_2^{2+}(aq)$ in 0.1 to 1 M $HClO_4$

$E^{o\prime} = 1.12$ V [25,27]

In the systems with pyridine ligands, the ammine complexes yielded

$Ru(NH_3)_5py^{3+}(aq) + e^- \rightarrow Ru(NH_3)_5py^{2+}(aq)$ $E^{o\prime} = 0.305$ V [25,29,30]

and in 1 M NaCl $E^{o\prime} = 0.490$ V [25]

and a series with pyridine derivatives gave [29]

$Ru(NH_3)_5py\text{-}4\text{-}CH(OH)_2^{3+}(aq) + e^- \rightarrow Ru(NH_3)_5py\text{-}4\text{-}CH(OH)_2^{2+}(aq)$

$E^{o\prime} = 0.300$ V

$Ru(NH_3)_5py\text{-}4Cl^{3+}(aq) + e^- \rightarrow Ru(NH_3)_5py\text{-}4Cl^{2+}(aq)$ $E^{o\prime} = 0.322$ V

$$Ru(NH_3)_5py\text{-}3\text{-}CONH_2^{3+}(aq) + e^- \rightarrow Ru(NH_3)_5py\text{-}3\text{-}CONH_2^{2+}(aq)$$

$E^{o\prime} = 0.353$ V

$$Ru(NH_3)_5py\text{-}3\text{-}Cl_3^{3+}(aq) + e^- \rightarrow Ru(NH_3)_5py\text{-}3\text{-}Cl_3^{2+}(aq) \quad E^{o\prime} = 0.361 \text{ V}$$

$$Ru(NH_3)_5py\text{-}4\text{-}CONH_2^{3+}(aq) + e^- \rightarrow Ru(NH_3)_5py\text{-}4\text{-}CONH_2^{2+}(aq)$$

$E^{o\prime} = 0.375$ V and 0.440 in 1 M KCl

$$Ru(NH_3)_5py\text{-}4\text{-}COOH^{3+}(aq) + e^- \rightarrow Ru(NH_3)_5py\text{-}4\text{-}COOH^{2+}(aq)$$

$E^{o\prime} = 0.382$ V

$$Ru(NH_3)_5py\text{-}4\text{-}COOCH_3^{3+}(aq) + e^- \rightarrow Ru(NH_3)_5py\text{-}4\text{-}COOCH_3^{2+}(aq)$$

$E^{o\prime} = 0.392$ V

$$trans\text{-}Ru(NH_3)_4py_2^{3+}(aq) + e^- \rightarrow trans\text{-}Ru(NH_3)_4py_2^{2+}(aq) \quad E^{o\prime} = 0.494 \text{ V}$$

$$cis\text{-}Ru(NH_3)_4py_2^{3+}(aq) + e^- \rightarrow cis\text{-}Ru(NH_3)_4py_2^{2+}(aq) \quad E^{o\prime} = 0.505 \text{ V}$$

In ammine-organonitrile complexes, electron-withdrawing substituents increase the oxidation potential so that the potentials were [29]

$$Ru(NH_3)_5NC\text{-}CH_3^{3+}(aq) + e^- \rightarrow Ru(NH_3)_5NC\text{-}CH_3^{2+}(aq) \quad E^{o\prime} = 0.426 \text{ V}$$

$$Ru(NH_3)_5NC\text{-}CH_2\text{-}CH_3^{3+}(aq) + e^- \rightarrow Ru(NH_3)_5NC\text{-}CH_2\text{-}CH_3^{2+}(aq)$$

$E^{o\prime} = 0.420$ V

$$Ru(NH_3)_5NC\text{-}CH\text{=}CH_2^{3+}(aq) + e^- \rightarrow Ru(NH_3)_5NC\text{-}CH\text{=}CH_2^{2+}(aq)$$

$E^{o\prime} = 0.512$ V

$$Ru(NH_3)_5NC\text{-}Ph^{3+}(aq) + e^- \rightarrow Ru(NH)_3)_5NC\text{-}PH^{2+}(aq) \quad E^{o\prime} = 0.485 \text{ V}$$

$$cis\text{-}Ru(NH_3)_4(NC\text{-}Ph)_2^{3+}(aq) + e^- \rightarrow cis\text{-}Ru(NH_3)_4(NC\text{-}Ph)_2^{2+}(aq)$$

$E^{o\prime} = 0.817$ V

where Ph = phenyl.

In the complex systems of pentaamine with other ammine ligands, cyclic voltammetry yielded [26]

$$Ru(NH_3)_5(NH_2\text{-}CH_2\text{-}C_6H_5)^{3+}(aq) + e^- \rightarrow Ru(NH_3)_5(NH_2\text{-}CH_2\text{-}C_6H_5)^{2+}(aq)$$

$E^{o\prime} = 0.120$ V

$Ru(NH_3)_5(NH=CH-C_6H_5)^{3+}(aq) + e^- \rightarrow Ru(NH_3)_5(NH=CH-C_6H_5)^{2+}(aq)$

$E^{o\prime} = 0.320$ V

$Ru(NH_3)_5NC-C_6H_5^{3+}(aq) + e^- \rightarrow Ru(NH_3)_5NC-C_6H_5^{2+}(aq)$ $\quad E^{o\prime} = 0.510$ V

$Ru(NH_3)_5NH_2-CH_3^{3+}(aq) + e^- \rightarrow Ru(NH_3)_5NH_2-CH_2^{2+}(aq)$ $\quad E^{o\prime} = 0.100$ V

$Ru(NH_3)_5NH_2C_6H_{11}^{3+}(aq) + e^- \rightarrow Ru(NH_3)_5NH_2C_6H_{11}^{2+}(aq)$ $\quad E^{o\prime} = 0.110$ V

$Ru(NH_3)_5NH-C_6H_{10}^{3+}(aq) + e^- \rightarrow Ru(NH_3)_5NH-C_6H_{10}^{2+}(aq)$ $\quad E^{o\prime} = 0.190$ V

In systems of ammine-Ru complexes of H_2S and related ligands, the following couples were measured at 25°C [27]:

$[Ru(NH_3)_5SH]^{2+}(aq) + H^+ + e^- \rightarrow [Ru(NH_3)_5H_2S]^{2+}(aq)$ in 1 M HCl at pH < 4

$E^{o\prime} = -0.05$ V

$[Ru(NH_3)_5SH]^{2+}(aq) + e^- \rightarrow [Ru(NH_3)_5SH]^+(aq)$ at pH = 7 in 1 M CH_3COONa

$E^{o\prime} = -0.29$ V

$cis\text{-}Ru(NH_3)_4(H_2O)S\text{-}(CH_3)_2^{3+}(aq) + e^- \rightarrow cis\text{-}Ru(NH_3)_4(H_2O)S\text{-}(CH_3)_2^{2+}(aq)$

$E^{o\prime} = 0.480$ V

$Ru(NH_3)_5(DMSO)^{3+}(aq) + e^- \rightarrow Ru(NH_3)_5(DMSO)^{2+}(aq)$ $\quad E^{o\prime} = 1.0$ V

where DMSO = dimethyl sulfoxide

$Ru(NH_3)_5(thiophene)^{3+}(aq) + e^- \rightarrow Ru(NH_3)_5(thiophene)^{2+}(aq)$

$E^{o\prime} = 0.61$ V

$Ru(NH_3)_5(S(CH_3)_2)^{3+}(aq) + e^- \rightarrow Ru(NH_3)_5(S(CH_3)_2)^{2+}(aq)$ $\quad E^{o\prime} = 0.5$ V

$Ru(NH_3)_5(S-C_2H_5)^{3+}(aq) + e^- \rightarrow Ru(NH_3)_5(S-C_2H_5)^{2+}(aq)$ $\quad E^{o\prime} = -0.43$ V

at pH 11

$trans\text{-}Ru(NH_3)_4(isn)(S\text{-}(CH_3)_2)^{3+}(aq) + e^- \rightarrow trans\text{-}Ru(NH_3)_4(isn)$

$(S\text{-}(CH_3)_2)^{2+}(aq)$ $\quad E^{o\prime} = 0.710$ V

where isn = isonicotinamide.

3. Potential Diagram for Ruthenium

```
     +8           +7           +6         +4      +3          +2        0
                              1.04
┌─────────────────────────────────────────┐
│RuO₄                                     │
│Ru(VIII)──0.99──Ru(VII)──0.591──Ru(VI)──2.0──Ru(IV)─Ru(III)──0.249──Ru(II)─Ru
│                              RuO₂·4H₂O  │   RuO₂·4H₂O
│                              └──1.4──┘  │   └──────0.68──────┘
└─────────────────────────────────────────┘
```

B. Osmium

1. Oxidation States

Osmium is the most chemically reactive of the platinum metals and its chemistry shows close analogies with that of ruthenium, especially in oxo compounds. As in the case of this metal, osmium takes up no less than nine oxidation states (from +VIII to 0) and displays a great tendency to form complexes with π-bonding ligands. In these complexes, the π-donor ligands help to stabilize the higher oxidation states and the π-acceptor ligands are more easily bonded to osmium of lower valence. The higher oxidation states are more stable in alkaline solutions, whereas the Os(IV) and Os(III) are more common in acid media, although OsO_4 is also stable in acidic solutions.

With the exception of a few important compounds (tetroxide, osmiamate, and pentacarbonyl), all osmium complexes in aqueous solutions are octahedrally coordinated. A selection of thermodynamic data on osmium compounds is given in Table 4.

2. Potentials of Osmium Systems

From Gibbs energy values and solubility products in water, Latimer [31] and later, Goldberg and Hepler [4] and Pourbaix et al. [32] have calculated

$$OsO_4(aq) + 8H^+(aq) + 8e^- \rightarrow Os(s) + 4H_2O \qquad E° = 0.84 \text{ V}$$

$$OsO_4(s) + 4H^+(aq) + 4e^- \rightarrow OsO_2(s) + 2H_2O \qquad E° = 1.005 \text{ V}$$

$$OsO_2(s) + 4H^+(aq) + 4e^- \rightarrow Os(s) + 2H_2O \qquad E° = 0.687$$

On the other hand, the potentiometric measurements of Cartledge [33] with Pt electrodes in 1.2×10^{-3} M H_2SO_4 or phosphate buffer solutions at pH between 3.5 and 6.5 at 25°C lead to

$$OsO_4(aq) + 4H^+(aq) + 4e^- \rightarrow OsO_2 \cdot 2H_2O(s) \qquad E° = 0.964 \pm 0.002 \text{ V}$$

Similar measurements with anhydrous OsO_2 gave a slightly lower value (0.95 V) for the OsO_2/OsO_4 couple. From the Gibbs energies deduced, the potential for the reaction

TABLE 4 Thermodynamic Data on Osmium[a]

Formula	Description	State	$\Delta H°$/kJ mol^{-1}	$\Delta G°$/kJ mol^{-1}	$S°$/J mol^{-1} K^{-1}
Os		c	0.	0.	32.6
Os		g	791.	745.	192.573
Os$^+$		g	1595.86		
Os^{2+}		g	3243.		
OsO$_3$		g	−283.7		
OsO$_3^+$		g	908.		
OsO$_4$	Yellow	c	−394.1	−304.9	143.9
OsO$_4$	White	c	−385.8	−303.7	167.8
OsO$_4$		g	−337.2	−292.8	293.8

OsO$_4$	Undiss.; std. state, $m=1$	aq	−378.2	−301.85	186.6
OsO$_4$	In CCl$_4$; std. state, $m=1$			−302.65	
OsO$_4^+$		g	860.6		
Os(OH)$_4$		amorph		−673.5	
OsF$_6$	Cubic	c			246.0
OsF$_6$		g			358.09
OsCl$_3$		c	−190.4		
OsCl$_4$		c	−254.8		
OsCl$_4$		g	−79.		
OsS$_2$		c	−146.0		

[a]National Bureau of Standards values are in italic.
Source: Data from Ref. 2.

$$OsO_2 \cdot 2H_2O(ppt) + 4H^+(aq) + 4e^- \rightarrow Os(s) + 4H_2O$$

has been calculated as $E° = 0.72$ V.

Earlier work by Anderson and Yost [34] and Sauerbrunn and Sandell [35] on the solubility of OsO_4 in alkaline media lead us to assume the formation of the perosmic acid H_2OsO_5, whose two dissociation constants are measured. However, recent spectrophotometric studies on aqueous OsO_4 solutions have induced Tridot and Bavay [36] to propose the formation of the osmenic acid $OsO_2(OH)_4$ with four ionization constants. This conclusion is supported by the extensive work of Griffith [37], who has determined by infrared spectrophotometry the structure $OsO_4(OH)_2^{2-}$, with the OH ligands in the *trans* position, for the alkali salts of this complex. The former results were used for further thermodynamic calculations, which must now be questioned on the basis of the new chemical work on this element.

Latimer [31] has estimated the potential for the half-reaction

$$OsCl_6^{2-}(aq) + e^- \rightarrow OsCl_3^{3-}(aq) \qquad E° = 0.85 \text{ V}$$

but this value differs considerably from that measured by potentiometry [38, 39], $E° = 0.45$ V, for both this couple and the system

$$OsBr_6^{2-}(aq) + e^- \rightarrow OsBr_6^{3-}(aq) \qquad E° = 0.45 \text{ V}$$

although it is possible that the halide species were hydrolyzed in the dilute solutions employed.

Determination of cell potentials for the $OsBr_6^{2-}/OsBr_6^{2-}$ couple in fairly concentrated HBr solutions (2 to 4 M) resulted in $E°' = 0.349$ V at 2.1 M HBr [40]; the potential is less positive at higher acid concentrations.

Opekar and Beran [41] have investigated the oxidation of Os(II) with a platinum rotating disk electrode in the system

$$Os(CN)_6^{3-}(aq) + e^- \rightarrow Os(CN)_6^{4-}(aq) \qquad E° = 0.634 \text{ V}$$

in neutral and alkaline solutions of KOH, KCN, HCl, and KNO_3; the standard potential was determined at zero ionic strength. Nevertheless, it was found that the half-wave potential of the oxidation reaction was dependent on the hydrogen concentration in acid solutions.

This potential value is quite different from that estimated by Meites [42], $E° = -0.75$ V (SCE) from polarographic data for the same system. The authors of a more recent work [43,44] suggest that Meites really measured the system

$$Os(OH)_2(CN)_4^{3-}(aq) + e^- \rightarrow Os(OH)_2(CN)_4^{4-}(aq)$$

which shows a formal potential of $E°' = -0.76$ V (SCE) in 1 M KOH + 0.1 M KCN solutions. These results also explain the unsuccessful attempts to oxidize Os(II) by polarography with a DME in cyanide or K_2SO_4 media [45].

Dwyer and co-workers [46] have measured the potential of 20 different cells involving Os(II) and Os(III) complexes with π-ligands vs. the SCE to determine the influence of the overall charge, conjugation coordinated halide, and substitution in the ligand on the stability of the complex cation. The potentiometric measurements were carried out with gold electrodes in aqueous solutions of halide with perchloric acid and the potential extrapolated to zero ionic strength.

The results show that a lower overall charge and/or a decrease in the ligand conjugation increases the stability of the Os(III) species. This is also achieved by halide coordination, particularly involving the Cl⁻ ligand. Substitution in the ligand seems to affect the stability of Os(III) favorably, but the effect on the redox potential is too small to draw reliable conclusions:

$Os(trpy)(bpy)(py)^{3+} + e^- \rightarrow Os(trpy)(bpy)(py)^{2+}$ $E° = 0.871$ V

$Os(trpy)(bpy)Cl^{2+} + e^- \rightarrow Os(trpy)(bpy)Cl^+$ $E° = 0.563$ V

$Os(py)_2(bpy)_2^{3+} + e^- \rightarrow Os(py)_2(bpy)_2^{2+}$ $E° = 0.834$ V

$Os(bpy)_2pyCl^{2+} + e^- \rightarrow Os(bpy)_2pyCl^+$ $E° = 0.484$ V

$Os(bpy)(py)_4^{3+} + e^- \rightarrow Os(bpy)(py)_4^{2+}$ $E° = 0.805$ V

$Os(bpy)(py)_3Cl^{2+} + e^- \rightarrow Os(bpy)(py)_3Cl^+$ $E° = 0.425$ V

$Os(bpy)_3^{3+} + e^- \rightarrow Os(bpy)_3^{2+}$ $E° = 0.885$ V

(b)$Os(bpy)_2(acac)^{2+} + e^- \rightarrow Os(bpy)_2(acac)^+$ $E° = 0.153$ V

$Os(trpy)_2^{3+} + e^- \rightarrow Os(trpy)_2^{2+}$ $E° = 0.987$ V

$Os(trpy)(py)_3^{3+} + e^- \rightarrow Os(trpy)(py)_3^{2+}$ $E° = 0.800$ V

$Os(trpy)(bpy)Br^{2+} + e^- \rightarrow Os(trpy)(bpy)Br^+$ $E° = 0.567$ V

$Os(trpy)(bpy)I^{2+} + e^- \rightarrow Os(trpy)(bpy)I^+$ $E° = 0.567$ V

$Os(bpy)_2pyI^{2+} + e^- \rightarrow Os(bpy)_2pyI^+$ $E° = 0.489$ V

$Os(bpy)_2pyBr^{2+} + e^- \rightarrow Os(bpy)_2pyBr^+$ $E° = 0.488$ V

$Os(bpy)(py)_3I^{2+} + e^- \rightarrow Os(bpy)(py)_3I^+$ $E° = 0.451$ V

$Os(bpy)(py)_3Br^{2+} + e^- \rightarrow Os(bpy)(py)_3Br^+$ $E° = 0.444$ V

$$\text{Os(trpy)(bpy)(3-CH}_3\text{py)}^{3+} + e^- \rightarrow \text{Os(trpy)(bpy)(3-CH}_3\text{py)}^{2+}$$

$$E^\circ = 0.867 \text{ V}$$

$$\text{Os(trpy)(bpy)(4-C}_3\text{H}_7\text{py)}^{3+} + e^- \rightarrow \text{Os(trpy)(bpy)(4-C}_3\text{H}_7\text{py)}^{2+}$$

$$E^\circ = 0.861 \text{ V}$$

$$\text{Os(trpy)(bpy)(4-C}_2\text{H}_5\text{py)}^{3+} + e^- \rightarrow \text{Os(trpy)(bpy)(4-C}_2\text{H}_5\text{py)}^{2+}$$

$$E^\circ = 0.859 \text{ V}$$

$$\text{Os(trpy)(bpy)(4-CH}_3\text{py)}^{3+} + e^- \rightarrow \text{Os(trpy)(bpy)(4-CH}_3\text{py)}^{2+}$$

$$E^\circ = 0.851 \text{ V}$$

where trpy = terpyridine, bpy = bipyridine, py = pyridine, and acac = acetyl acetone; (b) is a Os(II)/Os(I) couple.

3. *Potential Diagram for Osmium*

$$\text{Os(VIII)} \xrightarrow{1.005} \text{OsO}_2 \xrightarrow{0.687} \text{Os}$$

$$\underset{0.85}{\underline{\hspace{6cm}}}$$

with $+8$, $+4$, 0 oxidation states indicated above.

REFERENCES

1. G. T. Furukawa, M. L. Reilly, and J. S. Gallagher, J. Phys. Chem. Ref. Data, 3, 163 (1974).
2. D. D. Wagman, W. H. Evans, V. B. Parker, R. H. Schumm, I. Halow, S. M. Bailey, K. L. Churney, and R. L. Nuttall, *NBS Tables of Chemical Thermodynamic Properties*, J. Phys. Chem. Ref. Data, Vol. 11, Supplement 2 (1982).
3. D. R. Fredrickson and M. G. Chasanov, J. Chem. Eng. Data, 17, 21 (1972).
4. R. N. Goldberg and L. H. Hepler, Chem. Rev., 68, 229 (1968).
5. R. R. Buckley and E. E. Mercer, J. Phys. Chem., 70, 3103 (1966); Inorg. Chem., 4, 1692 (1965).
6. W. Bottcher, G. M. Brown, and N. Sutin, Inorg. Chem., 18, 1447 (1979).
7. D. D. Ford and A. W. Davidson, J. Am. Chem. Soc., 73, 1469 (1951).
8. G. Grube and G. Fromm, Z. Elektrochem., 46, 661 (1940); 47, 208 (1941).
9. J. R. Backhouse and F. P. Dwyer, J. Proc. R. Soc. N.S. Wales, 83, 138 (1949).
10. D. K. Atwood and J. T. De Vries, J. Am. Chem. Soc., 84, 2659 (1962).
11. L. W. Potts and H. S. Swofford, Anal. Chem., 47, 131 (1975).
12. J. E. Ferguson and A. Greenway, Aust. J. Chem., 31, 497 (1978).

13. M. Pourbaix, ed., *Atlas d'équilibres électrochimiques à 25°C*, Gauthier-Villars, Paris, 1963, p. 343.
14. P. Wehner and J. C. Hindman, J. Am. Chem. Soc., 72, 3911 (1950).
15. R. M. Wallace, J. Phys. Chem., 68, 2418 (1964).
16. P. Eichner, Bull. Soc. Chim. Fr., 2051 (1967).
17. W. M. Latimer, *Oxidation Potentials*, 2nd ed., Prentice-Hall, Englewood Cliffs, N.J., 1952, p. 228.
18. M. D. Silverman and H. A. Levy, J. Am. Chem. Soc., 76, 3319 (1954).
19. E. V. Luoma and C. H. Brubaker, Inorg. Chem., 5, 1637 (1966).
20. E. R. Connick and C. R. Hurley, J. Phys. Chem., 61, 1018 (1957).
21. K. W. Lam, K. E. Johnson, and D. G. Lee, J. Electrochem. Soc., 125, 1069 (1978).
22. E. R. Connick and C. R. Hurley, J. Am. Chem. Soc., 74, 5012 (1952).
23. T. J. Meyer and H. Taube, Inorg. Chem., 7, 2369 (1968).
24. D. K. Lavallee, C. Lavallee, J. C. Sullivan, and E. Deutsch, Inorg. Chem., 12, 570 (1973).
25. H. S. Lim, D. J. Barclay, and F. C. Anson, Inorg. Chem., 11, 1460 (1972).
26. S. E. Diamond, G. M. Tom, and H. Taube, J. Am. Chem. Soc., 97, 2661 (1975).
27. C. C. Kuehn and H. Taube, J. Am. Chem. Soc., 98, 689 (1976).
28. C. M. Elson, I. J. Itzkovich, J. McKenney, and J. A. Page, Can. J. Chem., 53, 2922 (1975).
29. T. Matsubara and P. C. Ford, Inorg. Chem., 15, 1107 (1976).
30. V. E. Alvarez, R. J. Allen, T. Matsubara, and P. C. Ford, J. Am. Chem. Soc., 96, 7686 (1974).
31. W. M. Latimer, *Oxidation Potentials*, 2nd ed., Prentice-Hall, Englewood Cliffs, N.J., 1952, p. 230.
32. M. Pourbaix, ed., *Atlas d'équilibres électrochimiques à 25°C*, Gauthier-Villars, Paris, 1963, p. 364.
33. G. H. Cartledge, J. Phys. Chem., 60, 1468 (1956).
34. L. H. Anderson and D. M. Yost, J. Am. Chem. Soc., 60, 1822 (1938).
35. R. D. Sauerbrunn and E. B. Sandell, J. Am. Chem. Soc., 75, 4170 (1953).
36. G. Tridot and J. Cl. Bavay, Bull. Soc. Chim. Fr., 2026 (1967); C.R. Acad. Sci., Paris, Ser. C, 276, 1025 (1973).
37. W. P. Griffith, Q. Rev. (Lond.), 19, 254 (1965); *The Chemistry of the Rare Pt Metals*, Wiley-Interscience, New York, 1967.
38. F. P. Dwyer, J. E. Humpoletz, and R. S. Nyholm, Proc. R. Soc. N.S. Wales, 80, 242 (1947).
39. F. P. Dwyer, H. A. McKenzie, and R. S. Nyholm, Proc. R. Soc. N.S. Wales, 80, 183 (1947).
40. R. W. Mertes, W. R. Crowell, and R. K. Brinton, J. Am. Chem. Soc., 72, 4218 (1950).
41. F. Opekar and P. Beran, J. Electroanal. Chem., 71, 120 (1976).
42. L. Meites, J. Am. Chem. Soc., 79, 4631 (1957).
43. F. Opekar, P. Beran, and Z. Samec, Electrochim. Acta, 22, 243 (1977).
44. F. Opekar and P. Beran, Electrochim. Acta, 22, 249 (1977).
45. J. B. Willis, J. Am. Chem. Soc., 67, 547 (1945).
46. D. A. Buckingham, F. D. Dwyer, and A. M. Sargeson, Inorg. Chem., 5, 1243 (1966).

15
Manganese, Technetium, and Rhenium

I. MANGANESE*

JAMES C. HUNTER and AKIYA KOZAWA   Union Carbide Corporation, Westlake, Ohio

There is a wealth of information on electrochemical properties of manganese compounds, both in aqueous and nonaqueous media. This chapter provides a brief summary of the aqueous electrochemistry of manganese compounds. A more complete discussion of the electrochemistry of these compounds, including nonaqueous electrochemistry, may be found in the *Encyclopedia of the Electrochemistry of the Elements* [1].

A. Oxidation States

Compounds and species are known with manganese in oxidation states ranging from Mn(−I) to Mn(VII). Mn(−I) is present in the ion $Mn(CO)_5^-$, while Mn(0) is found in the compound $Mn_2(CO)_{10}$. The metal itself is unstable in contact with water: its potential lies below that for hydrogen evolution throughout the entire pH range. In practice, however, the reaction with water is fairly slow. Mn(I) has been found in certain complexes, such as the $[Mn(CN)_6]^{5-}$ ion, but does not occur as a simple Mn^+ ion in solution. The Mn(II) state is generally the most stable state in aqueous solutions. In the acid region it exists in the form of the Mn^{2+} ion, which is stable over a wide range of potentials. In alkaline solutions Mn^{2+} is hydrolyzed to $Mn(OH)_2$, which has a somewhat narrower potential range of stability, being fairly easily oxidized to species containing Mn(III) or Mn(IV). Mn(III) species are strongly oxidizing in acidic solution, and tend to disproportionate to Mn(II) and Mn(IV) species; however, the Mn^{3+} ion has been studied in very strongly acidic solutions where the disproportionation is less favorable. Mn(III) forms a large number of complex species which have much greater stability than the Mn^{3+} ion. In alkaline solution one finds the stable Mn(III) compound Mn_2O_3 as well as the oxyhydroxide MnOOH. Mn(IV) is found in the oxide MnO_2, the sulfate $Mn(SO_4)_2$, the fluoride MnF_4, and in certain complex halides. However, only the oxide is stable in contact with water, and that is due mainly to its insolubility. MnO_2 finds wide use in the battery industry as a cathode material

---

Reviewer: M. Gross, Louis Pasteur University, Strasbourg, France.

for cells. Mn(V) is found in the manganate ion MnO_4^{3-}, which is very strongly oxidizing and unstable except in very highly alkaline solutions, tending to decompose to Mn(IV) and Mn(VI) or Mn(VII) species. Mn(VI) is found in the manganate ion MnO_4^{2-}, which, while being more stable than the manganate (V) ion, is still stable only in fairly strong alkaline solutions. Mn(VII) occurs in several compounds and species, of which the permanganates (MnO_4^-) are the most important. The other compounds of Mn(VII) are quite unstable, and in fact tend to decompose explosively. Permanganates find considerable use as oxidants in industrial processes, as well as in analytical chemistry. The MnO_4^- ion is thermodynamically unstable in aqueous solution; its potential lies above that for oxygen evolution throughout the entire pH range. In practice, however, the reaction of MnO_4^- with water in neutral solutions is slow, particularly if care is taken to protect the solution from light and to remove any oxidizable impurities from solution. In such cases, negligible decomposition occurs even after several months of storage. Thermodynamic data on manganese are given in Table 1.

B. Mn(II) Potentials

Numerous measurements and calculations of the potential of the Mn^{2+}/Mn couple have been made. We follow Liang [1] and select for the electrode reaction

$$Mn^{2+} + 2e^- \rightarrow Mn(c) \qquad E° = -1.18 \text{ V}$$

where $E°$ is calculated from the revised National Bureau of Standards (NBS) data [2] for the Gibbs energy of formation of the Mn^{2+} ion. Similarly, using the NBS values [2,3], one obtains for manganese (II) hydroxide:

$$Mn(OH)_2(c) + 2e^- \rightarrow Mn(c) + 2OH^- \qquad E° = -1.56 \text{ V}$$

For the reduction of the cyanide complex from Mn(II) to Mn(I), Treadwell and Raths [4] found a potential in 1.5 M NaCN,

$$[Mn(CN)_6]^{4-} + e^- \rightarrow [Mn(CN)_6]^{5-} \qquad E° = -1.06 \text{ V}$$

In contrast, Jakob and Senkowski [5] found the potential of

$$[Mn(CN)_5NO]^{2-} + e^- \rightarrow [Mn(CN)_5NO]^{3-} \qquad E° = 0.6 \text{ V}$$

which is also believed to involve reduction of Mn(II) to Mn(I) [6].

C. Mn(III) Potentials

From the work of Vetter and Manecke [7] and of Ciavatta and Grimaldi [8] on the potential of the Mn^{3+} ion measured in strongly acidic solutions, we take the potential for the reaction

$$Mn^{3+} + e^- \rightarrow Mn^{2+} \qquad E° = \text{ca. } 1.5 \text{ V}$$

Using this $E°$ value together with the NBS value for the Gibbs energy of formation of Mn^{2+}, we calculate the Gibbs energy of formation of Mn^{3+} to be approximately -83 kJ mol^{-1}. The Mn^{3+} ion is strongly oxidizing, and moreover it tends to disproportionate according to the reaction

TABLE 1 Thermodynamic Data on Manganese[a]

Formula	Description	State	$\Delta H°$/kJ mol^{-1}	$\Delta G°$/kJ mol^{-1}	$S°$/J mol^{-1} K^{-1}
Mn		g	280.7	238.5	173.7
Mn	α	c	0	0	32.01
Mn	γ	c	1.55	1.42	32.43
Mn$^+$		g	1004		
Mn^{2+}		g	2519		
Mn^{2+}		aq	−220.75	−228.1	−73.6
Mn^{3+}		aq	(−113)	(−83)	
MnO		g	124.22		
MnO		c	−385.22	−362.90	59.71
HMnO$_2^-$		aq		−507	
MnO$_2$	Pyrolusite, β	c	−520.03	−465.14	53.05
MnO$_2$	γ	c		−448.5	
MnO$_4^-$		aq	−541.4	−477.2	191.2
MnO$_4^{2-}$		aq	−653	−500.7	59

TABLE 1 (Continued)

Formula	Description	State	$\Delta H°$/kJ mol^{-1}	$\Delta G°$/kJ mol^{-1}	$S°$/J mol^{-1} K^{-1}
MnO_4^{3-}		aq		(−527)	
Mn_2O_3		c	−959.0	−881.1	110.5
MnOOH	γ	c		−563.2	
MnOOH	α	c		−557.7	
Mn_3O_4		c	−1387.8	−1283.2	155.6
$Mn(OH)_2$	Precipitated	amorph	−695.4	−615.0	99.2
MnF_2		c	−793	−751	92.26
$MnCl_2$		c	−481.29	−440.50	118.24
$MnCl_2 \cdot H_2O$		c	−789.9	−696.1	174.1
$MnBr_2$		c	−384.9	−371	(138)
MnI_2		c	−268.6	−271	(155)

MnS	Green	c	−214.2	−218.4	78.2
MnSO$_4$		c	−1065.25	−957.36	112.1
Mn(N$_3$)$_2$		c	385.8		
Mn$_5$N$_2$		c	−204.2		
Mn(NO$_3$)$_2$		c	−576.26		
Mn(NO$_3$)$_2$·6H$_2$O	Glassy	amorph	−2371.9		
Mn$_3$(PO$_4$)$_2$		c	−3116.7	−2900	(272)
Mn$_3$C		c	4.6	5.4	98.7
MnCO$_3$	Natural	c	−894.1	−816.7	85.8
MnSiO$_3$		c	−1320.9	−1240.5	89.1
Mn$_2$(CO)$_{10}$		c	−1677		

[a]National Bureau of Standards values are in italic; estimated values are in parentheses.

$$2Mn^{3+} + 2H_2O(\ell) \rightarrow Mn^{2+} + MnO_2(c) + 4H^+ \qquad \Delta G^\circ = -53 \text{ kJ mol}^{-1}$$

from which it is clear that appreciable concentrations of Mn^{3+} ion can be obtained only at high concentrations of acid and of Mn^{2+}.

For the reaction in alkaline solution, we write the reaction in terms of Mn_2O_3 and $Mn(OH)_2$, and calculate a potential from the Gibbs energies of formation:

$$Mn_2O_3(c) + 3H_2O(\ell) + 2e^- \rightarrow 2Mn(OH)_2(c) + 2OH^- \qquad E^\circ = -0.25 \text{ V}$$

For the reduction of the Mn(III) cyanide complex, Chadwick and Sharpe [6], on the basis of a review of experimental work, report

$$[Mn(CN)_6]^{3-} + e^- \rightarrow [Mn(CN)_6]^{4-} \qquad E^\circ = -0.24 \text{ V}$$

Comparing this value to that for reduction of the Mn^{3+} ion, it is apparent that complexing with CN^- stabilizes the Mn(III) to a much greater extent than Mn(II).

D. Mn(IV) Potentials

Numerous measurements of electrode potentials of Mn(IV) oxide materials have been made. Difficulties have often been encountered in obtaining these potentials, and the values obtained have not always agreed well with those calculated from thermochemical data. There are several reasons for this. First, Mn(IV) oxides can exist in a large number of structural forms, with differing degrees of thermodynamic stability. (For this reason we must specify in cell reactions the type of MnO_2 considered, except in cases where the most stable form, β-MnO_2, is involved.) Higher potentials are measured for reduction of the less stable forms, such as γ-MnO_2 and δ-MnO_2, than for the most stable form, β-MnO_2. Actually, the designation "MnO_2" is in many cases not accurate, since many of the structural forms contain additional water, other cations (e.g., Na, K, etc.), and lower-valent manganese. Only β-MnO_2 (pyrolusite) and ramsdellite (a mineral form of MnO_2) can be obtained as nearly stoichiometric MnO_2. However, it is the other less well defined Mn(IV) oxides that find the greatest use as battery cathode materials.

A second problem in obtaining potentials of Mn(IV) oxide materials arises from the fact that these oxides tend to undergo ion exchange, leading to changes in pH near the electrode and making establishment of steady potentials a difficult matter. This is more a problem with the less well defined oxides such as δ-MnO_2 and γ-MnO_2, but also does occur with β-MnO_2.

A third problem is that the actual reaction path may depend on the type of MnO_2 and the pH of the solution. Equilibrium may not be established between the MnO_2 and the thermodynamically stable reduction product. Rather, the potential measured may be between the MnO_2 and some metastable product. In some cases this reduction proceeds by a one-phase, solid-state reaction mechanism, so that the interpretation of measured potentials is different than that for a conventional two-phase redox system. Thus it is not surprising that there have been difficulties in measuring MnO_2 reduction potentials and in correlating them with thermochemical data.

In acidic solution, MnO_2 can be reduced to Mn^{2+}. Covington et al. [9] reviewed past work and performed a very thorough investigation of β-MnO_2.

They found that the ion-exchange capacity of the MnO$_2$ caused difficulty in obtaining steady potentials unless the solutions were buffered to maintain constant pH (of about 4.8) and constant Mn^{2+} concentration. Under these circumstances they measured potential values ranging from 1.233 to 1.241 V. The variation in potential appeared related to differences in nonstoichiometry of the β-MnO$_2$ samples, with the materials closest to stoichiometric MnO$_2$ giving the lowest potential values. If one calculates this potential using the NBS values for energy of formation of MnO$_2$(c) and Mn^{2+}, one obtains a value of 1.23 V, in reasonable agreement with the work of Covington et al. Thus

$$MnO_2(c) + 4H^+ + 2e^- \rightarrow Mn^{2+} + 2H_2O(\ell) \qquad E° = 1.23 \text{ V}$$

Similar work has been performed on other types of MnO$_2$. Ambrose et al. [10] measured potentials of 1.22 to 1.27 V for various preparations of α-MnO$_2$. Covington et al. [11] attempted to perform similar measurements on γ-MnO$_2$, the type most often used in batteries, but were unable to obtain steady potentials. In general, however, they observed that the potentials measured for γ-MnO$_2$ tended to be 40 to 90 mV higher than the poentials for β-MnO$_2$. This indicates that γ-MnO$_2$ is less stable than β-MnO$_2$ by 8 to 17 kJ mol^{-1}.

For the Mn(IV)/Mn(II) couple in alkaline solution we calculate from Gibbs energies of formation for the reaction

$$MnO_2(c) + 2H_2O(\ell) + 2e^- \rightarrow Mn(OH)_2(c) + 2OH^- \qquad E° = -0.05 \text{ V}$$

while for the reduction to the Mn(III) oxide we calculate

$$2MnO_2(c) + H_2O(\ell) + 2e^- \rightarrow Mn_2O_3(c) + 2OH^- \qquad E° = 0.15 \text{ V}$$

Thus one might expect the Mn(III) oxide Mn$_2$O$_3$ to be the stable product of MnO$_2$ reduction in alkaline solutions rather than Mn(OH)$_2$. In fact, it has been found that reduction of β-MnO$_2$ in alkaline solutions leads to the Mn(III) oxyhydroxide γ-MnOOH, for which Bode et al. [12] reported a Gibbs energy of formation of −563.2 kJ mol^{-1}. Thus one can write the reaction

$$MnO_2(c) + H_2O + e^- \rightarrow \gamma\text{-MnOOH}(c) + OH^- \qquad E° = 0.19 \text{ V}$$

For γ-MnO$_2$, Gabano [13] has reported a Gibbs energy of formation of −448.5 kJ mol^{-1}, a value consistent with the work of Covington et al. [9,11] in acidic solutions, described above. This value is included in Table 1; one should keep in mind, however, that γ-MnO$_2$ occurs with varying degrees of nonstoichiometry, and thus the $\Delta G°$ of a given material could deviate significantly from the value in the table. Using this value of $\Delta G°$ for γ-MnO$_2$ we can then calculate

$$2\gamma\text{-MnO}_2(c) + H_2O(\ell) + 2e^- \rightarrow Mn_2O_3(c) + 2OH^- \qquad E° = 0.32 \text{ V}$$

or

$$\gamma\text{-MnO}_2(c) + H_2O + e^- \rightarrow \gamma\text{-MnOOH}(c) + OH^- \qquad E° = 0.36 \text{ V}$$

However, it has been found that reduction of γ-MnO_2 in alkaline solution results in formation of α-MnOOH, *not* γ-MnOOH, and furthermore that the product is actually a solid solution of α-MnOOH and γ-MnO_2. One thus has a one-phase redox system, where the potential depends on the activities of Mn(III) and Mn(IV) in the solid solution, in a manner analogous to the Fe(III)/Fe(II) couple in aqueous solution [14]. The standard potential is then the potential measured when the activities of Mn(III) and Mn(IV) in the solid are equal. Using this principle, Kozawa and Powers [15] found an $E°$ of 0.103 V for reduction of γ-MnO_2, using the assumption that equal Mn(III) and Mn(IV) activities are obtained when the mole fractions in the solid are equal. Neumann and von Roda [16] noted the asymmetric shape of the discharge curve and concluded that the activities did not correspond exactly to the mole fractions of Mn(III) and Mn(IV) in the solid, and that equal activities are found when the mole fraction of Mn(IV) is 0.67; this assumption would lead to a somewhat higher $E°$ value. Using Gabano's [13] values of -448.5 kJ mol^{-1} for the Gibbs energy of formation of γ-MnO_2 and -557.7 kJ mol^{-1} for α-MnOOH, we can calculate

$$\gamma\text{-}MnO_2(c) + H_2O(\ell) + e^- \rightarrow \alpha\text{-}MnOOH(c) + OH^- \qquad E° = 0.30 \text{ V}$$

This potential is higher than that measured by Kozawa and Powers, and higher than expected even with the modification proposed by Neumann and von Roda. The reason for this discrepancy is not completely clear at present, but it may be due in part to uncertainties in the values of $\Delta G°$ for γ-MnO_2 and/or α-MnOOH, or it may indicate that our interpretation of the electrode potential data is not entirely correct. In any case, it appears that the potential for the actual reduction of γ-MnO_2 to α-MnOOH is lower than that for reduction to Mn_2O_3 or γ-MnOOH, and that because of mechanistic considerations, reduction of γ-MnO_2 to a metastable product is favored over reduction to the thermodynamically most stable product.

E. Mn(VII), Mn(VI), and Mn(V) Potentials

In acidic solutions, MnO_4^- is reduced to Mn^{2+}. From the Gibbs energies of formation one may calculate

$$MnO_4^- + 8H^+ + 5e^- \rightarrow Mn^{2+} + 4H_2O(\ell) \qquad E° = 1.51 \text{ V}$$

For reduction in alkaline solution, one may similarly calculate

$$MnO_4^- + 4H_2O(\ell) + 5e^- \rightarrow Mn(OH)_2(c) + 6OH^- \qquad E° = 0.34 \text{ V}$$

Unless there is an excess of reducing agent, the reduction of permanganate in alkaline solution tends to stop at Mn(IV), according to the reaction

$$MnO_4^- + 2H_2O(\ell) + 3e^- \rightarrow MnO_2(c) + 4OH^- \qquad E° = 0.60 \text{ V}$$

where the $E°$ is again calculated from Gibbs energies of formation. For the corresponding reaction in acidic solution we calculate

$$MnO_4^- + 4H^+ + 3e^- \rightarrow MnO_2(c) + 2H_2O(\ell) \qquad E° = 1.70 \text{ V}$$

These values are somewhat different from the values of 0.588 V and 1.679 V obtained from actual electrode potential measurements for the two reactions above [17,18]. However, it may be that some of the factors discussed in Section I.D affect the potential measured for this reaction. In particular, it is known that reduction of permanganate ion in alkaline solutions often results in δ-MnO_2, which is less stable than β-MnO_2, so one might expect to measure lower potentials than would be calculated on the basis of β-MnO_2. (Both studies cited above involved measurements in alkaline solution.)

For reduction of Mn(VII) to Mn(VI), Carrington and Symons [19] measured a potential of 0.558 V, in agreement with the value of 0.554 calculated from Gibbs energies of formation:

$$MnO_4^- + e^- \rightarrow MnO_4^{2-} \qquad E^\circ = 0.56 \text{ V}$$

For the further reduction of Mn(VI) to Mn(V), Thiele and Landsberg [20] reported a measured value of 0.26 V, Carrington and Symons [19] reported 0.285 V, and Schurig and Heusler [21] obtained a value of 0.274 V. We select for this reaction

$$MnO_4^{2-} + e^- \rightarrow MnO_4^{3-} \qquad E^\circ = \text{ca. } 0.27 \text{ V}$$

Using this E°, together with the Gibbs energy of formation of MnO_4^{2-}, we estimate the Gibbs energy of formation of MnO_4^{3-} to be -527 kJ mol^{-1}.

Finally, we use the Gibbs energies of formation to calculate the following additional potentials, which are included in the summary of manganese potentials:

$$MnO_4^{2-} + 4H^+ + 2e^- \rightarrow MnO_2(c) + 2H_2O(\ell) \qquad E^\circ = 2.27 \text{ V}$$

$$MnO_4^{2-} + 2H_2O(\ell) + 2e^- \rightarrow MnO_2(c) + 4OH^- \qquad E^\circ = 0.62 \text{ V}$$

$$MnO_4^{3-} + 4H^+ + e^- \rightarrow MnO_2(c) + 2H_2O(\ell) \qquad E^\circ = 4.27 \text{ V}$$

$$MnO_4^{3-} + 2H_2O + e^- \rightarrow MnO_2(c) + 4OH^- \qquad E^\circ = 0.96 \text{ V}$$

$$MnO_2 + 4H^+ + e^- \rightarrow Mn^{3+} + 2H_2O(\ell) \qquad E^\circ = 0.95 \text{ V}$$

F. Potential Diagrams for Manganese

Acid Solution

(See diagram on page 438.)

Alkaline Solution

(See diagram on page 438.)

F. Potential Diagrams for Manganese

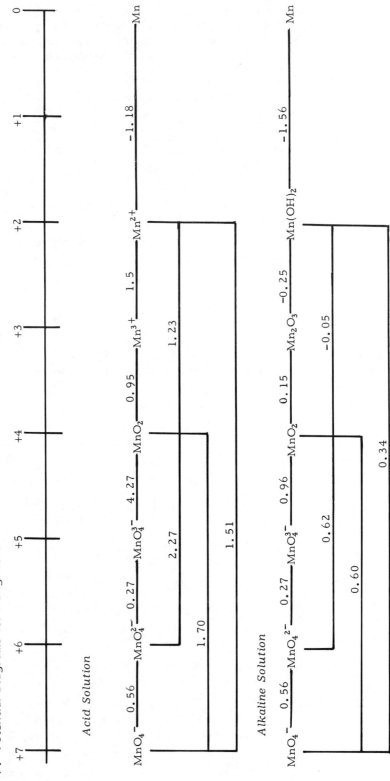

REFERENCES

1. C. C. Liang, in *Encyclopedia of the Electrochemistry of the Elements,* A. J. Bard, ed., Vol. 1, Marcel Dekker, New York, 1973, p. 349.
2. D. D. Wagman, W. H. Evans, V. B. Parker, R. H. Schumm, I. Halow, S. M. Bailey, K. L. Churney, and R. L. Nuttall, *NBS Tables of Chemical Thermodynamic Properties,* J. Phys. Chem. Ref. Data, Vol. 11, Supplement 2 (1982).
3. D. D. Wagman, W. H. Evans, V. B. Parker, I. Halow, S. M. Bailey, and R. H. Schumm, *Selected Values of Chemical Thermodynamic Properties,* Natl. Bur. Stand. Tech. Note 270-3, U.S. Government Printing Office, Washington, D.C., 1968; Tech. Note 270-4, May 1969.
4. W. D. Treadwell and W. E. Raths, Helv. Chim. Acta, 35, 2259 (1952).
5. W. Jakob and T. Senkowski, Rocz. Chem., 38, 1751 (1964).
6. B. M. Chadwick and A. G. Sharpe, in *Advances in Inorganic Chemistry and Radiochemistry,* H. J. Emeléus and A. G. Sharpe, eds., Vol. 8, Academic Press, New York, 1966, p. 83.
7. K. J. Vetter and G. Manecke, Z. Phys. Chem., *195,* 270 (1950).
8. L. Ciavatta and M. Grimaldi, J. Inorg. Nucl. Chem., *31,* 3071 (1969).
9. A. K. Covington, T. Cressey, B. G. Lever, and H. R. Thirsk, Trans. Faraday Soc., *58,* 1975 (1962).
10. J. Ambrose, A. K. Covington, and H. R. Thirsk, in *Power Sources 2,* D. H. Collins, ed., Pergamon Press, Oxford, 1968, p. 303.
11. A. K. Covington, P. K. Talukdar, and H. R. Thirsk, Electrochem. Technol., *5,* 523 (1967).
12. H. Bode, A. Schmier, and D. Berndt, Z. Elektrochem., *66,* 586 (1962).
13. J. P. Gabano, Compt. Rend., *264,* 262 (1967).
14. A. Kozawa and R. A. Powers, J. Chem. Educ., *49,* 587 (1972).
15. A. Kozawa and R. A. Powers, J. Electrochem. Soc., *113,* 870 (1966).
16. K. Neumann and E. von Roda, Ber. Bunseges. Phys. Chem., *69,* 347 (1965).
17. L. V. Andrews and D. J. Brown, J. Am. Chem. Soc., *57,* 254 (1935).
18. J. P. Gabano and J. P. Brenet, Electrochim. Acta, *1,* 242 (1959).
19. A. Carrington and M. C. R. Symons, J. Chem. Soc., 3373 (1956).
20. R. Thiele and R. Landsberg, Z. Phys. Chem., *236,* 95 (1967).
21. H. Schurig and K. E. Heusler, Fresenius' Z. Anal. Chem., *224,* 45 (1967).

II. TECHNETIUM AND RHENIUM*

ROBERT J. MAGEE AND HARRY BLUTSTEIN† La Trobe University, Melbourne, Victoria, Australia

A. Technetium

1. Oxidation States

Technetium occurs only in radioactive forms: the longest-lived isotopes are ^{97}Tc, ^{98}Tc ($t_{1/2} \cong 2 \times 10^6$ years) and ^{99}Tc ($t_{1/2} > 2 \times 10^5$ years).

Reviewer: M. Gross, Louis Pasteur University, Strasbourg, France.
†*Current affiliation*: Environment Protection Authority of Victoria, Melbourne, Victoria, Australia.

The latter is a weak β emitter and quite easy to handle. Technetium corresponding to the oxidation states +4, +5, +6, and +7 appears to have been well established in simple compounds such as $TcCl_4$, TcO_2(+4), TcF_5(+5), TcO_3(+6), and Tc_2O_7, Tc_2S_7(+7).

In solution, however, under certain conditions and usually by means of polarographic reduction techniques, oxidation states −1, +1, +2, +3, +4, +5, and +6 have been claimed. The lower oxidation states occur as intermediates, and the exact nature of the species present is often obscure. More recently, oxidation states +1 and +2 have been stabilized by nitrosyl and hydroxylamine ligands, after reduction of technetium (+4). The thermodynamic data are given in Table 2.

Latimer (in the first edition), commenting on the chemistry of technetium, felt that as sufficient quantities of the element became available, the chemistry would be studied in greater detail. Much work has been carried out on this element since Latimer made his prediction, but the resistance of the element to form cations has eliminated virtually all cationic aqueous chemistry. Attempts to prepare technetium complexes in low oxidation states usually start by reducing TcO_4^- or $TcCl_6^{2-}$ with suitable reducing agents and reacting with the ligand. From work of this kind the following states are known: $Tc_2(CO)_{10}$(0), $K_5Tc(CN)_6$(+1), $Tc(diars)_2Cl_2$(+2), and $[Tc(diars)_2Cl_2]^+$(+3).

The +5 and +6 oxidation states are relatively unknown and tend to disproportionate [e.g., $3Tc(V) = 2Tc(IV) + Tc(VIII)$]. Nitrosyl appears to be a special case in stabilizing the lower oxidation states +1 and +2: the stable compound $t - [(NH_3)_4Tc(NO)(OH_2)]Cl_2$ has been isolated. In a like manner, the diarsine ligand appears to stabilize the unstable +2, +3, and +5 states.

Technetium metal is not nearly as electropositive as manganese, and TcO_4^- and TcO_2 are much less powerful oxidizing agents than are the corresponding manganese compounds. It lies between rhenium and manganese in its ability to form metal-metal bonds in its halides and, with few exceptions, shows little similarity to manganese; but it does show considerable similarity to rhenium in its compounds.

The very stable +7 oxidation state is represented by the heptoxide, Tc_2O_7, and pertechnic acid, $HTcO_4$ (obtained by dissolving the heptoxide in water); the salts of the TcO_4^- ion; the heptasulfide, Tc_2S_7; the oxyhalides TcO_3F and TcO_3Cl; and the TcH_9^{2-} complexes. After the +7 oxidation state, the most stable is the +4 represented by TcO_2, TcS_2, and $TcCl_4$. Electric conductivity is observed in solid Tc_2O_7 but not in the melt; with Re_2O_7 the reverse is true. The crystal structures of Tc_2O_7 and Re_2O_7 are of different symmetries.

2. Technetium Potentials

The values of the electrode potentials of a number of oxidation states of technetium were first reported by Cobble et al. [1]. The heats of formation of Tc_2O_7(c) and $HTcO_4$(aq) were measured calorimetrically and, together with suitable experimental and estimated entropy values and the measured potential for the half-cell reaction

$$TcO_4^-(aq) + 4H^+ + 3e^- \rightarrow TcO_2(c) + 2H_2O$$

standard potentials were calculated. However, in later work, Cartledge and Smith [2] found that, although the TcO_2/TcO_4^- electrode is very sluggish

TABLE 2 Thermodynamic Data on Technetium[a]

Formula	Description	State	$\Delta H°$/kJ mol$^-$	$\Delta G°$/kJ mol$^-$	$S°$/J mol^{-1} K^{-1}
Tc	Metal	c	0.0	0.0	(1.8 ± 0.05)
Tc		g	38.7	—	(10.3 ± 0.0)
					10.3
TcO$_2$		c	(−24.1)	(−21.1)	(3.6 ± 0.1)
TcO$_3$		c	(−30.8 ± 1.2)	(−26.3)	(4.1 ± 0.1)
Tc$_2$O$_7$		c	−64.4 ± 0.4	(−53.2)	(10.9 ± 0.5)
			−63.6		
HTcO$_4$		c	−40.0 ± 0.3	33.8	(8.0 ± 0.5)
			39.9		
TcO$_4^-$		aq	(−41.3 ± 0.3)	−36.0	(11.3 ± 0.1)
KTcO$_4$		c	(−58.0 ± 0.4)	(−52.3 ± 0.4)	(9.5 ± 0.2)
K$_2$TcCl$_6$		c	—	—	(19.0 ± 0.2)
TcO$_3$Cl		c	—	—	18.1
Tc$^+$		g	79.1	—	

[a]National Bureau of Standards values are in italic; estimated values are in parentheses.
Source: Data from Refs. 1-3, 6-10.

at 278 K constant potential, values could be obtained in the absence of oxygen. The standard potential obtained in the absence of air differed by about 45 mV from that of Cobble et al.

The results obtained by Cartledge and Smith were used to revise the standard potentials for technetium and its compounds in acid solution. To calculate the Tc-TcO_2 and Tc-TcO_4^- potentials, the thermodynamic data used were $\Delta G°(TcO_2) = -369.5$ kJ mol^{-1}, $\Delta H°(TcO_2) = -421.3$ kJ mol^{-1}, and $\Delta G(TcO_4) = -630.1$ kJ mol^{-1}. De Zoubov and Pourbaix [3] used the values reported by Cartledge and Smith [2] and Latimer [4] together with estimated thermodynamic data to recalculate standard electrode potentials, especially those couples involving $Tc(+2)$. In more recent work by Cartledge [5], it was possible to measure reversible electrode potentials for couples containing technetium and lower oxides, which included hydrous oxides. Gibbs energies of formation were determined from technetium metal. Measurements of the potential of metallic technetium were carried out in sulfate and/or phthalate buffers, after various polarizations, with or without prior coating of the electrode with a film formed cathodically from pertechnetate. Other measurements were made on gold or platinum electrodes, after coating with a film and subjecting to various polarization procedures. Various stationary or metastable potentials were assigned to definite couples involving technetium in lower valence states by means of an empirical correlation between the valence of the cation and the free energy of formation of the oxides of a particular metal in different valence states. In this way, the couples involving the lower oxides or hydroxides of technetium were identified and the Gibbs energies of formation of $TcOH$, $Tc(OH)_2$, $Tc(OH)_3$, $Tc(OH)_4$, Tc_3O_4, and Tc_4O_7 determined. From the Gibbs energies of formation, the standard electrode potentials were calculated. The relevant standard potentials are listed in Table 3.

3. Summary of Technetium Potentials

The following potential diagram summarizes the technetium potentials.

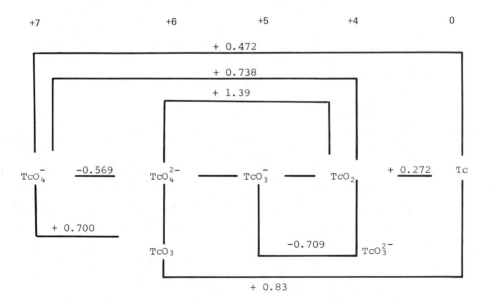

TABLE 3 Standard Potentials of Technetium

Half-reaction	Standard Potential, $E°/V$	Reference
$Tc^{2+} + 2e^- \rightarrow Tc(s)$	0.400	3
$TcO_2(c) + 4H^+ + 4e^- \rightarrow Tc(s) + 2H_2O$	0.272	2
$TcO_4^-(aq) + 4H^+ + 3e^- \rightarrow TcO_2(c) + 2H_2O$	0.738	2
$TcO_4^-(aq) + 2H^+ + e^- \rightarrow TcO_3(c) + H_2O$	0.700	1
$TcO_4^-(aq) + 8H^+ + 7e^- \rightarrow Tc(s) + 4H_2O$	0.472	1
$TcO_3(c) + 2H^+ + 2e^- \rightarrow TcO_2(c) + H_2O$	0.757	2
$TcO_2(c) + 4H^+ + 2e^- \rightarrow Tc^{2+} + 2H_2O$	0.144	3
$TcO_4^-(aq) + 8H^+ + 5e^- \rightarrow Tc^{2+} + 4H_2O$	0.500	3
$TcO_4^-(aq) + e^- \rightarrow TcO_4^{2-}$	−0.569	
$TcOH(s) + H^+ + e^- \rightarrow Tc(s) + H_2O$	0.031	5
$Tc(OH)_2(s) + 2H^+ + 2e^- \rightarrow Tc(s) + 2H_2O$	0.072	5
$Tc(OH)_2(s) + H^+ + e^- \rightarrow TcOH(s) + H_2O$	0.113	5
$Tc_3O_4(s) + 5H^+ + 5e^- \rightarrow 3TcOH(s) + H_2O$	0.161	5
$Tc(OH)_3(s) + 3H^+ + 3e^- \rightarrow Tc(s) + 3H_2O$	0.185	5
$Tc_3O_4(s) + 2H_2O + 2H^+ + 2e^- \rightarrow 3Tc(OH)_2(s)$	0.234	5
$Tc(OH)_3(s) + 2H^+ + 2e^- \rightarrow Tc(OH)(s) + 2H_2O$	0.262	5
$Tc(OH)_4(s) + 4H^+ + 4e^- \rightarrow Tc(s) + 4H_2O$	0.294	5
$Tc_4O_7(s) + 10H^+ + 10e^- \rightarrow 4TcOH(s) + 3H_2O$	0.338	5
$Tc(OH)_4(s) + 3H^+ + 3e^- \rightarrow TcOH(s) + 3H_2O$	0.382	5
$Tc(OH)_3(s) + H^+ + e^- \rightarrow Tc(OH)_2(s) + H_2O$	0.412	5
$Tc(OH)_4(s) + 2H^+ + 2e^- \rightarrow Tc(OH)_2(s) + 2H_2O$	0.518	5
$Tc(OH)_4(s) + H^+ + e^- \rightarrow Tc(OH)_3(s) + H_2O$	0.620	5
$3Tc(OH)_4(s) + 4H^+ + 4e^- \rightarrow Tc_3O_4(s) + 8H_2O$	0.657	5
$3Tc(OH)_3 + H^+ + e^- \rightarrow Tc_3O_4 + 5H_2O$	0.768	5

REFERENCES

1. J. W. Cobble, W. T. Smith, Jr., and G. E. Boyd, J. Am. Chem. Soc., 75, 5777 (1953).
2. G. H. Cartledge and W. T. Smith, Jr., J. Phys. Chem., 59, 1111 (1955).
3. N. de Zoubov and M. Pourbaix, *Comportement électrochimique du technétium. Diagramme d'équilibres tension-pH du systèm Tc-H_2O à 25°C*, Rap. tech. RT50 of Centre Belge d'étude de la corosion (CEBELCOR), 1957.
4. W. M. Latimer, *Oxidation Potentials*, 2nd ed., Prentice-Hall, New York, 1952, pp. 242-245; and "Recent References to Thermodynamic Data," University of California, Aug. 1954.
5. G. H. Cartledge, J. Electrochem. Soc., 118, 231 (1971).
6. G. E. Boyd, J. Chem. Educ., 36, 3 (1959).
7. Natl. Bur. Stand. Tech. Note 270-3, U.S. Government Printing Office, Washington, D.C., 1969.
8. D. D. Wagman, W. H. Evans, V. B. Parker, I. Halow, S. M. Bailey, and R. H. Shuman, *Selected Values of Chemical Thermodynamic Properties*, Natl. Bur. Stand. Tech. Note 270-4, U.S. Government Printing Office, Washington, D.C., 1969.
9. E. J. Baran, Z. Phys. Chem. Leipzig, 257, 829 (1976).
10. R. H. Busey, R. B. Bevan, Jr., and R. A. Gilbert, J. Chem. Thermodyn., 4, 77 (1972).

B. Rhenium

1. Oxidation States

Rhenium compounds and complexes have been prepared in all the oxidation states between +7 and −1. The aqueous chemistry of rhenium oxides is dominated by the very stable perrhenate anion ReO_4^-, to which lower oxides and some other compounds also readily hydrolyze. The perrhenate anion is stable over the entire pH range and acts as a weak oxidant and strong monobasic acid in aqueous solution. Other stable compounds form in oxidation states +3 and +4, while the lower oxidation states +1, 0, and −1 are stabilized in carbonyl, nitrosyl, σ-alkyl, σ-aryl, and π-arene complexes.

Rhenium +5 and +6 complexes have been prepared; however, they are quite reactive and have a tendency to disproportionate to more stable oxidation states. However, rhenium (+6) trioxide is inert and does not disproportionate. Simple halides form in the oxidation states +7 and +3 with a clear trend within the group, whereby iodide stabilizes the lower oxidation states, +4 to +1, and fluoride the higher oxidation states, +7 to +4.

Difficulties arise in the preparation of, and thermodynamic studies into, rhenium compounds (see Table 4) because reactions often yield a mixture of products. This occurs because it is easy to convert from one oxidation state to another under mild redox conditions.

Rhenium aqueous chemistry is confined almost exclusively to anion moieties; one feature is the formation of metal-metal bonds in compounds with oxidation states of +4 to −1. Rhenium +3 binuclear and trinuclear cluster compounds have been intensively studied because of their interesting chemistry. The reduction of perrhenate in aqueous hydrochloric acid with hypophosphite yields a dimeric species $[Re_2Cl_8]^{2-}$. X-ray crystallographic data of this anion show several unusual features. Each rhenium is approximately square-planar

TABLE 4 Thermodyanmic Data on Rhenium[a]

Formula	Description	State	$\Delta H°$/kJ mol^{-1}	$\Delta G°$/kJ mol^{-1}	$S°$/J mol^{-1} K^{-1}
Re	Metal	c	0	0	36.9
		g	769.9	724.7	188.83
Re$^-$	Std. state, $m = 1$	aq	46	10	230
Re$_2$O$_3$	Precipitated	c	−500	(−425)	(130)
ReO$_2$	Orthorhombic	c	−449	−391	47.8
ReO$_2$·2H$_2$O	Precipitated	c	−1010	−841	(130)
ReO$_3$		c	−589.1	−507.1	69.25
Re$_2$O$_7$		c	−1263	−1089	207.3
ReO$_4^-$		aq	−787	−695	201
HReO$_4$		c	−762	−656	158
		g	−665		
	Std. state, $m = 1$	aq	−762	−695	201
ReF$_6$		g	−1350		
ReF$_7$		g	−1410	(−1295)	(360)
ReCl$_3$		c	−264	−188	129

TABLE 4 (Continued)

Formula	Description	State	$\Delta H°$/kJ mol^{-1}	$\Delta G°$/kJ mol^{-1}	$S°$/J mol^{-1} K^{-1}
$ReCl_5$		c	−360		
		g	−370		
$ReCl_6^{2-}$	Std. state, $m = 1$	aq	−761	−590	
Re_3Cl_9		g	−573		
H_2ReCl_4		c	−636		
$ReBr_3$		c	−170		(150)
Re_3Br_9		g	−290		
ReI_3		c	−10		

ReO$_3$I	g	−444	
ReS$_2$	c	−190	(84)
Re$_2$S$_7$	c	−473	(200)
ReAs$_2$	c	5.4	
Re$_3$S	c	0	
Pb(ReO$_4$)$_2 \cdot$2H$_2$O	c	2230	*1900*	*310*
AgReO$_4$	c	736	*635.5*	*153*

[a]National Bureau of Standards values are in italic; estimated values are in parentheses.
Source: Data from Refs. 4-15.

coordinated; four chlorine atoms and two ReCl$_4$ groups are eclipsed. The rhenium-rhenium bond is very short at 0.224 nm. To explain this structure, Cotton and co-workers [1,2], using a molecular orbital approach, showed that a quadruple metal-metal interaction, in which a weak fourth bond forms from d_{xy}-d_{xy} orbital overlap, holds the chlorines in the eclipsed position. The [Re$_2$Br$_8$]$^{2-}$ ion can also be prepared [3]; however, the equivalent iodide and fluoride complexes have not been prepared. The Re-Re quadrupole bond length is relatively invariant and falls within the range 0.222 and 0.225 nm. Oxidation of these binuclear rhenium halides with chlorine and bromine gives the rhenium complexes [Re$_2$X$_9$]$^-$, via the intermediate [Re$_2$X$_9$]$^{2-}$; however, the structure of these complexes has not yet been elucidated. Trinuclear clusters [Re$_3$X$_{12}$]$^{3-}$ and Re$_3$X$_9$, in which the three rhenium atoms are in an equilateral triangle, and the quadrupole cluster Re$_4$X$_{12}$ are also known.

REFERENCES

1. F. A. Cotton, Inorg. Chem., 4, 334 (1965).
2. F. A. Cotton and C. B. Harris, Inorg. Chem., 6, 924 (1967).
3. F. A. Cotton, B. G. De Boer, and M. Jeremic, Inorg. Chem., 9, 2143 (1970).
4. Natl. Bur. Stand. Tech. Note 270-3, U.S. Government Printing Office, Washington, D.C., 1969.
5. H. Rabeneck, K. Rinke, and H. Schafer, Z. Anorg. Allg. Chem., 397, 112-116 (1973).
6. E. G. King, D. W. Richardson, and R. V. Meazek, U.S. Atom. Energy Comm., 1969 BM-RI-7323, 13 pp; Chem. Abstr., 72, 137193w (1970).
7. J. M. Stuve and M. J. Ferrante, U.S. Bureau of Mines, Rep. Invest. 1976, RI8199, 15 pp.
8. R. H. Busey, K. H. Gayer, R. A. Gilbert, and R. B. Bevan, J. Phys. Chem., 70, 2609 (1966).
9. J. P. King and J. W. Cobble, J. Am. Chem. Soc., 79, 1559 (1957).
10. J. E. McDonald and J. W. Cobble, J. Phys. Chem., 66, 791 (1962).
11. R. B. Bevan, Jr., R. A. Gilbert, and R. H. Busey, J. Chem. Phys., 70, 147 (1966).
12. J. Burgess, C. J. W. Fraser, I. Haigh, and R. D. Peacock, J. Chem. Soc., Dalton Trans., 501 (1973).
13. J. Burgess, J. Fawcett, R. D. Peacock, and D. Pickering, J. Chem. Soc., Dalton Trans., 1363 (1976).
14. R. H. Busey, E. D. Sprague, and R. B. Bevan, Jr., J. Phys. Chem., 73, 1039 (1969).
15. J. A. Connor, H. A. Skinner, and Y. Virmani, Faraday Symp. No. 8, High Temperature Chemistry, 1974.

2. Rhenium Potentials

On the basis of the free enrgy of formation of ReO$_4^-$, which was calculated from the standard potential for the ReO$_2$/ReO$_4^-$ couple in acidic solution [1] and the value for the standard Gibbs energy of formation for orthorhombic rhenium dioxide which was reported by Stuve and Ferrante [2], the standard potentials for the Re/ReO$_4^-$ couple in acidic and basic solutions were found to be

$$ReO_4^- + 8H^+ + 7e^- \rightarrow Re + 4H_2O \qquad E° = 0.34 \text{ V}$$

$$ReO_4^- + 4H_2O + 7e^- \rightarrow Re + 8OH^- \qquad E°_B = -0.604 \text{ V}$$

By direct cell measuements Hugus [1] measured the potentials of the ReO_2/ReO_4^- couple and found

$$ReO_4^- + 4H^+ + 3e^- \rightarrow ReO_2 + 2H_2O \qquad E° = 0.51 \text{ V}$$

$$ReO_4^- + 4H_2O + 3e^- \rightarrow ReO_2 \cdot 2H_2O + 4OH^- \qquad E°_B = -0.594 \text{ V}$$

A hydrated sesquioxide precipitate [3] is the outcome of the hydrolysis of rhenium trichloride in alkaline solution. In solution the sesquioxide is hydrated, but the exact composition has not been determined, and electrode potential calculations are based on the Re_2O_3 formulation. From the heat of formation of Re_2O_3 [3] of -500 kJ mol^{-1} and the entropy, estimated from the known entropies of lanthanum, bismuth, and antimony sesquioxides, of 130 J mol^{-1} K^{-1}, it is possible to calculate the Gibbs energy of Re_2O_3 as -425 kJ mol^{-1}. From this $E°_B$ can be calculated for the reactions

$$Re_2O_3 + 3H_2O + 6e^- \rightarrow 2Re + 6OH^- \qquad E°_B = -0.333 \text{ V}$$

$$2ReO_2 \cdot 2H_2O + 2e^- \rightarrow Re_2O_3 + 3H_2O + 2OH^- \qquad E°_B = -1.25 \text{ V}$$

From these values it is evident that Re_2O_3 is thermodynamically unstable in alkaline solution and disproportionates to rhenium metal and the hydrated rhenium dioxide. King and Cobble [4] measured the electrode potentials for the ReO_3/ReO_4^- couple, which were reversible over the pH range ca. 1.5 to 10.3.

$$ReO_4^- + 2H^+ + e^- \rightarrow ReO_3 + H_2O \qquad E° = 0.768 \pm 0.005 \text{ V}$$

$$ReO_4^- + H_2O + e^- \rightarrow ReO_3 + 2OH^- \qquad E°_B = 0.890 \pm 0.008 \text{ V}$$

From the enthalpy of formation of $ReCl_6^{2-}$ obtained from solution calorimetry [5] and using other available thermodynamic data, the electrode potentials for the half-reactions can be determined:

$$ReO_4^- + 8H^+ + 6Cl^- + 3e^- \rightarrow ReCl_6^{2-} + 4H_2O \qquad E° = 0.12 \text{ V}$$

$$ReCl_6^{2-} + 4e^- \rightarrow Re + 6Cl^- \qquad E° = 0.51 \text{ V}$$

On the basis of the claim by Lundell and Knowles [6] to have prepared rhenide (Re$^-$) in solution by the reduction of perrhenate with zinc in dilute sulfuric acid, Latimer [7] estimated $E° = -0.4$ for the Re/Re$^-$ couple. However, using more recent thermodynamic data provided by the National Bureau of Standards [8], in which the Gibbs energy of formation of the rhenide anion is given as 10 kJ mol^{-1}, it is possible to determine the standard potential of the Re$^-$/Re couple.

$$Re + e^- \rightarrow Re^- \qquad E° = -0.10 \text{ V}$$

3. Summary of Rhenium Potentials

Acidic Solution

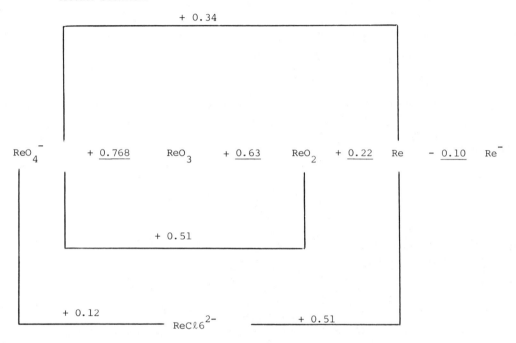

Basic Solution

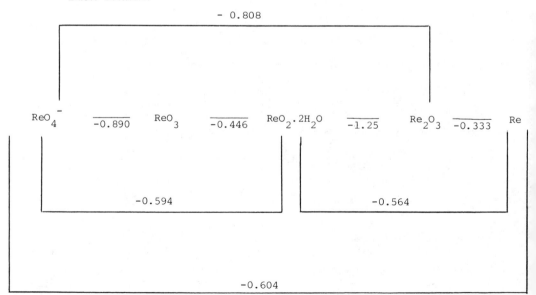

REFERENCES

1. Z. Z. Hugus, cited in *Oxidation Potentials*, 2nd ed., Prentice-Hall, Englewood Cliffs, N.J., 1952, p. 243.
2. J. M. Stuve and M. J. Ferrante, *Thermodynamic Properties of Rhenium Oxides, 8 to 1,400 K RI 8199*, Bureau of Mines, Washington, D.C., 1976.
3. R. H. Busey, E. D. Sprague, and R. B. Bevan, Jr., J. Phys. Chem., 73, 1039 (1969).
4. J. P. King and J. W. Cobble, J. Am. Chem. Soc., 79, 1559 (1957).
5. R. H. Busey, K. H. Grayer, R. A. Gilbert, and R. B. Bevan, Jr., J. Phys. Chem., 70, 2609 (1966).
6. G. E. F. Lundell and H. B. Knowles, Nat. Bur. Stand. J. Res., 18, 629 (1937).
7. W. M. Latimer, *Oxidation Potentials*, 2nd ed., Prentice-Hall, Englewood Cliffs, N.J., p. 243.
8. Nat. Bur. Stand. Tech. Note 270-4, U.S. Government Printing Office, Washington, D.C., May 1969.

16
Chromium, Molybdenum, and Tungsten

I. CHROMIUM*

KATSUMI NIKI   Yokohama National University, Yokohama, Japan

The most common and stable oxidation states of chromium are +2, +3, and +6. Several Cr(IV) and Cr(V) species are known as intermediates in oxidation-reduction reactions. These species are unstable with respect to disproportionation to Cr(III) and Cr(IV) or to reaction with solvents under most conditions. The most stable and important oxidation state is Cr(III). Cr(III) is amphoteric, forming compounds of the Cr(III) ion with acids and chromites with base, and forms a large number of relatively kinetically substitution inert complexes. Cr(II), known as the lowest oxidation state in aqueous solutions, is strongly reducing and easily oxidized to Cr(III) by oxygen. All Cr(VI) compounds are oxo compounds in aqueous solutions and are very strong oxidizing agents in acidic solutions.

The redox potentials of the Cr(III)/Cr and Cr(III)/Cr(II) systems have been investigated by direct cell measurements of many redox reactions of chromium species [1-3]. Latimer noticed from the disagreement between electrochemical data and thermal data that chromium electrodes are irreversible and relied entirely on the third-law calculation in the compilation of redox potentials. Dellien et al. [4] critically evaluated the previously reported thermodynamic properties, chemical equilibria, and standard potentials of chromium and its compounds by referring to the National Bureau of Standards (NBS) Technical Note 270-4 [5] (see Table 1).

A. Aqueous Cr(II) and Cr(III) Ions

The redox reaction of the Cr(II)/Cr couple is irreversible, so that the thermodynamic properties of Cr(II)(aq) ion rely entirely on the third-law calculation. On the other hand, there are some recent experimental data on the Cr(III)/Cr(II) couple [6] and Cr(III)(aq) ions [7].

Reviewer: L. Meites, Clarkson College of Technology, Potsdam, New York.

TABLE 1 Thermodynamic Data on Chromium[a]

Formula	Description[b]	State	$\Delta H°$/kJ mol^{-1}	$\Delta G°$/kJ mol^{-1}	$S°$/J mol^{-1} K^{-1}
Cr	Metal	c	0	0	23.7_7
Cr		g	$397._5$	$352._7$	174.3_9
Cr$^+$		g	$1056._5$		
Cr^{2+}		g	$2654._3$		
Cr^{3+}		g	5648		
CrO		g	222	192	222
Cr$_2$O$_3$		c	$-1139._7$	$-1058._1$	$81._7$
Cr(OH)$_3$	Hydrous	ppt	-984	-858	(105)
CrO$_2$		c	-598	-548	(54)
CrO$_3$		g	-272	-251	$266._1$
CrO$_3$		c	$-589._5$	-510	(67)
Ag$_2$CrO$_4$		c	-730.4_4	-640.6_1	$217._6$
Na$_2$CrO$_4$		c	$-1342._2$	$-1235._1$	176.6_2
Na$_2$Cr$_2$O$_7$		c	$-1979._0$	$-1800._4$	(268)
K$_2$CrO$_4$		c	$-1403._3$	$-1295._4$	$200._0$

$K_2Cr_2O_7$	c	−2062.3	−1882.8	291.2
$FeCr_2O_4$	c	−1444.7	−1343.9	146.0
$MgCr_2O_4$	c	−1786.1	−1671.5	106.0$_2$
CrO_2Cl_2	g	−538.1	−501.7	329.7
CrO_2Cl_2	ℓ	−579.5	−510.9	221.8
CrF_2	c	−778	−736	(84)
CrF_3	c	−1159	−1088	93.8$_9$
CrF_4	c	−1247	−1159	(138)
$CrCl_2$	c	−395.5	−356.1	115.3$_1$
$CrCl_3$	c	−556.5	−486.2	122.9$_3$
$CrBr_2$	c	−302.1	−289	(134)
CrI_2	c	−156.9	−163	(155)
CrI_3	c	−205	−205	(197)
Cr_3C_2	c	−80.8	−81.6	85.4$_4$
Cr_7C_3	c	−161.9	−166.9	200.8
Cr_2N	c	−126	−75	(50)
CrN	c	−121	−92	(25)

TABLE 1 (Continued)

Formula	Description[b]	State	$\Delta H°$/kJ mol^{-1}	$\Delta G°$/kJ mol^{-1}	$S°$/J mol^{-1} K^{-1}
Cr^{2+}		aq	−172	−174	(−100)
Cr^{3+}	$Cr(H_2O)_6^{3+}$	aq	−251	−215	(−293)
$Cr(OH)^{2+}$	i.s.	aq	−495	−430	−156
$Cr(OH)_2^+$	i.s.	aq	−748	−653	(−27)
$Cr(OH)_4^-$		aq	−1169	−1013	(238)
$CrCl^{2+}$	i.s.	aq	−397	−342	−181
$CrCl_2^+$	i.s.	aq	−544	−468	−74

$CrBr^{2+}$	i.s.	−351	−302	−194
$CrSCN^{2+}$	i.s.	−140	−183	−61
CrO_4^{2-}	aq	−881.1$_5$	−727.8	50.2$_1$
$HCrO_4^-$	aq	−878.2	−764.8	184.1
H_2CrO_4	aq	−841	−760.7	293
$Cr_2O_7^{2-}$	aq	−1490.3	−1301.2	262
CrO_4^{3-}	aq		−737.6	
CrO_3Cl^-	aq	−754.8	−664.8	205

[a] National Bureau of Standards [5] values are in italic; estimated values are in parentheses; some revised assignments are listed in J. Phys. Chem. Reference Data, *11*, Supplement 2 (1982).
[b] i.s., inner-sphere complex.

The standard potential of the Cr(III)/Cr(II) couple was first explored by Forbes and Richter [1] from the direct cell measurements and was found to be -0.402 ± 0.005 V. Grube and Schlechter [2] extended their work and obtained the value of -0.422 ± 0.002 V, which has been widely accepted. Biedermann and Romano [6] determined the formal potential of the Cr(III)/Cr(II) couple in 1 M NaCl by taking into account various factors affecting to the redox potential. The proposed value is -0.4295 ± 0.001 V. The standard potential of the Cr^{3+}(aq)/Cr^{2+}(aq) couple can be calculated by using the stability constants for $CrCl^{2+}$(aq,i.s.) (inner sphere) and $CrCl_2^+$(aq,i.s.), as will be discussed later, and by assuming that [Cr(II)] = Cr^{2+}. [Cr^{2+} concentration is assumed to be equal to the total concentration of Cr(II).]

$$Cr^{3+}(aq) + e^- \rightarrow Cr^{2+}(aq) \qquad E^\circ = -0.424 \text{ V}$$

Dellien and Hepler [7] determined the enthalpy of formation of Cr^{3+}(aq), by the calorimetric measurements, as $\Delta H^\circ = -251$ kJ mol^{-1}, which we adopted. We estimated the entropies of Cr^{2+}(aq) and Cr^{3+}(aq) by referring to those of various di- and trivalent aqueous ions and $S^\circ = -100$ J mol^{-1} K^{-1} for Cr^{2+}(aq) and -293 J mol^{-1} K^{-1} for Cr^{3+}(aq). The combination of the standard potential of the Cr^{3+}(aq)/Cr^{2+}(aq) couple and the enthalpy for Cr^{3+}(aq) with the estimated entropies leads to $\Delta H^\circ = -172$ kJ mol^{-1} for Cr^{2+}(aq), which is higher than the value -143.5 kJ mol^{-1} listed in NBS Technical Note 270-4. When we use the NBS Technical Note 270-4 value for the enthalpy of Cr^{2+}(aq), Dellien and Hepler's value for the enthalpy of Cr^{3+}(aq), the estimated entropy for Cr^{3+}(aq) and the standard redox potential of Cr^{3+}(aq)/Cr^{2+}(aq), the entropy of Cr^{2+}(aq) becomes -4.3 J mol^{-1} K^{-1}, which is unreasonably smaller than the estimated ones. We therefore choose $\Delta H^\circ = -172$ kJ mol^{-1} for Cr^{2+}(aq). Combining this enthalpy with the estimated entropy for Cr^{2+}(aq), we find $\Delta G^\circ = -174$ kJ mol^{-1} for Cr^{2+}(aq) and

$$Cr^{2+}(aq) + 2e^- \rightarrow Cr \qquad E^\circ = -0.90 \text{ V}$$

B. Cr(III) Hydroxides

The solubility product of precipitated chromium(III) hydroxide is given by

$$Cr(OH)_3(ppt) = Cr^{3+} + 3OH^- \qquad K_{sp} = 10^{-30}$$

and the combination of this K_{sp} with the selected ΔG° for Cr^{3+}(aq) leads to $\Delta G^\circ = -858$ kJ mol^{-1} for Cr(OH)$_3$(ppt). When we use ΔG° cited above and $\Delta H^\circ = -1064$ kJ mol^{-1} for Cr(OH)$_3$(ppt) listed in NBS Technical Note 270-4, the entropy of Cr(OH)$_3$(ppt) becomes negative, which is clearly unacceptable. We therefore choose an alternative approach to calculate ΔH° for Cr(OH)$_3$(ppt) by introducing an estimated entropy, $S^\circ = 105$ J mol^{-1} K^{-1}, for Cr(OH)$_3$(ppt). This estimation is made by referring to the entropies of various trivalent hydroxides.

It has been known that the Cr(OH)$_3$(ppt) and various Cr(III) species are converted by excess hydroxide ion to soluble chromite ion, the nature of which is not well understood. The formation of chromite ion is represented in a simplified fashion by

$$Cr(OH)_3(ppt) + OH^- = Cr(OH)_4^- \qquad K = 10^{-0.4}$$

By referring to the entropies of halide and hydroxide complexes ranged from $M^{3+}(aq)$ to MX_4^-, the entropies of the homologous series of Cr(III) hydroxides are assumed to be varied almost linearly with respect to the number of anions. The entropy of $Cr(OH)_4^-$ is estimated to be $S° = 238$ J mol^{-1} K^{-1}. The combination of this entropy with the stability constant of the reaction above enables us to calculate $\Delta G°$ for $Cr(OH)_4^-$.

Swaddle and Kong [8] have investigated the equilibrium of $Cr(H_2O)_6^{3+}$ represented by

$$Cr(H_2O)_6^{3+} \rightarrow Cr(H_2O)_5OH^{2+} + H^+ \qquad K = 1.6 \times 10^{-4}$$

in which hydrolytic species is represented by either $Cr(H_2O)_5OH^{2+}$ or by $Cr(OH)^{2+}$(aq,i.s.). This equilibrium leads to $\Delta G° = -430$ kJ mol^{-1}, $\Delta H° = -495$ kJ mol^{-1}, and $S° = -156$ J mol^{-1} K^{-1} for $Cr(OH)^{2+}$(aq,i.s.).

The second hydrolysis constant of Cr^{3+}(aq) is represented by

$$Cr(H_2O)_5OH^{2+} \rightarrow Cr(H_2O)_4(OH)_2^+ + H^+$$

Combining the second hydrolysis constant, $K = 10^{-7}$ [9], with $\Delta H° = 32.6$ kJ mol^{-1} for the reaction above, we obtain $S° = -111$ J mol^{-1} K^{-1} for $Cr(OH)_2^+$(aq, i.s.), which is quite unreasonable. The entropy of $Cr(OH)_2^+$(aq,i.s.) is considered to be $S° = -27$ J mol^{-1} K^{-1}. The combination of this estimated entropy and the enthalpy of the reaction above leads to $K = 2.3 \times 10^{-3}$.

In chloride solutions, Cr(III) ion forms chloride complexes and the stability constants were found [7,10] to be

$$Cr(H_2O)_6^{3+} + Cl^- = Cr(H_2O)_5Cl^{2+} + H_2O \qquad K = 0.20_2$$

The enthalpy of this reaction ranged from 21 to 26.3 kJ mol^{-1}, and we choose the first one to obtain $\Delta H° = -397$ kJ mol^{-1} and $S° = -181$ J mol^{-1} K^{-1} for $CrCl^{2+}$(aq,i.s.).

$$Cr(H_2O)_5Cl^{2+} + Cl^- \rightarrow Cr(H_2O)_4Cl_2^+ + H_2O \qquad K = 0.1$$

The enthalpy of this reaction is $\Delta H° = 21$ kJ mol^{-1}.

The equilibria with bromide ion and thiocyanide ion were given by

$$Cr(H_2O)_6^{3+} + Br^- \rightarrow Cr(H_2O)_5Br^{2+}$$

The stability constant and the enthalpy of this reaction are $K = 2 \times 10^{-3}$ and $\Delta H° = 21$ kJ mol^{-1}, respectively.

$$Cr(H_2O)_5^{3+} + SCN^- \rightarrow Cr(H_2O)_5SCN^{2+} + H_2O$$

The thermodynamic data of this reaction are $K = 1.2 \times 10^3$, $\Delta H° = -8.9$ kJ mol^{-1}, and $\Delta S° = 28.9$ J mol^{-1} K^{-1}.

The formal potential of the following systems has been determined.

$$Cr(CN)_6^{3-} + e^- \to Cr(CN)_6^{4-} \qquad E^{o'}(1\ M\ KCN) = -1.143\ V$$

$$[Cr(edta)(H_2O)]^- + e^- \to [Cr(edta)(H_2O)]^{2-} \qquad E^{o'}(0.1\ M\ KCl) = -0.99\ V$$

C. Chromate and Dichromate

The enthalpy of CrO_4^{2-} was determined directly from the enthalpy of solution of CrO_3 in excess hydroxide ion [11]. Combining ΔH° for CrO_4^{2-} with ΔG° for CrO_4^{2-} calculated from the solubility product of Ag_2CrO_4, we obtain S° for CrO_4^{2-}. These results lead to the values of $\Delta H^\circ = -881.15$ kJ mol^{-1} and S° between 46.4 and 52.7 J mol^{-1} K^{-1} for CrO_4^{2-}. The equilibrium of $HCrO_4^-$ is given by

$$HCrO_4^- \to CrO_4^{2-} + H^+ \qquad K = 3.3 \times 10^{-7}$$

$$2HCrO_4^- \to Cr_2O_7^{2-} + H_2O \qquad K = 33.9$$

which were obtained by several sets of investigators.

$$H_2CrO_4(aq) \to HCrO_4^- + H^+ \qquad K(I = 1.0\ M) = 6.3$$

$$HCrO_4^- + H^+ + Cl^- \to CrO_3Cl^- + H_2O \qquad K(I = 1.0\ M) = 17$$

Our selected ΔG° values for chromate and dichromate ions correspond to the following potentials:

$$HCrO_4^- + 7H^+ + 3e^- \to Cr^{3+} + 4H_2O \qquad E^\circ = 1.38\ V$$

$$CrO_4^{2-} + 4H_2O + 3e^- \to Cr(OH)_4^- + 4OH^- \qquad E^\circ = -0.13\ V$$

$$CrO_4^{2-} + 4H_2O + 3e^- \to Cr(OH)_3(ppt) + 5OH^- \qquad E^\circ = -0.11\ V$$

$$Cr_2O_7^{2-} + 14H^+ + 6e^- \to 2Cr^{3+} + 7H_2O \qquad E^\circ = 1.36\ V$$

D. Cr(IV) and Cr(V)

Bailey and Symons [12] have shown that when potassium chromate in potassium hydroxide-water melt were heated under nitrogen atmosphere, oxygen was evolved and chromate was converted into CrO_4^{3-}. In acid solution, CrO_4^{3-} quickly disproportionates to CrO_4^{2-} and Cr(III). In dilute alkaline solution, CrO_4^{2-} and chromite are formed. In the course of the oxidation of alcohols by Cr(IV), which was generated by the reaction of Cr(VI) with V(IV), it was found that Cr(IV) did not reduce Cr(VI) but rather oxidized the organic substrates. The redox potential of the Cr(VI)/Cr(V), Cr(V)/Cr(IV), and Cr(IV)/Cr(III) couples in acidic solutions was estimated [13].

$$Cr(VI) + e^- \rightarrow Cr(V) \qquad E^\circ = 0.55 \text{ V}$$
$$Cr(V) + e^- \rightarrow Cr(IV) \qquad E^\circ = 1.34 \text{ V}$$
$$Cr(IV) + e^- \rightarrow Cr(III) \qquad E^\circ = 2.10 \text{ V}$$

E. Potential Diagrams for Chromium

Acid Solution

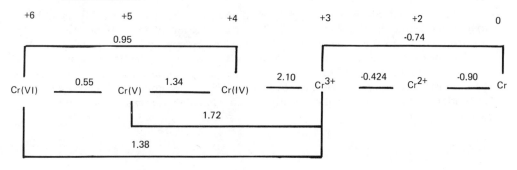

Basic Solution

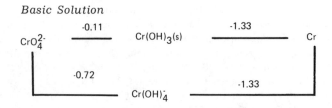

REFERENCES

1. G. S. Forbes and W. H. Richter, J. Am. Chem. Soc., *39*, 1140 (1917).
2. G. Grube and L. Schlechter, Z. Elektrochem., *32*, 178 (1926).
3. G. Grube and G. Breitinger, Z. Elektrochem., *33*, 112 (1927).
4. I. Dellien, F. M. Hall, and L. G. Hepler, Chem. Rev., *76*, 283 (1976).
5. D. D. Wagman, W. H. Evans, V. B. Parker, I. Halow, S. M. Bailey, and R. H. Schumm, *Selected Values of Chemical Thermodynamic Properties*, Natl. Bur. Stand. Tech. Note 270-4, U.S. Government Printing Office, Washington, D.C., 1969.
6. G. Biedermann and V. Romano, Acta Chem. Scand., *A29*, 615 (1975).
7. I. Dellien and L. G. Hepler, Can. J. Chem., *54*, 1383 (1976).
8. T. W. Swaddle and P. C. Kong, Can. J. Chem., *48*, 3223 (1970).
9. L. G. Sillén and A. E. Martell, *Stability Constants of Metal-Ion Complexes*, Spec. Publ. 17, Chemical Society, London (1964).
10. H. S. Gates and E. L. King, J. Am. Chem. Soc., *80*, 5011 (1958).
11. L. G. Hepler, J. Am. Chem. Soc., *80*, 6181 (1958).
12. N. Bailey and M. C. R. Symons, J. Chem. Soc., 203 (1957).
13. M. Rahman and J. Rocek, J. Am. Chem. Soc., *93*, 5455 (1971).

II. MOLYBDENUM*

NICHOLAS STOLICA Westfälische Wilhelms Universität, Münster, Westfalen, Federal Republic of Germany

A. Oxidation States

The most remarkable feature of molybdenum is its chemical versatility. It has oxidation states ranging from −II to VI, coordination numbers from 4 to 8, and the ability to form compounds with most inorganic and organic ligands. It also forms bi- and polynuclear compounds containing molybdenum-molybdenum bonds and/or bridging ligands. In aqueous solution the molybdenum VI oxidation state cannot be reduced below III. The aqueous chemistry of molybdenum involves oxidation states III to VI. The thermodynamic data are given in Table 2.

1. Mo(VI)

This is the most important oxidation state. The chemistry of this state is largely that of the oxide MoO_3, and its compounds, such as molybdates and dimolybdates. Molybdates form compounds with oxides of other metals. The molybdates $PbMoO_4$ and $MgMoO_4$ are widespread in nature. In solution Mo(VI) is found as colorless MoO_4^{2-} ion at pH > 6. Acid hydrolysis gives $Mo_7O_{24}^{6-}$ and $Mo_8O_{26}^{4-}$ from pH 6 to 1 [1]. At pH less than 1, $Mo_{36}O_{112}^{8-}$, $[HMoO_3]^+$, $[H_2Mo_2O_6]^{2+}$, and $[H_3Mo_2O_6]^{3+}$ are formed [2-4]. At higher acidities $MoO_3 \cdot H_2O$ and/or $MoO_3 \cdot 2H_2O$ are found [5]. Monomeric complexes of cis-MoO_2^{2+} appear in strong acids [1]. When molybdate solutions containing other oxanions (e.g., PO_4^{3-}, SiO_4^{4-}) are acidified, heteropolyanions are formed. In these anions 2 to 18 hexavalent Mo atoms surround one or more central atoms (heteroatoms). V, Ni, Ta, and transition metals can replace some of the Mo atoms in the heteropolystructure [6]. Heteropolyanions that have MoO_6 groups (octahedra) containing a single unshared oxygen can undergo reduction to mixed-valence species ("blues") without change of structure [7]. The oxidation-reduction of the following heteropolymolybdates has been studied in aqueous solutions: $\alpha[SiMo_{12}O_{40}]^{4-}$, $\beta[AsMo_{12}O_{40}]^{3-}$, $H_3[PMo_{12}O_{40}]$, $\beta\text{-}H_4[SiMo_{12}O_{40}]$, $H_4[PMo_{11}VO_{40}]$, $H_5[PMo_{10}V_2O_{40}]$, $[TeMo_6O_{24}]^{6-}$, $[CrMo_6O_{24}H_6]^{3-}$, $[AlMo_6O_{24}H_6]^{3-}$, $[FeMo_6O_{24}H_6]^{3-}$, $[IMo_6O_{24}]^{5-}$, $[Co_2Mo_{10}O_{38}H_4]^{6-}$, and $[P_2Mo_{18}O_{62}]^{6-}$. The $[Mo_{12}O_{40}]^{8-}$ species contain a central tetrahedral hole. Octahedral holes are formed by $[Mo_6O_{24}]^{12-}$ [1,8]. Mo(VI) forms many complexes. The oxide MoO_3 forms water-soluble dien-MoO_3, as well as 1:2 complexes with hexadentate mannitol and EDTA, and $[O_2MoOMoO_2]^{2+}$ with oxalate [1]. In HCl, $MoO_2Cl_2(H_2O)_2$ and $[MoO_2Cl_4]^{2-}$ are formed in 6 M and 12 M acid, respectively [5]. Bisulfate complexes formed by MoO_2^{2+} are $MoO_2(HSO_4)_4^{2-}$ and $MoO_2(HSO_4)_2$ [9]. The following complexes are formed by MoO_4^{2-}: $MoO_4(H_3cit)^{2-}$ [10], 1:2 complex with lactic acid, 1:1 with malic acid [11], and 1:2 and 1:1 with catechol at pH 5 to 7 and pH 0 to 1, respectively [1]. Polarographic reduction of molybdate ion takes

---

Reviewer: M. J. Sienko, Cornell University, Ithaca, New York.

place in two steps, Mo(VI) → Mo(V) and Mo(V) → Mo(III), in many media [12]. Molybdenum blue is formed on reduction of weakly acid (e.g., 0.5 to 1.5 M HCl) molybdate solutions by Sn, Zn, Al, SO_2, N_2H_4, or H_2S [5]. Electrolytic reduction also yields Mo blue [13]. The Mo in the blues has an oxidation number between V and VI. Typical formulas are $MoO_{2.0}(OH)$ and $MoO_{2.5}(OH)_{0.5}$ [5]. The Mo(VI) species undergoes substitution reactions. Bubbling H_2S through aqueous MoO_4^{2-} produces $[MoO_3S]^{2-}$, $[MoO_2S_2]^{2-}$, $[MoOS_3]^{2-}$, and $[MoS_4]^{2-}$ [1]. The following mixed-valent oxides have been prepared: Mo_4O_{11}, Mo_9O_{26}, $Mo_{17}O_{47}$, Mo_5O_{14}, Mo_8O_{23}, and $Mo_{18}O_{52}$ [14].

2. Mo(V)

The best characterized Mo(V) species in aqueous systems is $[MoOCl_5]^{2-}$. This green, monomeric, paramagnetic ion exists in HCl solutions of concentrations ⩾10 M [1]. It can be prepared by electrolytic or chemical (Hg) reduction of molybdates or MoO_3 in 12 M HCl, or by dissolving $MoCl_5$ in 12 M HCl [5]. As water is added dimerization takes place with maximum enhancement of the dimer spectrum at ca. 5 M HCl [1]. If the green ion $[MoOCl_5]^{2-}$ is dissolved in HCl of less than 2 M, it is converted into an orange-yellow diamagnetic species. This species can also be prepared by the electrolytic reduction of Mo(VI) in perchloric acid. The most recent structure proposed for this ion is $[Mo_2O_4]^{2+}$. A brown polynuclear species is formed from the orange Mo(V) aquo ion at pH > 1 [2]. Above 3 M HCl the complexes $[MoOCl]^{2+}$, $[MoOCl_2]^+$, and $MoOCl_3$ are formed with increasing acid concentration. At 5 M HCl $MoOCl_3$ is the dominant species [15]. In concentrated HBr (8.6 M), $[MoOBr_4^-]_2$ is the principal species, and at higher Mo(V) concentrations tetrameric species are formed Mo(V) complexes that have been isolated from aqueous acid contain the following groups: MoO^{3+}, $Mo_2O_3^{4+}$, $Mo_2O_4^{2+}$, and Mo_4O_{10} [1]. Some examples of monomeric complexes are $[MoO(NCS)_5]^{2-}$, $[MoOF_5]^{2-}$, $[MoOBr_5]^{2-}$, $[MoOCl_4]^-$, and $[Mo(CN)_8]^{3-}$ [1]. Molybdenum (V) has a great tendency to dimerize. Typical examples of dimeric complexes are: $M_4^I[Mo_2O_3(NCS)_8]$, $M_4^I[Mo_2O_4(NCS)_6]$, $Na_2[Mo_2O_4(EDTA)] \cdot H_2O$, $Na_2[Mo_2O_4(oxalate)_2(H_2O)_2]$, $Na_2[Mo_2O_4(cysteine)_2] \cdot 2H_2O$, and $Mo_2O_4L_2$ (L=HS–CH_2–CH_2–NH_2 and HS–CH_2–CH_2–COOH). Complexes containing the Mo_4O_{10} group have the composition $[Mo_4O_{10}(\ell)]$ where ℓ is glutathione or a related large hexadentate ligand derived from cysteine [1]. A new chloride oxide, MoO_2Cl, has been reported [16]. The dehydration of $MoO(OH)_3$ in three steps to yield Mo_2O_5 has been described [17].

3. Mo(IV)

A rose-colored Mo(IV) species can be prepared by heating an equimolar mixture of green Mo(III) and Mo(V) in aqueous HCl or H_2SO_4. The Mo(IV) does not undergo a disproportion reaction in aqeuous solution. It is very stable toward air oxidation. The rose-colored ion in aqueous HCl (>2 M) was formulated as $[MoOCl]^+$. In noncomplexing acids $[MoO]^{2+}$ is considered to be the dominant species. At lower acidities the rose-colored species is in equilibrium with a brown species $[MoO(OH)]^+$, which is the main species at about pH 1.6. This ion is considered to be polynuclear. At still lower acidities an insoluble hydroxide of Mo(IV) precipitates. A stable red Mo(IV) species is

TABLE 2 Thermodynamic Data on Molybdenum[a]

Formula	Description	State	$\Delta H°$/kJ mol^{-1}	$\Delta G°$/kJ mol^{-1}	$S°$/J mol^{-1} K^{-1}
Mo	Metal	c	0	0	28.66
Mo		g	658.1	612.5	181.950
Mo$^+$		g	1,349.30		
Mo^{2+}		g	2,914.24		
Mo^{3+}		g	5,541.7		
Mo^{4+}		g	10,025.		
Mo^{5+}		g	15,933.		
Mo^{6+}		g	22,506.		
Mo^{7+}		g	34,635.		
Mo^{8+}		g	48,543.		
Mo^{3+}		aq		(−57.7)	
MoO		g	389	355	(238)
MoO$_2$		c	−588.94	−533.01	46.28
MoO$_2$		g	13	0	(276)

Chromium, Molybdenum, and Tungsten

MoO_3		c	−745.09	−667.97	77.74
MoO_3		g	−356	−339	(280)
$(MoO_3)_2$		g	−1,176		
$(MoO_3)_3$		g	−1,925		
$(MoO_3)_4$		g	−2,619		
$(MoO_3)_5$		g	−3,310		
MoO_3		aq	−721.7		
MoO_3^+		g	837		
MoO_4		aq	−661.1	−580.3	167
MoO_5		aq	−584.1		
Mo_4O_{11}		c	−2,807	−2,512	290
Mo_9O_{26}		c	−6,531	−5,866	694
H_2MoO_4	White	c	−1,046.0	−912	(121)
H_2MoO_4		g	−849	−787	(356)
H_2MoO_4		aq	−1,007.5	−882.8	151
$H_2MoO_4 \cdot H_2O$	Yellow	c	−1,360		
MoO_4^{2-}		aq	−997.0	−838.5	38
Li_2MoO_4		c	−1,520.0	−1,410.4	(130)

TABLE 2 (Continued)

Formula	Description	State	$\Delta H°$/kJ mol^{-1}	$\Delta G°$/kJ mol^{-1}	$S°$/J mol^{-1} K^{-1}
Li$_2$MoO$_4$		g	−1,021		
Na$_2$MoO$_4$		c	−1,467.7	−1,353.9	159.4
Na$_2$MoO$_4$·2H$_2$O		c		−1,829.7	
Na$_2$Mo$_2$O$_7$		c	−2,245.1	−2,058.5	250.6
K$_2$MoO$_4$		c	−1,497.9	−1,384.9	(188)
K$_2$MoO$_4$		g	−1,138		
K$_2$Mo$_2$O$_7$		c	(−2,280)		
KHMoO$_4$		g	−1,013		
Rb$_2$MoO$_4$		c	−1,493.7	−1,387.0	(234)
Rb$_2$Mo$_2$O$_7$		c	(−2,276)		
Cs$_2$MoO$_4$		c	−1,514.6	−1,407.1	248.32
Cs$_2$MoO$_4$		g	−1,226		
Cs$_2$Mo$_2$O$_7$		c	−2,302.5	−2,122.1	(339.7)
BeMoO$_4$		c	(−1,323.4)	(−1,228.4)	129.3
MgMoO$_4$		c	−1,400.8	−1,295.8	118.8

MgMoO$_4$	0 K	g	−937	
CaMoO$_4$		c	−1,546.0	−1,439.3	122.6
CaMoO$_4$		g	−941		
SrMoO$_4$		c	−1,544	−1,435	(134)
SrMoO$_4$		g	−1,042		
BaMoO$_4$	Precipitated	c	−1,545.6	−1,439.	142.
BaMoO$_4$		g	−1,013		
MnMoO$_4$		c	−1,191.2		
FeMoO$_4$		c	−1,075	−975	129.3
Fe$_2$(MoO$_4$)$_3$		c	−2,937		
CoMoO$_4$		c	(−1,031.8)	(−937.2)	151.5
NiMoO$_4$		c	(−1,052.3)	(−947.7)	117.6
CuMoO$_4$		c	(−944.3)	(−848.1)	148.5
Ag$_2$MoO$_4$		c	−839.3	−750.6	226.
Ag$_2$MoO$_4$		aq	−786.6	−681.9	172.8
ZnMoO$_4$		c	−1,139.7	−1,043.5	133.72
CdMoO$_4$		c	−1,092.4	−997.0	142.26

TABLE 2 (Continued)

Formula	Description	State	$\Delta H°$/kJ mol^{-1}	$\Delta G°$/kJ mol^{-1}	$S°$/J mol^{-1} K^{-1}
Al$_2$(MoO$_4$)$_3$		c	−3,807	−3,485	286.60
Tl$_2$MoO$_4$		c		−948.9	
PbMoO$_4$		c	−1,051.9	−951.4	166.1
Mo$_7$O$_{24}^{6-}$		aq	−6,071	−5,251	301
MoF		g	272.0	240.2	237.2
MoF$_2$		g	−168.2	−179.9	270.7
MoF$_3$		g	−592.0	−582.8	301.2
MoF$_4$		g	−954.0	−920.1	320.1
MoF$_5$		c	−1,385.		
MoF$_5$		g	−1,241.4	−1,179.5	327.6
(MoF$_5$)$_2$		g			515.
(MoF$_5$)$_3$		g			707.
MoF$_6$		l	−1,585.40	−1,472.98	259.66
MoF$_6$		g	−1,557.66	−1,472.20	350.52
MoOF$_4$		c	−1,377.		

MoO_2F_2	g	−1,121	
$MoCl_2$	c	−285	−247	(126)
$MoCl_3$	c	−393.3	−466.1	118.16
$MoCl_4$	c	−477	−574.0	157.23
$MoCl_4$	g	−385	−356	(372)
$MoCl_5$	c	−527	−423	(238)
$MoCl_5$	g	−448	−393	(397)
$MoCl_6$	c	−523	−420.9	351
$MoCl_6$	g	−439	−379.5	494
$MoOCl_2$	c	−527.	−460	(130)
$MoOCl_3$	c	−623.08	−713.0	163.85
$MoOCl_3$	g	−515	−485	(368)
$MoOCl_4$	c	−652.16	−764.0	202.00
$MoOCl_4$	g	−586	−527	(381)
MoO_2Cl_2	c	−720.1	−815.9	135.65
MoO_2Cl_2	g	−628	−594	(343)
MoO_2Cl_2	aq	−796.6		

TABLE 2 (Continued)

Formula	Description	State	$\Delta H°$/kJ mol^{-1}	$\Delta G°$/kJ mol^{-1}	$S°$/J mol^{-1} K^{-1}
MoO$_2$Cl$_2$·H$_2$O		c	−1,028.8	−879	(180)
MoBr$_2$		c	−213.	−201	(138)
MoBr$_2$		g	−84	−100	(222)
MoBr$_3$		c	−259	−251	(218)
MoBr$_4$		c	−293	−285	(293)
MoBr$_4$		g	−151	−192	(460)
MoBr$_5$		c	−213	−170.3	(264)
MoOBr$_3$		c	−464	−414	(188)
MoOBr$_3$		g	−314.	−335	(423)
MoO$_2$Br$_2$		c	−636	−565	(142)
MoO$_2$Br$_2$		g	−527	−523	(360)
MoO$_2$Br$_2$		aq	−698.3		
MoI$_2$		c	(−105)	−110.5	(165.3)
MoI$_2$		g	134		
MoI$_3$		c	−113	−114.2	(208.4)

MoI₄	c	−75	(269.0)
MoI₅	c	−75	(326)
MoO₂I₂	g	−418	(360)
MgMoO₃	c	−1,182.8	91.04
CaMoO₃	c	−1,238	103.3
SrMoO₃	c	−1,280	
SrMoO₃	g	−590	
BaMoO₃	c	−1,234	
BaMoO₃	g	−653	
MoS₂	c	−275.3	62.59
MoS₃	c	−257.23	66.5
Mo₂S₃	c	−410	113
MoSe₂	c		89.08
MoTe₂	c		115.27
Mo₂N	c	−81.59	
MoC	c	−10.0	
Mo₂C	c	−45.6	65.86

MoI₄	c	−75	−77.4	(269.0)
MoI₅	c	−75	−78.2	(326)
MoO₂I₂	g	−418	−423	(360)
MgMoO₃	c	−1,182.8	−1,100.0	91.04
CaMoO₃	c	−1,238	−1,159	103.3
SrMoO₃	c	−1,280		
SrMoO₃	g	−590		
BaMoO₃	c	−1,234		
BaMoO₃	g	−653		
MoS₂	c	−275.3	−266.5	62.59
MoS₃	c	−257.23	−240.04	66.5
Mo₂S₃	c	−410	−397.9	113
MoSe₂	c			89.08
MoTe₂	c			115.27
Mo₂N	c	−81.59		
MoC	c	−10.0		
Mo₂C	c	−45.6	−46.4	65.86

TABLE 2 (Continued)

Formula	Description	State	$\Delta H°$/kJ mol^{-1}	$\Delta G°$/kJ mol^{-1}	$S°$/J mol^{-1} K^{-1}
Mo$_3$C$_2$		c	−64.0		
MoSi$_2$		c	−130	−130	67.8
Mo$_3$Si		c	−113.	−113.	106.3
Mo$_5$Si$_3$		c	−310.	−310.	207.5
Mo$_3$Ge		c	−21		
MoB		c	−88		
Mo$_2$B		c	−121		
KMoF$_6$		c	−2,075		

RbMoF$_6$	c	−2,084	
CsMoF$_6$	c	−2,092	
Na$_2$MoCl$_6$	c	−1,372	
K$_2$MoCl$_6$	c	−1,464	
Mo(CO)$_6$	c	−989.5	−884.9
Mo(CO)$_6$	g	−919.2	−864.0
Mo(CO)$_6^+$	g	−112.1	
Mo2(O$_2$CCH$_3$)$_4$	c	−1,976.5	
Mo(C$_5$H$_7$O$_2$)$_3$	c	−1,324.8	

	326.3
	429.9

aNational Bureau of Standards values are in italic; estimated values are in parentheses.

formed in aqueous solutions of $[Mo(H_2O)_6]^{3+}$ by oxidation by the solvent. It has the same absorption spectrum as the rose-colored ion. A binuclear structure $[MoO(H_2O)_4]_2^{4+}$ was proposed for the Mo(IV) ion based on the results of ion-exchange chromatography [2]. Sagi and Rao [18] obtained evidence for the existence of mixed-valence species having the oxidation numbers 3.5 and 4.5. A well-known Mo(IV) species in aqueous solution is $[Mo(CN)_8]^{4-}$. No evidence of protonation is seen up to 3 M $HClO_4$ [1]. Mo(IV) forms the thiocyanate complexes $[Mo_2O_2NCS]^{3+}$ and $[Mo_2O_2(NCS)_2]^{2+}$, depending on whether Mo(IV) or NCS^- is present in excess, respectively [2]. Molybdenite, MoS_2, is the most prevalent molybdenum mineral. A very important oxide is MoO_2.

4. Mo(III)

The pale yellow ion $[Mo(H_2O)_6]^{3+}$ can be prepared by aquation (about 2 days) of the red $[MoCl_6]^{3-}$ ion in aqueous HPTS (p-toluenesulfonic acid) or CF_3SO_3H. It is separated from the other products by ion-exchance chromatography. The hexaquo ion is easily oxidized by air to Mo(V). Under an inert atmosphere the ion is slowly oxidized by solvent water to Mo(IV). Green solutions of Mo(III) can be prepared by reducing acidic solutions of Mo(VI) in a Jones reductor (or Cd reductor). They can also be prepared by electrolytic reduction. The structure proposed for this ion is $[MoOH(H_2O)_4]_2^{4+}$. The green binuclear ion is stable in aqueous solution under an inert atmosphere but undergoes slow spectral changes. Air oxidizes it slowly to Mo(IV). Yellow solutions of $[Mo(H_2O)_6]^{3+}$ turn green after prolonged standing [2]. Salts of $[MoCl_6]^{3-}$ are stable in aqueous HCl > 9 M. At acid concentrations of 1 to 2 M, $[MoCl_6]^{3-}$ forms $[MoCl_5(H_2O)]^{2-}$ and undergoes substitution reactions with bidentate ligands such as oxalate, EDTA, in, σ-phen, and bipy. The $[Mo(NCS)_6]^{3-}$ ion and the cyanide complexes $K_2Mo(CN)_5$ and $K_4Mo(CN)_7(H_2O)_2$ are stable in aqueous solution [1].

5. Mo(II)

The red aquo ion $[Mo_2]^{4+}$ can be prepared by treating a solution of $K_4Mo_2(SO_4)_4$ in CF_3SO_3H with $Ba(CF_3SO_3)_2$. The $K_4Mo_2(SO_4)_4$ is prepared from $K_4Mo_2Cl_8$. There is evidence for the existence of the aquo ion even in solutions of $K_4Mo_2Cl_8$ in aqueous noncomplexing acids such as CF_3SO_3H or p-toluenesulfonic acid. The red ion in these solutions migrates to the cathode. The existence of the binuclear ion is also confirmed by Raman spectroscopy. The oxidation state of the $[Mo_2]^{4+}$ ion is quite stable in solution under an inert atmosphere. However, the adsorption spectrum changes within a few hours, indicating condensation of hydrolysis products of $[Mo_2]^{4+}$. Air oxidizes $[Mo_2]^{4+}$ to Mo(IV) and Mo(V). The species $[Mo_2(SO_4)_4]^{3-}$, having a stable oxidation state of 2.5+, exists in both the solid state and in solution. The aquo ion $[Mo_2]^{5+}$ disproportionates to Mo(II) and Mo(III) [2]. The metal cluster ion $[Mo_6Cl_8]^{4+}$ is stable in aqueous acid. Each Mo atom in the face of a cube is a binding site for the formation of $[MoCl_8L_6]^{n\pm}$. The ion $[(Mo_6Cl_8)Cl_6]^{2-}$ is stable in concentrated HCl. As the acidity of the solution is lowered, OH^- is substituted for Cl^- to make the complex $[(Mo_6Cl_8) \cdot (OH)_6]^{2-}$. The species $[(Mo_6Br_8)Br_4]$ is polymeric, whereas $[(Mo_6Br_8)(Br)_6]^{2-}$ is monomeric [1].

B. Gibbs Energy of Molybdate and Molybdic Acid

Dellien et al. [19] have adopted the value given in Table 2 as the best value for the Gibbs energy for formation of the molybdate ion on the basis of a critical review of existing data on different molybdates in the literature. They calculate for the following reaction given by Latimer [20],

$$MoO_4^{2-} + 4H_2O + 6e^- \rightarrow Mo + 8OH^- \qquad E° = -0.913 \text{ V}$$

Metallic molybdenum can be electrodeposited from aqueous solution only as an alloy with Fe, Co, or Ni [12,21]. Using the new thermodynamic data and the estimated value of the entropy of H_2MoO_4(aq) of Latimer (Table 2), we have recalculated the potential of the following reaction given by Latimer [20]:

$$H_2MoO_4(aq) + 6H^+ + 6e^- \rightarrow Mo + 4H_2O \qquad E° = 0.114 \text{ V}$$

The potentials of the following two reactions given by Pozdeeva et al. [22] have been recalculated. We obtained

$$H_2MoO_4(aq) + 2H^+ + 2e^- \rightarrow MoO_2 + 2H_2O \qquad E° = 0.646 \text{ V}$$

$$H_2MoO_4(aq) + 6H^+ + 3e^- \rightarrow Mo^{3+} + 4H_2O \qquad E° = 0.428 \text{ V}$$

We have calculated the oxidation-reduction potential of the Mo_4O_{11}/H_2MoO_4(aq) couple. The composition of Mo_4O_{11} is 1Mo(IV):3Mo(VI) [23,24].

$$4H_2MoO_4(aq) + 2H^+ + 2e^- \rightarrow Mo_4O_{11} + 5H_2O \qquad E° = 0.865 \text{ V}$$

The potential of the following reaction given by Pozdeeva et al. [22] has been recalculated. We give

$$MoO_4^{2-} + 2H_2O + 2e^- \rightarrow MoO_2 + 4OH^- \qquad E° = -0.780 \text{ V}$$

C. Reduction of Molybdic Acid to the V, IV, and III States

The electrolytic reduction of molybdic acid at hydrogen ion concentrations of 0.5 to 1.5 M results in the formation of molybdenum blues [13] (see above). At 2 M HCl Mo(VI) is reduced to Mo(V). At high HCl concentrations (e.g., 8 M) it is reduced to Mo(III). Mo(V) can also be reduced to Mo(III). Mo(VI) can be reduced to Mo(III) in a Jones reductor. Mo(IV) cannot be formed by the electrolytic reduction of Mo(VI) or Mo(V) [13]. However, Mo(IV) is formed on polarographic reduction of Mo(VI) according to Haight [25]. El-Shamy and El-Aggan [26] have given the following potential:

$$MoO_2^{2+} + 2H^+ + e^- \rightarrow MoO^{3+} + H_2O \qquad E° = 0.4826 \text{ V at } 30°C$$

This value was obtained from data in 0 to 4 M HCl solutions by plotting $E - (2RT/F) \ln [\text{HCl}]$ against $\sqrt{I}$ and extrapolating to zero ionic strength. They report the presence of two systems, one above 5.75 M and another below 3 M HCl. Other ionic species probably exist in between. Neisman and Mikhailovskaya [27] reported +0.50 V for the formal potential of the Mo(VI)/Mo(V) system in HCl solution.

Foerster and Fricke [13] determined the formal oxidation-reduction potentials of the Mo(VI)/Mo(V) and Mo(V)/Mo(III) systems in HCl solutions. Their results are given in Table 3. Karpova et al. [28] also determined the potentials for these two systems. Their values are given in Table 4. It is seen that agreement between the potentials in the two tables for the Mo(VI)/Mo(V) system is only approximate. Which one of the two sets of values is more accurate cannot be ascertained. It should be noted that the values of Foerster and Fricke [13] for the Mo(VI)/Mo(V) system extrapolate to 0.513 V at infinite dilution of HCl. This value is in reasonable agreement with that of El-Shamy and El-Aggan [26], and also with that of Neisman and Mikhailovskaya [27]. The value of -0.25 V given for the Mo(V)/Mo(III) (green) system by Foerster and Fricke is not very reliable according to the authors.

It is seen in Table 4 that the potentials of the two systems Mo(VI)/Mo(V) and Mo(V)/Mo(III) (green) do not depend on the hydrogen ion concentration at a chloride concentration of 10 M [28]. In another publication Karpova et al. [29] report that the current-potential curves of these two systems essentially do not depend on the hydrogen ion concentration at a total chloride concentration fo 10 M. On the basis of these results it can be said that the equation given above by El-Shamy and El-Aggan for the Mo(VI)/Mo(V) system does not hold for 10 M chloride. Similarly, the following equation for the Mo(V)/Mo(III) system is not valid in 10 M chloride solutions:

$$MoO^{3+} + 2H^+ + 2e^- \rightarrow Mo^{3+} + H_2O$$

El-Aggan et al. [30] determined the oxidation-reduction potentials of the Mo(V)/Mo(III) system, using a hydrogen electrode, over the range of hydrochloric acid concentration from 0.01 to 8 M:

$$MoO_2^+ + 4H^+ + 2e^- \rightarrow Mo^{3+} + 2H_2O \qquad E° = 0.089 \pm 0.02 \text{ V at } 30°C$$

This value was obtained by plotting $E - (2RT/F) \ln [HCl]$ against $\sqrt{I}$ and extrapolating to zero ionic strength. The expression $E - (2RT/F) \ln [HCl]$ was obtained from the cell reaction

TABLE 3 Dependence of the Formal Potentials in the Systems Mo(VI)/Mo(V), Mo(V)/Mo(III) (red), and Mo(V)/Mo(III) (green) on HCl Concentration at 18°C

[HCl]/M	Mo(VI)/Mo(V), $E^{o'}/V$	Mo(V)/Mo(III) (red), $E^{o'}/V$	Mo(V)/Mo(III) (green), $E^{o'}/V$
2	0.532	0.114	-0.25
4	0.551	0.176	
8	0.697	0.266	

TABLE 4 Dependence of the Formal Potentials in the Systems Mo(VI)/Mo(V) and Mo(V)/Mo(III) on HCl Concentration

[HCl]/M	Mo(VI)/Mo(V), $E^{o'}$/V		Mo(V)/Mo(III) (green), $E^{o'}$/V	
	No LiCl present	LiCl present [Cl⁻] = 10 M	No LiCl present	LiCl present [Cl⁻] = 10 M
0.1	0.375	0.636	0.031	0.235
1	0.392	0.631	0.099	0.233
2	0.426	0.634	0.123	0.231
4	0.501	0.624	0.129	0.223
6	0.534	0.618	0.173	0.222
8	0.612	0.619	0.215	0.222
10	0.618	0.618	0.220	0.220

$$MoO_2^+ + H_2 + 2H^+ \rightarrow Mo^{3+} + 2H_2O$$

Spectrophotometric studies showed the existence of two species of Mo(III), a green one at HCl < 2.23 M and a red one at HCl > 4.46 M. The shape of the curve of $E(V)$ vs. $\sqrt{I}$ also indicates the presence of two different species of tervalent molybdenum.

Sagi and Rao [18] measured the formal potentials of the Mo(V)/Mo(IV) and Mo(IV)/Mo(III) systems in HCl solutions. The values obtained are given in Fig. 1. The calculated potentails for the Mo(V)/Mo(III) system are also included in this figure. It is seen that if the potentials of the Mo(V)/Mo(III) (green) system are extrapolated to infinite dilution of HCl, the value obtained agrees reasonably well with the standard oxidation-reduction potential of 0.089 V determined by El-Aggan et al. Bergh and Haight [31] titrated Sn(II) with Mo(VI). From the titration curve they obtained the oxidation-reduction potential for the Mo(V)/Mo(III) system in 12 M HCl. They give the following equation:

$$[MoOCl_5]^{2-} + 2e^- + 2H^+ \rightarrow [Mo(OH_2)Cl_5]^{2-} \qquad E^{o\prime} = -0.38 \text{ V}$$

It should be pointed out that $[Mo(OH_2)Cl_5]^{2-}$ is known to exist only in 1 to 2 M HCl. In 12 M HCl $[MoCl_6]^{3-}$ is probably formed. From his studies on the polarography of Mo(VI) in HCl solutions, Haight [25] concluded that Mo(IV) is formed as a reduction product. He gives the following equation with an estimated oxidation-reduction potential:

$$[MoO_2Cl_4]^{2-} + 4H^+ + 2e^- \rightarrow [Mo(OH_2)_2Cl_4] \qquad E^{o\prime} = 0.15 \text{ V}$$

He states that it would be more accurate to give the half-wave potential in 1 M HCl for

$$Mo(VI) + 4H^+ + Cl^- + 2e^- \rightarrow Mo(IV) \qquad E^{o\prime} = 0.164 \text{ V}$$

The potentials would be ca. 0.40 V and ca. 0.70 V in 6 M and 10 M HCl, respectively [32].

Latimer [20] calculated the potential for the Mo/Mo(III) couple from his estimated value of the Gibbs energy of formation for the ion (Table 2):

$$Mo^{3+} + 3e^- \rightarrow Mo \qquad E^o = \text{ca.} -0.2 \text{ V}$$

1. Molybdenum Cyanides

Malik and Ali [33] reported the potential for the molybdo-molybdicyanide couple. For equimolar concentrations in neutral medium,

$$Mo(CN)_8^{3-} + e^- \rightarrow Mo(CN)_8^{4-} \qquad E^o = 0.725 \text{ V}$$

The measured potentials were plotted against the square root of the ionic strength and extrapolated to zero ionic strength. This result agrees quite well with the value of 0.726 V given by Kolthoff and Tomsicek [34] obtained in a similar manner.

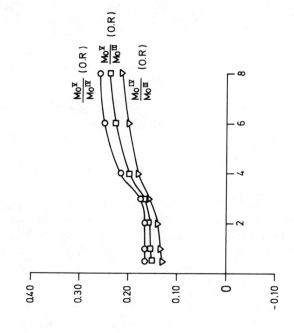
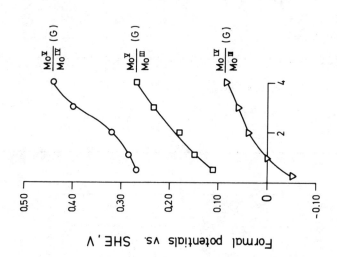

FIG. 1. Formal potentials of Mo(V)/Mo(III), Mo(V)/Mo(IV), and Mo(IV)/Mo(III) in media of varying concentrations of hydrochloric acid, using Mo(III) solutions as starting materials. G, green Mo(III); OR, orange-red Mo(III).

2. Molybdenum Dioxide

Molybdenum dioxide is insoluble in both acid and alkaline solutions [35]. The potentials of the following two reactions given by Pozdeeva et al. [22] and one by Pourbaix [36], respectively, have been recalculated using the thermodynamic data in Table 2. Our results are:

$$MoO_2 + 2H_2O + 4e^- \rightarrow Mo + 4OH^- \qquad E° = -0.980 \text{ V}$$

$$MoO_2 + 4H^+ + 4e^- \rightarrow Mo + 2H_2O \qquad E° = -0.152 \text{ V}$$

$$MoO_2 + 2H^+ + e^- \rightarrow Mo^{3+} + 2H_2O \qquad E° = -0.008 \text{ V}$$

3. Molybdenum Peroxide

From the heat of formation of MoO_4 in aqueous solution and the estimated entropy of Latimer [20] (Table 2), we have recalculated the potential of the following reaction given by Latimer:

$$MoO_4(aq) + 2H^+ + 2e^- \rightarrow H_2MoO_4(aq) \qquad E° = 1.567 \text{ V}$$

The value remains unchanged. This is due to the fact that the changes in the values of Gibbs energy of formation of $MoO_4(aq)$ and $H_2MoO_4(aq)$ cancel out.

D. Potential Diagrams for Molybdenum

Acid Solution

4. Silver Molybdate

Dellien et al. [19] calculated the potential of the Ag/Ag_2MoO_4 couple:

$$Ag_2MoO_4(c) + 2e^- \rightarrow 2Ag + MoO_4^{2-} \qquad E° = 0.455 \text{ V}$$

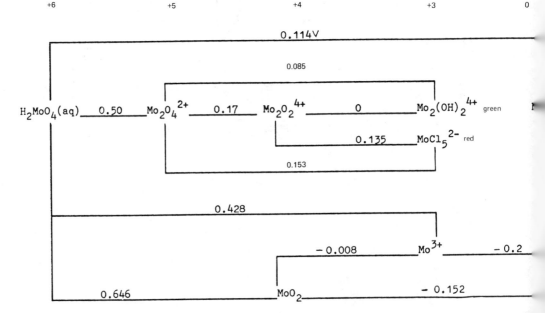

Basic Solution

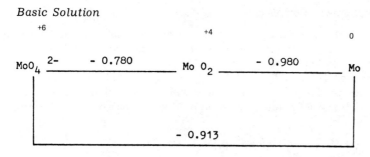

5. Molybdenum Carbide

Sundermann calculated the standard potential for the decomposition of molybdenum carbide [37]. We recalculated the potential using the new thermodynamic values in Table 2:

$$2MoO_3 + C + 12H^+ + 12e^- \rightarrow Mo_2C + 6H_2O \qquad E° = 0.116 \text{ V}$$

6. Molybdosilicate Ion

Launay et al. [38] give formal oxidation-reduction potentials for α-molybdosilicate, $[\alpha\text{-SiMo}_{12}O_{40}]^{4-}$, at pH 11.1 (Fig. 2). These results were obtained by polarography. Linear potential sweep voltammetry was also used to confirm the existence of the products. The Roman numerals in Fig. 2 denote the number of electrons taken up by one molybdosilicate ion during the reduction. (The α-molybdosilicate itself would be designated 0.) The molybdosilicate ion contains Mo(VI) atoms which are reduced to the Mo(V) state. It is seen that a maximum of 12 electrons can be taken up per molybdosilicate ion. The α-molybdosilicate ion can take up 2 to 8 electrons without changing its structure (upper series in Fig. 2). On further reduction, up to a maximum of 12 electrons, a new series having a different structure is formed. The species VI and VIII can exist in both series. The products VI', VIII, and VIII' are unstable. The species within one series are in equilibrium with each other, but not with members of the other series. It should be noted that polarography makes it possible to measure the formal oxidation-reduction potentials for pH values at which one of the components of a couple is stable only for a short period of time. (This would apply to both components if they are both intermediates.) The $E_{1/2}$ (half-wave potential) of the polarograms is taken to be the formal oxidation-reduction potential.

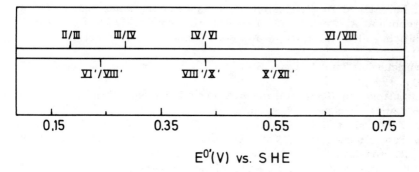

FIG. 2. Formal oxidation-reduction potentials at pH 11.1 for the two series of compounds formed by the reduction of α-molybdosilicate.

7. Molybdenum Complexes

Ott et al. [39] have measured the formal potentials of Mo(V) complexes with EDTA and cysteine using linear potential sweep voltammetry. The potentials given in Table 5 are the averages of the peak potentials for the cathodic and anodic sweeps [i.e., $E^{o'} = (1/2)(E_{pc} + E_{pa})$]. This procedure is valid for potential scan rates $\leqslant 1$ V s^{-1}. The complexes listed in Table 5 correspond to the species on the left side (i.e., the oxidized form) of the following equation:

$$Mo_2O_2X_2(L-L)^{2-} + 4H^+ + 4e^- \rightarrow Mo_2X_2(H_2O)_2(L-L)^{2-}$$

where L — L = EDTA or (cys)$_2$. The X's can both be either O or S, or they can be mixed, that is, one X can be O and one can be S. It is seen that the equation above corresponds to the oxidation-reduction couple Mo(V)$_2$/Mo(III)$_2$. The values of E_{pc} and E_{pa} depend on the scan rate, the bulk solution pH, the nature and concentration of the buffer species, and the nature and concentration of the supporting electrolyte cation.

TABLE 5 Formal Potentials of Mo(V) Complexes[a]

Complex	Acetate buffer 0.2 M HOAc/0.2 M NaOAc pH 4.70, $E^{o'}/V$	Borate buffer 0.1 M Na$_2$B$_4$O$_7$, pH 9.18 $E^{o'}/V$
Mo$_2$O$_4$(EDTA)$^{2-}$	−0.633	−0.815
Mo$_2$O$_3$S(EDTA)$^{2-}$	−0.453	−0.751
Mo$_2$O$_2$S$_2$(EDTA)$^{2-}$	−0.462	−0.741
Mo$_2$O$_4$(cys)$_2^{2-}$		−0.807
Mo$_2$O$_3$S(cys)$_2^{2-}$		−0.823

[a]Potentials vs. SHE; scan rate 0.1 V s^{-1}.

REFERENCES

1. G. P. Haight and D. R. Boston, J. Less-Common Metals, 36, 96 (1974).
2. M. Ardon and A. Pernick, J. Less-Common Metals, 54, 233 (1977).
3. A. Goiffon and B. Spinner, Bull. Soc. Chim. Fr., Nos. 11-12, 1081 (1977).
4. L. Krumenacker and J. Bye, Bull. Soc. Chim. Fr., No. 8, 3099, 3103 (1968).
5. F. A. Cotton and G. Wilkinson, *Advanced Inorganic Chemistry*, 3rd ed., Wiley-Interscience, New York, 1972.
6. G. A. Tsigdinos and G. Moh, *Aspects of Molybdenum and Related Chemistry*, Top. Curr. Chem., Vol. 76, Springer-Verlag, New York, 1978.
7. T. J. R. Weakley, Struct. Bonding, 18, 131 (1974).
8. G A. Tsigdinos and C. J. Hallada, J. Less-Common Metals, 36, 79 (1974).
9. A. Bali and K. C. Malhotra, Aust. J. Chem., 28, 481 (1975).
10. K. Ogura and Y. Enaka, Electrochim. Acta, 22, 833 (1977).

11. A. A. Fedorov and A. V. Pavlinova, Zh. Obshch. Khim., 42, 238 (1972).
12. A. J. Bard, ed., *Encyclopedia of the Electrochemistry of the Elements*, Vol. 5, Marcel Dekker, New York, 1976.
13. F. Foerster and E. Fricke, Z. Phys. Chem., 146, 81 (1930).
14. *Gmelin Handbuch der anorganischen Chemie*, Molybdenum, System No. 53, Part B1, Springer-Verlag, Berlin, 1975.
15. A. K. Babko and T. E. Get'man, Russ. J. Inorg. Chem., 4, 266
16. S. S. Eliseev and N. V. Gaidaenko, Khim. Tadzh., 225 (1973), P. M. Solozhenkin, ed., "Donish" Dushanbe, USSR; Chem. Abstr., 84, 53281h (1976).
17. K. A. Khaldoyanidi and Z. A. Grankina, Izv. Sib. Otd. Akad. Nauk SSSR, Ser. Khim. Nauk (1), 47 (1976); Chem. Abstr., 85, 13221c (1976).
18. S. R. Sagi and P. R. M. Rao, Talanta, 23, 427 (1976).
19. I. Dellien, F. M. Hall, and L. G. Hepler, Chem. Rev., 76, 283 (1976).
20. W. M. Latimer, *The Oxidation States of the Elements and Their Potentials in Aqueous Solutions*, 2nd ed., Prentice-Hall, Englewood Cliffs, N.J., 1952, p. 250.
21. A. T. Vas'ko, V. A. Kosenko, and V. N. Zaichenko, Tr. Ukr. Respub. Konf. Elektrokhim., 1st, 1973, Vol. 1, p. 238, A. V. Gorodyskii, ed., Kiev, USSR: Naukova Dumka, 1973; Chem. Abstr., 81, R 71680e (1974).
22. A. A. Pozdeeva, E. I. Antonovskaya, and A. M. Sukhotin, Zashch. Met. (Engl. transl.), 1, 15 (1965).
23. R. M. Peekema and M. W. Shafer, J. Electrochem. Soc., 123, 1779 (1976).
24. H. Kleykamp and A. Supawan, J. Less-Common Metals, 63, 237 (1979).
25. G. P. Haight, Jr., J. Inorg. Nucl. Chem., 24, 673 (1962).
26. H. K. El-Shamy and A. M. El-Aggan, J. Am. Chem. Soc., 75, 1187 (1953).
27. R. E. Neisman and M. I. Mikhailovskaya, Tr. Voronezh. Univ., 28, 47 (1953); Referat. Zh. Khim. 1956, Abstr. 35625; Chem. Abstr., 52, 16091i (1958).
28. L. A. Karpova, E. F. Speranskaya, and M. T. Kozlovskii, Zh. Anal. Khim. (Engl. transl.), 23, 7 (1968).
29. A. A. Karpova, E. F. Speranskaya, and M. T. Kozlovskii, Elektrokhimiya (Engl. transl.), 6, 1066 (1970).
30. A. M. El-Aggan, S. F. Sidarous, and M. A. El-Azmerlli, U.A.R. J. Chem., 14, 315 (1971).
31. A. A. Bergh and G. P. Haight, Jr., Inorg. Chem., 1, 688 (1962).
32. G. P. Haight, personal communication.
33. W. U. Malik and I. Ali, Indian J. Chem., 1, 374 (1963).
34. I. M. Kolthoff and W. J. Tomsieck, J. Phys. Chem., 40, 247 (1936).
35. H. Remy, *Lehrbuch der anorganischen Chemie*, Vol. II, 12/13 ed., Akademische Verlagsgesellschaft Geest und Portig, Leipzig, 1973, p. 221.
36. M. Pourbaix, ed., *Atlas of Electrochemical Equilibria in Aqueous Solutions*, Pergamon Press, Oxford, 1966, p. 272.
37. H. Sundermann, Mikrochim. Acta, Suppl. II, 30 (1967).
38. J. P. Launay, R. Massart, and P. Souchay, J. Less-Common Metals, 36, 139 (1974).
39. V. R. Ott, D. S. Swieter, and F. A. Schultz, Inorg. Chem., 16, 2538 (1977).

III. TUNGSTEN*

ALEXANDER T. VAS'KO Institute of General and Inorganic Chemistry, Ukrainian SSR Academy of Sciences, Kiev, USSR

A. Oxidation States. Some Peculiarities of the Chemical and Electrochemical Behavior of Tungsten

The chemical properties of tungsten, including its oxidation states, are determined first by the electron configuration of the atom of this metal. Therefore, we remind the reader of it: $[Xe]4f^{14}5d^46s^2$. The bond strength in compounds is determined, together with the energy parameters of the outer valence orbitals, by the extension of these orbitals and their overlapping. The radius values of the maximum electron density for tungsten are high in different oxidation states, giving rise to the ability of this metal to form metal-metal bonds in compounds. Tungsten exhibits a greater tendency to form such bonds than its analog, molybdenum, since the overlap of 5d orbitals is larger than that of 4d orbitals. The number of localized d electrons decreases with rising oxidation state of the metal, as a result of which the tendency to form metal-metal bonds decreases.

It has been reliably established that the following oxidation states are typical of tungsten in compounds: 0, I, II, III, IV, V, and VI. The consequence of such a large number of these states is variation of the redox and acid-base properties over a very wide range. The compounds corresponding to higher tungsten oxidation states exhibit oxidizing properties and those corresponding to lower oxidation states exhibit reducing properties.

Let us cite some compounds with different tungsten oxidation states. The metal with the oxidation state zero is found in tungsten carbonyl. The same oxidation state is also observed in tungsten compounds with some aromatic isonitriles. Oxidation state I is extremely unstable. Nevertheless, we were able to identify this state in the transition region of tungsten oxide near the metal-oxide boundary (i.e., in the solid phase) by x-ray photoelectron emission. Oxidation state II can be obtained only in the form of halides. In the objects of such kind, the ability of tungsten to form clusters of its atoms with metal-metal bond manifests itself most noticeably. For example, in dichloride, the structural unit is the $W_6Cl_8^{4+}$ cluster. The compounds in which the oxidation state of tungsten is II are very unstable. They are readily oxidizable even by water. Tungsten compounds with the oxidation state III are also unstable. This state can be obtained by reducing W(VI) with zinc powder in concentrated hydrochloric acid. As instances of stable compounds with the refractory metal oxidation state IV, we cite alkali hydroxotetracyanotungstates and tungsten dichalcogenides. High oxidation states of tungsten, V and especially VI, are stable. Compounds with such metal ion forms are very numerous and sufficiently well studied. When discussing the oxidation state of tungsten, one cannot but point out the peculiarity of tungsten to form nonstoichiometric compounds (e.g., with oxygen). It will be noted that within one and the same W-O system, one may observe a wide spectrum of electrical properties which are characteristic of metals, semiconductors, and dielectrics. This affects naturally the electrochemical behavior of such systems.

---

*Reviewer: J. Jordan, Pennsylvania State University, University Park, Pennsylvania.

Let us cite ionization potentials for tungsten:

	1st	2nd	3rd	4th	5th	6th
I/eV	7.98	17.7	(24)*	(35)*	(48)*	67.76

All of these potentials are high, and the first and the second are even higher than the corresponding potentials for other refractory metals. Hence ions of the W^{n+} type do not exist in aqueous solutions. In this connection we call attention to the high Gibbs energies of formation for many tungsten compounds with halogens, oxygen, sulfur, and nitrogen (see Section III.B), which indicate a considerable affinity of this metal for corresponding nonmetals. In contrast, tungsten does not form hydrides with hydrogen.

Tungsten trioxide, which is practically insoluble in water, exhibits an acidic nature when it interacts with alkalies. As a result, normal tungstates are formed. They exist both in the anhydrous state and as crystallohydrates. Normal ammonium, lithium, sodium, and potassium tungstates are easily soluble in water. Normal rubidium, cesium, and magnesium tungstates are also soluble. The other salts of this kind are practically insoluble.

At a pH below 7.0, hydrolysis of tungsten compounds takes place, resulting in the formation of polymeric species, paratungstates, the tendency of tungsten to polymerization being most pronounced as compared to other refractory metals. There are two kinds of paratungstate anions: $W_6O_{17}(OH)_7^{5-}$ (A) and $W_{12}O_{36}(OH)_{10}^{10-}$ (B). The former is formed by the acidification of normal-tungstate solution and the latter by the dissolution of crystalline salt. The paratungstate anion B is less reactive than the paratungstate anion A. The salts of paratungstic acid differ from normal tungstates by their lower solubility in water. From pH 4.5 on, the paratungstate anion A reacts with hydrogen ions to form the metatungstate $W_{12}O_{38}(OH)_2^{6-}$, which changes slowly to the ψ-metatungstate anion $W_{24}O_{72}(OH)_{12}^{12-}$. The charge and degree of polymerization of the latter are still to be determined accurately. Whereas the metatungstates are soluble, the ψ-metatungstates are slightly soluble (the exception is sodium salt). Such transformations take place if the concentration of tungstates exceeds 1×10^{-5} M. At lower concentrations, the acidification does not lead to polymerization. It is also important to note that in a strong acid medium (i.e., at an acid concentration over 1 M) cationic tungsten species exist (e.g., HWO_3^+ and WO_2^{2+}). For more detailed information on the state of ions, see Ref. 1.

It is typical of tungsten to form heteropoly acids with phosphoric, silicic, arsenic, boric, selenious, tellurous, germanic, zirconic, and other acids. These compounds are soluble in water, but they precipitate in the presence of NH_4^+, Cs(I), Rb(I), and Tl(I) ions. Mixed heteropolytungstates are also known, which are formed by a substitution or addition reaction. In the former case, tungsten may be substituted by molybdenum or vanadium. In the latter case, a large number of elements, especially Ni, Fe, and Zn, can act as components.

Thus there are many W(VI) compounds, and most of them are insoluble in aqueous solutions. It should be added to this that as the oxidation state of tungsten decreases, species are formed that manifest more and more alkaline nature, as a result of which they are only slightly soluble in alkaline solutions.

---

*Approximate values.

It is the formation of tungsten-medium interaction solid-phase products that may be a reason for the irreversibility of electrode processes involving tungsten in aqueous solutions [2].

B. Normal and Formal Potentials

Thermodynamic properties of tungsten and its compounds are given in Table 6. The choice of data was governed by the estimates contained in fundamental handbooks [3-7] and in the most recent critical review [8]. All the data, apart from those specified in the text, correspond to 298.15 K.

Normal electrode potentials for the processes of obtaining tungsten from its oxygen compounds in acid and alkali media can be evaluated from the thermodynamic data given in Table 6:

$$WO_3 + 6H^+ + 6e^- \rightarrow W + 3H_2O \qquad E^\circ = -0.090 \text{ V}$$

$$WO_4^{2-} + 4H_2O + 6e^- \rightarrow W + 8OH^- \qquad E^\circ = -1.074 \text{ V}$$

As follows from the information above, the tungsten formation potential is close to zero. In an alkaline medium, however, an appreciable negative potential must be attained to obtain the metal. During the electrolysis of tungsten-containing electrolytes, slightly soluble intermediates are formed on the cathode, which possess a high electron conductivity. The removal of oxygen from such compounds is rather hindered since high field strength cannot be maintained in low-resistance films. It should be added that tungsten intermediates act as hydrogen evolution catalysts. It follows that favorable conditions for a competing hydrogen evolution reaction are created on the surface of such compounds. Further reduction of tungsten ions, however, comes practically to an end. That is why elemental tungsten cannot be obtained under normal electrolysis conditions.

At ultrahigh current densities, though, one can obtain thin films of strongly reduced tungsten species, but this seems to be possible owing to the interaction of tungsten ions with electrochemically generated atomic hydrogen [2].

From the data given in Table 6, one can calculate standard potentials for many processes. In this way, E° for the process

$$Ag_2WO_4 + 2e^- \rightarrow 2Ag + WO_4^{2-} \qquad E^\circ = 0.457 \text{ V}$$

was calculated using thermodynamic data. However, the electrochemical measurements by Pan [9] gave the value $E^{\circ\prime} = 0.466$ V. In this connection it is worth stressing that it is rather difficult to realize in practice systems with tungsten and its compounds, which would behave reversibly. Therefore, the experimentally determined formal potential values should be regarded to a greater or lesser extent as tentative. The determined potential does not often correspond to the nominally symbolized system.

Geyer and Henze [10] determined formal potentials in 12.5 M HCl for the systems W(VI)/W(V) and W(V)/W(III):

$$WO_2Cl_3^- + e^- + 2H^+ + 2Cl^- \rightarrow WOCl_5^{2-} + H_2O \qquad E^{\circ\prime} = 0.36 \text{ V}$$

TABLE 6 Thermodynamic Data for Tungsten[a]

Formula	Description	State	ΔH_f°/kJ mol^{-1}	ΔG_f°/kJ mol^{-1}	S°/J mol^{-1} K^{-1}
W	Tungsten	c	0	0	32.7
W	Tungsten	g	(856.883)	(814.788)	173.84
W$^+$	Tungsten(I) cation	g	(1,632.92)	(1,576.63)	179.62
W$_3$O	Tritungsten monoxide	c	−316	(−281.48)	84
WO	Tungsten monoxide	g	(427)	(393)	(247)
WO$_2$	Tungsten dioxide	c	−589.5	(−533.728)	50.54
WO$_2$	Tungsten dioxide	g	(63)	(50)	(284)
WO$_2^{2+}$	Dioxotungsten(2+) ion	s		−502.1	
WO$_{2.72}$	Tungsten 2.72-oxide	c	−775.7	(−703.297)	68.6
WO$_{2.90}$	Tungsten 2.90-oxide	c	−818.4	(−741.844)	73.2
WO$_{2.96}$	Tungsten 2.96-oxide	c	−833.0	(−755.153)	74.9
WO$_3$	Tungsten trioxide	c	−842.7	(−763.865)	75.94
WO$_3$	Tungsten trioxide	g	−300	(−283.09)	285
WO$_4^{2-}$	Tetraoxotungstate(VI) ion	s	−1,073	−931.4	(97.52)

TABLE 6 (Continued)

Formula	Description	State	ΔH_f°/kJ mol^{-1}	ΔG_f°/kJ mol^{-1}	S°/J mol^{-1} K^{-1}
$(WO_3)_2$	Dimeric tungsten trioxide	g	$-1,167$	$(-1,088.34)$	416
W_3O_8	Tritungsten octaoxide	g	$(-1,711)$	$(-1,586)$	(494)
$(WO_3)_3$	Trimeric tungsten trioxide	g	$-2,025$	$(-1,871.20)$	504.6
$(WO_3)_4$	Tetrameric tungsten trioxide	g	$-2,803$	$(-2,590.48)$	605.0
WO_2OH^+	Dioxotungsten monohydroxide(1+) ion	s		-735.5	
HWO_4^-	Hydrogentetraoxotungstate(VI) ion	s		-952.7	
H_2WO_4	Tetraoxotungstic(VI) acid	c	$-1,130$	$-1,000$	(138)
H_2WO_4	Tetraoxotungstic(VI) acid	s	$-1,073$	-931.4	97.525
H_2WO_4	Tetraoxotungstic(VI) acid	g	-905.8		
$WO_2(OH)_2 \cdot 1.07H_2O$	Dioxotungsten dihydroxide 1.07-water	c	$-1,450$		
$HW_6O_{21}^{5-}$	Monohydrogenhexatungstate(5-) ion	s	$-5,843.4$	$-5,132.9$	360
$H_{0.18}WO_3$	0.18-Hydrogen tungsten trioxide (bronze)	c	-847.7		

Chromium, Molybdenum, and Tungsten

Formula	Name	state		
$H_{0.35}WO_3$	0.35-Hydrogen tungsten trioxide (bronze)	c	−852.3	
$Na_{0.53}WO_3$	0.53-Sodium tungsten trioxide (bronze)	c	−972.8	
$Na_{0.679}WO_3$	0.679-Sodium tungsten trioxide (bronze)	c	−1,007	−924.7	95.27
$(NH_4)_{0.25}WO_3$	0.25-Ammonium tungsten trioxide (bronze)	c	−938.9	
Li_2WO_4	Dilithium tetraoxotungstate	c	−1,603	
Li_2WO_4	Dilithium tetraoxotungstate	g	(−1,004)	
K_2WO_4	Dipotassium tetraoxotungstate	c	−1,581	
K_2WO_4	Dipotassium tetraoxotungstate	g	(−1,163)	
$KHWO_4$	Potassium hydrogentetraoxotungstate	g	(−1,075)	
Na_2WO_4	Disodium tetraoxotungstate	c	−1,547	−1,432	161
$Na_2W_2O_7$	Disodium ditungstate(2−)	c	−2,401	−2,213	254
$Na_2W_4O_{13}$	Disodium tetratungstate(2−)	c	−4,146	
$Na_6H_2W_{12}O_{40}$	Hexasodium dihydrogendodeca-tungstate(6−)	s	−13,054	
$Na_{10}H_2W_{12}O_{42}$	Decasodium hydrogendodeca-tungstate(10−)	s	−14,652	

TABLE 6 (Continued)

Formula	Description	State	ΔH_f°/kJ mol^{-1}	ΔG_f°/kJ mol^{-1}	S°/J mol^{-1} K^{-1}
BeWO$_4$	Beryllium tetraoxotungstate	c	(−1,513)	(−1,404.86)	(88.7)
MgWO$_4$	Magnesium tetraoxotungstate	c	−1,518	−1,406	101.2
MgWO$_4$	Magnesium tetraoxotungstate	g	(−900)		
CaWO$_4$	Calcium tetraoxotungstate	c	−1,641	−1,534	126.4
CaWO$_4$	Calcium tetraoxotungstate	g	(−946)		
SrWO$_4$	Strontium tetraoxotungstate	c	(−1,636)	−1,532	(138)
SrWO$_4$	Strontium tetraoxotungstate	g	(−1,046)		
BaWO$_4$	Barium tetraoxotungstate	c	−1,703		
3Sc$_2$O$_3$·WO$_3$	Scandium sesquioxide-tungsten trioxide (3/1)	c	−6,577.2	(−6,243.310)	351
Sc$_2$(WO$_4$)$_3$	Scandium(III) tetraoxotungstate	c	−4,657	(−4,316.156)	255
Sc$_2$(WO$_4$)$_3$	Scandium(III) tetraoxotungstate	c	(−4,448.51)	(−3,962)	234.61
Y$_2$O$_3$·WO$_3$	Yttrium sesquioxide-tungsten trioxide (1/1)	c	−2,929	(−2,755.33)	155
3Y$_2$O$_3$·WO$_3$	Yttrium sesquioxide-tungsten trioxide (3/1)	c	−6,694.0	(−6,352.111)	383

$Y_2(WO_4)_3$	Yttrium(III) tetraoxotungstate	c	−4,740	(−4,400.308)	383
$Y_2(WO_4)_3$	Yttrium(III) tetraoxotungstate	s	−4,666.4	(−4,171.473)	242.98
$La_2O_3 \cdot WO_3$	Lanthanum sesquioxide-tungsten trioxide (1/1)	c	−2,807	(−2,637.79)	192
$La_2O_3 \cdot 2WO_3$	Lanthanum sesquioxide-tungsten trioxide (1/2)	c	−3,707	(−3,454.62)	255
$3La_2O_3 \cdot WO_3$	Lanthanum sesquioxide-tungsten trioxide (3/1)	c	−6,412.0	(−6,074.373)	472.0
$La_2(WO_4)_3$	Lanthanum(III) tetraoxotungstate	c	−4,540	(−4,212.00)	343
$La_2(WO_4)_3$	Lanthanum(III) tetraoxotungstate	s	−4,633.8	(−4,159.85)	−147.58
$Ce(WO_4)_2$	Cerium(II) tetraoxotungstate	s	(−2,685)	(−2,369)	(99.692)
$Ce_2(WO_4)_3$	Cerium(III) tetraoxotungstate	c	−4,694		
$Ce_2(WO_4)_3$	Cerium(III) tetraoxotungstate	s	(−4,621.2)	(−4,144.80)	(126.66)
$3Nd_2O_3 \cdot WO_3$	Neodymium sesquioxide-tungsten trioxide (3/1)	c	−6,466.0	(−6,122.736)	542.2
$7Nd_2O_3 \cdot 4WO_3$	Neodymium sesquioxide-tungsten trioxide (7/4)	c	−16,737	(−15,819.25)	1,439
$Nd_2O_3 \cdot WO_3$	Neodymium sesquioxide-tungsten trioxide (1/1)	c	−2,831	(−2,660.55)	221

TABLE 6 (Continued)

Formula	Description	State	ΔH_f°/kJ mol^{-1}	ΔG_f°/kJ mol^{-1}	S°/J mol^{-1} K^{-1}
$Nd_2O_3 \cdot 2WO_3$	Neodymium sesquioxide-tungsten trioxide (1/2)	c	$-2,831$	$(-2,660.55)$	221
$Nd_2(WO_4)_3$	Neodymium(III) tetraoxotungstate	c	$-4,580.6$	$(-4,259.116)$	393
$Nd_2(WO_4)_3$	Neodymium(III) tetraoxotungstate	s	$(-4,610.35)$	$(-4,134.29)$	(124.99)
$Sm_2O_3 \cdot WO_3$	Samarium sesquioxide-tungsten trioxide (1/1)	c	$-2,825$	$(-2,660.85)$	238
$3Sm_2O_3 \cdot WO_3$	Samarium sesquioxide-tungsten trioxide (3/1)	c	$-6,508.6$	$-6,159.3$	(508.590)
$7Sm_2O_3 \cdot 4WO_3$	Samarium sesquioxide-tungsten trioxide (7/4)	c	$-16,863$	$(-15,947.40)$	1,416
$Sm_2(WO_4)_3$	Samarium(III) tetraoxotungstate	c	$-4,577$		
$Sm_2(WO_4)_3$	Samarium(III) tetraoxotungstate	c	$(-4,598.6)$	$(-4,117.6)$	(-145.91)
$3Eu_2O_3 \cdot WO_3$	Europium sesquioxide-tungsten trioxide (3/1)	c	$-5,977.7$	$-5,613.3$	(507.528)
$Eu_2(WO_4)_3$	Europium(III) tetraoxotungstate	s	$(-4,437.1)$	$(-3,949.72)$	(150.92)
$Gd_2O_3 \cdot WO_3$	Gadolinium sesquioxide-tungsten trioxide (1/1)	c	$-2,824$	$(-2,657.92)$	226

$Gd_2O_3 \cdot 2WO_3$	Gadolinium sesquioxide-tungsten trioxide (1/2)	c	−3,795	(−3,567.76)	361
$3Gd_2O_3 \cdot WO_3$	Gadolinium sesquioxide-tungsten trioxide (3/1)	c	−6,410.3	(−6,081.883)	569.0
$Gd_2(WO_4)_3$	Gadolinium(III) tetraoxotungstate	c	−4,607	(−4,284.834)	385
$Gd_2(WO_4)_3$	Gadolinium(III) tetraoxotungstate	s	(−4,587.3)	(−4,112.88)	(127.25)
$Dy_2O_3 \cdot WO_3$	Dysprosium sesquioxide-tungsten trioxide (1/1)	c	−2,866	(−2,692.74)	218
$3Dy_2O_3 \cdot WO_3$	Dysprosium sesquioxide-tungsten trioxide (3/1)	c	−6,543.0	(−6,191.236)	536.4
$7Dy_2O_3 \cdot 4WO_3$	Dysprosium sesquioxide-tungsten trioxide (7/4)	c	−16,945	(−15,892.6)	1,423
$Dy_2(WO_4)_3$	Dysprosium(III) tetraoxotungstate	c	−4,636	(−4,310.855)	389
$Ti(WO_4)_2$	Titanium(IV) bis(tetraoxotungstate)	c			245
ZrW_2O_8	Zirconium(IV) bis(tetraoxotungstate)	c			253
$Cr_2(WO_4)_3$	Chromium(III) tetraoxotungstate	s	−3,691	(−3,240.19)	−138.38
$MoO_4HWO_4^{3-}$	Tetraoxomolybdate hydrogentetraoxotungstate(3−) ion	s		−1,761	
$MnWO_4$	Manganese(II) tetraoxotungstate	c	−1,305	(−1,201.8)	133

TABLE 6 (Continued)

Formula	Description	State	ΔH_f°/kJ mol^{-1}	ΔG_f°/kJ mol^{-1}	S°/J mol^{-1} K^{-1}
FeWO$_4$	Iron(II) tetraoxotungstate	c	−1,184	(−1,084.05)	131.6
FeWO$_4$	Iton(II) tetraoxotungstate	s	(−1,160.3)	(−1,010.3)	(33.35)
Fe$_2$(WO$_4$)$_3$	Iron(III) tetraoxotungstate	s	(−3,312.3)	(−2,803.1)	(325.34)
CoWO$_4$	Cobalt(II) tetraoxotungstate	c	−1,138	(−1,036.59)	133
NiWO$_4$	Nickel(II) tetraoxotungstate	c	−1,159	(−1,035)	137
CuWO$_4$	Copper(II) tetraoxotungstate	c	−1,025	(−921.869)	129
CuWO$_4$	Copper(II) tetraoxotungstate	s	(−1,006)	(−865.80)	(4.786)
Ag$_2$WO$_4$	Silver(I) tetraoxotungstate	c	(−938.078)	−843.1	209
Ag$_2$WO$_4$	Silver(I) tetraoxotungstate	s	(−862.096)	(−777.103)	(242.77)
ZnWO$_4$	Zinc(II) tetraoxotungstate	c	(−1,228)	(−1,198.41)	130
ZnWO$_4$	Zinc(II) tetraoxotungstate	s	(−1,226.8)	(−1,078.52)	(−13.08)
CdWO$_4$	Cadmium(II) tetraoxotungstate	c	−1,177	(−1,071.9)	142.3
CdWO$_4$	Cadmium(II) tetraoxotungstate	s	(−1,148)	(−1,009.0)	(26.62)
HgWO$_4$	Mercury(II) tetraoxotungstate	c	−937	(−828.742)	154

Chromium, Molybdenum, and Tungsten

HgWO$_4$	Mercury(II) tetraoxotungstate	s	(−899.73)	(−766.68)	(72.371)
Hg$_2$WO$_4$	Mercury(I) tetraoxotungstate	c	−958	(−846.992)	222
Tl$_2$WO$_4$	Thallium(I) tetraoxotungstate	c	−1,159	(−1,059.80)	238
SnWO$_4$	Tin(II) tetraoxotungstate	c	−1,163	(−1,060.69)	151
SnWO$_4$	Tin(II) tetraoxotungstate	g	(−1,720)		
Sn$_2$WO$_5$	Ditin pentaoxotungstate	g	(−1,975)		
PbWO$_4$	Lead(II) tetraoxotungstate	c	−1,122	(−1,020.56)	168
PbWO$_4$	Lead(II) tetraoxotungstate	s	(−1,074.1)	(−955.75)	(110.54)
WF	Tungsten monofluoride	g	(388.69)	(353.93)	(251)
WF$_2$	Tungsten difluoride	g	(−105)		
WF$_3$	Tungsten trifluoride	g	(−531)		
WF$_4$	Tungsten tetrafluoride	g	−1,029		
WF$_5$	Tungsten pentafluoride	g	−1,447	(−1,329.65)	148
WF$_6$	Tungsten hexafluoride	ℓ	(−1,747.3)	(−1,635.12)	(264.61)
WF$_6$	Tungsten hexafluoride	g	−1,721.5	(−1,635.87)	354
WFO$_3^-$	Trioxofluorotungstate(1−) ion	s	−1,050		
WF$_2$O$_2$	Tungsten difluoride dioxide	g	−925		

TABLE 6 (Continued)

Formula	Description	State	ΔH_f°/kJ mol^{-1}	ΔG_f°/kJ mol^{-1}	S°/J mol^{-1} K^{-1}
WF_2O_2	Tungsten difluoride dioxide	s		(−1,057.5)	
WF_4O	Tungsten tetrafluoride monoxide	c	−1,394.4	(−1,298.13)	175
WF_4O	Tungsten tetrafluoride monoxide	ℓ	−1,405.2	(−1,297.36)	179
WF_4O	Tungsten tetrafluoride monoxide	g	−1,336.6	(−1,275.20)	334.6
WCl	Tungsten monochloride	g	(556.673)	(521.561)	262
WCl_2	Tungsten dichloride	c	−255	(−216.43)	126
WCl_2	Tungsten dichloride	g			309
WCl_4	Tungsten tetrachloride	c	−448	(−362.37)	192
WCl_4	Tungsten tetrachloride	g	−341	(−310.90)	379
WCl_5	Tungsten pentachloride	c	(−517.6)	(−410.23)	230
WCl_5	Tungsten pentachloride	ℓ	(−495.4)	(−393.55)	(248.50)
WCl_5	Tungsten pentachloride	g	−417	(−361.71)	404
$(WCl_5)_2$	Dimeric tungsten pentachloride	g	−862	−724	(711)
WCl_6	Tungsten hexachloride	c	−598.3	(−468.972)	268

WCl₆	Tungsten hexachloride	ℓ	(−569.400)	(−445.173)	(284.87)
WCl₆	Tungsten hexachloride	g	(−498.3)	(−413.6)	418
W₂Cl₁₀	Ditungsten decachloride	g	−977.8		
WCl₂O	Tungsten dichloride monoxide	c	615		(134)
WCl₂O₂	Tungsten dichloride dioxide	c	−791		(201)
WCl₂O₂	Tungsten dichloride dioxide	g	−678		(356)
WCl₂O₂	Tungsten dichloride dioxide	s	(−764.609)		
WCl₃O	Tungsten trichloride monoxide	c	−694.1	−607	(180)
WCl₄O	Tungsten tetrachloride monoxide	c	−721.3	−598	(172)
WCl₄O	Tungsten tetrachloride monoxide	ℓ	(−630.575)	(−532.916)	(254)
WCl₄O	Tungsten tetrachloride monoxide	g	−637.6	−577	385
WBr	Tungsten monobromide	g			272
WBr₅	Tungsten pentabromide	c	−313	(−271.27)	272
WBr₅	Tungsten pentabromide	ℓ	(−298.05)	(−262.31)	(293.35)
WBr₅	Tungsten pentabromide	g	(−223)	(−237.46)	460.2
WBr₆	Tungsten hexabromide	c	−345	(−292.44)	314
WBr₆	Tungsten hexabromide	g	−245	(−243.96)	486.6

TABLE 6 (Continued)

Formula	Description	State	ΔH_f°/kJ mol^{-1}	ΔG_f°/kJ mol^{-1}	S°/J mol^{-1} K^{-1}
WBr$_2$O	Tungsten dibromide monoxide	c	−557.3		
WBr$_2$O$_2$	Tungsten dibromide dioxide	c	−710.4		
WBr$_2$O$_2$	Tungsten dibromide dioxide	g	−556.5	−548	(360)
WBr$_2$O$_2$	Tungsten dibromide dioxide	s		(−710.267)	
WBr$_3$O	Tungsten tribromide monoxide	c	−527	−477	(192)
WBr$_4$O	Tungsten tetrabromide monoxide	c	−551.5	−494	(230)
WBr$_4$O	Tungsten tetrabromide monoxide	g	−225.9	−431	(439)
WI$_2$	Tungsten diiodide	c	(−42)		
WI$_3$	Tungsten triiodide	c	(−42)		
WIO$_2$	Tungsten monoiodide dioxide	c	−557.7		
WI$_2$O$_2$	Tungsten diiodide dioxide	c	(−613.8)		
WI$_2$O$_2$	Tungsten diiodide dioxide	g	−428.0	−435	(372)
WI$_2$O$_2$	Tungsten diiodide dioxide	s		(−605.7)	
WClF$_5$	Tungsten monoiodide pentafluoride	ℓ	−1,632		

WCl$_2$F$_4$	Tungsten dichloride tetrafluoride	ℓ	−1,473
LiWF$_6$	Lithium hexafluorotungstate(V)	c	−2,219
NaWF$_6$	Sodium hexafluorotungstate(V)	c	−2,211
KWF$_6$	Potassium hexafluorotungstate(V)	c	−2,224
RbWF$_6$	Rubidium hexafluorotungstate(V)	c	−2,236
CsWF$_6$	Cesium hexafluorotungstate(V)	c	−2,253
KWCl$_6$	Potassium hexachlorotungstate(V)	c	−996
K$_2$WCl$_7$	Dipotassium heptachlorotungstate(V)	c	−1,431
K$_3$W$_2$Cl$_9$	Tripotassium nonachloroditung-state(III)	c	(−1,218)
Rb$_3$W$_2$Cl$_9$	Trirubidium nonachloroditung-state(III)	c	(−2,423)
Cs$_3$W$_2$Cl$_9$	Tricesium nonachloroditung-state(III)	c	(−2,418)
WS$_2$	Tungsten disulfide	c	(−200)	(−192.85)	71.1
WO$_2$SO$_4$	Dioxotungsten tetraoxosulfate	s		(−1,257.73)	
WTe$_2$	Tungsten ditelluride	c	−38	(−24.77)	87.0
W$_2$N	Tungsten seminitride	c	−142	(−125.42)	105

TABLE 6 (Continued)

Formula	Description	State	ΔH_f°/kJ mol^{-1}	ΔG_f°/kJ mol^{-1}	S°/J mol^{-1} K^{-1}
$WO_2(NO_3)_2$	Dioxotungsten bis(trioxonitrate)	s		(−725.25)	
WB	Tungsten monoboride	c, α			55.2
W_2B	Tungsten semiboride	c	−87.0	(−100.99)	118
WC	Tungsten monocarbide	c, α	−40	(−39.49)	34
W_2C	Tungsten semicarbide	c, β	−26	(−29.50)	81.6
$W(CO)^+$	Monocarbonyltungsten(1+) ion	g	1,464		
$W(CO)_2^+$	Dicarbonyltungsten(1+) ion	g	1,105		
$W(CO)_3^+$	Tricarbonyltungsten(1+) ion	g	770		
$W(CO)_4^+$	Tetracarbonyltungsten(1+) ion	g	490		
$W(CO)_5^+$	Pentacarbonyltungsten(1+) ion	g	126		
$W(CO)_6$	Hexacarbonyltungsten	c	(−951.9)	(−847.507)	332
$W(CO)_6$	Hexacarbonyltungsten	g	(−875.04)	(−821.001)	501.2

$(WO_2)C_2O_4$	Dioxotungsten oxalate	s	$(-1,170.96)$
$H_2[WO_3C_2O_4]$	Oxalotungstic acid	s	$-1,442$
$WF_5(CH_3O)$	Pentafluoro(metoxo)tungsten	c	$-1,741$
$cis\text{-}WF_4(CH_3O)_2$	Cis-tetrafluorobis(metoxo)tungsten	liq	$-1,686$
$cis\text{-}WF_2(CH_3O)_4$	Cis-difluorotetrakis(metoxo)-tungsten	c	$-1,519$
$W(CH_3CN)(CO)_5$	(Acetonitrile)pentacarbonyl-tungsten	c	-750.6
$cis\text{-}W(CH_3CN)_2(CO)_4$	Cis-bis(acetonitrile)tetracarbonyl-tungsten	c	-555.2
$cis\text{-}W(CH_3CN)_3(CO)_3$	Cis-tris(acetonitrile)tricarbonyl-tungsten	c	-359
$W(CO)_3(C_5H_5N)$	Tricarbonyltris(pyridine)tungsten	c	-250
WSi_2	Tungsten disilicide	c	-92	-88	63
W_5Si_3	Pentatungsten trisilicide	c	-134	-138	230

[a]The values that agree with the values in the most recent handbooks [4,5,7] are in italic; estimated and approximate values are given in parentheses.

$$2WOCl_5^{2-} + 4e^- + 4H^+ \rightarrow W_2Cl_9^{3-} + 2H_2O + Cl^- \qquad E^{o\prime} = -.05 \text{ V}$$

It should be borne in mind that as was pointed out by Lingane and Small [11], three W(III) species—red (WCl_5^{2-}), colorless, and green ($W_2Cl_9^{3-}$)— may be formed in a hydrochloric acid medium. The formal potentials ($E^{o\prime}$) of the W(V)/W(III) system are, according to their data, approximately -0.2, -0.2, and $+0.1$ V, respectively.

Unlike the compounds of trivalent tungsten, it is peculiar to divalent tungsten halides that they, as has been pointed out, decompose water with hydrogen evolution. Therefore, $E^o_{W(III)/W(II)}$ is assumed to be still more negative than $E^o_{W(V)/W(III)}$ in the case of the most active red modification of trivalent tungsten ions.

According to the measurements of Boadsgaard and Treadwell [12]:

$$W^V(CN)_8^{3-} + e \rightarrow W^{IV}(CN)_8^{4-} \qquad E^{o\prime} = 0.457 \text{ V}$$

The redox properties of tetrahydroxotetracyanotungstate have been studied by Mikhalevich and Litvinchuk [13,14]. They [14] have determined the potential for the process (pH 13.7, 20°C)

$$W^V(OH)_4(CN)_4^{3-} + e^- \rightarrow W^{IV}(OH)_4(CN)_4^{4-} \qquad E^{o\prime} = -0.74 \text{ V}$$

The redox potential of this system depends not only on the oxidized species– reduced species ratio but also on the pH of the solution. The tetravalent tungsten ions under consideration possess pronounced reducing properties. Thus, using potassium tetrahydroxotetracyanotungstate, it is possible to reduce to metal the following cations: Cu(I), Ag(I), Cd(II), Hg(I), Pb(II), Sn(II), Bi(III), and Pd(II) [14].

Kabir-Ud-Din et al. [15] report the determination of $E^{o\prime}$ for the half-reaction

$$W^{VI}(OH)_4(CN)_4^{2-} + 2e^- \rightarrow W^{IV}(OH)_4(CN)_4^{4-} \qquad E^{o\prime} = -0.702$$

It is typical of tungsten to tend toward the formation of iso and heteropoly compounds. Their electrochemical behavior is dealt with in a number of papers. Let us cite experimentally determined redox potentials for the processes which may be conditionally considered to be reversible.

For the protonation of the $CoW_{12}O_{40}^{6-}$ anion, Pope and Varga [16] determined $E^{o\prime}$ in 1 M H_2SO_4 at 30°C:

$$CoW_{12}O_{40}^{6-} + 2H^+ + 2e^- \rightarrow H_2CoW_{12}O_{40}^{6-} \qquad E^{o\prime} = -0.046 \text{ V}$$

Lappina and Peacock [17] determined the following formal potentials in an acetate buffer solution (pH 4.50):

$$Cu^{II}W_{12}O_{40}^{6-} + e^- \rightarrow Cu^{I}W_{12}O_{40}^{7-} \qquad E^{o\prime} = -0.02 \text{ V}$$

In the copper(II) complex, which is obtained by this reaction, an intervalence charge transfer $Cu^I-W^{VI} \to Cu^{II}-W^V$ may take place to a certain degree. A similar charge transfer may take place in other polytungstates as well.

Shakhova and Motorkina [18] had shown long ago that the formal potential for the reduction of tungsten ion, if it forms part of germanotungsten heteropolyanion, is less positive than with tungstate anion. However, the addition of vanadium to germanotungstate leads to a considerable potential shift to more positive values. This is due to the fact that it is vanadium ion that is reduced but not tungsten ion as is the case in the vent of germanotungstate anion reduction.

When studying complex heteropolyanions, Weakley [19] determined the redox potential for the process

$$BCo^{III}H_2W_{11}O_{40}^{6-} + e^- \to BCo^{II}H_2W_{11}O_{40}^{7-} \qquad E^{o'} = 0.57 \text{ V}$$

Here cobalt ion is reduced.

The formal potentials for the processes involving 12-phosphorotungstate and 12-silicotungstate in 1 M H_2SO_4 at 30°C are taken from the paper by Pope and Varga [16]:

$$PW_{12}O_{40}^{3-} + e^- \to PW_{12}O_{40}^{4-} \qquad E^{o'} = 0.218 \text{ V}$$

$$PW_{12}O_{40}^{4-} + e^- \to PW_{12}O_{40}^{5-} \qquad E^{o'} = -0.039 \text{ V}$$

$$SiW_{12}O_{40}^{4-} + e^- \to SiW_{12}O_{40}^{5-} \qquad E^{o'} = 0.013 \text{ V}$$

$$SiW_{12}O_{40}^{5-} + e^- \to SiW_{12}O_{40}^{6-} \qquad E^{o'} = -0.192 \text{ V}$$

As can be seen from the data above, the potentials shift to more negative values with increasing heteropolytungstate charge.

According to the measurements by Pope and Papaconstantinou [20], the formal potentials for the transformations of heteropolymeric blues in 0.5 M H_2SO_4 at 30°C are*:

$$A-P_2W_{18}O_{62}^{6-} + e^- \to A-P_2W_{18}O_{62}^{7-} \qquad E^{o'} = 0.300 \text{ V}$$

$$A-P_2W_{18}O_{62}^{7-} + e^- \to A-P_2W_{18}O_{62}^{8-} \qquad E^{o'} = 0.160 \text{ V}$$

$$A-P_2W_{18}O_{62}^{8-} + 2e^- \to A-P_2W_{18}O_{62}^{10-} \qquad E^{o'} = -0.124 \text{ V}$$

$$B-P_2W_{18}O_{62}^{6-} + e^- \to B-P_2W_{18}O_{62}^{7-} \qquad E^{o'} = 0.286 \text{ V}$$

$$B-P_2W_{18}O_{62}^{7-} + e^- \to B-P_2W_{18}O_{62}^{8-} \qquad E^{o'} = 0.102 \text{ V}$$

---

*A is an isomer of $P_2W_{18}O_{62}^{6-}$; B is an isomer of $P_2W_{18}O_{62}^{6-}$.

$$B\text{-}P_2W_{18}O_{62}^{8-} + 2e \rightarrow B\text{-}P_2W_{18}O_{62}^{10-} \qquad E^{o\prime} = -0.159 \text{ V}$$

$$As_2W_{18}O_{62}^{6-} + e^- \rightarrow As_2W_{18}W_{62}^{7-} \qquad E^{o\prime} = 0.309 \text{ V}$$

$$As_2W_{18}O_{62}^{7-} + e^- \rightarrow As_2W_{18}O_{62}^{8-} \qquad E^{o\prime} = 0.147 \text{ V}$$

$$As_2W_{18}O_{62}^{8-} + 2e^- \rightarrow As_2W_{18}O_{62}^{10-} \qquad E^{o\prime} = -0.099 \text{ V}$$

From the comparison of data for less and more polymerized species, one may conclude that the potentials are more positive in the latter case.

The fact that the reversible $[NH_4]_6[P_2W_{18}O_{62}]$ reduction process, involving no decomposition, is a multistep one with attachment of one to six electrons, and the knowledge of formal potentials which correspond to single steps and increase as electrons are attached, allowed us to use the heteropoly compound above as an electron acceptor in determining approximate redox potentials of short-lived (lifetime of the order of microseconds) radicals—derivatives of carboxylic acids and glycol [21].

Flynn et al. [22] established reversible one-electron transfer in sulfuric acid and acetic acid media in the following processes:

$$H_2W_{11}V^VO_{40}^{7-} + e^- \rightarrow H_2W_{11}V^{IV}O_{40}^{8-} \text{ (pH 4)} \qquad E^{o\prime} = 0.50 \text{ V}$$

$$V^VW_5O_{19}^{3-} + e^- \rightarrow V^{IV}W_5O_{19}^{4-} \text{ (pH 2-4)} \qquad E^{o\prime} = 0.44 \text{ V}$$

Novosyolova and Barkovskii [23] determined the formal potential for a process involving vanadophosphate containing tungsten in a phosphoric acid-sulfuric acid medium:

$$PV^VW_{11}O_{40}^{6-} + e^- \rightarrow PV^{IV}W_{11}O_{40}^{5-} \qquad E^{o\prime} = 0.764 \text{ V}$$

Acerete et al. [24] determined formal potentials for α_1- and α_2-polytungstate species, containing vanadium (pH 4.7), from cyclic voltamograms:

$$\alpha_1\text{-}P_2W_{17}V^VO_{62}^{7-} + e^- \rightarrow \alpha_1\text{-}P_2W_{17}V^{IV}O_{62}^{8-} \qquad E^{o\prime} = 0.78 \text{ V}$$

$$\alpha_2\text{-}P_2W_{17}V^VO_{62}^{7-} + e^- \rightarrow \alpha_2\text{-}P_2W_{17}V^{IV}O_{62}^{8-} \qquad E^{o\prime} = 0.69 \text{ V}$$

The reader has surely already noticed that the formal potentials given relate to the systems where a rather slight decrease in tungsten oxidation state is attained as a result of a cathodic process, or where not a tungsten ion but some other metal (e.g., Co, V, and the like) is reduced in a complex ion. This is due to the fact that only these compounds behave reversibly or quasi-reversibly. In the case of strong reduction of tungsten ions, however, it does not appear possible to realize equilibrium conditions for reasons given in Section III.A Therefore, there are no corresponding reliably determined formal potentials in the literature.

C. Potential Diagrams for Tungsten

Acid Solution

$$W \xrightarrow{-0.119} WO_2 \xrightarrow{-0.031} W_2O_5 \xrightarrow{-0.029} WO_3$$
$$\underset{-0.090}{\underline{\hspace{5cm}}}$$

Basic Solution

$$W \xrightarrow{-0.982} WO_2 \xrightarrow{-1.259} WO_4^{2-}$$
$$\underset{-1.074}{\underline{\hspace{4cm}}}$$

REFERENCES

1. M. V. Mokhosoyev and N. A. Shevtsova, *Sostoyaniye ionov molibdena i volframa v vodnykh rastvorakh*, Buryatsk. Kn. Izdatelstvo, Ulan-Ude, 1977.
2. A. T. Vas'ko and S. K. Kovach, *Elektrokhimiya tugoplavkikh metallov*, Tekhnika, Kiev, 1983.
3. D. D. Wagmann, W. H. Evans, V. B. Parker, J. Halow, S. M. Bailey, and R. H. Schumm, *Selected Values of Chemical Thermodynamic Properties*, Natl. Bur. Stand. Tech. Note 270-4, U.S. Government Printing Office, Washington, D.C., 1969.
4. V. B. Parker, D. D. Wagmann, and W. H. Evans, *Selected Values of Chemical Thermodynamic Properties*, Natl. Bur. Stand. Tech. Note 270-6, U.S. Government Printing Office, Washington, D.C., 1971.
5. V. P. Glushko, ed., *Termicheskiye konstanty veshchestv (Handbook)*, AN SSSR, Moscow, 1974, Vol. VII, Parts I, II; 1978, Vol. VIII, Parts I, II.
6. *JANAF Thermochemical Tables*, 2nd ed., NSPDS-NBS 37, Cat. C13.48: 37, U.S. Government Printing Office, Washington, D.C., 1971.
7. M. W. Chase, J. L. Curnutt, A. T. Hu, H. Prophet, A. N. Syverud, and L. C. Walker, *JANAF Thermochemical Tables*, 1974 suppl., J. Phys. Chem. Ref. Data, 3, 311 (1974).
8. I. Dellien, F. M. Hall, and L. Hepler, Chem. Rev., 76(3), 283 (1976).
9. K. Pan, J. Chin. Chem. Soc. (Taipei), 1, 26 (1954).
10. R. Geyer and G. Henze, Wiss. Z. Tech. Hochsch. Chem. Leuna-Merseburg, 3, 261 (1960/1961).
11. S. S. Lingane and L. Y. Small, J. Am. Chem. Soc., 71, 973 (1949).
12. H. Boadsgaard and W. D. Treadwell, Helv. Chim. Acta, 38, 1669 (1955).
13. K. N. Mikhalevich and V. M. Litvinchuk, Nauchn. Zap. Lvovsk. Politekh. Inst., Ser. Khim. Tekh., 50, 16 (1958).
14. K. N. Mikhalevich and V. M. Litvinchuk, Zh. Neorgan. Khim., 9, 2391 (1964).
15. Kabir-Ud-Din, A. A. Khan, and M. A. Beg, J. Electroanal. Chem., 20, 239 (1969).

16. M. T. Pope and G. M. Varga, Jr., Inorg. Chem., 5, 1249 (1966).
17. A. G. Lappin and R. D. Peacock, Inorg. Chim. Acta, 46(4), L71 (1980).
18. Z. F. Shakhova and R. K. Motorkina, Zh. Obshch. Khim., 26, 2663 (1956).
19. T. J. R. Weakley, J. Chem. Soc., Dalton Trans., no. 3, 341 (1973).
20. M. T. Pope and E. Papaconstantinou, Inorg. Chem., 6, 1147 (1967).
21. E. Papaconstantinou, Z. Phys. Chem. N. F., 106, 283 (1977).
22. C. M. Flynn, Jr., M. T. Pope, and S. O'Donnell, Inorg. Chem., 13, 831 (1974).
23. I. M. Novosyolov and V. F. Barkovskii, in *Issledovaniye svoistv i primeneniye geteropolikislot v katalize*, Inst. Kataliza AN SSSR, Novosibirsk, 1978, p. 92.
24. R. Acerete, S. Harmalker, C. Hammer, M. T. Pope, and L. C. W. Baker, J. Chem. Soc., Chem. Commun., no. 17, 777 (1979).

17
Vanadium, Niobium, and Tantalum

I. VANADIUM*

YECHESKEL ISRAEL   IMI Institute for Research and Development, Haifa, Israel
LOUIS MEITES†   Clarkson College of Technology, Potsdam, New York

A. Introduction

The relative complexity of the chemistry of vanadium arises from the existence of the commonly known oxidation states II, III, IV, and V, and from the behavior of their ionic and molecular species in solutions. These oxidation states exist in solution in specific geometrical forms and in association with various ligands (including water, chloride, hydroxide, sulfate, and hydrogen sulfate), usually in the inner coordination sphere. Vanadium has a coordination number of 6 in most of its compounds and ions, although coordination numbers of 4, 5, and 8 have also been reported. In aqueous solutions, many of these compounds and ions undergo extensive hydrolysis and association, forming polymeric species. This is especially true of the higher oxidation states, of which some of the species participate in very sluggish equilibria.

Some features of the chemistry of vanadium(V) in aqueous solutions at intermediate pH values can be deduced from Latimer's *Oxidation Potentials* [1]. However, many more recent investigations, using different techniques, have established the existence of new complex isopolyvanadate(V) anions. Some of these predominate over others whose existences had been recognized previously, and some have ranges of existence that depend on both concentration and pH. It must be emphasized that the natures of some of the hydrolyzed species are still in doubt despite the large amount of research that has been devoted to their study. Furthermore, it has only recently been recognized that the lower oxidation states of vanadium can also exhibit marked hydrolysis and association. Uncertainties concerning the nature of all these species contribute to uncertainties in the electrochemical and thermochemical parameters that characterize the behavior of the element.

Since Latimer's book [1] was published, there have been a number of other works and compilations of data that are relevant to the standard potentials and thermochemistry of vanadium [2-10]. The thermodynamic data for vanadium are given in Table 1.

Reviewer: Loren Hepler, University of Lethbridge, Alberta, Canada.
†*Current affiliation*: George Mason University, Fairfax, Virginia.

TABLE 1 Thermodynamic Data for Vanadium and Some of Its Compounds and Ions at 298.15 K[a]

Formula	State	Description	ΔH_f°/kJ mol^{-1}	ΔG_f°/kJ mol^{-1}	S°/J mol^{-1} K^{-1}
V	c	Reference state	0.0	0.0	28.95
V	ℓ		17.87	15.61	36.53
V	g		515.5	469.8	182.2
V$^+$	g		1170.7		
V^{2+}	g		2590.4		
V^{2+}	aq	Standard state	(−226)	−218	(−130)
VO	c		−431.8	−404.2	39.0
VO	ℓ		−370.9	−352.3	69.14
VO	g		127.6	98.0	230.8
VBr$_2$	c		−365.3	−339	113
VBr$_2$	g		−155.2		
VCl$_2$	c		−452	−406	97.1
VCl$_2$	g		−256.5	−247	
VI$_2$	c		−251.5		134

Vanadium, Niobium, and Tantalum

VI_2	g	−4.2		
V^{3+}	g	5430.5		
V^{3+}	aq	(−259)	(−230)	Standard state
V_2O_3	c	−1219	−251.3	98.07
V_2O_3	ℓ	−1093	−1139	152.25
VBr_3	c	−433.5	−1029.5	167
VBr_3	g	−237.7	−418	
VCl_3	c	−580.7	−511.3	131
$VOCl$	c	−607	−556	75
VF_3	c			97.0
VI_3	c	−270.7	−268	184
VN	c	−217.2	−191.1	37.3
VN	g	523.0	490.6	233.5
V_2S_3	c	−950		
VO^+	aq		−451.8	Standard state
VOH^{2+}	aq		−471.9	Standard state
V_3O_5	c	−1946	−1816	163

TABLE 1 (Continued)

Formula	State	Description	ΔH_f°/kJ mol^{-1}	ΔG_f°/kJ mol^{-1}	S°/J mol^{-1} K^{-1}
V_4O_7	c		−2657	−2473	218
V^{4+}	g		9943.3		
VO_2	g		−232.6	−241.9	265.1
V_2O_4	c		−1427.2	−1318.6	103.5
V_2O_4	ℓ		−1332.2	−1244.4	173.5
VBr_4	g		−337		
$VBrCl_3$	g		−502		
VCl_4	ℓ		−569.4	−503.8	255
VCl_4	g		−525.5	−492.0	362.3
$VOCl_2$	c		−703	−636	130
VF_4	c		−1343	−1251	126
VI_4	g		−122.6		
$VOSO_4$	c		−1309	−1170	109
$VOSO_4$	aq	Standard state, undiss.		−1182	

VO^{2+}	aq	Standard state	−486.6	−446.4	−133.9
$VOOH^+$	aq	Standard state		−657	
$(VOOH)_2^{2+}$	aq	Standard state		−1331	
$V_4O_9^{2-}$	aq	Standard state		−2784	
$(VO)_3(PO_4)_2$	aq	Standard state		−3245	
$VOSCN^+$	aq	Standard state	−410	−360	34
V_6O_{13}	c		−4456.0	−4109	335
V^{5+}	g		16243		
V_2O_5	c		−1550.2	−1419.4	130.5
V_2O_5	ℓ		−1491.2	−1378.4	192.0
$V_2O_5 \cdot H_2O$	c		(−1845)	(−1657)	172
$V_2O_3(OH)_4$	g		−2049		
$VOCl_3$	ℓ		−734.7	−668.6	244
$VOCl_3$	g		−695.6	−659.3	344.2
VO_2Cl	c		−776.6	−702.0	96
VF_5	ℓ		−1480	−1373	176
VF_5	g		−1434	−1370	320.8

TABLE 1 (Continued)

Formula	State	Description	ΔH_f°/kJ mol^{-1}	ΔG_f°/kJ mol^{-1}	S°/J mol^{-1} K^{-1}
AgVO$_3$	c			−896.6	
Ag$_2$HVO$_4$	c			−1208	
Ag$_2$HVO$_4\cdot$AgOH	c			−1500	
Ca(VO$_3$)$_2$	c		−2329	−2170	179
Ca$_2$V$_2$O$_7$	c		−3083	−2893	220.5
Ca$_3$V$_2$O$_8$	c		−3778	−3561	275
Fe(VO$_3$)$_2$	c		−1899	−1750	197
Mg(VO$_3$)$_2$	c		−2202	−2039	161
Mg$_2$V$_2$O$_7$	c		−2836	−2645	200
Mn(VO$_3$)$_2$	c		−2000	−1849	198.7
NH$_4$VO$_3$	c		−1053.1	−888.3	140.6
NOVF$_6$	c		−1658	−1459	167
NaVO$_3$	c		−1146	−1064	114
Na$_3$VO$_4$	c		−1756	−1637	190

$Na_4V_2O_7$	c		−2917	−2720	318
$Pb_2V_2O_7$	c		*−2133*	−1946	274.1
$Pb_3(VO_4)_2$	c		*−2374*	−2161	354
$TlVO_3$	c			*−863.2*	
$Tl_4V_2O_7$	c			*−1955*	
VO_2^+	aq	Standard state	*−649.8*	*−587.0*	*−42.3*
VO_3^-	aq	Standard state	*−888.3*	*−783.7*	*50*
VO_4^{3-}	aq	Standard state		*−899.1*	
HVO_4^{2-}	aq	Standard state	*−1159.0*	*−974.9*	*17*
$H_2VO_4^-$	aq	Standard state	*−1174.0*	*−1020.9*	*121*
H_3VO_4	aq	Standard state		−1040.3	
$V_2O_7^{4-}$	aq	Standard state		−1720	
$HV_2O_7^{3-}$	aq	Standard state		*−1792*	
$H_3V_2O_7^-$	aq	Standard state		*−1864*	
$V_3O_9^{3-}$	aq	Standard state		−2356	
$V_4O_{12}^{4-}$	aq	Standard state		−3202	
$V_{10}O_{28}^{6-}$	aq	Standard state		−7675	

TABLE 1 (Continued)

Formula	State	Description	ΔH_f°/kJ mol^{-1}	ΔG_f°/kJ mol^{-1}	S°/J mol^{-1} K^{-1}
$HV_{10}O_{28}^{5-}$	aq	Standard state	−8694	−7708	1544
$H_2V_{10}O_{28}^{4-}$	aq	Standard state		−7729	
$VO_2 \cdot H_2O_2^+$	aq	Standard state		−746.4	
$VO \cdot H_2O_2^{3+}$	aq	Standard state		−523.4	
VAl_3	c		−109		
V_5Al_8	c		−293		
V_2C	c		−146	−146	51.0

$VC_{0.73}$	c	−92	−92	25
$VC_{0.88}$	c	−101.7	−99.2	26
$V(CO)_6$	g	−987		
$VN_{0.465}$	c	−132.2	−118.3	26.7
VSi_2	c	−305		
V_2Si	c	−155		
V_3Si	c	−109		
V_5Si_3	c	−393		

[a]National Bureau of Standards values are in italic; estimated values are in parentheses.
Source: Data from Refs. 5, 6, 9, 10, and 42.

B. Equilibrium Reactions

1. Vanadium(V)

The two most commonly known compounds of vanadium(V) are the pentoxide, V_2O_5, and ammonium metavanadate, NH_4VO_3. The orange pentoxide has a limited solubility (0.0043 M) in water and gives light yellow solutions that are acidic. As would be expected, V_2O_5 is more acidic than the oxides of vanadium in lower oxidation states, but may also be regarded as exhibiting basic properties when it reacts with, and consequently dissolves in, acidic aqueous solutions [5].

In very strongly acidic media there is some slight evidence for the existence of VO^{3+}(aq). At pH values between about 0 and 2, the principal species of vanadium(V) is the pervanadyl ion, VO_2^+(aq). Its formation in acidic media is described by the equation

$$V_2O_5(c) + 2H^+ \rightarrow 2VO_2^+ + H_2O(\ell) \tag{17.1}$$

whose equilibrium constant is equal to 3.42×10^{-2} [6]. At pH values between about 2 and 6.5, vanadium exists as a complex mixture of monomeric and polymeric anions in proportions that depend on pH and the total concentration of vanadium(V). It is suggested that the reader refer to the review by Pope and Dale [11], which contains a diagram outlining the ranges of existence of the various species as functions of these experimental variables. According to Schwarzenbach and Geier [12], the monomeric species $H_2VO_4^-$ (or VO_3^-), and possibly H_3VO_4(aq) [or HVO_3(aq)], may prevail if the concentration of vanadium(V) is lower than 10^{-4} mol dm^{-3}. The formation of H_3VO_4(aq) is described by the equation

$$VO_2^+ + 2H_2O(\ell) \rightarrow H_3VO_4(aq) + H^+ \tag{17.2}$$

whose equilibrium constant is equal to 6.1×10^{-4}, and its successive dissociation constants are given by $K_1 = 4.0 \times 10^{-4}$, $K_2 = 9.1 \times 10^{-9}$, and $K_3 = 1.0 \times 10^{-13}$.

The polyanions that form at higher concentrations of vanadium(V) are predominantly decanuclear species that have been represented by the formulas $H_2V_{10}O_{28}^{4-}$, $HV_{10}O_{28}^{5-}$, and $V_{10}O_{28}^{6-}$ [12-16]. Hexanuclear species were once thought to predominate at pH values from 2.0 to 6.5, but the best evidence now available seems to indicate that the decavanadates are considerably more important. The formation of $H_2V_{10}O_{28}^{4-}$ from the pervanadyl ion

$$10VO_2^+ + 8H_2O \rightarrow H_2V_{10}O_{28}^{4-} + 14H^+ \tag{17.3}$$

has an equilibrium constant, according to Rossotti and Rossotti [14], that is equal to 1.8×10^{-7}, and the product has acidic dissociation constants that are given by $K_5 = 2.5 \times 10^{-4}$ and $K_6 = 1.6 \times 10^{-6}$ in 1 M sodium perchlorate. As would be expected from the high charges on the ions involved, the values of these constants are sensitively dependent on the nature and composition of the ionic medium.

The nature(s) of the species formed by the polymerization of $H_2VO_4^-$, or by the depolymerization of the decavanadates at pH values between 6.5 and 8.2, is probably the most important of the unsovled problems. The so-called metavanadate ion predominates, but there is disagreement as to whether this is better represented as the trimer, $V_3O_9^{3-}$, or as the tetramer, $V_4O_{12}^{4-}$. Düllberg [17], supported by other investigators [18-21], postulated the formation of the trimer as follows:

$$3HVO_4^{2-} \rightarrow V_3O_9^{3-} + 3OH^- \quad (17.4)$$

$$3H_2VO_4^- \rightarrow V_3O_9^{3-} + 3H_2O(\ell) \quad (17.5)$$

$$3VO_3^- \rightarrow V_3O_9^{3-} \quad (17.6)$$

with equilibrium constants equal to 3.8×10^{-11}, 4.0×10^8, and 6.0×10^6, respectively. On the other hand, Jander and Jahr [22], supported by a good deal of evidence [15,23,24], have postulated the formation of the tetramer

$$4HVO_4^{2-} + 4H^+ \rightarrow V_4O_{12}^{4-} + 4H_2O(\ell) \quad (17.7)$$

with an equilibrium constant equal to 9.1×10^{43}. However, more recent evidence from studies of neutralization at 40°C (to hasten the attainment of equilibrium) has been interpreted as indicating that a trimeric and a tetrameric "metavanadate" are in equilibrium with each other, and other less likely species are also indicated [25]. Most of these studies have ignored the influence of the binding of counterions, although this has been shown to occur between sodium ion and several vanadate species [12] and is to be expected a priori when such highly charged ions are involved.

There seems to be conclusive evidence for the formation of the so-called pyrovanadate ion, $V_2O_7^{4-}$, at concentrations of vanadium above about 10^{-2} mol dm^{-3} and at pH values between about 8.5 and 13 [12,20]. There is also some evidence for the existence of other univalent, divalent, and trivalent monomeric and dimeric anions. The so-called orthovanadate ion, VO_4^{3-}, undoubtedly predominates in solutions having pH values above 13. The following equilibria are important in alkaline media:

$$2VO_4^{3-} + 2H^+ \rightarrow V_2O_7^{4-} + H_2O(\ell) \quad (17.8)$$

$$2HVO_4^{2-} \rightarrow V_2O_7^{4-} + H_2O(\ell) \quad (17.9)$$

and their equilibrium constants are equal to 1.1×10^{25} and 48, respectively [26]. Other thermodynamic data, which are given in Table 1, lead to a value of 6.0×10^{27} for the equilibrium constant of reaction (17.8) [6].

2. Vanadium(IV)

The blue divanadium tetroxide, V_2O_4, is amphoteric and about equally soluble in acidic and basic solutions [5]. In aqueous solutions vanadium(IV) exists mainly as monomeric, dimeric, and tetrameric forms and their hydrolysis products, depending on the pH and the concentration of vanadium(IV).

Acidic solutions contain the blue vanadyl ion, whose simplest formula is VO^{2+}. Several investigators [27-30] have provided strong evidence that the hydrolysis of VO^{2+} occurs in stepwise fashion, with $VOOH^+$ as the first product and with more complicated species as the products of further hydrolysis. The dimer $(VOOH)_2^{2+}$ [28] is favored at high concentrations of vanadium(IV) [6]. The reactions that occur in media containing sulfate are [27,28]

$$VO^{2+} + H_2O(\ell) \rightarrow VOOH^+ + H^+ \tag{17.10}$$

$$2VOOH^+ \rightarrow (VOOH)_2^{2+} \tag{17.11}$$

and their equilibrium constants are equal to 4.4×10^{-6} and 1.26×10^5, respectively; these values were obtained by ignoring the association of VO^{2+} and SO_4^{2-} to form the ion pair $VOSO_4$, for which the association constant is equal to 3.0×10^2 [31].

More nearly neutral solutions contain more complicated species that are poorly characterized. It has been speculated that $HV_2O_5^-$ (or VO_4^{4-}) is formed at pH values above about 4.5, and that, together with $(VOOH)_2^{2+}$, it undergoes condensation polymerization to yield $V_4O_9^{2-}$ in neutral or alkaline solutions containing high concentrations of vanadium(IV). Compounds having formulas such as $K_4V_4O_9 \cdot 7H_2O$ have been crystalized from alkaline solutions of vanadium(IV), although that is, of course, no proof that the anion exists at any substantial concentration in such solutions. The following equilibria have been reported:

$$HV_2O_5^- + 3H^+ \rightarrow (VOOH)_2^{2+} + H_2O(\ell) \tag{17.12}$$

$$2HV_2O_5^- \rightarrow V_4O_9^{2-} + H_2O(\ell) \tag{17.13}$$

$$V_4O_9^{2-} + 6H^+ \rightarrow 2(VOOH)_2^{2+} + H_2O(\ell) \tag{17.14}$$

with equilibrium constants equal to 1.83×10^{10}, 3.2_6, and 1.21×10^{20}, respectively.

3. *Vanadium(III)*

The black basic oxide V_2O_3 dissolves readily in acids to give green solutions of V^{3+}, or complexes of this ion [6,32]:

$$V_2O_3(c) + 6H^+ \rightarrow 2V^{3+} + 3H_2O(\ell) \tag{17.15}$$

$$V_2O_3(c) + 4H^+ \rightarrow 2VOH^{2+} + H_2O(\ell) \tag{17.16}$$

$$V_2O_3(c) + 2H^+ \rightarrow 2VO^+ + H_2O(\ell) \tag{17.17}$$

with equilibrium constants equal to 1.56×10^{13}, 2.14×10^7, and 2.0, respectively.

The slightly soluble hydrous oxide that is formed on adding alkali to V^{3+} is sometimes represented as $V(OH)_3$ [3]. Titrations of acidic solutions with alkalies reveals that hydrolysis, leading to the formation of a precipitate, begins at a pH value of about 3, depending on the concentration of vanadium(III). The investigations of Britton and Welford [13] show that solutions containing sulfate behave differently from those containing chloride. In sulfate media there is clear evidence for the formation of a basic sulfate complex, containing three atoms of vanadium and six hydroxide ions, in equilibrium with VOH^{2+}. There have been reports that the stable complexes $V_2(CH_3COO)_3(OH)_2^+$ and $H_3V(PO_4)_2$ are formed in acetate media and in solutions containing phosphoric acid, respectively, and there is evidence for complex formation in solutions containing ammonia [4], citrate, EDTA, tartrate, and many other common ligands. In noncomplexing media, the solution chemistry of vanadium(III) is confined to the acidic range (pH 4 or below) if the concentration exceeds about 10^{-4} M. Various products of the hydrolysis of vanadium(III) have been noted, including $V(OH)_2^+$ and $V_2(OH)_3^{3+}$ [32]. Meites [27] reported that the extent of hydrolysis increases on decreasing the acidity of a solution of vanadium(III): VOH^{2+} is formed if less than 0.6 mol of base is added per mole of vanadium(III) at pH values between 3.0 and 3.5 and at concentrations of vanadium(III) above about 10^{-3} M, but VO^+ is formed at lower concentrations and at pH values between 3.5 and 6.5:

$$V^{3+} + H_2O(\ell) \rightarrow VOH^{2+} + H^+ \tag{17.18}$$

$$VOH^{2+} \rightarrow VO^+ + H^+ \tag{17.19}$$

with equilibrium constants of 1.17×10^{-3} and 3×10^{-4}, respectively. Pajdowski [33], who studied the hydrolysis of $VCl_3 \cdot 6H_2O$ in both pure water and 3 M sodium chloride, reported values of 3.4×10^{-3} and 1.4×10^{-4}, respectively, for the same constants in water. In 3 M sodium chloride both constants, particularly that of reaction (17.19), had different values, because of the difference of ionic strength and possibly also because of some complexation with chloride ion, although polarographic data on the effect of chloride-ion concentration on the half-wave potentials of vanadium(III) and vanadium(II) [34] suggest that complexation is unlikely to be significant here.

Pajdowski [33] has also presented evidence for the existence of the dimer $V_2(OH)_2^{4+}$, which is formed, together with the two monomeric products, when the total concentration of vanadium(III) exceeds 6×10^{-3} M. The reaction that leads to its formation is

$$2V^{3+} + 2H_2O(\ell) \rightarrow V_2(OH)_2^{4+} + 2H^+ \tag{17.20}$$

and its equilibrium constant in 1 M sodium chloride is equal to 1.3×10^{-4}. Newton and Baker [35] have reported kinetic evidence for the existence of a hydrolytic dimer of vanadium(III), and Israel and Meites [4,36] have presented stirred-pool chronoamperometric evidence for the formation of species of vanadium(III) having a higher degree of polymerization, and possibly tetrameric, in solutions containing high concentrations of hydrogen sulfate ion.

4. *Vanadium(II)*

The black basic oxide having the ideal stoichiometric formula VO dissolves in nonoxidizing acids to yield violet solutions containing V^{2+} ions [6]:

$$VO(c) + 2H^+ \rightarrow V^{2+} + H_2O(\ell) \tag{17.21}$$

The equilibrium constant of this reaction is equal to 3.0×10^{10}. Padjowski [33] reported that V^{2+} is the only species in solution and that the solid oxide has the formula V_2O_2. Nonstoichiometric solids with a deficiency of either V or O are also known.

Although these statements suggest that the solution chemistry of vanadium(II) is rather simple, kinetic evidence has been presented for the formation of polymeric species in solutions containing high concentrations of hydrogen sulfate ion [4,36]. There is also evidence of complex formation in solutions containing ammonia [4], citrate, EDTA, and many other ligands.

C. Redox Behavior

In view of the complexity of the solution chemistry of vanadium, potential-pH diagrams can be useful in assessing the redox behavior of the different species containing vanadium in its various oxidation states. Post and Robins [6] have reconstructed the diagrams of Deltombe et al. [2], using critically selected data and more recent information concerning the identities of stable species and the conditions under which they exist.

Vanadium(IV) is the most stable oxidation state under many conditions. Under conditions that promote the formation of the complexes (e.g., in moderately concentrated solutions of phosphoric acid), however, it is a mild oxidizing agent and may oxidize iron(II); while in strongly alkaline solutions it is a fairly powerful reducing agent and readily undergoes oxidation by atmospheric oxygen. Metallic vanadium is a relatively strong reducing agent. Vanadium(II) is easily air-oxidized and is the least stable of the oxidation states of vanadium in aqueous solution; in solutions that contain such complexing agents as citrate and EDTA it is one of the strongest homogeneous reducing agents known. Vanadium(III) also undergoes slow oxidation by atmospheric oxygen. Vanadium(V) is a mild oxidizing agent under many conditions, but is capable of oxidizing chloride ion to chlorine in moderately concentrated solutions of hydrochloric acid.

1. Standard Potentials of Vanadium Couples

a. *Vanadium(V)-Vanadium(IV)*. Coryell and Yost [37] obtained data on the potential of this couple in the presence of hydrochloric acid, and Carpenter [38] combined their data with the following description of the half-reaction:

$$VO_2^+ + 2H^+ + e^- \rightarrow VO^{2+} + H_2O(\ell) \tag{17.22}$$

to calculate $E° = 1.000$ V. This value can, in turn, be used to calculate the standard potentials of other couples involving stable species of vanadium(V) and vanadium(IV). For example, it may be combined with equilibrium constants of reactions (17.3) and (17.10) to yield $E° = 0.723$ V for the half-reaction

$$H_2V_{10}O_{28}^{4-} + 24H^+ + 10e^- \rightarrow 10VOOH^+ + 8H_2O(\ell) \tag{17.23}$$

which in turn may be combined with the equilibrium constant of reaction (17.11) to yield $E° = 0.874$ V for the half-reaction

$$H_2V_{10}O_{28}^{4-} + 24H^+ + 10e^- \rightarrow 5(VOOH)_2^{2+} + 8H_2O(\ell) \tag{17.24}$$

Post and Robins [6] have calculated the standard changes of Gibbs energy for a number of half-reactions that involve vanadium(V)/vanadium(IV) couples.

Their values can be used to calculate the standard potentials of other such half-reactions, provided that these involve species that are well characterized and that reliable values of $E°$ and K are available. Some of the standard potentials quoted by Post and Robins are

$$V_2O_5(c) + 6H^+ + 2e^- \rightarrow 2VO^{2+} + 3H_2O(\ell) \qquad E° = 0.958 \text{ V} \qquad (17.25)$$

$$2HV_2O_7^{3-} + 8H^+ + 4e^- \rightarrow V_4O_9^{2-} + 5H_2O(\ell) \qquad E° = 0.997 \text{ V} \qquad (17.26)$$

$$HV_2O_7^{3-} + 4H^+ + 2e^- \rightarrow HV_2O_5^- + 2H_2O(\ell) \qquad E° = 0.991 \text{ V} \qquad (17.27)$$

b. *Vanadium(IV)/Vanadium(III)*. Jones and Colvin [39] assumed the half-reaction to be

$$VO^{2+} + 2H^+ + e^- \rightarrow V^{3+} + H_2O(\ell) \qquad (17.28)$$

and obtained $E° = 0.337$ V from potentiometric measurements made with solutions containing sulfuric acid. The value is somewhat uncertain because different values have been reported by other investigators [6,40]. The formal potential in solutions containing phosphoric acid becomes more positive as the concentration of the acid increases, because of complex formation [41], but the nature of the complex has not been elucidated. It may be noted that there is a corresponding change (toward more negative potentials) of the formal potential of the vanadium(III)/vanadium(II) couple [34].

Combining the value of the standard potential of reaction (17.28) with that of the equilibrium constant given by Meites [27] for the hydrolytic reaction (17.18) yields $E° = 0.164$ V for the half-reaction

$$VO^{2+} + H^+ + e^- \rightarrow VOH^{2+} \qquad (17.29)$$

and this in turn can be combined with the value of the equilibrium constant for the hydrolytic reaction (17.10) to yield $E° = 0.481$ V for the half-reaction

$$VOOH^+ + 2H^+ + e^- \rightarrow VOH^{2+} + H_2O(\ell) \qquad (17.30)$$

Post and Robins [6] reported values for the standard potentials of couples involving other substances that may predominate in neutral and alkaline media, including 0.536 V for the half-reaction

$$V_4O_9^{2-} + 6H^+ + 4e^- \rightarrow 2V_2O_3(c) + 3H_2O(\ell) \qquad (17.31)$$

and 0.542 V for the half-reaction

$$HV_2O_5^- + 3H^+ + 2e^- \rightarrow V_2O_3(c) + 2H_2O(\ell) \qquad (17.32)$$

c. *Vanadium(III)/Vanadium(II)*. Jones and Colvin [42] assumed the half-reaction to be

$$V^{3+} + e^- \rightarrow V^{2+} \tag{17.33}$$

and obtained $E° = -0.255$ V from potentiometric measurements made with solutions containing sulfuric acid. This value may be combined with that of the equilibrium constant for reaction (17.18) to yield $E° = -0.082$ V for the half-reaction

$$VOH^{2+} + H^+ + e^- \rightarrow V^{2+} + H_2O(\ell) \tag{17.34}$$

Values of some other standard potentials are

$$V_2O_3(c) + 6H^+ + 2e^- \rightarrow 2V^{2+} + 3H_2O(\ell) \qquad E° = 0.135 \text{ V} \tag{17.35}$$

$$V_2O_3(c) + 2H^+ + 2e^- \rightarrow 2VO(c) + H_2O(\ell) \qquad E° = -0.486 \text{ V} \tag{17.36}$$

d. *Vanadium(II)/Vanadium(0)*. Hill et al. [5] combined the standard potential of reaction (17.33) with the standard Gibbs energy of V^{3+}, -242.3 kJ mol^{-1}, to calculate the standard change of Gibbs energy for the reaction

$$V^{2+} + 2e^- \rightarrow V(c) \tag{17.37}$$

and the corresponding value $E° = -1.13$ V. A value of -1.175 V had been reported previously [2]. Combining the standard change of Gibbs energy for reaction (17.37) with the value of the equilibrium constant for reaction (17.21) yields $E° = -0.820$ V for the half-reaction

$$VO(c) + 2H^+ + 2e^- \rightarrow V(c) + H_2O(\ell) \tag{17.38}$$

e. *Potential Diagrams for Vanadium*. The data appearing in the preceding subsections are briefly summarized in the following diagrams. Values from the literature, in particular from Post and Robins [6], were recalculated to obtain the diagrams pertaining to basic solutions.

Strongly Acid Solutions

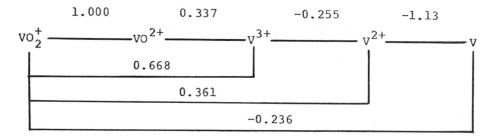

Weakly Acid Solutions, pH about 3.0-3.5

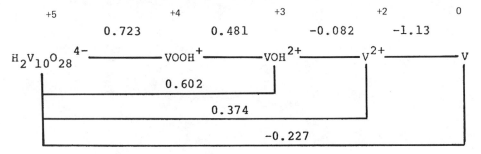

Neutral and Weakly Basic Solutions, pH about 6-12

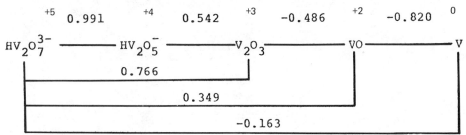

Strongly Basic Solutions

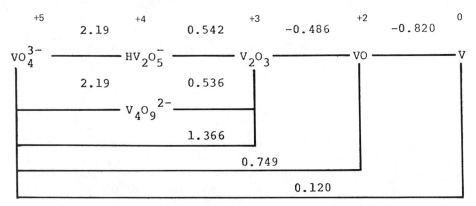

D. Thermochemistry

Table 1 is a collection of thermochemical data for vanadium and its compounds in the solid, liquid, and gaseous states, and also for ionic species containing the element in its various oxidation states in aqueous solutions. It was compiled from various sources. The most recent and reliable account of the thermochemical properties of vanadium and its oxides is that in the *JANAF Thermochemical Tables, 1975 Supplement* [10]. Data on the thermodynamic properties of vanadium and its compounds were assembled by Mah in U.S. Bureau of Mines Report of Investigations, No. 6727 [8], and included the experimental results obtained by a research group of the Bureau of Mines as

TABLE 2 Formal Potentials of Vanadium Couples in Various Media at 298.15 K

Couple	Medium	$E^{o'}/V$	References
V(V)/V(IV)	1 M HCl	1.02	43
	1 M HClO$_4$	1.02	43
	1 M H$_2$SO$_4$	1.0$_0$	43
V(IV)/V(III)	1 M H$_2$SO$_4$	0.36$_0$	43
V(III)/V(II)	Saturated citric acid	−0.329	34
	1 M HCl	−0.267	44
	12 M HCl	−0.29	34
	0.001-0.2 M EDTA, pH 5-8.3	−0.78	45
	Saturated hydrazine dihydrochloride	−0.21$_9$	34
	1 M K$_2$C$_2$O$_4$, pH 3.5-6.5	−0.895	46
	1 M HClO$_4$	−0.24	43,44
	7.3 M (42%) H$_3$PO$_4$	−0.49$_2$	34
	Saturated potassium hydrogen phthalate, pH 5.2	−0.62	47
	0.5 M H$_2$SO$_4$	−0.267	44
	0.1-1 M NH$_4$SCN	−0.22$_2$	47

as well as critically evaluated data from the literature. Wagman et al. compiled and tabulated thermochemical data on vanadium, its compounds, and its ionic species in solutions in NBS Technical Note 270-5 [9]. Other publications valuable for reference are that of Post and Robins [6] on thermodynamic diagrams for the vanadium-water system at 298.15 K, and the review by Hill et al. on the thermochemistry and oxidation potentials of vanadium [5].

The authors of each of the references above have evaluated the prior data critically and have readjusted them where necessary to accord with newer values that they added. Even so, in compiling the data of Table 1, we had to make further readjustments to ensure that they would be consistent with each other and with the values given in the text.

E. Formal Potentials

Many chemists regard formal potentials as having more practical utility than standard potentials because they are more directly related to experimental measurements. The formal potentials of a number of couples involving vanadium are therefore presented in Table 2. A few of them were obtained potentiometrically, but the great majority were obtained by polarographic or other voltammetric measurements with couples that satisfy rigorous criteria of reversibility. Although most of them were originally referred to the aqueous saturated calomel electrode, they are referred here to the standard hydrogen electrode. In making the conversions, the potential of the saturated calomel electrode was assumed to be equal to 0.241_4 V vs. SHE at 298.15 K.

ACKNOWLEDGMENT

The preparation of this manuscript was supported in part by Grant CHE-7727751 from the National Science Foundation.

REFERENCES

1. W. M. Latimer, *Oxidation Potentials*, 2nd ed., Prentice-Hall, Englewood Cliffs, N.J., 1952, pp. 258-261.
2. E. Deltombe, N. de Zoubov, and M. Pourbaix, in *Atlas d'équilibres électrochimiques à 25°C*, M. Pourbaix, ed., Gauthier-Villars, Paris, 1963, pp. 234-245.
3. G. Milazzo and S. Caroli, *Tables of Standard Electrode Potentials*, Chichester, West Sussex, England, 1978, pp. 214-219.
4. Y. Israel and L. Meites, in *Encyclopedia of Electrochemistry of the Elements*, A. J. Bard, ed., Vol. VII, Marcel Dekker, New York, 1976, pp. 293-466.
5. J. O. Hill, I. G. Worsley, and L. G. Hepler, Chem. Rev., *71*, 127 (1971).
6. K. Post and R. G. Robins, Electrochim. Acta, *21*, 401 (1976).
7. D. D. Perrin, *Dissociation Constants of Inorganic Acids and Bases in Aqueous Solutions*, Butterworth, London, 1969, pp. 210-212.
8. A. D. Mah, *Thermodynamic Properties of Vanadium and Its Compounds*, U.S. Bur. Mines Rep. Invest., No. 6727, U.S. Government Printing Office, Washington, D.C., 1961, 84 pp.
9. D. D. Wagman, W. H. Evans, V. B. Parker, I. Hallow, S. M. Bailey, R. H. Schumm, and K. L. Churney, *Selected Values of Chemical Thermodynamic Properties*, Natl. Bur. Stand. Tech. Note 270-5, U.S. Government Printing Office, Washington, D. C., 1971, pp. 1-5.
10. *JANAF Thermochemical Tables, 1975 Supplement*, Vol. 4, No. 1, pp. 117-175.
11. M. T. Pope and B. W. Dale, Q. Rev. Chem. Soc., *22*, 527 (1968).
12. G. Schwarzenbach and G. Geier, Helv. Chim. Acta, *46*, 906 (1963).
13. H. J. S. Britton and G. Welford, J. Chem. Soc., 764 (1940).
14. F. J. C. Rossotti and H. S. Rossotti, J. Inorg. Nucl. Chem., *2*, 201 (1956); Acta Chem. Scand., *10*, 957 (1956).
15. A. W. Naumann and C. J. Hallada, Inorg. Chem., *3*, 70 (1964).
16. G. Schwarzenbach, Pure Appl. Chem., *5*, 377 (1962).
17. P. Düllberg, Z. Phys. Chem., *45*, 129 (1903).
18. K. Schiller and E. Thillo, Z. Anorg. Chem., *310*, 261 (1961).
19. N. Ingri and F. Brito, Acta Chem. Scand., *13*, 1971 (1959).
20. F. Brito and N. Ingri, An. Real. Soc. Esp. Fiz. Quim., *B56*, 165 (1960).
21. R. A. Robinson and D. A. Sinclair, J. Chem. Soc., 642 (1934).

22. G. Jander and K. F. Jahr, Z. Anorg. Chem., 212, 1 (1933).
23. K. F. Jahr and L. Schoepp, Z. Naturforsch., 14b, 467 (1959).
24. K. F. Jahr, H. Schroth, and J. Fuchs, Z. Naturforsch., 18b, 1133 (1963).
25. F. Brito, An. Real. Soc. Esp. Fis. Quim., B62, 123 (1966); Acta Chem. Scand., 21, 1968 (1967).
26. Y. I. Sannikov, V. L. Zolotavin, and I. Y. Bezrukov, Russ. J. Inorg. Chem. (English trans.), 8, 474 (1963).
27. L. Meites, J. Am. Chem. Soc., 75, 6059 (1953).
28. F. J. C. Rossotti and H. S. Rossotti, Acta Chem. Scand., 9, 1171 (1955).
29. M. M. T. Khan and A. E. Martell, J. Am. Chem. Soc., 90, 6011 (1968).
30. L. G. Sillén and A. E. Martell, *Stability Constants of Metal Complexes*, Spec. Publ. 17, Chemical Society, London, 1964; R. M. Smith and A. E. Martell, eds., *Critical Stability Constants*, Vol. 4: *Inorganic Complexes*, Plenum, New York, 1976.
31. H. Strehlow and H. Wendt, Inorg. Chem., 2, 6 (1963).
32. J. Gandeboeuf and D. Souchay, J. Chim. Phys., 56, 358 (1959).
33. L. Pajdowski, Rocz. Chem., 37, 1351, 1363 (1963).
34. L. Meites, *Polarographic Techniques*, 2nd ed., Interscience, New York, 1965, pp. 630-632.
35. T. N. Newton and F. B. Baker, Inorg. Chem., 3, 569 (1964).
36. Y. Israel and L. Meites, J. Electroanal. Chem., 8, 99 (1964).
37. C. D. Coryell and D. M. Yost, J. Am. Chem. Soc., 55, 1909 (1933).
38. J. E. Carpenter, J. Am. Chem. Soc., 56, 1847 (1934).
39. G. Jones and J. H. Colvin, J. Am. Chem. Soc., 66, 1563 (1944).
40. E. H. Swift, *Introductory Quantitative Analysis*, Prentice-Hall, Englewood Cliffs, N.J., 1950, pp. 507-509.
41. G. Rao and L. S. A. Dikshitulu, Talanta, 10, 295 (1963).
42. G. Jones and J. H. Colvin, J. Am. Chem. Soc., 66, 1573 (1944).
43. E. H. Swift, *A System of Chemical Analysis*, Prentice-Hall, Englewood Cliffs, N.J., 1940, pp. 541-543.
44. J. J. Lingane and L. Meites, J. Am. Chem. Soc., 70, 2525 (1948).
45. R. L. Pecsok and R. S. Juvet, Jr., J. Am. Chem. Soc., 75, 1202 (1953).
46. J. J. Lingane and L. Meites, J. Am. Chem. Soc., 69, 1021 (1947).
47. J. J. Lingane and L. Meites, J. Am. Chem. Soc., 73, 2165 (1951).

II. NIOBIUM AND TANTALUM*

H. V. K. UDUPA† AND V. K. VENKATESAN Central Electrochemical Research Institute, Karaikudi, Tamil Nadu, India

M. KRISHNAN Thomas J. Watson Research Center, IBM Corporation, Yorktown Heights, New York

A. Niobium

Although the chemical properties of niobium (electron configuration Kr $4d^4 5s^1$) are in many respects, typical of second-row transition metals, at higher oxida-

---

*Reviewer: G. Nagasubramanian, JPL, Pasedena, California.
†*Current affiliation*: Titanium Equipment and Anode Manufacturing Company, Ltd., Vandalur, Madras, India.

tion states the properties of niobium are very similar to those of non-metals such as arsenic and phosphorus. For example, these elements form a considerable number of anionic species but few cationic species. The halides and oxyhalides are also similar in that bonding is largely covalent and the compounds are readily hydrolyzed [1]. These compounds are also volatile. At all pH values niobium is covered with a layer of oxide. Consequently, the electrochemical behavior depends on the intrinsic nature of the oxide layer and its stability in the various solutions considered [2].

The aqueous chemistry of niobium is rather limited becasue many of its compounds are either insoluble or react with water to form slightly soluble compounds. The thermodynamic data for niobium are given in Table 3.

1. Oxidation States

A number of oxidation states ranging from -1 to $+5$ and $+7$ have been mentioned by Pourbaix [2]. We shall be concerned only with the common oxidation states $+2$, $+4$, and $+5$, for which thermodynamic data exist. Because of the lack of thermodynamic data, the other possible oxidation states are not considered in detail [2].

2. Anhydrous Oxides

a. Niobium Monoxide (NbO). Niobium monoxide is a stoichiometric oxide with a very narrow range (few atom percent) of homogeneity; it has a metallic luster and excellent electrical conductivity. Even at 25°C, it is thermodynamically unstable in the presence of acid, neutral, and basic aqueous solutions. From a thermodynamic equilibrium standpoint, at all pH values it can theoretically decompose water with the evolution of hydrogen, while being oxidized to a higher oxide [2].

b. Niobium Dioxide (NbO_2). As in the case of monoxide, the range of homogeneity is small. NbO_2 is insoluble in water, but is a strong reducing agent in the dry state. It is thermodynamically unstable in water and aqueous solutions of any pH. Theoretically, it reduces water, evolving hydrogen and forming niobium pentoxide.

c. Niobium Pentoxide (Nb_2O_5). Niobium pentoxide is a dense white powder and is relatively inert chemically. It is attacked by concentrated hydrofluoric acid with the formation of fluorinated and oxyfluorinated compounds. When fused with alkali hydrogen sulfate or hydroxide, it can be brought into solution. Niobium pentoxide exhibits polymorphism. The principal polymorphs of stoichiometric Nb_2O_5 are obtained as the polycrystalline products of various thermal treatments of the pentoxide in air or oxygen [3]. A comprehensive review of the many stoichiometric and nonstoichiometric phases of niobium pentoxide has been published by Schäfer et al. [4,5]. According to the potential-pH diagram, niobium pentoxide can be reduced to a lower oxide and to metallic niobium. However, electrochemical reduction of Nb_2O_5 in aqueous solutions in the absence of any complexing agent appears to be difficult [2].

3. Hydrous Oxides

The insoluble hydrated pentoxide is obtained as a white precipitate with an indeterminate water content when water-soluble complexes of niobium are hydrolyzed or when solutions of niobates are acidified. The hydrous oxide has long been known as "niobic acid" even though its acidic properties are

TABLE 3 Thermodynamic Data for Niobium at 298 K

Substance	State	ΔH_f°/kJ mol^{-1}	ΔG_f°/kJ mol^{-1}	ΔS°/J mol^{-1} K^{-1}	References
Nb	c	0	0	36.40	21
Nb	g	725.9	681.1	186.256	21
NbO	c	-405.8	-378.6	48.1	21
NbO	g	213	184	238.97	21
NbO$_2$	c	-796.6	-740.9	54.52	22, 23
NbO$_2$	g	-214.6	-218.8	255.3	21
NbO$_3^-$	ao	—	-932.1	—	21
Nb$_2$O$_5$	c	-1901.6	-1768.1	137.24	24
Nb$_2$O$_5$	aq	-1919.6	—	—	21
Nb$_2$O$_5$	hyd ppt	-2004.14	—	—	25

Vanadium, Niobium, and Tantalum

$Nb(OH)_4^+$	ao	—	−1208.6	—	21
$Nb(OH)_5$	ao	—	−1448.3	—	21
NbF_5	c	−1813.8	−1699.1	160.2	26, 27
NbF_5	g	−1739.7	−1673.6	321.9	21
$NbCl_5$	c	−797.1	−682.8	210.9	28−30
$NbCl_4$	c	−694.5	−606.7	184.1	28
$NbCl_{3.13}$	c	−602.5	−531.4	150.62	31
$NbBr_5$	c	−556.2	−510.4	259.4	32
NbI_5 $NbCl_{3.13}$	c	−270.3	—	—	33
Nb_2N	c	−252.7	—	—	22
NbN	c	−235.9	−205.8	34.52	21, 34
Nb_2C	c	−196.6	−185.8	75.31	21
NbC	c	−138.9	−136.8	35.5	21, 35

not known [3]. It possesses no true solubility in water and its properties differ markedly depending on the method of preparation. Freshly precipitated niobic acid undergoes an "aging" process even at room temperature; its reactivity toward complexing agents therefore decrease on standing for a few days. The aging of niobic acid, together with its colloidal character, gives rise to many apparent anomalies and lack of reproducibility of its chemical behavior. For example, aging takes place even in the presence of moderately strong complexing agents which form soluble complexes with niobic acid. Therefore, unless a high concentration of complexing agent is maintained, irreversible hydrolysis of the complex results [6].

a. *Niobates.* In the dissolved form, niobium is present in the +5 state as niobate ions. When treated with molten alkali hydroxides or carbonates, niobium pentoxide is converted into the niobates of general formula $MNbO_3$, where M denotes a monovalent metal. The solutions are stable only at higher pH values. Precipitation occurs at pH values below 7. Most niobates are insoluble in water, the exceptions are the alkali metal niobates. Salts of other cations can be prepared by precipitation from solutions of potassium niobates, which are more soluble than the sodium salts. Of the known soluble niobates, the species that appear to be most stable in solution are the $[H_x Nb_6 O_{19}]^{(8-x)-}$ ions, $x = 0$, 1, or 2. Based on the analyses of absorption spectra, conductimetric titrations, and diffusion coefficient measurements, Jander and Eriel [7] conclude that in solutions of these salts the following pH-dependent reversible equilibrium exists.

$$H^+ + [Nb_6O_{19}(aq)]^{8-} \underset{}{\overset{pH \simeq 13}{\rightleftarrows}} [HNb_6O_{19}(aq)]^{7-}$$

There is also good evidence from electromotive force (emf) and spectroscopic studies that $(Nb_6O_{19})^{8-}$ is protonated in solution [8-11].

4. Mixed Oxides

In contrast to the niobates discussed above, which contain the metal atoms as discrete complex anions, the alkali metal metaniobates exist in the anhydrous state and are stable at high temperatures. These compounds and other mixed oxide systems, consisting of niobium and nonalkali metals, are of importance in connection with the high-temperature stability of niobium in oxidizing solutions. When alloys of niobium with various metals such as Ti, Cr, Al, Mo, Ta, W, and V are exposed to oxidizing conditions, the mixed oxides formed on the surface affect the electrode potentials and corrosion resistance of the alloys [12].

A number of nonstoichiometric mixed oxides of niobium in which the formal oxidation state of the metal less than 5+ are also known [3]. These contain an excess of the cometal with which they are associated. $Sr_{0.7}NbO_3$ and $Sr_{0.95}NbO_3$ are examples of this type which possess high electrical conductivity, variability of composition, and characteristic colors.

a. *Halogen Compounds.* Niobium forms pentahalides by the direct action of the halogens or gaseous hydrohalides on the metals. The fluoride dissolves vigorously in water with the evolution of heat to give clear solutions which precipitate insoluble hydrolysis products on long standing or boiling [13]. The other pentahalides are insoluble in water, but are freely soluble in organic solvents.

In addition to the pentahalides, a number of oxide-halides such as $NbOF_3$, $NbOCl_3$, and $NbOI_3$ are also known. While $NbOF_3$ is not fully characterized, $NbCl_3$ has been well studied and shown to be stable. $NbOBr_3$ and $NbOI_3$ are less stable than $NbOCl_3$ in that they are hydrolyzed on exposure to air. The other oxyhalides in which niobium is pentavalent are NbO_2F, NbO_2Br, and NbO_2I. By contrast with the oxide-trihalides, these compounds are stable on exposure to air and are not attacked by cold mineral acids.

The tetrahalides, such as NbF_4, $NbCl_4$, and $NbBr_4$, are also known. Niobium(IV) fluoride is a black, nonvolatile, and hygroscopic solid [14,15]. It hydrolyzes on exposure to air, forming NbO_2F. It reacts with water to give products of indeterminate composition. Niobium(IV) chloride is stable in air and is soluble in 2 M HCl [16].

The trihalides, in general, are formed by decomposition of higher halides. $NbCl_3$ is a stable, unreactive material. It is insoluble in water, mineral acids, and organic solvents, but is oxidized by concentrated nitric acid.

b. *Other Compounds.* Niobium forms a series of interstitial hydrides in which hydrogen atoms are accommodated within the expanded metal lattices. These hydrides are formed by absorption of hydrogen by the metal when it is used as a cathode in the electrolysis of dilute H_2SO_4 [3].

Niobium forms compounds with carbon, nitrogen, silicon, and boron. The available thermodynamic data are provided in Table 3.

5. Niobium Potentials

Latimer [17] has calculated the electrode potential $E°$ for the reaction

$$Nb_2O_5 + 10H^+ + 10e^- \rightarrow 2Nb + 5H_2O \qquad (17.39)$$

to be 0.65 V. This is based on a value of -1807.5 kJ mol^{-1} for the Gibbs energy of formation of Nb_2O_5.

Based on the emf measurements of Kiehl and Hart [18] for the Nb^{3+}/Nb^{5+} couple in H_2SO_4, Latimer has estimated the potential of the Nb/Nb^{3+} couple. If +5 niobium exists as a complex ion of the formula $NbO(SO_4)_2^-$ at higher H_2SO_4 concentrations (3.14 to 9.87 M), the following reaction can be written:

$$NbO(SO_4)_2^- + 2H^+ + 2e^- \rightarrow Nb^{3+} + H_2O + 2SO_4^{2-} \qquad E° \cong -0.1 \qquad (17.40)$$

From the solubility data for niobic oxide in sulfuric acid, Latimer estimates the potential for the $Nb/NbO(SO_4)_2^-$ couple:

$$NbO(SO_4)_2^- + 2H^+ + 5e^- \rightarrow Nb + H_2O + 2SO_4^{2-} \qquad E° \cong -0.63 \qquad (17.41)$$

If the potential for the $Nb/Nb(SO_4)_2^-$ couple is the same as that for $Nb/NbO(SO_4)_2^-$, the potential for the Nb/Nb^{3+} couple is estimated as -1.1 V. As pointed out by Latimer, in 3 M H_2SO_4, +3 niobium may not be the simple ion Nb^{3+}. Following Latimer [17], the electrode potential relationships among the various oxidation states can be summarized as follows:

$$Nb_2O_5 \xrightarrow{-0.1 \text{ V}} Nb^{3+} \xrightarrow{-1.1 \text{ V}} Nb$$
$$\underline{\; -0.65 \text{ V} \;}$$

TABLE 4 Thermodynamic Data for Tantalum at 298 K

Substance	State	ΔH_f°/kJ mol^{-1}	ΔG_f°/kJ mol^{-1}	ΔS°/J mol^{-1} K^{-1}	References
Ta	c	0	0	41.51	41, 21
TaO_2^+	ao	—	−842.6	—	21
Ta_2O_5	c	−2046.8	−1912.1	143.1	42, 43
Ta_2O_5	aq	−2078.2	—	—	21
TaF_5	c	−1903.7	−1790.8	171.5	44
TaF_5	ao	—	−1131.7	—	21
TaF_6^-	ao	—	−1431.7	—	21
TaF_7^{2-}	ao	—	−1729.5	—	21
$TaCl_3$	c	−581.6	−514.6	154.8	45

TaCl$_4$	c	−711.28	−623.4	192.6	45, 46
TaCl$_5$	c	−761.5	−707.1	422.6	45, 46
TaBr$_5$	c	−598.3	−552.3	271.9	32
NaTaCl$_6$	c	−1288.6	—	—	47
KTaCl$_6$	c	−1548.8	—	—	47
RbTaCl$_6$	c	−1363.9	—	—	47
CsTaCl$_6$	c	−1410.0	—	—	47
NH$_4$TaCl$_6$	c	−1133.0	—	—	21
TaC	c	−146.4	−144.8	42.3	21
TaN	c	−251.0	−221.8	41.8	34

Because of the assumptions involved regarding the nature of Nb^{3+}, Nb/Nb^{3+}, and Nb^{3+}/Nb^{5+}, these potentials are rather approximate. Grube and Grube [19] have determined the single cell potentials for Nb^{3+}/Nb^{5+} in concentrated HCl and H_2SO_4. They have suggested that the following equilibria may exist:

$$Nb^{5+} + 2e^- \rightarrow Nb^{3+} \tag{17.42}$$

$$NbO^{3+} + 2H^+ + 2e^- \rightarrow Nb^{3+} + H_2O \tag{17.43}$$

$$NbO_3^- + 6H^+ + 2e^- \rightarrow Nb^{3+} + 3H_2O \tag{17.44}$$

Using the Nernst relation and the percentage of Nb^{3+} oxidized, they have calculated the $E°$ value for Nb^{5+}/Nb^{3+} in 1.01 M HCl to be -0.395 V, and at 6.1 M HCl, $E°$ is -0.269 V. Thus, in HCl, a significant dependence on the acid concentration is observed. These values agree well with the $E°$ values computed from $\Delta G_f°$ for reaction (17.42), but not for (17.43) and (17.44). It may be inferred that these results only support the absence of NbO^{3+} and NbO_3^-. However, it is doubtful whether Nb^{5+} can exist as a free ion in such high concentrations of acids.

McCullough and Meites [20] have studied the electroreduction of Nb(V). They have concluded that in dilute solutions of niobium (<9 mM) in 10.75 F HCl, Nb(V) exists as a monomer, and at higher concentrations (>9 mM), it exists as a dimer. They have suggested that the dimeric species are probably joined by two oxygen bridges and an intermetallic bond which may permit the disproportionation of $[Nb(IV)]_2$ and $[Nb(IV) \cdot Nb(II)]$ and stabilize $[Nb(V) \cdot Nb(IV)]$, $[Nb(V) \cdot Nb(III)]$, and $[Nb(IV) \cdot Nb(III)]$ against dissociation.

B. Tantalum

1. Oxidation States

The chemical properties of tantalum and its compounds are, in general, similar to those of niobium [1]. Even though oxidation states ranging from -1 to $+7$ have been suggested by Pourbaix [2], only the +5, +4, and +3 states are known [3]. There is no clear evidence for the independent existence of tantalum dioxide and monoxide [37]. The important compounds of tantalum in the +5 state are the pentoxide, Ta_2O_5, and pentahalides. The existence of the +4 state in chloride, bromide, and iodide media is known. The oxyhalides and halide complexes with pyridine, γ-picoline, bipyridyl, and 1,10-phenanthroline are also known [36].

The tantalum oxide corresponding to the +3 state is not known, but the trihalides, except the iodide, have been prepared. At room temperature the trichloride is unreactive toward water or dilute mineral acids, but upon boiling, it dissolves in water to give blue-green solutions [36]. The thermodynamic data for tantalum are given in Table 4.

2. Tantalum Pentoxide and Fluoride

Tantalum pentoxide is thermodynamically stable in the presence of acid, neutral, and alkaline aqueous, noncomplexing solutions [2]. It dissolves very slightly in sulfuric acid and more readily in oxalic acid, forming complex sulfates and oxalate ions. With concentrated hydrofluoric acid, it forms fluorinated and oxyfluorinated complexes [17]. TaF_5 dissolves vigorously in water with the evolution of heat to give clear solutions that precipitate insoluble hydrolysis products on long standing or boiling [36]. The other halides are insoluble in water, but are soluble in organic solvents. However, the polynuclear halides of the formula Ta_6X_{12} (X = F, Cl, Br, I) are soluble in water [36].

3. Tantalum Potentials

From the free energy of the pentoxide, for the reaction

$$Ta_2O_5 + 10H^+ + 10e^- \rightleftharpoons 2Ta + 5H_2O$$

Pourbaix [2] has reported the value for $E° = -0.81$ at pH 1. The equilibrium potential for the reaction above is difficult to determine from experimental measurements, because the protective layer of Ta_2O_5 screens the metal surface from the solution.

Both crystalline solids and aqueous solutions of complex halides of tantalum have been reported [38]. Varga and Freund [39], based on ion-exchange and potentiometric measurements, have suggested that species of the type $TaF_n^{(5-n)+}$ (n = 4 to 9) are present in aqueous solutions and the potentials for $Ta/TaF_n^{(5-n)+}$ vary between -0.2 and -0.4 V. Haissinsky [40], based on potentiometric data in aqueous fluoride solutions, has suggested the presence of TaF_7^{2-} (aq) ion:

$$TaF_7^{2-} (aq) + 5e^- \rightleftharpoons Ta(c) + 7F^- (aq) \qquad E°(18°C) = -0.45 \text{ V}$$

The potential data and the $\Delta G_f^°$ of Ta_2O_5(c) are consistent with the hydrolysis of K_2TaF_7 in water and solubility of Ta_2O_5 in cold hydrofluoric acid.

$$Ta_2O_5 \xrightarrow{-0.81} Ta$$

REFERENCES

1. J. O. Hill, I. Worsley, and L. G. Hepler, Chem. Rev., 71, 127 (1971).
2. M. Pourbaix, *Atlas of Electrochemical Equilibria in Aqueous Solutions*, Pergamon Press, Elmsford, N.Y., 1966, pp. 246-250.
3. F. Fairbrother, *The Chemistry of Niobium and Tantalum*, Elsevier, London, 1967, p. 26.
4. H. Schäfer, R. Gruehn, and F. Schulte, Agnew. Chem., 28, 78 (1966).
5. R. Gruehn, J. Less-Common Metals, 11, 119 (1966).
6. F. Fairbrother, N. Ahmed, K. Edgar, and A. Thompson, J. Less-Common Metals, 4, 466 (1962).
7. G. Jander and D. Ertel, J. Inorg. Nucl. Chem., 14, 71 (1960).
8. G. Neuman, Acta Chem. Scand., 18, 278 (1964).
9. B. Spinner, Rev. Chim. Miner., 5, 839 (1968).
10. A. Goiffon, R. Granger, C. Bockel, and B. Spinner, Rev. Chim. Miner., 10, 487 (1973).
11. M. T. Pope, *Heteropoly and Isopoly Oxometalates*, Springer-Verlag, New York, 1983, p. 7.
12. E. J. Felton, J. Less-Common Metals, 9, 206 (1966).

13. F. Fairbrother, *The Chemistry of Niobium and Tantalum*, Elsevier, London, 1967, p. 79.
14. F. P. Gortesma and R. Didchenco, Inorg. Chem., *4*, 182 (1965).
15. C. H. Brubaker, Jr., and R. C. Young, J. Am. Chem. Soc., *74*, 3690 (1952).
16. H. Schäfer, E. Sibbing, and R. Gerkers, Z. Anorg. Allg. Chem., *307*, 163 (1961).
17. W. Latimer, *Oxidation Potentials*, 2nd ed., Prentice-Hall, Englewood Cliffs, N.J., 1952.
18. S. J. Kiehl and D. Hart, J. Am. Chem. Soc., *50*, 2337 (1928).
19. V. G. Grube and H. L. Grube, Z. Elektrochem., *44*, 771 (1938).
20. J. G. McCullough and L. Meites, J. Electroanal. Chem., *18*, 123 (1968).
21. D. D. Wagman, et al., "The NBS Tables of Chemical Thermodynamic Properties," J. Phys. Chem. Ref. Data, 2-207, Vol. 11, Suppl. 2 (1982).
22. A. D. Mah, J. Am. Chem. Soc., *80*, 3872 (1958).
23. J. Worrell, J. Phys. Chem., *68*, 952 (1964).
24. E. J. Huber, E. L. Head, C. E. Holley, E. K. Storms, and N. H. Krikoriau, J. Phys. Chem., *65*, 1846 (1961).
25. O. E. Myers and A. P. Brady, J. Phys. Chem., *64*, 591 (1960).
26. E. Greenberg, C. A. Natke, and W. H. Hubbard, J. Phys. Chem., *69*, (1965).
27. A. P. Brady, O. E. Myers, and J. K. Clauss, J. Phys. Chem., *64*, 588 (1960).
28. H. Schäfer and F. Kahlenberg, Z. Anorg. Allg. Chem., *305*, 291 (1960).
29. P. Gross, C. Hayman, D. L. Levi, and G. L. Wilson, Trans. Faraday Soc., *56*, 318 (1960).
30. F. J. Keneshea, D. Cubicciotti, G. Withers, and H. Eding., J. Phys. Chem., *72*, 1272 (1969).
31. H. Schäfer, Angew. Chem., *67*, 748 (1955); H. Schafer and K. D. Dohman, Z. Anorg. Allg. Chem., *300*, 1 (1965); A. Simon, H. G. Schnering, H. Wohrle, and H. Schäfer, Z. Anorg. Allg. Chem., *339*, 155 (1965).
32. P. Gross, C. Haymau, D. L. Levi, and G. L. Wilson, Trans. Faraday Soc., *58*, 890 (1962).
33. H. Schäfer and H. Heine, Z. Anorg. Allg. Chem., *352*, 258 (1967).
34. A. D. Mah and N. L. Giellert, J. Am. Chem. Soc., *78*, 3261 (1956).
35. L. B. Pankratz, W. W. Weller, and K. K. Kelley, U.S. Bur. Mines Rep. Invest., No. 6446, U.S. Government Printing Office, Washington, D.C., 1964.
36. F. Fairbrother, *The Chemistry of Niobium and Tantalum*, Elsevier, London, 1967, pp. 137-145.
37. R. J. Wasilewski, J. Am. Chem. Soc., *75*, 1001 (1953).
38. J. L. Hoard, W. J. Martin, M. E. Smith, and J. F. Whitney, J. Am. Chem. Soc., *76*, 3820 (1954).
39. L. P. Varga and H. Freund, J. Phys. Chem., *66*, 22, 187 (1962).
40. M. Haissinsky, A. Coche, and M. Cottin, J. Chim. Phys., *44*, 234 (1947).
41. K. K. Kelley and E. G. King, U.S. Bur. Mines Bull., No. 592, U.S. Government Printing Office, Washington, D.C., 1961.
42. G. L. Humphrey, J. Am. Chem. Soc., *76*, 976 (1954).

43. A. N. Kornilov, V. Y. Leonidov, and S. M. Skuratov, Dokl. Akad. Nauk SSSR, *144*, 355 (1962).
44. E. Greenberg, C. A. Natke, and W. N. Hubbard, J. Phys. Chem., *69*, 2089 (1965).
45. A. R. Kurbanov, A. V. Suvorov, S. A. Shchukarev, and G. I. Novikov, Zh. Neorg. Khim., *9*, 520 (1964).
46. H. Schafer and F. Kahlenberg, Z. Anorg. Allg. Chem., *305*, 178 (1960).
47. E. K. Smirnova, I. V. Vasilkova, and N. F. Kudryashova, Russ. J. Inorg. Chem., *9*, 268 (1964).

18
Titanium, Zirconium, and Hafnium

I. TITANIUM*

WILLIAM J. JAMES AND JAMES W. JOHNSON   University of Missouri—Rolla, Rolla, Missouri

A. Aqueous Solutions

The enthalpy of formation of TiO_2, $\Delta H^\circ_{f,298} = 14.46$ kJ mol^{-1} [1], approaches that of $(1/3)Al_2O_3$, 15.97 kJ mol^{-1}. This suggests that Ti is a very active metal with potentials comparable to those of Al. Accordingly, in aqueous solutions or in other oxidative media free of complexing agents, the surface of Ti reacts readily with its environment to form oxide films of varying composition and stability depending on the pH and the nature of the electrolyte.

Access of the electrolyte to the active metal surface is possible only via diffusion through the film barriers. Solution of the active metal surface proceeds by passage of ions through the barrier, a process equivalent to the passage of an anodic current. Hydrogen may also develop on the metal surface. It follows that Ti, in an aqueous medium, is not in a true state of equilibrium and its standard potentials therefore cannot be measured accurately. Accordingly, standard potentials of Ti are usually calculated from thermochemical data, some of which are tabulated in Table 1. The redox potentials are listed generally in order of increasing values except for those reactions that involve oxides in both anhydrous and hydrated forms (e.g., TiO_2, Ti_2O_3). Thermodynamic data are given in Table 2.

The standard potentials listed for Ti^{3+}/Ti^{2+} necessitate some explanation. The value of -0.368 V is based on the calculations of Forbes and Hall [2]. George and McClure [3], however, calculated a standard redox potential of about -2.3 V vs. SHE. Some experimental studies by other investigators [14,15] lend support to the calculations of George and McClure. In a very careful experimental approach, Oliver and Ross [14] dissolved about 2 mM of $Ti(H_2O)_6Cl_3$ in 0.1 M HCl and determined the polarographic behavior of the solution under a H_2 atmosphere. No reduction to Ti^{2+} was observed. Solu-

---

*Reviewer: J. Kuta (deceased), J. Heyrovsky Institute of Physical Chemistry and Electrochemistry, Prague, Czechoslovakia.

TABLE 1 Calculated Standard Potentials, $E°$, at 25°C for Ti in Aqueous Solution

Electrode reaction[a]	$E°$/V	References
$Ti^{3+} + e^- \rightarrow Ti^{2+}$	~−2.3	3
$Ti^{3+} + e^- \rightarrow Ti^{2+}$	−0.369	2
$Ti^{3+} + 3e^- \rightarrow Ti(s)$	−1.209	5
$Ti^{2+} + 2e^- \rightarrow Ti(s)$	−1.63	4,5,6,8
$Ti^{2+} + 2e^- \rightarrow Ti(s)$	−1.628	7
$TiO^{2+} + 2H^+ + 4e^- \rightarrow Ti(s) + H_2O$	−0.882	7
$TiO^{2+} + 2H^+ + 2e^- \rightarrow Ti^{2+} + H_2O$	−0.135	8
$TiO^{2+} + 2H^+ + e^- \rightarrow Ti^{3+} + H_2O$	0.100	6,8
$TiO_2^{2+} + H_2O + 2e^- \rightarrow HTiO_3^- + H^+$	1.303	8
$TiO_2^{2+} + 2H^+ + 2e^- \rightarrow TiO^{2+} + H_2O$	1.800	8
$TiO_2^{2+} + 4H^+ + 6e^- \rightarrow Ti(s) + 2H_2O$	0.012	8
$HTiO_3^- + 5H^+ + 2e^- \rightarrow Ti^{2+} + 3H_2O$	0.362	8
$TiO(s) + 2H^+ + 2e^- \rightarrow Ti(s) + H_2O$	−1.306	6
$Ti_2O_3(s) + 6H^+ + 2e^- \rightarrow 2Ti^{2+} + 3H_2O$	−0.478	8
$Ti_2O_3(h,s) + 6H^+ + 2e^- \rightarrow 2Ti^{2+} + 3H_2O$	−0.248	8
$Ti_2O_3(s) + 2H^+ + 2e^- \rightarrow 2TiO(s) + H_2O$	−1.123	8
$Ti_2O_3(h,s) + 2H^+ + 2e^- \rightarrow 2TiO(s) + H_2O$	−0.894	8
$2Ti_3O_5(s) + 2H^+ + 2e^- \rightarrow 3Ti_3O_3(h,s) + H_2O$	−1.178	8
$2Ti_3O_5(s) + 2H^+ + 2e^- \rightarrow 3Ti_2O_3(s) + H_2O$	−0.490	8
$TiO_2(s) + 4H^+ + 2e^- \rightarrow Ti^{2+} + 2H_2O$	−0.502	6,8

TABLE 1 (Continued)

Electrode reaction[a]	$E°/V$	References
$TiO_2(h,s) + 4H^+ + 2e^- \rightarrow Ti^{2+} + 2H_2O$	−0.169	6,8
$TiO_2(s) + 4H^+ + e^- \rightarrow Ti^{3+} + 2H_2O$	−0.666	8
$TiO_2(h,s) + 4H^+ + e^- \rightarrow Ti^{3+} + 2H_2O$	0.029	10
$2TiO_2(s) + 2H^+ + 2e^- \rightarrow Ti_2O_3(h,s) + H_2O$	−0.786	8
$2TiO_2(s) + 2H^+ + 2e^- \rightarrow Ti_2O_3(s) + H_2O$	−0.556	8
$2TiO_2(h,s) + 2H^+ + 2e^- \rightarrow Ti_2O_3(s) + H_2O$	−0.139	8
$2TiO_2(h,s) + 2H^+ + 2e^- \rightarrow Ti_2O_3(h,s) + H_2O$	−0.091	8
$3TiO_2(s) + 2H^+ + 2e^- \rightarrow Ti_3O_5(s) + H_2O$	−0.589	8
$3TiO_2(h,s) + 2H^+ + 2e^- \rightarrow Ti_3O_5(s) + H_2O$	−0.453	8
$Ti + H^+ + e^- \rightarrow TiH(s)$	−0.65	10, 13
$Ti + 2H^+ + 2e^- \rightarrow TiH_2(s)$	−0.45	10, 13
$Ti_2O_3(h,s) + 10H^+ + 10e^- \rightarrow 2TiH_2(s) + 3H_2O$	−0.785	9, 12
$TiO_2(h,s) + 6H^+ + 6e^- \rightarrow TiH_2(s) + 2H_2O$	−0.668	9, 12
$Ti_2O_3(h,s) + 10H^+ + 10e^- \rightarrow 2TiH_2(s) + 3H_2O$	−0.522	9, 12
$TiO_2(h,s) + 6H^+ + 6e^- \rightarrow TiH_2(s) + 2H_2O$	−0.450	9, 12
$TiF_6^{2-} + 4e^- \rightarrow Ti(s) + 6F^-$	−1.191	7
$TiF_4(s) + 4e^- \rightarrow Ti(s) + 4F^-$	−0.89	4
$TiCl_4(\ell) + 4e^- \rightarrow Ti(s) + 4Cl^-$	−0.39	11
$TiBr_4(s) + 4e^- \rightarrow Ti(s) + 4Br^-$	−0.52	11
$TiI_4(s) + 4e^- \rightarrow Ti(s) + 4I^-$	−0.58	11

[a](h), hydrated.

TABLE 2 Thermodynamic Data for Ti at 25°C[a]

Formula	Description	State	$\Delta H^\circ_{f,298}$/kJ mol^{-1}	$\Delta G^\circ_{f,298}$/kJ mol^{-1}	S°_{298}/J mol^{-1} K^{-1}	References
Ti		g	397.5	351.5	180.20	4
Ti	Metal	c	0	0	30.29	4
Ti$^+$		g	1063.1			4
Ti^{2+}		g	2384.9			4
Ti^{2+}		aq		(−314)		4,8
Ti^{3+}		g	5054.3			4
Ti^{3+}		aq		(−350)		4,8
TiO		g	108.8			4
TiO		c	−912.1	(−489.2)	34.77	4,8
TiO$_2$		c	−866.1	(−888.4)	50.25	4,8
TiO$_2$	Hydrated	c		(−821.3)		4,8
TiO$_2 \cdot$H$_2$O		c		(−1058.5)		4,8
TiO^{2+}		aq		−577.4		4,8
TiO$_2^{2+}$		aq		(−467.2)		8

Ti$_2$O$_3$		c	−1535.5	(−1432.2)	78.78	4,8
Ti$_2$O$_3$	Hydrated	c		−1388		8
Ti$_3$O$_5$		c	−2443	−2314	129.4	4,8
Ti(OH)$_3$		c		−1049.8		8
TiF$_2$		c	−828.4	−782.8	(80.3)	4
TiF$_3$		c	−1318	−1217.1	(91.2)	4
TiF$_4$		c	−1548	−1448.9	(124.7)	4
TiF$_6^{2-}$		aq	−2322.5	−2118.4	(83.7)	4
TiCl		g	439.3			4
TiCl$_2$		c	−477.0	−401.7	(108.8)	4
TiCl$_3$		c	−699.3	−619.2	(127.6)	4
TiCl$_4$		g			353.1	4
TiCl$_4$		ℓ	−750.2	−674.5	252.7	4
TiBr$_2$		c	−397.5	−382.4	(132.2)	4
TiBr$_3$		c	−552.3	−521.3	(154.0)	4
TiBr$_4$		c	−648.5	−610.9	(208.4)	4
TiI$_2$		c	−255.2	−257.7	(154.8)	4

TABLE 2 (Continued)

Formula	Description	State	$\Delta H^\circ_{f,\,298}$/kJ mol^{-1}	$\Delta G^\circ_{f,\,298}$/kJ mol^{-1}	S°_{298}/J mol^{-1} K^{-1}	References
TiI$_3$		c	−334.7	−320.1	(197.9)	4
TiI$_4$		c	−426.8	−426.8	(258.6)	4
TiN		c	−305.4	−276.6	30.1	4
TiC		c	−226	−222	24.3	4
TiH$_2$		c		−2.39		12
TiH$_2$		c		−4.92		13
TiH		c		−3.59		13
HTiO$_3^-$		aq		−955.9		10
FeTiO$_3$		c	−1207.1	−1125.1	105.9	4

^{a}Estimated values are in parentheses.

tions prepared by dissolving Ti, TiH_x, or TiO in mineral acids in the polarographic cell produced waves only for Ti^{3+}. Complexing of the Ti^{3+} ions with citrate and subjecting the system to a potential down to about -1.56 V (SHE) caused no reduction of Ti^{3+}. Based on their experimental results they estimated a reduction potential for Ti^{3+}/Ti^{2+} of about -2.1 V.

Although Latimer [4] cites the $E°$ for $Ti^{3+} + e^- \rightarrow Ti^{2+}$ to be -0.37 V, there can be little doubt, as suggested by Oliver and Ross [14], that the potential measured at zero current by Forbes and Hall [2] is a mixed potential limited on the positive side by oxidation of Ti(III), not Ti(II), and limited on the negative side by reduction of H^+.

Whereas the reduction of Ti^{3+} to Ti^{2+} does occur in dry acetonitrile solutions [16], all experimental evidence points to the high instability of Ti^{2+} in aqueous solutions, that is Ti appears to be oxidized directly to Ti^{3+} without any stable intermediates.

Experimental studies of the cathodic polarization of Ti in sulfate solutions containing Ti^{4+} and Ti^{3+} by Thomas and Nobe [15] indicate the reduction process to be the reduction of Ti^{4+} to Ti^{3+} and not of Ti^{3+} to Ti^{2+}, a further confirmation of the calculations of George and McClure.

For those cell reactions involving solid oxides of TiO_2 and Ti_2O_3, standard potentials are given for both stable anhydrous (a) and hydrated (h) forms of the oxides, $TiO_2 \cdot H_2O$ and $Ti(OH)_3$, respectively. For those reactions involving TiO_2, the standard potentials are computed from a Gibbs energy of formation for the anhydrous form of -5.080 kJ mol^{-1} and for the hydrated form -6.052 kJ mol^{-1}. The Gibbs energy of formation of the anhydrous form of Ti_2O_3 is -8.189 kJ mol^{-1} and of the hydrated form -12.01 kJ mol^{-1}.

The standard reduction potentials for the reactions

$$TiX_4(s) + 4e^- \rightarrow Ti(s) + 4X^- \qquad (18.1)$$

are taken from the studies by Beck [11] on the pitting of Ti metal in halide solutions. The values are based on thermodynamic data given by Latimer [4]. Beck also presents some experimental evidence which supports the calculated standard potential cited for $TiBr_4$.

Examination of Table 1 reveals the absence of Ti^+. However, the monovalent ion exists although its compounds are relatively unstable. Studies by Funaki et al. [17] reveal that the stability of the monohalides of Ti increases with increasing atomic weight of the halogen atom. Ti reacts violently with I_2 above 130°C; TiI_3, TiI_2, and TiI are formed in the presence of excess Ti. The monobromides decompose to form Ti and $TiBr_2$.

Barber et al. [18] have calculated enthalpies of formation of the solid mono-, di-, and trihalides of Ti, but no standard potential has been calculated for the reaction

$$TiX(s) + e^- \rightarrow Ti(s) + X^- \qquad (18.2)$$

1. Steady-State Potentials in Aqueous Solutions

It is evident that experimental verification of the standard potentials in Table 1 is rendered difficult by the presence of oxide or hydride films formed on the Ti surface, which causes it to behave as a noble metal. Ti dissolves rapidly only in HF or in acid media containing soluble fluorides or complexing agents. As the reactivity of Ti is closely tied to its potential, the measured potentials are not equilibrium values but rather are corrosion or dissolution potentials. As such, they depend on the metallurgical history of the metal; the surface preparation; nature, composition, and conductivity of surface layers; overpotential; pH; and current density. Accordingly, poor reproducibility of potential measurements is to be expected.

A partial listing of steady-state corrosion potentials of Ti in several aqueous solutions is given by James and Straumanis in *Encyclopedia of the Electrochemistry of the Elements*[19].

B. Potential Diagrams for Titanium

Acid Solution

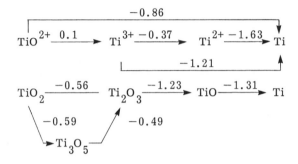

Basic Solution

$$TiO_2 \xrightarrow{-1.38} Ti_2O_3 \xrightarrow{-1.95} TiO \xrightarrow{-2.13} Ti$$

with $TiO_2 \xrightarrow{-1.42} Ti_3O_5 \xrightarrow{-1.32} Ti_2O_3$

ACKNOWLEDGMENT

The authors wish to thank Janet Thompson for typing the manuscript.

REFERENCES

1. O. Kubaschewski and E. L. L. Evans, *Metallurgical Thermochemistry*, Wiley, New York, 1965.
2. G. S. Forbes and I. P. Hall, J. Am. Chem. Soc., 46, 385 (1924).
3. P. George and D. S. McClure, *Progress in Inorganic Chemistry*, F. A. Cotton, ed., Vol. 1, Interscience, New York, 1959.
4. W. M. Latimer, *Oxidation Potentials*, 2nd ed., Prentice-Hall, Englewood Cliffs, N.J., 1952.
5. W. J. Hamer, M. S. Malmberg, and B. Rubin, J. Electrochem. Soc., 103, 8 (1956).

6. U. R. Evans, *The Corrosion and Oxidation of Metals*, Edward Arnold, London, 1960, p. 900.
7. A. J. de Béthune and N. A. Swendeman Loud, *Standard Aqueous Electrode Potentials and Temperature Coefficients at 25°C*, Hampel, Skokie, Ill., 1964.
8. J. Schmets, J. Van Muylder, and M. Pourbaix, in *Atlas of Electrochemical Equilibria in Aqueous Solutions*, M. Pourbaix, ed., translated from the French by J. A. Franklin, Pergamon Press, Oxford, 1966.
9. N. T. Thomas and K. Nobe, J. Electrochem. Soc., *117*, 622 (1970).
10. A. M. Sukhotin and L. I. Tungusiva, Protect. Met., *4*, 5 (1968).
11. T. R. Beck, J. Electrochem. Soc., *120*, 1310, 1317 (1972).
12. Y. V. Baymakov and O. A. Breganski, *U.S. Dept. of Commerce Rep.*, AD 613, 519, MT 64-317 (1966).
13. B. Stalinski and Z. Breganski, Bull. Acad. Pol. Sci., *10*, 247 (1962).
14. J. W. Oliver and J. W. Ross, Jr., J. Am. Chem. Soc., *85*, 2565 (1963).
15. N. T. Thomas and K. Nobe, J. Electrochem. Soc., *119*, 1450 (1972).
16. L. M. Kolthoff and F. G. Thomas, J. Electrochem. Soc., *111*, 1065 (1964).
17. K. Funaki, K. Uchimura, and Y. Kuniya, Kogyo Kagaku Zasshi, *64*, 1914 (1961); K. Funaki, K. Uchimura, and H. Matsunaga, Kogya Kagaku Zasshi, *64*, 129 (1961).
18. M. Barber, J. W. Linnett, and N. H. Taylor, J. Chem. Soc., 3323 (1961).
19. W. J. James and M. E. Straumanis, in *Encyclopedia of the Electrochemistry of the Elements*, A. J. Bard, ed., Marcel Dekker, New York, 1976, p. 317.

II. ZIRCONIUM AND HAFNIUM*

DENNIS G. TUCK   University of Windsor, Windsor, Ontario, Canada

Zirconium ($Z = 40$) and hafnium ($Z = 72$) are conveniently dealt with in a single short chapter, since there are many similarities, often extending beyong the qualitative level, in the chemistries of these two elements [1]. These are generally held to reflect the effects of the lanthanide contraction, which results in almost identical atomic radii [$r(Zr)$ 0.145 nm, $r(Hf)$ 0.144 nm], ionic radii [$r(Zr^{4+})$ 0.074 nm, $r(Hf^{4+})$ 0.075 nm], and charge densities, all of which may affect the chemical and physical properties of the elements and their compounds. These similarities extend to one area of crucial importance in electrochemistry, in that there are serious experimental difficulties hindering quantitative study of the aqueous solution chemistry of both elements. In consequence, much of the available information is the result of calculation rather than experimental determination.

Reviewer: William J. James, University of Missouri—Rolla, Rolla, Missouri.

A. Aqueous Solution Chemistry of Zirconium and Hafnium

The general chemistry of the elements has been discussed by a number of authors [1]. The highest oxidation state is +IV, but inorganic or organometallic compounds are also known for the states +I, +II, and +III. The aqueous solution chemistry is confined to compounds of M(IV), but some lower oxidation state species have been identified in nonaqueous or fused salt media. A number of complexes of both zirconium and hafnium have been prepared, and stability constant measurements have been reported in some cases [2].

A good review of early work on the aqueous solution chemistry of zirconium is given by Lister and McDonald [3] and more recent work on both elements has been discussed elsewhere [4-6]. In summary, there is no evidence for the presence of M^{4+}(aq) ions even in strongly acidic media, and those cationic species which do exist are formulated as oxocation derivatives, which exist only at $[H^+] < 0.5\ M$. Much of the evidence points to the ready formation of polymeric species, with oxo and hydroxo bridges. Hydrolysis is always a factor to be considered, and the hydroxides $M(OH)_4$ precipitate very readily, even in acidic solution; the solubility products are of the order of 10^{-56} for both compounds. In attempts to circumvent those problems, some workers have studied the behavior of aqueous solutions of the oxychlorides, but here again hydrolysis and oligomerization are of critical importance.

To compound the difficulties facing the experimentalist, zirconium and hafnium readily form a surface layer of oxide, so that passivity in solution may be significant. All in all, the dearth of reliable experimental results for the standard potential comes as no surprise.

B. Equilibrium Potentials

Any critical evaluation of the standard potentials for these two elements is rendered impossible by the absence of experimental data. As emphasized above, the nature of the aqueous solution chemistry of the elements prevents direct measurement of the electrode potentials, at least in the absence of complexing agents such as fluoride ion. In particular, the instability of the M^{4+}(aq) ion in aqueous solution, and the relatively narrow pH range in which solution chemistry can be studied, are daunting barriers to any experimental investigation.

The available thermodynamic results are collected in Tables 3 and 4 and one notes immediately that the data for hafnium are sparse, even relative to those available for zirconium. Many of the values are estimates rather than the result of direct measurement. A striking absence is the lack of reliable results for Hf^{n+}(g) ions. The $\Delta H°$ and $S°$ values for Hf(g) have apparently not been reported. The ionization energies have been estimated [12] as I_1 = ca. 7 eV (ca. 675 kJ mol^{-1}) and I_2 = 14.9 eV (1435 kJ mol^{-1}), values which are similar to those for zirconium, and it seems reasonable to assume that the $\Delta H°$ values for the formation of the Hf^+, Hf^{2+}, and so on, ions in the gas phase will also be close to those for the corresponding zirconium ions.

In his classical text, Latimer [9] considered the process

$$ZrO_2 + 4H^+ + 4e^- \rightarrow Zr + 2H_2O \qquad (18.3)$$

for which $E° = -1.45$ V (revised value, using data in Table 3). An estimate of $K = 0.006$ for the equilibrium

TABLE 3 Thermodynamic Data for Zirconium and Some of Its Compounds[a]

Formula	State	$\Delta H°$/kJ mol^{-1}	$\Delta G°$/kJ mol^{-1}	$S°$/J mol^{-1} K^{-1}
Zr	g	620.5 ± 14.5	578.0	181.3 ± 0.4
Zr	ℓ	(26.6)	(10.0)	(55.6)
Zr(α)	c,α	0.0	0.0	39.0
Zr(β)	c,β	7.14	4.84	46.7
Zr$^+$	g	1281 ± 17	1238 ± 16	183.53 ± 0.04
Zr^{2+}	g	2646	—	—
Zr^{3+}	g	4981	—	—
Zr^{4+}	g	8261	—	—
ZrO$_2$	c	−1097.5 ± 1.7	−1039.7	(50.4 ± 0.3)
ZrO$_2$·nH$_2$O	c	(−1186)	—	—
ZrO(OH)$_2$	c	(−1413)	—	—
Zr(OH)$_4$	c	(−1720)	—	—
ZrF	g	(83 ± 20)	(52)	(243 ± 8)
ZrF$_2$	c	(−960 ± 60)	(−910)	(75 ± 8)
ZrF$_3$	c	(−1400 ± 80)	(−1324)	(88 ± 8)

TABLE 3 (Continued)

Formula	State	$\Delta H°$/kJ mol^{-1}	$\Delta G°$/kJ mol^{-1}	$S°$/J mol^{-1} K^{-1}
ZrF$_4$	c	-1911 ± 1	-1813	117.3 ± 0.2
ZrCl	g	205 ± 21	(174)	(254 ± 8)
ZrCl$_2$	c	(-430 ± 40)	(-385)	(110 ± 12)
ZrCl$_3$	c	(-714 ± 60)	(-646)	(146 ± 12)
ZrCl$_4$	c	-979.8 ± 1.2	-889.0	181 ± 3
ZrBr	g	(300 ± 40)	(255)	(265 ± 8)
ZrBr$_2$	c	(-405 ± 40)	(-314)	(115 ± 12)
ZrBr$_3$	c	(-635)	(-606)	(172)
ZrBr$_4$	c	-760 ± 2.5	(-724)	(225)
ZrI	g	($591 + 40$)	(538)	(275)
ZrI$_2$	c	(-259)	(-275)	(150)
ZrI$_3$	c	(-397)	(-395)	(205)
ZrI$_4$	c	-489.9 ± 3.3	(-480.6)	(257)
ZrH	g	(516)	(483)	(216)
ZrO	g	60 ± 50	25	227 ± 8

[a]Parentheses indicate estimates quoted in Ref. 7. All uncertainties are quoted from the original source.

TABLE 4 Thermodynamic Data for Hafnium and Some of Its Compounds[a]

Formula	State	$\Delta H°$/kJ mol^{-1}	$\Delta G°$/kJ mol^{-1}	$S°$/J mol^{-1} K^{-1}
Hf	c	0.0	0.0	54.8
Hf^{4+}	aq	—	(−555)	—
HfO$_2$	c	−1136	(−1080)	(66)
HfF$_4$	c	−1930	—	—
HfCl$_2$	c	−545	—	—
HfCl$_3$	c	−780	—	—
HfCl$_4$	c	−992	—	—
HfBr$_2$	c	−450	—	—
HfBr$_4$	c	−835	—	—

[a] Parentheses indicate estimates quoted in Ref. 7. All uncertainties are quoted from the original source.

$$Zr(OH)_4 + 4H^+ \rightarrow Zr^{4+} + 4H_2O \tag{18.4}$$

then leads to $\Delta G°(Zr^{4+}_{(aq)}) = -590$ kJ mol^{-1}, dependent on $\Delta G°[Zr(OH)_4] = -1550$ kJ mol^{-1} (a value that is no longer quoted in Ref. 7). The resultant is

$$Zr^{4+} + 4e^- \rightarrow Zr \qquad E° = -1.55 \text{ V} \tag{18.5}$$

For hafnium, Latimer [11] estimates

$$HfO_2 + 4H^+ + 4e^- \rightarrow Hf + 2H_2O \qquad E° = -1.57 \text{ V} \tag{18.6}$$

The difference of 0.12 V from the corresponding zirconium arises almost entirely from the difference in the $\Delta G°$ values for the two oxides. The estimated value for the standard potential is

$$Hf^{4+} + 4e^- \rightarrow Hf \qquad E° = -1.70 \text{ V} \tag{18.7}$$

A recent calculation [10] gave $E° = -1.44$ V, obviously lower than Latimer's value, but closer to the zirconium potential.

C. Potential Diagrams for Zirconium and Hafnium

```
      +4           0
Zr⁴⁺ ──−1.55── Zr

Hf⁴⁺ ──−1.70── Hf
```

D. Conclusion

The aqueous solution chemistry of zirconium and hafnium presents formidable barriers to the experimental determination of standard electrode potentials. The values calculated from available thermochemical data are valid only within the limits of the assumptions made as to the species present in aqueous solution. The unambiguous identification of these species is a problem as yet unresolved.

REFERENCES

1. F. A. Cotton and G. Wilkinson, *Advanced Inorganic Chemistry*, 3rd ed., Wiley-Interscience, New York, 1972, p. 927 et seq.
2. L. G. Sillén and A. E. Martel, *Stability Constants of Metal-Ion Complexes*, Spec. Publ. 17 and 25, Chemical Society, London, 1964, 1971.
3. B. A. Lister and L. A. McDonald, *J. Chem. Soc.*, 4315 (1952).
4. D. C. Bradley and P. Thornton, *Comprehensive Inorganic Chemistry*, A. F. Trotman-Dickenson, exec. ed., Vol. 3, Pergamon Press, Oxford, 1973, p. 419.
5. P. Pascal, *Nouveau traité de chimie minérale*, Masson, Paris, 1963, Chap. 9.

6. J. J. Habeeb and D. G. Tuck, in *Encyclopedia of the Electrochemistry of the Elements*, A. J. Bard, ed., Vol. X, Marcel Dekker, New York, 19xx.
7. *JANAF Thermochemical Tables* (D. R. Stull and H. Prophet, directors), 2nd ed., National Bureau of Standards, Washington, D.C., 1971 (NSRDS-NBS 37).
8. *Selected Values of Chemical Thermodynamic Properties*, Natl. Bur. Stand. Circ. 500, U.S. Government Printing Office, Washington, D.C., 1952.
9. W. M. Latimer, *The Oxidation States of the Elements*, 2nd ed., Prentice-Hall, Englewood Cliffs, N.J., 1952, p. 270.
10. V. P. Vasil'ev, L. A. Kochergina, and A. I. Lytkin, Zh. Neorg. Khim., 20, 18 (1975).
11. W. M. Latimer, *The Oxidation States of the Elements*, 2nd ed., Prentice-Hall, Englewood Cliffs, N.J., 1952, p. 273.
12. E. Moore, ed., *Atomic Energy Levels*, Vol. III (rev.) National Bureau of Standards, Washington, D.C., 1971, pp. 143-147.

19
Boron, Aluminum, and Scandium

I. BORON*

SU-MOON PARK   University of New Mexico, Albuquerque, New Mexico

Boron has three valence electrons whose ground-state structure is $s^2p^1(^2P_{1/2})$, while scandium has the valence electronic structure of $s^2d^1(^2P_{3/2})$ in its ground state. Since boron has a very high ionization potential, the formation of cations is very difficult, and consequently their chemistry is not known. Covalently bound trivalent boron chemistry is quite well known, however. The thermodynamic data are given in Table 1.

A. Oxidation States

Boron has largely been understood by analogy to silicon, but boron has its own unique chemistry, as does any other first-row element. A series of investigations made by Stock [1] and his co-workers on boron hydrides has opened a whole new area. Boron chemistry has also contributed to the development of bonding theories and concepts due to its unique covalent bonding. That is, boron is an electron-deficient atom; thus there are not enough electrons to form conventional two-electron bonds between all adjacent pairs of atoms.

There are two isotopes of boron, ^{10}B (19.6%) and ^{11}B (80.4%). Pure boron, prepared in crystalline forms, has three allotropic forms: α-rhombohedral, β-rhombohedral, and tetragonal. Boron in its higher oxidation states is quite common, but crystalline boron is chemically also very stable. The oxide, BO, is well established in the gas phase, but its structure is unknown. The most important oxide is B_2O_3, which exists in both crystalline and glass forms. The suboxides, B_3O, B_4O_5, and $B_{6.6}O$ have been reported. A number of boron hydrides, such as B_2H_6, B_4H_9, and others, are well known and well characterized. The stoichiometry of the boranes ranges from the simplest B_2H_6 to the most complex $B_{20}H_{16}$. X-ray crystallographic and other studies have shown that the structures of boranes are different from those of hydrocarbons. The structure of diborane may best be described by Lipscomb's "semitopological" scheme [2]:

Reviewer: M. V. Plichon, Ecole Supérieure de Physique et de Chimie Industrielles, Paris, France.

TABLE 1 Thermodynamic Data on Boron[a]

Formula	Description	State	$\Delta H°$/kJ mol^{-1}	$\Delta G°$/kJ mol^{-1}	$\Delta S°$/J mol^{-1} K^{-1}
B		g	562.7	518.8	153.34
B	β-Rhombohedral	c	0	0	5.86
B	Amorph.	c	3.8		6.53
B$^+$		g	1,369.59		
B^{2+}		g	3,802.92		
B^{3+}		g	7,468.82		
B^{4+}		g	35,500.5		
B^{5+}		g	65,332.7		
B$_2$		g	830.5	774.0	201.79
BO		g	25	−4	203.43
BO$^+$		g	1,302.5		
BO$^-$		g	−300.8	(−360.1)	198.87
BO$_2$		g	−300.4	−305.9	229.45
BO$_2^+$		g	1,025.1		
BO$_2^-$		g	−666.5	(−730.8)	215.81

Boron, Aluminum, and Scandium

Formula	Description	State			
BO_2^-	Std. state, $m = 1$	aq	-772.37	-678.94	-37.2
B_2O_2		g	-454.8	-462.3	242.38
$B_2O_2^+$		g	894.5		
B_2O_3		g	843.79	831.99	279.70
B_2O_3		c	-1,272.78	-1,193.70	53.97
B_2O_3	Amorph.	c	-1,254.53	-1,182.4	77.8
$B_2O_3^+$		g	514.6		
$B_4O_7^{2-}$	Std. state, $m = 1$	aq		-2,605.0	
BH		g	449.61	419.29	171.75
BH_3	Borane	g	100.		
BH_4^-	Std. state, $m = 1$	aq	48.16	114.27	110.5
B_2H_6	Diborane	g	35.6	86.6	232.00
B_2H_6	In n-C_5H_{12}, std. state, $X_2 = 1$	g	25.1	96.2	164.8
B_4H_{10}	Tetraborane	g	66.1		
B_5H_9	Pentaborane	g	73.2	174.9	275.81
B_5H_9		ℓ	42.67	171.67	184.22

TABLE 1 (Continued)

Formula	Description	State	$\Delta H°/\text{kJ mol}^{-1}$	$\Delta G°/\text{kJ mol}^{-1}$	$\Delta S°/\text{J mol}^{-1}\text{ K}^{-1}$
B_5H_{11}		g	103.3		
B_5H_{11}		ℓ	73.2		
B_6H_{10}	Hexaborane	g	94.5		
B_6H_{10}		ℓ	56.27		
$B_{10}H_{14}$		g	31.55	216.14	353.1
$B_{10}H_{14}$	Decaborane	c	−45.2	192.0	176.56
HBO_2		g	−561.9	−551.0	239.95
HBO_2	Cubic	c	−804.03		
HBO_2	Monoclinic	c	−794.24	−723.4	38
HBO_2	Orthorhombic	c	−788.76	−721.7	50
HBO_2	Nonionized, std. state, $m = 1$	aq	−782.0		
$HB(OH)_2$		g		−523	
H_2BOH		g		−268	
$H_2BO_3^-$	Std. state, $m = 1$	aq	−1,053.5	−910.4	30.5

Boron, Aluminum, and Scandium

H_3BO_3		g	−994.1	
H_3BO_3		c	−1,094.32	−969.01	88.83
H_3BO_3	Nonionized, std. state, $m = 1$	aq	−1,072.32	−968.85	162.3
$B(OH)_4^-$	Std. state, $m = 1$	aq	−1,344.03	−1,153.32	102.5
$H_2BO_3 \cdot H_2O_2^-$	Std. state, $m = 1$	aq		−1,057.7	
$H_2BO_3 \cdot H_3BO_3 \cdot 2H_2O_2^-$	Std. state, $m = 1$	aq		−2,161.9	
$(BOH)_3$		g	−1,209		
$(HBO_2)_3$		g	−2,276		
$H_2B_4O_7^-$	Std. state, $m = 1$	aq		−2,720.0	
$HB_4O_7^-$	Std. state, $m = 1$	aq		−2,685.3	
BF		g	−122.1	−149.8	200.3
BF_3		g	−1,137.00	−1,120.35	254.01
BF_4^-	Std. state, $m = 1$	aq	−1,574.9	−1,487.0	180
B_2F_4		g	−1,440.1	−1,410.4	
BOF		g	−607		
$(BOF)_3$		g	−2,376		

TABLE 1 (Continued)

Formula	Description	State	$\Delta H°/\text{kJ mol}^{-1}$	$\Delta G°/\text{kJ mol}^{-1}$	$\Delta S°/\text{J mol}^{-1}\text{K}^{-1}$
HBF_4	In 14.67 HF + 58.72H_2O	aq	$-1,571.1$		
$BF_2(OH)_2^-$	Std. state, $m=1$	aq		$-1,341.0$	
BF_3OH^-	Std. state, $m=1$	aq	$-1,527.2$	$-1,414.6$	167
$B_3F_4O_3OH^{2-}$	Std. state, $m=1$	aq	$-3,152.6$		
BCl		g	149.49	120.91	213.13
BCl_3		g	-403.76	-388.74	289.99
BCl_3		ℓ	-427.2	-387.4	206.3
B_2Cl_4		ℓ	-490.4	-460.7	357.3
$BOCl$		g	-314		
$(BOCl)_3$		g	$-1,633.4$	$-1,550.2$	381
$(BOCl)_4$		g	$-2,100$		
$BClF_2$		g	-890.4	-876.1	272
BCl_2F		g	-645.2	-631.4	285
BBr		g	238.1	195.4	224.9
BBr_3		g	-205.64	-232.46	324.13

Boron, Aluminum, and Scandium

BBr$_3$	ℓ	−239.7		229.7
BBrF$_2$	g		−238.5	286.27
BBr$_2$F	g			309.86
BBrCl$_2$	g			310.29
BBr$_2$Cl	g			321.75
BI$_3$	g	71.13	20.75	349.07
BS	g	342.00	288.78	216.10
BS$_2$	g	121		
B$_2$S$_2$	g	151		
B$_2$S$_3$	g	67		
B$_2$S$_3$	c	−240.6		
SF$_4$·BF$_3$	c	−2,025		
BN	g	647.47	614.50	212.17
BN	c	−254.4	−228.4	14.8
B$_3$N$_3$H$_6$	g	−511.75	−390.12	288.57
B$_3$N$_3$H$_6$	ℓ	−541.0	−392.8	199.6

TABLE 1 (Continued)

Formula	Description	State	ΔH°/kJ mol^{-1}	ΔG°/kJ mol^{-1}	ΔS°/J mol^{-1} K^{-1}
$B_3N_3H_3Cl_3$	β-Trichloroborazole	g	−994.9		
$B_3N_3H_3Cl_3$	β-Trichloroborazole	ℓ	−1,067.0		
BP	Cubic	c	−79		
$B_{13}P_2$		c	−167		
B_4C		c	−71	−71	27.1
$B(CH_3)_3$		g	−124.3	−36.0	314.6
$B(CH_3)_3$		ℓ	−143.1	−32.2	238.9
$B(C_2H_5)_3$		g	−157.69	−15.06	437.6
$B(C_3H_5)_3$		ℓ	−194.6	9.2	336.69
$LiBH_4$		c	−184.72		
$NaBH_4$		c	−183.38		

[a]National Bureau of Standards values are in italic; estimated values are in parentheses.

carbons. The structure of diborane may best be described by Lipscomb's "semitopoligical" scheme [2]:

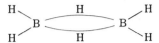

Owing both to the definite nonpolarity of the bonds and to the electron deficiency of the atoms, no oxidation numbers are assigned to boron in the boranes. The metal borides of compositions varying from M_4B to MB_{12} have been prepared and many of them show metallic conduction. The boron carbides are semiconductors, while the nitrides are electrical insulators. The B–N bonds in $B_3N_3H_6$, borazole, resemble the –C=C– bonds in aromatic compounds such as benzene. Borazine analogs of naphthalene and related hydrocarbons have been prepared by the pyrolysis of borazine or by other methods. Borohydrides such as $LiBH_4$ are well known as powerful and selective reductants, but their thermodynamic data are not available.

The most important oxidation state is +3, as seen in the oxide, B_2O_3, which forms metaboric acid, HBO_2, upon hydration. Further hydration of metaboric acid leads to boric acid, $B(OH)_3$. Boric acid is weakly acidic, similar to silicic acid, but not amphoteric, as $Al(OH)_3$ is. The boron halides excepting BF_3 are readily hydrolyzed. When small amounts of BF_3 are dissolved in water, fluoroborate, BF_4^-, is formed. Various sulfides and selenides, including B_2S_3 and B_2Se_3, have been well characterized.

B. Boric Acids and Their Salts

Boric acid, $B(OH)_3$, is moderately soluble in cold water, and the solubility increases markedly with temperature. Metal borates are of various types, the calcite type (e.g., $InBO_3$), the aragonite type (e.g., $NdBO_3$), and the vaterite type (e.g., $EuBO_3$). Many other meta- and pyroborates are known. Many borates occur naturally in the hydrated form.

In solution, boric acid is a very weak monobasic acid, behaving as a Lewis acid, not as a proton donor:

$$B(OH)_3(aq) + H_2O(\ell) \rightarrow B(OH)_4^- + H^+ \qquad K = 1 \times 10^{-9}$$

The calculated dissociation constant from the standard Gibbs energy data is 1.8×10^{-9}. At concentrations lower than $0.025\ M$, $B(OH)_3$ and $B(OH)_4^-$ are in equilibrium, but at higher concentrations, acidity increases, and the formation of polymeric species,

$$3B(OH)_3 \rightarrow B_3O_3(OH)_4^- + H^+ + 2H_2O \qquad K = 1.4 \times 10^{-7}$$

is consistent with pH measurements [3,4]. The polyborate formed appears to have three boron atoms with one negative charge. The equilibrium constant for the above polymerization reaction has been reported to be 110 and 145 by several investigators [3,4]. Other polyanions in solution such as mononegative tetramers, pentamers, and hexamers as well as the corresponding dinegative species have been postulated, but the data are not conclusive enough to justify any of these species.

The metaborate ion, $B_3O_6^{3-}$, hydrates very rapidly and spontaneously forms $B(OH)_3$ in the presence of water. No evidence for a neutral polymer $(HBO_2)_x$ has been found. It is therefore believed that neutral, trinegative forms are unstable with respect to their monomeric forms, trigonal $B(OH)_3$ and tetrahedral $B(OH)_4^-$, respectively.

The fluoroborate ion is hydrolyzed easily in water:

$$BF_4^- + H_2O(\ell) \rightarrow [BF_3OH]^- + HF(aq) \qquad K = 2.3 \times 10^{-3}$$

This species is also hydrolyzed in basic media,

$$BF_4^- + 4OH^- \rightarrow H_2BO_3^- + 4F^- + H_2O(\ell) \qquad \Delta G^\circ = -146.7 \text{ kJ mol}^{-1}$$

Boron trifluoride is one of the strongest Lewis acids known and readily reacts with many electron pair donors, such as water, ethers, alcohols, amines, phosphines, and others, to form adducts. Other boron trihalides have similar properties and are stronger Lewis acids than BF_3.

C. Boron Potentials

The reduction potential of boric acid is calculated from Gibbs energy data in acidic media,

$$B(OH)_3(aq) + 3H^+ + 3e^- \rightarrow B(c) + 3H_2O \qquad E^\circ = -0.890 \text{ V}$$

In basic media, we calculate

$$B(OH)_4^- + 3e^- \rightarrow B(c) + 4OH^- \qquad E^\circ = -1.811 \text{ V}$$

Temperature coefficients are calculated to be 0.55 and -1.45 mV K^{-1}, respectively, from entropy data.

The potential of BH_4^-,

$$BO_2^-(aq) + 6H_2O + 8e^- \rightarrow BH_4^- + 8OH^- \qquad E^\circ = -1.241 \text{ V}$$

can also be calculated. Gardiner and Collat [5] have reported two polarographic waves, one at $E_{1/2} = -0.11$ V vs. SCE (0.13 V vs. SHE) and one at -0.64 V vs. SCE (-0.40 V vs. SHE). The first polarographic wave at -0.11 V vs. SCE was assigned to the oxidation of BH_4^- and the second at -0.64 V vs. SCE to the oxidation of a hydrolysis product of BH_4^-, BH_4OH^-. From a controlled potential coulometry experiment, it was found that eight electrons are involved in the first wave and three in the second. The postulated oxidation of BH_3OH^- in the second wave is

$$BO_2^- + (3/2)H_2(aq) + 2H_2O(\ell) + 3e^- \rightarrow BH_3OH^- + 3OH^-$$

The following is a list of calculated standard electrode potentials for the reactions shown [6]:

$$BF_4^- + 3e^- \rightarrow B(c) + 4F^- \qquad E^\circ = -1.284 \text{ V}$$

$$2H_3BO_3(aq) + 12H^+ + 12e^- \rightarrow B_2H_6(g) + 6H_2O \qquad E^\circ = -0.519 \text{ V}$$

Boron, Aluminum, and Scandium

$$B_4O_7^{2-} + 26H^+ + 24e^- \to 2B_2H_6(g) + 7H_2O \qquad E° = -0.483 \text{ V}$$

$$2B(c) + 6H^+ + 6e^- \to B_2H_6(g) \qquad E° = -0.150 \text{ V}$$

The precision for calculated potentials in this chapter reflect those of original literature.

D. Organoborates and Organoboric Acids

Organoborates of the general structure shown below are easily prepared by the reaction of orthoboric acid with alcohols or of metal alkoxides with trialkyl orthoborates.

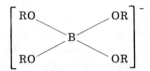

The reactions of various diols and polyols with sodium borate result in cyclic monoborates (from 1,2- and 1,3-diols), borospiranes (from pinacols), or dicyclic bis-borates (from pentaerythritols).

Following are equilibrium constants originally listed by Latimer [7]:

$$B(OH)_3 + \text{glycerine} = B(OH)_3 \cdot \text{glycerine} \qquad K = 0.9$$

$$B(OH)_3 \text{ glycerine} = H^+ + A^- \qquad K = 3 \times 10^{-7}$$

Phenylboric acid	$K = 137 \times 10^{-10}$	
o-Tolylboric acid	$K = 18.1 \times 10^{-10}$	
Benzylboric acid	$K = 75.5 \times 10^{-10}$	
β-Phenylethylboric acid	$K = 10.0 \times 10^{-10}$	
n-Butylboric acid	$K = 1.82 \times 10^{-10}$	

E. Potential Diagram for Boron

$$\overset{+3}{B(OH)_3} \xrightarrow{-0.890} \overset{0}{B}$$

REFERENCES

1. A. E. Stock, *Hydrides of Boron and Silicon*, Cornell University Press, Ithaca, N.Y., 1933.
2. W. N. Lipscomb, *Boron Hydrides*, W. A. Benjamin, New York, 1963.
3. J. O. Edwards, J. Am. Chem. Soc., 75, 6154 (1953).
4. N. Ingri, G. Laberström, M. Frydman, and L. G. Sillén, Acta Chem. Scand., 11, 1034 (1957).
5. J. A. Gardiner and J. W. Collat, Inorg. Chem., 4, 1208 (1965).

6. For a more extensive list, see M. I. Bellavance and B. Miller, *Encyclopedia of the Electrochemistry of the Elements*, A. J. Bard, ed., Vol. 2, Marcel Dekker, New York, 1974.
7. W. M. Latimer, *The Oxidation States of the Elements and Their Potentials in Aqueous Solutions*, Prentice-Hall, Englewood Cliffs, N.J., 1952.

II. ALUMINUM*

GEORGES G. PERRAULT   Laboratoire D'Electrochimie Interfaciale du CNRS, Meudon, France.

A. Oxidation States

The +3 oxidation state is the only one of importance. However, lower oxidation states and several of the corresponding compounds are known in the gaseous state. Some evidence has been obtained for an unstable +1 state as an intermediate in nonaqueous solutions. Al^+ ions have been postulated several times as an intermediate in electrochemical reactions at the aluminum electrode in aqueous solutions, but there is no convincing evidence of its possible existence and no values can be selected for the thermodynamical functions of the subvalent ions in aqueous solutions, although several estimated values have been proposed for them.

The hydroxide is amphoteric. With acids it forms compounds of the ion Al^{3+} and with bases, aluminates. Fluoride is acidic and many fluoaluminates are known.

B. Thermodynamic Data

The ΔH_f°, ΔG_f°, and S° values adopted by the National Bureau of Standards [NBS 270-3 (1968), NSRDS-NBS 37(1971)] have been accepted in most cases. Other values have been selected from various origins. More recent CODATA [1] and U.S. Geological Survey [2,3] recommended values are preferred against other selected values. See Table 2. Error intervals are given from original papers.

C. Electrode Potential

From the values selected in Table 2, the standard potentials can be calculated:

$$Al^{3+} + 3e^- \rightarrow Al \qquad E^\circ = -1.676 \text{ V}$$

$$Al(OH)_3(s) + 3e^- \rightarrow Al + 3OH^- \qquad E^\circ = -2.300 \text{ V}$$

$$Al(OH)_4^- + 3e^- \rightarrow Al + 4OH^- \qquad E^\circ = -2.310 \text{ V}$$

Isothermal coefficients for the Al/Al^{3+} and $Al/Al(OH)_3$ equilibria have also been determined [4].

---

*Reviewer: M. V. Plichon, Ecole Supérieure de Physique et de Chimie Industrielles, Paris, France.

TABLE 2 Thermodynamic Data on Aluminum[a]

Formula	Description	State	$\Delta H°/\text{kJ mol}^{-1}$	$\Delta G°/\text{kJ mol}^{-1}$	$S°/\text{J mol}^{-1}\text{ K}^{-1}$
Al		c	0	0	28.35 ± 0.08*
Al		g	329.7 ± 4.0*	285.8	164.44 ± 0.03*
Al$^+$		g	910.10		
Al^{2+}		g	2,732.9		
Al^{3+}		g	5,483.9		
Al^{3+}	Std. state, $m = 1$	aq	−531	−485	−322
AlO		g	91.2	65.3	218.3
AlO$^+$	Std. state, $m = 1$	aq		−654.2	
Al$_2$O		g	−131.4 ± 29.3	−159	259.4
(AlO)$_2$		g	−405.8 ± 29.3		(266.5)
AlO$_2$		g	(−184 ± 84)	(−147)	(245.6 ± 16.7)
AlO$_2^-$	Std. state, $m = 1$	aq	−918.8	−823.4	−20.9
AlO$_2$H		g	(−460 ± 63)	(−424)	(254.4 ± 6.3)
AlO$_2$H		c		−908.13	
AlOH		g	(−180 ± 12.5)	(−163)	(216.3 ± 2.1)

TABLE 2 (Continued)

Formula	Description	State	$\Delta H°/\text{kJ mol}^{-1}$	$\Delta G°/\text{kJ mol}^{-1}$	$S°/\text{J mol}^{-1}\text{K}^{-1}$
Al_2O_3	α-Corundum	c	$-1,675.7 \pm 1.3$*	$-1,582.3$	50.92 ± 0.10*
Al_2O_3	δ-Corundum	c	$-1,665$		
Al_2O_3	γ-Corundum	c	$-1,653$		(52.5)
Al_2O_3		amorph	$-1,632$		
$Al_2O_3 \cdot H_2O$	Boëhmite	c	$-1,975.4$	$-1,825.4$	96.88
$Al_2O_3 \cdot H_2O$	Diaspore	c	$-2,000$	$-1,841$	70.54
$Al(OH)_3$	Gibbsite	c	$-1,293.1 \pm 1.2$	$-1,154.9 \pm 1.2$	68.44 ± 0.14
$Al(OH)_3$	Bayerite	c	$-1,276.3$		
AlH	Monomer	g	259.2	231.2	187.8
AlH_3		g	75 ± 42	91 ± 42	170 ± 2
AlH_3	Electrochem.			103.8 ± 11.7	
AlH_3	Polymer α	c	-11.42 ± 0.84	46.48 ± 0.96	30.04
AlH_3	Polymer γ	c	-180 ± 0.63	50.41	48.9
AlH^{2+}	Electrochem.			-365.6 ± 12.6	

$AlOH^{2+}$	Std. state, *m* = 1	aq		*−694.1*	
$Al(OH)_3$		amorph	*−1,276*	−1,137.6	448.1
$Al(OH)_3$	Std. state, *m* = 1	aq		−1,094.6	
$Al(OH)_4^-$	Std. state, *m* = 1	aq	*−1,490.3*	*−1,297.8*	*117*
$Al(OH)_3 \cdot H_2O$		c		−1,376.42	
$Al(OH)_3 \cdot 3H_2O$		c		−1,850.4	
$BeAl_2O_4$		c	−2,280 ± 21		66.3
$MgAl_2O_4$		c	*−2,306.2 ± 7.9*		*80.62 ± 0.42*
$NaAlO_2$		c	*−1,133.1 ± 0.7*		*70.40*
$LiAlO_2$		c	*−1,189.6 ± 8.4*		53.35
AlF		g	*−265.3 ± 3.4*	−284	215.05 ± 0.04
AlF_2		g	−477		
AlF_2^+	Std. state, *m* = 1	aq		*−803.3*	
AlF^{2+}	Std. state, *m* = 1	aq	−1,113		
AlF_3	Std. state, *m* = 1, nonionized	aq	*−1,519*	*−1,414*	*−25*
AlF_3	Std. state, *m* = 1, ionized	aq	*−1,531*	*−1,322*	*−363.2*

TABLE 2 (Continued)

Formula	Description	State	$\Delta H°$/kJ mol^{-1}	$\Delta G°$/kJ mol^{-1}	$S°$/J mol^{-1} K^{-1}
AlF$_3$		g	−1,204.5	−1,188	277
AlF$_3$		c	−1,510.4 ± 1.3*	−1,431	66.5 ± 0.4*
Al$_2$F$_6$		g	−2,633.6 ± 16.8	−2,551	387 ± 13
AlOF		g	−586.6 ± 10.9		(234.24)
H$_3$AlF$_6$	In 800 HF +8000 H$_2$O		−2,489		
AlF$_6^{3-}$	Std. state, $m = 1$	aq	−2,522	−2,271.2	
Na$_3$AlF$_6$	Cryolite	c	−3,309.5 ± 4.2		238.4 ± 1.7
Li$_3$AlF$_6$		c	−3,383.6 ± 4.6		187.9 ± 0.2
LiAlF$_4$		c	−1,853.5 ± 8.4		(328.3 ± 8.4)
NaAlF$_4$		c	−1,868.5 ± 9.6		(340.0 ± 8.4)
AlF$_3$·(1/2)H$_2$O		c	−1,494	−1,395	117
AlF$_3$·3H$_2$O		c	−2,296	−2,051	209
AlF$_3$·2NH$_4$F·(3/2)H$_2$O		c	−2,817		
AlCl		g	−51.46 ± 4.18	−77.6	227.83 ± 0.08
AlCl$_2$		g	−330		

Boron, Aluminum, and Scandium 571

$AlCl_3$	g	−584.5 ± 2.9	−570.2	314.3 ± 4.2
$AlCl_3$	c	−705.6 ± 0.8	−628.8	109.3 ± 4.2
$AlCl_3$	aq	−1,033	−878	−152.3
		Std. state, $m = 1$		
Al_2Cl_4	g	−811 ± 21		376 ± 21
Al_2Cl_6	g	−1,295.5 ± 3.3	−1,220	475.5 ± 4.2
$AlOCl$	c	−793.3 ± 1.3		(54 ± 4)
$AlOCl$	g	−339		
$AlCl_3 \cdot 6H_2O$	c	−2,691		
$KAlCl_4$	c	−1,196.6 ± 10.5		(197 ± 8)
$NaAlCl_4$	c	−1,142 ± 4		(188 ± 8)
K_3AlCl_6	c	−2,092 ± 4		(377 ± 8)
Na_3AlCl_6	c	−1,979 ± 4		(347 ± 8)
$K_3Al_2Cl_9$	c	−2,860.1 ± 12.6		(469)
$AlBr$	g	15.1 ± 20.9		239.51
$AlBr_2$	g	−33		
$AlBr_3$	g	−423 ± 17	−452	352.1
$AlBr_3$	c	−527 ± 13	−504	180.2 ± 1.0

TABLE 2 (Continued)

Formula	Description	State	$\Delta H°$/kJ mol^{-1}	$\Delta G°$/kJ mol^{-1}	$S°$/J mol^{-1} K^{-1}
AlBr$_3$	Std. state, $m = 1$	aq	−895	−799	−74.5
Al$_2$Br$_6$		g	−968.8 ± 29.3	−978	547.3
AlI		g	59.8 ± 8.4		(247.87)
AlI$_2$		g	96		
AlI$_3$		g	−205.0 ± 7.5	(−253)	(363.07)
AlI$_3$		c	−309.2 ± 6.3	(−305)	(189.5 ± 8.4)
AlI$_3$	Std. state, $m = 1$	aq	−699	−640	12.1
Al$_2$I$_6$		g	−516.7		
(AlI$_3$)$_3$		g	−502.3 ± 12.6		584.12
2AlI$_3$·3PbI$_2$		c	−979		
2AlI$_3$·3PbI$_2$·10H$_2$O		c	−4,694		
AlS		g	200 ±84	150	230.5
Al$_2$S		g	21		
Al$_2$S$_3$		c	−724		

$Al_2(SO_4)_3$		c	-3,440.8	-3,100.1	239.3
$Al_2(SO_4)_3$	Std. state, $m = 1$	aq	-3,791	-3,205	-583.2
$Al_2(SO_4)_3 \cdot 6H_2O$		c	-5,311.5	-4,622.4	469.0
$Al_2(SO_4)_3 \cdot 8H_2O$		c	-8,878.8		
AlSe		g	347		
Al_2Se		g	104		
Al_2Se_3		c	-565		
AlN		c	-318.0		17.514
AlN		g	435 ± 84		(211.59)
$Al(NO_3)_3$	Std. state, $m = 1$	aq	-1,154.7	-820	117.6
$Al(NO_3)_3 \cdot 6H_2O$		c	-2,850.4	-2,203.8	467.6
$Al(NO_3)_3 \cdot 9H_2O$		c	-3,757.0		
$NH_4Al(SO_4)_2$		c	-2,352.2	-2,033.4	216.3
$NH_4Al(SO_4)_2$	Std. state, $m = 1$	aq	-2,481	-2,054.3	-168.2
$NH_4Al(SO_4)_2 \cdot 12H_2O$		c	-5,942.5	-4,937.9	697.1
$(NH_4)_2O \cdot 3Al_2O_3 \cdot 4SO_3 \cdot 6H_2O$		c	-10,078.0	-8,892.2	687.0
$(NH_4)_2O \cdot 3Al_2O_3 \cdot 5SO_3 \cdot 9H_2O$		c	-11,544.9		

TABLE 2 (Continued)

Formula	Description	State	ΔH°/kJ mol^{-1}	ΔG°/kJ mol^{-1}	S°/J mol^{-1} K^{-1}
AlBO$_2$		g	-548 ± 17		(259.4)
AlPO$_4$	Berlinite	c	$-1,733.8$	$-1,617.9$	90.79
H$_6$(NH$_4$)$_3$Al$_5$(PO$_4$)$_8 \cdot$18H$_2$O	Ammonium tara-kanite	c	$-18,747.7$	$-16,100.6$	1,422.1
(CH$_3$CO$_2$)$_3$Al	Triacetate	c	$-1,892.4$		
Al(BH$_4$)$_3$		ℓ	-16.3	144.7	289
LiAlH$_4$		c	-117.1 ± 8.4		(87.8 ± 8.4)
NaAlH$_4$		c	-112.9		
KAlH$_4$		c	-166.4		
CsAlH$_4$		c	-164.8		
Mg(AlH$_4$)$_2$		c	-96.6		
Al$_2$SiO$_5$	Andalusite	c	$-2,590.8 \pm 2.1$		93.17 ± 0.42
Al$_2$SiO$_5$	Kyanite	c	$-2,593.0 \pm 2.1$		83.77 ± 0.42
Al$_2$SiO$_5$	Sillimanite	c	$-2,587.8 \pm 2.1$		96.13 ± 0.42
Al$_2$Si$_2$O$_5$(OH)$_4$	Halloysite	c	$-4,101.5 \pm 2.7$	$-3,780.8 \pm 2.7$	203.3 ± 1.3

Kaolinite	$Al_2Si_2O_5(OH)_4$	c	$-4,120.1 \pm 2.6$	$-3,799.4 \pm 2.7$	203.0 ± 1.3
Mullite	$Al_6Si_2O_{13}$	c	$-6,815.8 \pm 6.3$		274.8 ± 2.5
Dickite	$Al_2Si_2O_5(OH)_4$	c	$-4,118.8 \pm 2.7$	$-3,796.3 \pm 2.7$	197.1 ± 1.3
Anorthite	$CaAl_2Si_2O_8$	c	$-4,243.0 \pm 3.1$	$-4,017.3 \pm 3.1$	193.3 ± 0.3
Hexagonal anorthite	$CaAl_2Si_2O_8$	c	$-4,222.6 \pm 3.3$	$-4,001.4 \pm 3.3$	214.8 ± 1.3
	$CaAl_2Si_2O_8$	Glass	$-4,171.3 \pm 3.3$	$-3,956.8 \pm 3.3$	273.3 ± 2.5
Lawsonite	$CaAl_2Si_2O_7(OH)_2 \cdot H_2O$	c	$-4,879.1 \pm 3.7$	$-4,525.6 \pm 3.9$	237.6 ± 2.1
Gehlenite	$Ca_2Al(Si_{0.5}Al_{0.5})_2O_7$	disordered	$-4,007.6 \pm 2.8$	$-3,808.7 \pm 2.9$	209.8 ± 1.6
Leonhardite	$Ca_2Al_4Si_8O_{24} \cdot 7H_2O$	c	$-14,246.5 \pm 9.6$	$-13,197.1 \pm 10.2$	922.2 ± 10.9
Low albite	$NaAlSi_3O_8$	c	$-3,935.1 \pm 3.4$	$-3,711.1 \pm .4$	207.4 ± 0.4
Analbite	$NaAlSi_3O_8$	c	$-3,924.2 \pm 3.6$	$-3,706.5 \pm 3.6$	226.4 ± 0.4
	$NaAlSi_3O_8$	glass	$-3,875.5 \pm 3.7$	$-3,665.3 \pm 3.7$	251.9 ± 1.8
Nepheline	$NaAlSiO_4$	c	$-2,110.3 \pm 2.0$	$-1,995.7 \pm 2.1$	124.3 ± 1.3
Analcime	$NaAlSi_2O_6 \cdot H_2O$	c	$-3,309.8 \pm 2.3$	$-3,091.7 \pm 2.5$	234.4 ± 2.5
Microcline	$KAlSi_3O_8$	c	$-3,967.7 \pm 3.4$	$-3,742.3 \pm 3.4$	214.2 ± 0.4
High sanidine	$KAlSi_3O_8$	c	$-3,959.5 \pm 3.4$	$-3,739.7 \pm 3.4$	232.9 ± 0.5

TABLE 2 (Continued)

Formula	Description	State	ΔH°/kJ mol^{-1}	ΔG°/kJ mol^{-1}	S°/J mol^{-1} K^{-1}
$KAlSi_3O_8$		glass	−3,914.7 ± 3.4	−3,703.5 ± 3.5	261.6 ± 2.3
$KAlSiO_4$	Kaliophilite	c	−2,121.9 ± 1.4	−2,006.0 ± 1.5	133.3 ± 1.3
$KAl_2(AlSi_3O_{10})\cdot(OH)_2$	Muscovite	c	−5,976.7 ± 3.2	−5,600.7 ± 3.3	306.4 ± 0.6
$KAlSi_2O_6$	Leucite	c	−3,038.7 ± 2.8	−2,875.9 ± 2.9	200.2 ± 1.7
$LiAlSiO_4$	Eucryptite	c	−2,123.3 ± 2.0	−2,009.2 ± 2.0	103.8 ± 0.8
$LiAlSi_2O_6$	α-Spodumene	c	−3,053.5 ± 2.8	−2,880.2 ± 2.8	129.3 ± 0.8
$LiAlSi_2O_6$	β-Spodumene	c	−3,025.3 ± 2.8	−2,859.5 ± 2.8	154.4 ± 1.2

[a]National Bureau of Standards recommended values are in italic; CODATA values are indicated by an asterisk; estimated values are in parentheses.

	$dE°/dT_{isoth}$/mV K^{-1}
$Al^{3+} + 3e^- \rightarrow Al$	0.504
$Al(OH)_3 + 3e^- \rightarrow Al + 3OH^-$	−0.93

Values close to these standard potentials have been observed in a few transitory experiments in which freshly cut or vapor-deposited aluminum surfaces were exposed to acid solutions:

$Al^{3+} + 3e^- \rightarrow Al$ $E = -1.67$ V   pH 3.2 [5]

$E = -1.60$ V $2.17 < $ pH $ < 5.20$ [6]

Potential-pH diagrams considering these equilibria have been calculated several times, as have the variations introduced by temperature changes [7]. However, in stationary experimental measurements, numerous investigations have shown that the equilibrium potentials of the aluminum electrode in aqueous solutions do not correspond to these standard potentials and E-pH diagrams. At very low partial pressure of oxygen, they can instead be correlated with equilibria involving aluminum hydrides [8].

Neutral trihydride AlH$_3$ and an ionized form AlH^{2+} have been considered and values of the $\Delta G_f^°$ of these two compounds obtained by potentiometric determination. Several values have been proposed for the $\Delta G_f^°$ of AlH$_3$ and it has been shown that they can be correlated to the polymerization degree of the hydride, the hydride electrochemically formed in alkaline solutions appearing intermediate between gaseous monomer and chemically prepared polymeric trihydride [9].

With the selected values, the standard potentials of the equilibria involving the hydrides can be calculated:

$Al^{3+} + H^+ + 2e^- \rightarrow AlH^{2+}$ $E° = -0.620$ V

$Al(OH)_{3am} + 4H^+ + 2e^- \rightarrow 3H_2O + AlH^{2+}$ $E° = -0.313$ V

$Al(OH)_4^- + H_2O + 2e^- \rightarrow AlH^{2+} + 5OH^-$ $E° = -1.984$ V

$Al(OH)_4^- + 3H_2O + 6e^- \rightarrow AlH_{3elec} + 7OH^-$ $E° = -1.748$ V

$Al(OH)_{3am} + 3H^+ + 6e^- \rightarrow 3OH^- + AlH_{3elec}$ $E° = -1.329$ V

The corresponding E-pH diagram has been observed experimentally [8] and several potential measurements obtained in alkaline solutions are close to these values.

pH 14 $E = -1.780$ V [10]

pH 14.2 $E = -1.730$ V   [11]

pH 14 $E = -1.748_2$ V   [9]

Other calculations of E-pH diagrams have been made with consideration of H$^-$ ions [12], but no experimental data can, so far, be correlated with the corresponding calculated equilibria. Several other assumptions, including the presence of other ionized hydrides and of hydroxyhydrides, have been considered, but none seems to correspond to the experimental results except possibly AlH$_2$OH [13].

The equilibrium potential of Al in fluorhydric solutions can also be calculated:

$$AlF_6^{3-} + 3e^- \rightarrow Al + 6F^- \qquad E° = -2.067 \text{ V}$$

and its isothermal coefficient has been determined [4]:

$$(dE°/dT)_{isoth} = -0.20 \text{ mV K}^{-1}$$

D. Equilibria in Solutions

From the selected values we can calculate the dissociation constants for the species in solution. The hydroxide dissociation,

$$Al(OH)_3 \rightarrow Al^{3+} + 3OH^-$$

corresponds to

$$\log[Al^{3+}] = -\Sigma \Delta G°/RT - 3 \log K_{H_2O} - 3 \text{ pH}$$

This equilibrium has been studied many times and shows variations with the various crystallographic forms.

For the amorphous hydroxide we obtain a value for the solubility product K_S:

$$\log K_S = -31.657$$

Other hydrolysis equilibria can also be calculated:

$$Al^{3+} + H_2O \rightarrow Al(OH)^{2+} + H^+$$

$$\log K = -4.923 \qquad K = 1.19 \times 10^{-5}$$

while Bjerrum [14] reported

$$K = 1.4 \times 10^{-5}$$

$$Al^{3+} + 4OH^- \rightarrow Al(OH)_4^- \qquad \log K = -32.165$$

which corresponds to a pH value for $[Al^{3+}] = [Al(OH)_4^-]$ of 5.96, while a recent determination [15], gives

$$5.95 < pH < 6.25$$

$$Al(OH)_3 + OH^- \rightarrow Al(OH)_4^- \qquad \log K = -0.509$$

$$\log[Al(OH)_4^-] = 13.491 - pH$$

Generally the hydrolysis is complicated by the formation of complexes such as in sulfate solutions [16]:

$$Al^{3+} + H_2O + SO_4^{2-} \rightarrow AlOHSO_4 + H^+ \qquad K = 1.25 \times 10^{-2}$$

$$2AlSO_4^+ + 2H_2O \rightarrow (AlOHSO_4)_2 + 2H^+ \qquad K = 1.25 \times 10^{-8}$$

Acid dissociation constant of the aluminic acid can also be calculated,

$$Al(OH)_3 \rightarrow AlO_2^- + H^+ + H_2O \qquad K = 3.24 \times 10^{-14}$$

while various investigators have found values ranging from 6×10^{-12} to 1.3×10^{-14} [17-19].

In fluorohydric solutions, the successive constants for the equilibria between Al^{3+} and F^- have also been determined [20]:

$$AlF^{2+} \rightarrow Al^{3+} + F^- \qquad K = 7.4 \times 10^{-7}$$

$$AlF_2^+ \rightarrow Al^{3+} + 2F^- \qquad K = 7.1 \times 10^{-12}$$

$$AlF_3 \rightarrow Al^{3+} + 3F^- \qquad K = 1.0 \times 10^{-15}$$

$$AlF_4^- \rightarrow Al^{3+} + 4F^- \qquad K = 1.8 \times 10^{-18}$$

$$AlF_5^{2-} \rightarrow Al^{3+} + 5F^- \qquad K = 4.3 \times 10^{-20}$$

$$AlF_6^{3-} \rightarrow Al^{3+} + 6F^- \qquad K = 1.44 \times 10^{-20}$$

The last value has been used to calculate ΔG_f° for AlF_6^{3-}.

E. Potential Diagrams for Aluminum

Acid Solution

$$\overset{+3}{Al^{3+}} \underset{}{\xrightarrow{-1.676}} \overset{0}{Al}$$

Alkaline Solution

$$\text{Al(OH)}_3 \xrightarrow{-2.300} \text{Al}$$
$$\text{Al(OH)}_4^- \xrightarrow{-2.310}$$

+3 above Al(OH)₃/Al(OH)₄⁻, 0 above Al.

REFERENCES

1. CODATA recommended key values for thermodynamics, 1977, J. Chem. Thermodyn., *10*, 903 (1978).
2. B. S. Hemingway and R. A. Robie, J. Res. U.S. Geol. Surv., 5(4), 413 (1977).
3. B. S. Hemingway and R. A. Robie, Geochim. Cosmochem., *41*, 1402 (1977).
4. A. J. de Bethune, T. S. Licht, and N. Swendeman, J. Electrochem. Soc., *106*, 616 (1959).
5. T. Hagyard, W. B. Earl, K. J. Kirkpatrick, J. G. Watson, J. Electrochem. Soc., *113*, 962 (1966).
6. T. Hagyard and J. H. Williams, Trans. Faraday Soc., *57*, 2288 (1961).
7. P. A. Malachowski, in *Encyclopedia of the Electrochemistry of the Elements*, A. J. Bard, ed., Vol. 6, Marcel Dekker, New York, 1977, Chap. 3.
8. G. G. Perrault, J. Electrochem. Soc., *126*, 199, 808 (1979).
9. G. G. Perrault, ECS Meet., St. Louis, Mo., May 1980.
10. Yu. F. Fateev, G. G. Vrjosek, and L. I. Antropov, Dvoinoi Sloi. Absorb. Tverd. Elektrod. Mater. Symp. III, Tartu, 1972, p. 281.
11. I. M. Kolthoff and C. J. Sambucetti, Anal. Chim. Acta, *21*, 17 (1959).
12. L. I. Antropov, G. G. Vrjosek, and Yu. F. Fateev, Zashch. Met., *11*(3), 300 (1975).
13. G. G. Perrault, UNESCO Discuss. Meet. Electrochem. Light Metals, Belgrade, Sept, 1982; Bull. Soc. Chim. Beograd, *48*, Supp. 155 (1983).
14. N. Bjerrum, Z. Phys. Chem., *59*, 350 (1907).
15. H. Schott, J. Pharm. Sci., *66*, 1548 (1977).
16. Guiter, C. R. Acad. Sci., *226*, 1092 (1948).
17. A. Maffei, Gass. Ital. Chem., *64*, 149 (1934).
18. R. Fricke, Koll. Z., *49*, 41 (1929).
19. I. M. Kolthoff, Z. Anorg. Allg. Chem., *112*, 185 (1920).
20. C. Brosset and J. Orring, Sven. Kem. Tidsk., *55*, 101 (1943).

III. SCANDIUM*

SU-MOON PARK   University of New Mexico, Albuquerque, New Mexico

A. Oxidation States

The only known chemistry of scandium has to do with the Sc(III) oxidation state. The coordination chemistry of scandium has been well studied recently [1]. The hydroxide, $Sc(OH)_3$, is more basic than aluminum hydroxide, but

---

Reviewer: M. V. Plichon, Ecole Supérieure de Physique et de Chimie Industrielles, Paris, France.

TABLE 3 Thermodynamic Data on Scandium[a]

Formula	Description	State	$\Delta H°$/kJ mol^{-1}	$\Delta G°$/kJ mol^{-1}	$\Delta S°$/J mol^{-1} K^{-1}
Sc		c	0	0	34.76
Sc		g	381.2	339.3	174.68
Sc$^+$		g	1,010.4		
Sc^{2+}		g	2,245.5		
Sc^{3+}		g	4,634.6		
Sc^{3+}	Std. state, $m = 1$	aq	−614.2	−586.6	−255.2
Sc^{4+}		g	11,723.1		
Sc$_2$		g	611.	556.	255.
ScO		g	−54.0	−79.9	224.47
Sc$_2$O		g	−22.2		
Sc$_2$O$_3$		c	−1,908.7	−1,819.2	77.0
Sc(OH)$^{2+}$	Std. state, $m = 1$	aq		−799	
Sc(OH)$_3$	Amorph.	c		−1,226	
ScF		g	−135.6	−161.1	222.2
ScF^{2+}	Std. state, $m = 1$	aq		−904	

TABLE 3 (Continued)

Formula	Description	State	$\Delta H°$/kJ mol^{-1}	$\Delta G°$/kJ mol^{-1}	$\Delta S°$/J mol^{-1} K^{-1}
ScF_2		g	−649	661	276
ScF_2^+	Std. state, $m = 1$	aq		−1,217	
ScF_3		c	−1,648	−1,573	92
ScF_3		g	−1,280	−1,268	293
ScF_3	Undiss., std. state, $m = 1$	aq		−1,523	
ScF_4^-	Std. state, $m = 1$ (ionic strength = 0.5)	aq		−1,816	
NH_4ScF_4		c	−2,163	−1,971	151
$(NH_4)_3ScF_6$		c	−3,092	−2,678	318
$ScCl$		g	115.9	89.1	234.30
$ScCl^{2+}$	Std. state, $m = 1$ (ionic strength = 0.5)	aq		−728.0	
$ScCl_2^+$	Std. state, $m = 1$	aq		−869.0	
$ScCl_3$		c	−925.1	−846.0	105
$ScCl_3$		g	−657	640	320.1

Boron, Aluminum, and Scandium 583

$ScCl_3 \cdot 6H_2O$		c	−2,798.7	
ScOI		c	−970	(59)
$Sc(OH)_2Cl$	Amorph.	c		−912
$Sc_2(OH)_5Cl$	Amorph.	c		−1,155
$Sc(ClO_3)^{2+}$	Std. state, m = 1	aq		−2,389
$ScBr^{2+}$	Std. state, m = 1	aq		−589.9
$ScBr_3$		c	−743.1	−812.1
$Sc(BrO_3)^{2+}$	Std. state, m = 1	aq		−588.7
$Sc(BrO_3)_2^{+}$		aq		−587.4
ScS		g	181.2	130.9	235.6
$Sc(SO_4)^{+}$	Std. state, m = 1	aq	−1,497.0	−1,454.4	71.1
$Sc(SO_4)_2^{-}$	Std. state, m = 1	aq	−2,389.9	−2,108.3	38
$Sc(SeO_4)^{+}$	Std. state, m = 1	aq		−1,038.5
$Sc(SeO_4)_2^{-}$		aq		−1,484.1
$Sc(NO_3)^{2+}$	Std. state, m = 1	aq		−704.6
ScAs		c	−268	

[a]National Bureau of Standards values are in italic; estimated values are in parentheses.

more acidic than hydroxides of yttrium and lanthanum. As a result, scandiumates are generally unstable.

The hydroxide and fluoride have a low solubility, but the fluoride shows an appreciable solubility in higher concentration of F^- due to formation of the complex, ScF_6^{3-}. Other salts with weak acid anions generally have low solubilities. The hydroxide cannot be completely precipitated from ammonium solution probably because of complex formation. The thermodynamic data on scandium are given in Table 3.

B. Scandium Potentials

The redox potential of the Sc^{3+}/Sc redox pair is calculated to be

$$Sc^{3+} + 3e^- \rightarrow Sc \qquad E^\circ = -2.03 \text{ V}$$

Sevoyukov and Hussan [2] obtained a polarographic half-wave potential of -1.745 V vs. SCE (-1.503 V vs. SHE) for the reduction above in 0.35 M LiCl solution (pH 5.7). The fact that this potential is less negative than the calculated one could be connected to a possible prewave due to hydrogen ion reduction.

Since the hydroxide has a low solubility in water, $K_{sp} = 4 \times 10^{-31}$, the potential for the reduction of scandium hydroxide must be calculated:

$$Sc(OH)_3 + 3e^- \rightarrow Sc(c) + 3OH^- \qquad E^\circ = -2.60 \text{ V}$$

Likewise, for the oxide the reduction potential is

$$Sc_2O_3(c) + 3H_2O + 3e^- \rightarrow 2Sc(c) + 6OH^- \qquad E^\circ = -2.74 \text{ V}$$

with an equilibrium constant of 4×10^{-37} for the reaction

$$Sc_2O_3(c) + 3H_2O \rightarrow 2Sc^{3+}(aq) + 6OH^-$$

Following are the redox potentials calculated for the reactions

$$ScF^{2+} + 3e^- \rightarrow Sc(c) + F^- \qquad E^\circ = -2.16 \text{ V}$$

$$ScF_2^+ + 3e^- \rightarrow Sc(c) + 2F^- \qquad E^\circ = -2.28 \text{ V}$$

$$ScF_3(aq) + 3e^- \rightarrow Sc(c) + 3F^- \qquad E^\circ = -2.37 \text{ V}$$

Some reported equilibrium constants are also listed below for possible use in the calculation of standard electrode potentials of appropriate redox pairs.

$$Sc(OH)_2Cl(c) \rightarrow Sc^{3+}(aq) + 2OH^- + Cl^- \qquad K_{sp} = 2.2 \times 10^{-22}$$

$$Sc(OH)_5Cl \rightarrow 2Sc^{3+}(aq) + 5OH^- + Cl^- \qquad K_{sp} = 4.6 \times 10^{-53}$$

$$Sc^{3+}(aq) + F^- \rightarrow ScF^{2+}(aq) \quad K = 1.2 \times 10^7$$

$$ScF^{2+}(aq) + F^- \rightarrow ScF_2^+(aq) \quad K = 6.5 \times 10^5$$

$$ScF_2^+ + F^- \rightarrow ScF_3(aq) \quad K = 3.0 \times 10^4$$

$$ScF_3(aq) + F^- \rightarrow ScF_4^-(aq) \quad K = 7 \times 10^2$$

For more extensive listings of equilibrium constants, the reader should consult the recently published review article [3].

C. Potential Diagram for Scandium

$$\overset{+3}{Sc^{3+}} \xrightarrow{-2.03} \overset{0}{Sc}$$

REFERENCES

1. G. A. Melson and R. W. Stotz, Coord. Chem. Rev., 7, 133 (1971).
2. N. N. Sevoyukov and M. Z. Hassan, Zh. Anal. Khim., 22, 1487 (1967).
3. J. G. Travers, I. Dellien, and L. G. Hepler, Thermochim. Acta, 15, 89 (1976).

20
Yttrium, Lanthanum, and the Lanthanide Elements *

LESTER R. MORSS Argonne National Laboratory, Argonne, Illinois

The chemistry of yttrium, lanthanum, and the lanthanide elements is, both in solids and in solution, primarily that of a group of classical tripositive ions. The stability of the +3 state for yttrium, lanthanum, and lutetium is obvious, since the energetics for oxidation and reduction of the completed-subshell +3 ions are unfavorable. To oxidize any of these elements from the +3 to the +4 state, the energy required to remove the outermost electron (the fourth ionization energy, I_4) is very large since the electron must be removed from a closed subshell (4p for yttrium, 5p for lanthanum, and 4f for lutetium). This energy expenditure is not recovered in the lattice energy in a solid, or in the solvation energy in a solution, for the +4 ion. To reduce any of these elements from the +3 to the +2 state, the energy gained in adding an electron ($-I_3$) is relatively small because the electron enters an empty subshell. This small energy gain is more than counterbalanced by the loss of lattice or hydration energy from a +3 to a +2 ion. Simply stated, the fourth ionization energies are much greater than the third (Table 1).

The predominant (although not exclusive) stability of the +3 ions for the lanthanides cerium through ytterbium is striking, and the explanation for this stability is somewhat more subtle. Since the +3 state of these elements is the most stable, the energetics of oxidation and reduction of the tripositive ions are unfavorable. With the exception of reduction of Gd^{3+}, the electron being lost in oxidation, or gained in reduction, of each of these tripositive ions is a 4f electron (Table 1), shielded by the outer (5p) filled subshell and participating only marginally in bonding. As is the case for yttrium, lanthanum, and lutetium, the +3 ions are the most stable because $I_4 \gg I_3$, but not for the same reason: For these elements $I_4 \gg I_3$ because it is the deeply buried 4f electrons that are being ionized.

Much interest in the redox chemistry of the lanthanides arises from perturbations to this dominant behavior. There are three principal causes of these perturbations. The first cause is the systematic increase of ionization energies I_3 and I_4 through the f^7 configuration, the sudden drop at f^8, and the systematic increase through f^{14}. Theoretical reasons for these trends have been well elucidated [2-5a]. These trends favor the reductions $f^6 \rightarrow f^7$

Reviewers: J. Fuger, University of Liège, Liège, Belgium; F. David, Nuclear Physics Institute, Orday, France.

TABLE 1 Electronic Configurations and Ionization Energies for Neutral Through Triply Ionized Lanthanides

Element	Neutral (M)		Singly ionized (M$^+$)		Doubly ionized (M^{2+})		Triply ionized (M^{3+})		Quadruply Ionized (M^{4+})
	Config.	I.E./kJ mol^{-1}	Config.	I.E./kJ	Config.	I.E./kJ mol^{-1}	Config.	I.E./kJ mol^{-1}	Config.
Y	$4d5s^2$	615.6	$5s^2$	1181.	$5s$	1980.	$4p^6$	5963	$4p^5$
La	$5d6s^2$	538.1	$5d^2$ / $5d6s$	1067.1	$5d$	1850.4	$5p^6$	4819.5	$5p^5$
Ce	$4f5d6s^2$	534.3	$4f5d^2$	1047.	$4f^2$	1948.8	$4f$	3546.7	$5p^6$
Pr	$4f^3 6s^2$	528.1	$4f^3 6s$	1018.	$4f^3$	2086.4	$4f^2$	3761.1	$4f$
Nd	$4f^4 6s^2$	533.1	$4f^4 6s$	1035.	$4f^4$	2132.4	$4f^3$	3899.	$4f^2$
Pm	$4f^5 6s^2$	538.6	$4f^5 6s$	1052.	$4f^5$	2152.	$4f^4$	3966.	$4f^3$
Sm	$4f^6 6s^2$	544.5	$4f^6 6s$	1068.	$4f^6$	2258.	$4f^5$	3995.	$4f^4$

Eu	$4f^7 6s^2$	547.1	$4f^7 6s$	1084.6	$4f^7$	2404.	$4f^6$	4110.	$4f^5$
Gd	$4f^7 5d 6s^2$	593.4	$4f^7 5d 6s$	1166.	$4f^7 5d$	1990.	$4f^7$	4245.	$4f^6$
Tb	$4f^9 6s^2$	565.8	$4f^9 6s$	1122.	$4f^9$	2114.	$4f^8$	3839.	$4f^7$
Dy	$4f^{10} 6s^2$	573.0	$4f^{10} 6s$	1126.	$4f^{10}$	2200.	$4f^9$	4001.	$4f^8$
Ho	$4f^{11} 6s^2$	581.0	$4f^{11} 6s$	1138.	$4f^{11}$	2204.	$4f^{10}$	4101.	$4f^9$
Er	$4f^{12} 6s^2$	589.3	$4f^{12} 6s$	1151.	$4f^{12}$	2194.	$4f^{11}$	4115.	$4f^{10}$
Tm	$4f^{13} 6s^2$	596.7	$4f^{13} 6s$	1163.	$4f^{13}$	2285.	$4f^{12}$	4119.	$4f^{11}$
Yb	$4f^{14} 6s^2$	603.4	$4f^{14} 6s$	1175.6	$4f^{14}$	2415.	$4f^{13}$	4220.	$4f^{12}$
Lu	$4f^{14} 5d 6s^2$	523.5	$4f^{14} 6s^2$	1341.	$4f^{14} 6s$	2022.3	$4f^{14}$	4360.	$4f^{13}$

Source: Ref. 1, which also gives estimated standard deviation. Data for yttrium from C. E. Moore, NSRDS-NBS 34, Natl. Bur. Stand., U.S. Dept. of Commerce, Washington, D.C., 1970.

and $f^{13} \to f^{14}$ (explaining in part the relative stability of Eu^{2+} and Yb^{2+}) and the oxidations $f^1 \to f^0$ and $f^8 \to f^7$ (explaining the relative stability of Ce^{4+} and Tb^{4+}). Less stable, but also observed as a result of these changes in ionization energies, are Sm^{2+}, Tm^{2+}, Pr^{4+}, Nd^{4+}, and Dy^{4+}. Systematic treatments have been presented by Johnson [6-8] and Morss [9].

The second cause of perturbations in redox behavior of the lanthanides is the metallic divalency of europium and ytterbium. This effect is illustrated effectively in Fig. 2 of Ref. 10, in which $E°(Ln^{3+}/Ln)$ are plotted. Since elemental Eu and Yb exist as divalent rather than trivalent metals, the divalent state must be unusually stable for these metals. These stable divalent metals cause $E°(Eu^{3+}/Eu)$ and $E°(Yb^{3+}/Yb)$ to be less negative than other $E°(Ln^{3+}/Ln)$ values; that is, the divalent metals destablize Eu^{3+} and Yb^{3+}.

The third and most subtle cause explains the existence of saline halides of dysprosium, since neither effect above is able to stabilize this divalent ion. Halides such as $DyCl_2$ can exist because $\Delta G° < 0$ for

$$Dy(s) + 2DyCl_3(s) = 3DyCl_2(s) \qquad (20.1)$$

By breaking down reaction (20.1) in terms of a Born-Haber cycle, Johnson [8] showed that both I_3 and, to a lesser extent, ΔH (sublimation) for Dy are responsible for its divalency. Thus the irregular behavior of ΔH (sublimation) across the lanthanides is partly responsible for the marginal divalency of some of the lanthanides, and contributes to that of samarium and thulium. This behavior of ΔH(sublimation) has been explained in terms of the divalency of the corresponding gaseous atoms [11,12].

An additional perturbation is attributed to the "tetrad effect" in enthalpies of hydration [10,13]. Although this effect has theoretical justification, its magnitude has variously been calculated as ca. 40 kJ mol^{-1} [14], ca. 20 kJ mol^{-1} [10], 5 to 10 kJ mol^{-1} [8], or as insignificant [15] with respect to uncertainties in ionization energies.

I. TRIVALENT IONS

A. $E°(M^{3+}/M)$ in Acid Solution

Herman and Rairden [16] have thoroughly reviewed polarographic reduction behavior of the trivalent lanthanides in various solvents. Although it was for some time believed that polarographic reductions of Ln^{3+} to $Ln(Hg)$ in two steps could be observed, it is now known that only the reductions M^{3+}/M^{2+} (Sm, Eu, Yb) can be measured polarographically. Therefore, values of $E°(M^{3+}/M)$ must be calculated from enthalpy and entropy determinations.

For all the Ln(III) aquo ions except Pm^{3+}, reliable enthalpies of formation and standard entropies [17] have been determined calorimetrically and have been critically reviewed [9,18,19,19a]. These data, and relevant recent thermodynamic data for other important lanthanide species, are reported in Table 2 [17-37]. From the aquo ion values (Table 2) we have calculated $\Delta G_f°(M^{3+},aq)$ and $E°(M^{3+}/M)$ as was earlier done by Guillaumont and David [10] and by Morss [9]. The less negative $\Delta G_f°(Eu^{3+},aq)$ and $\Delta G_f°(Yb^{3+},aq)$ and therefore the less negative $E°(Eu^{3+}/Eu)$ and $E°(Yb^{3+}/Yb)$ are caused by the divalency of Eu and Yb metals as explained in the preceding section. We note that a number of very recent surveys of lanthanide chemistry [16,38,39] ignore recent experimental data, causing their $E°(M^{3+}/M)$ tabulations to be obsolete

TABLE 2 Thermodynamic Data on Yttrium, Lanthanum, and the Lanthanide Elements[a]

Formula	Description	State	ΔH_f°/kJ mol^{-1}	ΔG_f°/kJ mol^{-1}	S°/J mol^{-1} K^{-1}
Yttrium					
Y		c	0	0	44.4
Y		g	421.3		
Y$^+$		g	1043		
Y^{2+}		g	2230		
Y^{3+}		g	4216		
Y^{3+}	Std. state, m = 1	aq	−715	−686	−249
Y$_2$O$_3$	Cubic	c	−1905.6	−1817.0	99
Y(OH)$_3$		c	−1435	−1291	99.2
YF$_3$		c	−1719	−1645	100
YCl$_3$		c	−996		
YCl$_3$	Std. state, m = 1	aq	−1216		

TABLE 2 (Continued)

Formula	Description	State	ΔH_f°/kJ mol^{-1}	ΔG_f°/kJ mol^{-1}	S°/J mol^{-1} K^{-1}
			Lanthanum		
La		c	0	0	57.0
La		g	431.0		
La$^+$		g	975.3		
La^{2+}		g	2048		
La^{3+}		g	3905		
La^{3+}	Std. state, $m=1$	aq	−709.4	−688	−209
La$_2$O$_3$	Hexag.	c	−1794.2	−1706.5	127.3
La$_2$O$_3$	Cubic	c	−1799		
La(OH)$_3$		c	−1410	−1278	117.8
LaF$_3$		c	−1699		
LaF$_3 \cdot$H$_2$O		c	−1987		

LaCl$_3$	Hexag.	c	−1073.2	−998	137.6
LaCl$_3$		g	−731		
LaCl$_3$	Std. state, $m = 1$	aq	−1210.9	−2715	463
LaCl$_3 \cdot$ 7H$_2$O		c	−3181		
LaOCl		c	−1011		
LaBr$_3$		c	−907.2		
LaI$_3$		c	−667		
CERIUM					
Ce		c	0	0	72.0
Ce		g	420.1		
Ce$^+$		g	960.6		
Ce^{2+}		g	2013		
Ce^{3+}		g	3969		
Ce^{3+}	Std. state, $m = 1$	aq	−700.4	−676	−205

TABLE 2 (Continued)

Formula	Description	State	ΔH_f°/kJ mol^{-1}	ΔG_f°/kJ mol^{-1}	S°/J mol^{-1} K^{-1}
Ce^{4+}		g	7522		
Ce^{4+}	1M HClO$_4$	aq	−580	−510	−419
CeO_2		c	−1088.7	−1024.6	62.3
Ce_2O_3	Hexag.	c	−1796.2	−1706.2	150.6
CeF_3		c	−1703	−1625	115
$CeF_3 \cdot H_2O$		c	−1977		
$CeCl^{2+}$	Std. state, $m = 1$	aq	−845	−803	−88
$CeCl_3$	Hexag.	c	−1058.0	−982	(150.5)
$CeCl_3$		g	−727		
$CeCl_3$	Std. state, $m = 1$	aq	−1201.9		
$CeCl_3 \cdot 7H_2O$		c	−3174		

CeOCl	c	−1004		
CeI$_3$	c	−650		
PRASEODYMIUM				
Pr	c	0	0	73.2
Pr	g	356.9		
Pr$^+$	g	891.2		
Pr^{2+}	g	1915		
Pr^{3+}	g	4008		
Pr^{3+}	aq	−706.2	−681	−207
Pr^{4+}	g	7776		
Pr^{4+}	aq	(−439)	(−371)	(−412)
PrO$_2$	c	−958		
Pr$_2$O$_3$	c	−1823.4	(−1734)	(154)
Pr$_2$O$_3$	c	−1827.6		

Std. state, m = 1
1 M HClO$_4$
Hexag.
Cubic

TABLE 2 (Continued)

Formula	Description	State	ΔH_f°/kJ mol^{-1}	ΔG_f°/kJ mol^{-1}	S°/J mol^{-1} K^{-1}
Pr_6O_{11}		c	−5686.1	−5360.6	
$Pr(OH)^{2+}$		aq		−863	
$Pr(OH)_2^+$		aq		−1076	
$Pr(OH)_3$		c	−1419	−1286	131.7
PrF_3		c	−1689		
$PrF_3 \cdot H_2O$		c	−1983		
$PrCl^{2+}$		aq		−817	
$PrCl_3$	Hexag.	c	−1059.0	−983	153.3
$PrCl_3$		g	−731		
$PrCl_3$	Std. state, $m = 1$	aq	−1207.7		
$PrCl_3 \cdot 6H_2O$		c	−2883		

PrCl$_3 \cdot$ 7H$_2$O		c	-3180	
PrOCl		c	-1014	
PrBr$_3$		c	-891.4	
PrI$_3$		c	-654	
NEODYMIUM				
Nd		c	0	0	71.5
Nd		g	326.9		
Nd$^+$		g	866.2		
Nd^{2+}		g	1907		
Nd^{2+}	Std. state, $m = 1$	aq	(-402)	-419	(-2)
Nd^{3+}		g	4045		
Nd^{3+}	Std. state, $m = 1$	aq	-696.6	-672	-206
Nd^{4+}		g	7951		
Nd^{4+}	1 M HClO$_4$	aq	(-263)	(-197)	(-413)

TABLE 2 (Continued)

Formula	Description	State	ΔH_f°/kJ mol^{-1}	ΔG_f°/kJ mol^{-1}	S°/J mol^{-1} K^{-1}
Nd$_2$O$_3$	Monocl.	c	−1807.9	−1721	158.6
Nd(OH)$_3$		c	−1679		129.9
NdF$_3$		c	−1973		
NdF$_3 \cdot$ H$_2$O		c	−706.9		
NdCl$_2$		c	−1041.8	−966	153.4
NdCl$_3$	Hexag.	c	−715		
NdCl$_3$		g	−1197.8		
NdCl$_3$	Std. state, $m = 1$	aq	−2875.2	−2461	417
NdCl$_3 \cdot$ 6H$_2$O		c	−1001		
NdOCl		c	−873.2		
NdBr$_3$		c	−639		
NdI$_3$		c			

PROMETHIUM

Pm	c	0	0	
Pm	g	(318)		(71)
Pm⁺	g	(863)		
Pm²⁺	g	(1920)		
Pm²⁺	aq	(−390)	(−406)	(−8)
Pm³⁺	g	(4079)		
Pm³⁺	aq	(−688)	(−663)	(−209)
PmCl₃	c			(1.55)

Std. state, $m = 1$

Std. state, $m = 1$

SAMARIUM

Sm	c	0	0	69.6
Sm	g	206.7		
Sm⁺	g	757.4		
Sm²⁺	g	1831		
Sm²⁺	aq	(−504)	(−514)	(−26)

Std. state, $m = 1$

TABLE 2 (Continued)

Formula	Description	State	ΔH_f°/kJ mol^{-1}	ΔG_f°/kJ mol^{-1}	S°/J mol^{-1} K^{-1}
Sm^{3+}		g	4095		
Sm^{3+}	Std. state, $m = 1$	aq	−691.1	−667	−207
Sm_2O_3	Monocl.	c	−1823.6	−1735.5	151.0
Sm_2O_3	Cubic	c	−1827.4	−1737.4	(145)
SmF_3		c	−1669		
$SmF_3 \cdot H_2O$		c	−1968		
$SmCl^{2+}$		aq		−800	
$SmCl_2$		c	−802.5		
$SmCl_3$		c	−1026.0	−950	150.1
$SmCl_3$	Hexag.	g	Unknown		
$SmCl_3$	Std. state, $m = 1$	aq	−1192.6		

SmCl$_3 \cdot$ 6H$_2$O		c	−2870	−2456	414
SmOCl		c	−992		
SmBr$_3$		c	−857.3		
EUROPIUM					
Eu		c	0	0	77.8
Eu		g	177.4		
Eu$^+$		g	730.7		
Eu^{2+}		g	1821		
Eu^{2+}	Std. state, $m = 1$	aq	−527.8	−541	−10
Eu^{3+}		g	4232		
Eu^{3+}	Std. state, $m = 1$	aq	−605.6	−576	−216
EuO		c	−592.0	−556.9	83.6
Eu$_2$O$_3$	Monocl.	c	−1651.4	(−1556)	(143)
Eu$_2$O$_3$	Cubic	c	−1662.7	(−1565)	(137)
Eu$_3$O$_4$		c	−2272	−2142	205

TABLE 2 (Continued)

Formula	Description	State	ΔH_f°/kJ mol^{-1}	ΔG_f°/kJ mol^{-1}	S°/J mol^{-1} K^{-1}
Eu(OH)$_3$		c	−1332	−1195	119.88
EuF$_3$		c	−1571		
EuF$_3 \cdot$H$_2$O		c	−1874		
EuCl^{2+}		aq	−773	−711	−151
EuCl$_2^+$		aq		−838	
EuCl$_2$		c	−824		
EuCl$_3$	Hexag.	c	−936.5	−856	144.1
EuCl$_3$		g	651		
EuCl$_3 \cdot$6H$_2$O	Std. state, $m = 1$	aq	−1107.1		
EuCl$_3 \cdot$6H$_2$O		c	−2785	−2366	407
EuBr$_3$		c	−779		

GADOLINIUM

Gd	c	0	0	68.1
Gd	g	397.5		
Gd^+	g	997.1		
Gd^{2+}	g	2170		
Gd^{3+}	g	4167		
Gd^{3+}	aq	−687.0	−660	−219
Gd_2O_3	Monocl.	c	−1815.6	(−1730)	(157)
Gd_2O_3	Cubic	c	−1826.9	−1739.6	151
$Gd(OH)_3$		c			126.6
GdF^{2+}		aq		−964	
GdF_3		c	−1699		
$GdF_3 \cdot H_2O$		c	−1954		
$GdCl_3$	Hexag.	c	−1007.6	−933	151.4

TABLE 2 (Continued)

Formula	Description	State	ΔH_f°/kJ mol^{-1}	ΔG_f°/kJ mol^{-1}	S°/J mol^{-1} K^{-1}
GdCl$_3$		g	−696		
GdCl$_3$	Std. state, $m = 1$	aq	−1188.5		
GdCl$_3 \cdot$ 6H$_2$O		c	−2865	−2451	408.2
GdOCl		c	−979		
GdBr$_3$		c	−828.8		
GdI$_3$		c	−594		
TERBIUM					
Tb		c	0	0	73.2
Tb		g	388.7		
Tb$^+$		g	960.7		
Tb^{2+}		g	2078		
Tb^{3+}		g	4198		

Yttrium, Lanthanum, and the Lanthanide Elements

Tb^{3+}	Std. state, m = 1	aq	−698	−668	−224
Tb^{4+}		g	8043		
Tb^{4+}	1M HClO$_4$	aq	(−443) ?	(−369) ?	(−436)
TbO$_2$		c	−971.5		
Tb$_2$O$_3$	Cubic	c	−1865.2	(−1776.9)	(158)
Tb(OH)$_3$		c			128.4
TbO$_{1.817}$		c	−962.3		
TbF$_3$		c	−1707		
TbCl$_3$	PuBr$_3$	c	−1007		
TbCl$_3$		g	−703		
TbCl$_3$	Std. state, m = 1	aq	−1199.5		
TbCl$_3 \cdot$ 6H$_2$O		c	−2869	−2441	403
TbOCl		c	−985		

TABLE 2 (Continued)

DYSPROSIUM

Formula	Description	State	ΔH_f°/kJ mol^{-1}	ΔG_f°/kJ mol^{-1}	S°/J mol^{-1} K^{-1}
Dy		c	0	0	74.8
Dy		g	290.4		
Dy$^+$		g	869.6		
Dy^{2+}		g	2002		
Dy^{2+}	Std. state, $m = 1$	aq	(−418)	(−430)	(−15)
Dy^{3+}		g	4208		
Dy^{3+}	Std. state, $m = 1$	aq	−695.6	−664	−229
Dy^{4+}		g	8215		
Dy^{4+}		aq	(−307)	(−233)	(−435)
Dy$_2$O$_3$	Cubic	c	−1863.2	−1771.6	149.8
DyF^{2+}		aq		−966	

Yttrium, Lanthanum, and the Lanthanide Elements

DyF$_3$		c	−1692	
DyCl$_2$		c	−693	
DyCl$_3$	PuBr$_3$	c,γ	−977	
DyCl$_3$	YCl$_3$	c,β	−989.1	
DyCl$_3$		g	−678	
DyCl$_3$	Std. state, m = 1	aq	−1197.0	
DyCl$_3 \cdot 6H_2O$		c	−2861	−2443	401.7
DyOCl		c	−978	
DyBr$_3$		c	−833.8	
DyI$_3$		c	−607	

HOLMIUM

Ho		c	0	0	75.3
Ho		g	300.6	
Ho$^+$		g	887.8	

TABLE 2 (Continued)

Formula	Description	State	ΔH_f°/kJ mol^{-1}	ΔG_f°/kJ mol^{-1}	S°/J mol^{-1} K^{-1}
Ho^{2+}		g	2032		
Ho^{3+}		g	4242		
Ho^{3+}	Std. state, $m = 1$	aq	−707	−675	−228
Ho$_2$O$_3$	Cubic	c	−1880.9	−1791.6	158.2
Ho(OH)$_3$		c			130.03
HoF$_3$		c	−1698		
HoCl$_3$		c	−995		
HoCl$_3$		g	−675		
HoCl$_3$	Std. state, $m = 1$	aq	−1208		
HoCl$_3 \cdot$ 6H$_2$O		c	−2878	−2460	406.2
HoOCl		c	−991		
HoI$_3$		c	−623		

ERBIUM

Er	c	0	0	77.8
Er	g	311.9		
Er$^+$	g	907.4		
Er^{2+}	g	2065		
Er^{3+}	g	4265		
Er^{3+}	aq	−708.1	−673	−235
Er$_2$O$_3$	c	−1897.8	−1808.9	155.6
		Cubic		
ErF^{2+}	aq	−1694	−976	
ErF$_3$	c	−1694		
ErCl$_3$	c	−994.5		
ErCl$_3$	g	−669		
ErCl$_3$	aq	−1209.6		
		Std. state, $m = 1$		
ErCl$_3 \cdot$6H$_2$O	c	−2877	−2457	398.7

Std. state, $m = 1$

TABLE 2 (Continued)

Formula	Description	State	ΔH_f°/kJ mol^{-1}	ΔG_f°/kJ mol^{-1}	S°/J mol^{-1} K^{-1}
ErOCl		c	−991		
ErBr$_3$		c	−838.9		
ErI$_3$		c	−613		
THULIUM					
Tm		c	0	0	74.01
Tm		g	232.2		
Tm$^+$		g	835.1		
Tm^{2+}		g	2004		
Tm^{2+}	Std. state, $m = 1$	aq	(−442)	(−450)	(−28)
Tm^{3+}		g	4295		
Tm^{3+}	Std. state, $m = 1$	aq	−705.2	−671	−236
Tm$_2$O$_3$	Cubic	c	−1888.7	−1797.4	149

TmF$_3$	c	−1656	(−1579)	(120)
TmCl$_2$	c	−709.1		
TmCl$_3$	c	−991.0		
TmCl$_3$	g	−670		
TmCl$_3$	aq, Std. state, $m = 1$	−1206.7		
TmCl$_3 \cdot 6H_2O$	c			399.6
TmOCl	c	−991		
TmI$_3$	c	−602		
YTTERBIUM				
Yb	c	0	0	59.9
Yb	g	155.6		
Yb$^+$	g	765.2		
Yb^{2+}	g	1947		
Yb^{2+}	aq, Std. state, $m = 1$	(−537)	(−544)	(−47)

TABLE 2 (Continued)

Formula	Description	State	ΔH_f°/kJ mol^{-1}	ΔG_f°/kJ mol^{-1}	S°/J mol^{-1} K^{-1}
Yb^{3+}		g	4368		
Yb^{3+}	Std. state, $m = 1$	aq	−674.5	−643	−241
Yb_2O_3	Cubic	c	−1814.5	−1729.3	142
YbF^{2+}		aq		−948	
YbF_3		c	−1570		
$YbCl_2$		c	−800		
$YbCl_3$		c	−960.0		
$YbCl_3$		g	−639		
$YbCl_3$	Std. state, $m = 1$	aq	−1176.0		
$YbCl_3 \cdot 6H_2O$		c	−2846	−2429	396
$YbOCl$		c	−962		

LUTETIUM

Lu	c	0	0	51.0
Lu	g	427.6		
Lu$^+$	g	957.3		
Lu^{2+}	g	2328		
Lu^{3+}	g	4356		
Lu^{3+}	aq	−702.6	−667	−264	Std. state, $m = 1$
Lu$_2$O$_3$	c	−1878.2	−1788.9	110	Cubic
LuF^{2+}	aq		−969		
LuF$_3$	c	−1701			
LuCl$_3$	c	−985.7			
LuCl$_3$	g	−666			
LuCl$_3$	aq	−1204.1			Std. state, $m = 1$
LuCl$_3 \cdot$ 6H$_2$O	c	−2869	−2449	376	

TABLE 2 (Continued)

Formula	Description	State	ΔH_f°/kJ mol^{-1}	ΔG_f°/kJ mol^{-1}	S°/J mol^{-1} K^{-1}
LuOCl		c	−990		
LuI$_3$		c	−548		

[a]Estimated values are given in parentheses.

and to be particularly erroneous for Eu and Yb. Recommended $E°$ values are listed in the potential diagrams at the end of the chapter.

B. $E°[M(OH)_3/M]$ in Basic Solution

It has long been known that the Ln^{3+} ions decrease in size from La through Lu. The resultant increase in charge-to-volume ratio for these ions is reflected in a decreasing basicity as one proceeds from $La(OH)_3$ to $Lu(OH)_3$, that is, increasing hydrolysis from $La^{3+}(aq)$ to $Lu^{3+}(aq)$. Although there is general agreement that the equilibrium constants (solubility product constants, K_{sp}, or K^b_{s10} in the nomenclature of Baes and Mesmer [40]) for the reactions

$$Ln(OH)_3(s) = Ln^{3+}(aq) + 3OH^-(aq) \tag{20.2}$$

$$K_{sp} = K^b_{s10} = [Ln^{3+}][OH^-]^3 \gamma_\pm^4 \tag{20.3}$$

decrease irregularly from lanthanum through lutetium, there is by no means agreement on values for these equilibrium constants. The precipitates that form initially yield larger K_{sp} [41] (log K_{sp} less negative) than aged precipitates, because the basic polymers that generate precipitates grow and equilibrate slowly. We have utilized K_{sp} values for aged precipitates because they represent most nearly the thermodynamically stable hydroxides. [Note that Latimer (1952) used K_{sp} values for fresh precipitates, perhaps because they simulate the conditions in a reversible cell under basic conditions.] Graphical presentations of $E°$ vs. pH for lanthanide species are available [42]. Values for K_{sp} have been tabulated by Aksel'rud [43] and in the *Gmelin Handbook* [44] and assessed by Baes and Mesmer [40]. In Table 3 we have generally accepted Baes and Mesmer's assessment. Our $E°_B$ values (basic solution, pH 14) have been calculated from the $E°_A$ values (acid solution) and the K_{sp} values of Table 3.

Two additional important equilibria in near-neutral or basic solution are the hydrolysis of Ln^{3+} (written here as a weak acid hydrolysis, using modified* Baes and Mesmer"s hydrolysis nomenclature) [40]:

$$Ln^{3+}(aq) + H_2O = Ln(OH)^{2+}(aq) + H^+(aq) \quad K = K_{1,1} \tag{20.4}$$

and the hydrolysis of $Ln(OH)_3$ itself acting as a weak acid:

$$Ln(OH)_3(s) + H_2O = Ln(OH)_4^-(aq) + H^+(aq) \quad K = K_{s1,4} \tag{20.5}$$

We quote the values of Baes and Mesmer in Table 3 for these equilibria. We note that recently Schmidt et al. [48] obtained $K_{1,1}$ values for Eu^{3+} and Yb^{3+} about 10 times smaller than those selected by Baes and Mesmer. Values of $K_{s1,4}$ are so small that they have negligible effects on K_{sp}.

---

*We use symbols such as $K_{1,1}$ instead of K_{11} to avoid confusion with numbered equations.

TABLE 3 Important Lanthanide Equilibria in Basic Solutions[a]

Ln	M^{3+} I.R./nm	Unit cell vol. [44,46,47] $M(OH)_3$ (nm^3)	log K_{sp} [40,43,44]	log $K_{1,1}$ [40]	log $K_{s1,4}$ [40]
Y	0.0900	0.1203	−24.5	−7.7	−36.5
La	0.1032	0.1431	−21.7	−8.5	
Ce	0.101	0.1398	−22.1	−8.3	
				−8.1 [41]	
Pr	0.099	0.1359	−22.1	−8.1	
Nd	0.0983	0.1335	−23.1	−8.0	−37.1
Pm	0.097	0.1301	(−24)		
Sm	0.0958	0.1293	−25.2	−7.9	
Eu	0.0947	0.1278	−26.5	−7.8	
Gd	0.0938	0.1260	−26.9	−8.0	−34.4

Tb	0.0923	0.1244	−26.3	−7.9	
Dy	0.0912	0.1224	−25.9	−8.0	−33.5
Ho	0.0901	0.1208	−26.5	−8.0	
Er	0.0890	0.1190	−26.6	−7.9	−32.6
Tm	0.0880	0.1178	−26.6	−7.7	
Yb	0.0868	0.1173	−26.6	−7.7	−32.7
Lu	0.0861	Unknown	−27.0	−7.6	

[a] Estimated values are given in parentheses. $K_{sp} = [M^{3+}][OH^-]^3 \gamma_\pm^4 (= K_{s10}^b$ in notation of Baes and Mesmer [40]; $K_{1,1} = [M(OH)^{2+}][H^+]/[M^{3+}]$ (corrected for activities); $K_{s1,4} = [M(OH)_4^-][H^+]$ (corrected for activities).

[b] From Ref. 45, coordination number 6. It is recognized that this is not the true coordination number in aqueous solution, but these radii parallel the ions' basicity.

[c] For $Ce(OH)_4$, $CeO_2 \cdot nH_2O$, or CeO_2 (CeO_2 is assumed to be the most stable and to be the equilibrium solid phase in contact with basic solution), log $K_{sp} = -63.1$ (see the text). This value was also used for $Pr(OH)_4$, $Nd(OH)_4$, and $Tb(OH)_4$.

II. DIVALENT AND TETRAVALENT LANTHANIDES

Reversible cell potentials are measurable only for the Ce^{4+}/Ce^{3+} and the Eu^{3+}/Eu^{2+} couples. These values, and polarographic measurements for several Ln^{3+}/Ln^{2+} couples, have been tabulated by Herman and Rairden [16] and in the *Gmelin Handbook* [39] and are discussed below in sections on specific elements. For some couples of interest, electrochemical measurements must be supplemented by calorimetric and spectroscopic techniques.

"Calorimetric" estimation of electrode potentials of species that are not stable in aqueous solution requires the determination of the enthalpy of formation of a solid compound containing the element in the desired oxidation state (e.g., MO_2 or MCl_2). A thermochemical cycle is then devised that utilizes enthalpies of solution of isostructural compounds and entropy estimates to yield the desired $E°$. Despite all the approximations involved, $E°$ estimates of this sort have error limits typically no greater than ±0.2 V.

To utilize spectroscopic correlations, Nugent et al. [49,50] studied the electron-transfer and f-d spectra of the lanthanide(III) ions in solution complexes and in solids. They compared these measurements with available $E°$ values and with the systematics of the refined electron spin-pairing theory [51,52]. They calculated $E°$ values systematically for all the M^{4+}/M^{3+} and M^{3+}/M^{2+} couples. Their values have been compared with calorimetric and electrochemical measurements and thermochemical cycles [7,9] and are cited in the oxidation-state potential diagrams when there is no experimental thermodynamic value. A recent thermochemical assessment [37] has confirmed these earlier systematic treatments [7,9].

III. INDIVIDUAL ELEMENTS

A. Yttrium

The chemistry of yttrium is that of a classical trivalent element. Most of its properties are those of a highly electropositive metal which is thermodynamically unstable with respect to its trivalent ion in aqueous solution. The trivalent ion is moderately acidic; its ionic radius is nearly the same as that of dysprosium or holmium and its solution, complex formation, hydrolysis, and crystal chemistry parallels dysprosium and the heavier rare earths.

For illustrative purposes, we calculate entries in the oxidation-state potential diagram for yttrium from entries in Tables 2 and 3. Beginning with $\Delta H_f°$ and $S°$ data from Table 2, $\Delta G_f°(Y^{3+},aq) = -685$ kJ mol^{-1} was calculated from equation (1.22) and $E°(Y^{3+}/Y)$ was calculated from Equation (1.8): $E° = -685/(3F) = -2.37$ V. To calculate $E_B°$, we sum

$$Y^{3+}(aq) + 3e^- \rightarrow Y(s) \qquad \Delta G° = 685 \text{ kJ mol}^{-1}$$

$$Y(OH)_3(s) \rightarrow Y^{3+}(aq) + 3OH^-(aq) \qquad \Delta G° = -RT \ln K_{sp} = +140 \text{ kJ mol}^{-1}$$

$$Y(OH)_3(s) + 3e^- \rightarrow Y(s) + 3OH^-(aq) \qquad \Delta G° = 825 \text{ kJ mol}^{-1}$$

(20.6)

using K_{sp} from Table 3. Then $E_B° = -2.85$ V.

As a check on self-consistency, we calculate from ΔG_f° entries in Table 2:

$$1.5H_2O(\ell) + (1/2)Y_2O_3(s) + 3e^- \rightarrow Y(s) + 3OH^-(aq) \qquad (20.7)$$

$\Delta G_7^\circ = +792$ kJ mol^{-1}. Subtracting equation (20.7) from Equation (20.6), we obtain

$$Y(OH)_3(s) \rightarrow (1/2)Y_2O_3(s) + 1.5H_2O(\ell) \qquad (20.8)$$

for which $E_8^\circ = -0.11$ V, so that $\Delta G_8^\circ = +32$ kJ mol^{-1}. The Gibbs energy data of Table 2 yield $\Delta G_8^\circ = +27$ kJ mol^{-1}. Considering the uncertainties in the experimental data, these two values of ΔG_8° are essentially consistent.

B. Lanthanum

Lanthanum is more electropositive than any of the lanthanides. This property is principally due to the ease of ionizing La(g) to La^{3+}(g), but is counteracted somewhat by the high stability of La metal and by the small enthalpy of hydration of La^{3+}. In basic solution, however, the heavy lanthanides are more electropositive than lanthanum because of the lower solubility of their hydroxides.

C. Cerium(IV)

This strong oxidant is used extensively in analytical and organic chemistry. Its oxidizing power is affected by complexing, by hydrolysis, and by polymerization. Its redox behavior was studied many years ago by Smith and Getz [53], who published the data in Table 4. Their data showed the effect of complexing of Ce(IV) and were essentially confirmed by more recent data in H$_2$SO$_4$ and in HClO$_4$ [54,55]. Sherrill et al. [56] recognized that hydrolysis of Ce(IV) is extensive even in acid solution. They characterized the hydrolysis products Ce(OH)$^{3+}$ and Ce(OH)$_2^{2+}$. Hardwick and Robertson [57] identified a dimeric species and several sulfate complexes. Wadsworth et al. [58] summarized the literature data on the Ce^{4+}/Ce^{3+} couple in 1957 and recommended a formal potential $E_f = 1.74$ V at an ionic strength of 2 mol L^{-1}.

More recent evidence includes potentiometric studies that extrapolate Ce species in perchlorate media [59,60] and confirm the potentials in sulfuric acid [61]. Hydrolysis has been studied spectrophotometrically [61-64] and turbidimetrically [65]. There remains poor consistency among these data, conflicting interpretations of the data, and even disagreement with respect to the reaction taking place. Baes and Mesmer [40] have attempted to rationalize these data and they recommend $E^\circ = 1.77$ V and $Q_{1,1} = 13$ ($K_{1,1}$ uncorrected for activities) in 3 M perchlorate for reaction (9):

$$Ce^{4+}(aq) + H_2O(\ell) \rightarrow Ce(OH)^{3+}(aq) + H^+(aq) \qquad K = K_{1,1} \qquad (20.9)$$

There is agreement that extrapolation to infinite dilution yields [57,64] $K_{1,1} = 6$. (Symbols such as $K_{1,1}$ and $Q_{1,1}$ are modifications of those used by Baes and Mesmer [40].)

TABLE 4 Formal Potentials $E^{o'}/V$ for $Ce^{4+} + e^- \to Ce^{3+}$

[Acid] /mol L^{-1}	1	2	4	6	8
HNO$_3$	1.61	1.62	1.61	-	1.56
HCl	1.28	-	-	-	-
HClO$_4$	1.70	1.71	1.75	1.82	1.87
(1/2)H$_2$SO$_4$	1.44	1.44	1.43	-	1.42

Other recent assessments of $E°(Ce^{4+}/Ce^{3+})$ yield 1.74 V [19,60] and 1.76 V [9,61]. We proceed to reassess calorimetric data because of the extensive hydrolysis [reaction (20.9)] that accompanies all electrochemical experiments.

Several solution-calorimetry reactions have been studied that include the Ce(IV)/Ce(III) aqueous couple. We restrict our consideration to perchloric acid solutions (0.5 M unless otherwise noted), and we indicate all Ce(IV) by Ce^{4+}, with correction for hydrolysis to follow.

For the reaction with $H_2O_2(aq)$,

$$Ce^{4+}(aq) + (1/2)H_2O_2(aq) \to Ce^{3+}(aq) + H^+(aq) + (1/2)O_2(aq) \tag{20.10}$$

Evans [66] used H_2O_2 in excess and reported $\Delta H_{10} = -69.9$ kJ mol^{-1}. To this datum we assign an uncertainty of ±0.8 kJ mol^{-1}. The same reaction was studied by Fitzgibbon et al. [67] with Ce^{4+} in excess; they found $\Delta H_{10} = -71.5 \pm 1.2$ kJ mol^{-1}. Using literature data for $H_2O_2(aq)$, we calculate for the reaction

$$Ce^{4+}(aq) + (1/2)H_2(g) \to Ce^{3+}(aq) + H^+(aq) \tag{20.11}$$

$\Delta H_{11} = -166.0 \pm 1.0$ kJ mol^{-1}. For the reaction

$$Ce^{4+}(aq) + Fe^{2+}(aq) \to Ce^{3+}(aq) + Fe^{3+}(aq) \tag{20.12}$$

Evans [66] found -123.8 ± 2.3 kJ mol^{-1}, Fontana [68] found -118.5 ± 0.5 kJ mol^{-1}, and Fitzgibbon et al. [67] found -126.19 ± 0.42 kJ mol^{-1}. Rejecting the value of Fontana because of the systematic discrepancies of Fontana's measurements as compared with those of Evans and Fitzgibbon et al., we estimate from this reaction and literature data [67,69] $\Delta H_{11} = -166.7 \pm 0.8$ kJ mol^{-1}.

Finally, for the reaction

$$H_2O(\ell) + Ce^{4+}(aq) + (1/2)U^{4+}(aq) \rightarrow Ce^{3+}(aq) + (1/2)UO_2^{2+}(aq) + 2H^+(aq)$$

(20.13)

Fontana [68] measured -89.4 ± 0.2 kJ mol^{-1} and Fitzgibbon et al. [67] measured -96.82 ± 0.59 kJ mol^{-1}. Literature data [69,70] combined with Fitzgibbon's measurements yields $\Delta H_{11} = -167.4 \pm 0.8$ kJ mol^{-1}.

From these data we consider those of Fitzgibbon et al. to be the most consistent and we accept $\Delta H_{11} = -166.7 \pm 1.0$ kJ mol^{-1}. Accepting[9] $\Delta H_f^\circ(Ce^{3+},aq) = -700.4 \pm 2.1$ kJ mol^{-1} as valid in 0.5 M HClO$_4$, we calculate $\Delta H_f^\circ(Ce^{4+},aq) = -534 \pm 3$ kJ mol^{-1}. This value must be corrected for hydrolysis.

We accept [64] $\Delta H_9 = 67 \pm 5$ kJ mol^{-1} and $K_{1.1} = 6.4 \pm 0.4$, whence $[Ce(OH)^{3+}]/[Ce^{4+}] \cong 12.8$ in 0.5 M HClO$_4$, so that 93% of Ce(IV) is hydrolyzed under these conditions. The correction to convert ΔH_{11} for hydrolusis is $0.93 \times 67 = 62$ kJ mol^{-1}. Therefore, $\Delta H_{11}^\circ = -166.7 + 62 = -105 \pm 12$ kJ mol^{-1}. From the temperature coefficient of the electromotive force (emf) of reaction (20.11), Conley [60] derived $\Delta H_{11} = -121 \pm 3$ kJ mol^{-1} and Morss [9] recalculated ΔH_{11} from Conley's data as -124 kJ mol^{-1}. We select $\Delta H_{11} = -120 \pm 10$ kJ mol^{-1}, whence $\Delta H_f^\circ(Ce^{4+},aq) = \Delta H_f^\circ(Ce^{3+},aq) - \Delta H_{11} = -700.4 - (-120) = -580 \pm 10$ kJ mol^{-1}. Using $\Delta S_{11}^\circ = 149 \pm 10$ J mol^{-1} K^{-1} from recalculation of Conley's data [9] and other entropy data from Table 2, we calculate $\Delta G_{11}^\circ = -165 \pm 20$ kJ mol^{-1} and E° (Ce^{4+}/Ce^{3+}) = 1.71 ± 0.02 V.

Temperature-coefficient emf measurements of Conley [60] in 1 M HClO$_4$ lead to $\Delta S_{11} = -26.7 \pm 1.7$ J K^{-1}, so that $\Delta G_{11} = -166.7 - 0.29815 (-26.7) = -159 \pm 3$ kJ mol^{-1}. Correcting for hydrolysis, we calculate E° (Ce^{4+}/Ce^{3+}) = $-(-159/96.487) - 0.05916 \log (1 - 0.93) = 1.72 \pm 0.03$ V. Thus enthalpy and entropy calculations are consistent with measurements of cell potentials and with recent assessments. We recommend E°(Ce^{4+}/Ce^{3+}) = 1.72 ± 0.02 V and $\Delta G_f^\circ(Ce^{4+},aq) = \Delta G_f^\circ(Ce^{3+},aq) - F(1.72) = -510$ kJ mol^{-1} (Table 2).

In basic solution, we consider CeO$_2$ to be the stable solid species in equilibrium with aqueous solution, a conclusion also reached by de Zoubov and Van Muylder [42]. Since we know that $\Delta G_f^\circ(CeO_2,s) = -1025.4$ kJ/mol [19, 20], we calculate for

$$CeO_2(s) + 2H_2O(\ell) \rightarrow Ce^{4+}(aq) + 4OH^-(aq)$$

(20.14)

$\Delta G_{14}^\circ = -510 + 4(-157.34) - (-1025.4) - 2(-237.19) = 360$ kJ mol^{-1}, and $K_{sp} = 1 \times 10^{-63}$. This value has been used in Table 3 (log $K_{sp} = -63.1 \pm 0.5$) and has also been used to calculate M^{4+}/M^{3+} potentials in basic solution for Ce, Pr, Nd, Tb, and Dy.

D. Praseodymium(IV)

Although there are early unconfirmed reports of Pr(IV) in complexed aqueous solution (evidence for and against having been summarized by Johnson [8]), there are a few Pr(IV) compounds that are stable and isostructural with corresponding Ce(IV) compounds. Of these, the most appropriate to use for estimation of E°(Pr^{4+}/Pr^{3+}) is PrO$_2$.

Morss and Fuger [32] compared ΔH° for the reaction

$$MO_2(s) + 4H^+(aq) \rightarrow M^{4+}(aq) + 2H_2O(\ell)$$

(20.15)

(for all dioxides with the fluorite structure) with the unit cell length of the dioxides. For Pr^{4+}(aq) they selected ΔH_f^o = -439 kJ/mol calculated from the $E°(Pr^{4+}/Pr^{3+})$ = 3.2 ± 0.2 V selected by Nugent et al. [50] and $S°$ values selected by Morss [9]. The value of $\Delta H_f^o(Pr^{4+}$,aq) correlates very well with a linear plot of ΔH_{15} against unit cell length of the dioxides. Thus we accept $E°(Pr^{4+}/Pr^{3+})$ = 3.2 ± 0.2 V. Other estimates (2.9 V by Johnson [8] and 3.9 V by Morss [9]) reflect alternative thermodynamic cycles as well as uncertainties in the fourth ionization energies of Pr and in ΔH_{15}.

Hobart et al. [71] were able to oxidize Pr^{3+}(aq) to Pr(IV) in 5.5 M K_2CO_3 solution at a potential of 1.4 V vs. SHE. Since the $E°$ for the Ce^{4+}/Ce^{3+} in this medium is shifted 1.7 V in 5.5 M K_2CO_3, one may conclude that $E°(Pr^{4+}/Pr^{3+}) \cong 3.1$ V.

E. Terbium(IV)

Nugent et al. [50] selected $E°(Tb^{4+}/Tb^{3+})$ = 3.1 V and Johnson [8] calculated $E°$ = 2.7 V. Hobart et al. [71] oxidized Tb^{3+} to Tb(IV) in 5.5 M K_2CO_3 solution at + 1.3 V. Assuming that the carbonato complexes of Tb(IV) are as stable as those of Ce(IV) and Pr(IV), one estimates $E°(Tb^{4+}/Tb^{3+}) \cong 3.0$ V. The correlation of Morss and Fuger [32] leads to $\Delta H_{15}^o = 3 \pm 20$ kJ mol^{-1} and, with $\Delta H_f^o(TbO_2,s)$ = -972 kJ mol^{-1} [19,20], we calculate $\Delta H_f^o(Tb^{4+}$,aq) = -383 ± 20 kJ mol^{-1}. With entropies from Table 2, $E°(Tb^{4+}/Tb^{3+})$ = 3.6 ± 0.3 V. Unfortunately, this value conflicts with those of Nugent and Johnson. Because of these divergent values, we estimate $E°(Tb^{4+}/Tb^{3+})$ = 3.1 ± 0.5 V. Further research is clearly needed on this system.

F. Other Tetravalent Lanthanides

The only other lanthanide(IV) ions observed in compounds (mixed fluorides and oxides) are Nd^{4+} and Dy^{4+}. For these ions, the $E°(M^{4+}/M^{3+})$ selections of Nugent et al. [50] have been accepted, with estimated error limits of ± 0.5 V.

G. Europium and Other Divalent Lanthanides

The Eu^{3+}/Eu^{2+} reduction has been studied both calorimetrically and electrochemically. The data from different observers, different methods of measurements, and independent assessments are all consistent with $E°$ = -0.35 ± 0.03 V [8,9,16,19,33,39,71-74].

Divalent samarium and ytterbium have been observed as unstable species in aqueous solution, in several compounds, and in polarographic experiments. Polarographic results led Johnson [7] to recommend $E°(Sm^{3+}/Sm^{2+})$ = -1.50 ± 0.20 V and $E°(Yb^{3+}/Yb^{2+})$ = -1.10 ± 0.10 V. These values are consistent with $\Delta G_f^o(M^{2+}$,aq) and $\Delta G_f^o(M^{3+}$,aq) and $E°(M^{3+}/M^{2+})$ selected by Morss (-1.56 V and -1.04 V) [9] but not with ΔG_f^o values selected by Schumm et al. [19]. Herman and Rairden [16] tabulated nonaqueous polarographic data. If we use a correction from $E_{1/2}$ to $E°$ for the Eu^{3+}/Eu^{2+} couple, and apply this correction to the polarographic values in five solvents [16], we calculate averaged values of $E°(Sm^{3+}/Sm^{2+})$ = -1.51 ± 0.05 V and $E°(Yb^{3+}/Yb^{2+})$ = -1.05 ± 0.03 V. We recommend -1.55 ± 0.05 V and -1.05 ± 0.05 V for these two couples.

Unfortunately, other lanthanides show only +3 → 0 reduction waves, so to estimate $E°(M^{3+}/M^{2+})$ for Nd, Dy, and Tm we must turn to spectroscopic and thermochemical measurements. Spectroscopic potentials have been derived by Nugent et al. [50] and thermochemical estimates have been made by

Johnson [7], by Morss and Fahey [24], and by Kim and Oishi [37]. Values in Table 2 represent assessments for Nd, Sm, Dy, Tm, and Yb that are consistent (within error limits) with polarographic, spectroscopic, and thermochemical determinations.

By co-crystallization of trace amounts of trivalent lanthanides (Ln = Ce, Nd, Gd, Tb, Dy, Er, and Tm) with PrOCl from a melt of $LnCl_2 + LnCl_3$ in equilibrium with $PrCl_{2+x}$ or $HoCl_{2+x}$, Mikheev and co-workers [75-78] have calculated $E°(Ln^{3+}/Ln^{2+})$ for ln = Ce, Pr, Gd, Tb, Ho, and Er to lie in the region −2.7 to −2.9 V. Mikheev [79] attributes the observation of divalency in these molten salts to the existence of $f^{\bar{n}}d$ electron configurations in the divalent ions.

In basic solution the hydroxide $Eu(OH)_2 \cdot H_2O$ can be prepared, but it is not sufficiently stable to reach equilibrium with $Eu^{2+}(aq)$. It is isostructural with $Sr(OH)_2 \cdot H_2O$ and $Ba(OH)_2 \cdot H_2O$ [80] and one might hypothesize similar solubility equilibria of Eu and Sr hydroxides. Because of the speculative nature of such an estimate, we have not considered any bivalent lanthanides in basic solution. Here we follow de Zoubov and Van Muylder [42], who considered $Ln^{2+}(aq)$ only in their potential-pH diagrams because any solid dihydroxides would be unstable with respect to the trihydroxides.

It is worth noting that monoxides of lanthanides other than Eu have recently been prepared by carrying out the reaction

$$Ln(s) + Ln_2O_3(s) \rightarrow 3LnO(s) \tag{20.16}$$

at high pressure [81], which generates a favorable $P\Delta V$ terms to make ΔG negative. In the case of NdO, high pressure also appears to stabilize the $4f^3 5d$ configuration of Nd^{2+}, which yields the metallic monoxide $Nd^{3+}(e^-)O^{2-}$ (the $5d$ electron participates in bonding). Thermodynamic calculations [82] should be confirmed by direct calorimetric measurement.

Gordon et al. [83] have measured rate constants for the pulse-radiolysis reaction

$$Ln^{3+}(aq) + e^-_{aq} \rightarrow Ln^{2+}(aq)$$

for most of the lanthanides. One group (Eu^{3+}, Yb^{3+}, and Sm^{3+}) is easily reduced and has rate constants of the order of 10^{10} M^{-1} s^{-1}. A second group (Tm^{3+}, Gd^{3+}, Ho^{3+}, and Er^{3+}) can be reduced with difficulty, and has rate constants of 10^7 to 10^8 M^{-1} s^{-1}. Other lanthanides showed much smaller rate constants and were considered relatively unreducible. For the first six ions listed above, $Ln^{2+}(aq)$ spectra were obtained with a streak camera-TV scanning instrument. Except for Dy^{3+}, which should be in the second group, and Gd^{3+}, which should be in the last group, these rate data are consistent with thermochemical, electrochemical, and spectroscopic determinations of $E°(Ln^{3+}/Ln^{2+})$.

This work was performed under the auspices of the Office of Basic Energy Sciences, Division of Chemical Sciences, U.S. Department of Energy, under contract W-31-109-ENG-38. The author appreciates helpful suggestions from reviewers and from D. A. Johnson (Open University, U.K.), J. C. Sullivan (Argonne), and G. R. Choppin (Florida State University).

IV. POTENTIAL DIAGRAMS FOR THE GROUP

Acid Solution

$$+4 \qquad\qquad +3 \qquad\qquad +2 \qquad\qquad 0$$

$Y^{3+} \xrightarrow{-2.37} Y$

$La^{3+} \xrightarrow{-2.38} La$

$Ce^{4+} \xrightarrow{1.72} Ce^{3+} \xrightarrow{-2.34} Ce$

$Pr^{4+} \xrightarrow{3.2} Pr^{3+} \xrightarrow{-2.35} Pr$

$Nd^{4+} \xrightarrow{4.9} Nd^{3+} \xrightarrow{-2.6} Nd^{2+} \xrightarrow{-2.2} Nd$; overall $Nd^{3+} \to Nd$: -2.32

$Pm^{3+} \xrightarrow{-2.29} Pm$

$Sm^{3+} \xrightarrow{-1.55} Sm^{2+} \xrightarrow{-2.67} Sm$; overall $Sm^{3+} \to Sm$: -2.30

$Eu^{3+} \xrightarrow{-0.35} Eu^{2+} \xrightarrow{-2.80} Eu$; overall $Eu^{3+} \to Eu$: -1.99

$Gd^{3+} \xrightarrow{-2.28} Gd$

$Tb^{4+} \xrightarrow{3.1} Tb^{3+} \xrightarrow{-2.31} Tb$

$Dy^{4+} \xrightarrow{5.7} Dy^{3+} \xrightarrow{-2.5} Dy^{2+} \xrightarrow{-2.2} Dy$; overall $Dy^{3+} \to Dy$: -2.29

$Ho^{3+} \xrightarrow{-2.33} Ho$

$Er^{3+} \xrightarrow{-2.32} Er$

$Tm^{3+} \xrightarrow{-2.3} Tm^{2+} \xrightarrow{-2.3} Tm$; overall $Tm^{3+} \to Tm$: -2.32

$Yb^{3+} \xrightarrow{-1.05} Yb^{2+} \xrightarrow{-2.8} Yb$; overall $Yb^{3+} \to Yb$: -2.22

$Lu^{3+} \xrightarrow{-2.30} Lu$

Basic Solution

+4		+3		0
		$Y(OH)_3$	$\xrightarrow{-2.85}$	Y
		$La(OH)_3$	$\xrightarrow{-2.80}$	La
CeO_2	$\xrightarrow{-0.7}$	$Ce(OH)_3$	$\xrightarrow{-2.78}$	Ce
PrO_2	$\xrightarrow{0.8}$	$Pr(OH)_3$	$\xrightarrow{-2.79}$	Pr
$[NdO_2]$	$\xrightarrow{2.5}$	$Nd(OH)_3$	$\xrightarrow{-2.78}$	Nd
		$Pm(OH)_3$	$\xrightarrow{-2.76}$	Pm
		$Sm(OH)_3$	$\xrightarrow{-2.80}$	Sm
		$Eu(OH)_3$	$\xrightarrow{-2.51}$	Eu
		$Gd(OH)_3$	$\xrightarrow{-2.82}$	Gd
TbO_2	$\xrightarrow{0.9}$	$Tb(OH)_3$	$\xrightarrow{-2.82}$	Tb
$[DyO_2]$	$\xrightarrow{3.5}$	$Dy(OH)_3$	$\xrightarrow{-2.80}$	Dy
		$Ho(OH)_3$	$\xrightarrow{-2.85}$	Ho
		$Er(OH)_3$	$\xrightarrow{-2.84}$	Er
		$Tm(OH)_3$	$\xrightarrow{-2.83}$	Tm
		$Yb(OH)_3$	$\xrightarrow{-2.74}$	Yb
		$Lu(OH)_3$	$\xrightarrow{-2.83}$	Lu

REFERENCES

1. W. C. Martin, L. Hagan, J. Reader, and J. Sugar, J. Phys. Chem. Ref. Data, 3, 771-779 (1974); E. F. Worden and J. G. Conway, in *Lanthanide and Actinide Chemistry and Spectroscopy*, N. M. Edelstein, ed., A. Chem. Soc. Symp. Ser. 131 (1980), pp. 381-425.
2. D. A. Johnson, J. Chem. Soc. A, 1525-1528 (1969).
3. K. L. Vander Sluis and L. J. Nugent, J. Opt. Soc. Am., 64, 687-695 (1974).
4. K. L. Vander Sluis and L. J. Nugent, J. Chem. Phys., 60, 1927-1930 (1974).
5. C. K. Jørgensen, in *Gmelin Handbuch der anorganischen Chemie*, 8th ed., Seltenerdelemented, Part B1, Springer-Verlag, Berlin, 1976.
5a. D. A. Johnson, *Some Thermodynamic Aspects of Inorganic Chemistry*, 2d ed., Cambridge University Press, Cambridge, UK, 1982.
6. D. A. Johnson, J. Chem. Soc. A, 1528-1530 (1969).
7. D. A. Johnson, J. Chem. Soc., Dalton Trans., 1671-1675 (1974).
8. D. A. Johnson, Adv. Inorg. Chem. Radiochem., 20, 1-132 (1977).
9. L. R. Morss, Chem. Rev., 76, 827-841 (1976).
10. R. Guillaumont and F. David, Radiochim. Radioanal. Lett., 17, 25-39 (1974).
11. L. Brewer, J. Opt. Soc. Am., 61, 1101-1111 (1971).
12. L. J. Nugent, J. L. Burnett, and L. R. Morss, J. Chem. Thermodyn., 5, 665-678 (1973).
13. L. J. Nugent, J. Inorg. Nucl. Chem., 32, 3485-3491 (1970).
14. F. David, Rev. Chim. Miner., 10, 565-576 (1973).
15. S. Goldman and L. R. Morss, Can. J. Chem., 53, 2695-2700 (1975).
16. H. B. Herman and J. R. Rairden, in *Encyclopedia of the Electrochemistry of the Elements*, A. J. Bard, ed., Vol. 6, Marcel Dekker, New York, 1976, Chap. 2.
17. S. L. Bertha and G. R. Choppin, Inorg. Chem., 8, 613-617 (1969).
17a. R. J. Hinchey and J. W. Cobble, Inorg. Chem., 9, 917-921 (1970).
18. D. D. Wagman, W. H. Evans, V. B. Parker, I. Halow, S. M. Bailey, R. H. Schumm, and K. L. Churney, *Selected Values of Chemical Thermodynamic Properties*, Nat. Bur. Stand. Tech. Note 270-5, U.S. Government Printing Office, Washington, D.C., 1971; reissued in SI units as J. Phys. Chem. Ref. Data, 11, Suppl. 2 (1982).
19. R. H. Schumm, D. D. Wagman, S. Bailey, W. H. Evans, and V. B. Parker, *Selected Values of Chemical Thermodynamic Properties*, Nat. Bur. Stand. Tech. Note 270-7, U.S. Government Printing Office, Washington D.C., 1973; reissued in SI units as J. Phys. Chem. Ref. Data, 11, Suppl. 2 (1982). For $LnCl_3(g)$, vaporization data were taken from C. E. Meyers and D. T. Graves, J. Chem. Eng. Data, 22, 440-445 (1977); M. H. Hannay and C. E. Meyers, J. Less-Common Metals, 66, 145-150 (1979); C. E. Meyers and M. H. Hannay, J. Less-Common Metals, 70, 15-24 (1980).
19a. F. H. Spedding, J. A. Rard, and A. Habenschuss, J. Phys. Chem., 81, 1069-1074 (1977).
20. K. A. Gschneider, Jr., N. Kippenhan, and O. D. McMasters, Thermochemistry of the Rare Earths, IS-RIC-6 Rare Earth Information Center, Iowa State University, Ames, Iowa, 1973.
21. G. K. Johnson, R. G. Pennell, K.-Y. Kim, and W. N. Hubbard, J. Chem. Thermodyn., 12, 125-136 (1980).

22. C. Hurtgen, J. Fuger, and D. Brown, J. Chem. Soc. Dalton Trans., 70-75 (1980).
23. L. R. Morss and M. C. McCue, Inorg. Chem., 14, 1624-1627 (1975).
24. L. R. Morss and J. A. Fahey, Proc. 12th Rare Earth Res. Conf., Vol. 1, Denver Research Institute, Denver, Colo., 1976, pp. 443-450.
25. Y.-C. Kim, J. Oishi, and S.-H. Kang, J. Chem. Thermodyn., 9, 973-977 (1977).
26. O. Greis and J. M. Haschke, in *Handbook on the Physics and Chemistry of Rare Earths*, K. A. Gschneidner and L. Eyring, eds., Vol. 5, North-Holland, Amsterdam, 1982, Chap. 45.
27. Y.-C. Kim, J. Oishi, and S.-H. Kang, J. Chem. Thermodyn., 10, 975-981 (1978).
28. Y.-C. Kim, M. Misumi, H. Yano, and J. Oishi, J. Chem. Thermodyn., 11, 657-662 (1979).
29. J. Fuger, L. R. Morss, and D. Brown, J. Chem. Soc., Dalton Trans., 1076-1078 (1980).
30. R. W. Mar and R. G. Bedford, J. Less-Common Metals, 71, 317 (1980).
31. Y.-C. Kim, M. Kanazawa, and J. Oishi, J. Chem. Thermodyn., 12, 811-816 (1980).
32. L. R. Morss and J. Fuger, J. Inorg. Nucl. Chem., 43, 2059-2064 (1981).
33. L. R. Morss and H. O. Haug, J. Chem. Thermodyn., 5, 513-524 (1973).
34a. R. D. Chirico and E. F. Westrum, Jr., J. Chem. Thermodyn., 12, 71-85 (1980); (b) J. Warmkessel, J. Chem. Thermodyn., 11, 835-850 (1979).
35. E. F. Westrum, Jr., J. Chem. Thermodyn., 15, 305-325 (1983).
36. O. D. McMasters, K. A. Gschneidner, Jr., E. Kaldis, and G. Sampietro, J. Chem. Thermodyn., 6, 845-857 (1974).
37. Y.-C. Kim and J. Oishi, J. Less-Common Metals, 65, 199-210 (1979).
38. T. Moeller, in *Comprehensive Inorganic Chemistry*, A. F. Trotman-Dickenson, exec. ed., Vol. 4, Pergamon Press, Oxford, 1973, Chap. 44.
39. *Gmelin Handbuch der anorganischen Chemie*, 8th ed., Seltenerdelemente, System No. 39, Part B7, Springer-Verlag, Berlin, 1979, pp. 119, 128-145.
40. C. F. Baes, Jr., and R. E. Mesmer, *The Hydrolysis of Cations*, Wiley-Interscience, New York, 1976.
41. J. Kragten and L. G. Decnop-Weever, Talanta, 25, 147-150 (1978).
42. N. De Zoubov and J. Van Muylder, in *Atlas of Electrochemical Equilibria in Aqueous Solutions*, M. Pourbaix, ed., Pergamon Press, Oxford, 1966, pp. 183-197.
43. N. V. Aksel'rud, Russ. Chem. Rev., 32, 353-366 (1963).
44. *Gmelin Handbuch der anorganischen Chemie*, Seltenerdelemente, System No. 39, Part C2, Springer-Verlag, Berlin, 1974.
45. R. D. Shannon, Acta Crystallogr., $A32$, 751-767 (1976).
46. D. R. Dillin, W. O. Milligan, and R. J. Williams, J. Appl. Crystallogr., 6, 492-494 (1973).
47. G. W. Beall, W. O. Milligan, and H. A. Wolcott, J. Inorg. Nucl. Chem., 39, 65-70 (1977).
48. K. H. Schmidt, J. C. Sullivan, S. Gordon, and R. C. Thompson, Inorg. Nucl. Chem. Lett., 14, 429-434 (1978).
49. L. J. Nugent, R. D. Baybarz, J. L. Burnett, and J. L. Ryan, J. Inorg. Nucl. Chem., 33, 2503-2530 (1971).
50. L. J. Nugent, R. D. Baybarz, J. L. Burnett, and J. L. Ryan, J. Phys. Chem., 77, 1528-1539 (1973).

51. C. K. Jørgensen, Mol. Phys., 5, 271-277 (1962).
52. J. L. Ryan and C. K. Jørgensen, Mol. Phys., 7, 27-29 (1963).
53. G. F. Smith, and C. A. Getz, Ind. Eng. Chem. Anal. Ed., 10, 191-195 (1938).
54. F. Verbeek and Z. Eeckhaut, Bull. Soc. Chim. Belges, 67, 204-224 (1958).
55. J. R. Stokely, Jr., R. D. Baybarz, and W. D. Schults, Inorg. Nucl. Chem. Lett., 5, 877-884 (1969).
56. M. S. Sherrill, C. B. King, and R. C. Spooner, J. Am. Chem. Soc., 65, 170-179 (1943).
57. T. J. Hardwick and E. Robertson, Can J. Chem., 29, 818-837 (1951).
58. E. Wadsworth, F. R. Duke, and C. A. Goetz, Anal. Chem., 29, 1824-1825 (1957).
59. F. B. Baker, T. W. Newton, and M. Kahn, J. Phys. Chem., 64, 109-112 (1960).
60. H. L. Conley, M. S. thesis, University of California, Lawrence Radiation Lab. Rep. UCRL-9332, Berkeley, Calif., 1960.
61. V. I. Kravtsov, L. Ya. Smirnova, and O. A. Babintseva, Elektrokhimiya, 13, 1487-1492 (1978).
62. H. G. Offner and D. A. Skoog, Anal. Chem., 38, 1520-1521 (1966).
63. Z. Amjad and A. McAuley, J. Chem. Soc., Dalton Trans., 2521-2526 (1974).
64. K. G. Everett and D. A. Skoog, Anal. Chem., 43, 1541-1547 (1971).
65. K. P. Louwrier and T. Steemers, Inorg. Nucl. Chem. Lett., 12, 185-189 (1976).
66. M. W. Evans, in *The Transuranium Elements*, G. T. Seaborg, J. J. Katz, and W. M. Manning, eds., National Nuclear Energy Series, Part IV, Vol. 14B, McGraw-Hill, New York, 1949, pp. 282-294.
67. G. C. Fitzgibbon, T. W. Newton, and C. E. Holley, Jr., 30th Calorimetry Conf., Seattle, Wash., July 16-19, 1975.
68. B. J. Fontana, in *Chemistry of Uranium—Collected Papers*, J. J. Katz and E. Rabinowitch, eds., TID-5290, Book 1, U.S. Atomic Energy Commission, Oak Ridge, Tenn., 1958, pp. 279-288.
69. D. D. Wagman, W. H. Evans, V. B. Parker, I. Halow, S. M. Bailey, and R. H. Schumm, Natl. Bur. Stand. Tech. Notes 270-3 (1968) and 270-4 (1969), U.S. Government Printing Office, Washington, D.C.; reissued in SI units as J. Phys. Chem. Ref. Data, 11, Suppl. 2 (1982).
70. J. Fuger and F. L. Oetting, *The Chemical Thermodynamics of Actinide Elements and Compounds*, Part 2: *The Actinide Aqueous Ions*, International Atomic Energy Agency, Vienna, 1976.
71. D. E. Hobart, K. Samhoun, J. P. Young, V. E. Norvell, G. Mamantov, and J. R. Peterson, Inorg. Nucl. Chem. Lett., 16, 321-328.
72. L. B. Anderson and D. J. Macero, J. Phys. Chem., 67, 1942 (1963).
73. G. Biedermann and G. S. Terjosin, Acta Chem. Scand., 23, 1896-1902 (1969).
74. G. Biedermann and H. B. Silber, Acta Chem. Scand., 27, 3761-3768 (1973).
75. N. B. Mikheev, B. G. Korshunov, L. N. Auerman, I. A. Rumer, and A. J. Galushko, Radiokhimiya, 23, 624-627 (1981).
76. R. A. D'yachkova, L. N. Auerman, I. A. Rumer, and N. B. Mikheev, Radiokhimiya, 24, 112-114 (1982); Sov. Radiochem., 23, 98-100 (1982).
77. N. B. Mikheev, Radiochim. Acta, 32, 69-80 (1983).
78. N. B. Mikheev, L. N. Auerman, and I. A. Rumer, Zh. Neorg. Khim., 28, 1329-1331.

79. N. B. Mikheev, submitted to *J. Less-Common Metals*.
80. H. Barnighausen, Z. Anorg. Allg. Chem., *342*, 233-239 (1966).
81. J. M. Leger, N. Yacoubi, and J. Loriers, Inorg. Chem., *19*, 2252-2254 (1980).
82. J. M. Leger, N. Yacoubi, and J. Loriers, J. Solid State Chem., *36*, 261-270 (1981).
83. S. Gordon, J. C. Sullivan, W. A. Mulac, D. Cohen, and K. H. Schmidt, *Proceedings of the 4th Symposium on Radiation Chemistry*, Akadémiai Kiadó, Budapest, 1976, pp. 753-760.

21
The Actinides*

L. MARTINOT and J. FUGER   University of Liège, Liège, Belgium

I. THE ACTINIDE ELEMENTS

The actinides are the 14 elements that follow actinium ($Z = 89$) in the periodic table, just as the lanthanides are the 14 elements that follow lanthanum ($Z = 59$). The two groups of elements are characterized by the progressive filling of the $5f$ and $4f$ electronic subshells, respectively. However, while the $4f$ electrons fall well below the $5d$ electrons at the beginning of the lanthanide series, the $5f$ and the $6d$ electrons in the actinides have similar energy levels, especially at the beginning of the series. The stabilization of the $5f$ electrons, compared to that of the $6d$ electrons, appears only progressively with increasing Z and, in fact, at the beginning of the series, the $6d$ electrons are still involved in chemical bonding. The electronic structures of the gaseous atoms in the ground state of the lanthanides and actinides are shown for comparison in Table 1.

Because the energy difference between the $5f$ and $6d$ electrons is small at the beginning of the series, electronic shifts may occur when changing the chemical or even the physical state of an actinide species. Discussion of this topic is beyond the scope of the present compilation. In addition, the $5f$ electrons of the actinides are more shielded than the $4f$ electrons of the lanthanides. These features lead to a large number of valency states for the elements in the first part of the actinide series. Table 2 compares the oxidation states of the lanthanides and of the actinides. Oxidation states up to +7 have been identified for neptunium, plutonium, and americium; in that valency state neptunium exhibits the closed-shell structure of radon. These heptavalent species can be obtained in very basic media; there is a marked decrease in stability when going from neptunium to americium. Heptavalent compounds of neptunium and plutonium often display isomorphism with the corresponding rhenium compounds.

The multiplicity of the valency states of some actinides leads to very complicated redox behavior. The case of plutonium, for which four oxidation states can coexist at equilibrium in solution, is unique in the periodic table.

Reviewers: James R. Stokely of Oak Ridge National Laboratory, Oak Ridge, Tennessee; Lester R. Morss and James C. Sullivan, Argonne National Laboratory, Argonne, Illinois.

TABLE 1 Electronic Structures of the Gaseous Atoms in the Ground State of the Lanthanides and the Actinides

Lanthanides				Actinides			
La		$5d$	$6s^2$	Ac		$6d$	$7s^2$
Ce	$4f^2$		$6s^2$	Th		$6d^2$	$7s^2$
Pr	$4f^3$		$6s^2$	Pa	$5f^2$	$6d$	$7s^2$
Nd	$4f^4$		$6s^2$	U	$5f^3$	$6d$	$7s^2$
Pm	$4f^5$		$6s^2$	Np	$5f^4$	$6d$	$7s^2$
Sm	$4f^6$		$6s^2$	Pu	$5f^6$		$7s^2$
Eu	$4f^7$		$6s^2$	Am	$5f^7$		$7s^2$
Gd	$4f^7$	$5d$	$6s^2$	Cm	$5f^7$	$6d$	$7s^2$
Tb	$4f^9$		$6s^2$	Bk	$5f^9$		$7s^2$
Dy	$4f^{10}$		$6s^2$	Cf	$5f^{10}$		$7s^2$
Ho	$4f^{11}$		$6s^2$	Es	$5f^{11}$		$7s^2$
Er	$4f^{12}$		$6s^2$	Fm	$5f^{12}$		$7s^2$
Tm	$4f^{13}$		$6s^2$	Md	$5f^{13}$		$7s^2$
Yb	$4f^{14}$		$6s^2$	No	$5f^{14}$		$7s^2$
Lu	$4f^{14}$	$5d$	$6s^2$	Lr	$5f^{14}$	$6d$	$7s^2$

Source: Adapted from L. Brewer, J. Opt. Soc. Am., *61*, 1101-1111 (1971).

TABLE 2 Oxidation States of Lanthanides and Actinides[a]

Lanthanides

Z	57	58	59	60	61	62	63	64	65	66	67	68	69	70	71
Elements	La	Ce	Pr	Nd	Pm	Sm	Eu	Gd	Tb	Dy	Ho	Er	Tm	Yb	Lu
Oxidation states				(2)		2	2			(2)	(2)		(2)	2	
	3	3	3	3	3	3	3	3	3	3	3	3	3	3	3
		4	4	(4)					4	(4)					

Actinides

Z	89	90	91	92	93	94	95	96	97	98	99	100	101	102	103
Elements	Ac	Th	Pa	U	Np	Pu	Am	Cm	Bk	Cf	Es	Fm	Md	No	Lr
Oxidation states							(2)			(2)	2	2	2	2	
	3	(3)	(3)	3	3	3	3	3	3	3	3	3	3	3	3
		4	4	4	4	4	4	4	4						
			5	5	5	5	5								
				6	6	6	6								
					7	7									

[a]Values in parentheses were not observed in aqueous media. Italic type indicates the most stable species in dilute mineral acid solution.

With the exception of thorium and protactinium all actinides display the +3 oxidation state, which becomes progressively more stable with increasing Z. At the levels of americium, trivalency becomes predominant, and there is a close analogy between heavy actinides and the corresponding lanthanides. A trend toward divalency arises with the heavier actinides from californium to nobelium. Divalent nobelium is the most stable aqueous species of this element, reflecting the special stability of the filled $5f^{14}$ shell. The last actinide lawrencium, is strictly trivalent, as this state also represents the filled $5f$ shell. The tendency to achieve the half-filled $5f^7$ shell is apparent in the relative stability in solution of tetravalent berkelium, whereas its counterpart, tetravalent terbium, can be obtained only in the solid state. On the other hand, divalent americium has been stabilized only in solids, whereas its homolog, divalent europium, is rather easily prepared in aqueous solution.

The progressive filling of the $5f$ inner shell together with the increase in nuclear charge results in a contraction of the electronic orbitals analogous to that observed in the lanthanide series; this contraction is reflected in the ionic radii. Table 3 displays that effect for the +3 and +4 ions. In particular, ionic sizes of the actinide ions compared to those of the corresponding lanthanides are the basis for the estimation of the solubility products of the hydroxides of the +3 transplutonium ions, for which experimental data are lacking. In turn, these solubility products are used in the calculation of the potentials in basic media and are listed in Table 4. The thermodynamic data are derived from calorimetry, entropy determinations, and electromotive force measurements according to the data available for each element.

Few direct determinations of the entropies of the actinide ions exist and most of the entropy data are derived values. The entropy of Pu^{3+}(aq) has been calculated from the enthalpy and Gibbs energy of solution of $PuCl_3 \cdot 6H$ $PuCl_3 \cdot 6H_2O$(c) according to the reaction

$$PuCl_3 \cdot 6H_2O(c) \rightarrow Pu^{3+}(aq) + 3Cl^-(aq) + 6H_2O(\ell)$$

and from the entropy of $PuCl_3 \cdot 6H_2O$(c), which itself was estimated from that of the isostructural salt $SmCl_3 \cdot 6H_2O$ (aq). The entropies of other trivalent actinide ions are calculated by assuming that the variation of entropies with ionic radii is similar to that observed for the trivalent lanthanide ions. No entropy determination exists for any other trivalent actinide ion. The only experimental value available for a tetravalent actinide is that of Th^{4+}(aq); the entropies adopted for all the other tetravalent ions result from combining existing enthalpies of formation with potential data and with accepted entropies for the other ions. The entropy of NpO_2^+(aq) has been obtained from the value of the entropy of NpO_2^+(aq) with the knowledge of the thermodynamic parameters associated with the reduction of NpO_2^{2+}(aq) to NpO_2^+(aq); the entropies of all the other actinyl V ions are derived values. The entropies of the actinyl VI ions rest on experimental data of UO_2^{2+}(aq) and NpO_2^{2+}(aq). The value for the latter ion rests, however, on an estimation of the entropy of $NpO_2(NO_3)_2 \cdot 6H_2O$(c).

Unless otherwise specified, all thermodynamic data used in this compilation are based on parts, I, II, III, VIII, X, and XII of the following critical assessment.

TABLE 3 Effective Ionic Radii (nm) of Lanthanides and Actinides for Coordination Number $n = 6$

M^{3+}	M^{4+}	M^{3+}		M^{4+}
La 0.1032		Ac	0.112	
Ce 0.101	0.087	Th		0.094
Pr 0.099	0.085	Pa	0.104	0.090
Nd 0.0983		U	0.1025	0.089
Pm 0.097		Np	0.101	0.087
Sm 0.0958		Pu	0.100	0.086
Eu 0.0947		Am	0.0975	0.085
Gd 0.0938		Cm	0.097	0.085
Tb 0.0923	0.076	Bk	0.096	0.083
Dy 0.0912		Cf	0.095	0.082
Ho 0.0901		Es	0.0935[a]	
Er 0.0890		Fm	0.0925[a]	
Tm 0.0880		Md	0.0915[a]	
Yb 0.0868		No	0.0905[a]	
Lu 0.0861		Lr	0.090[a]	

[a] As estimated by L. R. Morss, in *The Chemistry of the Actinide Elements*, J. J. Katz, G. T. Seaborg, and L. R. Morss, eds., Chap. 17, 2nd ed. (to be published).
Source: Adapted from R. D. Shannon, Acta Crystallogr. *A 32*, 751-767 (1976).

The Chemical Thermodynamics of Actinide Elements and Compounds, V. Medvedev, M. H. Rand, E. F. Westrum, Jr., Editors (F. L. Oetting, Executive Editor), International Atomic Energy Agency, Vienna.

 Part I: *The Actinide Elements*, F. L. Oetting, M. H. Rand, and R. J. Ackerman (published in 1976).
 Part II: *The Actinide Aqueous Ions*; J. Fuger and F. L. Oetting (published in 1976).
 Part III: *Miscellaneous Actinide Compounds*, E. H. P. Cordfunke and P. A. G. O'Hare (published in 1978).
 Part IV: *The Actinide Chalcogenides (Excluding Oxides)*; F. Grønvold, J. Drowart, and E. F. Westrum, Jr. (published in 1984).
 Part V: *The Actinide Binary Alloys*; P. Chiotti, V. V. Akhachinskij, I. Ansara, and M. H. Rand (published in 1981).

Part VI: *The Actinide Carbides*; C. E. Holley, M. H. Rand, and E. K. Storms (published in 1984).

Part VII: *The Actinide Pnictides*; P. E. Potter, K. C. Mills, and Y. Takahashi (in preparation).

Part VIII: *The Actinide Halides*; J. Fuger, V. B. Parker, W. N. Hubbard, and F. L. Oetting (published in 1983). In addition, use was made of the report NBSIR 80-2029, *The Thermochemical Properties of the Uranium-Halogen Containing Compounds*, V. B. Parker, Washington, D.C., 1980.

Part IX: *The Actinide Hydrides*; H. E. Flotow, J. M. Haschke, and S. Yamauchi (published in 1984).

Part X: *The Actinide Oxides*; M. H. Rand, R. J. Ackerman, F. Grønvold, F. L. Oetting, and A. Pattoret (in preparation).

Part XI: *Selected Ternary Systems*; P. E. Potter, H. Holleck, and K. E. Spear (in preparation).

Part XII: *The Actinide Aqueous Inorganic Complexes*; J. Fuger, I. L. Khodakovskij, V. A. Medvedev, and J. D. Navratil (in preparation).

TABLE 4 Activity Products K_{sp} Used to Calculate Potentials in Basic Media; $K_{sp} = [M^{n+}][OH^-]^n \gamma_{\pm}^{n+1}$

M	$M(OH)_3$ ($-\log K_{sp}$)	MO_2 ($-\log K_{sp}$)
Ac	20.9	
Th		49.4
Pa		
U	22.2	57.8
Np	22.4	60.0
Pu	22.6	63.3
Am	23.3	64
Cm	24.0	64
Bk	25.0	65
Cf	25.5	65
Es	26.0	

For U to Am: $MO_2OH \rightarrow MO_2^+ + OH^-$; $\log K_{sp} \cong -10$

$MO_2(OH)_2 \rightarrow MO_2^{2+} + 2OH^-$; $\log K_{sp} \cong -22$

Source: Adapted from L. Morss, in *The Chemistry of the Actinide Elements*, J. J. Katz, G. T. Seaborg, and L. R. Morss, eds., Chap. 17, 2nd ed. (to be published); and based on C. F. Baes, Jr., and R. E. Mesmer, *The Hydrolysis of Cations*, Wiley-Interscience, New York, 1976.

The Actinides

Part XIII: *The Gaseous Actinide Ions*; D. L. Hildenbrand, L. V. Gurvich, and V. S. Yungman (published in 1985).
Part XIV: *The Actinide Aqueous Organic Complexes*; P. A. Baisden, G. R. Choppin, B. F. Myasoedev, and J. D. Navratil (in preparation).

This critical assessment is based on auxiliary values recommended by "CODATA; Recommended Key values for Thermodynamics 1977," J. Chem. Thermodyn. *10*, 903-906 (1978).

Description of the hydrolytic behavior of the various actinide ions is based on the critical assessment of C. F. Baes, Jr., and R. E. Mesmer; *The Hydrolysis of Cations*, Wiley-Interscience, New York, 1976.

In the assessments above, some thermodynamic data were reported without uncertainty limits. Where feasible, these limits have been evaluated in the present chapter. To minimize as much as possible rounding-off errors when converting data from calories to joules, and vice versa, uncertainty limits have purposely been reported with two significant figures.

A comprehensive review of reduction-potential literature data has appeared in the *Encyclopedia of the Electrochemistry of the Elements*, A. J. Bard; ed., Marcel Dekker, New York, Volume 8, Chapter II, "Actinides," L. Martinot, 1978.

II. ACTINIUM

A. Oxidation States

The trivalent actinium ion is the only actinium species that exists in aqueous solutions.

The longest-lived isotope ^{227}Ac has a half-life of only 21.8 years. A number of short-lived β^- and γ emitters among the decay products of ^{227}Ac makes working with weighable amounts of actinium very difficult even a few days after thorough purification; the only experimental work utilizes radiopolarographic techniques. The thermodynamic data are given in Table 5.

B. Gibbs Energies and Potentials

From the analysis of the polarographic first-wave amalgamation potential, Nugent [1] concludes that $E°$ (Ac^{3+}/Ac) equals -2.18 ± 0.07 V, which is consistent with the analysis of David et al. [2], who propose -2.13 ± 0.12 V. The enthalpy of formation of Ac^{3+}(aq) is based on systematics correlating the thermodynamic properties of lanthanides and actinides; the entropies of Ac(c,α) and Ac^{3+}(aq) are also based on correlations.

C. Potential Diagrams for Actinium

Acid Solution

$$\text{Ac}^{3+} \xrightarrow{-2.13} \text{Ac}$$

Basic Solution

$$\text{Ac(OH)}_3 \xrightarrow{-2.6} \text{Ac}$$

TABLE 5 Thermodynamic Data on Actinium[a]

Formula	Description	State	ΔH_f°/kJ mol^{-1}	ΔG_f°/kJ mol^{-1}	S°/J mol^{-1} K^{-1}
Ac	Metal, α	c	0	0	62.8 ± 4.2
Ac^{3+}	Std. state, $m = 1$	aq	−653 ± 25	−640 ± 25	−180 ± 17
Ac(OH)$_3$		c	(−1371)	(−1231)	(96)

[a]Estimated values are given in parentheses.

REFERENCES

1. L. J. Nugent, J. Inorg. Nucl. Chem., 37, 1767-1770 (1975).
2. F. David, K. Samhoun, R. Guillaumont, and L. J. Nugent, Proc. Joint Session, 4th Int. Symp. Transplutonium Elements—5th Int. Conf. Plutonium and Other Actinides, Baden-Baden, 1975, C. Lindner and W. Müller, eds., North-Holland, Amsterdam, 1976, pp. 97-105; and also J. Inorg. Nucl. Chem., 40, 69-74 (1978).

III. THORIUM

A. Oxidation States

Tetravalent thorium is the only ionic species that has been characterized in solution. It must be pointed out that $Th^{4+}(aq)$ is the only 4+ ion for which the standard entropy is known directly from experimental measurements:

$$S°(Th^{4+}, aq) = -422.6 \pm 16.7 \text{ J mol}^{-1} \text{ K}^{-1}$$

The hydrolysis of Th^{4+} leads to the formation of several complex species, such as $Th(OH)^{3+}$, $Th(OH)_2^{2+}$, and $Th(OH)_4(aq)$, and also to polymeric species $Th_2(OH)_2^{6+}$, $Th_4(OH)_8^{8+}$, and $Th_6(OH)_{15}^{9+}$. $ThO_2(c)$ is considered to be the only thermodynamically stable species of hydrolyzed thorium in the solid state. The thermodynamic data are given in Table 6.

B. Gibbs Energies and Potentials

Using the Gibbs energy of formation of $Th^{4+}(aq)$, we find $E°$ $(Th^{4+}/Th) = -1.83 \pm 0.01$ V, while in basic media we calculate $E° = -2.56$ V for the reaction

$$ThO_2 + 2H_2O + 4e^- \rightarrow Th + 4OH^-$$

C. Potential Diagrams for Thorium

Acid Solution

$$\overset{+4}{Th^{4+}} \xrightarrow{-1.83} \overset{0}{Th}$$

Basic Solution

$$ThO_2 \xrightarrow{-2.56} Th$$

IV. PROTACTINIUM

A. Oxidation States

Protactinium exists in aqueous solutions in the tetravalent and pentavalent oxidation states. The pentavalent species is the most stable as tetravalent

TABLE 6 Thermodynamic Data on Thorium

Formula	Description	State	ΔH_f°/kJ mol^{-1}	ΔG_f°/kJ mol^{-1}	S°/J mol^{-1} K^{-1}
Th		g	597.1 ± 2.1	556.4 ± 2.1	190.04 ± 0.40
Th	Metal, α	c	0	0	53.39 ± 0.40
Th^{4+}	Std. state $m = 1$	aq	−769.0 ± 2.5	−704.9 ± 5.4	−422.6 ± 16.7
Th^{4+}	1 M HCl	aq	−769.9 ± 1.7		
Th^{4+}	6 M HCl	aq	−759.4 ± 2.1		
ThO$_2$		c	−1226.4 ± 1.7	−1168.8 ± 1.7	65.23 ± 0.08
Th(OH)$_4$	Undiss.	aq		(−1588)	
Th(OH)$^{3+}$	1 M NaClO$_4$	aq	−1030.1 ± 2.5	−918.5 ± 5.6	−348.9 ± 16.7[a]
Th$_2$(OH)$_2^{6+}$	1 M NaClO$_4$	aq	−2047.7 ± 5.0	−1857.7 ± 11.2	−586.5 ± 33.5[a]
Th(OH)$_2^{2+}$	1 M NaClO$_4$	aq	−1282.5 ± 2.5	−1134.9 ± 5.6	−236.7 ± 16.7[a]
Th$_4$(OH)$_8^{8+}$	1 M NaClO$_4$	aq	−5121.2 ± 10.0	−4608.8 ± 22.3	−684.8 ± 66.9[a]

Compound					
$Th_6(OH)_{15}^{9+}$	1 M NaClO$_4$	aq	-8447.5 ± 15.0	-7577.4 ± 33.5	-667.6 ± 100.4^a
ThF_4		c	-2097.9 ± 8.4	-2003.3 ± 8.4	142.0 ± 0.2
$ThOF_2$		c	-1669 ± 10	-1592 ± 10	101.2 ± 8.4
$ThCl_4$		c	-1186.2 ± 1.7	-1094.1 ± 2.5	190.4 ± 4.2
$ThOCl_2$		c	-1232.6 ± 8.4	-1169.8 ± 8.5	114.6 ± 4.2
$ThBr_4$	β	c	-964.4 ± 2.1	-925.5 ± 3.3	228.0 ± 8.4
$ThOBr_2$		c	-1130 ± 13	-1077 ± 14	130 ± 13
ThI_4		c	-670.7 ± 2.1	-661.5 ± 4.4	255 ± 13
$ThOI_2$		c	-996.6 ± 2.5	-958.6 ± 2.8	144.8 ± 4.2
$Th(SO_4)_2$		c	-2536 ± 21	-2306 ± 22	167 ± 21
$Th(NO_3)_4$		c	-1446 ± 13		
$Th(NO_3)_4 \cdot 5H_2O$		c	-3007.9 ± 4.2	-2325.5 ± 4.2	543.1 ± 0.4
$Th(C_2O_4)_2$		c	-1339 ± 42		

[a] Uncertainty limits predicted on estimates for the entropy of the ion Th^{4+}.

protactinium is rapidly oxidized by air. In very acidic media ($[H^+] \geq 2\,M$) the predominant tetravalent species is the ion Pa^{4+}, which appears to be less easily hydrolyzable than pentavalent protactinium.

In the +5 oxidation state the protactinium species is different from corresponding uranium, neptunium, plutonium, and americium species which form the well-known MO_2^+ ion. The predominant species at pH < 0 in noncomplexing media is $PaOOH^{2+}$; in very acidic media PaO^{3+} may possibly exist. In 3 M HCl, the predominant species of pentavalent protactinium is $Pa\,OOHCl^+$, while in fluoride media numerous oxyfluorinated and fluorinated ions up to PaF_8^{3-} have been characterized. In 4 to 8 M HF, PaF_7^{2-} is the predominant species.

The hydrolytic behavior of protactinium in both the +4 and +5 oxidation states is very complex and has been studied only in a fragmentary fashion. Hydrolysis and polymerization of pentavalent protactinium occurs even in complexing media, such as HCl, H_2SO_4, and HNO_3, with the precipitation of a hydrous oxide as the ultimate step. It must be pointed out that at concentrations higher than 10^{-5} or 10^{-4} M, stable solutions of pentavalent protactinium can be obtained only in the presence of fluoride ions. The thermodynamic data are given in Table 7.

B. Gibbs Energies and Potentials

From the Gibbs energy, we find in 1 M HCl,

$$Pa^{4+} + 4e^- \rightarrow Pa \qquad E = -1.46 \pm 0.06\text{ V}$$

By comparison with the potentials of the couple V^{3+}/V^{2+}, Haïssinsky and Pluchet [1] report the value -0.29 ± 0.03 V for the formal potential of the system Pa^{5+}/Pa^{4+} in 6 M HCl. In 1 M HCl, Fried and Hindman [2] estimate this potential as -0.1 ± 0.1 V. In the absence of other data we accept Fried and Hindman's value for the formal potential in 1 M HCl with the assumption that oxidation proceeds according to the reaction $Pa^{4+} + H_2O \rightarrow PaOOH^{2+} + 2H^+ + (1/2)H_2O$.

C. Potential Diagram for Protactinium (in 1 M HCl)

$$\begin{array}{ccc} +5 & +4 & 0 \end{array}$$

$$PaOOH^{2+} \xrightarrow{-0.1} Pa^{4+} \xrightarrow{-1.46} Pa$$

$$\underline{\qquad -1.19 \qquad}$$

TABLE 7 Thermodynamic Data on Protactinium[a]

Formula	Description	State	ΔH_f°/kJ mol^{-1}	ΔG_f°/kJ mol^{-1}	S°/J mol^{-1} K^{-1}
Pa		g	607 ± 21	563 ± 21	197.9 ± 1.7
Pa	Metal, α	c	0	0	51.9 ± 1.7
Pa^{4+}	1 M HCl	aq	−620 ± 13	−564 ± 17	−397 ± 42
Pa^{4+}	6 M HCl	aq	−608 ± 13		
Pa^{4+}	12 M HCl-0.05 M HF	aq	−666 ± 13		
PaOOH^{2+}	1 M HCl	aq	−1113 ± 21	−1050 ± 21	−21 ± 21
Pa^{5+}	12 M HCl-0.05 M HF	aq	−672 ± 15		
PaF$_4$		c	−1945 ± 17	−1853 ± 17	146.9 ± 4.2

TABLE 7 (Continued)

Formula	Description	State	$\Delta H°$/kJ mol^{-1}	$\Delta G°$/kJ mol^{-1}	$S°$/J mol^{-1} K^{-1}
PaCl$_4$		c	−1044 ± 13	−954 ± 13	193.7 ± 4.2
PaCl$_5$		c	−1143 ± 15	−1032 ± 15	238.5 ± 8.4
PaOCl$_2$		c	−1093.7 ± 8.8	−1018.4 ± 9.1	125.5 ± 8.4
PaBr$_4$		c	−826 ± 13	−794 ± 13	233.5 ± 8.5
PaBr$_5$		c	−862 ± 16	−820 ± 17	289 ± 17
PaOBr$_2$		c	−1001 ± 15	−952 ± 16	142 ± 17
PaI$_4$		c	−524 ± 15	−516 ± 16	259 ± 13
PaO$_2$		c	−1109 ± 15[a]	−1054 ± 15	74.0 ± 5.0

[a] J. Fuger, "Thermodynamic Properties of the Actinides: Current Perspectives," Actinide 81, Pacific Grove, Calif., September 1981.

The Actinides

REFERENCES

1. M. Haissinsky and E. Pluchet, J. Chim. Phys., 59, 608-610 (1962).
2. S. Fried and J. C. Hindman, J. Am. Chem. Soc., 76, 4863-4864 (1954).

V. URANIUM

A. Oxidation States

Uranium exists in aqueous solutions in the +3, +4, +5, and +6 oxidation states. The ion UO_2^{2+} is the most stable and is difficult to reduce. The lowest oxidation state is slowly oxidized by water to U^{4+}; in the presence of air, oxidation of U^{3+} to U^{4+} is very rapid. U^{4+} is stable to water but is slowly oxidized by air to UO_2^{2+}. UO_2^+ has a very limited range of stability; it is the most stable at pH 2.5 but even at that pH it slowly disproportionates to U^{4+} and UO_2^{2+} according to the reaction.

$$2UO_2^+ + 4H^+ \rightarrow U^{4+} + UO_2^{2+} + 2H_2O$$

The hydrolysis of U^{4+} begins at pH < 1, and with increasing pH several species, from UOH^{3+} to $U(OH)_4$, form progressively. The thermodynamically stable solid species is $UO_2(c)$. Solubility measurements of $UO_2(c)$ have also indicated the formation of $U(OH)_5^-$. The hydrolysis of UO_2^{2+} begins at about pH 3. In all the media investigated (NaCl, NaClO$_4$, KNO$_3$), $UO_2(OH)^+$ is the first species that is produced and the dimerized ion $(UO_2)_2(OH)_2^{2+}$ rapidly becomes the principal hydrolysis product. $(UO_2)_3(OH)_5^+$ forms on further hydrolysis. There is evidence for the occurrence of $(UO_2)_3(OH)_4^{2+}$ only in NaCl solutions. The crystalline compound $UO_3 \cdot H_2O$ is the solid that appears upon precipitation, but the thermodynamically stable phase is $UO_3 \cdot 2H_2O(c)$. The thermodynamic data are given in Table 8.

B. Gibbs Energies and Potentials

From the Gibbs energies we obtain the following values for the standard potentials:

$$U^{4+} + 4e^- \rightarrow U \qquad E^\circ = -1.38 \pm 0.01 \text{ V}$$

$$U^{3+} + 3e^- \rightarrow U \qquad E^\circ = -1.66 \pm 0.02 \text{ V}$$

$$U^{4+} + e^- \rightarrow U^{3+} \qquad E^\circ = -0.52 \pm 0.05 \text{ V}$$

The potential of the U^{4+}/U^{3+} couple in 1 M HClO$_4$ has been reported by Kritchevsky and Hindman [1] as -0.631 ± 0.005 V. Using the correction to standard conditions for the corresponding plutonium couple (see Section VII), we obtain $E^\circ = -0.607 \pm 0.007$ V. According to Ahrland et al. [2], the correction to standard conditions leads to $E^\circ = -0.596$ V. It is obvious that the various values for the standard potential of the U^{4+}/U^{3+} couple are in marginal agreement. Only reasons of coherence with the accepted thermodynamic data [i.e., measured $\Delta H_f^\circ[U^{3+}(aq)]$ and $\Delta H_f^\circ[U^{4+}(aq)]$, as well as estimated $S^\circ[U^{3+}(aq)]$ and $S^\circ[U^{4+}(aq)]$ consistent with accepted entropies for other

TABLE 8 Thermodynamic Data on Uranium[a]

Formula	Description	State	ΔH_f°/kJ mol^{-1}	ΔG_f°/kJ mol^{-1}	S°/J mol^{-1} K^{-1}
U		g	531.4 $^{+8.4}_{-4.2}$	486.8 $^{+8.4}_{-4.2}$	199.7 ± 0.4
U	Metal, α	c	0	0	50.21 ± 0.12
U^{3+}		aq	−489.1 ± 3.8	−480.7 ± 4.6	−174.9 ± 8.4
U^{4+}		aq	−591.2 ± 3.3	−530.9 ± 2.1	−414 ± 21
U^{4+}	0.5 M HClO$_4$	aq	−588.3 ± 3.7		
U^{4+}	1 M HClO$_4$	aq	−589.9 ± 3.7		
U^{4+}	1 M HCl	aq	−587.4 ± 3.3		
U^{4+}	6 M HCl	aq	−569.9 ± 3.3		
UO$_2$		c	−1084.9 ± 0.9	−1031.8 ± 0.9	77.03 ± 0.21
UO$_2^+$		aq	−1032.6 ± 5.9	−968.6 ± 5.4	−25.1 ± 8.4
UO$_2^{2+}$		aq	−1018.8 ± 1.7	−952.7 ± 2.1	−97.1 ± 3.8
UO$_3$	Orth., γ	c	−1223.8 ± 1.3	−1145.8 ± 1.3	96.11 ± 0.40

$UO_3 \cdot H_2O$	c, β	-1533.4 ± 0.8	(-1397.0 ± 2.5)	(133.9 ± 8.4)
U_3O_8	c	-3574.8 ± 2.5	-3369.6 ± 2.5	282.5 ± 0.5
$U(OH)^{3+}$	aq	-828.1 ± 4.2	-765.4 ± 7.5	-193 ± 21
UO_2OH^+	aq, 0.5 M KNO_3	-1258.6 ± 8.5	-1158.2 ± 3.3	19 ± 21
$(UO_2)_2(OH)_2^{2+}$	aq, 3 M $NaClO_4$	-2569.5 ± 3.4	-2345.3 ± 4.2	-36.3 ± 7.7
$(UO_2)_3(OH)_5^+$	aq, 3 M $NaClO_4$	-4384.5 ± 5.5	-3949.2 ± 6.3	83 ± 24
UF_3	c	-1502.0 ± 5.0	-1433.4 ± 5.0	123.4 ± 0.4
UF_4	c	-1914.2 ± 4.2	-1823.4 ± 4.2	151.7 ± 0.2
UF_5	c, α	-2075.2 ± 5.9	-1968.7 ± 7.1	199.6 ± 13
UF_5	c, β	-2083.2 ± 6.3	-1970.7 ± 7.4	179.5 ± 13
UF_6	c	-2197.0 ± 1.7	-2068.6 ± 1.7	227.6 ± 1.3
UOF_2	c	-1504.6 ± 6.3	-1434.3 ± 6.4	119.2 ± 4.2
UO_2F_2	c	-1653.5 ± 1.3	-1557.3 ± 1.3	135.6 ± 0.4

TABLE 8 (Continued)

Formula	Description	State	ΔH_f°/kJ mol^{-1}	ΔG_f°/kJ mol^{-1}	S°/J mol^{-1} K^{-1}
UCl_3		c	−866.1 ± 4.2	−798.7 ± 3.8	159.0 ± 1.3
UCl_4		c	−1018.8 ± 2.5	−929.7 ± 2.5	197.1 ± 0.8
UCl_5		c	−1059 ± 13	−950 ± 13	242.7 ± 8.4
UCl_6		c	−1092 ± 13	−962 ± 13	285.8 ± 1.7
$UOCl_2$		c	−1066.9 ± 2.5	−996.2 ± 2.5	138.3 ± 0.2
$UOCl_3$		c	−1163 ± 21	1071 ± 21	171.5 ± 8.4
UO_2Cl_2		c	−1243.6 ± 1.3	−1145.9 ± 1.3	150.5 ± 0.2
UBr_3		c	−698.7 ± 4.2	−673.2 ± 4.8	192.5 ± 8.4
UBr_4		c	−802.1 ± 2.5	−767.5 ± 3.5	238.5 ± 8.4
UBr_5		c	−810.4 ± 8.4	−769.4 ± 9.2	293 ± 13

UOBr$_2$	c	−973.6 ± 8.4	−929.7 ± 8.4	−157.6 ± 0.3
UOBr$_3$	c	−954 ± 21	−900 ± 21	205 ± 13
UO$_2$Br$_2$	c	−1137.4 ± 1.3	−1066.5 ± 1.8	169.4 ± 4.2
UI$_3$	c	−467.3 ± 4.2	−466.5 ± 4.9	221.7 ± 8.4
UI$_4$	c	−518.8 ± 2.9	−513.2 ± 3.8	263.6 ± 8.4
U(SO$_4$)$_2$	c	−2310 ± 13	−2085 ± 15	180 ± 21
UO$_2$SO$_4$	c, β	−1845.1 ± 0.8	−1685.7 ± 2.5	163.2 ± 8.4
UO$_2$(NO$_3$)$_2$	c	−1351.4 ± 2.1	−1106.4 ± 4.1	240 ± 12
UO$_2$(NO$_3$)$_2$·6H$_2$O	c	−3168.1 ± 2.1	−2585.3 ± 2.1	505.6 ± 2.5
UO$_2$(NO$_3$)$_2$	aq, m = 1	−1433.4 ± 2.5	−1174.9 ± 2.5	196.6 ± 4.6
U(C$_2$O$_4$)$_2$	c	−1351 ± 42		

[a] Estimated values are given in parentheses.

actinide ions] lead to our use, in the summary of potentials, of the value $E° = -0.52 \pm 0.05$ V for the U^{4+}/U^{3+} couple.

The standard Gibbs energy change for the reaction

$$UO_2^{2+} + 2H^+ + H_2 \rightarrow U^{4+} + 2H_2O$$

is $\Delta G° = -52.68 \pm 0.18$ kJ mol^{-1} and thus $E° = 0.273 \pm 0.005$ V.

The latter value is in perfect agreement with the value reported by Wagman et al. [3], $E° = 0.273 \pm 0.005$ V, based on a Debye-Hückel extrapolation of the data of Sobkowski and Minc [4] and Sobkowski [5] in perchloric solution. The agreement with the standard potential obtained by Nikolaeva [6], $E° = 0.272$ V, is also excellent.

The formal potential of the couple UO_2^{2+}/UO_2^+ in 1 M HClO$_4$ has been determined by different authors and the results are in good agreement. The value of Kritchevsky and Hindman, 0.063 ± 0.004 V, is accepted; to reach the standard state we use the same correction as that proposed by Brand and Cobble [7] for the corresponding Np couple (i.e., 0.1 V) and obtain $E°(UO_2^{2+}/UO_2^+) = 0.163 \pm 0.05$ V.

C. Potential Diagrams for Uranium

Acid Solution

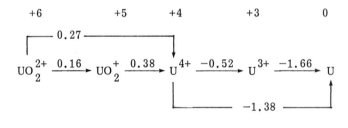

Basic Solution

$$UO_2(OH)_2 \xrightarrow{-0.3} UO_2 \xrightarrow{-2.6} U(OH)_3 \xrightarrow{-2.10} U$$

REFERENCES

1. E. S. Kritchevsky and J. C. Hindman, J. Am. Chem. Soc., 71, 2096-2102 (1949).
2. S. Ahrland, J. Liljenzin, and J. Rydberg, in *Comprehensive Inorganic Chemistry*, Pergamon Press, Elmsford, N.Y., 1973.
3. D. D. Wagman, W. H. Evans, V. B. Parker, R. H. Schumm, I. Halow, S. M. Bailey, K. L. Churney, and R. L. Nuttal, *NBS Tables of Chemical Thermodynamic Properties*, J. Phys. Chem. Ref. Data, Vol. 11, Supplement 2 (1982).
4. J. B. Sobkowski and S. Minc, J. Inorg. Nucl. Chem., 19, 101-106 (1961).
5. J. B. Sobkowski, J. Inorg. Nucl. Chem., 23, 81-90 (1961).
6. N. M. Nikolaeva, Izv. Sib. Otd. Akad. Nauk SSSR, Ser. Khim., 1, 14-17 (1974).
7. J. R. Brand and J. W. Cobble, J. Inorg. Chem., 9, 912-917 (1970).

The Actinides

VI. NEPTUNIUM

A. Oxidation States

Neptunium exists in solution in the +3, +4, +5, +6, and +7 oxidation states. The most stable species is NpO_2^+, which disproportionates only at high acidities and at moderate or high Np concentrations by the reaction

$$2NpO_2^+ + 4H^+ \rightarrow Np^{4+} + NpO_2^{2+} + 2H_2O$$

Np^{3+} is stable in water but is quickly oxidized by air to Np^{4+}; Np^{4+} is slowly oxidized by air to NpO_2^+. Hexavalent neptunium is stable in acidic solutions as NpO_2^{2+} but is easy to reduce to NpO_2^+. Heptavalent neptunium is a strong oxidizing agent in acid media; in 1 M $HClO_4$ or 1 M HNO_3 the usually accepted heptavalent species is NpO_3^+, which is reduced by water to NpO_2^{2+}. In basic media (0.2 to 14 M NaOH) heptavalent neptunium slowly reduces at the rate of a few percent per month. It is believed by many authors to exist as NpO_5^{3-}, which we shall adopt for reason of convenience, but species such as $NpO_2(OH)_6^{3-}$ or $NpO_4(OH)_2^{3-}$ have also been suggested. References concerning the various possible species of Np(VII) can be found in Part II of the IAEA assessment, which serves as the basis for our data selection. The hydrolysis behavior of Np^{4+} in solution is believed to be similar to that of U^{4+} and Pu^{4+}, but only the first hydrolytic species $NpOH^{3+}$ has been well characterized. The thermodynamically stable solid species is $NpO_2(c)$. The known hydrolyzed species of NpO_2^+ are $NpO_2OH(aq)$ and $NpO_2OH(am)$. The hydrolysis products of NpO_2^{2+} are NpO_2OH^+, $(NpO_2)_2(OH)_2^{2+}$, $(NpO_2)_3(OH)_5^+$, and $NpO_2(OH)_2(am)$, and are thus analogous to those of UO_2^{2+} and PuO_2^{2+}.

The first hydrolysis step of NpO_3^+ corresponds to NpO_3OH, whose solubility product has been determined as 10^{-24}. The thermodynamic data are given in Table 9.

B. Gibbs Energies and Potentials

From the Gibbs energies we calculate the three following standard potentials:

$$Np^{4+} + 4e^- \rightarrow Np \qquad E° = -1.30 \pm 0.02 \text{ V}$$

$$Np^{3+} + 3e^- \rightarrow Np \qquad E° = -1.79 \pm 0.01 \text{ V}$$

$$Np^{4+} + e^- \rightarrow Np^{3+} \qquad E° = 0.15 \pm 0.03 \text{ V}$$

The best values of the formal potential for the couple Np^{4+}/Np^{3+} are equal to 0.137 ± 0.005 V in 1 M HCl [1] and 0.155 ± 0.001 V in 1 M $HClO_4$ [2]. Using for the latter value the same correction to the standard conditions as that discussed in the case of the Pu^{4+}/Pu^{3+} couple, we obtain and adopt the value $E° = 0.179 \pm 0.005$ V, which is in agreement with that deduced from the Np^{3+}/Np and Np^{4+}/Np couples.

The most precise value for the formal potential of the couple NpO_2^{2+}/NpO_2^+ in 1 M $HClO_4$ is 1.13638 ± 0.00016 V obtained by Sullivan et al. [3], which is consistent with the value reported by Cohen and Hindman [2a]; $E° = 1.1373 \pm 0.001$ V. Brand and Cobble [4] derived $E° = 1.236 \pm 0.010$ V by using a Debye-Hückel extrapolation of their experimental results in perchloric medium. In this case an accurate value is therefore available for the difference

TABLE 9 Thermodynamic Data on Neptunium[a]

Formula	Description	State	ΔH_f°/kJ mol^{-1}	ΔG_f°/kJ mol^{-1}	S°/J mol^{-1} K^{-1}
Np		g	464.8 ± 4.2	420.9 ± 4.2	197.61 ± 0.25
Np	Metal, α	c	0	0	50.46 ± 0.25
Np^{3+}		aq	−527.2 ± 2.1	−517.1 ± 3.3	−179.1 ± 8.4
Np^{3+}	1 M HCl	aq	−526.8 ± 1.7		
Np^{4+}		aq	−556.1 ± 4.2	502.9 ± 7.5	−389 ± 21
Np^{4+}	1 M HClO$_4$	aq	−555.6 ± 4.2		
Np^{4+}	1 M HCl	aq	−551.9 ± 1.7		
NpO$_2^+$		aq	−978.2 ± 4.6	−915.0 ± 5.4	−20.9 ± 8.4
NpO$_2^{2+}$		aq	−860.6 ± 4.6	−795.8 ± 5.4	−92.0 ± 8.4
NpF$_3$		c	−1528.8 ± 5.4	−1460.3 ± 5.5	124.7 ± 4.2
NpF$_4$		c	−1874 ± 13	−1784 ± 13	152.7 ± 4.2

NpCl$_3$	c	−898.3 ± 3.3	−831.7 ± 4.1	161.5 ± 8.4
NpCl$_4$	c	−984.1 ± 1.7	−896.2 ± 2.1	201.7 ± 4.2
NpOCl$_2$	c	−1038 ± 13	−968 ± 13	141.0 ± 8.4
NpBr$_3$	c	−730.5 ± 2.1	−706.0 ± 5.5	197 ± 17
NpBr$_4$	c	−771.1 ± 1.7	−737.6 ± 4.3	243 ± 13
NpOBr$_2$	c	−946 ± 13	−903 ± 13	158.2 ± 8.4
NpI$_3$	c	−512.9 ± 2.1	513.4 ± 5.5	226 ± 17
NpO$_2$	c	−1074.0 ± 2.5	−1021.8 ± 2.5	80.3 ± 0.4
NpO$_3 \cdot$H$_2$O	c	−1379.0 ± 4.6	−1246.3 ± 8.4	146 ± 33
NpO$_2$(NO$_3$)$_2$	c	(−1209)		
NpO$_2$(NO$_3$)$_2 \cdot$6H$_2$O	c	−3009.1 ± 5.0	−2429.2 ± 5.2	516.3 ± 4.6
Np(C$_2$O$_4$)$_2$	c	−1975 ± 21		

[a]Estimated values are given in parentheses.

between the standard and the formal potentials; Brand and Cobble recommend that the difference of 0.1 V be applied similarly to the couples UO_2^{2+}/UO_2^+, PuO_2^{2+}/PuO_2^+, and AmO_2^{2+}/AmO_2^+, where no experimental data at infinite dilution are available.

In various media we adopt the following formal potentials for the couple NpO_2^{2+}/NpO_2^+ [1]:

1 M HCl $E^{\circ\prime} = 1.14$ V

1 M HNO$_3$ $E^{\circ\prime} = 1.14$ V

1 M H$_2$SO$_4$ $E^{\circ\prime} = 1.082$ V

Cohen and Hindman [2] have measured the formal potential of the couple NpO_2^+/Np^{4+} in 1 mol kg^{-1} HClO$_4$ as 0.7391 ± 0.0010 V; we accept this value to be the same in 1 M HClO$_4$. We correct this potential to standard conditions using the same correction as that used for the corresponding plutonium couple; this gives $E^{\circ}(NpO_2^+/Np^{4+}) = 0.670 \pm 0.060$ V. As ΔG° values yield $E^{\circ}(NpO_2^+/Np^{4+}) = 0.65 \pm 0.08$ V, we adopt the average value of 0.66 ± 0.06 V for the standard potential. In 1 M HCl and in 1 M H$_2$SO$_4$, respectively, we accept for the formal potentials of the same couple the values 0.737 ± 0.006 V and 0.99 V [1].

Np(VII) exists as an unstable species in acid solutions and the formal potential corresponding to the reaction $NpO_3^+ + 2H^+ + e^- \rightarrow NpO_2^{2+} + H_2O$ has been reported as 2.04 ± 0.03 V in 1 M HClO$_4$ [5] and > 2.00 V in 1 M HNO$_3$ [6]. In basic solutions ($\cong$ 0.2 M NaOH to 14 M NaOH) the species NpO_5^{3-} is stable and the electrochemical reduction to Np(VI) is described by several authors as

$$NpO_5^{3-} + H_2O + e^- \rightarrow NpO_4^{2-} + 2OH^-$$

The results of Zielen and Cohen [7] in 1 M NaOH appear to be the most precise; they report $E^{\circ}(NpO_5^{3-}/NpO_4^{2-}) = 0.5821 \pm 0.00045$ V. In view of the value of (0.600 V) reported by Simakin and Matyashchuk [8] for that couple in the same medium, we suggest the value 0.58 ± 0.01 V for the formal potential in 1 M NaOH.

C. Potential Diagrams for Neptunium

Acid Solution

```
   +7           +6          +5          +4          +3           0
           2.04        1.24        0.66        0.18       -1.79
NpO_3^+  ------>  NpO_2^{2+} ----> NpO_2^+ ----> Np^{4+} ----> Np^{3+} ----> Np
                     |_____ 0.95 _____|    |_____ -1.30 _____|
```

Basic Solution

```
           0.58              0.6            0.3          -2.1          -2.2
NpO_5^{3-} ----> NpO_2(OH)_2 ----> NpO_2OH ----> NpO_2 ----> Np(OH)_3 ----> Np
```

REFERENCES

1. J. Hindman, L. Magnuson, and T. Lachapelle, J. Am. Chem. Soc., 71, 687-693 (1949).
2. (a) D. Cohen and J. Hindman, J. Am. Chem. Soc., 74, 4679-4682 (1952); (b) D. Cohen and J. Hindman, J. Am. Chem. Soc., 74, 4682-4685 (1952).
3. J. C. Sullivan, J. C. Hindman, and A. J. Zielen, J. Am. Chem. Soc., 83, 3373-3378 (1961).
4. J. R. Brand and J. W. Cobble, Inorg. Chem., 9, 912-917 (1970).
5. C. Musikas, F. Couffin and M. Marteau, J. Chim. Phys., 71, 641-648 (1974).
6. J. Sullivan and A. Zielen, Inorg. Nucl. Chem. Lett., 5, 927-931 (1969).
7. A. Zielen and D. Cohen, J. Phys. Chem., 74(2), 394-405 (1970).
8. G. A. Simakin and I. V. Matyashchuk, Radiokhimiya, 11, 481-482 (1969).

VII. PLUTONIUM

A. Oxidation States

As for neptunium, plutonium exhibits all oxidation states from +3 through +7. Tetravalent plutonium is the most stable species in relatively concentrated acid media (e.g., 6 M HNO_3), but at lower acidities, it disproportionates at a slow rate to essentially Pu^{3+} and PuO_2^{2+} [1]. For instance, a 0.025 M Pu^{4+} solution in 0.344 M HNO_3 was found to reach equilibrium in ca. 17 h, at which time it contained 12% Pu^{3+}, 66% Pu^{4+}, and 22% PuO_2^{2+}, while in 0.5 M HCl small amounts of PuO_2^+ were also found at equilibrium. At pH 2 or above, hydrolysis of Pu^{4+} becomes important. Pu^{3+} is not oxidized by water or by air at room temperature in HCl or $HClO_4$ solutions, but is oxidized slowly by O_2 in HCl, or H_2SO_4-HCl solutions.

The closeness of the potentials of the couples PuO_2^{2+}/PuO_2^+, PuO_2^{2+}/Pu^{4+}, PuO_2/Pu^{4+}, and Pu^{4+}/Pu^{3+} and their pH dependence makes it possible to obtain all four oxidation states simultaneously in solution. In addition, the stability of the various species depends markedly on the presence of complexing agents and the rates of oxidation-reduction and disproportionation reactions.

PuO_2^+ rapidly disproportionates below pH 1.5 in $HClO_4$ media. At low plutonium concentrations essentially no disproportionation occurs at pH 2.5. In view of hydrolysis phenomena, the maximum stability range of PuO_2^+ lies between pH 1.5 and 5. PuO_2^{2+} is a stronger oxidizing agent than UO_2^{2+} but less oxidizing than NpO_2^{2+}; PuO_2^{2+} is a fairly stable ion. The plutonium potential-pH diagram in $HClO_4$ media at pH up to 3 has been particularly well studied by Silver [2].

Heptavalent plutonium is unstable in acidic media and quantitative information on its reducing behavior is lacking; in basic media ($[OH^{-1}] > 0.2 M$) the heptavalent species, which is believed to be PuO_5^{3-}, begins to decompose as soon as oxidation is stopped.

The hydrolysis of Pu^{4+} gives rise to the following species: $Pu(OH)^{3+}$, $Pu(OH)_2^{2+}$, and $Pu(OH)_3^+$, leading to the formation of $Pu(OH)_4(aq)$; $PuO_2(c)$ is considered to be the thermodynamically stable species of hydrolyzed tetravalent plutonium in the solid state. The anionic species $Pu(OH)_5^-$ has also been postulated by analogy with tetravalent uranium.

The first hydrolyzed species of PuO_2^+ is $PuO_2OH(aq)$, which appears at about pH 8, followed by the precipitation of $PuO_2(OH)$ (am). It should be

TABLE 10 Thermodynamic Data on Plutonium

Formula	Description	State	ΔH_f°/kJ mol^{-1}	ΔG_f°/kJ mol^{-1}	S°/J mol^{-1} K^{-1}
Pu		g	345.2 ± 4.2	309.1 ± 4.2	177.1 ± 0.4
Pu	Metal, α	c	0	0	56.1 ± 0.4
Pu^{3+}		aq	−592.0 ± 2.1	−578.6 ± 3.3	−184.5 ± 8.4
Pu^{3+}	6 M HCl	aq	−592.0 ± 1.7		
Pu^{4+}		aq	−536.4 ± 3.3	−481.6 ± 3.3	−389 ± 21
Pu^{4+}	1 M HClO$_4$	aq	−535.6 ± 3.3		
PuO$_2^+$		aq	−914.6 ± 5.9	−849.8 ± 7.5	−20.9 ± 8.4
PuO$_2^{2+}$		aq	−822.2 ± 6.7	−756.9 ± 7.1	−87.9 ± 8.4
PuO 1.515	bcc	c	−836 ± 11[a]	−769 ± 11[a]	72 ± 10

Compound	State	$\Delta H°_f$	$\Delta G°_f$	$S°$
$PuO_{1.500}$	hex	-828 ± 10[b]	-788 ± 10[b]	71 ± 10
PuO_2	c	-1056.2 ± 0.7	-988.0 ± 0.7	66.13 ± 0.26
PuF_3	c	-1585.7 ± 2.9	-1515.5 ± 3.0	126.11 ± 0.38
PuF_4	c	-1846 ± 21	-1753 ± 21	147.23 ± 0.29
PuF_6	c	-1862 ± 29	-1730 ± 34	221 ± 17
$PuCl_3$	c	-959.8 ± 1.7	-892.1 ± 2.1	163.6 ± 4.2
$PuCl_3 \cdot 6H_2O$	c	-2773.6 ± 2.1	-2365.2 ± 2.7	420.1 ± 5.8
$PuCl_4$	Unstable	-963.6 ± 7.5		
$PuOCl$	c	-930.9 ± 1.7	-881.4 ± 2.3	104.2 ± 5.4
$PuBr_3$	c	-792.9 ± 2.1	-768.0 ± 5.4	201 ± 17
$PuOBr$	c	-872.8 ± 4.6	-838.9 ± 6.8	121 ± 17
PuI_3	c	-579.9 ± 2.5	-579.8 ± 5.7	230 ± 17

[a] Values adjusted to be compatible with adopted data for $PuO_{1.500}$ (c,hex).
[b] T. M. Besmann and T. B. Lindemer, J. Am. Chem. Soc., 66, 782-785 (1983).

pointed out that at such a pH, disproportionation of PuO_2^+ (see above) is the dominant phenomenon.

The hydrolysis of PuO_2^{2+} follows the same steps as those of UO_2^{2+} and NpO_2^{2+}: $PuO_2(OH)^+$, $(PuO_2)_2(OH)_2^{2+}$, and $(PuO_2)_3(OH)_5^+$. The solubility product of $PuO_2(OH)_2$ is ill defined, for it increases from 1 M KOH solutions to 10 M KOH solutions with the formation of a more hydrolyzed ion, $PuO_2(OH)_3^-$. The thermodynamic data are given in Table 10.

B. Gibbs Energies and Potentials

Using the Gibbs energies of formation of $Pu^{4+}(aq)$ and $Pu^{3+}(aq)$, we calculate

$$Pu^{4+} + 4e^- \rightarrow Pu \qquad E^\circ = -1.25 \pm 0.01 \text{ V}$$

$$Pu^{3+} + 3e^- \rightarrow Pu \qquad E^\circ = -2.00 \pm 0.01 \text{ V}$$

$$Pu^{4+} + e^- \rightarrow Pu^{3+} \qquad E^\circ = 1.01 \pm 0.01 \text{ V}$$

In 1 M $HClO_4$ we adopt the formal potential 0.9821 ± 0.0005 V for the couple Pu^{4+}/Pu^{3+} from the virtually identical results of Connick and McVey [3] and Rabideau and Lemons [4]. We take $E^\circ = 1.006 \pm 0.003$ V from the data in 0.1 M $HClO_4$ by Schwabe and Nebel [5], who estimated the values for the activity coefficients. In this case the difference between the formal and the standard potentials is well established and we have applied the same difference, 0.024 V, to the couples U^{4+}/U^{3+} and Np^{4+}/Np^{3+}.

From the numerous results for the Pu^{4+}/Pu^{3+} couple in various media, we select the following formal potentials:

1 M HCl	$E^{\circ\prime} = 0.9703 \pm 0.0005$ V	[4]
1 M HNO_3	$E^{\circ\prime} = 0.916 \pm 0.006$ V	[6]
1 M H_2SO_4	$E^{\circ\prime} = 0.74 \pm 0.02$ V	[7]
1 M HF	$E^{\circ\prime} = \cong 0.50$ V	[7]

On the basis of the Gibbs energy change for the reaction

$$Pu^{3+} + 2H_2O \rightarrow PuO_2^{2+} + H^+ + (3/2)H_2$$

in 1 M $HClO_4$, 295.9 ± 0.6 kJ mol^{-1} deduced from several concordant data, the PuO_2^{2+}/Pu^{3+} potential in that medium becomes 1.022 ± 0.0016 V. Accordingly, the PuO_2^{2+}/Pu^{4+} potential in 1 M $HClO_4$ becomes 1.042 ± 0.001 V, while the corresponding standard potential based on Gibbs energies of formation of the ions is accepted as 1.032 ± 0.037 V.

The formal potential of the couple PuO_2^{2+}/PuO_2^+ in 1 M $HClO_4$ is given by Rabideau [6] as 0.9164 ± 0.0002 V; after a correction to standard states analogous to that obtained by Brand and Cobble for the corresponding neptunium couple, we reach $E^\circ = (1.016 \pm 0.05)$ V.

In 0.1 M HNO_3 we adopt $E(PuO_2^{2+}/PuO_2^+) = 0.917$ V [8], while in 1 M HCl the formal potential of the same couple is found to be 0.91 V [9].

The formal potential in 1 M NaOH corresponding to the reaction

$$PuO_5^{3-} + H_2O + e^- \rightarrow PuO_4^{2-} + 2OH^-$$

is accepted as 0.95 ± 0.15 V from the measurements of Krot et al. [10]. Musikas [11] recommends the value 0.94 V and proposes the reaction

$$PuO_5^{3-} + 3H_3O + e^- \rightarrow PuO_2(OH)_3^- + 3OH^-$$

C. Potential Diagrams for Plutonium

Acid Solution

$$PuO_2^{2+} \xrightarrow{1.02} PuO_2^+ \xrightarrow{1.04} Pu^{4+} \xrightarrow{1.01} Pu^{3+} \xrightarrow{-2.00} Pu$$

with $PuO_2^{2+} \xrightarrow{1.03} Pu^{4+}$ and $Pu^{4+} \xrightarrow{-1.25} Pu$

Basic Solution

$$PuO_5^{3-} \xrightarrow{0.95} \left\{ \begin{array}{c} PuO_2(OH)_2 \\ PuO_2(OH)_3^- \end{array} \right\} \xrightarrow{0.3} PuO_2(OH) \xrightarrow{0.9} PuO_2 \xrightarrow{-1.4} Pu(OH)_3 \xrightarrow{2.46} Pu$$

REFERENCES

1. J. M. Cleveland, in *Plutonium Handbook*, O. J. Wick, ed., Gordon and Breach, London, 1967, p. 405.
2. (a) G. L. Silver, Radiochem. Radioanal. Lett., 7, 1-5 (1971); (b) G. L. Silver, J. Inorg. Nucl. Chem., 43, 2997-2999 (1981).
3. R. E. Connick and W. H. McVey, J. Am. Chem. Soc., 73, 1798-1804 (1951).
4. S. W. Rabideau and J. F. Lemons, J. Am. Chem. Soc., 73, 2895-2899 (1951).
5. K. Schwabe and D. Nebel, Z. Phys. Chem., 220, 339-354 (1962).
6. S. W. Rabideau, J. Chem. Soc., 78, 2705-2707 (1956).
7. J. Howland, J. Hindman, and R. Kraus, U.S. Atomic Energy Comm. U.S. Rep. C. K. 1371.
8. P. Artyukhin, A. Gel'man, and V. Medvedoskii, Dokl. Akad. Nauk. SSSR, 120, 98-101 (1958).
9. M. Kasha, Natl. Nuclear Energy Ser. IV, 14B, Chap. 8.
10. N. N. Krot, A. D. Gel'man, and M. P. Mefod'eva, Vestn. Akad. Nauk. SSSR, 40, 57-61 (1970).
11. C. Musikas, *Gmelin Handbuch der anorganischen Chemie*, Transurane, System No. 55, Vol. 20, Part D, Springer-Verlag, Berlin, 1975.

VIII. AMERICIUM

A. Oxidation States

Americium can exist in aqueous solutions in the +3, +4, +5, +6, and +7 oxidation states. Am^{3+}, which is difficult to oxidize, is the most stable species. Am^{4+} is not stable in noncomplexing aqueous solution; this species completely and rapidly disproportionates according to the scheme

$$2Am^{4+} + 2H_2O \rightarrow Am^{3+} + AmO_2^+ + 4H^+ \qquad \log K = 21$$

It can be calculated that at equilibrium the Am^{4+} concentration relative to the other americium species is only 10^{-10}. However, an appreciable stability appears in concentrated fluoride and carbonate media in which americium exists as negatively charged complexes; tetravalent americium has also been stabilized in concentrated phosphate and phosphotungstate solutions. The status of the stability of the quadrivalent state of americium in complexing media has been reviewed rather recently [1].

AmO_2^+ disproportionates to Am^{3+} and AmO_2^{2+} at pH values below ca. 1.5. At equilibrium AmO_2^+ should never be the predominant species; nevertheless, slow kinetic processes allow the preparation of rather pure AmO_2^+ solutions either by chemical or electrochemical oxidation of Am^{3+} or by self-reduction of AmO_2^{2+} (see below).

AmO_2^{2+} can be stabilized in the presence of strong oxidizing agents, but when using the usual ^{241}Am isotope (α-emitter, half-life 433 years), the self-reduction under the influence of the radiolytic products of the medium plays an important role. In 1 M $HClO_4$ the self-reduction of AmO_2^{2+} is about 4 to 5% per hour, yielding AmO_2^+, which itself reduces to Am^{3+} at about half that rate. These rates vary largely with the medium and the presence of scavengers for the radiolysis products. The use of the long-life isotope ^{243}Am (α-emitter, half-life 7900 years) will minimize these problems, but with the exception of recent thermodynamic data on $AmO_2(c)$ [2], results involving ^{243}Am are essentially nonexistent.

Although this matter has been controversial for some time, the formation of heptavalent americium has been announced in 3 to 5 M NaOH by oxidation of hexavalent americium using γ irradiation in the presence of N_2O or $K_2S_2O_8$. No quantitative data exist on this species, which is reported to be a stronger oxidizing agent than heptavalent plutonium [3]. Transient formation ($t_{1/2} \cong 5 \times 10^{-6}$ s) of Am(II) in perchlorate media has been observed in pulse radiolysis experiments involving a streak camera technique to characterize the spectrum of the transient species [4].

Experimental data for the hydrolytic behavior of Am^{3+} as well as for the other heavier trivalent actinides are scanty and possibly subject to systematic error. Quantitative data on the hydrolytic behavior of these elements in other valency states are simply nonexistent. We therefore believe it preferable to use comparisons with the data accepted for uranium, neptunium, and plutonium, and in the case of the trivalent state, comparisons with the properties of +3 lanthanide ions, as described in the introduction to this chapter. We thus take $Am(OH)_3(c)$, $AmO_2(c)$, $AmO_2OH(c)$, and $AmO_2(OH)_2/(c)$ for the various thermodynamically stable entities existing in contact with a basic solution. We must also mention that $Am(OH)_4(c)$ (probably hydrated AmO_2) has also been prepared by oxidation of $Am(OH)_3(c)$ [5]. The thermodynamic data are given in Table 11.

TABLE 11 Thermodynamic Data on Americium[a]

Formula	Description	State	ΔH_f°/kJ mol^{-1}	ΔG_f°/kJ mol^{-1}	S°/J mol^{-1} K^{-1}
Am		g	284.1 ± 4.0	242.3 ± 4.0	194.4 ± 2.5
Am	Metal, α	c	0	0	54.5 ± 2.5
Am^{3+}		aq	−616.7 ± 1.3	−599.1 ± 3.8	−201 ± 13
Am^{3+}	1 M HCl	aq	−616.3 ± 0.9		
Am^{3+}	6 M HCl	aq	−613.0 ± 1.3		
Am^{4+}		aq	−406.0 ± 6.0[b]	−346.6 ± 8.6	−406 ± 21
AmO$_2^+$		aq	−805.0 ± 4.6	−741.0 ± 5.4	−20.9 ± 8.4
AmO$_2^{2+}$		aq	−651.9 ± 2.1	−587.4 ± 3.3	−88 ± 33
AmO$_2$		c	−932.2 ± 2.7[b]	(−880)	(86)

Am$_2$O$_3$	hex.	−1690.4 ± 7.9[c]	(−1613)	(158)
AmF$_3$	c	−1588 ± 12	−1523 ± 13	127.6 ± 4.2
AmF$_4$	c	−1720 ± 29	−1628 ± 30	148.5 ± 4.2
AmCl$_3$	c	−977.8 ± 1.7	−910.9 ± 2.5	164.8 ± 6.3
AmOCl	c	−953.5 ± 2.5	−904.2 ± 3.3	102.5 ± 7.1
AmBr$_3$	c	−809.6 ± 7.1	−786.6 ± 8.7	205 ± 17
AmOBr	c	−897.5 ± 8.4	−862.7 ± 6.8	117 ± 17
AmI$_3$	c	−612.1 ± 7.1	−613.8 ± 9.4	234 ± 21

[a]Estimated values are given in parentheses.
[b]L. R. Morss and J. Fuger, J. Inorg. Nucl. Chem., *43*, 2059-2064 (1981).
[c]L. R. Morss and D. C. Sonnenberger, Proc. IUPAC Conf. on Chem. Thermodyn., Symposium on Thermodynamics of Nuclear Materials, Hamilton, Canada, August 1985; J. Nucl. Matl., *130*, 266-272 (1985).

B. Gibbs Energies and Potentials

From the Gibbs energies of formation we obtain the following standard potentials:

$$Am^{4+} + 4e^- \rightarrow Am \quad E° = -0.90 \pm 0.02 \text{ V}$$

$$Am^{3+} + 3e^- \rightarrow 3e^- \rightarrow Am \quad E° = -2.07 \pm 0.01 \text{ V}$$

$$Am^{4+} + e^- \rightarrow Am^{3+} \quad E° = 2.62 \pm 0.09 \text{ V}$$

The status of the Am^{4+}/Am^{3+} couple prior to 1980 has been reviewed in a paper (2) dealing with the determination of the enthalpy of formation of $AmO_2(c)$. Using auxiliary data these measurements yield a value of 2.62 ± 0.09 V. This result is in agreement with a value deduced from spectroelectrochemical studies in 2 M Na_2CO_3-$NaHCO_3$, for which a formal potential of 0.92 ± 0.01 V was found; correction to standard conditions using a shift potential of ca. 1.7 V based on similar lanthanide IV/III potentials yielded $E°(Am^{4+}/Am^{3+}) = 2.62$ V [6]. Earlier data based on calorimetric measurements on $AmO_2(c)$ as well as spectroscopic and electrochemical data yielded values scattered between 2.2 and 2.9 V [1]. Since fewer assumptions are involved in the most recent values, we adopt $E°(Am^{4+}/Am^{3+}) = 2.62 \pm 0.09$ V.

Penneman and Asprey [7] made direct measurements of the cell potential of the couple AmO_2^{2+}/AmO_2^+. They report a formal potential of (1.600 ± 0.005) V in 1 M $HClO_4$. After a correction to the standard state analogous to that obtained by Brand and Cobble for the corresponding neptunium couple, we reach $E° = 1.70 \pm 0.05$ V. Nevertheless, if the latter value is adopted, its combination with accepted enthalpies of formation for the ions would result in an entropy difference between AmO_2^{2+} and AmO_2^+ which would appear somewhat large considering the accepted difference between other actinyl ions. The enthalpy data are more accurate than the potential measurements of Penneman and Asprey because of the adverse effects of radiolysis and H_2O_2 production in ^{241}Am solutions and we shall therefore take the value $E°(AmO_2^{2+}/AmO_2^+) = 1.59 \pm 0.06$ V, which is fully consistent with enthalpy data and with the expected entropy difference between $AmO_2^{2+}(aq)$ and $AmO_2^+(aq)$.

The remaining standard potentials may be calculated from the Gibbs energies of formation of the various ions based on calorimetric measurements and the accepted entropies since no direct electrochemical determinations are available. We thus obtain $E°(AmO_2^+/Am^{3+}) = (1.72 \pm 0.03$ V. Using the latter value we also calculate

$$E°(AmO_2^+/Am^{4+}) = 0.82 \pm 0.03 \text{ V}$$

$$E°(AmO_2^{2+}/Am^{4+}) = 1.20 \pm 0.03 \text{ V}$$

$$E°(AmO_2^{2+}/Am^{3+}) = 1.67 \pm 0.01 \text{ V}$$

Nugent et al. [8] have estimated -2.3 V as the standard potential $E°(Am^{3+}/Am^{2+})$ from spectroscopic data on hexahalogeno complexes.

In basic media, Penneman et al. [5] indicated that $E°(AmO_2/Am(OH)_3)$ is at least 0.5 V. Nevertheless, we calculate a lower value, 0.22 V, from $\Delta G°$ data in acid solution and from the activity products quoted in Table III. In 1 M NaOH, Nikolaevsky et al. [9] have estimated $E°(Am(VII)/Am(VI)) \cong$ 1.05 V.

Potential Diagrams for Americium

Acid Solution

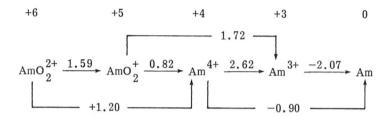

Basic Solution

$$AmO_2(OH)_2 \xrightarrow{0.9} AmO_2OH \xrightarrow{0.7} AmO_2 \xrightarrow{0.22} Am(OH)_3 \xrightarrow{-2.53} Am$$

REFERENCES

1. A. S. Saprykin, V. P. Shilov, V. I. Spitsyn, and N. N. Krot, Dokl. Akad. Nauk SSSR, 226, 853-856 (1976).
2. L. R. Morss, and J. Fuger, J. Inorg. Nucl. Chem., 43, 2059-2064 (1981).
3. N. N. Krot, V. P. Shilov, V. B. Nikolaevsky, A. K. Pikaev, A. D. Gelman, and V. I. Spitsyn, Akad. Nauk SSSR, 217(3), 589-592 (1974).
4. (a) J. C. Sullivan, S. Gordon, W. A. Mulac, K. A. Schmidt, D. Cohen, and R. Sjoblom, Inorg. Nucl. Chem. Lett., 12, 599-601 (1976); (b) S. Gordon, W. A. Mulac, K. A. Schmidt, R. Sjoblom, and J. C. Sullivan, Inorg. Chem., 17, 294-296 (1978).
5. R. A. Penneman, J. C. Coleman, and T. K. Keenan, J. Inorg. Nucl. Chem., 17, 138-145 (1961).
6. D. E. Hobart, K. Samhoun, and J. R. Peterson, Radiochim. Acta, 31, 139-145 (1982).
7. R. A. Penneman and L. B. Asprey, Rep. AECU-936, 1950.
8. L. J. Nugent, R. D. Baybarz, J. L. Burnett, and J. L. Ryan, J. Phys. Chem., 77, 1528-1537 (1973).
9. V. B. Nikolaevsky, V. P. Shilov, N. N. Krot, and V. F. Peretrukhin, Radiokhimiya, 17, 426-430 (1975).

IX. CURIUM

A. Oxidation States

Curium exists in solution in the +3 and the +4 oxidation states. Cm^{3+} is a stable species, while tetravalent curium has only been observed in concentrated fluoride [1] and phosphotungstate [2] media as complexes which, how-

The Actinides

ever, undergo rapid reduction by the radiolysis products of these media (^{242}Cm: α-emitter; half-life 163.0 days; ^{244}Cm: α-emitter; half-life 18.1 years). The increased availability of ^{248}Cm (α-emitter; half-life 3.4×10^5 years), for which radiation effects are comparatively negligible, will be welcomed by experimentalists.

We consider $Cm(OH)_3(c)$ and $CmO_2(c)$ for the calculation of potentials in basic solutions. The thermodynamic data are given in Table 12.

B. Free Energies and Potentials

The standard potential of the couple Cm^{3+}/Cm is -2.06 ± 0.03 V as calculated from the free energy of formation of $Cm^{3+}(aq)$. No direct determination of potentials by electrochemical techniques is available. From the cutoff in the ultraviolet absorption spectrum of tetravalent curium in 15 M aqueous CsF solutions, Nugent et al. [3] estimate 3.1 ± 0.2 V as the standard potential of the couple Cm^{4+}/Cm^{3+}, whereas a value of 3.2 V was deduced from the wavelength of the first f-d absorption band of hexahalogenocomplexes [1]. This very positive value indicates indeed that the tetravalent ion can be stabilized only in very complexing media.

A standard potential of -4.4 ± 0.3 V for the Cm^{3+}/Cm^{2+} [3,4] couple has been proposed by Nugent et al. from spectroscopic studies. More recently, the value -3.7 V has been proposed by the same authors [5] to take into consideration the half-filled shell effect in the hydration energy of $Cm^{3+}(aq)$. Such negative values lead to the conclusion that Cm^{2+} cannot exist even transiently in aqueous solution. However, some evidence for the transient existence ($t_{1/2} \cong 12 \times 10^{-6}$ s) of Cm(II) has been put forward, based on pulse radiolysis studies in perchlorate media [6].

Amalgamation studies using Cm^{3+} solutions led Nugent [3] to the conclusion that only a three-electron process is involved.

C. Potential Diagrams for Curium

Acid Solution

$$\overset{+4}{Cm^{4+}} \xrightarrow{3.2} \overset{+3}{Cm^{3+}} \xrightarrow{-2.06} \overset{0}{Cm}$$

Basic Solution

$$CmO_2 \xrightarrow{0.7} Cm(OH)_3 \xrightarrow{-2.5} Cm$$

TABLE 12 Thermodynamic Data on Curium[a]

Formula	Description	State	ΔH_f°/kJ mol^{-1}	ΔG_f°/kJ mol^{-1}	S°/J mol^{-1} K^{-1}
Cm		g	387.4 ± 2.1	350.1 ± 3.6	197 ± 10
Cm	Metal, α	c	0	0	72.0 ± 2.5
Cm^{3+}		aq	−615.0 ± 5.0	−595.8 ± 6.3	−188 ± 13
Cm^{3+}	1 M HCl	aq	−614.6 ± 4.6		
CmO$_2$		c	−911.0 ± 6.0[b]	(−854)	(87)
Cm$_2$O$_3$	Mono.	c	−1682 ± 12[c]	(−1595)	(161)
CmCl$_3$		c	−974.4 ± 5.4	−902.5 ± 6.0	166.1 ± 8.4
CmOCl		c	−963.2 ± 2.5	−909.2 ± 9.2	105 ± 10

[a]Estimated values are given in parentheses.
[b]Cited by J. Fuger in *Actinides in Perspective*, N. M. Edelstein, ed. (Pergamon, Oxford), 1981, 409-439, based on results by L. R. Morss, J. Fuger, J. Goffart, and R. G. Haire, Actinide 81, LBL Report 12441, p. 263.
[c]L. R. Morss, J. Fuger, J. Goffart, and R. G. Haire, Inorg. Chem., 22, 1993-1996 (1983).

REFERENCES

1. T. K. Keenan, J. Am. Chem. Soc., 83, 3719-3720 (1961).
2. (a) A. S. Saprykin, V. P. Shilov, V. I. Spitsyn, and N. N. Krot, Dokl. Akad. Nauk SSSR, 226, 853-856 (1976); (b) V. N. Kosyakov, G. A. Timofeev, E. A. Erin, V. I. Andreev, V. V. Kopytov, and G. A. Simakin, Radiokhimiya, 19, 511-517 (1977).
3. L. J. Nugent, R. D. Baybarz, J. L. Burnett, and J. L. Ryan, J. Inorg. Nucl. Chem., 33, 2503-2530 (1976).
4. L. J. Nugent, J. Inorg. Nucl. Chem., 37, 1767-1770 (1975).
5. L. J. Nugent, in *International Review of Science; Inorganic Chemistry Series 2*, Vol. 7, K. W. Bagnall, vol. ed., Butterworth, London University Park Press, Baltimore, 1975, p. 195.
6. J. C. Sullivan, S. Gordon, V. Mulac, K. Schmidt, D. Cohen, and R. Sjoblom, Inorg. Nucl. Chem. Lett., 12, 599-601 (1976).

X. BERKELIUM

A. Oxidation States

Berkelium has been characterized in aqueous solutions in the trivalent and tetravalent states. Except in very complexing media, Bk^{3+} is the stable species. The oxidation of Bk^{3+} by $Cr_2O_7^{2-}$ or BrO_3^- leads to Bk^{4+}, which is fairly stable, reflecting the influence of its f^7 configuration. Nevertheless, in view of the short half-life of the usual isotope (^{249}Bk: β-emitter; half-life 314 days), radiolytic reduction occurs in a matter of hours even though the decay energy $[E(\beta^-_{max}) = 0.11$ MeV] is rather low. On the contrary, in $2\,M\,K_2CO_3$, radiolytically induced oxidation of Bk^{3+} to Bk^{4+} is observed with a half-oxidation time of 2.8 h. In these solutions Bk^{4+} is strongly complexed by the carbonate ions.

^{249}Cf, which forms in freshly purified ^{249}Bk solutions at the rate of 0.2% per day, is an unavoidable source of chemical contamination. In the past few years ^{249}Bk has become available in multimilligram quantities and experiments in the submilligram range are possible. The thermodynamic data are given in Table 13.

B. Gibbs Energies and Potentials

Weaver and Fardy [1] and Weaver and Stevenson [2] measured the potential difference between the Ce^{4+}/Ce^{3+} and the Bk^{4+}/Bk^{3+} couples as a function of acidity in $HClO_4$ and HNO_3 media. Extrapolation of their results to standard conditions indicates that the standard potential of the Ce^{4+}/Ce^{3+} couple is approximately 60 to 70 mV more positive than the Bk^{4+}/Bk^{3+} couple. Using the accepted value of $E°(Ce^{4+}/Ce^{3+}) = 1.74$ V [3], we adopt $E°(Bk^{4+}/Bk^{3+}) = 1.67 \pm 0.07$ V. Combination of this value with the more recent [4] experimental value for the enthalpy of formation of Bk^{3+}(aq) and the estimated entropies of Bk^{3+}(aq) and $Bk(c,\alpha)$ yields the other thermodynamic data.

From the data we obtain

$$Bk^{4+} + 4e^- \to Bk \qquad E° = -1.05 \pm 0.07 \text{ V}$$

$$Bk^{3+} + 3e^- \to Bk \qquad E° = -2.01 \pm 0.03 \text{ V}$$

TABLE 13 Thermodynamic Data on Berkelium

Formula	Description	State	ΔH_f°/kJ mol^{-1}	ΔG_f°/kJ mol^{-1}	S°/J mol^{-1} K^{-1}
Bk		g	310 ± 48[a]	—	—
Bk	Metal, α	c	0	—	78.2 ± 1.3[a]
Bk^{3+}		aq	−601.0 ± 5.0	−581.0 ± 7.0	−188 ± 17
Bk^{4+}		aq	−483 ± 16	−420 ± 10	−406 ± 42

[a]J. W. Ward, J. Less-Common Metals, 93, 279-292 (1983).

The value for the standard potential of the couple Bk^{3+}/Bk^{2+} has been estimated as −2.8 ± 0.2 V by Nugent et al. [5]; this very negative value indicates that Bk^{2+} cannot be characterized in aqueous media.

The formal potential of the Bk^{4+}/Bk^{3+} as measured by direct potentiometry in 1 M HClO$_4$ is given as 1.54 ± 0.1 V at 23°C [6]. A value of 1.60 V is also reported [7] in the same medium. In 1 M H$_2$SO$_4$ the formal potential of the same couple is 1.37 V [8,9] (derived also by direct potentiometry), while in 0.05 M H$_2$SO$_4$ the corresponding value is 1.43 V [8,10]. In 1 M HNO$_3$ the formal potential is reported as 1.54 V [6] or as 1.56 V [7]. The relative values of these formal potentials are in line with the complexation of the berkelium species in the various media.

C. Potential Diagram for Berkelium in Acid Solution

$$\text{Bk}^{4+} \xrightarrow{1.67} \text{Bk}^{3+} \xrightarrow{-2.01} \text{Bk}$$

with overall -1.05 from +4 to 0.

REFERENCES

1. B. Weaver and J. J. Fardy, Inorg. Nucl. Chem. Lett. 5, 145-146 (1969).
2. B. Weaver and J. N. Stevenson, J. Inorg. Nucl. Chem., 33, 1877-1881 (1971).
3. D. D. Wagman, W. H. Evans, V. B. Parker, R. H. Schumm, I. Halow, S. M. Bailey, K. L. Churney, and R. L. Nuttall, NBS Tables of Chemical Thermodynamic Properties, J. Phys. Chem. Ref. Data, Vol. 11, Supplement 2 (1982).
4. J. Fuger, R. G. Haire, and J. R. Peterson, J. Inorg. Nucl. Chem., 43, 3209-3212 (1981).
5. L. J. Nugent, R. D. Baybarz, J. L. Burnett, and J. L. Ryan, J. Phys. Chem., 77, 1528-1539 (1973).
6. J. R. Stokely, R. D. Baybarz, and J. R. Peterson, J. Inorg. Nucl. Chem., 34, 392-393 (1972).

7. G. A. Simakin, V. N. Kosyakov, A. A. Baranov, E. A. Erin, V. V. Kopytov, and G. A. Timofeev, Radiokhimiya, 19, 336-372 (1977).
8. J. R. Stokely, R. D. Baybarz, and W. Shults, Inorg. Nucl. Chem. Lett., 5, 877-886 (1969).
9. Yu. M. Kulyako, Y. A. Frenkel, I. A. Lebedev, T. I. Trofimov, B. F. Myasoedov, and A. N. Mogilev'ski, Radiochim. Acta, 28, 119-122 (1981).
10. R. Propst and M. Hyder, J. Inorg. Nucl. Chem., 32, 2205-2216 (1970).

XI. CALIFORNIUM

A. Oxidation States

Californium is stable in aqueous solutions only in the trivalent state. It appears from amalgamation experiments that divalent californium may exist as a transient species. Spectral evidence for the oxidation of californium to the tetravalent state in phosphotungstate solutions, using persulfate as oxidant, has been presented [1], even in these highly complexing media Cf(IV) is reduced by water in a matter of minutes. The relatively long-lived ^{249}Cf (α-emitter; half-life 360 years) is available for experiments in the submilligram range.

B. Gibbs Energies and Potentials

Mikheev et al. [2] reported that trivalent californium was reduced and co-crystallized with $SmCl_2$ in the reduction of $SmCl_3$ to $SmCl_2$ in aqueous-alcoholic solutions. Evidence for the formation of Cf^{2+} in acetonitrile medium has been established by Friedman et al. [3] by taking in account the similarity of the Cf^{3+}/Cf^{2+} couple with the corresponding samarium couple: the half-wave potential for californium is 0.2 V less negative than the samarium couple. From the above-mentioned data, the value $E°(Cf^{3+}/Cf^{2+}) = -1.6 \pm 0.1$ V is accepted by Nugent et al. [4] in their systematic analysis of the subject.

A preliminary value of -617 ± 11 kJ mol^{-1} was reported by Raschella et al. [5] for $\Delta H_f°(Cf^{3+},aq)$. Very recently, Fuger et al. [6] reported $\Delta H_f°(Cf^{3+},aq) = -577 \pm 5$ kJ mol^{-1}, using well-characterized and analyzed metal sample. The latter result leads to $E°(Cf^{3+}/Cf) = -1.92 \pm 0.03$ V, using in the calculation the same entropy change as for the corresponding berkelium system. This result is slightly less negative than the values of -2.01 ± 0.05 V [7] (interpolated from data on other actinides), -2.01 ± 0.05 V [8] and -2.03 ± 0.04 V [9] obtained from radiopolarographic data. The value derived from thermochemical measurements shall be adopted here, as the evaluation of amalgamation potentials used in radiopolarography may involve larger uncertainties.

The use of $E°(Cf^{3+}/Cf) = -1.93 \pm 0.03$ V together with $E°(Cf^{3+}/Cf^{2+}) = -1.6 \pm 0.1$ V [4] leads to $E°(Cf^{2+}/Cf) = -2.1 \pm 0.1$ V in agreement with the value proposed by Nugent [7] from a review of amalgamation experiments. All these data indicate the possible formation of a +2 transient state in aqueous solution.

C. Potential Diagram for Californium in Acid Solution

$$Cf^{3+} \xrightarrow{-1.6} Cf^{2+} \xrightarrow{-2.1} Cf$$
$$\underline{\qquad -1.93 \qquad}$$

REFERENCES

1. V. N. Kosyakov, G. A. Timofeev, E. A. Erin, V. I. Andreev, V. V. Kopytov, G. A. Simakin, Radiokhimiya, 19, 511-517 (1977).
2. N. B. Mikheev, V. I. Spitsyn, A. N. Kamenskaya, N. A. Rosenkevitch, I. A. Rumer, L. N. Auermann, Radiokhimiya, 14, 486-487 (1972).
3. H. A. Friedman, J. R. Stokely, R. D. Baybarz, Inorg. Nucl. Chem. Lett., 8, 433-441 (1972).
4. L. J. Nugent, R. D. Baybarz, J. L. Burnett, J. L. Ryan, J. Phys. Chem., 77, 1528-1536 (1973).
5. D. L. Raschella, R. G. Haire, J. R. Peterson, Radiochim. Acta, 30, 41-43 (1982).
6. J. Fuger, R. G. Haire, J. R. Peterson, J. Less-Common Metals, 98, 315-321 (1984).
7. L. J. Nugent, J. Inorg. Nucl. Chem., 37, 1767-1770 (1975).
8. F. David, K. Samhoun, R. Guillaumont, N. Edelstein, J. Inorg. Nucl. Chem., 40, 69-74 (1978).
9. K. Samhoun, F. David, J. Inorg. Nucl. Chem., 41, 357-363 (1979).

XII. EINSTEINIUM

A. Oxidation States

Trivalent einsteinium is a stable species in aqueous solution. Similarly to californium, divalent einsteinium has been identified as a transient species from radiopolarography experiments.

^{253}Es (α-emitter; half-life 20.47 days) is available for experiments at the microgram level. Radiolysis is also a source of porblems when experiments are carried out with nontracer amounts of ^{253}Es.

B. Gibbs Energies and Potentials

Quantitative experimental data on the einsteinium oxidoreduction behavior are rather scanty. From radiopolarographic measurements, Samhoun [1a] and Samhoun and David [1b] obtains $E°(Es^{3+}/Es) = -1.98 \pm 0.05$ V while considering an amalgamation potential of 0.52 ± 0.04 V. For the Es^{3+}/Es^{2+} potential, Mikheev et al. [2] deduced a value of $E°(Es^{3+}/Es^{2+})$ -1.51 ± 0.04 V from the similarity of the reduction behavior of einsteinium and californium as studied by their co-crystallization with $SmCl_2$. Spectroscopic studies on halogeno complexes of lanthanides and trivalent actinides, together with a generalized treatment of the f electron energetics [3,4], led Nugent to accept a value of $E°(Es^{3+}/Es^{2+}) = -1.3 \pm 0.2$ V, which is more in line with the progressive stabilization of the +2 oxidation state.

In view of the large uncertainties on the potentials, we shall not speculate in estimating the thermodynamic properties.

C. Potential Diagram for Einsteinium in Acid Solution

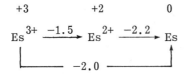

REFERENCES

1. (a) K. Samhoun, Thesis, University of Paris-Sud (Orsay), ref. IPNO-T-76-12, 1976; (b) K. Samhoun and F. David, J. Inorg. Nucl. Chem., *41*, 357-363 (1979).
2. (a) N. B. Mikheev, V. I. Spitsyn, A. N. Kamenskaya, N. A. Rozenkevitch, I. A. Rumer, and L. N. Auermann, Radiokhimiya, *14*, 486-487 (1972); (b) N. B. Mikheev, A. N. Kamenskaya, I. A. Rumer, V. I. Spitsyn, R. A. Diatchkova, and N. A. Rozenkevitch, Radiochim. Radioanal. Lett., *9*, 247-254 (1972).
3. L. J. Nugent, R. D. Baybarz, J. L. Burnett, and J. L. Ryan, J. Phys. Chem., *77*, 1528-1536 (1973).
4. L. J. Nugent, J. Inorg. Nucl. Chem., *37*, 1767-1770 (1975).

XIII. FERMIUM

A. Oxidation States

As for californium and einsteinium, the Fm^{3+} ion is stable in solution. Amalgamation experiments and radiopolarography also show the existence of divalent fermium. Only tracer amounts of ^{255}Fm (α-emitter; half-life 20.1 h) and ^{257}Fm (α-emitter; half-life 100.5 days) are presently available and will be available in the foreseeable future because of the extremely low production yields of fermium isotopes.

B. Potentials

Chemical reduction studies of Fm^{3+} to Fm^{2+}, using magnesium followed by cocrystallization with $SmCl_2$ in aqueous alcoholic solutions, led Mikheev et al. [1] to conclude that $E°(Fm^{3+}/Fm^{2+})$ is within 0.2 V that of Yb^{3+}/Yb^{2+}. They report $E°(Fm^{3+}/Fm^{2+}) = -1.15$ V. The systematic relationship for the calculation of the 3+/2+ potentials of the actinides (see Section XII) gives -1.1 ± 0.2 V [2].

From the radiopolarographic reduction of Fm^{3+}, Samhoun and David [3] assumed that divalent fermium is present at the surface of the cathode and that the electrochemical reduction corresponds to $Fm^{2+} + 2e^- + Hg \rightarrow Fm(Hg)$; they obtained $E°(Fm^{2+}/Fm) = -2.37$ V, assuming an amalgamation potential of 0.9 V. This value is in reasonable agreement with a value predicted earlier, $E°(Fm^{2+}/Fm) = -2.48 \pm 0.05$ V [4].

C. Potential Diagram for Fermium in Acid Solution

$$Fm^{3+} \xrightarrow{-1.15} Fm^{2+} \xrightarrow{-2.37} Fm$$
$$\underbrace{\qquad\qquad -1.96 \qquad\qquad}$$

REFERENCES

1. N. B. Mikheev, V. I. Spitsyn, A. N. Kamenskaia, N. A. Konovalova, I. A. Rumer, L. N. Auermann, and A. M. Podorozhnyi, Inorg. Nucl. Chem. Lett., *13*, 651-656 (1977).
2. L. J. Nugent, R. D. Baybarz, J. L. Burnett, and J. L. Ryan, J. Phys. Chem., *77*, 1528-1536 (1973).

3. K. Samhoun and F. David, in *Transplutonium Elements*, W. Müller and R. Lindner, eds., North-Holland, Amsterdam, 1976, p. 297.
4. L. J. Nugent, J. Inorg. Nucl. Chem., 37, 1767-1770 (1975).

XIV. MENDELEVIUM

A. Oxidation States

Md^{3+} is a stable ion; however, it can be reduced rather easily to Md^{2+}, which can be maintained in solution in absence of oxygen. Experiments with ^{256}Md (α-emitter; E.C.; half-life 77 min) are so far restricted to a maximum of 10^6 atoms.

B. Potentials

$E°(Md^{3+}/Md^{2+})$ has been reported as about -0.1 V [1], about -0.2 V [2], and greater than -0.3 V [3] from studies involving coprecipitation and extraction of the chemically reduced mendelevium. We adopt a value of -0.15 ± 0.15 V. For the couple Md^{2+}/Md, Samhoun et al. [4] and David et al. [5] obtain $E°(Md^{2+}/Md) = -2.4$ V if an amalgamation potential of 0.9 V is taken into consideration.

The results of the above-mentioned authors demonstrate that mendelevium and fermium have quite similar behavior but that the stabilization of the +2 oxidation state is more pronounced in the case of mendelevium.

C. Potential Diagram for Mendelevium in Acid Solution

$$\begin{array}{ccc} +3 & +2 & 0 \end{array}$$

$$Md^{3+} \xrightarrow{-0.15} Md^{2+} \xrightarrow{-2.4} Md$$

$$\underset{-1.7}{\longleftarrow \qquad \longrightarrow}$$

REFERENCES

1. J. Maly and B. B. Cunningham, Inorg. Nucl. Chem. Lett., 3, 445-451 (1967).
2. E. K. Hulet, R. W. Lougheed, J. D. Brady, R. E. Stone, and M. S. Coops, Science, 158, 486-487 (1967).
3. R. J. Silva, R. L. Hahn, M. L. Mallory, C. F. Bemis, P. F. Dittner, O. L. Keller, J. R. Stokely, and K. S. Toth, in *Uses of Cyclotrons in Chemistry, Metallurgy and Biology*, C. B. Amphlett, ed., Butterworth, London, 1970, p. 42.
4. K. Samhoun, F. David, R. L. Hahn, G. D. O'Kelley, J. R. Tarrant, and D. E. Hobart, as cited by E. K. Hulet, Rep. UCRL 82642, 1979.
5. F. David, K. Samhoun, E. K. Hulet, P. A. Baidsen, R. Dougan, J. H. Landrum, R. W. Lougheed, J. F. Wild, and G. D. O'Kelley, as cited by E. K. Hulet, Rep. UCRL 82642, 1977; and *Lanthanide-Actinide Chemistry and Spectroscopy*, ACS Symposium Series 131, N. Edelstein, ed., American Chemical Society, Washington, D.C., 1980.

XV. NOBELIUM

At best, experiments involving up to 1000 atoms of nobelium have been carried out so far. Cathodic deposition, ion exchange, separation, coprecipitation, and solvent extraction studies [1,2] at the level of 50 to 100 atoms per experiment with ^{255}No (α-emitter; half-life c. 3 min) indicate that this element behaves as a +2 ion in aqueous solutions; rather strong oxidizing agents are required to obtain $No^{3+}(aq)$. These experiments lead to approximately 1.4 to 1.5 V for $E°(No^{3+}/No^{2+})$. The behavior of nobelium in aqueous solutions reflects the stability of the $5f^{14}$ electronic shell. By a modified radiopolarographic technique, Meyer et al. [3] obtain $E°(No^{2+}/No) = -2.5$ V assuming an amalgamation potential of 0.9 V.

A. Potential Diagram for Nobelium in Acid Solution

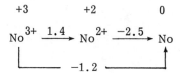

REFERENCES

1. J. Maly, T. Sikkeland, R. J. Silva, and A. Ghiorso, Science, *160*, 1114-1115 (1968).
2. R. J. Silva, T. Sikkeland, M. Nurmia, A. Chiorso, and E. K. Hulet, J. Inorg. Nucl. Chem., *31*, 3405-3409 (1969).
3. R. E. Meyer, W. J. McDowell, P. F. Dittner, R. J. Silva, and J. R. Tarrant, J. Inorg. Nucl. Chem., *38*, 1171-1173 (1976).

XVI. LAWRENCIUM

Only fast solvent extraction experiments with a few atoms ^{255}Lr (α-emitter; half-life 35 s) have been carried out [1]; they indicate that only the trivalent state exists in solution, as expected for the last element of the actinides series. The generalized f electron energetics treatment by Nugent [2] leads to $E°(Lr^{3+}/Lr) = -2$ V, while a value of -2.06 ± 0.12 V is obtained by David et al. [3] from correlations between the thermodynamic properties of lanthanides and actinides.

REFERENCES

1. R. J. Silva, R. H. Hahn, M. L. Mallory, C. E. Bemis, P. F. Dittner, O. L. Keller, J. R. Stokely, and K. S. Toth, *Uses of Cyclotrons in Chemistry, Metallurgy and Biology*, C. B. Amphlett, ed., Butterworth, London, 1970, p. 42.
2. L. J. Nugent, J. Inorg. Nucl. Chem., *37*, 1767-1770 (1975).
3. F. David, K. Samhoun, R. Guillaumont, and N. Edelstein, J. Inorg. Nucl. Chem., *40*, 69-74 (1978).

22
Beryllium, Magnesium, Calcium, Strontium, Barium, and Radium

I. BERYLLIUM*

RICHARD E. PANZER, Consultant, El Sobrante, California

The general chemistry of beryllium is influenced drastically by its small ionic radius, 0.031 nm, and its correspondingly high charge-to-radius ratio, $z/r = 64.5$ nm^{-1}, the highest of any element. This may be compared to the z/r ratios of the other light metals given in Table 1. From these values, the similarity of Be to Al can be seen; but Be resists oxidation in spite of its high heat of oxidation, 609.6 kJ mol^{-1}. This is a kinetic effect which arises because the volume of the oxide is so much larger than the metal atom from which it is formed that the oxide surface film resists removal and protects the metal from further attack. It is this aspect which allows the use of beryllium metal in various spacecraft applications where heat resistance is required.

In solutions, a consequence of the high charge-to-radius ratio of the Be^{2+} ion is that it is a strong order-producing ion. That is, it strongly orients solvent molecules in its vicinity, bringing to them a greater degree of order than they normally possess [1]. The high charge density on the Be^{2+} ion polarizes the surrounding solvent molecules, causing them to orient with the negative ends of the dipoles toward the Be^{2+} ion. Any positive species present, protons, and so on, are repelled to such an extent that thermal energy eventually transfers the positive species to a more distance molecule. The result is formation of a polynuclear species such as that known for BeCl$_2$, BeF$_2$, and other compounds. The polymeric structures are not planar but are difficult to show on the printed page.

$$\diagdown_{\diagup}Be\diagup^{Cl}_{\diagdown Cl}Be\diagup^{Cl}_{\diagdown Cl}Be\diagup^{Cl}_{\diagdown Cl}Be\diagdown^{\diagup} \quad \text{etc.}$$

Because of the polymeric structures, the identity and activity of charged species of beryllium in solution are unpredictable. Additionally, Be is amphoteric and the formation of many of its compounds is a direct result of

---

Reviewer: M. Chemla, Pierre and Marie Curie University, Paris, France.

TABLE 1 Comparison of Some Charge-to-Radius Ratios for Light Metal Ions

Element	$(Z/r)/\text{nm}^{-1}$
Be^{2+}	64.5
Al^{3+}	60.0
Mg^{2+}	30.7
Zn^{2+}	24.1
Ca^{2+}	20.2
Li^{1+}	16.7

this aspect of its chemistry. Also noteworthy is the melting point of Be metal, 1285°C, the highest of the light metals and a direct result of the close-packed hexagonal structure of the metal combined with the small atomic radius, 0.112 nm, resulting in a very stable structure that is not easily disrupted.

Beryllium exists in two oxidation states, +1 and +2. At the time of Latimer's publication (1952) there was some question of the existence of the lower state, but extensive investigations made since that time have definitely proven its existence. Spectroscopic evidence continues to be published, as exemplified by the work on a new predissociated state of the beryllium monochloride molecule [2]. Additionally, the corrosion film compounds occurring on beryllium have been analyzed by Kolot et al. using mass spectroscopy [3]. They found that the Be^{1+} ions were always formed, even though in their experiments the metal had been mechanically cleaned prior to exposure.

A peculiar phenomenon occurring when Be metal is anodically oxidized in aqueous chloride solutions was reported by Laughlin et al [4]. When an anode of beryllium is subjected to potentials of about 1.25 V (vs. SHE) at apparent current densities of around 0.03 A cm^{-2}, a gelatinous white precipitate forms in the electrode compartment. After a delay period of some minutes, depending on the CD, fine black particles are observed to form throughout the anode compartment. There is a delay time from the initiation of the electrolysis until the rather sudden formation of black particles, which have been found to be beryllium metal. Although Straumanis and Mathis [5] have objected to proposed mechanisms for the formation of the beryllium particles, calling it a "chunk effect," no further evidence has been presented to explain why the particulates form in the body of the solution away from the anode. Epelboin et al. [6] have shown, using EtOH with 0.1 g L^{-1} of H_2O present, that at potentials greater than 2.5 V, the simple Be^{1+} ion forms next to the anode in the solvent poor film, and eventually it oxidizes to the Be^{2+} ion. For completely aqueous solutions, it is most likely that the initial ion forming is similar to the cuprous ion, Cu_2^{2+} (i.e., Be_2^{2+}). Diffusing away from the anode, this ion undergoes a disproportionation reaction, as proposed by Laughlin et al.:

$$Be_2^{2+} \rightarrow Be(s) + Be^{2+} \tag{22.1}$$

The fact that no chlorine or oxygen is formed during anodization of beryllium metal in NaCl(aq) is additional evidence that monovalent Be ions are indeed formed. Confirmation of many of these observations was obtained by Panzer [7] during the course of investigations in both aqueous and nonaqueous electrochemistry of beryllium. Finally, the work of Aida et al. [8] appears to provide additional verification of the existence of the Be^{1+} ions.

A. Gibbs Energies for Beryllium

From the extensive investigations conducted in the course of the nuclear and space programs, a large number of thermodynamic values for beryllium compounds have accumulated. All the values in Table 1 and 2 were taken from the publication of the National Bureau of Standards (NBS) [9]. Values for many beryllium compounds utilized in nonaqueous systems are available from the same sources, but these have been omitted from the tables here. Additionally, other values are available, such as those obtained in nonaqueous solvents and fused salts by Plambeck [10].

B. Beryllium Potentials

For the beryllium standard state, Be^{2+}, $m = 1$, one calculates $E°$ using the value of the Gibbs energy from Table 2 for the reaction

$$Be^{2+} + 2e^- \rightarrow Be \qquad E° = -1.97 \text{ V} \qquad (22.2)$$

Attempts to measure beryllium potentials in aqueous systems have not been successful in producing reversible or very reproducible values in most cases. For example, Bodforss [11] obtained values around -0.6 V. *Gmelin Handbuch* [1] lists numerous potentials obtained for beryllium in aqueous solutions by various investigators. However, polarographers have obtained results indicating those techniques are useful for quantitative determination of beryllium.

Venkataratnam and Rao [12] obtained two well-defined waves at $E_{1/2}$ of -1.79 and -2.00 V in solutions of 0.15 M LiCl and at pH 2.38. At pH 3.3 and below, the diffusion current at the first stage of the reduction remains proportional to the concentration of Be up to 8×10^{-3} mol L^{-1}. The two beryllium waves have been attributed to the reaction

$$Be(H_2O)^{2+} \rightarrow Be(OH)_2 + 2H^+ + 2H_2O \qquad 2H^+ + 2e^- \rightarrow H_2 \qquad (22.3)$$

which takes place by stages with formation of intermediate hydroxy complexes. With more acidic solutions, there is a shift of the half-wave potential, but the diffusional current at $E_{1/2} = -1.79$ V is independent of the pH. The diffusion current of the second wave increases between pH 2.66 and 3.33 with increase of pH and becomes constant at 3.33, possibly due to formation of hydroxy complexes of beryllium.

Mendyntsev and Dyatlova [13] studied the decomposition and polarography of beryllium hydroxy complexes on the DME. Their studies indicated that around pH 4.0 the limiting current falls sharply as hydrolysis becomes the principal reaction. Generally, the rate of discharge of Be ions on Hg(DME) is determined by the preceding decomposition of hydroxy complexes of Be in the double layer.

In their book on the analysis for beryllium, in the section on polarographic methods, Novoseleva and Batsanova [14] indicate that sharp waves

TABLE 2 Thermodynamic Data on Beryllium and Its Compounds[a]

Formula	Description	State	$\Delta H°$/kJ mol^{-1}	$\Delta G°$/kJ mol^{-1}	$S°$/J mol^{-1} K^{-1}
Be		c	0	0	9.50
Be		g	324.30	286.6	136.16
Be$^+$		g	1,229.95		
Be^{2+}		g	2,993.2		
Be^{2+}	Std. state, $m = 1$	aq	−382.8	−379.7	−129.7
Be^{+3}		g	+17,824.8	17,847.6	
Be^{+4}		g	+38,860.2		
BeO		c	−609.61	−580.32	14.14
BeO$_2^{2-}$	Std. state, $m = 1$	aq	−790.77	−640.15	−158.99
Be$_2$O		g	−83.68		
(BeO)$_2$		g	−430.95		
(BeO)$_3$		g	−1087.84		
(BeO)$_4$		g	−1,635.9		
(BeO)$_5$		g	−2,167.3		

Group IIA Elements

(BeO)$_6$			g	−2,723.8		
BeH			g	−316.3	−285.77	176.61
BeH$_2$			c	−19.25		
Be(OH)$_2$	α		c	−902.49	−815.55	51.88
Be(OH)$_2$	β		c	−905.84	−817.55	50.2
	Freshly precipitated	amorph	−897.89		
		g	−661.07		
Be(OH)$_2$·(3/4)H$_2$O		amorph	−1,116.7		
Be$_3$(OH)$_3^{3+}$	Std. state, $m = 1$	aq		−1,801.6	
BeF			g	−174.89	−203.3	205.64
BeF$_2$	α, quartz	c	−1,026.75	−979.47	53.35
	β, cristobalite	c	−1,023.8		
	glassy	amorph	−1,022.15		
		g	−793.7		
		aq	−1,046.0		
	In 100(HF + 1.5H$_2$O)	aq	−1,051.86		

TABLE 2 (Continued)

Formula	Description	State	$\Delta H°/\text{kJ mol}^{-1}$	$\Delta G°/\text{kJ mol}^{-1}$	$S°/\text{J mol}^{-1}\text{ K}^{-1}$
	In 100(HF + 2.0H$_2$O)	aq	−1,053.45		
	In 100(HF + 3.0H$_2$O)	aq	−1,055.96		
	In 100(HF + 4.0H$_2$O)	aq	−1,057.88		
	In 100(HF + 5.0H$_2$O)	aq	−1,059.19		
	In 100(HF + 6.0H$_2$O)	aq	−1,059.89		
	In 100(HF + 7.0H$_2$O)	aq	−1,060.28		
	In 100(HF + 8.0H$_2$O)	aq	−1,060.48		
Be$_2$OF$_2$		g	−1,221.73		
BeCl		g	58.58	29.29	217.57
BeCl$_2$	α	c	−490.36	−445.60	82.68
	β	c	−495.8	−448.94	75.81
		g	−358.57		
	Dilute	aq	−715.88		
	In 45.8(HCl + 7.25H$_2$O)	aq	−677.8		
	In 38.3(HCl + 9.72H$_2$O)	aq	−687.4		

Group IIA Elements

$BeCl_2 \cdot 4H_2O$	In 6.38(HCL + 10.65H$_2$O)	aq	−684.5	
Be_2Cl_4		c	−1,808.3	
$BeBr_2$		g	−845.17	
		c	−353.55	
	In 6.30 m HCl	aq	−586.60	
BeI_2		c	−192.46	
	In 6.30 m HCl	aq	−453.96	
BeS		c	−234.30	
$BeSO_4$	α, tetragonal	c	−1,205.2	−1,093.87	77.91
	Std. state, m = 1	aq	−1,292.0	−1,124.3	−109.6
$BeSO_4$	In 15 H$_2$O	aq	−1,275.2	
	In 20 H$_2$O	aq	−1,277.5	
	In 25 H$_2$O	aq	−1,278.47	
	In 50 H$_2$O	aq	−1,281.8	
	In 75 H$_2$O	aq	−1,282.86	
	In 100 H$_2$O	aq	−1,283.48	
	In 200 H$_2$O	aq	−1,284.74	

TABLE 2 (Continued)

Formula	Description	State	$\Delta H°/\text{kJ mol}^{-1}$	$\Delta G°/\text{kJ mol}^{-1}$	$S°/\text{J mol}^{-1}\text{ K}^{-1}$
	In 400 H_2O	aq	−1,285.91		
	In 500 H_2O	aq	−1,286.29		
	In 1000 H_2O	aq	−1,287.54		
	In 2000 H_2O	aq	−1,288.67		
	In 3000 H_2O	aq	−1,289.5		
	In 4000 H_2O	aq	−1,289.76		
	In 5000 H_2O	aq	−1,290.22		
	In 100(HCl + 55.5H_2O)	aq	−1,234.28		
	200(HCl + 27.75H_2O)	aq	−1,232.61		
	300(HCl + 18.5H_2O)	aq	−1,230.5		
	400(HCl + 13.88H_2O)	aq	−1,229.26		
$BeSO_4 \cdot H_2O$		c	−1,523.8		
$BeSO_4 \cdot 2H_2O$		c	−1,823.14	−1,598.2	163.22
$BeSO_4 \cdot 4H_2O$	Tetragonal	c	−2,423.75	−2,080.66	232.97

Group IIA Elements

BeBr$_2 \cdot$2H$_2$S		c	−469.03
BeI$_2 \cdot$2H$_2$S		c	−292.04
BeSeO$_4$		c	−890.36
	Std. state, m = 1	aq	−981.98	−820.9	−75.7
	In 800 H$_2$O	aq	−979.47
BeSeO$_4 \cdot$2H$_2$O		c	−1,507.50
BeSeO$_4 \cdot$4H$_2$O		c	−2,112.92
Be$_3$N$_2$	Cubic	c	−588.27
	Hexagonal	c	−571.1
Be(NO$_3$)$_2$		aq	−799.14
BeCl$_2 \cdot$2NH$_3$		c	−843.9
BeCl$_2 \cdot$4NH$_3$		c	−1,089.5
BeCl$_2 \cdot$6NH$_3$		c	−1,247.67
BeCl$_2 \cdot$12NH$_3$		c	−1,715.0
BeBr$_2 \cdot$4NH$_3$		c	−983.66
BeBr$_2 \cdot$6NH$_3$		c	−1,148.1
BeBr$_2 \cdot$10NH$_3$		c	−1,461.0

TABLE 2 (Continued)

Formula	Description	State	$\Delta H^\circ/\text{kJ mol}^{-1}$	$\Delta G^\circ/\text{kJ mol}^{-1}$	$S^\circ/\text{J mol}^{-1} \text{K}^{-1}$
$BeI_2 \cdot 4NH_3$		c	−868.60		
$BeI_2 \cdot 6NH_3$		c	−1,034.28		
$BeI_2 \cdot 13NH_3$		c	−1,583.23		
Be_2C		c	−117.15		
$BeCO_3$		c	−1,025.08		
Be_2SiO_4		c	−2,149.3	−1,919.62	64.31
$Be(BO_2)_2$		g	−1,359.8		
$Be_3B_2O_6$		c	−3,139.26		

BeOAl		g	−20.92
BeO·Al$_2$O$_3$	Chrysoberyl	c	−2,300.78	−2,178.6	66.27
BeO·Al$_2$O$_3$		c	−5,627.06	−5,320.37	175.73
BeMoO$_4$		c	−1,372.35		
NbBe$_2$		c	−61.09		
NbBe$_8$		c	−83.68		
NbBe$_{12}$		c	−125.52		
UBe$_{13}$		c	−163.18		

[a] All values are from the National Bureau of Standards; all values are for the temperature 298.15 K. Concentrations of solutions are expressed in terms of the number of moles of solvent associated with one mole of solute; if no concentration is indicated, the solution is assumed to be "dilute."

for beryllium are obtained in acid solutions with LiCl, Li_2SO_4, or tetraethylammonium iodide as the background electrolyte. The half-wave potential is -1.85 V. With increased pH the diffusion current decreases and beryllate solutions give no polarographic waves. Efforts continue to measure redox couples of beryllium in aqueous media, but because of the considerations presented above, attempts to measure exact potentials of beryllium in aqueous solutions are doomed to failure.

A great body of literature has grown during the past 20 years concerning the formation of beryllium complexes in aqueous and nonaqueous solutions. Additionally, many studies have been conducted in fused systems and extensive references are available on spectroscopic investigations unique to interest in beryllium and its structure. This mass of data provides eloquent testimony to the facts of the element's unique chemistry, its high charge-to-radius ratio, and its distinctive physical structure. Indeed, in all these aspects, beryllium tends to be more like a transition metal than like alkaline earth metal as would be expected solely by its position on the periodic table.

C. Potential Diagram for Beryllium

$$Be^{2+} \xrightarrow{-1.97} Be$$
$$\phantom{Be^{2+}}+2 \phantom{\xrightarrow{-1.97}} 0$$

REFERENCES

1. R. Gurney, *Ionic Processes in Solution*, McGraw-Hill, New York, 1953.
2. M. Carleer, B. Burtin, and R. Colin, Can. J. Phys., 55(6), 582-588 (1977).
3. V. Ya. Kolot, V. F. Rybalko, M. Fogel, and G. F. Tikhinskii, Zasch. Met., 3(6), 723-730 (1967).
4. B. Laughlin, J. Kleinberg, and A. W. Davidson, J. Am. Chem Soc., 78, 559 (1956).
5. M. E. Straumanis, and D. L. Mathis, J. Electrochem. Soc., 109, 434 (1962).
6. I. Epelboin, M. Froment, and M. Garreau, Corrosion (Rueil-Malmaison, Fr.), 18(7), 433-440 (1970).
7. R. E. Panzer, Final Rep. Under Contract Request ER-9085 from NASA Electronics Res. Lab to Naval Weapons Center, Corona, Calif., 1969.
8. H. Aida, I. Epelboin, and M. Garreau, J. Electrochem. Soc., 118, 243-248 (1971).
9. V. B. Parker, D. D. Wagman, and W. H. Evans, Natl. Bur. Stand. Tech. Note 270-6, U.S. Government Printing Office, Washington, D.C., November 1971.
10. J. A. Plambeck, Can. J. Chem., 47(8), 1401-1410 (1969).
11. S. Bodforss, Z. Phys. Chem., 124, 68 (1926); 130, 85-89 (1927).
12. G. Venkataratnam, and B. S. V. R. Rao, J. Sci. Ind. Res., 17B, 360-362 (1958).
13. V. V. Mendyntsev, and N. M. Dyatlova, Tr. Vses. Nauchno-Issled. Inst. Khim. Reaktiv. Osoba Chist. Khim. Veschestv., 30, 318-328 (1967); Chem. Abstr., 68, 35280 (1968).
14. A. V. Novoseleva, and L. R. Batsanova, *Analytical Chemistry of Beryllium*, trans. from the Russian, Israel Program for Scientific Translation, Jerusalem, 1968.

II. MAGNESIUM*

GEORGES G. PERRAULT Laboratoire d'Electrochimie Interfaciale du CNRS, Meudon, France

A. Oxidation States

The +2 oxidation state is the only one of importance. However, the +1 state has been considered several times and a certain number of corresponding compounds are known in the gaseous state. Ions with oxidation state different from +2 are known in the gaseous state and probably exist in nonaqueous solutions, but there is no evidence of their possible existence in aqueous solutions. The enthalpy of formation of aqueous Mg^+ has been obtained by theoretical calculations [1], and it has been shown that it has no stability domain in aqueous solutions [2]. However, calculations show that it might be considered as an intermediate species in the anodic dissolution of magnesium [3].

B. Thermodynamic Data

The thermodynamic data on magnesium are given in Table 3. The ΔH_f°, ΔG_f°, and S° values adopted by the National Bureau of Standard [NBS 270-6 (1971), NSRDS-NBS 37 (1971)] have been accepted in most cases, and some of them are different from those accepted in the *Encyclopedia of the Electrochemistry of the Elements* [2]. Other values have been selected from various origins. More recent CODATA [4] and U.S. Geological Survey [5] recommended values are preferred against other selected values. Error intervals are given from original papers.

C. Electrode Potential

From the values selected in Table 3 the standard potentials of Mg/Mg^+ and Mg/Mg^{2+} can be calculated:

$$Mg^{2+} + 2e^- \to Mg \qquad E^\circ = -2.356 \text{ V}$$

$$Mg(OH)_2(s) + 2e^- \to Mg + 2OH^- \qquad E^\circ = -2.687 \text{ V}$$

$$Mg^+ + e^- \to Mg \qquad E^\circ = -2.657 \text{ V}$$

Isothermal coefficients for the Mg/Mg^{2+} and $Mg/Mg(OH)_2$ equilibria have also been determined [5] (see p. 698):

---

*Reviewer: M. Chemla, Pierre and Marie Curie University, Paris, France.

TABLE 3 Thermodynamic Data on Magnesium[a]

Formula	Description	State	$\Delta H°$/kJ mol^{-1}	$\Delta G°$/kJ mol^{-1}	$S°$/J mol^{-1} K^{-1}
Mg	Metal	c	0	0	32.68*
Mg		g	147.10 ± 0.80	113.08	148.585 ± 0.020
Mg$^+$		g	884.9 ± 0.4		154.23
Mg$^+$		aq	−230	−256.4	51.8
Mg^{2+}		g	2347.3		
Mg^{2+}	Std. state, $m = 1$	aq	−466.62	−454.6	−138
MgO	Periclase	c	−601.5 ± 0.3	−569.2	26.95 ± 0.15
MgO	Microcrystal	c	−597.69	−565.69	27.89
MgH		g	172.2 ± 48.1	142	(193)
MgH$_2$		c	−76.1 ± 9.2	−35.9	31.07 ± 0.84
MgD$_2$		c	−73.10	−32.6	41.4
MgOH$^+$	Std. state, $m = 1$	aq		−626.5	
MgOH		g	(217 ± 83)		(221)

Group IIA Elements

Mg(OH)$_2$		c	−924.2 ± 2.1	−833.2	63.15
Mg(OH)$_2$	Precipitate	amorph	−920.0		
Mg(OH)$_2$	Std. state, $m=1$	aq	−926.4	−769.1	−148.9
MgF		g	−222.1 ± 5.4	−247.6	220.8
MgF$_2$		c	−1124.2 ± 1.2	−1069.7	57.2 ± 0.4
MgF$_2$		g	−723.5	−730.2	258.2
(MgF$_2$)$_2$		g	−1722.5 ± 14.2		(312.1)
MgF$_2$		g	−43.5 ± 41.8	−67	233.18
MgCl		c	−641.30 ± 0.46	−591.5	89.586
MgCl$_2$			−800.76	−716.8	−25.1
MgCl$_2$	Std. state, $m=1$	aq			
MgCl$_2 \cdot$H$_2$O		c	−966.16	−861.40	137.2
MgCl$_2 \cdot$2H$_2$O		c	−1279.1	−1117.5	179.8
MgCl$_2 \cdot$4H$_2$O		c	−1898.1	−1622.7	263.9
MgCl$_2 \cdot$6H$_2$O		c	−2497.8	−2113.9	365.9
Mg(ClO$_4$)$_2$		c	−568.62		
Mg(ClO$_4$)$_2$	Std. state, $m=1$	aq	−725.15	−471.72	225.8

TABLE 3 (Continued)

Formula	Description	State	$\Delta H°/\text{kJ mol}^{-1}$	$\Delta G°/\text{kJ mol}^{-1}$	$S°/\text{J mol}^{-1}\text{ K}^{-1}$
$Mg(ClO_4)_2 \cdot 2H_2O$		c	−1218.2		
$Mg(ClO_4)_2 \cdot 4H_2O$		c	−1836.3		
$Mg(ClO_4)_2 \cdot 6H_2O$		c	−2444.3	−1862.2	520.6
$MgO \cdot MgCl_2$		c	−1306.8		
$MgO \cdot MgCl_2 \cdot 6H_2O$		c	−3113.0		
$MgO \cdot MgCl_2 \cdot 16H_2O$		c	−6034.2		
$MgOHCl$		c	−799.2	−731.4	83.6
$3Mg(OH)_2 \cdot MgCl_2$		c	−3462.2		
$3Mg(OH)_2 \cdot MgCl_2 \cdot 4H_2O$		c	−4699.3		
$3Mg(OH)_2 \cdot MgCl_2 \cdot 7H_2O$		c	−5593.8		
$3Mg(OH)_2 \cdot MgCl_2 \cdot 8H_2O$		c	−5891.6		
$5Mg(OH)_2 \cdot MgCl_2 \cdot 8H_2O$		c	−7727.0		
$MgBr$		g	−54 ± 3		244.6 ± 2.09
$MgBr_2$		c	−518.6 ± 4.2	−468.5	(119.2 ± 8.4)
$MgBr_2$	Std. state, $m=1$	aq	−709.60	−662.42	26.8

Group IIA Elements

MgBr$_2$·6H$_2$O		c	−2408.8	−2055.0	397
MgI$_2$		c	−363.8	−358.0	130.11
MgI$_2$	Std. state, $m=1$	aq	−576.94	−557.9	84.5
MgIO$_3^+$	Std. state, $m=1$	aq		−586.7	
MgS		c	−345.8	−341.7	50.31
MgSO$_3$		c	−1007.8		
MgSO$_3$·3H$_2$O		c	−1930.8		
MgSO$_3$·6H$_2$O		c	−2816.1		
MgSO$_4$		c	−1261.2 ± 20.9	−1170.1	91.35 ± 0.84
MgSO$_4$	Std. state, $m=1$	aq	−1375.4	−1198.9	117.9
MgSO$_4$·H$_2$O		c	−1601.2	−1428.1	126.3
MgSO$_4$·H$_2$O		amorph	−1574.1	−1404.3	138.0
MgSO$_4$·2H$_2$O		c	−1895.2		
MgSO$_4$·4H$_2$O		c	−2495.3		
MgSO$_4$·6H$_2$O		c	−3085.4	−2630.8	347.9

TABLE 3 (Continued)

Formula	Description	State	$\Delta H°/\text{kJ mol}^{-1}$	$\Delta G°/\text{kJ mol}^{-1}$	$S°/\text{J mol}^{-1}\text{K}^{-1}$
$MgSO_4 \cdot 7H_2O$		c	−3387.0	−2870.5	372
$MgS_2O_3 \cdot 3H_2O$		c	−1947.1		
$MgS_2O_3 \cdot 6H_2O$		c	−2847.1		
$3Mg(OH)_2 \cdot MgSO_4 \cdot 8H_2O$		c	−6454.0		
$MgSeO_3$		c	−899.75		
$MgSeO_3$		amorph	−892.22		
$MgSeO_3 \cdot 6H_2O$		c	−2704.4		
$MgSeO_4$		c	−968.04		
$MgSeO_4$	Std. state, $m=1$	aq	−1065.5	−895.8	−84.06
$MgSeO_4 \cdot H_2O$		c	−1294.4		
$MgSeO_4 \cdot 4H_2O$		c	−2188.8		
$MgSeO_4 \cdot 6H_2O$		c	−2694.0		
Mg_3N_2		c	−460.8 ± 4.2		(88 ± 8)
$Mg(NO_3)_2$		c	−790.21	−589.2	163.9

$Mg(NO_3)_2$	Std. state, $m=1$	aq	-881.14	-677.1	154.7
$Mg(NO_3)_2 \cdot 2H_2O$		c	-1468.4		
$Mg(NO_3)_2 \cdot 6H_2O$		c	-2612.0	-2079.7	463
$Mg(NH_3)_2^{2+}$		aq	-621.4		
$Mg(NH_3)_2Cl_2$		aq	-956.0		
$Mg(NH_3)_2SO_4$		aq	-1530.1		
$3MgSO_3 \cdot (NH_4)_2SO_3 \cdot 6H_2O$		c	-5770.7		
$3MgSO_3 \cdot (NH_4)_2SO_3 \cdot 18H_2O$		c	-9325.0		
MgN		g	(288 ± 25)		(224.6)
$MgP_2O_7^{2-}$	Std. state, $m=1$	aq	-2724.5	-2413.8	-79
$Mg_2P_2O_7$	α	c			154.8
$Mg_3(PO_4)_2$		c	-3743.3 ± 10.5	-3537	(188.2)
$MgHPO_4$		aq	-1755.6		
$MgNH_4PO_4 \cdot 6H_2O$		c	-3680.1		
$Mg_3(AsO_4)_2$		c	-3091.3		

TABLE 3 (Continued)

Formula	Description	State	$\Delta H°$/kJ mol^{-1}	$\Delta G°$/kJ mol^{-1}	$S°$/J mol^{-1} K^{-1}
$Mg(H_2AsO_4)_2$		aq	−2281.2		
$MgHAsO_4$		aq	−1362.9		
$MgNH_4AsO_4 \cdot 6H_2O$		c	−3315.0		
$MgCO_3$	Magnesite	c	−1111.1 ± 8.4	−1011.6	65.82
$MgCO_3 \cdot 3H_2O$	Nesquehonite	c		−1725.4	
$MgCO_3 \cdot 5H_2O$	Lansfordite	c		−2198.4	
MgC_2O_4		c	−1268.4		
MgC_2O_4	Std. state, $m=1$	aq	−1291.4	−1128.3	−92.4
$Mg(C_2O_4)_2^{2-}$	Std. state, $m=1$	aq		−1827.1	
$MgC_2O_4 \cdot 2H_2O$		c		−1632.6	
$MgHCO_3^+$	Std. state, $m=1$	aq		−1046.7	
$Mg(CO_2H)_2$	Formate	c			143.9
$Mg(C_2H_3O_2)^+$	Std. state, $m=1$	aq		−830.9	

Group IIA Elements

Formula	Name	State			
$Mg(C_2H_3O_3)_2$	Glycollate	c		−1749.7	
$Mg(C_2H_3O_3)_2$		aq		−1768.1	
$Mg(C_2H_3O_3)_2 \cdot 2H_2O$		c		−2346.9	
$3MgCO_3, Mg(OH)_2 \cdot 3H_2O$		c	−4601.4		
$MgCN_2$	Cyanamide	c		−252.2	
$Mg(CN)_2$		aq		−163.9	
$Mg(NH_2CH_2CO_2)^+$	Std. state, $m = 1$	aq	−789.1		
$MgSiO_3$	Clinoenstatite	c	−1460.9 ± 1.2	−1547.8 ± 1.2	67.9 ± 0.4
Mg_2SiO_4	Forsterite	c	−2051.3 ± 1.3	−2170.4 ± 1.3	95.2 ± 0.8
$Mg_3Si_2O_5(OH)_4$	Chrysotile	c	−4034.0 ± 2.9	−4361.7 ± 2.8	221.3 ± 1.7
$Mg_3Si_2O_5(OH)_4$	Antigonite	c			222.5
$Mg_3Si_4O_{10}(OH)_2$	Talc	c	−5536.4 ± 4.3	−5915.9 ± 4.3	160.8 ± 0.6
$PbI_2 \cdot 2MgI_2$		c		−863.6	
$MgO \cdot Al_2O_3$		c	−2187.6	−2312.2	80.59
$Mg_2Al_4Si_5O_{18}$	Cordierite	c	−8594	−9104	406.9
$MgHgBr_4$		aq		−879.9	

TABLE 3 (Continued)

Formula	Description	State	$\Delta H°$/kJ mol^{-1}	$\Delta G°$/kJ mol^{-1}	$S°$/J mol^{-1} K^{-1}
Mg(HgBr$_3$)$_2$		aq	−1041.7		
2MgBr$_2$·HgBr$_2$		aq	−1594.1		
4MgBr$_2$·HgBr$_2$		aq	−3016.0		
MgHg(CN)$_4$		aq	81.1		
Mg[Hg(CN)$_3$]$_2$		aq	341.2		
2Hg(CN)$_2$·MgCl$_2$		aq	−238.8		
2Hg(CN)$_2$·MgCl$_2$·6H$_2$O		c	−1992.7		
2Hg(CN)$_2$·MgBr$_2$		aq	−150.6		
2Hg(CN)$_2$·MgBr$_2$·8H$_2$O		c	−2500.4		
2Hg(CN)$_2$·MgI$_2$		aq	−35.1		
2Hg(CN)$_2$·MgI$_2$·8H$_2$O		c	−2401.3		
MgFe$_2$O$_4$		c	−1427.7	−1316.4	123.8
MgFe(CN)$_6^-$	Std. state, $m = 1$	aq		−1199.3	
MgFe(CN)$_6^{2-}$	Std. state, $m = 1$	aq		−1170.9	

Compound		Std. state, $m = 1$			
$Mg_2Fe(CN)_6$		aq	−478.0	−214.5	−181.1
$MgCrO_4$		c	−1342.8		
$MgCrO_4$		aq	−1354.9		
$MgCr_2O_4$		c	−1782.7	−1668.1	105.97
$MgMoO_4$		c	−1400.1	−1295.1	118.8
$MgWO_4$		c	−1515.1 ± 33.5	−1420.2	101.12 ± 0.84
MgV_2O_6	Metavanadate	c	−2200.5	−2038.4	160.6
$Mg_2V_2O_7$	Pyrovanadate	c	−2834.5	−2644.0	200.3
$MgTiO_3$	Metatitanate	c	−1571.8 ± 6.3	−1483.7	74.52 ± 0.42
$MgTi_2O_5$		c	−2508.1 ± 10.5	−2367.7	135.5 ± 6.3
Mg_2TiO_4	Orthotitanate	c	−2163.3 ± 4.2	−2045.8	95.09 ± 0.84
$2La(NO_3)_3 \cdot 3Mg(NO_3)_2 \cdot 24H_2O$		c			2195

[a] National Bureau of Standards values are in italic; CODATA values are underlined; estimated values are given in parentheses.

	$(dE°/dT)_{isoth}$/mV K^{-1}
$Mg^{2+} + 2e^- \rightarrow Mg$	0.103
$Mg(OH)_2(s) + 2e^- \rightarrow Mg + 2OH^-$	−0.945

Experimental measurements of the rest potential of the magnesium electrode do not correspond to these standard potentials and have been correlated, in neutral and alkaline solutions at very low partial pressure of oxygen, with the Mg/MgH_2 equilibrium for which the standard potentials can be calculated [2]:

$$Mg^{2+} + 2H^+ + 4e^- \rightarrow MgH_2 \qquad E° = -1.084 \text{ V}$$

$$Mg(OH)_2(s) + 2H_2O + 4e^- \rightarrow MgH_2 + 4OH^- \qquad E° = -1.663 \text{ V}$$

In acid solutions instability of the hydride leads to mixed potentials between metal dissolution and hydrogen evolution.

D. Equilibrium in Aqueous Solutions

The dissociation equilibrium for the hydroxide can be calculated and corresponds to

$$Mg(OH)_2(s) \rightarrow Mg^{2+} + 2OH^- \qquad K = 8.9 \times 10^{-12}$$

$$\log[Mg^{2+}] = 16.95 - 2 \text{ pH}$$

The solubility equilibrium has been studied many times and shows that the values obtained are slightly different with the various forms of the hydroxide.
The second dissociation equilibrium can also be calculated:

$$Mg(OH)_2(s) \rightarrow MgOH^+ + OH^- \qquad K_2 = 2.4 \times 10^{-9}$$

$$\log[MgOH^+] = 5.35 - \text{pH}$$

However, it has been shown [3] that $MgOH^+$ has no thermodynamical stability domain.
The $MgOH^+$ ion shows condensation and the thermodynamic function of the reaction has been determined [7]:

$$4MgOH^+ \rightarrow Mg_4(OH)_4^{4+}$$

with

$\Delta G° = -53.1$ kJ mol^{-1}

$\Delta H° = 48.9$ kJ mol^{-1}

$\Delta S° = 347$ J mol K^{-1}

E. Potential Diagrams for Magnesium

Acid Solution

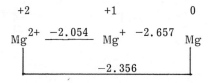

Basic Solution

+2   0

Mg(OH)$_2$ $\underline{-2.687}$ Mg

REFERENCES

1. S. Pobedinski, G. A. Krestov, and L. I. Kuzmin, Izv. Vyssh. Ucheb. Zaved. Khim. Khim. Tekhnol., 65, 768 (1963).
2. G. G. Perrault, in *Encyclopedia of the Electrochemistry of the Elements*, A. J. Bard, ed., Vol. 8, Marcel Dekker, New York, 1979, Chap. 4.
3. G. G. Perrault, UNESCO Discuss. Meet. Electrochem. Light Metals, Belgrade, Yugoslavia, Sept. 1982; Bull. Soc. Chim. Beograd, 48 suppl., 155 (1983).
4. CODATA recommended key values for thermodynamics, 1977, J. Chem. Thermodyn., 10, 903 (1978).
5. B. S. Hemingway and R. A. Robie, J. Res. U.S. Geol. Surv., 5(4), 413 (1977).
6. A. J. de Béthune, T. S. Licht, and N. Svendeman, J. Electrochem. Soc., 106, 616 (1959).
7. K. A. Burkov, E. A. Bus'ko, L. A. Garmash, and G. V. Khonin, Zh. Neorg. Khim., 23, 1767 (1978).

III. CALCIUM, STRONTIUM, BARIUM, AND RADIUM*

SHINOBU TOSHIMA[†] Tohoku University, Sendai, Miyagi, Japan

A. Calcium

1. Oxidation States

The only stable oxidation state in aqueous solution is +2, although other valence states, +1 and from +3 to +15, were found spectroscopically in relation to electron configuration, as well as the Ca_2 molecule. Calcium is very close to magnesium with respect to electrochemical properties, but its hydroxide is more basic than magnesium hydroxide and the solubilities of the sulfate and oxalate are much lower [1]. The thermodynamic data are given in Table 4.

2. Calcium Gibbs Energies and Potentials

For the standard potential of the Ca^{2+}/Ca couple, Latimer computed −2.87 V from the enthalpy of dissolution of Ca in acid and the entropies of the hydrogen and calcium ions [2].

Although several values varying from −2.87 to −2.758 V have been reported, the most reliable would be [3,4]

$$Ca^{2+} + 2e^- \rightarrow Ca \qquad E° = -2.84 \pm 0.01 \text{ V}$$

The standard potentials of the alkali earth metals have been measured using the standard potential of the amalgams in aqueous solutions, combined with the potential of a nonaqueous cell which has the amalgam as the positive electrode and the pure metal as the negative electrode. Difficulties about these measurements originate from the rapid corrosion and passivation of the metal [4].

Longhi et al. [5] determined the standard potentials of the calcium amalgam electrode over the temperature range 283.15 to 343.15 K, from electromotive force (emf) measurements of the cell

$$\text{Pt} \mid Ca_x Hg_{1-x} \mid CaCl_2(aq) \mid AgCl \mid Ag \mid \text{Pt}$$

yielding

$$E°[Ca(Hg)] = -1.727256 - 1.321118 \times 10^{-3}T + 1.322877 \times 10^{-6}T^2 \text{ V}$$

The $E°$ value at 25°C is −2.003 V, which is slightly different from the value −1.996 ± 0.002 V found by Butler [6].

Taking into account the fact that the emf of the cell

$$-Ca(s) \mid CaI_2(\text{pyridine}) \mid Ca(Hg)$$

---

*Reviewer: M. Chemla, Pierre and Marie Curie University, Paris, France.
[†]Current affiliation: Akita University, Akita, Japan.

was greater than 0.813 V [7], the highest experimental value of the standard potential of the Ca^{2+}/Ca couple may be evaluated to be $-2.003 - 0.813 = -2.816$ V.

If the emf of the cell

$$Pb \mid PbCO_3 \mid CaCO_3 \mid CaCl_2 \cdot H_2O \mid Hg_2Cl_2 \mid Hg$$

were combined with thermodynamic data, the standard potential might be -2.868 V [6].

For the couple in alkaline solution [8],

$$Ca(OH)_2 + 2e^- \rightarrow Ca + 2OH^- \qquad E° = -3.026 \text{ V}$$

The corresponding value for the solubility product of the hydroxide

$$Ca(OH)_2 = Ca^{2+} + 2OH^-$$

is $K = 2.03 \times 10^{-6}$. Isothermal temperature coefficients of the standard potentials at 25°C, $(dE°/dT)_{isoth}$, are

$$Ca^{2+} + 2e^- \rightarrow Ca \qquad -0.175 \text{ mV K}^{-1}$$

$$Ca(OH)_2 + 2e^- \rightarrow Ca + 2OH^- \qquad -0.965 \text{ mV K}^{-1}$$

respectively [8]. The values of the potentials for other half-reactions are [9,10]

$$CaO(hyd) + 2H^+ + 2e^- \rightarrow Ca + H_2O \qquad E° = -2.189 \text{ V}$$

$$Ca^{2+} + 2H^+ + 4e^- \rightarrow CaH_2 \qquad E° = -1.045 \text{ V}$$

$$CaO(hyd) + 4H^+ + 4e^- \rightarrow CaH_2 + H_2O \qquad E° = -0.706 \text{ V}$$

$$Ca + 2H^+ + 2e^- \rightarrow CaH_2 \qquad E° = 0.776 \text{ V}$$

$$CaO_2 + 2H^+ + 2e^- \rightarrow CaO(hyd) + H_2O \qquad E° = 1.547 \text{ V}$$

$$CaO_2 + 4H^+ + 2e^- \rightarrow Ca^{2+} + 2H_2O \qquad E° = 2.224 \text{ V}$$

The polarographic half-wave potential of calcium was found to be -1.974 V by Kimura [11] and by Zlotowski and Kolthoff [12] in tetraethylammonium salts as supporting electrolytes. A procedure for suppressing pronounced maxima was developed.

TABLE 4 Thermodynamic Data on Calcium

Formula	Description	State	ΔH°/kJ mol^{-1}	ΔG°/kJ mol^{-1}	S°/J mol^{-1} K^{-1}
Ca		g	178.2	144.3	154.884
	Metal	c	0.0	0.0	41.42
Ca$^+$		g	774.25		
Ca^{2+}		g	1,925.9		
	Std. state, $m=1$	aq	−542.83	−553.54	−53.1
CaO		c	−635.09	−604.04	39.75
CaO$_2$		c	−652.7		
CaH		g	228.9	200.4	201.63
CaH$_2$		c	−186.2	−147.3	42.0
Ca(OH)$^+$	Std. state, $m=1$	aq		−718.4	
Ca(OH)$_2$		c	−986.09	−898.56	83.39
	Std. state, $m=1$	aq	−1,002.82	−868.14	−74.5
CaF		g	−272.0	−298.0	229.3
CaF$_2$		c	−1,219.64	−1,167.34	68.87

CaCl	Std. state, $m=1$	aq	−1,208.1	−1,111.2	−80.75
		g	−97.9	−124.3	241.42
CaCl$_2$		c	−795.8	−748.1	104.6
	Std. state, $m=1$	aq	−877.13	−816.05	59.8
CaCl$_2 \cdot$ H$_2$O		c	−1,109.2		
CaCl$_2 \cdot$ 2H$_2$O		c	−1,402.9		
CaCl$_2 \cdot$ 4H$_2$O		c	−2,009.6		
CaCl$_2 \cdot$ 6H$_2$O		c	−2,607.9		
Ca(OCl)$_2$		aq	−754.4		
Ca(ClO$_4$)$_2$		c	−736.76		
	Std. state, $m=1$	aq	−801.49	−570.78	310.87
Ca(ClO$_4$)$_2 \cdot$ 4H$_2$O		c	−1,948.9	−1,476.83	433.5
CaBr$_2$		c	−682.8	−663.6	130.
	Std. state, $m=1$	aq	−785.92	−761.49	111.7
CaBr$_2 \cdot$ 6H$_2$O		c	−2,506.2	−2,153.1	410.
CaI$_2$		c	−533.5	−528.9	142.
	Std. state, $m=1$	aq	−653.21	−656.72	169.5

TABLE 4 (Continued)

Formula	Description	State	$\Delta H°/\text{kJ mol}^{-1}$	$\Delta G°/\text{kJ mol}^{-1}$	$S°/\text{J mol}^{-1} \text{K}^{-1}$
$Ca(IO_3)_2$		c	$-1{,}002.5$	-839.3	230.
$Ca(IO_3)_2 \cdot 6H_2O$		c	$-2{,}780.7$	$-2{,}267.7$	452.
CaS		c	-482.4	-477.4	56.5
$CaSO_4$	Insol., anhydrite	c	$-1{,}434.11$	$-1{,}321.85$	106.7
	Std. state, $m = 1$	aq	$-1{,}452.10$	$-1{,}298.17$	-33.0
	Soluble α	c	$-1{,}425.24$	$-1{,}313.48$	108.4
	Soluble β	c	$-1{,}420.80$	$-1{,}309.05$	108.4
$CaSO_4 \cdot (1/2)H_2O$	Macro; α	c	$-1{,}576.74$	$-1{,}436.83$	130.5
	Micro; β	c	$-1{,}574.65$	$-1{,}435.87$	134.3
$CaSO_4 \cdot 2H_2O$	Selenite	c	$-2{,}022.63$	$-1{,}797.45$	194.1
CaSe		c	-368.2	-363.2	67.
$CaSeO_4$		c	$-1{,}109.8$		
$CaSeO_4 \cdot 2H_2O$		c	$-1{,}706.6$	$-1{,}487.0$	222.
$Ca(N_3)_2$	Azide	c	14.6		
$Ca(NO_3)_2$		c	-938.39	-743.20	193.3

$Ca(NO_3)_2 \cdot 2H_2O$	Std. state, $m=1$	aq	-957.55	-776.22	239.7
$Ca(NO_3)_2 \cdot 2H_2O$		c	-1,540.76	-1,229.34	269.4
$Ca(NO_3)_2 \cdot 3H_2O$		c	-1,838.0	-1,471.9	319.2
$Ca(NO_3)_2 \cdot 4H_2O$		c	-2,132.33	-1,713.47	375.3
$Ca_2P_2O_7$	β	c	-3,338.8	-3,132.1	189.24
$Ca_3(PO_4)_2$	β, low temp.	c	-4,120.8	-3,884.8	236.0
	α, high temp.	c	-4,109.9	-3,875.6	240.91
	Std. state, $m=1$	aq	-4,183.2	-3,698.2	-603.
$CaHPO_4$		c	-1,814.39	-1,681.26	111.38
	Std. state, $m=1$	aq	-1,834.98	-1,642.81	-86.6
$CaHPO_4 \cdot 2H_2O$		c	-2,403.58	-2,154.76	189.45
$Ca(H_2PO_4)_2$		c	-3,104.70		
	Std. state, $m=1$	aq	-3,135.41	-2,814.33	127.6
$Ca(H_2PO_4) \cdot H_2O$		c	-3,409.67	-3,058.42	259.8
$Ca_{10}(PO_4)_6(OH)_2$	Hydroxyapatite	c	-13,477.	-12,678.	780.7
	Std. state, $m=1$	aq	-13,552.4	-11,962.9	-1,883.
$Ca_{10}(PO_4)_6F_2$	Fluorapatite	c	-13,744.	-12,982.	775.7

TABLE 4 (Continued)

Formula	Description	State	$\Delta H°$/kJ mol^{-1}	$\Delta G°$/kJ mol^{-1}	$S°$/J mol^{-1} K^{-1}
$Ca_3(AsO_4)_2$		c	−3,298.7	−3,063.1	226.
CaC_2		c	−59.8	−64.9	69.96
$CaCO_3$	Calcite	c	−1,206.92	−1,128.84	92.9
	Aragonite	c	−1,207.13	−1,127.80	88.7
	Std. state, $m = 1$	aq	−1,219.97	−1,081.44	−110.0
CaC_2O_4		c	−1,360.6		
	Std. state, $m = 1$	aq	−1,367.7	−1,227.46	−7.5
$CaC_2O_4 \cdot H_2O$		c	−1,674.86	−1,513.98	156.5
$Ca(C_2H_3O_2)_2$		c	−1,479.5		
	Std. state, $m = 1$	aq	−1,514.86	−1,292.35	120.1
$CaO \cdot SiO_2$	Wollastonite	c	−1,634.94	−1,549.71	81.92
	Pseudowollastonite	c	−1,628.4	−1,544.7	87.36
$2CaO \cdot SiO_2$	β	c	−2,307.5	−2,192.8	127.74
	γ	c	−2,317.9	−2,201.2	120.79
$3CaO \cdot SiO_2$		c	−2,929.2	−2,784.0	168.6

Group IIA Elements

$3CaO \cdot 2SiO_2$	Rankinite	c	−3,961.0	−3,761.4	210.79
$CaO \cdot B_2O_3$		c	−2,030.96	−1,924.10	104.85
$CaO \cdot 2B_2O_3$		c	−3,360.25	−3,167.12	134.7
$2CaO \cdot B_2O_3$		c	−2,734.41	−2,596.67	145.10
$3CaO \cdot B_2O_3$		c	−3,429.08	−3,259.92	183.7
$CaO \cdot Al_2O_3$		c	−2,326.3	−2,208.7	114.22
$CaO \cdot 2Al_2O_3$		c	−3,977.7	−3,770.6	177.82
$3CaO \cdot Al_2O_3$		c	−3,587.8	−3,411.6	205.8
$CaO \cdot Al_2O_3 \cdot 2SiO_2$	Anorthite, hexag.	c	−4,222.5	−3,997.8	202.5
$CaO \cdot Al_2O_3 \cdot 2SiO_2$	Anorthite, hexag.	c	−4,201.9	−3,973.9	191.6
$CaO \cdot Al_2O_3 \cdot 2SiO_2 \cdot 2H_2O$	Lawsonite	c	−4,858.5	−4,505.3	237.7
$2CaO \cdot Al_2O_3 \cdot SiO_2$	Gehlenite	c	−3,985.7	−3,783.6	198.3
$2CaO \cdot 2Al_2O_3 \cdot 8SiO_2 \cdot 7H_2O$	Leonhardite	c	−14,211.4	−13,165.4	922.2
$CaO \cdot Fe_2O_3$		c	−1,520.34	−1,412.81	145.35
$2CaO \cdot Fe_2O_3$		c	−2,139.28	−2,001.79	188.78
$CaFe(CN)_6^-$	Std. state, $m = 1$	aq		159.4	

TABLE 4 (Continued)

Formula	Description	State	$\Delta H°$/kJ mol^{-1}	$\Delta G°$/kJ mol^{-1}	$S°$/J mol^{-1} K^{-1}
CaFe(CN)$_6^{2-}$	Std. state, $m = 1$	aq		119.7	
Ca$_2$Fe(CN)$_6$	Std. state, $m = 1$	aq	−630.1	−412.17	−11.3
Ca$_3$[Fe(CN)$_6$]$_2$	Std. state, $m = 1$	aq	−504.6	−202.1	381.2
CaMoO$_4$		c	−1,541.4	−1,434.7	122.6
	Std. state, $m = 1$	aq	−1,540.5	−1,389.9	−25.9
CaWO$_4$		c	−1,645.15	−1,538.50	126.40
	Std. state, $m = 1$	aq	−1,618.4		
CaO·V$_2$O$_5$		c	−2,329.27	−2,169.70	179.1
2CaO·V$_2$O$_5$		c	−3,083.36	−2,893.19	220.5
3CaO·V$_2$O$_5$		c	−3,777.94	−3,561.09	274.9

CaO·TiO$_2$	Perovskite	c	−1,660.6	−1,575.3	93.63
3CaO·2TiO$_2$		c	−3,950.5	−3,751.4	234.7
CaTiSiO$_5$	Sphene	c	−2,603.3	−2,461.9	129.20
CaZrO$_3$		c	−1,766.9	−1,681.1	100.08
CaMg$_2$	γ	c	−40.2	−39.3	104.27
CaMgC$_2$O$_6$	Dolomite	c	−2,326.3	−2,163.5	155.18
CaO·MgO·SiO$_2$	Monticellite	c	−2,263.08		
CaO·MgO·2SiO$_2$	Diopside	c	−3,206.2	−3,032.1	142.93
2CaO·MgO·2SiO$_2$	Ankermanite	c	−3,877.19	−3,679.95	209.2
3CaO·MgO·2SiO$_2$	Mervinite	c	−4,567.7	−4,340.5	253.1
2CaO·5MgO·8SiO$_2$·H$_2$O	Tremolite	c	−12,359.	−11,631.	548.9

Source: Ref. 13; 1982 values are in italic.

3. Potential Diagrams for Calcium

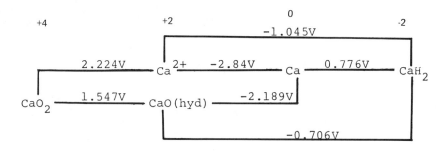

REFERENCES

1. *Gmelin Handbuch der anorganischen Chemie*, 8th ed., Calcium, System No. 28, Verlag Chemie, Weinheim, West Germany, 1957.
2. W. M. Latimer, J. Phys. Chem., *31*, 1267-1269 (1927).
3. J. O'M. Bockris and J. F. Herringshaw, Discuss. Faraday Soc., *1*, 328-334 (1947).
4. A. J. Bard, ed., *Encyclopedia of Electrochemistry of the Elements*, Vol. 1, Marcel Dekker, New York, 1973 pp. 405-466.
5. P. Longhi, T. Mussini, and E. Vaghi, Chim. Ind. (Milan), *57*, 169-172 (1975).
6. J. N. Butler, Electroanal. Chem., *17*, 309-317 (1968).
7. F. L. E. Shibata, J. Sci. Hiroshima Univ., Al 147-157 (1931).
8. A. J. de Béthune, T. S. Licht, and N. Swendeman, J. Electrochem. Soc., *106*, 616-622 (1959).
9. M. Pourbaix, *Atlas of Electrochemical Equilibria*, Pergamon Press, Oxford, 1966, pp. 147-148.
10. G. Milazzo and S. Caroli, *Tables of Standard Electrode Potentials*, Wiley, New York, 1978, pp. 70-75.
11. G. Kimura, Coll. Czech. Chem. Commun., *4*, 492 (1932).
12. I. Zlotowski and I. M. Kolthoff, J. Phys. Chem., *49*, 386-405 (1945).
13. Natl. Bur. Stand. Tech. Note 270-6, U.S. Government Printing Office, Washington, D.C., Nov. 1971, pp. 30-56; D. D. Wagman, W. H. Evans, V. B. Parker, R. H. Schumm, I. Halow, S. M. Bailey, K. L. Churney, and R. L. Nuttall, *NBS Tables of Chemical Thermodynamic Properties*, J. Phys. Chem. Ref. Data, Vol. 11, Supplement 2 (1982).

B. Strontium

1. Oxidation States

The only stable oxidation state in aqueous solution is +2, although other valence states, +1 and from +3 to +14, were found spectroscopically in relation to electron configuration, as well as the Sr_2 molecule. Monohalides, monohydride, and monoxide may exist under spectroscopic observation. The

Group IIA Elements

hydroxide is more basic than calcium, and the solubilities of the sulfate and chromates are less than those of the calcium salts [1]. The thermodynamic data are given in Table 5.

2. Strontium Gibbs Energies and Potentials

The most reliable value of the standard potential of the Sr^{2+}/Sr couple would be [2]

$$Sr^{2+} + 2e^- \rightarrow Sr \qquad E° = -2.89 \pm 0.01 \text{ V}$$

Longhi et al. [3] determined the standard potentials of the strontium amalgam electrode over the temperature range 283.15 to 343.15 K, from emf measurements of the cell

$$Pt \mid St_x Hg_{1-x} \mid SrCl_2(aq) \mid AgCl \mid Ag \mid Pt$$

$$E°[Sr(Hg)] = -1.448013 - 2.289353 \times 10^{-3}T + 2.582182 \times 10^{-6}T^2 \text{ V}$$

The $E°$ value at 25°C is -1.901 V.

For the potential of the strontium couple in alkaline solution [4],

$$Sr(OH)_2 + 2e^- \rightarrow Sr + 2OH^- \qquad E° = -2.88 \text{ V}$$

The values for other half-reactions are [5,6]

$$SrO(hyd) + 2H^+ + 2e^- \rightarrow Sr + H_2O \qquad E° = -2.047 \text{ V}$$

$$Sr^{2+} + 2H^+ + 4e^- \rightarrow SrH_2 \qquad E° = -1.085 \text{ V}$$

$$SrO(hyd) + 4H^+ + 4e^- \rightarrow SrH_2 + H_2O \qquad E° = -0.665 \text{ V}$$

$$Sr + 2H^+ + 2e^- \rightarrow SrH_2 \qquad E° = 0.718 \text{ V}$$

$$SrO_2 + 2H^+ + 2e^- \rightarrow SrO(hyd) + H_2O \qquad E° = 1.492 \text{ V}$$

$$SrO_2 + 4H^+ + 2e^- \rightarrow Sr^{2+} + 2H_2O \qquad E° = 2.333 \text{ V}$$

Isothermal temperature coefficients of the standard potentials at 25°C, $(dE°/dT)_{isoth}$, are

$$Sr^{2+} + 2e^- \rightarrow Sr \qquad -0.191 \text{ mV K}^{-1}$$

$$Sr(OH)_2 + 2e^- \rightarrow Sr + 2OH^- \qquad -0.96 \text{ mV K}^{-1}$$

respectively [4].

The polarographic half-wave potential of the strontium was found to be -1.864 V by Zlotowski and Kolthoff [7] in tetraethylammonium salts as supporting electrolytes.

TABLE 5 Thermodynamic Data on Strontium

Formula	Description	State	$\Delta H°$/kJ mol^{-1}	$\Delta G°$/kJ mol^{-1}	$S°$/J mol^{-1} K^{-1}
Sr		g	164.4	131.0	164.51
	Metal	c	0	0	52.3
Sr$^+$		g	720.11		
Sr^{2+}		g	1790.58		
	Std. state, m = 1	aq	−545.80	−559.44	−32.6
SrO		c	−592.0	−561.9	54.4
SrO$_2$		c	−633.4		
SrH		g	217.	188.	212.55
SrH$_2$		c	−180.3		
SrOH$^+$	Std. state, m = 1	aq		−721.3	
Sr(OH)$_2$		c	−959.0		
	In 800 H$_2$O	aq	−1004.6		
SrF		g	−288.7	−314.2	239.78
SrF$_2$		c	−1216.3	−1164.8	82.13
SrCl		g	−126.4	−152.7	251.88

Group IIA Elements

$SrCl_2$	α	c	−828.8	−781.2	114.85
	Std. state, $m = 1$	aq	−880.10	−821.95	80.3
$SrCl_2 \cdot H_2O$		c	−1136.8	−1036.4	172.
$SrCl_2 \cdot 2H_2O$		c	−1438.0	−1282.0	218.
$SrCl_2 \cdot 6H_2O$		c	−2623.8	−2241.2	390.8
$Sr(OCl)_2$		aq	−755.2		
$Sr(ClO_4)_2$		c	−762.79		
	Std. state, $m = 1$	aq	−804.46	−576.68	331.4
$Sr(ClO_4)_2 \cdot 4H_2O$		c	−1962.3		
$SrBr_2$		c	−717.6	−697.1	135.10
	Std. state, $m = 1$	aq	−788.89	−767.74	132.2
$SrBr_2 \cdot H_2O$		c	−1031.4	−954.4	180.
$SrBr_2 \cdot 6H_2O$		c	−2531.3	−2174.4	406.
$Sr(BrO_3)_2 \cdot H_2O$		c	−1105.	−791.2	280.
SrI_2		c	−558.1		
	Std. state, $m = 1$	aq	−656.18	−662.62	190.0
$Sr(IO_3)_2$		c	−1019.2	−855.2	234.

TABLE 5 (Continued)

Formula	Description	State	$\Delta H°$/kJ mol^{-1}	$\Delta G°$/kJ mol^{-1}	$S°$/J mol^{-1} K^{-1}
Sr(IO$_3$)$_2$·H$_2$O		c	−1310.4	−1089.5	276.
Sr(IO$_3$)$_2$·6H$_2$O		c	−2789.9	−2274.8	456.
SrS		c	−472.4	−467.8	68.2
SrSO$_4$		c	−1453.1	−1341.0	117.
	Std. state, $m = 1$	aq	−1455.07	−1304.07	−12.6
SrSe		c	−385.8		
SrSeO$_4$		c	−1142.6		
Sr(N$_3$)$_2$		c	8.8		
Sr(NO$_3$)$_2$		c	−978.2	−780.15	194.56
	Std. state, $m = 1$	aq	−960.52	−782.12	260.2
Sr(NO$_3$)$_2$·4H$_2$O		c	−2154.8	−1730.71	369.0
SrHPO$_4$		c	−1821.7	−1688.7	121.
Sr$_3$(AsO$_4$)$_2$		c	−3317.1	−3080.3	255.

SrC_2			c	−75.
$SrCO_3$	Strontianite		c	−1220.1	−1140.1	97.1
	Std. state, $m = 1$		aq	−1222.9	−1087.34	−89.5
SrC_2O_4			c	−1370.7		
	Std. state, $m = 1$		aq	−1370.7	−1233.4	8.8
$SrC_2H_3O_2^+$	Std. state		aq		−935.1	
$SrSiO_3$			c	−1633.9	−1549.8	96.6
Sr_2SiO_4			c	−2304.5	−2191.2	153.1
$SrFe(CN)_6^-$	Std. state, $m = 1$		aq		153.5	
$Sr_2Fe(CN)_6$	Std. state, $m = 1$		aq	−636.0	−424.0	29.7
$Sr_3[Fe(CN)_6]_2$	Std. state, $m = 1$		aq	−513.8	−219.7	442.7
$SrWO_4$			c	−1639.7	−1531.	138.
$SrTiO_3$			c	−1672.39	−1588.41	108.8
Sr_2TiO_4			c	−2287.4	−2178.6	159.0
$SrZrO_3$			c	−1767.3	−1682.8	115.1

Source: Ref. 8; 1982 values are in italic.

3. Potential Diagram for Strontium

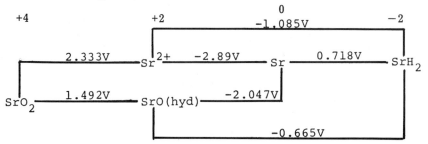

REFERENCES

1. *Gmelin Handbuch der anorganischen Chemie*, 8th ed., Strontium (System No. 29), Verlag Chemie, Weinheim, West Germany, 1960.
2. J. O'M. Bockris and J. F. Herringshaw, Discuss. Faraday Soc., *1*, 328-334 (1947).
3. P. Longhi, T. Mussini, and E. Vaghi, J. Chem. Thermodyn., *7*, 767-776 (1975).
4. A. J. de Béthune, T. S. Licht, and N. Swendeman, J. Electrochem. Soc., *106*, 616-622 (1959).
5. M. Pourbaix, ed., *Atlas of Electrochemical Equilibria*, Pergamon Press, Oxford, 1966, p. 148.
6. G. Milazzo and S. Caroli, *Tables of Standard Electrode Potentials*, Wiley, New York, 1978, pp. 80-81.
7. I. Zlotowski and I. M. Kolthoff, J. Am. Chem. Soc., *66*, 1431-1435 (1944), J. Phys. Chem., *49*, 386-405 (1945).
8. Nat. Bur. Stand. Tech. Note 270-6, U.S. Government Printing Office, Washington, D.C., Nov. 1971, pp. 57-77; D. D. Wagman, W. H. Evans, V. B. Parker, R. H. Schumm, I. Halow, S. M. Bailey, K. L. Churney, and R. L. Nuttall, *NBS Tables of Chemical Thermodynamic Properties*, J. Phys. Chem. Ref. Data, Vol. 11, Supplement 2 (1982).

C. Barium

1. Oxidation States

The only stable oxidation state in aqueous solution is +2, although other valence states, +1, +3, and +4, were found spectroscopically in relation to electron configuration. Monohalides, monohydride, and monoxide may exist under spectroscopic observation. The solubilities of the sulfate and chromate are less than those of the lighter elements of the family [1]. The thermodynamic data are given in Table 6.

2. Barium Gibbs Energies and Potentials

The most reliable value of the standard potential of the Ba^{2+}/Ba couple would be [2]

$$Ba^{2+} + 2e^- \rightarrow Ba \qquad E° = -2.92 \pm 0.01 \text{ V}$$

Group IIA Elements

Ardizzone et al. [3] determined the standard potentials of the barium amalgam electrode over the temperature range 283.15 to 343.15 K from emf measurements of the cell

$$\text{Pt} \mid \text{Ba}_x\text{Hg}_{1-x} \mid \text{BaCl}_2(\text{aq}) \mid \text{AgCl} \mid \text{Ag} \mid \text{Pt}$$

yielding

$$E°[\text{Ba(Hg)}] = -1.160117 - 2.765854 \times 10^{-3}T + 3.025048 \times 10^{-6}T^2 \text{ V}$$

The value at 25°C is −1.717 V.

For the couple in alkaline solution [4],

$$\text{Ba(OH)}_2 + 2e^- \rightarrow \text{Ba} + 2\text{OH}^- \qquad E° = -2.81 \text{ V}$$

The values of the potentials for other half-reactions are

$$\text{Ba(OH)}_2 \cdot 8\text{H}_2\text{O} + 2e^- \rightarrow \text{Ba(s)} + 2\text{OH}^- + 8\text{H}_2\text{O} \qquad E° = -2.99 \text{ V [4,6]}$$

$$\text{BaO(hyd)} + 2\text{H}^+ + 2e^- \rightarrow \text{Ba} + \text{H}_2\text{O} \qquad E° = -2.166 \text{ V [5]}$$

$$\text{Ba}^{2+} + 2\text{H}^+ + 4e^- \rightarrow \text{BaH}_2 \qquad E° = -1.110 \text{ V [5,6]}$$

$$\text{BaO(hyd)} + 4\text{H}^+ + 4e^- \rightarrow \text{BaH}_2 + \text{H}_2\text{O} \qquad E° = -0.741 \text{ V [5]}$$

$$\text{Ba} + 2\text{H}^+ + 2e^- \rightarrow \text{BaH}_2 \qquad E° = -0.685 \text{ V [5,6]}$$

$$\text{BaO}_2(\text{hyd}) + 2\text{H}^+ + 2e^- \rightarrow \text{BaO(hyd)} + \text{H}_2\text{O} \qquad E° = 1.626 \text{ V [5]}$$

$$\text{BaO}_2(\text{hyd}) + 4\text{H}^+ + 2e^- \rightarrow \text{Ba}^{2+} + 2\text{H}_2\text{O} \qquad E° = 2.365 \text{ V [5]}$$

Isothermal temperature coefficients of the standard potentials at 25°C, $(dE°/dT)_{\text{isoth}}$, are

$$\text{Ba}^2 + 2e^- \rightarrow \text{Ba} \qquad -0.395 \text{ mv K}^{-1}$$

$$\text{Ba(OH)}_2 + 2e^- \rightarrow \text{Ba} + 2\text{OH}^- \qquad -0.93 \text{ mV K}^{-1}$$

$$\text{Ba(OH)}_2 \cdot 8\text{H}_2\text{O} + 2e^- \rightarrow \text{Ba(s)} + 2\text{OH}^- + 8\text{H}_2\text{O} \qquad 0.38 \text{ mV K}^{-1}$$

respectively [4].

The polarographic half-wave potential of barium was found to be −1.694 V by Zlotowski and Kolthoff [7] in tetraethylammonium salts as supporting electrolytes.

TABLE 6 Thermodynamic Data on Barium

Formula	Description	State	$\Delta H°$/kJ mol^{-1}	$\Delta G°$/kJ mol^{-1}	$S°$/J mol^{-1} K^{-1}
Ba		g	180.	146	170.134
	Metal	c	0	0	62.8
Ba$^+$		g	688.98		
Ba^{2+}		g	1660.46		
	Std. state, $m = 1$	aq	−537.64	−560.74	9.6
BaO		c	−553.5	−525.1	70.42
BaO$_2$		c	−634.3		
BaH		g	222.	197.	218.78
BaH$_2$		c	−178.7		
BaOH$^+$	Std. state, $m = 1$	aq	−944.7	−730.5	
Ba(OH)$_2$		c	−995.4	−875.3^a	−8.4^a
	In 500 H$_2$O	aq			
Ba(OH)$_2$·8H$_2$O		c	−3342.2	−2793.2	427.
BaF		g	−326.4	−350.6	246.0

Group IIA Elements

BaF$_2$		c	−1207.1	−1156.9	96.36
	Std. state, $m = 1$	aq	−1202.90	−1118.38	−18.0
BaCl		g	−167.		258.6
BaCl$_2$		c	−858.6	−810.4	123.68
	Std. state, $m = 1$	aq	−871.95	−823.24	122.6
BaCl$_2 \cdot$ H$_2$O		c	−1160.6	−1055.71	166.9
BaCl$_2 \cdot$ 2H$_2$O		c	−1460.13	−1296.45	202.9
Ba(OCl)$_2$		aq	−750.2		
Ba(ClO$_2$)$_2$		c	−680.3	−531.4	197.
Ba(ClO$_4$)$_2$		c	−800.0		
	In 900 H$_2$O	aq	−795.8		
Ba(ClO$_4$)$_2 \cdot$ 3H$_2$O		c	−1961.6	−1270.7	393.
BaBr$_2$		c	−757.3	−736.8	146.
	Std. state, $m = 1$	aq	−780.73	−768.68	174.5
BaBr$_2 \cdot$ 2H$_2$O		c	−1366.1	−1230.5	226.
Ba(BrO$_3$)$_2$		c	−752.66	−577.4	243.
Ba(BrO$_3$)$_2 \cdot$ H$_2$O		c	−1054.8	−824.62	292.5

TABLE 6 (Continued)

Formula	Description	State	$\Delta H°$/kJ mol^{-1}	$\Delta G°$/kJ mol^{-1}	$S°$/J mol^{-1} K^{-1}
BaI$_2$		c	−602.1		
	Std. state, m = 1	aq	−648.02	−663.92	232.2
Ba(IO$_3$)$_2$		c	−1027.2	−864.8	249.4
	Std. state, m = 1	aq	−980.3	−816.7	246.4
Ba(IO$_3$)$_2$·H$_2$O		c	−1322.1	−1104.2	297.
BaS		c	−460.	−456.	78.2
BaSO$_4$		c	−1473.2	−1362.3	132.2
	Std. state, m = 1	aq	−1446.91	−1305.37	29.7
BaSe		c	−372.		
BaSeO$_3$		c	−1040.6	−968.2	167.
BaSeO$_4$		c	−1146.4	−1044.7	176.
BaN$_2$		c	−172.		
Ba(N$_3$)$_2$·H$_2$O		c	−308.4	−105.0	188.
BaNO$_3$	Std. state, m = 1	aq	−735.5	−666.72	172.
Ba(NO$_3$)$_2$		c	−992.07	−796.72	213.8

Group IIA Elements

BaHPO$_4$	Std. state, $m = 1$	aq	−952.36	−783.41	302.5
Ba(H$_2$PO$_2$)$_2$		c	−1814.6		
	In 400 H$_2$O	c	−1762.3		
Ba$_3$(AsO$_4$)$_2$	Hydrated precipitate	aq	−1763.1		
			−3427.		
BaC$_2$		c	−75.		
BaCO$_3$	Witherite	c	−1216.3	−1137.6	112.1
	Std. state, $m = 1$	aq	−1214.78	−1088.63	−47.3
BaC$_2$O$_4$		c	−1368.6		
Ba(HCO$_3$)$_2$	Std. state, $m = 1$	aq	−1921.63	−1734.44	192.0
Ba(C$_2$H$_3$O$_2$)$_2$		c	−1484.5		
	Std. state, $m = 1$	aq	−1509.67	−1299.55	182.8
BaO·SiO$_2$		c	−1623.60	−1540.26	109.6
BaO·2SiO$_2$		c	−2548.1	−2410.8	153.1
2BaO·SiO$_2$		c	−2287.8	−2174.8	176.1
2BaO·3SiO$_2$		c	−4184.8	−3963.1	258.2
BaSiF$_6$		c	−2952.2	−2794.1	163.

TABLE 6 (Continued)

Formula	Description	State	$\Delta H°$/kJ mol^{-1}	$\Delta G°$/kJ mol^{-1}	$S°$/J mol^{-1} K^{-1}
BaO·Al$_2$O$_3$		c	−2326.		
Ba$_2$Fe(CN)$_6$		aq	−609.6		
Ba$_3$[Fe(CN)$_6$]$_2$	Std. state, $m = 1$	aq	−489.1	−223.8	569.4
BaMnO$_4$		c		−1119.2	
Ba(ReO$_4$)$_2$·4H$_2$O		c	−3368.	−2918.3	377.
BaCrO$_4$		c	−1445.9	−1345.28	158.6
BaMoO$_4$		c	−1548.	−1439.7	138.
BaTiO$_3$		c	−1659.8	−1572.3	107.9
Ba$_2$TiO$_4$		c	−2243.0	−2133.0	196.6
BaZrO$_3$		c	−1779.4	−1694.5	124.7
BaSrTiO$_4$		c	−2276.1	−2167.7	191.6

[a]National Bureau of Standards, Selected Values of Chemical Thermodynamic Properties, Part I, 1961, p. 414.
Source: Ref. 8.

Group IIA Elements

3. Potential Diagram for Barium

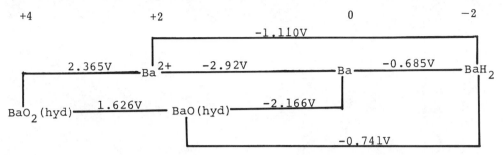

REFERENCES

1. *Gmelin Handbuch der anorganischen Chemie*, 8th ed., Barium (System No. 30), Verlag Chemie, Weinheim, West Germany, 1960.
2. J. O'M. Bockris and J. F. Herringshaw, Discuss. Faraday Soc., 1, 328-334 (1947).
3. S. Ardizzone, P. Longhi, T. Mussini, and S. Rondinini, J. Chem. Thermodyn., 8, 557-564 (1976).
4. A. J. de Béthune, T. S. Licht, and N. Swendeman, J. Electrochem. Soc., 106, 616-622 (1959).
5. M. Pourbaix, ed., *Atlas of Electrochemical Equilibria*, Pergamon Press, Oxford, 1966, p. 149.
6. G. Milazzo and S. Caroli, *Tables of Standard Electrode Potentials*, Wiley, New York, 1978, pp. 84-89.
7. I. Zlotowski and I. M. Kolthoff, J. Am. Chem. Soc., 66, 1431-1435 (1944); J. Phys. Chem., 49, 386-405 (1945).
8. Natl. Bur. Stand. Tech. Note 270-6, U.S. Government Printing Office, Washington, D.C., Nov. 1971, pp. 78-102.

D. Radium

1. Oxidation States

The only stable oxidation state in aqueous solution is +2. Ra^{2+} is colorless and most Ra salts are white. All Ra compounds having contact with air are pale blue luminescent; the sulfate glows red. The chemistry resembles closely that of Ba^{2+}, but RaO is highly aggressive. The sulfate is slightly less soluble than $BaSO_4$. The hydroxide is the most basic of the group [1]. The thermodynamic data are given in Table 7.

2. Radium Gibbs Energies and Potentials

For the potential of the Ra^{2+}/Ra couple, the calculated value from the free energy of the ion is [2-4],

$$Ra^{2+} + 2e^- \rightarrow Ra \qquad E° = -2.916 \text{ V}$$

The isothermal temperature coefficient of the standard potential at 25°C, $(dE°/dT)_{isoth}$, is [4]

TABLE 7 Thermodynamic Data on Radium

Formula	Description	State	$\Delta H°$/kJ mol^{-1}	$\Delta G°$/kJ mol^{-1}	$S°$/J mol^{-1} K^{-1}
Ra		g	159.	130.	176.36
	Metal	c	0	0	71.
Ra$^+$		g	674.54		
Ra^{2+}		g	1659.80		
	Std. state, $m = 1$	aq	−527.6	−561.5	54.
RaO		c	−523.		
RaCl$_2$		c			139.
	Std. state, $m = 1$	aq	−861.9	−823.8	167.
RaCl$_2 \cdot$ 2H$_2$O		c	−1464.	−1302.9	213.
Ra(IO$_3$)$_2$		c	−1026.8	−868.6	272.
RaSO$_4$		c	−1471.1	−1365.6	138.
	Std. state, $m = 1$	aq	−1436.8	−1306.2	75.
Ra(NO$_3$)$_2$		c	−992.	−796.2	222.
	Std. state, $m = 1$	aq	−942.2	−784.1	347.

Source: Ref. 5.

$Ra^{2+} + 2e^- \rightarrow Ra$ -0.59 mV K^{-1}

We have as the value of the potential of the following reaction,

$RaO + 2H^+ + 2e^- \rightarrow Ra + H_2O$ $E° = -1.319$ V [2,3]

3. Potential Diagram for Radium

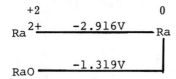

REFERENCES

1. *Gmelin Handbuch der anorganischen Chemie*, 8th ed., Radium (System No. 31), Springer-Verlag, Berlin, 1977.
2. M. Pourbaix, ed., *Atlas of Electrochemical Equilibria*, Pergamon Press, Oxford, 1966, pp. 156-157.
3. G. Milazzo and S. Caroli, *Tables of Standard Electrode Potentials*, Wiley, New York, 1978, pp. 90-93.
4. A. J. de Béthune, T. S. Licht, and N. Swendeman, J. Electrochem. Soc., *106*, 616-622 (1959).
5. Natl. Bur. Stand. Tech. Note 270-6, Nov. 1971, p. 103.

23
Lithium, Sodium, Potassium, Rubidium, Cesium, and Francium*

HIROYASU TACHIKAWA Jackson State University, Jackson, Mississippi

I. CHARACTERISTICS OF THE GROUP

The elements of this group have one s electron outside a noble gas core and the state of the monoatomic gas atoms is $s^1(^2S_{1/2})$. Because of the similarity and simplicity of the chemistry, the various thermodynamic properties will be discussed for the group as a whole, with the exception of francium.

The ionization energy, particularly the first ionization energy, of the elements of this group is very low. The second ionization energies are approximately 10 times higher than the first; thus the chemistry of the alkali metals is almost entirely the chemistry of their +1 cations. Because of the strong electropositive nature of the alkali metals, the elements behave as the strongest reducing agents and react to form compounds with a very high degree of ionic bonding. The cations have low charge and large size, so that the lattice energies of their salts are relatively low. Consequently, most of the salts are dissociated completely in aqueous solution. In fact, salts with a solubility of less than $0.01\ M$ are very rare. The hydroxides are the strongest bases available. The standard thermodynamic states of the elements are the solid metals. At low concentrations of the metal vapors, the gases are monoatomic because of the weak bonding between atoms. At higher concentrations appreciable formation of the diatomic molecules occurs.

The properties of alkali metal compounds vary systematically, particularly the properties of K, Rb, and Cs. Sodium does not deviate from the regular trends nearly as much as Li does. In many of its properties, Li is quite different from the other alkali metals. Some compounds of lithium, such as fluoride, carbonate, and phosphate, are relatively insoluble, and those of the other alkali metals are reasonably soluble. The solubility of these compounds of lithium increases in the presence of acid by forming acid salts (FHF^-, HCO_3^-, $H_2PO_4^-$). However, adding acid to a solution of sodium carbonate precipitates $NaHCO_3$. Lithium is the only alkali metal that oxidizes in air to form Li_2O; all other alkali metals form the peroxides M_2O_2. Because of their smallness and high charge density, the lithium compounds, having significantly higher lattice energies, are much more stable than the corresponding sodium compounds. Lithium hydride is stable at atmospheric pressure to 800 to

---

Reviewer: M. Chemla, Pierre and Marie Curie University, Paris, France.

900°C, whereas sodium hydride decomposes at 350°C. Lithium is the only alkali metal that reacts with N_2 gas. Li_3N is very stable, whereas Na_3N is unstable at room temperature. Lithium carbide is the only alkali metal carbide formed readily by direct reaction. The electrode potential of Li is the most negative among the standard potentials of the alkali metals. The ionization and sublimation energies decrease regularly down through the alkali metal family. However, Li is a highly active reducing agent because of its very high hydration energy. The energy released in the hydration of the small Li^+ ion more than compensates for its higher ionization and sublimation energies. Lithium chlorides have exceptionally high solubilities compared with corresponding salts of other alkali metals because of their high heat of hydration.

II. FRANCIUM

The most stable francium isotope is ^{223}Fr, which is formed from ^{227}Ac by alpha decay. Its half-life is only 21 min, and consequently its chemistry is known only from experiments using less than microgram amounts. The results of such experiments have confirmed that the chemistry of Fr resembles that of Cs. Considering the great similarity between Rb and Cs, all of its thermodynamic values should be very close to the corresponding values for cesium.

III. ALKALI ELEMENT GIBBS ENERGIES AND POTENTIALS

The thermodynamic data for alkali elements are given in Table 1. The standard potentials for the metal-metal ion couples are calculated [1] from the molar Gibbs energy change in Table 1 and are shown in Table 2 together with the measured values for the metal-metal ion couples. Lewis and his co-workers [2-4] were the first to measure the potentials of alkali metals successfully. The reaction of the alkali metals with water makes it difficult to obtain the equilibrium conditions for direct measurements of their electrode potentials. They have measured the potentials of alkali ions against the alkali metal-amalgam electrodes at such low concentrations of the metal that decomposition of water did not occur. They then measured the electromotive force (emf) of the amalgam against the pure metal in nonaqueous solvents.

Their first measurement was made for sodium metal potential in which the following cell system was used:

Na(s) | NaI(in ethylamine) | Na(0.2062% amalgam)

which they combined with a measurement on the aqueous cell

Na(0.2062% amalgam) | NaOH(0.2 M)/NaCl(0.2 M)/KCl(0.2 M)/KCl(1 M), HgCl(s) | Hg(l)

These investigations were carried out with remarkable skill and ingenuity and their $E°$ values for sodium and potassium are probably accurate to 1 mV. However, the experimental difficulties in the case of lithium and rubidium were somewhat greater and the results were not as accurate.

TABLE 1 Thermodynamic Data on the Alkali Metals and Their Compounds[a]

Formula	Description	State	$\Delta H°$/kJ mol^{-1}	$\Delta G°$/kJ mol^{-1}	$S°$/J mol^{-1} K^{-1}
		Lithium			
Li		g	160.71	128.03	138.67
Li		ℓ	2.38	0.933	33.95
Li	Metal	c	0.000	0.000	29.10
Li$^+$		g	687.16	649.99	132.91
Li$^+$		aq	−278.46	−293.80	14.23
Li^{2+}		g	7,983.9		
Li^{3+}		g	19,802		
Li$_2$		g	210.87	169.49	196.85
Li$_2$O		g	−166.94	−187.31	229.00
Li$_2$O		ℓ	−552.83	−521.33	55.06
Li$_2$O		c	−598.73	−562.11	37.89
Li$_2$O$_2$		g	−242.67	−245.72	273.45
Li$_2$O$_2$		c	−632.62	−570.98	56.48
LiO		g	84.10	60.47	210.85

TABLE 1 (Continued)

Formula	Description	State	$\Delta H°$/kJ mol^{-1}	$\Delta G°$/kJ mol^{-1}	$S°$/J mol^{-1} K^{-1}
LiOF		g	(−41.84)	(−45.80)	246.27
LiON		g	179.9	174.6	245.2
LiO$^-$		g	−66.9	−80.83	199.1
LiNaO		g	−104.6	−126.4	256.4
LiH		g	134.31	111.52	170.81
LiH		ℓ	−63.16	−49.39	48.18
LiH		c	−90.65	−68.48	20.03
LiHF$_2$		c	−938.05	−870.69	71.00
LiOH		g	−241.42	−248.33	220.08
LiOH		ℓ	−477.19	−432.87	48.26
LiOH		c	−487.85	−441.91	42.81
LiOH		aq	−508.40	−451.12	3.77
LiOH·H$_2$O		c	−789.81	−683.25	74.06
Li$_2$(OH)$_2$		g	−708.77	−666.07	250.61
LiF		g	−332.63	−353.42	200.17

LiF	ℓ	−588.05	−566.26	57.35
LiF	c	−612.96	−584.70	35.66
LiF	aq	−607.56	−570.28	4.60
Li$_2$F$_2$	g	−927.17	−928.68	265.93
Li$_3$F$_3$	g	−1,514.61	−1,486.58	297.34
LiCl	g	−195.72	−27.26	212.82
LiCl	ℓ	−390.76	−372.23	78.43
LiCl	c	−408.27	−384.03	59.30
LiCl	aq	−445.92	−424.97	69.45
LiCl·H$_2$O	c	−712.58	−632.62	103.76
LiClO	g	−14.23	−18.17	256.33
LiClO$_4$	ℓ	−358.50	−243.28	164.20
LiClO$_4$	c	−380.7	−253.99	125.5
Li$_2$Cl$_2$	g	−598.55	−600.79	288.67
Li$_2$ClF	g	−753.96	−753.04	267.93
Li$_3$Cl$_3$	g	−1,004.7	−979.06	335.65
LiBr	g	−153.97	−189.45	224.23

TABLE 1 (Continued)

Formula	Description	State	$\Delta H°$/kJ mol^{-1}	$\Delta G°$/kJ mol^{-1}	$S°$/J mol^{-1} K^{-1}
LiBr		ℓ	−338.23	−332.08	84.60
LiBr		c	−350.91	−341.62	74.06
LiBr		aq	−399.36	−396.60	94.98
Li$_2$Br$_2$		g	−501.66	−531.79	314.43
LiI		g	−79.91	−123.14	232.13
LiI		ℓ	−260.95	−259.82	83.37
LiI		c	−271.08	−266.92	(75.73)
LiI		aq	−334.39	−345.47	123.43
Li$_2$I$_2$		g	−342.17	−387.15	325.19
Li$_2$S		c	−445.60		
Li$_2$SO$_3$		c	−1,169.0		
Li$_2$SO$_4$		c,II	−1,434.4	−1,324.7	(113)
Li$_2$SO$_4$		aq	−1,464.4	−1,329.6	45.6
Li$_2$Se		c	381.16		
Li$_3$N		c	−198.7	−155.4	37.66

LiN	g	334.72	309.87	208.13
LiN$_3$	c	10.8	77.4	71.8
LiNO$_2$	c	−404.2	−332.6	(89.1)
LiNO$_3$	c	−482.33	−389.5	(105)
LiNO$_3$	aq	−485.03	−404.3	160.7
LiNH$_2$	c	−182.0		
Li$_3$As	c	−340		
Li$_3$Sb	c	−180		
Li$_3$Bi	c	−230		
Li$_2$C$_2$	s	−59.4	−56.13	58.6
Li$_2$CO$_3$	ℓ	−119.79	−1,105.54	127.28
Li$_2$CO$_3$	c	−1,215.6	−1,132.4	90.37
Li$_2$CO$_3$	aq	−1,233.2	−1,115.7	−24.7
Li$_2$SiO$_3$	ℓ	−1,625.2	−539.2	96.06
Li$_2$SiO$_3$	c	−1,649.5	−1,558.8	80.29
Li$_2$Si$_2$O$_5$	ℓ	−2,513.1	−2,379.4	160.2
Li$_2$Si$_2$O$_5$	c	−2,560.9	−2,416.9	125.5

TABLE 1 (Continued)

Formula	Description	State	$\Delta H°$/kJ mol^{-1}	$\Delta G°$/kJ mol^{-1}	$S°$/J mol^{-1} K^{-1}
Li$_2$SiF$_6$		c	−2,882.4		
LiBH$_4$		c	−190.46	−124.78	75.8
LiBO$_2$		g	−671.04	−676.50	258.33
LiBO$_2$		ℓ	−999.22	−947.11	65.23
LiBO$_2$		c	−1,017.92	−961.78	51.67
Li$_2$B$_4$O$_7$		ℓ	−3,317.1	−3,130.5	173.9
Li$_2$B$_4$O$_7$		c	−3,362.3	−3,170.2	155.6
Li$_2$B$_6$O$_{10}$		c	−4,659.7	−4,382.3	188.3
Li$_2$B$_8$O$_{13}$		c	−5,914.5	−5,564.7	265.3
LiAlF$_4$		g	−1,853.5	−1,813.3	328.3
Li$_3$AlF$_6$		ℓ	−3,317.7	−3,175.2	245.9
Li$_3$AlF$_6$		c	−3,383.6	−3,223.8	187.9
LiAlH$_4$		c	−117	−48.4	87.9
LiAlO$_2$		ℓ	−1,143.5	−1,089.2	80.3
LiAlO$_2$		c	−1,189.6	−1,127.3	53.4

LiBeF$_3$	g	−887	−864.54	267.31
LiBeF$_3$	c	−1,632.6	−1,557.2	89.1
Li$_2$BeF$_4$	ℓ	−2,222.5	−2,130.5	164.4
Li$_2$BeF$_4$	c	−2,255.2	−2,151.3	124.7
Li$_2$TiO$_3$	ℓ	−1,566.4	−1,492.3	147.93
Li$_2$TiO$_3$	c	−1,670	−1,579.8	91.76
LiReO$_4$	c	−1,060.2		

Sodium

Metal

Na	g	107.68	77.30	153.61
Na	ℓ	2.41	0.498	57.85
Na	c	0.000	0.000	51.45
Na$^+$	g	609.84	574.88	147.85
Na$^+$	aq	−239.66	−261.87	60.25
Na^{2+}	g	5,180.2		
Na^{3+}	g	12,099.7		
Na$_2$	g	137.53	99.61	230.10
NaO	g	83.68	61.30	229.02

TABLE 1 (Continued)

Formula	Description	State	ΔH°/kJ mol^{-1}	ΔG°/kJ mol^{-1}	S°/J mol^{-1} K^{-1}
NaO$_2$		c	−260.7	−218.7	115.9
Na$_2$O		ℓ	−372.8	−338.9	91.59
Na$_2$O		c	−415.89	−377.08	75.27
Na$_2$O$_2$		c	−513.21	−449.65	94.88
NaO$^-$		g	−121.3	−134.0	217.4
NaH		g	124.27	102.94	188.3
NaH		c	−56.44	−33.57	40.02
NaOH		g	−231.96	−237.58	238.10
NaOH		ℓ	−419.88	−376.48	73.72
NaOH		c,II	−427.77	−381.53	64.18
NaOH		aq	−469.60	−419.17	49.79
NaOH·H$_2$O		c	−732.91	−623.42	84.52
NaF		g	−293.30	−312.59	217.50
NaF		ℓ	−543.44	−520.11	74.55
NaF		c	−573.63	−543.36	51.30

NaF	aq	−568.77	−538.36	50.63
Na$_2$F$_2$	g	−822.16	−820.96	301.61
NaF·HF	c	−906.25	−807.51	90.92
NaCl	g	−181.42	−201.32	229.69
NaCl	ℓ	−385.92	−365.68	95.06
NaCl	c	−411.12	−384.04	72.12
NaCl	aq	−407.11	−393.04	115.48
NaClO$_2$	c	−303.97	−227.82	(111.7)
NaClO$_3$	c	−358.69	(−274.89)	(129.7)
NaClO$_4$	c	−382.75	−254.32	142.3
NaClO$_4$	aq	−371.08	−272.63	242.25
Na$_2$Cl$_2$	g	−566.1	−565.94	325.34
NaBr	g	−143.9	−177.78	241.11
NaBr	ℓ	−339.34	−332.42	104.35
NaBr	c	−361.41	−349.26	86.82
NaBr	aq	−360.58	−364.69	141.0
NaBrO$_3$	c	−342.8	−252.6	130.5

TABLE 1 (Continued)

Formula	Description	State	$\Delta H°$/kJ mol^{-1}	$\Delta G°$/kJ mol^{-1}	$S°$/J mol^{-1} K^{-1}
Na$_2$Br$_2$		g	−486.35	−514.30	390.7
NaI		g	−87.61		248.9
NaI		ℓ	−266.5	−269.9	120.76
NaI		c	−287.9	−284.6	98.50
NaI		aq	−295.6	−296.8	169.5
Na$_2$S		ℓ	−367.7	−357.6	101.0
Na$_2$S		c	−372.4	−361.4	97.9
Na$_2$S$_2$		c	−412.1		
Na$_2$S$_2$		aq	−437.6		
Na$_2$SO$_3$		c	−1,090.4	−1,002.1	146.0
Na$_2$SO$_4$		ℓ	−1,356.6	−1,247.5	179.0
Na$_2$SO$_4$		c,I	−1,381.1	−1,265.7	157.9
Na$_2$SO$_4$		c,III	−1,384.9	−1,268.2	153.3
Na$_2$SO$_4$		c,V	−1,387.2	−1,269.3	149.6
Na$_2$SO$_4$		aq	−1,386.8	−1,265.7	137.7

$Na_2S_2O_3$	c	−1,117.1	
$Na_2S_2O_5$	c	−1,460.6	
$Na_2S_2O_6$	c	−1,673.2	(−1,531.8)
NaHS	c,II	−236.4	(−213.4)
$NaHSO_4$	c	−1,126.3	(−1,021.1)
Na_2SeO_3	c	−963.6	
Na_2SeO_3	aq	−989.9	
Na_2SeO_4	c	−1,093.4	(−970.3) (117)
NaHSe	c,II	−116.3	
Na_2Te	c,II	−351.5	−345 (131)
Na_2TeO_4	c	−1,309	77.4
NaN_3	c	21.3	99.4 70.5
$NaNO_2$	c,II	−359.4	−283.7 (106)
$NaNO_3$	c,II	−424.8	−365.9 116.3
$NaNH_3$	c	−118.8	−59.0 76.9
$NaPO_3$	c	−1,207.5	
Na_3PO_4	c	−1,924	−1,802 (194)

TABLE 1 (Continued)

Formula	Description	State	$\Delta H°$/kJ mol^{-1}	$\Delta G°$/kJ mol^{-1}	$S°$/J mol^{-1} K^{-1}
$Na_4P_2O_7$		c	−3,183.2	(−3,001)	
NaH_2PO_3		c	−1,210.9		
Na_2HPO_3		c	−1,414.2	(−1,315)	
Na_2HPO_4		c	−1,746.4	(−1,624)	
Na_3AsO_4		c	−1,527		
Na_2C_2	Sodium carbide	c	17.2		
Na_2CO_3		ℓ	−1,109.6	−1,031.8	138.7
Na_2CO_3		c	−1,130.9	−1,048.1	138.8
Na_2CO_3		aq		−1,051.9	
$Na_2C_2O_4$	Sodium oxalate	c	−1,315.0	−1,289	(155)
$NaCHO_2$	Sodium formate	c	−648.7		
$NaHCO_3$		c	−947.7	−851.9	102.1
$NaHCO_3$		aq	−930.9	−848.9	
$NaHC_2O_4 \cdot H_2O$		c	−1,381.6		(155)
$NaC_2H_3O_2$	Sodium acetate	c	−710.4		

NaCN		g	94.27	67.32	243.3
NaCN		ℓ	−84.35	−76.19	125.5
NaCN		c	−90.71	−80.42	118.4
Na$_2$CN$_2$		g	−8.79	−21.9	347.0
NaCNO	Sodium cyanate	c	−400.0	−359.4	119.2
NaCO$_2$NH$_2$	Sodium carbamate	c	−750.6		
NaCNS		c	−174.6	(−153)	(112)
Na$_2$SiO$_3$		ℓ	−1,510.9	−1,427.7	150.5
Na$_2$SiO$_3$		c	−1,561.4	−1,467.4	113.8
Na$_2$SiO$_3$		glass	−1,506	−1,426	
Na$_2$SiO$_5$		c			163
Na$_2$Si$_2$O$_5$		ℓ	−2,438.5	−2,300	188.7
Na$_2$Si$_2$O$_5$		c	−2,470.1	−2,324.2	164.1
Na$_4$SiO$_4$		c			195.8
Na$_2$SiF$_6$		c	−2,832	−2,553.9	(215)
Na$_2$SnO$_3$		c	−1,154	(−1,079)	
Na$_2$PbO$_3$		c	−857	(−787)	

TABLE 1 (Continued)

Formula	Description	State	$\Delta H°$/kJ mol^{-1}	$\Delta G°$/kJ mol^{-1}	$S°$/J mol^{-1} K^{-1}
$NaBO_2$		g	−656.9	−659.8	271.5
$NaBO_2$		ℓ	−959.4	−907.1	86.2
$NaBO_2$		c	−979.1	−923.0	73.5
$NaBO_3$		c	−342.7	−252.7	130.5
$Na_2B_4O_7$		ℓ	−3,256.8	−3,064.4	198.7
$Na_2B_4O_7$		c	−3,276.9	−3,081.5	189.5
$Na_2B_4O_7 \cdot 10H_2O$		c	−6,264.3		
$Na_2B_6O_{10}$		c	−4,580.6	−4,302.8	232.2
$NaBH_4$	Sodium borohydride	c,I	−191.8	−127.1	101.4
$NaAlO_2$		c	−1,133.0	−1,069.4	70.3
$NaAlH_4$		c	−113	−48.5	124
$NaAlF_4$		g	−1,868.6	−1,826.7	345.5
Na_3AlF_6		ℓ	−3,238.8	−3,088.2	286.8
Na_3AlF_6		c,II	−3,309.5	−3,144.7	238.4
$NaAlCl_4$		c	−1,142	−1,042	188

Na_2AlCl_6	c	−1,979.0	−1,828.8	347.3
Na_2ZnO_2	c	−787		
Na_2PtCl_6	c	−1,144.7	(−989)	(312)
$Na_2PtBr_6 \cdot 6H_2O$	c	−2,720.0		
Na_2PtI_6	aq	−711.7		
Na_2IrCl_6	c	−976.5		
Na_2OsCl_6	c	−1,237.6		
$NaReO_4$	c	−1,043.5		
Na_3RhCl_6	c	−1,555.2		
Na_2FeO	c	−1,050		
Na_2MnO_4	c	−1,146		
Na_2CrO_4	c	−1,360.2	(−1,232)	(174)
$Na_2Cr_2O_7$	c	−1,962		
Na_2MoO_4	c	−1,540	−1,354	159
$Na_2Mo_2O_7$	c	−2,244.3		
Na_2WO_4	c	−1,544.7	−1,429.8	160.3
$Na_2W_2O_7$	c	−2,485		

TABLE 1 (Continued)

Formula	Description	State	$\Delta H°$/kJ mol^{-1}	$\Delta G°$/kJ mol^{-1}	$S°$/J mol^{-1} K^{-1}
Na$_2$W$_4$O$_{13}$		c	−4,311		
Na$_3$VO$_4$		c	−1,757		
Na$_2$UO$_4$		c	−2,096	(−1,987)	(197)
Potassium					
K		g	89.16	60.92	160.2
K		ℓ	2.28	0.26	71.46
K		c	0.000	0.000	64.68
K$^+$		g	514.2	481.2	154.5
K$^+$		aq	−251.2	−282.3	103
K^{2+}		g	3,590.5		
K^{3+}		g	8,004		
K$_2$		g	127.1	91.21	249.7
KO		g	71.1	50.04	237.9
KO$_2$		c	−283	−249	117
KO$_3$		c	−260	(−190)	

K_2O	c	−363	−322	94.1
K_2O_2	c	−495.8	−429.7	113
K_2O_3	c	−523	−418	(83.3)
K_2O_4	c	−561	−417	(93.7)
KO^-	g	−138	−149	226.2
KH	g	123	103	198
KH	c	−57.7	−34.1	50.2
KOH	g	−228	−233	246
KOH	ℓ	−413	−373	97.9
KOH	c	−425.9	−380	79.5
KOH	aq	−481.2	−439.7	92.0
KF	g	−326	−344	226
KF	ℓ	−547.9	−521.0	75.7
KF	c	−567.4	−537.7	66.6
$KF \cdot 2H_2O$	c	−1,159	−1,016	150
K_2F_2	g	−859.4	−856.5	323
KHF_2	ℓ	−912.9	−854.4	135

TABLE 1 (Continued)

Formula	Description	State	$\Delta H°/\text{kJ mol}^{-1}$	$\Delta G°/\text{kJ mol}^{-1}$	$S°/\text{J mol}^{-1}\text{ K}^{-1}$
KHF_2		c	−928.4	−860.6	105
KCl		g	−214.7	−233.4	239
KCl		ℓ	−421.8	−395.1	86.7
KCl		c	−436.7	−408.8	82.6
KCl		aq	−418.7	−413.5	157.7
K_2Cl_2		g	−617.6	−617.8	352.8
$KClO_3$		c	−391.2	−289.9	143.0
$KClO_3$		aq	−349.5	−284.9	266
$KClO_4$		c	−430.1	−300	151
$KClO_4$		aq	−382.6	−293.0	285
KBr		g	−180.1	−212.8	250.4
KBr		ℓ	−376.5	−366.0	105.5
KBr		c	−393.8	−380.4	95.94
KBr		aq	−372.1	−385.1	183
K_2Br_2		g	−540.6	−566.5	376

KBrO$_3$	c	−332	−244	149.2
KBrO$_3$	aq	−291	−237	265
KI	g	−129.9	−170.4	258.8
KI	ℓ	−309.6	−310.0	124.3
KI	c	−327.6	−324.1	110.8
KI	aq	−307.1	−334.0	212
KI$_3$	c	−320.5	−307.5	(195)
KIO$_3$	c	−508.4	−425.5	151.5
KIO$_3$	aq	−481.2	−418.0	218
KIO$_3$	c	−472.0	(−395.4)	(159)
K$_2$S	c	−418	−404	(111)
K$_2$S$_4$	c	−472.8	−463.6	(161)
K$_2$SO$_4$	c,II	−1,433.7	−1,316.4	176
K$_2$S$_2$O$_5$	c	−1,517		
K$_2$S$_2$O$_6$	c	−1,731		
K$_2$S$_2$O$_8$	c	−1,918	−1,693	
K$_2$S$_4$O$_6$	c	−1,766		

TABLE 1 (Continued)

Formula	Description	State	$\Delta H°$/kJ mol^{-1}	$\Delta G°$/kJ mol^{-1}	$S°$/J mol^{-1} K^{-1}
KHS		c	−264		
KHSO$_4$		c	−1,158	(−1,043)	
KHSe		c	−150		
KTeO$_4$		c	−1,014.8	−917.34	178
KN$_3$		c	−1.38	75.48	85.98
KNO$_2$		c	−370	−282	(117)
KNO$_3$		c	−492.71	−393.1	132.9
KNO$_3$		aq	−457.77	−392.8	291
KH$_2$PO$_4$		c	−1,569	−1,419	(144)
KH$_2$AsO$_4$		c	−1,136	−991.6	155.1
K$_2$CO		aq		−1,093	
K$_2$CO$_3$		ℓ	−1,130.6	−1,049.4	170.4
K$_2$CO$_3$		c	−1,150.2	−1,064.6	155.5
K$_2$C$_2$O$_4$	Potassium oxalate	c	−1,342	−1,241	(169)

KCHO₂	Potassium formate	c	−661.1	−591.2	(105)
KHCO₃		c	−959.4	−860.6	(111)
KCN		g	79.5	53.6	253.0
KCN		ℓ	−104.1	−94.68	134.3
KCN		c	−113.5	−102.0	127.8
K₂C₂N₂		g	−8.4	−21	373.0
KCNO		c	−412	−374	
KCNS		c	−203.4		
K₂SiF₆		c	−2,807	−2,724	(229)
K₂SiO₃		c	−262		138
K₂SnCl₆		c	−1,518	−1,369	(314)
KBH₄		c	−226.9	−159.8	106.6
KBF₄		g	−1,552	−1,504.4	315.3
KBF₄		ℓ	−1,869	−1,772.3	150.1
KBF₄		c	−1,887	−1,785	134
K₂B₄O₇		ℓ	−3,289	−3,100.3	237
K₂B₄O₇		c	−3,334	−3,136	208

TABLE 1 (Continued)

Formula	Description	State	$\Delta H°$/kJ mol^{-1}	$\Delta G°$/kJ mol^{-1}	$S°$/J mol^{-1} K^{-1}
$K_2B_6O_{10}$		c	-4,633.4	-4,353.66	251
$K_2B_8O_{13}$		ℓ	-5,872.54	-5,511.6	298
$K_2B_8O_{13}$		c	-5,945.13	-5,582.79	293
$KAlH_4$		c	-167	-99.6	(129)
K_3AlF_6		c	-3,326		285
$K_3Al_2F_6$		c	(-3,320)	(-3,164)	285
$KAlCl_4$		c	-1,197	-1,095	197
K_3AlCl_6		c	-2,092	-1,938	380
$K_3Al_2Cl_9$		c	-2,860	-2,626	468.6
$KAl(SO_4)_2$		c	-2,465	-2,236	205
$KAl(SO_4)_2 \cdot 12H_2O$		c	-6,057.34	-5,137.1	687.4
$KAg(CN)_2$		c	-16	36	(142)
K_2PtCl_4		c	-1,064	-958.1	(283)
K_2PtCl_6		c	-1,259	-1,109	334
K_2PtBr_4		c	-923.4	-877.4	(341)

K₂PtBr₆	c	−1,040	−930.5	(323)
K₂IrCl₆	c	−1,196	−1,021	(346)
K₃IrCl₆	c	−1,565	−1,382	(382)
K₂OsCl₆	c	−1,172	−1,015	(345)
KReO₄	c	−1,105	−997.9	168
K₂PdCl₄	c	−1,095	−996.2	(281)
K₂PdCl₆	c	−1,187	−1,037	(325)
K₂PdBr₄	c	−908.8	−885.3	(331)
K₃RhCl₆	c	−1,435	−1,287	(394)
K₃Fe(CN)₆	c	−173	−51.9	420.0
K₄Fe(CN)₆	c	−523.4	−351	(361)
K₃FeCO(CN)₅	c	−574.9		
KMnO₄	c	−813.4	−713.8	174.2
K₂CrO₄	g	−618.8	−200	353
K₂CrO₄	c	−1,414	−1,299	(187)
K₂Cr₂O₇	c	−2,030	−1,864	291
KCr(SO₄)₂	c	−2,351	−2,134	(240)

TABLE 1 (Continued)

Formula	Description	State	$\Delta H°$/kJ mol^{-1}	$\Delta G°$/kJ mol^{-1}	$S°$/J mol^{-1} K^{-1}
KCr(SO$_4$)$_2$·12H$_2$O		c	−5,786.9	−4,870	(707.9)
KVO$_4$		c	−1,146	−1,026	
	Rubidium				
Rb		g	85.81	55.86	169.99
Rb	Metal	c,l	0.000	0.000	69.5
Rb$^+$		g	494.955		164
Rb$^+$		aq	−246	−282.2	124
Rb^{2+}		g	3,154.5		
Rb$_2$		g	123	78.7	272
RbO$_2$		c	−285	(−220)	
Rb$_2$O		c	−330	−291	(110)
Rb$_2$O$_2$		c	−425.5	−350	(104)
Rb$_2$O$_3$		c	−488.3	−387	(106)
Rb$_2$O$_4$		c	−528.0	−396	(116)
RbH		g	138	(115)	208

RbOH	c,II	−413.8	−364	(70.7)
RbOH	aq	−476.6	−439.53	114
RbF	g			239
RbF	c	−549.28	−520.1	(72.8)
RbF	aq	−575.7	−558.69	115
RbHF$_2$	c	−909.2		120
RbCl	g	−234		249
RbCl	c	−430.58	−405	(94.6)
RbCl	aq	−414	−413	179
RbClO$_3$	c	−392	−292	152
RbClO$_3$	aq	−345	−285	287
RbClO$_4$	c,II	−434.59	−306.2	161
RbClO$_4$	aq	−377.8	−293.0	306
RbBr	g	−184		260
RbBr	c	−389.2	−378.2	108.3
RbBr	aq	−367	−385.0	205
RbI	g	−134		267

TABLE 1 (Continued)

Formula	Description	State	$\Delta H°$/kJ mol^{-1}	$\Delta G°$/kJ mol^{-1}	$S°$/J mol^{-1} K^{-1}
RbI		c	−328	−326	118.0
RbI		aq	−303	−333.9	233
Rb$_2$S		c	−348	−337	(134)
Rb$_2$SO$_4$		c	−1,424.7	−1,309	(192)
Rb$_2$SO$_4$		aq	−1,400	−1,306.4	141
RbHS		c	−261		
RbNO$_3$		c,II	−489.70	−390	(141)
RbNO$_3$		aq	−453.1	−392.7	271
RbNH$_2$		c	−108		
Rb$_2$CO$_3$		c	−1,128	−1,043	(97.5)
Rb$_2$CO$_3$		aq		−1,093	
RbHCO$_3$		c	−956.0	−855.2	(123)
RbReO$_3$		c	−1,075		
Rb$_2$UO$_4$		c	−2,038		

Cesium

Metal

Cs	g	78.78	51.21	175.49
Cs	ℓ	2.09	0.025	92.0
Cs	c	0.000	0.000	85.14
Cs$^+$	g	452.3	420.87	169.7
Cs$^+$	aq	−248	−282	133
Cs^{2+}	g	2,730		
Cs$_2$	g	106	72.4	283.9
Cs$_2$O	g	−92.0	−105.5	317.9
Cs$_2$O	c	−318	−274	(124)
CsO	g	62.8	42.55	255.4
Cs$_2$O$_2$	c	−402.5	−327	(118)
Cs$_2$O$_3$	c	−465.3	−360	(120)
CsO$_2$	c	−290	(−211)	
Cs$_2$O$_4$	c	−519.7	−387	(131)
CsH	g	121	102	214.4
CsOH	c,II	−407	−355	(77.8)

TABLE 1 (Continued)

Formula	Description	State	$\Delta H°$/kJ mol^{-1}	$\Delta G°$/kJ mol^{-1}	$S°$/J mol^{-1} K^{-1}
CsOH		aq	−477.8	−439.3	123
CsF		g	−356.5	−373.3	243.1
CsF		ℓ	−543.84	−515.09	90.08
CsF		c	−554.67	−525.38	88.28
CsF		aq	−568.6	−558.6	123
Cs$_2$F$_2$		g	−890.10	−883.91	352.2
CsHF$_2$		c	−904.2		135
CsCl		g	−240.2	−257.9	255.96
CsCl		ℓ	−434.47	−406.18	101.7
CsCl		c,II	−433.0	−404	(97.5)
CsCl		aq	−415	−371.4	188
Cs$_2$Cl$_2$		g	−659.8	−656.85	383
CsClO$_3$		c	−386		(161)
CsClO$_4$		c	−434.7	−306.6	175.3
CsClO$_4$		aq	−379	−292.8	315

CsBr	g	(−201)		266
CsBr	c	−395	−383	121
CsBr	aq	−369	−384.8	214
CsBrO$_3$	c	−374	−285	162
CsI	g	−138		273
CsI	c	−337	−333	130
CsI	aq	−304	−333.8	242
CsIO$_3$	c	−526.3		
Cs$_2$S	c	−339	−325	(148)
Cs$_2$SO$_3$	c	−1,103		
Cs$_2$SO$_4$	c,II	−1,420.0	−1,300	(206)
Cs$_2$SO$_4$	aq	−1,403	−1,306.1	283
Cs$_2$S$_2$O$_5$	c	−1,521		
CsHS	c	−263		
CsNO$_3$	c,II	−494.2	−393	(148)
CsNO$_3$	aq	−454.4	−392.5	363
Cs$_2$CO$_3$	c	−1,119	−1,019	(177)

TABLE 1 (Continued)

Formula	Description	State	$\Delta H°$/kJ mol^{-1}	$\Delta G°$/kJ mol^{-1}	$S°$/J mol^{-1} K^{-1}
Cs_2CO_3		aq		$-1{,}092$	
$CsHCO_3$		c	-955.6	-831.8	(130)
Cs_2SiF_6		c	$-2{,}801$	$-2{,}634$	(266)
Cs_2SiO_3		c	$-1{,}530$		
$CsBO_3$		c	-374	-285	162
$CsAlH_4$		c	-165	-98.3	(151)
$CsAl(SO_4)_2 \cdot 12H_2O$		c	$-6{,}064.7$	$-5{,}098.2$	686

Francium

| Fr | | g | 72.80 | 46.65 | 181.9 |

Fr	c		94.14
Fr$_2$O	c	(−339)	(157)
FrF	g	346	251
FrF	c	(−523)	98.7
FrCl	g	233	263
FrCl	c	(−435)	(113)
FrBr	g	186	274
FrBr	c	(−397)	(125)
FrI	g	142	282
FrI	c	(−347)	(137)

[a]National Bureau of Standards values are in italic; values from other sources are in roman; estimated values are in parentheses.

TABLE 2 Standard Potentials of the Alkali Metals Versus Standard Hydrogen Electrode in Aqueous System

Metal	$E°/V$ (calculated value)	$(dE°/dT)/mV\ K^{-1}$	$E°/V$ (measured value)
$Li^+ + e^- \rightarrow Li$	−3.045	−0.534	−3.040
$Na^+ + e^- \rightarrow Na$	−2.714	−0.772	−2.713
$K^+ + e^- \rightarrow K$	−2.925	−1.080	−2.924
$Rb^+ + e^- \rightarrow Rb$	−2.925	−1.245	−2.924
$Cs^+ + e^- \rightarrow Cs$	−2.923	−1.197	−2.923

The standard potentials measured by Lewis et al. [2-4,7,9] are Li^+/Li, -3.027 V; Na^+/Na, -2.715 V; K^+/K, -2.925 V; Rb^+/Rb, -2.924 V. The measured value for lithium by Lewis et al. [3,4] showed a large discrepancy from the calculated value compared with the differences for the other alkali metals. The value reported by Huston and Butler [5] is -3.040 V and is closer to the calculated value. They made measurement by eliminating the liquid junction potentials, which may have caused a large discrepancy between the calculated value and the value measured by Lewis and co-workers. Smith and Taylor [6] later remeasured the standard potential of sodium in which dimethylamine, which reacts with sodium much less readily than does ethylamine, was used as the solvent in the nonaqueous cell. They reported the potential of -2.7132 V vs. SHE.

For the standard potential of potassium metal Armbruster and Crenshaw [8] reported the potential of -2.9243 V after remeasuring the potential of the corresponding aqueous system without a liquid junction. Popovych and co-workers [12] measured the potential of potassium electrode to be -2.9230 V by using dropping amalgam electrodes containing approximately 0.02 wt % potassium. The standard potential of cesium has been measured to be -2.923 V by Bent et al. [10] after they found a solvent, dimethylamine, which was suitable for the determination of its potential.

The temperature coefficient of standard potentials for alkali metals have been calculated by de Béthune et al. [11] and shown in the third column of Table 2. A value of $dE°/dT$ for sodium metal reported by Smith and Taylor is -0.7532 mV K^{-1} [6], slightly different from the calculated value given in the table.

One frequently wishes to know a Gibbs energy or enthalpy of solution of a group I salt. For that reason the data for the aqueous solution of many salts have been included [1] in Table 3. This is done for convenience, since the values can always be calculated as the sum of the properties of the individual ions. To illustrate trends within the group, a summary of the Gibbs energies of solution of the more important compounds is given in Table 3.

There are not many slightly soluble compounds in this group. The solubility products for a number of the less soluble compounds have been calculated from the Gibbs energy data:

TABLE 3 Summary of Gibbs Energies of Solution of Certain Alkali Compounds (ΔG_{sol} kJ mol^{-1} at 25°C)

	OH^-	Cl^-	SO_4^{2-}	CO_3^{2-}
Li^+	-9.21	-40.94	-4.9	16.49
Na^+	-37.64	-9.00	-0.08	-3.8
K^+	-59.7	-4.65	9.87	-23.85
Rb^+	-75.53	-8.38	2.35	-49.79
Cs^+	-84.3	32.6	-6.11	-72.8

$$Li_2CO_3 \rightarrow 2Li^+ + CO_3^{2-} \quad K_{sp} = 3.1 \times 10^{-1}$$

$$NaHCO_3 \rightarrow Na^+ + HCO_3^- \quad K_{sp} = 1.2 \times 10^{-3}$$

$$K_2PtCl_6 \rightarrow 2K^+ + PtCl_6^{2-} \quad K_{sp} = 1.4 \times 10^{-6}$$

$$KClO_4 \rightarrow K^+ + ClO_4^- \quad K_{sp} = 8.9 \times 10^{-3}$$

$$RbClO_4 \rightarrow Rb^+ + ClO_4^- \quad K_{sp} = 3.8 \times 10^{-3}$$

$$CsClO_4 \rightarrow Cs^+ + ClO_4^- \quad K_{sp} = 3.2 \times 10^{-3}$$

IV. POTENTIAL DIAGRAMS FOR THE GROUP

+1   0

$Li^+ \xrightarrow{-3.040} Li$

$Na^+ \xrightarrow{-2.713} Na$

$K^+ \xrightarrow{-2.924} K$

$Rb^+ \xrightarrow{-2.924} Rb$

$Cs^+ \xrightarrow{-2.923} Cs$

REFERENCES

1. W. M. Latimer, *Oxidation Potentials*, 2nd ed., Prentice-Hall, Englewood Cliffs, N.J., 1952.
2. G. N. Lewis and C. A. Kraus, J. Am. Chem. Soc., *32*, 1459 (1910).
3. G. N. Lewis and F. G. Keyes, J. Am. Chem. Soc., *35*, 340 (1913).
4. G. N. Lewis and W. L. Argo, J. Am. Chem. Soc., *37*, 1983 (1915).
5. R. Huston and J. N. Butler, J. Phys. Chem., *72*, 4263 (1968).
6. E. R. Smith and J. K. Taylor, J. Res. Natl. Bur. Stand., *25*, 731 (1940).
7. G. N. Lewis and F. G. Keyes, J. Am. Chem. Soc., *34*, 119 (1912).
8. M. H. Armbruster and J. L. Crenshaw, J. Am. Chem. Soc., *56*, 2525 (1934).
9. G. N. Lewis and M. Randall, *Thermodynamics and the Free Energy of Chemical Substances*, McGraw-Hill, New York, 1923.
10. H. E. Bent, G. S. Forbes, and A. F. Forziatti, J. Am. Chem. Soc., *61*, 709 (1939).
11. A. J. dé Bethune, T. S. Licht, and N. Swendeman, J. Electrochem. Soc., *106*, 616 (1959).
12. A. J. Dill, L. M. Itzkowitz, and O. Popovych, J. Phys. Chem., *72*, 4580 (1968).

24
Inert Gases*

BRUNO JASELSKIS Loyola University of Chicago, Chicago, Illinois

The elements of this family in the ground state have completely filled s and p orbitals, corresponding to the spectroscopic notation of $^1S_0$. All these elements are quite inert and only a limited number of stable compounds can be found with the heavier elements, such as krypton, xenon, and radon. The ease of combination and formation of new compounds increases with decreasing ionization potential, as shown in Table 1. In general, these compounds can be formed only with the most electronegative elements, such as fluorine and to some extent with chlorine, and indirectly with oxygen. Krypton and radon form compounds only of +2 oxidation state, while xenon with fluorine forms compounds with oxidation states +2, +4, and +6. Some of these xenon fluorides can be transformed to relatively stable oxyfluorides and oxides, such as XeO_3 and XeO_4.

Helium and argon do not form stable compounds; however, metastable transitory compounds have been reported to be formed in solid low-temperature matrices and have been observed spectroscopically.

Inasmuch as the reactions of noble gas compounds are highly irreversible, the oxidizing power can only be measured thermodynamically by using the following methods:

1. Studying noble gas-fluorine equilibria as a function of temperature, and relating the change of enthalpy or heat of reaction to the equilibrium constant by the van't Hoff equation

$$d(\ln K)/dT = \Delta H°_{react}/RT^2$$

   as described by Weinstock et al. [1]

2. Measuring the direct calorimetric reaction of the noble gas compound with hydrogen as illustrated by Stein and Plurien [2]:

$$XeF_4(g) + 2H_2(g) \rightarrow Xe(g) + 4HF(g)$$

$$\Delta H_{react} = 4\Delta H°_f(HF) - \Delta H_f(XeF_4)$$

Reviewer: Gordon Ewing, New Mexico State University, Las Cruces, New Mexico.

TABLE 1 Ionization Potentials of Noble Gases

Element	Configuration of outer electrons	Atomic number	First I.P./eV	ΔH_{vap}/kJ mol^{-1}	Promotion energy/eV $ns^2np^6 \rightarrow ns^2np^5(n+1)s$
Helium	$1s^2$	2	24.58	0.092	—
Neon	$2s^2sp^6$	10	21.56	1.84	16.6
Argon	$3s^23p^6$	18	15.76	6.28	11.5
Krypton	$4s^24p^6$	36	14.00	9.66	9.9
Xenon	$5s^25p^6$	54	12.13	13.68	8.3
Radon	$6s^26p^6$	86	10.75	17.99	6.8

3. Measuring enthalpies of the reaction between noble gas compound and hydrogen iodide as described by Gunn and Williamson [3], O'Hare et al. [4], and Johnson [4a]
4. Measuring calorimetric heats of combustion or decomposition as determined by Pepekin et al. [5]
5. Establishing a very approximate value of Gibbs energy by using various strength oxidizing agents as applied in the study of radon difluoride by Stein [6]
6. Measuring mass spectroscopic appearance potential as described by Svec and Flesch [7]
7. Utilizing photoionization mass spectroscopic measurements which provide quite accurate values comparable to those observed by the calorimetric measurements [8]

The Gibbs energy is calculated using experimentally determined enthalpies. Enthalpy and entropy values are tabulated in the National Bureau of Standards (NBS) Circular 500 and Technical Note 270-3. The electrochemical potentials then can be calculated from the Gibbs energy.

Yet the exact determination of bond strengths is in question as compared to the values determined from heats of formation of gaseous xenon disfluoride, xenon tetrafluoride, and xenon hexafluoride by calorimetry and photoionization mass spectroscopy [9]. The calorimetric values are always higher than corresponding photoionization results and are outside the quoted experimental errors. The differences in the enthalpies of formation determined by the two methods for XeF_2, XeF_4, and XeF_6 are 13 ± 3, 42.0 ± 8, and 109 ± 33 kJ mol^{-1}, respectively [9]. Consequently, at best, some of the values presented contain some of these uncertainties in the final analysis.

I. KRYPTON

A. Oxidation States

Krypton forms only relatively stable krypton difluoride, which is apparently thermodynamically unstable [10,11]. It decomposes slowly at room temperature and rather fast at higher temperatures. Higher-oxidation-state krypton fluorides have not been confirmed. Krypton difluoride in nonaqueous systems forms adducts with antimony, tantalum, arsenic, platinum, and gold fluorides of the type: $KrF \cdot MF_5$, $2KrF_2 \cdot MF_5$ (where M is Ta, Nb, Sb, Tl), $KrF \cdot PtF_6$, $KrF \cdot AsF_6$, $KrF \cdot AuF_6$, $Kr_2F_3 \cdot SbF_6$, and $KrF_2 \cdot BrF_5$ as reported by various groups [12-15].

B. Thermodynamic Studies

Thermodynamic studies involving krypton difluoride have been reported by Weinstock et al. [1] and by Gunn [16,17]. In addition, molecular energies of krypton difluoride have been estimated from the appearance potentials as determined by mass spectrometric methods by Sessa and McGee [18]. The enthalpy of sublimation has been determined from the vapor pressure vs. temperature variation from the slope of logarithm of the vapor pressure vs. $1/T$ as reported by Gunn [16,17]. Krypton difluoride is a volatile solid and has vapor pressures of 1.5, 10, 29, and 73 Torr at -40, -15.5, 0, and $+15°C$, respectively. The following thermodynamic values have been reported:

$$\Delta H^{\circ}_{\text{sublim}}[\text{KrF}_2(c)] = 41.4 \text{ kJ mol}^{-1} \text{ [17]}$$

$$\Delta H^{\circ}_{f}[\text{Kr}(g) + \text{F}_2(g)] = 15.5 \text{ kJ mol}^{-1}$$

$$\Delta H_f[\text{KrF}_2(g)] = 60.2 \text{ kJ mol}^{-1} \text{ [16]}$$

The appearance potentials as reported by Sessa and McGee [18], using 1(Ar) = 15.76 eV as a check are

$$A(\text{Kr}+,\text{KrF}_2)^{84}\text{Kr} = 13.21 \pm 0.25 \text{ eV}$$

$$A(\text{KrF}+,\text{KrF}_2)\text{KrF}^+ = 13.71 \pm 0.20 \text{ eV}$$

The estimated bond dissociation energy for KrF, $D_{av}(\text{KrF})$, is 48.9 kJ mol^{-1}, compared to the calculated value from the calorimetrically determined ΔH_f being 79.1 kJ mol^{-1} or 0.82 eV [16]. The reduction of krypton difluoride by the reaction

$$\text{KrF}_2 \rightarrow \text{Kr}^{2+} + \text{F}_2 + 2e^-$$

yields a predicted value of 13.4 eV.

Krypton difluoride in water decomposes to yield krypton, oxygen, and hydrofluoric acid. Thus the aqueous solution chemistry involving krypton difluoride is rather limited. However, krypton difluoride is soluble in anhydrous HF and can be used as a powerful fluorinating agent [19].

II. XENON

A. Oxidation States and Chemistry

Bartlett reported the first chemical compound of xenon, XePtF$_6$, in 1962 [20]. He had previously oxidized molecular oxygen using platinum hexafluoride and had reasoned that xenon might be similarly oxidized because xenon and molecular oxygen have essentially identical first ionization potentials (12.13 and 12.2 eV, respectively). Immediately following Bartlett's paper numerous xenon compounds were prepared and characterized, notably by the Argonne National Laboratory group, as well as others [1,21-24].

The principal oxidation states of xenon in various compounds are +2, +4, +6, and +8. Xenon yields relatively stable fluorine-containing compounds such as XeF$_2$, XeF$_4$, and XeF$_6$ and oxyfluorides. In general, the direct preparation of xenon fluorides is readily possible at elevated temperatures and high pressures by adding fluorine in excess of xenon; however, the preparation of oxygen-containing xenon compounds cannot be achieved by direct reaction of oxygen with xenon. Oxygen-containing xenon compounds can be prepared indirectly by the hydrolysis of xenon tetrafluoride or xenon hexafluoride or by the reaction of these fluorides with silica. In general, solid xenon oxides such as XeO$_3$ and XeO$_4$ are shock and heat sensitive and must be handled with utmost care. However, the salts of Xe(VIII) are quite stable, while most of the Xe(VI) salts are unstable with the exception of MXeO$_3$F (where M is Cs or Rb).

Inert Gases

The chemistry of xenon in various oxidation states is considerably more complex than that for krypton or radon. Xenon fluorides form, under anhydrous conditions, various adduct salts with metal fluorides, such as AsF_5, PtF_6, TeF_6, SbF_5, TaF_5, and others, but also undergo other chemical reactions. In aqueous solutions, xenon fluorides undergo hydrolysis: xenon tetra- and hexafluorides yield aqueous xenon trioxide, while xenon difluoride yields oxygen and hydrofluoric acid.

$$XeF_2 + H_2O \rightarrow Xe(g) + (1/2)O_2(g) + 2HF(aq)$$

$$3XeF_4 + 6H_2O \rightarrow XeO_3(aq) + 2Xe(g) + 12HF(aq) + (3/2)O_2(g)$$

$$XeF_6 + 3H_2O \rightarrow XeO_3(aq) + 6HF(aq)$$

Thus reduction potentials for Xe(VI), Xe(VIII), and xenon difluoride in aqueous solutions have been calculated indirectly from thermodynamic measurements.

III. XENON FLUORIDES AND THEIR CHEMICAL PROPERTIES

Xenon fluorides are readily prepared by the reaction of xenon with fluorine in nickel or a Monel container on heating and elevated pressure. Equilibrium and optimum conditions for these reactions have been reported by Weinstock et al. [1]. Xenon fluorides are thermodynamically quite stable and can be considered as belonging to a class of "electron-rich" compounds comparable to ICl_2^-, ICl_4^-, and IF_6^-.

A. Xenon Difluoride

Xenon difluoride is prepared from xenon and fluorine mixtures using rather mild conditions, by heating a mixture containing an excess of xenon in a nickel container or by reacting xenon with fluorine in a silica or pyrex container placed in sunlight [24-27]. Xenon difluoride in anhydrous hydrofluoric acid dissolves to an extent of 1672 g per 1000 g of HF at 30°C. Solutions of XeF_2 in anhydrous HF are clear and show no electrical conductance, indicating the presence of molecular XeF_2 rather than ionic species [36]. These solutions are indefinitely stable. Also, XeF_2 dissolves in acetonitrile and NOF·3HF without apparent decomposition [28,29].

Xenon difluoride is not only a good fluorinating agent in vapor as well as in liquid phases, but it is also a powerful oxidizer. Xenon fluoride with xenon tetrafluoride forms a well-defined crystalline compound, $XeF_2 \cdot XeF_4$ [30], which originally was thought to be XeF_3 [31]. Furthermore, XeF_2 with $HOTeF_5$, HSO_3F, and $HClO_4$ forms the adducts $F-Xe-OTeF_5$, $Xe(OTeF_5)_2$ [32], $F-Xe-SO_3F$, $Xe(SO_3F)_2$, $F-Xe-ClO_4$, $Xe(ClO_4)$, and others [33]. Also, adducts of XeF_2 with MF_5 (where M is As, Os, Ir, or Pt) have been reported [34]. Redox reactions of XeF_2 with PtF_x and PdF_x have been reported by Bartlett et al. [35].

B. Xenon Tetrafluoride

Xenon tetrafluoride is commonly prepared by mixing xenon and fluorine in a 1:3 ratio in a nickel container and heating at 400°C [20]. At higher fluorine ratios production of xenon hexafluoride is increased. Xenon tetrafluoride is colorless and is somewhat soluble in anhydrous HF [36] and NOF·3HF [28]. Xenon tetrafluoride is stable and can be stored in a metal container. It reacts rapidly with hydrogen and at 400°C with xenon gas to yield xenon difluoride [22]. It also reacts with fluorine and O_2F_2 between 140 and 195 K to yield xenon hexafluoride. In anhydrous HF xenon tetrafluoride reacts with platinum to yield platinum tetrafluoride and xenon gas. Xenon tetrafluoride fluorinates organic molecules such as ethylene and reacts with tetrahydrofuran, dioxane, and ethanol. However, xenon tetrafluoride dissolves in trifluoroacetic anhydride without reaction. An unstable $Xe(CF_3COO)_4$ solid has been observed in the reaction of trifluoroacetic acid with xenon tetrafluoride [37].

Reactions of xenon tetrafluoride have been studied in various solvents, such as AsF_3, IF_5, NH_3, SO_3, CCl_3F, $CCl_2FCCl_2F_2$, $(CH_3)_2SO$, C_5H_5N, and acetonitrile [38]. Adducts of xenon tetrafluoride with PF_5, AsF_5, and SbF_5 in BrF_3 have been reported and isolated [25].

C. Xenon Hexafluoride

Xenon hexafluoride is easily prepared by heating xenon and fluorine in a 1:20 ratio to a total pressure of 50 atm (5.07 MPa) and temperature of approximately 250°C, yielding better than 95% XeF_6. Separation of xenon hexafluoride from XeF_4 can be achieved by the fact that XeF_6 forms a complex with sodium fluoride, whereas XeF_4 does not [39].

Xenon hexafluoride is very soluble in anhydrous HF [36]. Xenon hexafluoride is thermodynamically stable and can be stored in nickel or Monel containers. It attacks quartz, producing xenon oxyfluorides such as $XeOF_4$, XeO_2F_2, and highly explosive XeO_3. Xenon hexafluoride reacts at room temperature with hydrogen, yielding xenon gas and hydrogen fluoride. It reacts with mercury to liberate xenon gas and mercuric fluoride. Xenon hexafluoride with cyclopentane yields perfluorocyclopentane. It reacts with HCl to yield chlorine, hydrogen fluoride, and xenon gas; with NH_3 it yields ammonium fluoride, nitrogen, and xenon gases. In anhydrous HF it yields ions such as XeF_5^+ and HF_2^-. Xenon hexafluoride forms a number of complex adducts with Lewis acids: $2XeF_6 \cdot SbF_5$, $XeF_6 \cdot SbF_5$, $XeF_6 \cdot 2SbF_5$, $2XeF_6 \cdot VF_5$, and others [40-42], and with NOF at room temperature XeF_6 yields a solid $XeF_6 \cdot 2NOF$ [43]. Xenon hexafluoride also forms stable adducts with alkali fluorides of the type $2MF \cdot XeF_6$ (where M is Cs, Rb, K, or Na) [44].

Xenon hexafluoride with silica yields xenon oxide tetrafluoride, $XeOF_4$ [45]. This xenon oxide tetrafluoride reacts with hydrogen at 300°C to yield xenon gas, HF, and water. It also forms adducts with other fluorides. In general, this oxide is less reactive than XeF_6. Adducts of SbF_5, RbF, and CsF have been reported [46]. The xenon dioxide difluoride is prepared by the reaction of xenon trioxide with xenon oxide tetrafluoride [47]. The chemistry of this compound is rather limited. Xenon trioxide difluoride, XeO_3F_2, is generated by the reaction of XeF_6 with solid sodium perxenate [48,49].

Thermodynamic and physical properties of xenon fluorides have been determined by various methods: (1) calorimetric methods based on the reaction of xenon fluoride with hydrogen or with phosphorus trifluoride, (2) mass spectroscopic methods, (3) measurement of equilibrium constants at different temperatures, (4) reaction with oxygen and polyethylene, and (5) photoionization measurements. All of these measurements yield useful information; however, in some cases the agreement is not the best. Some of the physical and thermodynamic values are summarized in Tables 2, 3, and 4. Tabulated values of thermodynamic properties are based primarily on the more recent publications. A comparison of these values as obtained by various methods yields a

TABLE 2 Physical and Thermodynamic Properties of Xenon Fluorides

Property			
Melting point, T/°C	129	117.1	49.5
Vapor pressure[a] at 25°C, p/Torr	4.6	2.5	28.9
ΔH of sublimation/ kJ mol^{-1}	55.2 ± 0.8 [47] 52.3 ± 0.8 [50] 55.7 ± 0.1 [55]	60.6 ± 0.2 [47] 64.0 ± 0.8 [50] 60.9 [54]	59.1 [51] 62.3 [58]
ΔH of combustion/ kJ mol^{-1}	176.2 [5]		
Density of solid at 25°C, ρ/g cm^{-3}	4.32 [52]	4.04	3.56 [53]
Bond energy/ kJ mol^{-1}	129.7 [1] 125.3 ± 0.8 [56]	129.3 [1] 130.3 ± 0.6 [56]	124.2 [1] 132.3 ± 0.8 [56]

[a]Vapor pressure of the solid is calculated using the following expressions:

$$\log(p/\text{Torr})\text{XeF}_2(c) = -3057.67/T - 1.23521 \log T + 13.969736$$

(T range 0 to 115°C) [47]

$$\log(p/\text{Torr})\text{XeF}_4(c) = 3226.21/T - 0.43434 \log T + 12.301738$$

(T range 2 to 117°C) [47]

$$\log(p/\text{Torr})\text{XeF}_6(c) = 3093.9/T + 11.8379$$

(T range 18 to 50°C) [51]

substantial difference for the heats of formation of xenon fluorides, as has been reported: ΔH_f — 104.6 to −188 kJ mol^{-1} for XeF$_2$, −201 to −241 kJ mol^{-1} for XeF$_4$, and −294 to −347 kJ mol^{-1} for XeF$_6$.

IV. AQUEOUS CHEMISTRY OF XENON COMPOUNDS AND THEIR OXIDATION POTENTIALS

A. Aqueous Divalent Xenon

Xenon difluoride dissolves in water to the extent of 2.5 g per 100 mL at 0°C and at this temperature decomposes with a half-life of 7 h to yield xenon gas, oxygen, and hydrogen fluoride [59].

$$\text{XeF}_2(c) + \text{H}_2\text{O}(\ell) \rightarrow \text{Xe}(g) + 2\text{HF}(\ell) + (1/2)\text{O}_2(g)$$

TABLE 3 Thermodynamic Properties of Xenon Fluorides at 298.15 K[a]

	Compound					
	$XeF_2(g)$	$XeF_2(c)$	$XeF_4(g)$	$XeF_4(c)$	$XeF_6(g)$	$XeF_6(c)$
$S°/J\,mol^{-1}\,K^{-1}$	259.4 ± 0.1 [57]	115.1 ± 0.1 [55]	323.9 ± 0.4 [54]	166.9 ± 0.4 [54]	381.5 ± 4 [51]	210.4 ± 0.4 [51]
$\Delta S°_f/J\,mol^{-1}\,K^{-1}$	-112.9 ± 0.1 [55]	-257.2 ± 0.1 [55]	-250.9 ± 1. [54]	-407.9 ± 1 [54]	-396.2 ± 4 [51]	-567.2 ± 4.5 [51]
$\Delta H°_f/kJ\,mol^{-1}$	-107.0 ± 0.9 [56]	-162.8 ± 0.9 [55]	-206.2 ± 0.9 [54,56]	-267.1 ± 0.9 [54,56]	-279.0 ± 2.5 [56]	-338.2 ± 2.2 [56]
$\Delta G°_f/kJ\,mol^{-1}$	-73.4 ± 0.9 [55]	-86.1 ± 0.9 [55]	-131.4 ± 0.9 [54]	-145.5 ± 0.9 [54]	-152.0 ± 3. [51]	-169.0 ± 2.2 [51]

[a]Auxiliary data: $S°_{Xe,g}/J\,mol^{-1}\,K^{-1}$ at 298.15 K is 169.573 and $S°_{F_2,g}/J\,mol^{-1}\,K^{-1}$ at 298.15 K is 202.689.
Source: Data taken in part from Refs. 47, 51, and 54-56.

TABLE 4 Ideal Gas Thermodynamic Properties of Xenon Fluorides at 298.15 K

	C_p°/J mol^{-1} K^{-1}	S°/J mol^{-1} K^{-1}	$[(H^\circ - H_0^\circ)/T]$/ J mol^{-1} K^{-1}	$[(G^\circ - H_0^\circ)/T]$/ J mol^{-1} K^{-1}	References
XeF$_2$(c)	75.60 ± 0.08	115.09 ± 0.12	51.17 ± 0.5	−63.92 ± 0.06	55
XeF$_2$(g)	54.14	259.45	42.13	−217.3	57
XeF$_4$(c)	118.4 ± 0.12	167.0 ± 0.17	77.24 ± 0.08	−89.76 ± 0.09	54
XeF$_4$(g)	89.96	323.98	63.89	−260.09	54
XeF$_6$(g)	126.2	374.0	80.92	−293.1	57
XeF$_6$(c)	171.6	210.4	104.0	−106.4	51

Measurement of the specific conductance of aqueous XeF_2 solutions indicates that the electrolytic dissociation of XeF_2 cannot be detected since the molar conductance of XeF_2 in the concentration range 5×10^{-3} to 2×10^{-2} M is 0.4 Ω^{-1} cm^2 mol^{-1} [60,61].

Xenon difluoride is not only a good fluorinating agent, but also a strong oxidizer capable of oxidizing chloride to chlorine, Ce(III) to Ce(IV), Co(II) to Co(III), Ag(I) to Ag(II), iodate to periodate, xenate to perxenate [59,62], and bromate to perbromate [63]. In water and, in particular, in alkaline solutions, xenon difluoride produces transient slightly yellow solutions, presumably due to the XeO species [59,64]. The decomposition of xenon difluoride in water is catalyzed by hydroxide (in 0.007 M NaOH at 25°C the pseudo-first-order rate constant, k, is 5×10^{-2} s^{-1}) [64]. The decomposition of xenon difluoride in water is also catalyzed to a lesser extent by fluoride, bicarbonate, carbonate, dihydrogen phosphate, monohydrogen phosphate, phosphate, and by cations which form strong fluoride complexes, such as thorium, lanthanum, beryllium, zirconium [59,65], and hydrogen fluoride [66]. Hydrolysis reaction is inhibited by potassium nitrate and sodium perchlorate [65].

Reduction potentials of XeF_2 in aqueous solutions reported in the literature vary from 2.18 to 2.64 V [59,67-69]. The wide variation in these values can be attributed directly to the estimated values of Gibbs energy of formation for aqueous XeF_2. Furthermore, xenon difluoride as an oxidizing agent can oxidize substances by two-electron transfer in one step or in two steps, as was observed in some of the fluorination reactions. The limiting step of the reduction process in aqueous solution appears to depend on the hydrolysis reaction of xenon difluoride, which may produce an intermediate, XeO, as an effective oxidizing species or partially hydrolyzed intermediate such as XeF^+ involving one electron transfer. These reactions have been investigated by Goncharov et al. [67,67a] and Lempert et al. [67b] and corresponding one-electron-transfer potentials have been calculated.

The electrode potential of aqueous xenon difluoride has been calculated from the Gibbs energy of formation of -74.7 kJ mol^{-1} $XeF_2(g)$, vapor pressure of 4.5 Torr at 25°C [1], and its solubility, which has been estimated to be 0.4 mol kg^{-1}. The Gibbs energy of formation for $XeF_2(aq)$ is then calculated to be 85.1 kJ mol^{-1}, yielding a reduction potential value of 2.64 V for the reaction [68]

$$XeF_2(aq) + 2H^+(aq) + 2e^- \rightarrow Xe(g) + 2HF(aq)$$

Goncharov and co-workers have redetermined the Gibbs energy of formation for the $XeF_2(aq)$ according to the reactions

$$XeF_2(aq) + H_2(g) \rightarrow Xe(g) + 2HF(aq)$$

$$XeF_2(aq) + H_2(g) \rightarrow Xe(g) + 2H^+(aq) + 2F^-(aq)$$

to be 144.8 kJ mol^{-1}. Using this value, reduction potentials calculated for the reactions above are 2.32 and 2.14 V, respectively. Thermodynamic properties for the aqueous xenon difluoride are summarized in Table 5.

Calculations of reduction potentials for a single electron transfer by xenon difluoride yielding an intermediate, XeF(aq), can be achieved from the

TABLE 5 Summary of Thermodynamic and Physical Properties for XeF_2 in Aqueous Solutions at 298.15 K

	Value	Reference
$\Delta G_f^\circ [XeF_2(aq)]/kJ\ mol^{-1}$	−85.1	68
$\Delta G_f^\circ [XeF_2(aq)]/kJ\ mol^{-1}$	−124.7	67
$\Delta G_f^\circ [XeF_2(aq)]/kJ\ mol^{-1}$	−144.8	69
$\Delta G_f^\circ [XeF_2(aq)]/kJ\ mol^{-1}$	−145.2	67
ΔH(dissolution)$/kJ\ mol^{-1}$	−38 ± 13	69
$\Delta H^\circ [XeF_2(aq)]/kJ\ mol^{-1}$	−209 ± 4	69
$S^\circ [XeF_2(aq)]/J\ mol^{-1}\ K^{-1}$	85.3	69
$S^\circ [XeF_2(aq)]/J\ mol^{-1}\ K^{-1}$	157.3	67
Molal solubility $S/\ mol\ kg^{-1}$	−0.4	68

estimated values of ΔH_f°, ΔS°, and ΔG° at 298 K for XeF(aq). Reduction potentials for the reactions

$$XeF_2(aq) + e^- \rightarrow XeF(aq) + F^-(aq)$$

$$XeF(aq) + e^- \rightarrow Xe(g) + F^-(aq)$$

are calculated by using Gibbs energies ΔG_f° at 298.15 K for aqueous species of XeF_2(aq), F^-(aq), and XeF(aq) as −145.2, −78.8, and 46 kJ mol^{-1}, yielding the potential values of 0.9 and 3.37 V, respectively. Thus xenon difluoride in aqueous solution as a one-electron oxidizing reagent has an oxidizing power somewhat greater than that of Fe(III)/Fe(II) couple, and in the reactions where two electrons are transferred, the oxidizing power is greater than 2.1 V. Yet, when a XeF(aq) species acts as an oxidizing agent, the reduction potential is 3.37 V, which is by far the highest value yet reported for aqueous solutions. Thermodynamic values for XeF in aqueous solutions are summarized in Table 6.

B. Aqueous Solutions of Xenon Trioxide

Solutions of xenon trioxide, xenic acid, are prepared from the hydrolysis of xenon hexafluoride, xenon tetrafluoride, or xenon oxygen tetrafluoride, $XeOF_4$ [70-72]. However, xenon hexafluoride is preferable since the hydrolysis in water yields exclusively aqueous xenon trioxide as indicated:

$$XeF_6(g) + 3H_2O \rightarrow XeO_3(aq) + 6HF(aq)$$

TABLE 6 Summary of Thermodynamic Values for XeF in Aqueous Solutions at 298.15 K[a]

	Value	Reference
$\Delta H_f^\circ[\text{XeF(g)}]/\text{kJ mol}^{-1}$	69 ± 2.2	67
$\Delta H_f^\circ[\text{XeF(g)} \rightarrow \text{XeF(aq)}]/\text{kJ mol}^{-1}$	−59	67
$\Delta G_f^\circ[\text{XeF(g)}]/\text{kJ mol}^{-1}$	74 ± 2.2	67
$\Delta G_f^\circ[\text{XeF(aq)}]/\text{kJ mol}^{-1}$	46.0	67
$\Delta G_f^\circ[\text{XeF(g)} \rightarrow \text{XeF(aq)}]/\text{kJ mol}^{-1}$	−26.8	67
$S^\circ[\text{XeF(g)}]/\text{J mol}^{-1}\text{K}^{-1}$	254	67
$S^\circ[\text{XeF(aq)}]/\text{J mol}^{-1}\text{K}^{-1}$	146	83

[a]Auxiliary data used for the calculation of ΔG° of the reaction taken in part from Refs. 66 and 67:

$\Delta H_{298}(\text{H}\cdots\text{F}$ bonding in $\text{H}_2\text{O}) = 29.3$ kJ mol^{-1} ΔH(sublimation, XeF_2) = 55.7 kJ mol^{-1}

$\Delta H_{298}[(\text{F}\cdots\text{Xe}\cdots\text{F}\cdots\text{H})$ bonding] = 33.5 kJ mol^{-1} ΔH(fusion, XeF_2) = 4.2 kJ mol^{-1}

ΔH vaporization of XeF_2 = 51.5 kJ mol^{-1}

The hydrolysis reaction can be violent and in general must be carried out by passing xenon hexafluoride vapor with inert gas into water [73]. Hydrofluoric acid from the hydrolyzate may be eliminated by treatment of solution with calcium carbonate or magnesium oxide [74,75], or small amounts can be prepared by dry-freeze evaporation of water and HF. Solid xenon trioxide is shock sensitive and decomposes explosively according to the equation

$$\text{XeO}_3 \rightarrow \text{Xe} + (3/2)\text{O}_2$$

Xenon trioxide has very low vapor pressure at room temperature and has been observed in the vapor by mass spectrometry [76], and small amounts of XeO_3 have been purified by vacuum sublimation.

Xenon trioxide may be dissolved in pure t-butyl alcohol. In the absence of impurities these solutions are stable and can be titrated as monoprotic acid by adding 1 mol of alkali per mole of XeO_3 to yield a salt approximating the empirical composition $\text{XeO}_3\cdot\text{M}-\text{OC}_4\text{H}_9\cdot\text{C}_4\text{H}_9\text{OH}$. These salts are easily decomposed but are not explosive [77].

Solutions of pure XeO_3 are slightly acidic and can be titrated as a very weak acid. The equilibrium constant for the reaction

$$XeO_3 + OH^- \rightarrow HXeO_4^-$$

has been determined by Appelman and Malm to have a value of 1.5×10^3 [75].

Aqueous xenon trioxide is a strong oxidizer; it oxidizes readily Hg(0) to Hg(II), Fe(II) to Fe(III), Mn(II) to Mn(IV), chloride to chlorine, bromide to bromine, Pu(III) to Pu(IV), Np(V) to Np(VI), and organic alcohols, carboxylic acids, aldehydes, and amines [31,67b,78,79]. Organic compounds apparently are oxidized via a free-radical route in which impurities play an important role. In alkaline solutions, xenon trioxide disproportionates to yield xenon gas and perxenate.

Most of the xenate salts are unstable; only the alkali salts containing fluoride, such as $MXeO_3F$ (where M is K, Rb, or Cs) have comparable stability to that of perxenate. Cesium bromo- and chloroxenates have been prepared and have been characterized; however, both of these are unstable [80].

The enthalpy of formation $\Delta H_f^\circ(298)[XeO_3(c)]$ is 401 kJ mol^{-1}, and the enthalpy of formation of aqueous xenon trioxide at 298.15 K was found by O'Hare et al. [4] to be 418 ± 1 kJ mol^{-1}, or as recently reported the value is 413 kJ mol^{-1} [4a]. This value was used to calculate the thermodynamic oxidizing power of aqueous xenon trioxide. Since the reactions of XeO_3 are irreversible, the oxidation potential can only be measured thermodynamically. The potentials of the XeO_3/Xe couple in acidic and of the $HXeO_4^-/Xe$ couple in basic solutions have been calculated from experimental data by O'Hare et al. [4] as 2.1 ± 0.02 and 1.24 ± 0.02 V, respectively, using the thermodynamic data summarized in Table 7. The potential in acid media for the reaction

$$XeO_3(aq) + 6H^+ + 6e^- \rightarrow Xe(g) + 3H_2O(\ell)$$

is calculated from the ΔG_f° value at 298.15 K using available thermodynamic data. The estimation of the Gibbs energy of formation of aqueous xenon trioxide is achieved by determining enthalpy and entropy for aqueous xenon trioxide solutions:

1. The heat of formation of aqueous XeO_3 is determined from calorimetric measurements of the enthalpies of reaction between $XeO_3(aq)$ and HI(aq):

$$XeO_3(aq) + 6H^+ + 9I^-(aq) \rightarrow 3I_3^-(aq) + Xe(g) + 3H_2O(\ell)$$

to yield a value of 418.1 or 413 kJ mol^{-1} using the new value of ΔG_f° for $XeO_3(aq)$.

2. The heat of solution for the reaction

$$XeO_3(c) \rightarrow XeO_3(aq)$$

$$\Delta H_{solution} = \Delta H_f^\circ[XeO_3(aq)] - \Delta H_f[XeO_3(c)] = 16 \text{ kJ mol}^{-1}$$

TABLE 7 Thermodynamic Properties of Xenon Trioxide at 298.15 K

	Value	References
$\Delta H_f^\circ[XeO_3(c)]$/kJ mol^{-1}	402 ± 8	81
$\Delta H_f^\circ(XeO_3, H_2O)$/kJ mol^{-1}	418 ± 1	4
ΔH (dissolution)/kJ mol^{-1}	16 ± 8	81
$S^\circ[XeO_3(c)]$/J mol^{-1} K^{-1}	79.5	81
$S^\circ[XeO_3(g)]$/J mol^{-1} K^{-1}	288.3	82
	289.1	57
S^{int}/J mol^{-1} K^{-1}	112.9	82,83
$\bar{S}^\circ[XeO_3(aq)]$/J mol^{-1} K^{-1}	184 ± 17	82
$\Delta S_f^\circ[Xe(g) + 3/2O_2(g) \rightarrow XeO_3(aq)]$/J mol^{-1} K^{-1}	−293 ± 17	84,4
$\Delta G_f^\circ[XeO_3(aq)]$/kJ mol^{-1}	505 ± 5	5
$\Delta G_f^\circ[Xe(g) + 3H_2O(\ell) \rightarrow XeO_3(aq) + 6H^+ + 6e]$/kJ mol^{-1}	1212 ± 5	84,84a
$\Delta G_f^\circ[H_2O(\ell)]$/kJ mol^{-1}	−237.2	84,84a
$\Delta G_f^\circ[OH^-(aq)]$/kJ mol^{-1}	−157.5	84
$C_p^\circ(XeO_3)$/J mol^{-1} K^{-1}	289.1	57
$(H^\circ - H_0^\circ)/T$/J mol^{-1} K^{-1}	46.3	
$(G^\circ - H_0^\circ)/T$/J mol^{-1} K^{-1}	242.8	

3. The standard partial molar entropy is estimated by the expression

$$\bar{S}^\circ[XeO_3(aq)] = S^{int} - 1.5 R \ln M + [41.84 - 0.920 (V_m/cm^3 mol^{-1})]$$

$$J\ mol^{-1}\ K^{-1}$$

in which S^{int} is calculated by the equation

$$S^{int} = S^\circ[XeO_3(g)] - 7.5 R \ln M - 2.5 R \ln T + 9.67\ J\ mol^{-1}\ K^{-1}$$

using the following values: $V_m = 39.4$ cm^3 mol^{-1}, $M = 179.298$ g, $T = 298.15$ K, and $S°[XeO_3(g)] = 288.3$ J mol^{-1} K^{-1}: thus the calculated values of S^{int} 114 J mol^{-1} K^{-1} and $\bar{S}°[XeO_3(aq)] = 184 \pm 17$ J mol^{-1} K^{-1}, respectively.

4. The standard entropy of formation, $\Delta S_f°$, for the reaction

$$Xe(g) + (3/2)O_2(g) \rightarrow XeO_3(aq)$$

may be calculated using $S°$ values of 169.6 and 205.0 J mol^{-1} K^{-1} for $Xe(g)$ and $O_2(g)$ [84].

$$\Delta S_f° = 184.1 \pm 16.7 - 170.41 - (3/2)(205) = 293 \pm 17 \text{ J mol}^{-1} \text{ K}^{-1}$$

5. The standard Gibbs energy of formation for $XeO_3(aq)$ is obtained by

$$\Delta G_f° = \Delta H_f° - T \Delta S_f° = 505.4 \pm 5 \text{ kJ mol}^{-1}$$

6. The $\Delta G°$ for the reaction

$$XeO_3(aq) + 6H^+(aq) + 6e^- \rightarrow Xe(g) + 3H_2O(\ell)$$

is calculated using the values of $\Delta G_f°[XeO_3(aq)]$ as either 505 or 500 and of $\Delta G_f°[H_2O(\ell)]$ as -237.2 kJ mol^{-1}, respectively. Thus, taking the calculated value of $\Delta G_f°$,

$$\Delta G° = 3\Delta G_f°[H_2O(\ell)] - \Delta G_f°[XeO_3(aq)] = -1217 \text{ or } -1212 \text{ kJ mol}^{-1}$$

the reaction potential for the XeO_3/Xe couple in acid media corresponds to 2.10 V.

The Gibbs energy $\Delta G°$ at 298.15 K in the basic media for the reaction

$$HXeO_4^-(aq) + 3H_2O(\ell) + 6e^- \rightarrow Xe(g) + 7OH^-(aq)$$

is calculated by the summation of the Gibbs energies of formation of the aqueous species

$$\Delta G° = 7\Delta G_f°[OH^-(aq)] - 3\Delta G_f°[H_2O(\ell)] - \Delta G_f°[HXeO_4^-(aq)]$$

to yield the value -717.8 kJ mol^{-1}, corresponding to 1.24 V. The calculated value of the Gibbs energy of formation, $\Delta G_f°[HXeO_4^-(aq)]$ is based on the reaction

$$HXeO_4^-(aq) \rightarrow OH^-(aq) + XeO_3(aq)$$

using the equilibrium constant of 6.7×10^{-4} [75]. Uncertainties in the estimation of $\Delta G°$ values for aqueous XeO_3 are on the order of 1.5%; thus the calculated $E°$ values are known to be ±0.02 V at best.

C. Aqueous Perxenate

Aqueous perxenate in strongly alkaline solution is formed by the hydrolysis of XeF_6 or by disproportionation reaction of xenate [85-87]. Sodium perxenate, $Na_4XeO_6 \cdot xH_2O$, slightly soluble salt, is commonly obtained by using sodium hydroxide. This salt is only slightly soluble in water having solubility at 25°C, approximately 0.003 M in 0.1 M NaOH [85]. Potassium hydroxide yields an insoluble salt $K_4XeO_6 \cdot 2XeO_3$ [75]. Alkali and many other metal perxenates are stable; however, silver perxenate is unstable and decomposes violently [88].

Sodium perxenate in concentrated sulfuric acid yields xenon tetroxide [89]. Sodium perxenate in water is basic due to the following reactions:

$$Na_4XeO_6 + H_2O \rightarrow 4Na^+ + HXeO_6^- + OH^-$$

$$H_2XeO_6^{2-} + OH^- \rightarrow HXeO_6^{3-} + H_2O$$

and yields an equilibrium constant at 25°C of about 4×10^3 [72]. The formation of $H_3XeO_6^-$ and of H_4XeO_6 cannot be observed because of the instability of perxenate in acid solutions. Measurements of perxenate in alkaline solutions suggest the presence of following equilibrium:

$$HXeO_6^{3-} + OH^- \rightarrow XeO_6^{4-} + H_2O$$

having an equilibrium constant of approximately 3 [90].

Aqueous perxenate is a strong oxidizer. It rapidly oxidizes iodate to periodate in both alkaline and acidic solutions [75]; chlorate is oxidized to perchlorate only in acidic media [91]. In basic solutions perxenate oxidizes Co(II) to Co(III) and Ce(III) to Ce(IV) [75]. Perxenate oxidizes quantitatively hydrazine to nitrogen, Cr(III) to Cr(VI), Mn(II) to Mn(VII), and Ce(III) to Ce(IV) [92]. It also oxidizes Ag(I) to Ag(III) [88] and Cu(II) to Cu(III).

The reaction potential for the perxenate reaction is calculated by using the enthalpy of formation of $XeO_4(g)$, entropy, and the Gibbs energy of dissolution of $XeO_4(g)$. Thermodynamic values for the perxenate are summarized in Table 8.

The potential value of 1.18 ± 0.03 V for the perxenate-xenon couple in alkaline solution:

$$HXeO_6^{3-}(aq) + 5H_2O(\ell) + 8e^- \rightarrow Xe(g) + 11OH^-(aq)$$

is calculated using ionization constants of 10^{-2} and 10^{-6} for the reactions

$$H_4XeO_6 \rightarrow H^+ + H_3XeO_6^-$$

$$H_3XeO_6^- \rightarrow H^+ + H_2XeO_6^{2-}$$

TABLE 8 Summary of Thermodynamic Properties at 298.15 K for XeO_4 and Perxenate

	Value	Reference
$\Delta H_f[XeO_4(g)]/kJ\ mol^{-1}$	642.24	93
$S°[XeO_4(g)]/J\ mol^{-1}\ K^{-1}$	276.14 294.64	93, 93a
$\Delta G[XeO_4(g) + 2H_2O(\ell) \to H_4XeO_6(aq)]/kJ\ mol^{-1}$	0	93
$C_p°[XeO_4(g)]/J\ mol^{-1}\ K^{-1}$	76.86 78.15	93a, 57
$(H° - H_0°)/T/J\ mol^{-1}\ K^{-1}$	53.55 54.62	93a, 57
$-(G° - H_0°)/T/J\ mol^{-1}\ K^{-1}$	241.1 242.1	93a, 57
$\Delta G°[Xe(g) + 6H_2O(\ell) \to H_4XeO_6 + 8H^+ + 8e^-]/kJ\ mol^{-1}$	1682.7	
$-\Delta H_f[Na_4XeO_6(c)]/kJ\ mol^{-1}$	836.3	100
$-\Delta H_f[Na_4XeO_6(c) + 2H_2SO_4 \xrightarrow{pH\ 1.3} 2Na_2SO_4(aq) + XeO_3(aq)]/kJ\ mol^{-1}$	287.1 ± 1	100
$\Delta H[Na_4XeO_6(c) \to Na_4XeO_6 \cdot xH_2O(c)]/kJ\ mol^{-1}$	55.0 ± 3	100

respectively, as reported by Appelman and Malm [75]. Also, the potential of the perxenate-xenon couple in acidic medium for the reaction

$$H_4XeO_6 + 8H^+ + 8e^- \rightarrow Xe(g) + 6H_2O$$

is calculated to be 2.18 V. Combination of these results with Xe(VI)/Xe couple yields potential values of approximately 2.42 ± 0.06 and 0.99 ± 0.06 V in acidic and basic media, respectively, for the reactions

$$H_4XeO_6 + 2H^+ + 2e^- \rightarrow XeO_3 + 3H_2O$$

$$HXeO_6^{3-} + 2H_2O + 2e^- \rightarrow HXeO_4^- + 4OH^-$$

The calculated potentials are consistent with perxenate oxidation reactions in which Ag(I) is oxidized to Ag(II) or Ag(III), and iodate to periodate. The ability of ozone to oxidize Xe(VI) in alkaline solutions to perxenate is also in agreement with the calculated values.

V. RADON

A. Oxidation States and Chemistry

Radon, Rn, is not only highly radioactive, but also quite scarce, since it is formed from the nuclear decay of ^{238}Th and ^{232}Th and decomposes with a half-life of 3.8 days to yield polonium and other daughters. Unlike argon, krypton, and xenon, which differ from each other by 18 electrons, radon has 32 electrons more than xenon. The interval between xenon and radon is much longer than that between lighter inert gases, due to the filling of $4f$ and $5d$ orbitals. Thus radon is much heavier than xenon and more metallic. It forms compounds having more ionic character than the fluorides of xenon or krypton. Also, the ionization potential of radon, 10.75 eV, is quite comparable to that of iodine, 10.45 eV. Consequently, radon difluoride, RnF_2, the only known fluoride, dissociates to yield ionic species and is also not as volatile as the xenon or krypton difluoride.

Radon difluoride can be prepared either by heating the reaction mixture of fluorine and radon or by increasing the radon concentration, which produces intense alpha radiation, causing the formation of large numbers of ions and excited atoms [6,94]. Attempts to prepare radon compounds other than radon difluoride by various methods have been unsuccessful; hydrolysis of radon difluoride have not yielded an oxide or oxyfluoride. Similarly, the reactions of radon with chlorine produced no reaction products. In aqueous solutions radon difluoride reacts to produce hydrogen fluoride and oxygen [95,96]:

$$RnF_2 + H_2O \rightarrow Rn + (1/2)O_2 + 2HF$$

In anhydrous hydrogen fluoride and in other halogen fluorides radon difluoride dissolves and forms ionic solutions. Radon difluoride has been prepared by spontaneous reaction of radon with ClF, ClF_3, ClF_5, BrF_3, BrF_5, and IF_7 at room temperature [97]. Electromigration studies in HF and HF-BrF_3

Inert Gases

solutions indicate that radon is present as cationic as well as anionic species such as Rn^{2+}, RnF^+, and RnF_3^-.

B. Thermodynamic Studies

Determination of thermodynamic functions for radon difluoride poses almost insurmountable problems because of the scarcity of radon and its high radioactivity. The Gibbs energy values can only be inferred from the reaction of radon with various oxidizing agents. Since the free energies of formation for most of the halogen fluorides are known [98,99], the oxidizing power of these reagents can be ranked as follows:

$$ClF_5 > ClF > IF_7 > ClF_3 > BrF_5 > BrF_3 > AsF_5 > IF_5$$

Upper and lower limits for the formation of radon difluoride can be deduced from the fact that radon is oxidized by all of the halogen fluorides above except arsenic and iodine pentafluorides. Thus the Gibbs energy for formation of radon difluoride must lie between the ΔG_f° values for the reduction of BrF_3 to BrF and AsF_5 to AsF_3, which when corrected to a standard state and concentration of 1.0 mol L^{-1} should correspond to the standard Gibbs energy of formation of dissolved radon difluoride to ΔG_f° values between -121.3 and -213.4 kJ mol^{-1}. Admittedly, this spread is large and probably a better value of ΔG_f might be obtained by using oxidation reagents having oxidizing power between BrF_3 and AsF_5; however, this provides only a "ballpark" value [6].

VI. SUMMARY OF ELECTRODE POTENTIALS FOR AQUEOUS XENON COMPOUNDS

Aqueous xenon difluoride, xenon trioxide, and perxenate are powerful oxidants. The potentials calculated from the available thermodynamic data are consistent with chemically observed reactions and are summarized in Table 9.

A. Potential Diagrams for Aqueous Xenon Compounds

Acid Solution

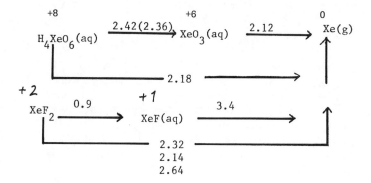

TABLE 9 Summary of Potentials for the Reactions

Medium	$E°/V$
Acid	
$XeO_3(aq) + 6H^+(aq) + 6e^- \rightarrow Xe(g) + 3H_2O(\ell)$	2.10 ± 0.02
$H_4XeO_6(aq) + 8H^+(aq) + 8e^- \rightarrow Xe(g) + 6H_2O(\ell)$	2.18 ± 0.03
$H_4XeO_6(aq) + 2H^+(aq) + 2e^- \rightarrow XeO_3(aq) + 3H_2O(\ell)$	2.42 ± 0.03
$XeF_2(aq) + 2H^+ + 2e^- \rightarrow Xe(g) + 2HF(aq)$	2.64
$XeF_2(aq) + H_2(g) \rightarrow Xe(g) + 2HF(aq)$	2.32
$XeF_2(aq) + H_2(g) \rightarrow Xe(g) + 2H^+(aq) + 2F^-(aq)$	2.14
$XeF(aq) + e^- \rightarrow XeF(aq) + F^-(aq)$	0.90
$XeF(aq) + e^- \rightarrow Xe(g) + F^-(aq)$	3.4
Basic	
$HXeO_6^{3-} + 5H_2O(\ell) + 8e^- \rightarrow Xe(g) + 11OH^-(aq)$	1.18 ± 0.03
$HXeO_4^-(aq) + 3H_2O(\ell) + 6e^- \rightarrow Xe(g) + 7OH^-(aq)$	1.24 ± 0.03
$HXeO_6^{3-}(aq) + 2H_2O(\ell) + 2e^- \rightarrow HXeO_4^-(aq) + 4OH^-(aq)$	0.99 ± 0.06

Basic Solution

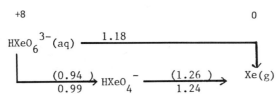

REFERENCES

1. B. Weinstock, E. E. Weaver, and C. P. Knop, Inorg. Chem., 5, 2189 (1966).
2. G. Stein and P. Plurien, in *Noble Gas Compounds*, H. H. Heyman, ed., University of Chicago Press, Chicago, 1963, p. 147.
3. S. R. Gunn and S. M. Williamson, Science, *140*, 177 (1963).
4. P. A. G. O'Hare, G. K. Johnson, and E. H. Appleman, Inorg. Chem., *9*, 332 (1970).
4a. G. K. Johnson, J. Chem. Thermodyn., *9*, 835 (1977).
5. V. I. Pepekin, Yu. A. Lebedev, and A. Ya. Apin, Zh. Fiz. Khim., *43*, 1564 (1969).
6. L. Stein, Yale Sci. Mag., *44*(8), 2 (1970).

7. H. J. Svec and G. D. Flesch, Science, *142*, 954 (1963).
8. J. Berkowitz, W. A. Chupka, P. M. Guyoi, J. H. Holloway, and R. Spohr, J. Phys. Chem., *75*, 1461 (1971).
9. J. F. Liebman, in *Inorganic and Nuclear Chemistry: Herbert H. Hyman Memorial Volume*, J. J. Katz and I. Shift, eds., Pergamon Press, Oxford, 1976, p. 155.
10. A. V. Grosse, A. D. Kirshenbaum, A. G. Streng, and L. V. Streng, Science, *139*, 1047 (1963).
11. F. Scheiner, J. G. Malm, and J. C. Hindeman, J. Am. Chem. Soc., *87*, 25 (1965).
12. V. N. Prusakoc and V B. Sokolov, Russ. J. Phys. Chem., *45*, 1673 (1971).
13. B. Frlec and J. H. Holloway, J. Chem. Soc. Commun., *11*, 370 (1973).
14. R. J. Gillespie and R. J. Schrobilgen, Inorg. Chem., *15*, 22 (1976).
15. J. H. Holloway, G. J. Schrobilgen, S. Bukshpan, W. Hilbrants, and H. DeWaard, J. Chem. Phys., *66*, 2627 (1977).
16. S. R. Gunn, J. Phys. Chem., *71*, 2934 (1967).
17. S. R. Gunn, J. Am. Chem. Soc., *88*, 5924 (1966).
18. P. A. Sessa and H. A. McGee, Jr., J. Phys. Chem., *73*, 2078 (1969).
19. H. H. Claassen, G. L. Goodman, J. G. Malm, and F. Schreiner, J. Chem. Phys., *42*, 1229 (1965).
20. N. Bartlett, Proc. Chem. Soc., *218* (1962).
21. H. H. Hyman, ed., *Noble Gas Compounds*, University of Chicago Press, Chicago, 1963, pp. 47, 106.
22. H. H. Claassen, H. Selig, and J. G. Malm, J. Am. Chem. Soc., *84*, 3593 (1962).
23. S. Maricic and Z. Weksli, Croat. Chem. Acta, *34*, 189, 191 (1962).
24. J. G. Malm and C. L. Chernick, Inorg. Synth., *8*, 254 (1967).
25. L. V. Streng and A. G. Streng, Inorg. Chem., *4*, 1370 (1965).
26. J. H. Holloway, Chem. Commun., 22 (1966).
27. S. M. Williamson, Inorg. Syn., *11*, 147 (1968).
28. A. V. Nikolaev, A. S. Nazarov, A. A. Opalovskii, and A. F. Trippel, Dokl. Akad. Nauk SSSR, *186*, 1331 (1969).
29. H. Meinert and S. Ruediger, Z. Chem., *9*, 239 (1967).
30. P. Allamagny, M. Langignard, and P. Dognin, C.R. Acad. Sci. Paris, Ser. C, *266*, 711 (1968).
31. S. Siegl and E. Gebert, J. Am. Chem. Soc., *85*, 240 (1963).
32. F. Sladky, Angew. Chem., *81*, 330 (1969).
33. N. Bartlett, M. Wechsberg, F. O. Sladky, P. A. Vulliner, G. R. Jones, and R. D. Burbank, Chem. Commun., *703* (1969).
34. N. Bartlett and F. O. Sladky, J. Am. Chem. Soc., *90*, 5316 (1968).
35. N. Bartlett, B. Zemva, and L. Graham, J. Fluorine Chem., *7*, 301 (1976).
36. H. H. Hyman and L. A. Quarterman, *Noble Gas Compounds*, University of Chicago Press, Chicago, 1963, p. 275.
37. A. Iskraut, R. Taubenest, and E. Schumacher, Chimia, *18*, 188 (1964).
38. H. Meinert, G. Kauschka, and S. Ruediger, Z. Chem., *7*, 111 (1967).
39. I. Sheft, T. M. Spittler, and F. H. Martin, Science, *145*, 701 (1967).
40. H. Selig, Science, *144*, 537 (1964).
41. K. E. Pullen and G. H. Cady, Inorg. Chem., *6*, 2267 (1967).
42. G. J. Moody and H. Selig, J. Inorg. Nucl. Chem., *28*, 2429 (1966).
43. G. J. Moody and H. Selig, Inorg. Nucl. Chem. Lett., *2*, 319 (1966).
44. J. G. Malm, F. Schreiner, and D. W. Osborne, Inorg. Nucl. Chem. Lett., *1*, 97 (1965).

45. D. F. Smith, Science, *140*, 899 (1963).
46. H. Selig, Inorg. Chem., *5*, 183 (1966).
47. F. Schreiner, G. N. McDonald, and C. L. Chernick, J. Phys. Chem., *72*, 1162 (1968).
48. J. F. Huston, J. Phys. Chem., *45*, 3334 (1967).
49. J. F. Huston, Inorg. Nucl. Chem. Lett., *4*, 29 (1968).
50. J. Jortner, E. G. Wilson, and S. A. Rice, J. Am. Chem. Soc., *85*, 814 (1963).
51. F. Schreiner, D. W. Osborne, J. G. Malm, and G. McDonald, J. Chem. Phys., *51*, 4838 (1969).
52. S. Siegel and E. Gebert, J. Am. Chem. Soc., *85*, 240 (1963).
53. P. A. Agron, C. K. Johnson, and H. A. Levy, Inorg. Nucl. Chem. Lett., *1*, 145 (1965).
54. D. W. Osborne, F. Schreiner, H. E. Flatow, and J. G. Malm, J. Chem. Phys., *57*, 3401 (1972).
55. D. W. Osborne, H. E. Flatow, and J. G. Malm, J. Chem. Phys., *57*, 4670 (1972).
56. G. K. Johnson, J. G. Malm, and W. N. Hubbard, J. Chem. Thermodyn., *4*, 879 (1972).
57. S. A. Kudchadker and A. P. Kudchadker, Proc. Indian Acad. Sci. Sect. A, *73*(5), 261 (1971).
58. H. H. Claassen, *The Noble Gases*, Heath, Boston, 1966, p. 51.
59. E. H. Appelman and J. G. Malm, J. Am. Chem. Soc., *86*, 2297 (1964).
60. I. Feher and M. Semptey, Magy. Kem. Foly., *76*, 954 (1963).
61. E. H. Appelman, Inorg. Chem., *6*, 1268 (1967).
62. I. S. Kirin, V. K. Isupov, V. I. Tikhonov, N. V. Ivannikova, Yu. K. Gusev, and G. G. Selikhov, Zh. Neorg. Khim., *12*, 1088 (1967).
63. E. H. Appelman, J. Am. Chem. Soc., *90*, 1900 (1968).
64. E. J. Appelman, Inorg. Chem., *6*, 1305 (1967).
65. M. T. Beck and L. Dozsa, J. Am. Chem. Soc., *89*, 5713 (1967).
66. A. V. Nikolaev, A. A. Opalovskii, and A. S. Nazarov, Dokl. Akad. Nauk SSSR, *181*, 361 (1968).
67. A. A. Goncharov, Yu. N. Kozlov, and A. P. Purmal, Zh. Fiz. Khim., *53*, 2685 (1979).
67a. A. A. Goncharov, Yu. N. Kozlov, and A. P. Purmal, Zh. Fiz. Khim., *55*, 1607 (1981).
67b. S. A. Lempert, D. B. Lempert, N. N. Aleinikov, and P. K. Agasyan, J. Gen. Chem. USSR, *49*, 1907 (1979).
68. J. G. Malm and E. H. Appelman, Atom. Energy Rev., *7*, 3 (1969).
69. V. D. Klimov, V. D. Prusakov, and V. B. Sokolov, Radiochimya, *13*, 725 (1971).
70. F. B. Dudley, G. Gard, and G. H. Cady, Inorg. Chem., *2*, 228 (1963).
71. D. F. Smith, J. Am. Chem. Soc., *85*, 816 (1963).
72. S. M. Williamson and C. W. Koch, Science, *139*, 1046 (1963).
73. B. Jaselskis, T. M. Spittler, and J. L. Huston, J. Am. Chem. Soc., *88*, 2149 (1966).
74. A. D. Kirshenbaum and A. B. Grosse, Science, *142*, 580 (1963).
75. E. H. Appelman and J. G. Malm, J. Am. Chem. Soc., *86*, 2141 (1964).
76. M. H. Studier and J. L. Huston, J. Phys. Chem., *71*, 457 (1967).
77. B. Jaselskis and J. P. Warriner, J. Am. Chem. Soc., *91*, 210 (1969).
78. B. Jaselskis and S. Vas, J. Am. Chem. Soc., *86*, 2078 (1964).
79. H. J. Rhodes and M. I. Blake, J. Pharm. Sci., *56*, 1352 (1967).

80. B. Jaselskis, T. M. Spittler, and J. L. Huston, J. Am. Chem. Soc., *89*, 2770 (1967).
81. S. R. Gunn, in *Noble Gas Compounds*, H. H. Hyman, ed., University of Chicago Press, Chicago, 1963, p. 149.
82. G. Narajan, Bull. Soc. Chim. Belges, *73*, 665 (1964).
83. R. E. Powell and W. M. Latimer, J. Chem. Phys., *19*, 1139 (1951).
84. D. D. Wagman, W. H. Evans, V. B. Parker, I. Halow, S. M. Bailey, and R. H. Schumm, Natl. Bur. Stand. Tech. Note 270-3, U.S. Government Printing Office, Washington, D.C., 1968.
84a. CODATA, Recommended Key Values for Thermodynamics, 1975; CODATA Bulletin, *17* (1976).
85. J. G. Malm, B. D. Holt, and R. W. Bane, in *Noble Gas Compounds*, H. H. Hyman, ed., Univeristy of Chicago Press, Chicago, 1963, p. 167.
86. E. H. Appelman and J. G. Malm, *Preparative Inorganic Reactions*, Vol. 2, Interscience, New York, 1965, p. 348.
87. E. H. Appelman, Inorg. Synth., *9*, 205 (1968).
88. D. M. Gruen, in *Noble Gas Compounds*, H. H. Hyman, ed., University of Chicago Press, Chicago, 1963, p. 174.
89. J. L. Huston, M. H. Studier, and E. N. Sloth, Science, *143*, 1161 (1964).
90. J. L. Peterson, H. H. Claassen, and E. H. Appelman, Inorg. Chem., *9*, 619 (1970).
91. E. H. Appelman and M. Anbar, Inorg. Chem., *4*, 1066 (1965).
92. R. W. Bane, Analyst, *90*, 756 (1965).
93. S. R. Gunn, J. Am. Chem. Soc., *87*, 2290 (1965).
93a. R. S. McDowell and B. Asprey, J. Chem. Phys., *57*, 3062 (1972).
94. P. R. Fields, L. Stein, and M. H. Zirkin, J. Am. Chem. Soc., *84*, 4164 (1962).
95. M. W. Hazeltine and H. C. Moser, J. Am. Chem. Soc., *89*, 2497 (1967).
96. K. Flohr and E. H. Appelman, J. Am. Chem. Soc., *90*, 3584 (1968).
97. L. Stein, J. Am. Chem. Soc., *91*, 5396 (1969).
98. L. Stein, in *Halogen Chemistry*, V. Gutmann, ed., Vol. 1, Academic Press, London, 1967, pp. 133-224.
99. P. A. G. O'Hare and W. N. Hubbard, J. Phys. Chem., *69*, 4358 (1965).
100. Yu. N. Matyushin, T. S. Kon'kova, E. A. Miroshnichenko, V. K. Isupov, N. N. Aleinikov, B. L. Korsunskii, F. I. Dubovitskii, and Yu. A. Lebedev, Proc. Acad. Sci. USSR (English), *239*, 401 (1978).

Appendix: Synopsis of Standard Potentials

Selected half-reaction potentials are listed in order of increasingly positive (anodic) assignments, providing a convenient guide to redox couples in a given range of standard potentials. The synopsis consists of two tables, one each for acid and basic solutions, which have been advisedly restricted to selected half-reactions. Values for additional couples are given in the text.

TABLE A Standard Potentials in Acid Solutions

Couple	$E°/V$
$(3/2)N_2 + H^+ + e^- \longrightarrow HN_3$	-3.10
$Li^+ + e^- \longrightarrow Li$	-3.045
$K^+ + e^- \longrightarrow K$	-2.925
$Rb^+ + e^- \longrightarrow Rb$	-2.925
$Cs^+ + e^- \longrightarrow Cs$	-2.923
$Ba^{2+} + 2e^- \longrightarrow Ba$	-2.92
$Ra^{2+} + 2e^- \longrightarrow Ra$	-2.916
$Sr^{2+} + 2e^- \longrightarrow Sr$	-2.89
$Ca^{2+} + 2e^- \longrightarrow Ca$	-2.84
$Na^+ + e^- \longrightarrow Na$	-2.714
$No^{2+} + e^- \longrightarrow No$	-2.5
$Md^{2+} + 2e^- \longrightarrow Md$	-2.4
$Fm^{2+} + 2e^- \longrightarrow Fm$	-2.37
$La^{3+} + 3e^- \longrightarrow La$	-2.37
$Y^{3+} + 3e^- \longrightarrow Y$	-2.37
$Ce^{3+} + 3e^- \longrightarrow Ce$	-2.34

TABLE A (Continued)

Couple	$E°/V$
$Nd^{3+} + 3e^- \longrightarrow Nd$	−2.32
$Sm^{3+} + 3e^- \longrightarrow Sm$	−2.30
$Gd^{3+} + 3e^- \longrightarrow Gd$	−2.29
$Mg^{2+} + 2e^- \longrightarrow Mg$	−2.356
$Lu^{3+} + 3e^- \longrightarrow Lu$	−2.30
$1/2H_2 + e^- \longrightarrow H^-$	−2.25
$Cf^{2+} + 2e^- \longrightarrow Cf$	−2.2
$Es^{2+} + 2e^- \longrightarrow Es$	−2.2
$Am^{3+} + 3e^- \longrightarrow Am$	−2.07
$AlF_6^{3-} + 3e^- \longrightarrow Al + 6F^-$	−2.067
$Cm^{3+} + 3e^- \longrightarrow Cm$	−2.06
$Sc^{3+} + 3e^- \longrightarrow Sc$	−2.03
$Bk^{3+} + 3e^- \longrightarrow Bk$	−2.01
$Cf^{3+} + 3e^- \longrightarrow Cf$	−2.0
$Es^{3+} + 3e^- \longrightarrow Es$	−2.0
$Be^{2+} + 2e^- \longrightarrow Be$	−1.97
$Fm^{3+} + 3e^- \longrightarrow Fm$	−1.96
$Th^{4+} + 4e^- \longrightarrow Th$	−1.83
$Np^{3+} + 3e^- \longrightarrow Np$	−1.79
$Md^{3+} + 3e^- \longrightarrow Md$	−1.7
$Zr^{4+} + 4e^- \longrightarrow Zr$	−1.70
$Al^{3+} + 3e^- \longrightarrow Al$	−1.67
$U^{3+} + 3e^- \longrightarrow U$	−1.66
$Ti^{2+} + 2e^- \longrightarrow Ti$	−1.63
$Hf^{4+} + 4e^- \longrightarrow Hf$	−1.56

TABLE A (Continued)

Couple	$E°/V$
$No^{3+} + 3e^- \longrightarrow No$	−1.2
$SiF_6^{2-} + 4e^- \longrightarrow Si + 6F^-$	−1.2
$TiF_6^{2-} + 4e^- \longrightarrow Ti + 6F^-$	−1.191
$Mn^{2+} + 2e^- \longrightarrow Mn$	−1.18
$V^{2+} + 2e^- \longrightarrow V$	−1.13
$Nb^{3+} + 3e^- \longrightarrow Nb$	−1.1
$H_3BO_3 + 3H^+ + 3e^- \longrightarrow B + 3H_2O$	−0.890
$SiO_2(vit) + 4H^+ + 4e^- \longrightarrow Si + 2H_2O$	−0.888
$TiO^{2+} + 2H^+ + 2e^- \longrightarrow Ti + H_2O$	−0.882
$Ta_2O_5 + 10H^+ + 10e^- \longrightarrow 2Ta + 5H_2O$	−0.81
$Zn^{2+} + 2e \longrightarrow Zn$	−0.7626
$TlI + e^- \longrightarrow Tl + I^-$	−0.74
$Te + 2H^+ + 2e^- \longrightarrow H_2Te$	−0.740
$TlBr + e^- \longrightarrow Tl + Br^-$	−0.658
$Nb_2O_5 + 10H^+ + 10e^- \longrightarrow 2Nb + 5H_2O$	−0.65
$TlCl + e^- \longrightarrow Tl + Cl^-$	−0.5568
$Ga^{3+} + 3e^- \longrightarrow Ga$	−0.529
$U^{4+} + e^- \longrightarrow U^{3+}$	−0.52
$Sb + 3H^+ + 3e^- \longrightarrow SbH_3(g)$	−0.510
$H_3PO_2 + H^+ + e^- \longrightarrow P(w) + 2H_2O$	−0.508
$H_3PO_3 + 2H^+ + 2e^- \longrightarrow H_3PO_2 + H_2O$	−0.499
$Fe^{2+} + 2e \longrightarrow Fe$	−0.44
$Cr^{3+} + e^- \longrightarrow Cr^{2+}$	−0.424
$Cd^{2+} + 2e \longrightarrow Cd$	−0.4025
$Ti^{3+} + e^- \longrightarrow Ti^{2+}$	−0.37

TABLE A (Continued)

Couple	$E°/V$
$PbI_2 + 2e^- \longrightarrow Pb + 2I^-$	-0.365
$PbSO_4 + 2e^- \longrightarrow Pb + SO_4^{2-}$	-0.3505
$Eu^{3+} + e^- \longrightarrow Eu^{2+}$	-0.35
$In^{3+} + 3e^- \longrightarrow In$	-0.3382
$Tl^+ + e^- \longrightarrow Tl$	-0.3363
$PbBr_2 + 2e^- \longrightarrow Pb + 2Br^-$	-0.280
$Co^{2+} + 2e^- \longrightarrow Co$	-0.277
$H_3PO_4 + 2H^+ + 2e^- \longrightarrow H_3PO_3 + H_2O$	-0.276
$PbCl_2 + 2e^- \longrightarrow Pb + 2Cl^-$	-0.268
$Ni^{2+} + 2e^- \longrightarrow Ni$	-0.257
$V^{3+} + e^- \longrightarrow V^{2+}$	-0.255
$2SO_4^{2-} + 4H^+ + 4e^- \longrightarrow S_2O_6^{2-} + 2H_2O$	-0.253
$SnF_6^{2-} + 4e^- \longrightarrow Sn + 6F^-$	-0.25
$N_2 + 5H^+ + 4e^- \longrightarrow N_2H_5^+$	-0.23
$As + 3H^+ + 3e^- \longrightarrow AsH_3$	-0.225
$Mo^{3+} + 3e^- \longrightarrow Mo$	-0.2
$CuI + e^- \longrightarrow Cu + I^-$	-0.182
$CO_2 + 2H^+ + 2e \longrightarrow HCOOH(aq)$	-0.16
$AgI + e^- \longrightarrow Ag + I^-$	-0.1522
$Si + 4H^+ + 4e^- \longrightarrow SiH_4$	-0.143
$Sn^{2+} + 2e^- \longrightarrow Sn$	-0.136
$Pb^{2+} + 2e^- \longrightarrow Pb$	-0.1251
$WO_3(c) + 6H^+ + 6e^- \longrightarrow W + 3H_2O$	-0.090
$P(w) + 3H^+ + 3e^- \longrightarrow PH_3$	-0.063
$O_2 + H^+ + e \longrightarrow HO_2$	-0.046

Standard Potentials

TABLE A (Continued)

Couple	$E°/V$
$Hg_2I_2 + 2e^- \longrightarrow 2Hg + 2I^-$	−0.0405
$Se + 2H^+ + 2e^- \longrightarrow H_2Se$	−0.028
$GeO_2 + 4H^+ + 4e^- \longrightarrow Ge(hex) + 2H_2O$	−0.009
$2H^+ + 2e \longrightarrow H_2$	0.000
$CuBr + e^- \longrightarrow Cu + Br^-$	0.033
$HCOOH(aq) + 2H^+ + 2e^- \longrightarrow HCHO(aq) + H_2O$	0.056
$AgBr + e^- \longrightarrow Ag + Br^-$	0.0711
$TiO^{2+} + 2H^+ + e^- \longrightarrow Ti^{3+} + H_2O$	0.100
$CuCl + e^- \longrightarrow Cu + Cl^-$	0.121
$C + 4H^+ + 4e^- \longrightarrow CH$	0.132
$Hg_2Br_2 + 2e^- \longrightarrow 2Hg + 2Br^-$	0.13920
$S + 2H^+ + 2e \longrightarrow H_2S$	0.144
$Np^{4+} + e^- \longrightarrow Np^{3+}$	0.15
$Sn^{4+} + 2e^- \longrightarrow Sn^{2+}$	0.15
$Sb_4O_6 + 12H^+ + 12e^- \longrightarrow 4Sb + 6H_2O$	0.1504
$SO_4^{2-} + 2H^+ + 2e^- \longrightarrow H_2SO_3 + H_2O$	0.158
$Cu^{2+} + e^- \longrightarrow Cu^+$	0.159
$UO_2^{2+} + e^- \longrightarrow UO_2^+$	0.16
$BiOCl + 2H^+ + 3e^- \longrightarrow Bi + H_2O + Cl^-$	0.1697
$2H_2SO_3^- + 3H^+ + 2e^- \longrightarrow HS_2O_4^- + 2H_2O$	0.173
$ReO_2 + 4H^+ + 4e^- \longrightarrow Re + 2H_2O$	0.22
$AgCl + e^- \longrightarrow Ag + Cl^-$	0.2223
$HCHO(aq) + 2H^+ + 2e^- \longrightarrow CH_3OH(aq)$	0.232
$(CH_3)_2SO_2 + 2H^+ + 2e^- \longrightarrow (CH_3)_2SO + 2H_2O$	0.238
$HAsO_2(aq) + 3H + 3e^- \longrightarrow As + 2H_2O$	0.248

TABLE A (Continued)

Couple	$E°/V$
$UO_2^{2+} + 4H^+ + 2e^- \longrightarrow U^{4+} + 2H_2O$	0.27
$HCNO + H^+ + e^- \longrightarrow 1/2 C_2N_2 + H_2O$	0.330
$VO^{2+} + 2H^+ + e^- \longrightarrow V^{3+} + H_2O$	0.337
$ReO_4^- + 8H^+ + 7e^- \longrightarrow Re + 4H_2O$	0.34
$Cu^{2+} + 2e^- \longrightarrow Cu$	0.340
$AgIO_3 + e^- \longrightarrow Ag + IO_3^-$	0.354
$Fe(CN)_6^{3-} + e^- \longrightarrow Fe(CN)_6^{4-}$	0.3610
$C_2N_2 + 2H^+ + 2e^- \longrightarrow 2HCN(aq)$	0.373
$UO_2^+ + 4H^+ + e^- \longrightarrow U^{4+} + 2H_2O$	0.38
$H_2N_2O_2 + 6H^+ + 4e^- \longrightarrow 2NH_3OH^+$	0.387
$2H_2SO_3 + 2H^+ + 4e^- \longrightarrow S_2O_3^{2-} + 3H_2O$	0.400
$Ag_2CrO_4 + 2e^- \longrightarrow 2Ag + CrO_4^{2-}$	0.4491
$Ag_2MoO_4 + 2e^- \longrightarrow 2Ag + MoO_4^{2-}$	0.486
$PdBr_4^{2-} + 2e^- \longrightarrow Pd + 4Br^-$	0.49
$RhCl_6 + 3e^- \longrightarrow Rh + 6Cl^-$	0.5
$H_2SO_3 + 4H^+ + 4e^- \longrightarrow S + 3H_2O$	0.500
$2H_2SO_3 + 4H^+ + 6e^- \longrightarrow S_4O_6^{2-} + 6H_2O$	0.507
$ReO_4^- + 4H^+ + 3e^- \longrightarrow ReO_2 + 2H_2O$	0.51
$Cu^+ + e^- \longrightarrow Cu$	0.520
$TeO_2(c) + 4H^+ + 4e^- \longrightarrow Te + 2H_2O$	0.53
$I_2 + 2e^- \longrightarrow 2I^-$	0.5355
$I_3^- + 2e^- \longrightarrow 3I^-$	0.536
$AgBrO_3 + e^- \longrightarrow Ag + BrO_3^-$	0.546
$Cu^{2+} + Cl^- + e^- \longrightarrow CuCl$	0.559
$TeOOH^+ + 3H^+ + 4e^- \longrightarrow Te + 2H_2O$	0.559

TABLE A (Continued)

Couple	$E°/V$
$H_3AsO_4 + 2H^+ + 2e^- \longrightarrow HAsO_2 + 2H_2O$	0.560
$MnO_4^- + e^- \longrightarrow MnO_4^{2-}$	0.56
$S_2O_6^{2-} + 4H^+ + 2e^- \longrightarrow 2H_2SO_3$	0.569
$CH_3OH(aq) + 2H^+ + 2e^- \longrightarrow CH_4 + H_2O$	0.59
$Sb_2O_5 + 6H^+ + 4e^- \longrightarrow 2SbO^+ + 3H_2O$	0.605
$Au(SCN)_4^- + 3e^- \longrightarrow Au + 4SCN^-$	0.636
$PdCl_4^{2-} + 2e^- \longrightarrow Pd + 4Cl^-$	0.64
$AgC_2H_3O_2 + e^- \longrightarrow Ag + C_2H_3O_2^-$	0.643
$Cu^{2+} + Br^- + e^- \longrightarrow CuBr$	0.654
$Ag_2SO_4 + 2e^- \longrightarrow 2Ag + SO_4^{2-}$	0.654
$NpO_2^+ + 4H^+ + e^- \longrightarrow Np^{4+} + 2H_2O$	0.66
$O_2 + 2H^+ + 2e^- \longrightarrow H_2O_2$	0.695
$HN_3 + 11H^+ + 8e^- \longrightarrow 3NH_4^+$	0.695
$PtBr_4^{2-} + 2e^- \longrightarrow Pt + 4Br^-$	0.698
$2NO + 2H^+ + 2e^- \longrightarrow H_2N_2O_2$	0.71
$H_2SeO_3 + 4H^+ + 4e^- \longrightarrow Se + 3H_2O$	0.739
$PtCl_4^{2-} + 2e^- \longrightarrow Pt + 4Cl^-$	0.758
$Rh^{3+} + 3e^- \longrightarrow Rh$	0.76
$(SCN)_2 + 2e^- \longrightarrow 2SCN^-$	0.77
$Fe^{3+} + e^- \longrightarrow Fe^{2+}$	0.771
$Hg_2^{2+} + 2e^- \longrightarrow 2Hg$	0.7960
$Ag^+ + e^- \longrightarrow Ag$	0.7991
$2NO_3^- + 4H^+ + 2e^- \longrightarrow N_2O_4 + 2H_2O$	0.803
$IrBr_6^{2-} + e^- \longrightarrow IrBr_6^{3-}$	0.805
$AmO_2^+ + 4H^+ + e^- \longrightarrow Am^{4+} + 2H_2O$	0.82

TABLE A (Continued)

Couple	$E°/V$
$OsO_4(c) + 8H^+ + 8e^- \longrightarrow Os + 4H_2O$	0.84
$AuBr_4^- + 3e^- \longrightarrow Au + 4Br^-$	0.854
$2HNO_2 + 4H^+ + 4e^- \longrightarrow H_2N_2O_2$	0.86
$IrCl_6^{3-} + 3e^- \longrightarrow Ir + 6Cl^-$	0.86
$Cu^{2+} + I^- + e^- \longrightarrow CuI$	0.861
$IrCl_6^{2-} + e^- \longrightarrow IrCl_6^{3-}$	0.867
$2Hg^{2+} + 2e^- \longrightarrow Hg$	0.9110
$Pd^{2+} + 2e^- \longrightarrow Pd$	0.915
$NO_3^- + 3H^+ + 2e^- \longrightarrow HNO_2 + H_2O$	0.94
$NO_3^- + 4H^+ + 3e^- \longrightarrow NO + 2H_2O$	0.957
$AuBr_2^- + e^- \longrightarrow Au + 2Br^-$	0.960
$PtO + 2H^+ + 2e^- \longrightarrow Pt + H_2O$	0.980
$HNO_2 + H^+ + e^- \longrightarrow NO + H_2O$	0.996
$AuCl_4^- + 3e^- \longrightarrow Au + 4Cl^-$	1.002
$Pu^{4+} + e^- \longrightarrow Pu^{3+}$	1.01
$PuO_2^{2+} + e^- \longrightarrow PuO_2^+$	1.02
$PuO_2^{2+} + 4H^+ + 2e^- \longrightarrow Pu^{4+} + 2H_2O$	1.03
$N_2O_4 + 4H^+ + 4e^- \longrightarrow NO + 2H_2O$	1.039
$PuO_2^+ + 4H^+ + e^- \longrightarrow Pu^{4+} + 2H_2O$	1.04
$Sb_2O_5 + 2H^+ + 2e^- \longrightarrow Sb_2O_4 + H_2O$	1.055
$Br_2(l) + 2e^- \longrightarrow 2Br^-$	1.065
$ICl_2^- + e^- \longrightarrow 2Cl^- + 1/2 I_2$	1.07
$N_2O_4 + 2H^+ + 2e^- \longrightarrow 2HNO_3$	1.07
$Cu^{2+} + 2CN^- + e^- \longrightarrow Cu(CN)_2^-$	1.12
$H_2O_2 + H^+ + e^- \longrightarrow OH + H_2O$	1.14

Standard Potentials

TABLE A (Continued)

Couple	$E°/V$
$SeO_4^{2-} + 4H^+ + 2e^- \longrightarrow H_2SeO_3 + H_2O$	1.151
$ClO_3^- + 3H^+ + 2e^- \longrightarrow HClO_2 + H_2O$	1.181
$ClO_2 + H^+ + e^- \longrightarrow HClO_2$	1.188
$S_2Cl_2 + 2e^- \longrightarrow 2S + 2Cl^-$	1.19
$IO_3^- + 6H^+ + 5e^- \longrightarrow 1/2 I_2 + 3H_2O$	1.195
$ClO_4^- + 2H^+ + 2e^- \longrightarrow ClO_3^- + H_2O$	1.201
$O_2 + 4H^+ + 4e^- \longrightarrow 2H_2O$	1.229
$MnO_2 + 4H^+ + 2e^- \longrightarrow Mn^{2+} + 2H_2O$	1.23
$N_pO_2^{2+} + e^- \longrightarrow NpO_2^+$	1.24
$N_2H_5^+ + 3H^+ + 2e^- \longrightarrow 2NH_4^+$	1.275
$PdCl_6^{2-} + 2e^- \longrightarrow PdCl_4^{2-} + 2Cl^-$	1.288
$2HNO_2 + 4H^+ + 4e^- \longrightarrow N_2O + 3H_2O$	1.297
$NH_3OH^+ + 2H^+ + 2e^- \longrightarrow NH_4^+ + H_2O$	1.35
$Cl_2 + 2e^- \longrightarrow 2Cl^-$	1.3583
$Cr_2O_7^{2-} + 14H^+ + 6e^- \longrightarrow 2Cr^{3+} + 7H_2O$	1.36
$2NH_3OH^+ + H^+ + 2e^- \longrightarrow N_2H_5^+ + 2H_2O$	1.41
$HO_2 + H^+ + e^- \longrightarrow H_2O_2$	1.44
$PbO_2(\alpha) + 4H^+ + 2e^- \longrightarrow Pb^{2+} + 2H_2O$	1.468
$BrO_3^- + 6H^+ + 5e \longrightarrow 1/2 Br_2 + 3H_2O$	1.478
$Mn^{3+} + e^- \longrightarrow Mn^{2+}$	1.5
$MnO_4^- + 8H^+ + 5e^- \longrightarrow Mn^{2+} + 4H_2O$	1.51
$Au^{3+} + 3e^- \longrightarrow Au$	1.52
$AmO_2^{2+} + e^- \longrightarrow AmO_2^+$	1.59
$NiO_2 + 4H^+ + 2e^- \longrightarrow Ni^{2+} + 2H_2O$	1.593
$H_5IO_6 + H^+ + 2e^- \longrightarrow IO_3^- + 3H_2O$	1.603

TABLE A (Continued)

Couple	$E°/V$
$HBrO + H^+ + e^- \longrightarrow 1/2 Br_2 + H_2O$	1.604
$HClO + H^+ + e^- \longrightarrow 1/2 Cl_2 + H_2O$	1.630
$Bk^{4+} + e \longrightarrow Bk^{3+}$	1.67
$AmO_2^{2+} + 4H^+ + 3e^- \longrightarrow Am^{3+} + H_2O$	1.67
$HClO_2 + 2H^+ + 2e^- \longrightarrow HClO + H_2O$	1.674
$PbO_2(\alpha) + SO_4^{2-} + 4H^+ + 2e^- \longrightarrow PbSO_4 + 2H_2O$	1.698
$MnO_4^- + 4H^+ + 3e^- \longrightarrow MnO_2 + 2H_2O$	1.70
$Ce^{4+} + e \longrightarrow Ce^{3+}$	1.72
$AmO_2^+ + 4H^+ + 2e^- \longrightarrow Am^{3+} + 2H_2O$	1.72
$H_2O_2 + 2H^+ + 2e^- \longrightarrow 2H_2O$	1.763
$Au^+ + e^- \longrightarrow Au$	1.83
$Co^{3+} + e^- \longrightarrow Co^{3+}$	1.92
$HN_3 + 3H^+ + 2e^- \longrightarrow NH_4^+ + N_2$	1.96
$S_2O_8^{2-} + 2e^- \longrightarrow 2SO_4^{2-}$	1.96
$Ag^{2+} + e^- \longrightarrow Ag^+$	1.980
$O_3 + 2H^+ + 2e^- \longrightarrow O_2 + H_2O$	2.075
$F_2O + 2H^+ + 4e^- \longrightarrow 2F^- + H_2O$	2.153
$OH + H^+ + e^- \longrightarrow H_2O$	2.38
$O(g) + 2H^+ + 2e^- \longrightarrow H_2O$	2.430
$Am^{4+} + e^- \longrightarrow Am^{3+}$	2.62
$F_2 + 2e^- \longrightarrow 2F^-$	2.87
$F_2 + 2H^+ + 2e^- \longrightarrow 2HF(aq)$	3.053

TABLE B Standard Potentials in Basic Solutions

Couple	$E°/V$
$Ca(OH)_2 + 2e^- \longrightarrow Ca + 2OH^-$	−3.026
$Ba(OH)_2 + 2e^- \longrightarrow Ba + 2OH^-$	−2.99
$Sr(OH)_2 + 2e^- \longrightarrow Sr + 2OH^-$	−2.88
$Y(OH)_3 + 3e^- \longrightarrow Y + 3OH^-$	−2.85
$Ho(OH)_3 + 3e^- \longrightarrow Ho + 3OH^-$	−2.85
$Er(OH)_3 + 3e^- \longrightarrow Er + 3OH^-$	−2.84
$Tm(OH)_3 + 3e^- \longrightarrow Tm + 3OH^-$	−2.83
$Lu(OH)_3 + 3e^- \longrightarrow Lu + 3OH^-$	−2.83
$Gd(OH)_3 + 3e^- \longrightarrow Gd + 3OH^-$	−2.82
$Tb(OH)_3 + 3e^- \longrightarrow Tb + 3OH^-$	−2.82
$La(OH)_3 + 3e^- \longrightarrow La + 3OH^-$	−2.80
$Sm(OH)_3 + 3e^- \longrightarrow Sm + 3OH^-$	−2.80
$Dy(OH)_3 + 3e^- \longrightarrow Dy + 3OH^-$	−2.80
$Pr(OH)_3 + 3e^- \longrightarrow Pr + 3OH^-$	−2.79
$Ce(OH)_3 + 3e^- \longrightarrow Ce + 3OH^-$	−2.78
$Nd(OH)_3 + 3e^- \longrightarrow Nd + 3OH^-$	−2.78
$Pm(OH)_3 + 3e^- \longrightarrow Pm + 3OH^-$	−2.76
$Yb(OH)_3 + 3e^- \longrightarrow Yb + 3OH^-$	−2.74
$Mg(OH)_2 + 2e^- \longrightarrow Mg + 2OH^-$	−2.687
$Sc(OH)_3 + 3e^- \longrightarrow Sc + 3OH^-$	−2.60
$ThO_2 + 2H_2O + 4e^- \longrightarrow Th + 4OH^-$	−2.56
$Am(OH)_3 + 3e^- \longrightarrow Am + 3OH^-$	−2.53
$Cm(OH)_3 + 3e^- \longrightarrow Cm + 3OH^-$	−2.5
$Pu(OH)_3 + 3e^- \longrightarrow Pu + 3OH^-$	−2.46
$Al(OH)_4^- + 3e^- \longrightarrow Al + 4OH^-$	−2.310

TABLE B (Continued)

Couple	$E°/V$
$Al(OH)_3(g) + 3e^- \longrightarrow Al + 3OH^-$	-2.300
$Np(OH)_3 + 3e^- \longrightarrow Np + 3OH^-$	-2.2
$TiO + H_2O + 2e^- \longrightarrow Ti + 2OH^-$	-2.13
$U(OH)_3 + 3e^- \longrightarrow U + 3OH^-$	-2.10
$H_2PO_2^- + e^- \longrightarrow P + 2OH^-$	-2.05
$Ti_2O_3 + H_2O + 2e^- \longrightarrow 2TiO + 2OH^-$	-1.95
$B(OH)_4^- + 3e^- \longrightarrow B + 4OH^-$	-1.811
$SiO_3^{2-} + 3H_2O + 4e^- \longrightarrow Si + 6OH^-$	-1.7
$HPO_3^{2-} + 2H_2O + 2e^- \longrightarrow H_2PO_2^- + 3OH^-$	-1.57
$Mn(OH)_2 + 2e^- \longrightarrow Mn + 2OH^-$	-1.56
$ZnS + 2e^- \longrightarrow Zn + S^{2-}$	-1.44
$PuO_2 + 2H_2O + e^- \longrightarrow Pu(OH)_3 + OH^-$	-1.4
$2TiO_2 + H_2O + 2e^- \longrightarrow Ti_2O_3 + 2OH^-$	-1.38
$Zn(CN)_4^{2-} + 2e^- \longrightarrow Zn + 4CN^-$	-1.34
$Cr(OH)_3 + 3e^- \longrightarrow Cr + 3OH^-$	-1.33
$Cr(OH)_4^- + 3e^- \longrightarrow Cr + 4OH^-$	-1.33
$Zn(OH)_4^{2-} + 2e^- \longrightarrow Zn + 4OH^-$	-1.285
$CdS + 2e^- \longrightarrow Cd + S^{2-}$	-1.255
$Zn(OH)_2 + 2e^- \longrightarrow Zn + 2OH^-$	-1.248
$H_2GaO_3^- + H_2O + 3e^- \longrightarrow Ga + 4OH^-$	-1.22
$Te + 2e^- \longrightarrow Te^{2-}$	-1.14
$PO_4^{3-} + 2H_2O + 2e^- \longrightarrow HPO_3^{2-} + 3OH^-$	-1.12
$WO_4^{2-} + 4H_2O + 6e^- \longrightarrow W + 8OH^-$	-1.074
$ZnCO_3 + 2e^- \longrightarrow Zn + CO_3^{2-}$	-1.06
$Zn(NH_3)_4^{2+} + 2e^- \longrightarrow Zn + 4NH_3$	-1.04

TABLE B (Continued)

Couple	$E°/V$
$HGeO_3^- + 2H_2O + 4e^- \longrightarrow Ge + 5OH^-$	−1.03
$MnO_2 + 2H_2O + 4e^- \longrightarrow Mn + 4OH^-$	−0.980
$CNO^- + H_2O + 2e^- \longrightarrow CN^- + 2OH^-$	−0.97
$Cd(CN)_4^{2-} + 2e^- \longrightarrow Cd + 4CN^-$	−0.943
$SO_4^{2-} + H_2O + 2e^- \longrightarrow SO_3^{2-} + 2OH^-$	−0.94
$PbS + 2e^- \longrightarrow Pb + S^{2-}$	−0.923
$MoO_4^{2-} + 4H_2O + 6e^- \longrightarrow Mo + 8OH^-$	−0.913
$Sn(OH)_6^- + 2e^- \longrightarrow HSnO_2^- + H_2O + 3OH^-$	−0.91
$P + 3H_2O + 3e^- \longrightarrow PH_3 + 3OH^-$	−0.89
$2H_2O + 2e^- \longrightarrow H_2 + 2OH^-$	−0.828
$Cd(OH)_2 + 2e^- \longrightarrow Cd + 2OH^-$	−0.824
$VO + H_2O + 2e^- \longrightarrow V + 2OH^-$	−0.820
$HFeO_2^- + H_2O + 2e^- \longrightarrow Fe + 3OH^-$	−0.8
$MoO_4^{2-} + 2H_2O + 2e^- \longrightarrow MoO_2 + 4OH^-$	−0.780
$CdCO_3 + 2e^- \longrightarrow Cd + CO_3^{2-}$	−0.734
$Co(OH)_2 + 2e^- \longrightarrow Co + 2OH^-$	−0.733
$Ni(OH)_2 + 2e^- \longrightarrow Ni + 2OH^-$	−0.72
$CrO_4^{2-} + 4H_2O + 3e^- \longrightarrow Cr(OH)_4^- + 4OH^-$	−0.72
$Ag_2S + 2e^- \longrightarrow Ag + S^{2-}$	−0.691
$FeO_2^- + H_2O + e^- \longrightarrow HFeO_2^- + OH^-$	−0.69
$AsO_2^- + 2H_2O + 3e^- \longrightarrow As + 4OH^-$	−0.68
$Se + 2e^- \longrightarrow Se^{2-}$	−0.67
$AsO_4^{3-} + 2H_2O + 2e^- + AsO_2^- + 4OH^-$	−0.67
$Sb(OH)_4^- + 3e^- \longrightarrow Sb + 4OH^-$	−0.639
$Cd(NH_5)_4^{2+} + 2e^- \longrightarrow Cd + 4NH_3$	−0.622

TABLE B (Continued)

Couple	$E°/V$
$ReO_4^- + 2H_2O + 7e^- \longrightarrow Re + 8OH^-$	−0.604
$ReO_4^- + 2H_2O + 3e^- \longrightarrow ReO_2 + 4OH^-$	−0.594
$2SO_3^{2-} + 3H_2O + 4e^- \longrightarrow S_2O_3^{2-} + 6OH^-$	−0.58
$ReO_2 + H_2O + 4e^- \longrightarrow Re + 4OH^-$	−0.564
$Cu_2S + 2e^- \longrightarrow 2Cu + S^{2-}$	−0.542
$HPbO_2^- + H_2O + 2e^- \longrightarrow Pb + 3OH^-$	−0.502
$Ni(NH_3)_6^{2+} + 2e^- \longrightarrow Ni + 6NH_5$	−0.476
$Sb(OH)_6^- + 2e^- \longrightarrow Sb(OH)_4^- + 2OH^-$	−0.465
$Bi_2O_3 + 3H_2O + 6e^- \longrightarrow Bi + 6OH^-$	−0.452
$S + 2e^- \longrightarrow S^{2-}$	−0.45
$NiCO_3 + 2e^- \longrightarrow Ni + CO_3^{2-}$	−0.45
$Cu(CN)_2^- + e^- \longrightarrow Cu + 2CN^-$	−0.44
$TeO_3^{2-} + 3H_2O + 4e^- \longrightarrow Te + 6OH^-$	−0.42
$Cu_2O + H_2O + 2e^- \longrightarrow 2Cu + 2OH^-$	−0.365
$SeO_3^{2-} + 3H_2O + 4e^- \longrightarrow Se + 6OH^-$	−0.36
$Tl(OH) + e^- \longrightarrow Tl + OH^-$	−0.343
$O_2 + e^- \longrightarrow O_2^-$	−0.33
$Ag(CN)_2^- + e^- \longrightarrow Ag + 2CN^-$	−0.31
$Cu(SCN) + e^- \longrightarrow Cu + SCN^-$	−0.310
$CuO + H_2O + 2e^- \longrightarrow Cu + 2OH^-$	−0.29
$Mn_2O_3 + 2H_2O + 2e^- \longrightarrow 2Mn(OH)_2 + 2OH^-$	−0.25
$2CuO + H_2O + 2e^- \longrightarrow Cu_2O + 2OH^-$	−0.22
$Cu(NH_3)_2^+ + e^- \longrightarrow Cu + 2NH_3$	−0.100
$O_2 + H_2O + 2e^- \longrightarrow HO_2^- + OH^-$	−0.0649
$Tl(OH)_3 + 2e^- \longrightarrow TlOH + 2OH^-$	−0.05

TABLE B (Continued)

Couple	$E°/V$
$MnO_2 + H_2O + 2e^- \longrightarrow Mn(OH)_2 + 2OH^-$	−0.05
$AgCN + e^- \longrightarrow Ag + CN^-$	−0.017
$NO_3^- + H_2O + 2e^- \longrightarrow NO_2^- + 2OH^-$	0.01
$SeO_4^{2-} + H_2O + 2e^- \longrightarrow SeO_3^{2-} + 2OH^-$	0.03
$Co(NH_3)_6^{3+} + e^- \longrightarrow Co(NH_3)_6^{2+}$	0.058
$TeO_4^{2-} + H_2O + e^- \longrightarrow TeO_3^{2-} + 2OH^-$	0.07
$HgO(red) + H_2O + 2e^- \longrightarrow Hg + 2OH^-$	0.0977
$N_2H_4 + 2H_2O + 2e^- \longrightarrow 2NH_5(aq) + 2OH^-$	0.1
$VO_4^{3-} + 4H_2O + 5e^- \longrightarrow V + 8OH^-$	0.120
$Co(OH)_3 + e^- \longrightarrow Co(OH)_2 + OH^-$	0.17
$HO_2^- + H_2O + e^- \longrightarrow OH + 2OH^-$	0.184
$O_2^- + H_2O + e^- \longrightarrow HO_2^- + OH^-$	0.20
$PbO_2(\beta) + H_2O + e^- \longrightarrow HPbO_2 + OH^-$	0.208
$PbO_2(\beta) + H_2O + 2e^- \longrightarrow PbO(red) + 2OH^-$	0.247
$IO_5^- + 3H_2O + 6e^- \longrightarrow I^- + 6OH^-$	0.257
$ClO_3^- + H_2O + 2e^- \longrightarrow ClO_2^- + 2OH^-$	0.295
$PuO_2(OH)_2 + e^- \longrightarrow PuO_2OH + OH^-$	0.3
$PbO_3^{2-} + 2H_2O + 2e^- \longrightarrow HPbO_2^- + 3OH^-$	0.330
$Ag_2O + H_2O + 2e^- \longrightarrow 2Ag + 2OH^-$	0.342
$Ag(NH_3)_2^+ + e^- \longrightarrow Ag + NH_3$	0.373
$ClO_4^- + H_2O + 2e^- \longrightarrow ClO_3^- + 2OH^-$	0.374
$O_2 + 2H_2O + 4e^- \longrightarrow 4OH^-$	0.401
$NH_2OH + H_2O + 2e^- \longrightarrow NH_3 + 2OH^-$	0.42
$Ag(SO_3)_2^{3-} + e^- \longrightarrow Ag + 2SO_3^{2-}$	0.43
$Ag_2CO_3 + 2e^- \longrightarrow 2Ag + CO_3^{2-}$	0.47

TABLE B (Continued)

Couple	$E°/V$
$IO^- + H_2O + 2e^- \longrightarrow I^- + 2OH^-$	0.472
$NiO_2 + 2H_2O + 2e^- \longrightarrow Ni(OH)_2 + 2OH^-$	0.490
$FeO_4^{2-} + 2H_2O + 3e^- \longrightarrow FeO_2^- + 4OH^-$	0.55
$BrO_3^- + 3H_2O + 6e^- \longrightarrow Br^- + 6OH^-$	0.584
$RuO_4^- + e^- \longrightarrow RuO_4^{2-}$	0.593
$MnO_4^{2-} + 2H_2O + 2e^- \longrightarrow MnO_2 + 4OH^-$	0.62
$2AgO + H_2O + 2e^- \longrightarrow Ag_2O + 2OH^-$	0.640
$H_3IO_6^{2-} + 2e^- \longrightarrow IO_3^- + 3OH^-$	0.656
$PbO_4^{4-} + 3H_2O + 2e^- \longrightarrow HPbO_2^- + 5OH^-$	0.680
$ClO_2^- + H_2O + 2e^- \longrightarrow ClO^- + 2OH^-$	0.681
$Ag_2O_3 + H_2O + 2e^- \longrightarrow 2AgO + 2OH^-$	0.739
$BrO^- + H_2O + 2e^- \longrightarrow Br^- + 2OH^-$	0.766
$HO_2^- + H_2O + 2e^- \longrightarrow 3OH^-$	0.867
$ClO^- + H_2O + 2e^- \longrightarrow Cl^- + 2OH^-$	0.890
$ClO_2 + e^- \longrightarrow ClO_2^-$	1.041
$O_3 + H_2O + 2e^- \longrightarrow O_2 + 2OH^-$	1.246
$OH + e^- \longrightarrow OH^-$	1.985

Author Index

Numbers in parentheses are reference numbers and indicate that an author's work is referred to although the name may not be cited in text. Underscored numbers give the page on which the complete reference is listed.

Aaron, A., 145(29), 161
Abel, E., 45(15), 47
Acerete, R., 504(24), 506
Acree, S. F., 147(40), 162
Aditya, S., 279(46), 284
Agafonov, I. L., 164(11b), 165(11b), 171
Agafonova, A. L., 164(11b), 165(11b), 171
Agasyan, P. K., 772(67b), 775(67b), 784
Ageeva, E. D., 358(37), 364
Ageno, F., 331(30), 332(30), 338
Agron, P. A., 769(53), 784
Ahlberg, J. E., 88(18), 91
Ahluwahlia, J. C., 276(18), 283
Ahmed, N., 530(6), 535
Ahrland, S., 181(10), 185(10), 187, 645(2), 650
Aida, H., 677(8), 686
Aksel'rud, N. V., 615(43), 627
Aleinikov, N. N., 772(67b), 775(67b), 777(100), 784, 785
Ali, I., 469(33), 475
Allamagny, P., 767(30), 783
Allen, P. L., 101(4), 124
Allen, R. J., 418(30), 427
Allen, T. L., 86(7), 90
Altynov, V. I., 228(12), 235
Alvarez, V. E., 418(30), 427
Ambrose, J., 435(10), 439

Amira, M. F., 141(58), 162
Amjad, Z., 619(63), 628
Amosse, J., 383(6), 390
Anaorg, Z., 239(4), 246
Andersen, T. N., 197(4,5), 200
Anderson, L. B., 622(72), 628
Anderson, L. H., 424(34), 427
Anderson, T. F., 154(51), 162
Andreev, E. N., 120(29), 125
Andreev, V. I., 664(2b), 667, 669(1), 670
Andrews, L. V., 437(17), 439
Andronov, E. A., 362(41), 365
Angstadt, R. T., 230(24), 231(24, 28), 235
Anson, F. C., 354(24), 364 417(25), 418(25), 427
Antonovskya, E. I., 466(22), 471(22), 474
Antropov, L. I., 577(10), 578(12), 580
Appelman, E. H., 78(1), 83(1), 83, 91(5), 92(6), 92, 763(4), 765(4), 769(59), 772(59,61,63, 64,68), 773(68,75), 774(75), 775(4,75), 776(4), 778(75,86, 87), 779(61), 780(75,96), 782, 784, 785
Ardizzone, S., 717(3), 724
Ardon, M., 453(2), 464(2), 465(2), 474

Argo, W. L., 728(4), 761(4), 762
Ariya, S. M., 163(3), 171
Armbruster, M. H., 761(8), 762
Arnett, E. M., 103(6), 124
Artyukhin, P., 658(8), 659
Arvia, A. J. 329(21), 331(21), 338
Asprey, L. B., 663(7), 664
Atwood, D. K., 416(10), 426
Auerman, L. N., 623(75,76,78), 628, 670(2), 670, 671(24), 671, 672(1), 672
Awad, S. A., 119(25), 125, 218(8), 220, 228(16), 235
Awtrey, A. D., 104(13), 124

Babaeva, A. V., 345(21), 364
Babintseva, O. A., 619(61), 620(61), 628
Babko, A. K., 464(15), 474
Backhouse, J. R., 416(9), 426
Baeckström, S., 79(2), 83
Baes, C. F., 391(6), 407(6), 411, 615(40), 617(40), 619(40), 627
Bagotskii, V. S., 57(10,12,13), 66
Baidsen, P. A., 672(5), 673
Bailey, N., 482(12), 483
Bailey, S., 40(2), 46
Bailey, S. H., 163(16), 171(16), 172
Bailey, S. M., 94(1), 100(1), 105(1), 113(1), 117(1), 122(1), 124, 127(4), 137(4), 139, 159(59), 162, 249(1), 256, 257(6), 262(6), 264(6), 264, 266(11,12), 275(11,12), 276(11), 278(11), 279(11,12), 280(11), 281(11), 282(11), 283, 287(3,12), 293, 391(3), 409(3), 411, 430(2,3), 439, 441(8), 444, 475(5), 479(5), 483, 486(3), 505, 507(9), 515(9), 525, 590(18), 620(19,69), 621(19,69), 622(19), 626, 628, 776(84), 785
Baker, C. T., 254(8), 256
Baker, F. B., 519(35), 526, 619(59), 628
Baker, L. C. W., 504(24), 506
Bali, A., 464(9), 474
Bane, R. W., 778(85,92), 785
Baran, E. J., 441(9), 444
Baranov, A. A., 669(7), 669
Barber, M., 545(18), 547

Barbier, M. J., 383(6), 390
Barclay, D. J., 417(25), 418(25), 427
Bard, A. J., 77(4), 77, 237(1), 239(1), 240(5), 242(5), 243(21), 245(21), 246(21), 246, 247, 464(12), 466(12), 474, 700(4), 711
Barin, I., 194(13,14), 200, 203(9,10), 206, 216(15,16), 220, 226(39,40), 235, 391(4), 411
Barkovskii, V. F., 504(23), 506
Barnes, H. L., 282(57), 285
Barnighausen, H., 623(80), 628
Bartelt, H., 371(3), 381
Bartholomew, C. H., 377(22), 382
Bartelt, H., 374(15), 382
Bartlett, N., 766(20), 767(20,33,34,35), 783
Bary, R., 304(18), 312
Basolo, F. G., 44(14), 47
Basu, A. K., 279(46), 284
Bates, R. G., 42(10), 46, 147(39,40), 162, 277(25), 284, 294(9,10), 304(9,18,28), 305(9,10,28), 308(28), 311, 312
Batsanova, L. R., 677(14), 686
Baum, R. L., 254(5), 256
Baur, E., 181(11), 187
Bavay, J. Cl., 424(36), 427
Baybarz, R. D., 618(49,50), 619(55), 622(50), 627, 628, 663(8), 664, 665(3), 667, 669(5,6,8), 669, 670(4), 670, 671(3), 671, 672(2), 672
Bayer, G., 343(12), 364
Baymakov, Y. V., 541(12), 544(12), 546
Bazhin, N. M., 63(31), 66
Beall, G. W., 616(47), 627
Beck, M. T., 772(65), 784
Beck, T. R., 541(11), 545(11), 546
Beck, W. W., 230(23), 231(23), 232(23), 235
Beg, M. A., 502(15), 505
Behar, D., 60(20), 66
Behl, W. K., 380(36), 382
Bell, R. P., 86(6), 90, 243(22), 247
Bell, W. E., 343(9), 364
Bemis, C., 672(3), 673, 674(1), 674
Bennett, R. J., 254(9), 257
Bent, H. E., 761(10), 762

Beran, P., 424(41,43,44), 427
Berdnikov, V. M., 63(31), 66
Berecki-Biedermann, C., 257(7), 264
Bergh, A. A., 469(31), 475
Berka, A., 231(29), 235
Berkowitz, J., 765(8), 783
Berndt, D., 435(12), 439
Berne, E., 304(15), 312
Bertocci, U., 292(14), 29
Bertram, J., 280(49), 284
Beukenkamp, J., 145(25), 161
Bevan, R. B., Jr., 441(10), 444, 447(8,11,14), 448, 449(3,5), 451
Bezurkov, I. Y., 517(26), 526
Bhagwat, W. V., 141(14), 161
Biedermann, G., 242(15), 247, 257(7), 264, 475(6), 480(6), 483
Bielski, B. H. J., 60(21,21a), 61(21, 21a), 66
Biernat, R. H., 409(31), 412
Bindra, P., 266(17), 283
Bjerrum, J., 261(13), 265, 313(1), 316(1), 317(13), 318(1), 319, 329(22), 330(22), 338
Bjerrum, N., 141(8), 161, 578(14), 580
Blackburn, I., 279(48), 280(48), 284
Blake, M. I., 773(79), 775(79), 784
Blanc, E., 118(24), 125
Blanc, J. C. L., 391(8), 407(8), 412
Blomgren, G. E., 254(4), 256
Bloom, M. C., 114(19), 125
Boadsgaard, H., 502(12), 505
Bockris, J. O'M., 207(3), 211(3), 212(3), 213(3), 213, 700(3), 711(2), 711, 717(2), 717, 724
Bockel, C., 530(10), 535
Bode, H., 435(12), 439
Bodforss, S., 677(11), 686
Boldin, R., 373(12), 380(35), 381, 382
Bolzan, J. A., 329(21), 331(21), 338
Bond, A. M., 186(12), 187
Bondarev, N. K., 279(45), 280(45), 284
Bone, S. J., 230(25), 235, 280(49), 284

Bonilla, C. F., 230(20), 235, 326(12), 329(12), 338
Bonk, J. F., 310(45), 312
Borisowa, L. M., 373(12), 381
Bosch, W., 145(26), 161
Boston, D. R., 453(1), 464(1), 465(1), 474
Bottcher, W., 416(6), 426
Bounsall, E. J., 389(21), 390
Bourgault, P. L., 335(39), 337(39), 339
Bowen, R. J., 57(14), 66
Bower, V. E., 294(10), 304(10), 305(10), 311
Bowersox, D. F., 88(16), 91
Boyd, G. E., 440(1), 441(1,6), 443(1), 444
Brackett, T. E., 277(28), 284
Bradley, D. C., 548(4), 552
Brady, A. P., 528(25), 529(27), 536
Brady, J. D., 672(2), 673
Brand, J. R., 650(7), 650, 651(4), 655
Bratu, E., 45(15), 47
Breeley, R. S., 294(11), 304(11), 311
Breganski, O. A., 541(12), 544(12), 546
Breganski, Z., 541(13), 544(13), 547
Breitinger, G., 475(3), 483
Brenet, J. P., 437(18), 439
Brewer, L., 590(11), 626
Brinton, R. K., 424(40), 427
Brito, F., 517(20,25), 525, 526
Britton, H. J. S., 164(11a), 165(11a), 171, 516(13), 518(13), 525
Brodd, R. J., 249(2), 254(9), 255(2), 256, 257
Brodsky, A. N., 278(34), 280(34), 284
Broier, A. F., 84(3), 90
Bronsted, J. N., 227(7), 235
Brosset, C., 579(20), 580
Brown, A. P., 266(17), 283
Brown, D. J., 437(17), 439
Brown, G. M., 416(6), 426
Brown, I. D., 265(5), 283
Brown, M. G., 89(22), 91
Brown, R. A., 178(11), 179

Brubaker, C. H., 417(19), 427, 531(15), 536
Brzezinska, M. B., 316(12), 319
Buckingham, D. A., 425(46), 427
Buckley, R. R., 413(5), 426
Buehrer, T. F., 316(15), 319
Buist, G. J., 89(21), 91
Bukshpan, S., 765(15), 783
Bulakh, A. A., 336(52), 337(52), 339
Burbank, J., 233(36), 235
Burbank, R. D., 767(33), 783
Burgess, J., 447(12,13), 448
Burkov, K. A., 698(7), 699
Burnett, J. L., 590(12), 618(49,50), 622(50), 626, 627, 663(8), 664, 665(3), 667, 669(5), 669, 670(4), 670, 671(3), 671, 672(2), 672
Burnett, M. G., 377(26), 378(26), 382
Burtin, B., 676(2), 686
Busey, R. H., 441(10), 444, 447(8, 11,14), 448, 449(3,5), 451
Bus'ko, E. A., 698(7), 699
Butler, E. A., 88(16), 91
Butler, J. N., 700(6), 710(6), 711, 761(5), 762
Bye, J., 453(4), 474

Cady, G. H., 768(41), 773(70), 783, 784
Cahan, B. D., 231(28), 235
Cann, J. Y., 228(13,14), 235
Carasso, J. I., 204(4), 205(4), 206(4), 206, 207(2), 211(2), 212(2), 213(2), 213
Cardi, S., 306(36), 307(36), 312
Carleer, M., 676(2), 686
Carmody, W. R., 221(1), 227(1), 234
Caroli, S., 243(23), 247, 257(8), 261(8), 262(8), 264, 266(14), 282(14), 283, 326(8), 327(8), 329(8), 338, 371(6), 381, 409(33), 412, 507(3), 518(3), 525, 710(10), 711, 716(6), 717, 723(6), 724(3), 724, 726(3), 726
Carpenter, J. E., 520(38), 526
Carr, D. S., 326(12), 329(12), 338
Carrett, A. B., 310(45), 312
Carrington, A., 437(19), 439

Cartledge, G. H., 421(33), 427, 440(2), 441(2), 442(2,5), 443(2,5), 444
Cason, D. L., 86(10), 90
Cavoli, S., 67(2), 69
Cerutti, P., 326(14), 329(14), 338, 391(5), 407(5), 411
Chaberek, S., 330(24), 338
Chadwick, B. M., 430(6), 439
Chalkin, V. A., 91(3), 92
Chang, J. C., 389(20), 390
Charlot, G., 282(54), 285
Chartier, P., 226(10), 227(10), 232(10), 235
Chasanov, M. G., 415(3), 426
Chase, M. W., 486(7), 501(7), 505
Chateau, H., 304(17), 312, 313(4), 318(4), 319
Chauchard, J., 278(42), 279(42), 281(42), 284
Chemla, M., 42(9), 46
Chernick, C. L., 766(24), 767(24), 768(47), 769(47), 770(47), 783, 784
Chernyaev, I. I., 355(34), 356(34), 364
Chevalet, J., 62(24), 66
Chizhikov, D. M., 336(53), 337(53), 339
Chomutov, N. E., 139(1), 154(1), 161, 166(14), 167(14), 171
Choudhary, B. K., 279(44), 284
Christensen, J. J., 147(41), 162, 339(3), 343(3,6), 363, 364, 377(22), 382
Christiane, V. F., 159(57), 162
Chupka, W. A., 765(8), 783
Churney, K. L., 524(9), 525, 590(18), 626
Ciavatta, L., 430(8), 439
Claassen, H. H., 766(19,22), 768(22), 783
Clark, W. M., 110(15), 124
Clauss, J. K., 529(27), 536
Cleveland, J. M., 655(1), 659
Cobble, J. W., 276(18), 283, 409(32), 412, 440(1), 441(1), 443(1), 444, 447(9,10), 448, 590(17), 626, 650(7), 650, 651(4), 655
Coche, A., 535(40), 536
Cohen, D., 278(36), 284, 623(83), 629, 645(2a,2b), 651(2a,2b),

Author Index

[Cohen, D.,]
  654(2a,2b,7), 655, 660(4a), 664, 655(6), 667
Cole, L. G., 87(11), 90
Coleman, J. C., 660(5), 664(5), 664
Colin, R., 676(2), 686
Collat, J. W., 564(5), 565
Collumea, A., 282(54), 285
Colombier, I., 329(17), 338
Colvin, J. H., 515(42), 521(39), 522(42), 526
Conley, H. L., 619(60), 620(60), 621(60), 628
Connick, R. E., 104(13), 124, 417(20,22), 427, 658(3), 659
Connolly, P. J., 377(26), 378(26), 382
Connor, J. A., 447(15), 448
Coops, M. S., 672(2), 673
Conway, B. E., 41(4), 45(4b,4c,16, 17), 46, 47, 54(4), 66, 335(39, 40), 337(39,40), 339
Corey, E. J., 65(35), 66
Coryell, C. D., 520(37), 526
Cossi, D., 385(8), 390
Coté, P. A., 287(11), 293
Cottin, M., 535(40), 536
Cotton, F. A., 49(1), 65, 200(1), 206, 207(1), 213, 321(1), 338, 383(3), 390, 440(1), 448(1,2,3), 448, 453(5), 464(5), 474, 547(1), 548(1), 552
Couffin, F., 654(5), 655
Courtney, R. C., 330(24), 338
Couturier, Y., 287(6), 293
Covington, A. K., 231(27), 235, 240(6), 246, 276(21), 279(43), 280(51), 283, 284, 434(9), 435(9,10,11), 439
Cowling, R. D., 380(34), 382
Cowperthwaite, I. A., 229(18), 235
Craig, D. N., 233(34,35), 235
Crenshaw, J. L., 761(8), 762
Cressey, T., 434(9), 435(9), 439
Criss, C. M., 409(32), 412
Crouthamel, C. E., 89(20), 91
Crowell, W. R., 424(40), 427
Crowther, J. P., 145(28), 161
Cubicciotti, D., 180(4), 187, 529(30), 536
Cunningham, B. B., 672(1), 673
Curley-Fiorino, M. E., 319(20), 320

Curnutt, J. L., 486(7), 501(7), 505
Cuta, F., 330(26), 338
Cutforth, B. D., 265(4,5), 283
Czapski, G., 60(20), 62(25,26), 66

Dale, B. W., 516(11), 525
Darland, W. G., 254(9), 257
Dash, U. N., 306(34), 307(34), 312
David. F., 590(10,14), 626, 637(2), 639, 672(3,4,5), 672, 673
Davidson, A. W., 416(7), 426, 676(4), 686
Davies, C. G., 265(4,5), 283
Davies, G., 373(13), 381
Davis, H. J., 163(2), 166(2), 171
Davis, C. W., 144(22), 161
Dean, P. A. W., 265(4), 283
de Béthune, A. J., 54(5), 66, 67(3), 69(3), 69, 109(16), 116(16), 121(16), 124, 189(3), 195(3), 196(3), 197(3), 198(3), 199(3), 200, 203(5), 204(5), 205(5), 206, 207(4), 211(4), 213, 217(6), 218(6), 219(6), 220, 227(6), 228(6), 229(6), 232(6), 234, 257(9), 261(9,15), 262(9,15), 263(15), 264(9,15), 264, 265, 282(53), 284, 305(25), 306(25), 307(25), 310(25), 312, 326(9), 327(9), 328(9), 329(9), 338, 540(7), 541(7), 546, 566(4), 578(4), 580, 710(8), 711, 716(4), 717, 723(4), 724(4), 724, 726, 761(11), 762
DeBoer, B. G., 448(3), 448
Decnop-Weever, L. G., 615(41), 627
de Kay Thompson, M., 333(48), 339
Delahay, P., 227(11), 230(11), 232(11), 233(11), 235
Delbert, J., 343(6), 364
Dellien, I., 466(19), 472(19), 474, 475(4,7), 480(7), 481(7), 483, 486(8), 505, 585(3), 585
Deltombe, E., 198(7), 199(7), 200, 220(12), 220, 507(2), 520(2), 522(2), 525
Demitras, G. C., 321(2), 338
Denham, H. G., 330(23), 338
Denny, T. O., 327(15), 338

Dergacheva, M. B., 266(7), 283
Desideri, P. G., 84(2), 87(2), 90(2), 90
Deutsch, E., 417(24), 427
DeVries, T., 228(15), 235, 278(36), 284, 377(24), 382, 416(10), 426
DeWaard, H., 765(15), 783
deZoubov, N., 127(2), 129(2), 130(2), 131(2), 132(2), 133(2), 139, 172(5), 177(5), 179, 220(12), 220, 441(3), 442(3), 443(3), 444, 507(2), 520(2), 522(2), 525, 615(42), 621(42), 623(42), 627
Dhar, N. R., 141(14), 161
Diamond, S. E., 417(26), 418(26), 420(26), 427
Diatchkova, R. A., 671(2b), 671
Dibrov, I. A., 337(36,37), 339
Didchenco, R., 531(14), 536
Dikshitulua, L. S. A., 521(41), 526
Dill, A. J., 761(12), 762
Dillin, D. R., 616(46), 627
Dirkse, T. P., 254(3), 256, 309(43), 312
Distefano, N., 158(55), 159(55), 162
Dittner, P. F., 672(3), 673(3), 673, 674(1), 674
Divisek, J., 61(23), 66
Dkhara, S. Ch., 354(29,30), 355(29,30,31,32), 357(29,36), 358(29,36), 359(29), 360(30,32, 40), 361(29,32,40), 362(29,31, 36), 363(31,32), 364, 365
Dobos, D., 94(2), 102(2), 119(2), 124
Dobrowolski, J., 294(1), 311
Dobson, J. V., 42(8), 46, 231(27), 235, 276(21), 283
Dognin, P., 767(30), 783
Doody, F. G., 45(15), 47
Dorfman, L. M., 59(17), 60(17,20), 66
Dougan, R., 672(5), 673
Dozsa, L., 772(65), 784
Dracka, O., 371(4), 381
Dreyer, I., 91(3), 92
Dreyer, R., 91(3), 92
Druce, J. G. F., 180(3), 187
Drucker, C., 94(3), 124
Dubovitskii, F. I., 779(100), 785
Duby, P., 287(5), 293

Dudley, F. B., 773(70), 784
Dudova, I., 91(1), 92
Duisman, J. A., 232(32), 233(32), 235
Duke, F. R., 619(58), 628
Dullberg, P., 516(17), 525
Dutkiewicz, E., 316(12), 319
Duus, H. C., 159(56), 162
Dvorak, V., 231(29), 235
Dwyer, F. P., 389(23), 390, 410(25), 412, 416(9), 424(39, 40), 425(46), 426, 427
D'yachkova, N. N., 114(18), 119(18), 125
D'yachkova, R. A., 623(76), 628
Dyatlova, N. M., 677(13), 686

Earl, W. B., 577(5), 580
Eatough, D., 339(3), 343(3,6), 363, 364
Edgar, K., 530(6), 535
Eding, H., 529(30), 536
Edwards, J. O., 104(14), 124, 563(3), 565
Eeckhaut, Z., 619(54), 627
Egan, E. P., 141(10), 144(20), 161
Eichner, P., 416(16), 427
Eiverum, G. W., 85(11), 90
El-Aggan, A. M., 466(26), 467(26, 30), 469(26), 474, 475
El-Awady, A. A., 389(21), 390
El-Azmerlli, M. A., 467(30), 475
Elhady, Z. A., 228(16), 235
Eliseev, S. S., 464(16), 474
El Sayed, J. A., 217(5), 220
El-Shamy, H. K., 466(26), 467(26), 469(26), 474
Elson, C. M., 417(28), 418(28), 427
El Wakkad, S. E. S., 217(5), 220, 333(47), 339
Emara, S. E., 333(47), 339
Enaka, Y., 464(10), 474
Epelboin, I., 676(6), 677(8), 686
Erenburg, A. M., 313(8), 317(8, 17), 318(8,17), 319, 320
Erin, E. A., 664(2b), 667, 669(1, 7), 669, 670
Erofeev, B. V., 278(41), 284
Ertel, D., 530(7), 535

Evans, D. H., 313(5), 316(5), 317(5), 318(5), 319
Evans, E. L., 194(11), 200, 203(8), 206, 226(37), 235, 325(6), 338, 539(1), 546
Evans, M. W., 620(66), 628
Evans, U. R., 540(6), 541(6), 546
Evans, W. H., 40(2), 46, 94(1), 100(1), 105(1), 113(1), 117(1), 122(1), 124, 127(4), 137(4), 139, 159(59), 162, 163(16), 171(16), 172, 249(1), 256, 257(6), 261(11), 262(6), 264(11), 264, 266(11,12), 275(11,12), 276(11), 277(11,12), 278(11), 279(11,12), 280(11), 281(11), 282(11), 283, 287(3, 12), 293, 391(3), 409(3), 411, 430(2,3), 439, 441(8), 444, 475(5), 479(5), 483, 486(3,4), 501(4), 505, 507(9), 515(9), 525, 590(18), 620(19,69), 621(19,69), 622(19), 626, 628, 650(3), 650, 677(9), 686, 776(84), 785
Everett, K. G., 619(64), 621(64), 628
Eyring, H., 197(4,5), 200

Fabyan, C., 59(15), 66
Fahey, J. A., 623(24), 626
Fairbrother, F., 527(3), 530(3,6, 13), 534(3,36), 535(36), 535, 536
Faita, G., 79(3,4), 83(4), 83
Faktor, M. M., 204(4), 205(4), 206(4), 206, 207(2), 211(2), 212(2), 213(2), 213
Fardy, J. J., 667(1), 669
Farina, R., 375(17), 382
Farkas, J., 242(10), 246
Farr, T. D., 149(46), 162
Farrar, T., 85(23), 91
Farrell, J., 19(6), 37
Fasman, A. B., 343(15), 344(15), 364
Fateev, Yu. F., 577(10), 578(12), 580
Faull, J. H., 86(9), 87(9), 90
Fawcett, J., 447(13), 448
Fedorov, A. A., 464(11), 474
Fee, J. A., 61(21b,22), 62(21b,

[Fee, J. A.,] 22), 66
Feher, I., 772(60), 784
Felin, M., 355(33), 364
Felton, E. J., 530(12), 535
Fenwick, F., 176(8), 179
Ferguson, J. E., 416(12), 426
Ferrante, M. J., 447(7), 448(2), 448, 451
Ferrell, D. T., Jr., 279(48), 280(48), 284
Ferris, L. M., 257(3), 264
Fields, P. R., 780(94), 785
Fischer, W., 332(35), 337(35), 339
Fischerova, E., 371(4), 381
Fitzgerald, M. E., 257(2), 264
Fitzgibbon, G. C., 620(67), 628
Flatow, H. E., 769(55), 770(54, 55), 771(54,55), 784
Fleischmann, M., 42(7), 46, 232(31), 235, 266(17), 283
Flesch, G. D., 765(7), 783
Flis, I. E., 75(4), 77
Flohr, K., 780(96), 785
Flynn, C. M., Jr., 504(22), 506
Foerster, F., 337(43,44,45), 339, 464(13), 466(13), 467(13), 474
Fogel, M., 676(3), 686
Foner, S. N., 64(33), 66
Fontana, B. J., 620(68), 621(68), 628
Foote, C. S., 64(32), 66
Forbes, G. S., 475(1), 480(1), 483, 539(2), 540(2), 545(2), 546, 761(10), 762
Ford, D. D., 416(7), 426
Ford, P. C., 417(29), 418(29,30), 419(29), 427
Ford-Smith, M. H., 186(14), 187
Fordyce, J. S., 254(5), 256
Fortunatov, N. S., 336(54), 339
Forziatti, A. F., 761(10), 762
Frandsen, M., 407(12), 412
Fraser, C. J. W., 447(12), 448
Fredrickson, D. R., 415(3), 426
Frei, V., 141(19), 161
Frenkel, Y. A., 669(9), 669
Freund, H., 535(39), 536
Fricke, E., 464(13), 466(13), 467(13), 474
Fricke, R., 579(18), 580
Fridman, Ya. D., 178(12), 179
Fried, S., 642(2), 645

Friedman, H. A., 670(3), 670
Fritts, D. H., 139(5), 139
Frlec, B., 765(13), 783
Froment. M., 676(6), 686
Fromm, G., 416(8), 426
Frydman, M., 563(4), 565
Fuchs, J., 517(24), 526
Fuger, J., 621(32,70), 622(32), 627, 628, 660(2), 664, 667(4), 669
Fujita, S., 265(2,3), 283
Funaki, K., 545(17), 547
Furukawa, G. T., 342(1), 352(1), 363, 387(1), 390, 415(1), 423(1), 426

Gabano, J. P., 435(13), 436(13), 437(18), 439
Gadet, M. C., 313(4,7), 316(9,10), 317(9,10), 318(4,7,9), 319
Gaglioti, V., 327(32), 332(32), 338
Gaidaenko, N. V., 464(16), 474
Gall, B. L., 59(17), 60(17), 66
Gallagher, J. S., 415(1), 423(1), 426
Gallagher, L. S., 387(1), 390
Galloway, W. J., 276(19), 283
Galus, Z., 217(1), 218(11), 220
Galushko, A. J., 623(75), 628
Gandeboeuf, J., 518(32), 519(32), 526
Garber, H. K., 326(14), 329(14), 338
Gard, G., 773(70), 784
Gardiner, J. A., 564(5), 565
Garmash, L. A., 698(7), 699
Garner, C. S., 343(4), 363, 389(20,21), 390
Garreau, M., 676(6), 607(8), 686
Garret, A. B., 329(18), 330(18), 331(18), 338
Garvin, D., 287(13), 293
Gastinger, F., 242(18), 247
Gates, H. S., 481(10), 483
Gauguin, R., 198(8,9,10), 199(9), 200
Gauthier, J., 278(42), 279(42), 281(42), 284
Gave, G. C. B., 305(30), 312
Gayer, K. H., 329(18), 330(18,27), 331(18), 338, 447(8), 448, 449(5), 451
Gebert, E., 767(31), 769(52),

[Gebert, E.,]
 775(31), 783, 784
Gedansky, L. M., 287(2), 293, 294(4), 305(4), 306(4), 307(4), 311
Geier, G., 516(12), 517(12), 525
Gelfman, M. J., 357(36), 358(36), 360(40), 361(40), 362(36), 364, 365
Gelles, F., 86(6), 90
Gelman, A. D., 658(8), 659(10), 659, 660(3), 664
George, J. H. B., 243(22), 247
George, P., 388(13), 390, 410(26), 412, 539(3), 540(3), 546
Georgi, K., 333(46), 339
Gerasimov, A. D., 114(18), 119(18), 125
Gerber, H. K., 391(5), 407(5), 411
Gerke, R. H., 221(2), 228(2), 229(2), 234, 316(14), 319
Gerkers, R., 531(16), 536
Gertman, E. M., 164(10(, 171
Gessner, H., 331(31), 332(31), 338
Getman, F. H., 119(26), 125
Get'man, T. E., 464(15), 474
Getz, C. A., 619(53), 627
Geyer, R., 486(10), 505
Ghiorso, A., 673(1,2), 673, 674
Giauque, W. F., 87(12), 90, 163(5), 171, 232(32), 233(32), 235, 277(29), 284
Gibson, N. A., 410(25), 412
Giellert, N. L., 529(34), 536
Gierst, L., 62(24), 66
Giggenbach, W., 282(55), 285
Gilbert, R. A., 441(10), 444, 447(8,11), 448, 449(5), 451
Gileadi, E., 335(39,40), 337(39, 40), 339
Gillespie, R. J., 63(27), 66, 265(4, 5), 283, 765(14), 783
Giner, J., 42(6), 46
Ginstrup, D., 353(22,23), 354(22, 23), 364
Gladyshev, V. P., 266(6), 283
Glasstone, S., 232(30), 235
Glavin, S. R., 181(7), 187, 227(9), 235
Glushko, V. P., 100(9), 113(9), 117(9), 122(9), 124, 294(7),

[Glushko, V. P.,]
  303(7), 311, 486(5), 501(5), 505
Gmelin, H., 409(35), 412
Goetz, C. A., 619(58), 628
Göhr, H., 381(37), 382
Goiffon, A., 453(3), 474, 530(10), 535
Goldberg, R. N., 343(5), 344(5), 353(5), 354(5), 364, 383(4), 385(4), 390, 413(4), 421(4), 426
Goldfinger, P., 262(16), 265
Golding, M., 305(24), 312
Goldman, S., 590(15), 626
Goldschmidt, V., 41(3), 44(3), 46
Golendzinowski, M., 257(1), 264
Gomer, R., 19(8), 37
Goncharov, A. A., 772(67,67a), 773(67), 774(67), 784
Goodman, G. L., 766(19), 783
Gordon, B. M., 309(41), 312
Gordon, J. E., 103(7), 124
Gordon, S., 59(16), 66, 615(48), 623(83), 627, 629, 660(4a,4b), 664, 665(6), 667
Gortesma, F. P., 531(14), 536
Graham, L., 767(35), 783
Granger, R., 530(10), 535
Grankina, Z. A., 464(17), 474
Grechina, N. K., 186(13), 187
Greeley, R. S., 304(21), 312
Green, G., 154(49), 159(49), 162
Greenberg, E., 529(26), 532(44), 536, 537
Greenway, A., 416(12), 426
Grenthe, I., 181(10), 185(10), 187
Grieb, M. W., 371(2), 381
Griffith, R. O., 141(17), 154(48), 161, 162
Griffith, W. P., 377(25), 382, 424(37), 427
Grigorieva, T. V., 337(36), 339
Grimaldi, M., 430(8), 439
Grinberg, A. A., 357(36), 358(36), 362(36), 364, 345(20), 353(25), 354(25), 364, 388(14), 390
Gritzner, G., 391(1), 411
Gross, P., 529(29,32), 536
Grosse, A., 765(10), 773(74), 774(74), 783, 784
Grube, G., 177(9), 179, 409(35), 412, 416(8), 426, 475(2,3), 480(2), 483

Grube, H. L., 534(19), 536
Grube, V. G., 534(19), 536
Gruen, D. M., 778(88), 779(88), 785
Gruehn, R., 527(4,5), 535
Grzybowsky, A. K., 141(5), 161
Gschneidner, K. A., Jr., 621(20), 622(20), 626
Gubeli, A. O., 254(6), 256, 287(11), 293
Guillaumont, R., 590(10), 626, 637(2), 639
Guiter, C. R., 579(16), 580
Gunn, S. R., 154(4), 159(49), 162, 163(8), 171, 763(3), 765(3,16,17), 766(16), 773(81), 776(81), 778(81), 782, 783, 785
Gupta, S. R., 276(22), 283
Gurnett, J. L., 257(4), 264
Gurney, R., 675(1), 677(1), 686
Gusev, K., 772(62), 784
Guthrie, J. P., 140(3), 141(3), 143(3), 161
Guyoi, P. M., 765(8), 783
Gyarfas, E. C., 410(25), 412

Habeeb, J. J., 186(14), 187, 548(6), 553
Hagan, L., 589(1), 626
Hagesawa, H., 84(5), 90
Hagyard, T., 577(5,6), 580
Hahn, R., 672(3,4), 674(1), 673, 674
Haigh, I., 447(12), 448
Haight, G. P., 453(1), 464(1), 465(1), 466(25), 469(31,32), 474, 475
Haire, R. G., 667(4), 669, 670(5), 670
Haissinsky, M., 535(40), 536, 642(1), 645
Hakeem, M. A., 240(6), 246
Hall, F. M., 466(19), 472(19), 474, 475(4), 483, 486(8), 505
Hall, I. P., 539(2), 540(2), 545(2), 546
Hallada, C. J., 453(8), 474, 516(15), 517(15), 525
Halow, I., 40(2), 46, 94(1), 100(1), 105(1), 113(1), 117(1), 122(1), 124, 127(4), 137(4), 139, 159(59), 162, 163(16),

[Halow, I.,]
171(16), 172, 249(1), 256, 257(6), 262(6), 264(6), 264, 266(11), 275(11), 276(11), 277(11), 278(11), 279(11), 283, 287(3,12), 293, 430(23), 439, 441(8), 444, 475(5), 479(5), 483, 486(3), 505, 507(9), 515(9), 525, 590(18), 620(69), 621(69), 626, 628, 776(84), 785

Halow, J., 391(3), 409(3), 411
Halpern, J., 376(21), 382
Hamer, W. J., 229(19), 230(19,22), 233(19), 235, 277(30), 284, 310(44), 312, 540(5), 546
Hammer, C., 504(24), 506
Hampson, N. A., 257(14), 261(14), 262(14), 265
Hampton, W. H., 407(11), 412
Hanania, G. I. H., 388(13), 390, 410(26,29,30), 412
Hanzlik, J., 376(20), 377(20), 382
Hardwick, T. J., 619(57), 628
Haring, M. M., 217(2), 220, 221(5), 234, 278(31), 284, 326(11), 329(11), 338
Harmalker, S., 504(24), 506
Harned, H. S., 229(19), 230(19), 233(19), 235, 257(2), 264, 277(30), 284
Harris, C. B., 448(2), 448
Harrison, J. H., 57(14), 66
Hart, D., 531(18), 536
Hassan, M. Z., 584(2), 585
Hatfield, M. R., 221(5), 234
Haug, H. O., 622(33), 627
Häussermann, W., 304(22), 305(22), 308(22), 312
Haydon, E., 59(18), 66
Hayes, A. M., 89(20), 91
Hayman, C., 529(29,32), 536
Hayward, P., 154(47), 162
Hazeltine, M. W., 780(95), 785
Head, A. J., 141(9), 161
Head, E. L., 528(24), 536
Hébert, G., 287(11), 293
Heckner, K. -H., 63(30), 66
Heidt, L. J., 59(19), 66
Heimler, D., 84(2), 87(2), 90(2), 90
Heine, H., 529(33), 536
Hejtmanek, M., 330(25,26), 331(25), 338

Hemingway, B. S., 566(2,3), 580, 687(5), 699
Henze, G., 486(10), 505
Hepler, L. G., 242(12), 246, 266(10), 276(10), 277(10), 279(10), 281(10), 282(10), 283, 287(2), 293, 294(4), 305(4), 306(4), 307(4), 311, 378(27), 382, 391(5), 407(5), 411, 466(19), 472(19), 474, 475(4,7), 480(7), 481(7), 482(11), 483, 486(8), 505, 507(5), 515(5), 516(5), 517(5), 522(5), 524(5), 525, 527(1), 534(1), 535, 585(3), 585
Hepler, L. H., 343(5), 344(5), 353(5), 354(5), 364, 383(4), 388(4), 390, 413(4), 421(4), 426
Herman, H. B., 590(16), 618(16), 622(16), 626
Herringshaw, J. F., 700(3), 711(2), 711, 717(2), 717, 724
Hervier, B., 304(17), 312
Herzberg, G., 287(8), 293
Hetzer, H. B., 304(28), 305(28), 308(28), 312
Heusler, K. E., 407(10), 408(10, 19), 412, 437(21), 439
Hickling, A., 63(28), 66, 101(4), 124, 337(38), 339
Hiddlestone, J. N., 42(7), 46
Hieber, W., 378(28), 382
Higginson, W. C. E., 374(16), 375(16), 382
Hilbrants, W., 765(15), 783
Hill, J. O., 507(5), 515(5), 516(5), 517(5), 522(5), 524(5), 525, 527(1), 534(1), 535
Hill, S., 63(28), 66
Hills, G. J., 276(20,22), 278(40), 283, 284
Hinchey, R. J., 590(17), 626
Hindeman, J. C., 765(11), 783
Hindman, J. C., 416(14), 427, 642(2), 645(1,2a,2b), 645, 650, 651(1,2a,2b,3), 654(1, 2a,2b), 655
Hin-Fat, L., 374(16), 375(16), 382
Hippel, A. T., 114(19), 125
Hipperson, W. C. P., 89(21), 91
Hoar, T. P., 218(9), 220
Hoard, J. L., 294(2), 311, 535(38),

[Hoard, J. L.,]
536
Hoare, J. P., 55(6,7,8), 56(7), 66, 343(14), 345(14), 364, 383(7), 385(11), 390
Hobart, D. E., 622(71), 628, 663(6), 664, 672(4), 673
Holley, C. E., 528(24), 536
Holley, C. E., Jr., 620(67), 628
Holloway, J. H., 765(8,13,15), 767(26), 783
Holmes, W. S., 141(11), 161
Holt, B. D., 778(85), 785
Hopkins, T. E., 277(28), 284
Horiguchi, G., 84(5), 90
Horii, H., 265(2,3), 283
Hornung, E. W., 277(28), 284
Howell, O. R., 332(34), 339
Hu, A. T., 486(7), 501(7), 505
Hubbard, A. T., 354(24), 364
Hubbard, W., 529(26), 532(44), 536, 537, 769(56), 770(56), 781(99), 784, 785
Hübel, W., 378(28), 382
Huber, E. J., 528(24), 536
Huber, K. P., 287(8), 293
Huchital, D. H., 372(9), 381
Hudson, R. L., 64(33), 66
Huey, C. S., 218(10), 220
Hugel, R., 287(7), 293
Hughes, W. S., 163(7), 164(7), 171
Hugus, Z. Z., 242(12), 246 448(1), 449(1), 451
Hulet, E. K., 672(2,5), 673(2), 673, 674
Hume, D. H., 331(29), 332(29), 338
Hume, D. N., 305(30), 312
Humphrey, G. L., 532(42), 536
Humpoletz, J. E., 424(38), 427
Hurlen, T., 408(16,17,18), 412
Hurley, C. R., 417(20,22), 427
Huston, J., 768(48), 773(73,76,80), 774(73,76), 775(80), 777(80), 778(89), 784, 785
Huston, R., 761(5), 762
Hyder, M., 669(10), 669
Hyman, H. H., 766(21), 767(36), 768(36), 783

Ilan, Y. A., 62(25), 66
Immerwahr, G., 278(39), 284
Ingri, N., 517(20), 525, 563(4), 565
Inyard, R. E., 343(9), 364
Irani, R. R., 145(30,34,36), 146(34,38), 147(30), 161, 162
Ireland, P. R., 265(4,5), 283
Irvine, D. H., 388(13), 390, 410(26,28,29,30), 412
Irving, R. J., 141(12), 161
Iskraut, A., 768(37), 783
Israel, Y., 507(4), 519(4,36), 520(4,36), 525, 526
Isupov, V. K., 772(62), 779(100), 784, 785
Itzkovich, I. J., 417(28), 418(28), 427
Itzkowitz, L. M., 761(12), 762
Iuff, B. B., 144(20), 161
Ivakin, A. A., 164(10), 171
Ivannikova, N. V., 772(62), 784
Ives, D. J. G., 276(20,22), 278(40), 283, 284
Ivett, R. W., 228(15), 235
Izatt, R. M., 147(41), 162, 339, (3), 343(3,6), 363, 364, 377(22), 382

Jackson, E., 343(7), 344(7), 364, 388(19), 389(19), 390
Jackson, P., 164(11a), 165(11a), 171
Jaffe, I., 261(11), 264(11), 264
Jahr, K. F., 517(22,23,24), 526
Jain, D. V. S., 145(35), 162
Jakob, W., 430(5), 439
James, W. J., 545(19), 547
Jander, G., 517(22), 526, 530(7), 535
Janz, G., 305(26), 312
Jaselskis, B., 773(73,77,78,80), 774(73,77), 775(78,80), 777(80), 784, 785
Jauregui, E. A., 329(21), 331(21), 338
Jeremic, M., 448(3), 448
Jesser, R. A., 378(27), 382
Jeunehomme, M., 262(16), 265

Johnson, D. A., 371(8), 372(8), 381, 587(2), 590(6,7,8), 618(8), 621(8), 622(7,8), 623(7), 626
Johnson, G. K., 763(4,4a), 765(4, 4a), 769(53,56), 770(56), 775(4, 4a), 776(4), 782, 784
Johnson, G. L., 91(1), 92
Johnson, K. E., 417(21), 427
Johnson, P. R., 313(2), 316(2), 318(2), 319
Jolly, W. L., 211(7), 212(7,8), 213
Jones, G., 79(2), 83, 515(42), 521(39), 522(42), 526
Jones, G. R., 767(33), 783
Jonte, J. H., 304(12), 311
Jorgensen, C. K., 587(5), 618(51, 52), 626, 627
Jortner, J., 769(50), 784
Jowett, M., 229(17), 235
Juvet, R. S., Jr., 524(46), 526

Kabanov, B. N., 409(34), 412
Kabir-Ud-Din 502(15), 505
Kahlenberg, F., 529(28), 533(46), 536, 537
Kahn, M., 619(59), 628
Kamalov, N. N., 359(39), 363(42), 365
Kamenskaya, A. N., 670(2), 670, 671(2a,2b), 671, 672(1), 672
Kanevski, E., 17(4), 37
Kangro, W., 242(14), 247
Kao, Y. K., 139(5), 139
Kappenstein, C., 287(7), 293
Karpova, A. A., 467(29), 475
Karpova, L. A., 467(28), 474
Kasha, M., 65(34), 66, 658(9), 659
Kassab, A., 218(8), 220
Kastening, B., 61(23), 66
Kauschka, G., 768(38), 783
Kazakov, V. A., 178(10), 179, 313(6), 316(6,11), 318(6), 319
Kazakova, G. D., 278(41), 284
Kazarnovskii, I. A., 309(42), 312
Kazarnovsky, I., 116(21), 118(21), 125
Keefer, R. M., 86(7), 90
Keenan, T. K., 660(5), 644(1,5), 664, 665(1), 667
Keller, O. L., 672(3), 674(1), 673, 674

Kelley, K. K., 163(6), 171, 529(35), 532(41), 536
Kemball, C., 377(26), 378(26), 382
Keneshea, F. J., 529(30), 536
Kepler, L. O., 326(14), 329(14), 338
Kern, D. M. H., 57(9), 66
Keyes, F. G., 728(3), 761(3,7), 762
Khaldoyanidi, K. A., 464(17), 474
Khalkin, V. A., 91(2), 92
Khan, A. A., 502(15), 505
Khan, A. U., 65(34), 66
Khan, M. M. T., 517(29), 526
Khan, O. A., 336(52), 337(52), 339
Khananova, E. Y., 345(21), 364
Khidzhazi, A., 356(35), 364
Khonin, G. V., 698(7), 699
Khvorostukhina, N. A., 242(19), 247
Kiehl, S. J., 531(18), 536
Kim, J. J., 371(5), 381
Kim, Y. C., 618(37), 623(37), 627
Kimura, G., 710(11), 711
King, C. B., 619(56), 628
King, E. G., 163(6), 171, 287(1), 293, 447(6), 448, 532(41), 536
King, E. L., 481(10), 483
King, J. P., 447(9), 448
Kippenhan, N., 621(20), 622(20), 626
Kirin, I. S., 772(62), 784
Kirkpatrick, K. J., 577(5), 580
Kirshenbaum, A. D., 765(10), 773(74), 774(74), 783, 784
Kisova, L., 257(1), 264
Kiss, L., 242(10), 246
Kita, H., 43(11), 46
Klimov, V. D., 772(69), 773(69), 784
Kleinberg, J., 676(4), 686
Klemm, W., 239(4), 246
Kleykamp, H., 343(13), 364, 466(24), 474
Knacke, O., 194(13,14), 200, 203(9,10), 206, 216(15,16), 220, 226(39,40), 235, 391(4), 411
Knowles, H. B., 449(6), 451
Kochergina, L. A., 552(10), 553

Koch, C. W., 773(72), 778(72), 784
Koepp, H. M., 378(29), 382, 410(24), 412
Kolbin, N. I., 387(10), 390
Kolet, V. Ya., 676(3), 686
Kolthoff, I. M., 45(15), 47, 141(13), 145(26), 161, 331(29), 332(29), 338, 410(27), 412, 471(34), 475, 545(16), 547, 578(11), 579(19), 580, 710(12), 711, 716(7), 717, 723(7), 724
Komandenko, V. M., 119(27), 120(30), 125
Kong, P. C., 481(8), 483
Kon'kova, T. S., 779(100), 785
Konopik, N., 141(6), 161, 165(12), 171
Konovalova, N. A., 672(1), 672
Konrad, D., 367(1), 371(1), 381
Kopytov, V. V., 664(2b), 667, 669(1,7), 669, 670
Korablina, L. S., 355(34), 356(34), 364
Kornfiel, F., 335(50), 339
Kornilov, A. N., 532(43), 537
Korshunov, B. G., 623(75), 628
Korsunskii, B. L., 779(100), 785
Kortüm, G., 304(22), 305(22), 308(22), 312
Kosenko, V. A., 466(21), 474
Kossiakoff, A., 309(39), 312
Kosyakov, V. N., 664(2b), 667, 669(7), 669
Kovach, L., 88(15), 91
Kovach, S. K., 486(2), 505
Kovonova, S. V., 176(7), 177(7), 179
Kozawa, A., 436(14,15), 439
Kozin, L. F., 266(7), 283
Kozlov, Yu. N., 772(67,67a), 773(67), 774(67), 784
Kozlovskii, M. T., 266(6), 283, 467(28,29), 474, 475
Kragten, J., 615(41), 627
Krasova, G. N., 362(41), 365
Kraus, C. A., 88(17), 91, 728(2), 761(2), 762
Krause, E. P., 57(11), 66
Kravtsov, V. I., 343(16), 354(26, 27), 364, 388(16,17), 389(22), 390, 619(61), 620(61), 628
Krestov, G. A., 687(1), 699
Krikoriau, N. H., 528(24), 536

Kritchevsky, E. S., 645(1), 650
Kroo, E., 91(2), 92
Krot, N. N., 664(2), 659(10), 659, 660(1,3), 663(1), 664(9), 664, 667
Krumenacker, L., 453(4), 474
Ksandr, Z., 330(25,26), 331(25), 338
Kubaschewski, O., 194(11,14), 200, 203(8,10), 206, 216(16), 220, 226(37,40), 235, 325(6), 338, 391(4), 411, 539(1), 546
Kudchadker, A. P., 770(57), 771(57), 776(57), 784
Kudchadker, S. A., 770(57), 771(57), 776(57), 784
Kudryashova, N. F., 533(47), 537
Kuehn, C. C., 417(27), 418(27), 420(27), 427
Kukushkin, Y. N., 354(29,30), 355(29,30,31,32), 357(29), 358(29,37,38), 359(29), 360(30, 32,40), 361(29,32,40), 362(29, 31,41), 363(31,32), 364, 365
Kullberg, L., 292(15), 293
Kullgren, C., 330(28), 338
Kulyako, Yu. M., 669(9), 669
Kuniya, Y., 545(17), 547
Kurbanov, A. R., 532(45), 533(45), 537
Kuta, J., 391(1), 411
Kutyukov, C. G., 343(15), 344(15), 364
Kuzmin, L. I., 687(1), 699

Lagerström, G., 204(6), 206, 563(4), 565
Laitinen, H. A., 371(2), 381
Lam, K. W., 417(21), 427
Lamb, A. B., 407(15), 412
Lambert, J. P., 62(24), 66
Lamberts, S. M., 145(31), 147(31), 162
LaMer, V. K., 87(13), 90, 257(5), 264
Lamoreaux, R. H., 87(12), 90
Landi, V. R., 59(19), 66
Landrum, J. H., 672(5), 673
Landsberg, R., 437(20), 439
Lange, E., 15(1), 37, 45(15), 47
Langignard, M., 767(30), 783

Langmuir, D., 408(21), 412
Lansberg, R., 63(30), 66
Lappin, A. G., 502(17), 506
Lapshin, A. I., 316(11), 319
Larson, J. W., 326(14), 329(14), 338, 391(5), 407(5), 411
Larson, W. D., 278(37), 279(47), 284
Latham, R. J., 257(14), 261(14), 262(14), 265
Latilagic, N., 319(18), 320
Latimer, W. M., 67(1), 69, 84(1), 85(1), 88(1), 90, 91(4), 92, 100(11), 113(11), 116(11), 117(11), 119(11), 122(11), 124, 127(1), 129(1), 130(1), 131(1), 132(1), 133(1), 134(1), 139, 154(54), 159(54), 162, 163(1), 164(1), 165(1), 171(1), 171, 172(2), 177(2), 178(2), 179, 180(2), 181(8), 187, 189(1), 194(1), 196(1), 197(1), 198(1), 199, 203(2), 204(2), 205(2), 206, 211(7), 212(7), 213, 216(4), 217(4), 218(4), 219(4), 220, 226(8), 227(8), 228(8), 229(8), 230(8), 231(8), 232(8), 235, 239(3), 242(3,12), 243(3), 245(3), 246(3), 246, 261(12), 262(12), 264(12), 264, 277(26), 280(26), 282(26), 284, 305(23), 306(23), 307(23), 310(23), 312, 325(4), 326(4), 328(4), 329(4), 330(4), 331(4), 338, 416(17), 427, 421(31), 424(31), 427, 442(4), 444, 449(7), 451, 466(20), 469(20), 474, 507(1), 525, 531(17), 534(17), 536, 539(4), 540(4), 541(4), 542(4), 543(4), 544(4), 545(4), 546, 548(9), 552(11), 553, 565(7), 566, 700(2), 711, 728(1), 761(1), 762, 774(83), 785
Lattmann, W., 181(11), 187
Laughlin, B., 676(4), 686
Launay, J. P., 472(38), 475
Lavallee, C., 417(24), 427
Lavallee, D. K., 417(24), 427
Lavrentiev, V. N., 353(25), 354(25), 364
Lawicki, W., 316(12), 319
Lebedev, I. A., 669(9), 669
Lebedev, Yu. A., 765(5), 769(5),
[Lebedev, Yu. A.,] 776(5), 779(100), 782, 785
Lebedj, V. J., 279(45), 280(45), 284
Leberl, O., 141(6), 161, 165(12), 171
Leden, I., 304(15), 312, 353(23), 354(23), 364
Lee, D. G., 417(21), 427
Leger, J. M., 623(81,81,82), 628
Leger, V. Z., 249(2), 255(2), 256
Leiniger, R. F., 91(1), 92
Lemmonds, T. J., 304(20), 312
Lemons, J. F., 658(4), 659
Lempert, D. B., 772(67b), 775(67b), 784
Lempert, S. A., 772(67b), 775(67b), 784
Leonidov, V. Y., 532(43), 537
Lepri, L., 84(2), 87(2), 90(2), 90
Leuschke, W., 276(23), 284
Lever, B. G., 434(9), 435(9), 439
Levi, D. L., 529(29,32), 536
Levine, S., 261(11), 264(11), 264
Levy, H. A., 416(18), 417(18), 427, 769(53), 784
Lewinsohn, M. H., 87(13), 90
Lewis, G. B., 141(9), 161
Lewis, G. N., 45(15), 47, 728(2, 3,4), 761(2,3,4,7,9), 762
Lewis, J. D., 89(21), 91
Liang, C. C., 429(1), 430(1), 439
Licht, T. S., 257(9), 261(9), 262(9), 264(9), 264, 566(4), 578(4), 580, 710(8), 711, 716(4), 717, 723(4), 724(4), 724, 726, 761(11), 762
Liebhafsky, H. A., 79(5), 83, 88(19), 91
Liebman, J. F., 765(9), 783
Lietzke, M. H., 294(11), 304(11, 19,20,21), 311, 312
Liler, M., 232(31), 235
Liljenzin, J., 645(2), 650
Lim, H. S., 417(25), 418(25), 427
Linderbaum, S., 145(25), 161
Lindsay, W. L., 391(9), 407(9), 412
Lingane, J. J., 221(4), 234, 313(3, 5), 316(3,5), 317(3,5), 318(3, 5), 319, 524(45,47,48), 526
Lingane, S. S., 502(11), 505

Linnett, J. W., 545(18), 547
Lipkowski, J., 257(1), 264
Lipscomb, W. N., 555(2), 565
Lister, B. A., 548(3), 552
Litvinchuk, V. M., 502(13,14), 505
Longhi, P., 70(1), 74(1), 77, 700(5), 711, 716(3), 717(3), 717, 724
López-López, M. A., 326(13), 329(13), 338
Loriers, L., 623(81,82), 628
Losev, V. V., 242(8,13), 246, 247, 409(34), 412
Loud, N. A. S., 67(3), 69(3), 69, 109(16), 116(16), 121(16), 124, 189(3), 195(3), 196(3), 197(3), 198(3), 199(3), 200, 203(5), 204(5), 205(5), 206, 207(4), 211(4), 213, 217(6), 218(6), 219(6), 220, 227(6), 228(6), 229(6), 232(6), 234, 282(53), 284, 305(25), 306(25), 307(25), 310(25), 312, 326(9), 327(9), 328(9), 329(9), 338
Lougheed, R. W., 672(2,5), 673
Louwrier, K. P., 619(65), 628
Lovrecek, B., 207(3), 211(3), 212(3), 213(3), 213
Luca, C., 254(10), 257
Luckman, A., 242(16), 247
Luff, B. B., 141(10), 161
Luk'yanycheva, V. I., 57(10), 66
Lundberg, W. O., 88(18), 91
Lundell, G. E. F., 449(6), 451
Luoma, E. V., 417(19), 427
Luther, R., 309(37), 310(37), 312
Lyalikov, Yu. S., 304(16), 312
Lytkin, A. I., 552(10), 553

Macaskill, J. B., 294(9), 305(9), 311
Macero, D. J., 622(72), 628
Maffei, A., 579(17), 580
Magnuson, L., 645(1), 651(1), 654(1), 655
Mah, A. D., 287(1), 293, 507(8), 525, 528(22), 529(22,34), 536
Maki, N., 375(18), 382
Malachowski, P. A., 577(7), 580
Malhotra, K. C., 145(35), 162, 464(9), 474
Malik, W. U., 469(33), 475

Mallory, M. L., 672(3), 673, 674(1), 674
Malm, J. G., 765(11), 766(19,22, 24), 767(24), 768(22,44), 769(51,56,59), 770(51,54,56), 771(51,54,55), 772(59,68), 773(68,75), 774(75), 775(75), 778(75,85,86), 780(75), 783, 784, 785
Malmberg, M. S., 540(5), 546
Maloy, J. T., 139(5), 139
Maly, J., 672(1), 673(1), 673
Mamantov, G., 622(71), 628
Manecke, G., 430(7), 439
Marchon, M. J. C., 282(54), 285
Marcovic, T., 319(18), 320
Marcus, Y., 266(15), 276(15), 283
Margearu, V., 254(10), 257
Maricic, S., 766(23), 783
Markosyan, G. N., 242(11), 246
Marteau, M., 654(5), 655
Martell, A. E., 261(10), 262(10), 264, 278(33), 284, 304(14), 312, 330(24), 338, 481(9), 483, 517(29,30), 526, 548(2), 552
Martin, D. S., 89(20), 91, 304(12), 311
Martin, F. H., 768(40), 783
Martin, W. C., 589(1), 626
Martin, W. J., 535(38), 536
Massart, R., 472(38), 475
Mathis, D. L., 676(5), 686
Mathur, P. B., 306(35), 312
Matsubara, T., 417(29), 418(29, 30), 419(29), 427
Matyashchuk, I. V., 654(8), 655
Matyushin, Yu. N., 779(100), 785
McAuley, A., 619(63), 628
McCarty, R. D., 39(1), 46
McClure, D. S., 539(3), 540(3), 546
McCoubrey, J. C., 85(8), 90
McCullough, J. G., 534(20), 536
McDonald, G. N., 768(47), 769(47, 51), 770(47,51), 771(51), 784
McDonald, J. E., 447(10), 448
McDonald, L. A., 548(3), 552
McDowell, W. J., 673(3), 674
McGavey, J. J., 59(18), 66
McGee, H. A., Jr., 765(18), 766(18), 783
McGilvery, J. D., 145(28), 161
McKenney, J., 417(28), 418(28),

[McKenney, J.,] 427
McKenzie, H. A., 389(23), 390, 424(39), 427
McKeown, A., 141(17), 154(48), 161, 162
McKerrell, H., 141(12), 161
McLendon, G., 379(33), 382
McMasters, O. D., 621(20), 622(20), 626
McMillan, J. A., 294(5), 303(5), 309(40), 311, 312
McTigue, P., 19(6), 37
McVey, W. H., 658(3), 659
Meazek, R. V., 447(6), 448
Medvedoskii, V., 658(8), 659
Mefod'eva, M. P., 659(10(, 659
Meinert, H., 767(29), 768(38), 783
Meisel, D., 62(25), 66
Meites, L., 424(42), 427, 507(4), 517(27), 519(4,27,34,36), 520(4,36), 521(27,34), 524(34, 45,47,48), 525, 526, 534(20), 536
Mellor, J. W., 145(23), 161
Meloun, M., 371(4), 381
Melson, G. A., 580(1), 585
Mendyntsev, V. V., 677(13), 686
Mercer, E. E., 85(23), 91, 413(5), 426
Merlin, J. C., 149(45), 162
Mertes, R. W., 424(40), 427
Mesmer, R. E., 615(40), 617(40), 619(40), 627
Meyer, R. E., 673(3), 674
Meyer, T. J., 417(23), 427
Michallides, M. S., 410(30), 412
Mikhailovskaya, M. I., 467(27), 474
Mikhailovskaya, V. I., 336(54), 337(54), 339
Mikhalevich, K. N., 502(13,14), 505
Mikheev, N. B., 623(75,76,77,78, 79), 628, 670, 671(2a,2b), 671
Milazzo, G., 67(2), 69, 166(15), 171, 243(23), 247, 257(8), 261(8), 262(8), 264, 266(14), 282(14), 283, 306(36), 307(36), 312, 326(8), 327(8), 329(8), 338, 371(6), 381, 409(33), 412, 507(3), 518(3), 525, 710(10), 711, 716(6), 717, 723(6), 724(3), 724, 726(3), 726

Miller, B., 242(9), 246
Milligan, W. O., 616(46,47), 627
Minc, S., 650(4), 650
Minzl, E., 343(10), 364
Miroshnichenko, E. A., 779(100), 785
Mischenko, K., 15(1), 37, 75(3), 77
Mitra, R. P., 145(35), 162
Moeller, T., 590(38), 627
Mogilev'ski, A. N., 669(9), 669
Moh, G., 453(6), 474
Mokhosoyev, M. V., 485(1), 505
Molodov, A. I., 242(11), 246
Monatsch, J., 147(43), 162
Monk, C. B., 141(58), 144(22), 145(24), 161, 162, 327(15), 338
Moody, G. J., 768(42,43), 783
Mooney, W. F., 379(33), 382
Moore, E., 548(12), 553
Mori, T., 265(3), 283
Morrow, J. C., 411(36), 412
Morss, L. R., 590(9,12,15), 618(9), 620(9), 621(9,32), 622(9,32,33), 623(24), 626, 660(2), 664
Morton, C., 141(15), 161
Moser, H. C., 265(1), 283, 780(95), 785
Motorkina, R. K., 503(18), 506
Motov, D. L., 57(12), 66
Moussa, A. A., 181(9), 187
Mulac, V., 623(83), 629, 660(4a, 4b), 665(6), 664, 667
Munson, R. A., 145(37), 162
Murate, F., 326(10), 329(10), 338
Muraveiskaya, G. S., 355(34), 356(34), 364
Murray, R. C., 408(23), 412
Musikas, C., 654(5), 655, 659(11), 659
Muss, J., 145(27), 161
Mussini, T., 70(1), 74(1), 77, 79(3, 4), 83(4), 83, 700(5), 711, 716(3), 717
Myasoedov, B. F., 669(9), 669
Myers, O. E., 528(25), 529(27), 536
Mykytiuk, D. P., 159(56), 162

Nachtrieb, N., 237(2), 246
Nagai, T., 354(28), 364

Naito, T., 75(2), 77
Nall, D. S., 304(19), 312
Naqvi, S. M. A., 306(35), 312
Narajan, G., 773(82), 776(82), 785
Narakidze, I. G., 178(10), 179
Näsänen, R., 329(20), 330(20), 338
Nash, Ch. P., 282(58), 285
Natke, C. A., 529(26), 532(44), 536, 537
Naumann, A. W., 516(15), 517(15), 525
Nazarov, A. S., 767(28), 768(28), 772(66), 774(66), 783, 784
Neiding, A. B., 309(42), 312
Neisman, R. E., 467(27), 474
Nemec, I., 231(29), 235
Neuman, G., 530(8), 535
Neumann, H. N., 86(10), 90
Neumann, K., 436(16), 439
Nevsky, O. B., 114(18), 119(18), 125
Newman, G. H., 254(4), 256
Newton, T. N., 519(35), 526
Newton, T. W., 619(59), 620(67), 628
Nigmetova, R. Sh., 266(7), 283
Nikolaev, A. V., 767(28), 768(28), 772(66), 774(66), 783, 784
Nikolaeva, N. M., 344(18), 364, 650(6), 650
Nikolaevsky, V. B., 660(3), 664(9), 664
Nobe, K., 539(15), 541(9), 545(15), 546, 547
Norseev, Yu. V., 91(3), 92
Norvell, V. E., 622(71), 628
Novikov, G. I., 532(45), 533(45), 537
Novoseleva, A. V., 677(14), 686
Novosyolov, I. M., 504(23), 506
Noyes, A. A., 294(2), 309(39), 311, 312
Nugent, L. J., 587(3,4), 590(12, 13), 618(49,50), 622(50), 626, 627, 637(1,2), 639, 663(8), 664, 665(3,4,5), 667, 669(5), 669, 670(4,6), 670, 671(3,4), 671, 672(2,4), 672, 674(2), 674
Nunez, L., 287(10), 293
Nurmia, M., 673(2), 674
Nuttal, R. L., 650(3), 650
Nyholm, R. S., 389(23), 390, 424(38), 424(39), 427

Nylem, P., 141(16), 161

O'Donnell, S., 504(22), 506
Oetting, F. L., 621(70), 628
Offner, H. G., 619(62), 628
Oglaza, J., 294(1), 311
Ogura, K., 464(10), 474
O'Hare, P. A. G., 763(4), 765(4), 775(4), 776(4), 781(99), 782, 785
Oishi, J., 618(37), 623(37), 627
O'Kelley, G. D., 672(4,5), 673
Olin, A., 180(6), 181(6), 187
Oliver, J. W., 539(14), 547
Olofson, G., 266(10), 276(10), 277(10), 279(10), 281(10), 282(10), 283
Opalovskii, A. A., 767(28), 768(28), 772(66), 774(66), 783, 784
Opekar, F., 424(41,43,44), 427
Orring, J., 579(20), 580
Osborne, D. W., 768(44), 769(51, 55), 770(51,54,55), 771(51,54, 55), 783, 784
Osman-Zade, Sh. D., 114(20), 125
Ostannii, N. I., 278(41), 284
Ott, V. R., 473(39), 475

Padhi, M. C., 306(34), 307(34), 312
Page, J. A., 388(18), 390, 417(28), 418(28), 427
Paik, W., 197(4), 200
Pajdowski, L., 519(33), 526
Palanisamy, T., 139(5), 139
Pan, K., 486(9), 505
Pankratz, L. B., 287(1), 293, 529(35), 536
Pantani, F., 385(8), 390
Pantony, D. A., 343(7), 344(7), 364, 388(19), 389(19), 390
Panzer, R. E., 677(7), 686
Papaconstantinou, E., 503(20), 504(21), 506
Papadopoulos, M. N., 277(29), 284
Paris, R. A., 149(45), 162
Parker, H. C., 88(17), 91
Parker, V. B., 40(2), 46, 94(1), 100(1), 105(1), 113(1), 117(1), 122(1), 124, 127(4), 137(4), 139, 159(59), 162, 163(16), 171(16), 172, 249(1), 256, 257(6),

[Parker, V. B.,]
  262(6), 264(6), 264, 266(11, 12), 275(11,12), 276(11), 277(11,12), 278(11), 279(11,12), 280(11), 281(11), 282(11), 283, 287(3,12,13), 293, 391(3), 409(3), 411, 430(2,3), 439, 441(8), 444, 475(5), 479(5), 483, 486(3,4), 501(4), 505, 507(9), 515(9), 525, 590(18), 620(19,69), 621(19,69), 622(19), 626, 628, 650(3), 650, 677(9), 686, 776(84), 785

Parks, W. G., 257(5), 264
Pascal, P., 548(5), 552
Passmore, J., 63(27), 66
Pavlinova, A. V., 464(11), 474
Pchelnikov, A. P., 242(8), 246
Peacock, R. D., 447(12,13), 448, 502(17), 506
Pearson, R. G., 44(14), 47
Pecherskaya, A. G., 220(13), 220
Pecsok, R. L., 524(46), 526
Peekema, R. M., 466(23), 474
Penneman, R. A., 660(5), 663(7), 664(5), 664
Pepekin, V. I., 765(5), 769(5), 776(5), 782
Peretrukhin, V. F., 664(9), 664
Perkampus, H. H., 103(8), 124
Pernick, A., 453(2), 464(2), 465(2), 474
Perrault, G. G., 577(8,9), 578(9,13), 580, 687(2,3), 698(2,3), 699
Perrin, D. D., 204(7), 206, 212(9), 213, 278(38), 284, 378(30), 382, 507(7), 525
Peshchevitskii, B. I., 313(6,8), 316(6,11), 317(8,17), 318(6,8,17), 319, 320
Peterson, J. R., 622(71), 628, 663(6), 664, 667(4), 669(6), 669, 670(5), 670
Petitfaux, C., 287(6), 293
Petrova, G. M., 388(16), 390
Pfanhauser, W., 334(49), 335(49), 339
Pickering, D., 447(13), 448
Pierre, D., 65(36), 66
Pikaev, A. K., 660(2), 664
Pilcher, G., 287(10), 293
Piskunova, V. N., 304(16), 312

Pitman, A. L., 172(5), 177(5), 179
Pitzer, K. S., 140(2), 141(2), 146(2), 161, 294(2), 311
Plambeck, J. A., 677(10), 686
Pletcher, D., 266(17), 283
Plieth, W. J., 127(3), 128(3), 129(3), 130(3), 131(3), 132(3), 133(3), 134(3), 139
Pluchet, E., 642(1), 645
Plurien, P., 763(2), 782
Pobedinski, S., 687(1), 699
Podlaha, J., 141(19), 161
Podlahova, J., 141(19), 161
Podorozhnyi, A. M., 672(1), 672
Pokorny, F., 309(37), 310(37), 312
Popa, G., 254(10), 257
Pope, M. T., 502(16), 503(16,20), 504(22,24), 506, 516(11), 525, 530(11), 535
Popov, A. I., 84(4), 90
Popov, L. V., 355(33), 364
Popovych, O., 761(12), 762
Post, K., 507(6), 515(6), 516(6), 517(6), 518(6), 519(6), 520(6), 521(6), 522(6), 524(6), 525
Potter, R. L., 158(55), 159(55), 162
Potts, L. W., 416(11), 426
Pouradier, J., 313(4,7), 316(9,10), 317(9,10), 318(4,7,9), 319
Pourbaix, M., 67(4), 69(4), 69, 117(22), 125, 127(2), 129(2), 130(2), 131(2), 132(2), 133(2), 139, 165(13), 166(13), 167(13), 171, 172(5), 177(5), 179, 189(2), 194(2), 195(2), 196(2), 198(7), 199(2,7), 199, 200, 203(3), 204(3), 206(3), 206, 212(10), 213(10), 213, 216(7), 217(7), 218(7), 219(7), 220(7), 220, 226(21), 227(11), 230(11,21), 232(11), 233(11), 235, 308(38), 310(38), 312, 319(19), 320, 326(7), 327(7), 328(7), 331(7), 333(7), 337(7), 338, 343(8), 344(8), 345(8), 353(8), 364, 383(5), 385(5), 388(12), 390, 408(22), 409(22), 412, 416(13), 421(32), 426, 427, 441(3), 442(3), 443(3), 444, 471(36), 475, 507(2), 520(2), 522(2), 525, 527(2), 534(2), 535(2), 535, 540(8), 541(8), 546, 710(9),

Author Index

[Pourbaix, M.,]
 711, 716(5), 717, 723(5),
 724(2), 724, 726(2), 726
Powell, R. E., 305(31), 309(31),
 312, 774(83), 785
Powers, R. A., 254(9), 257, 436(14,
 15), 439
Pozdeeva, A. A., 466(22), 471(22),
 474
Prasad, B., 279(44), 284
Pratt, J. M., 313(2), 316(2),
 318(2), 319, 377(23), 382
Price, H. I., 229(17), 235
Prokopcikas, A., 327(16), 328(16),
 333(16), 338
Prophet, H., 44(12), 47, 266(13),
 275(13), 283, 486(7), 501(7),
 505
Propst, R., 669(10), 669
Prusakoc, V. N., 765(12), 783
Prusakov, V. D., 772(69), 773(69),
 784
Prytz, M., 218(11), 220
Ptitsyn, B. V., 228(12), 235,
 353(25), 354(25), 364, 388(15),
 389(15), 390
Pullen, K. E., 768(41), 783
Purmal, A. P., 772(67,67a),
 773(67), 774(67), 784

Quarterman, L. A., 767(36),
 768(36), 783
Quist, A. S., 141(7), 161

Rabani, J., 60(20), 66
Rabeneck, H., 447(5), 448
Rabideau, S. W., 658(4,6), 659
Rahman, M., 482(13), 483
Rairden, J. R., 590(16), 618(16),
 622(16), 626
Ramette, R., 304(13), 311
Randall, M., 221(3), 234, 407(12),
 412, 761(9), 762
Randin, J. P., 197(6), 200
Randles, J. E. B., 19(5,7), 35(5),
 37
Rao, B. S. V. R., 677(12), 686
Rao, G., 521(41), 526
Rao, K. V., 57(11), 66
Rao, P. R. M., 465(18), 469(18),
 474

Raschella, D. L., 670(5), 670
Raths, W. E., 430(4), 439
Ratkovskii, I. A., 242(20), 247
Ray, P., 294(3), 309(3), 311
Reader, J., 589(1), 626
Reardon, J. F., 280(50), 284
Redlich, O., 45(15), 47
Reichinstein, D., 120(28), 125
Reid, W. E., Jr., 207(5), 213
Reilly, M. L., 342(1), 352(1), 363,
 387(1), 390, 415(1), 423(1),
 426
Remy, H., 471(35), 475
Rhodes, H. J., 773(79), 775(79),
 784
Rice, S. A., 769(50), 784
Richardson, D. W., 447(6), 448
Richardson, H. K., 333(48), 339
Richter, W. H., 475(1), 480(1),
 483
Riddiford, A. C., 380(34), 382
Rieman, W., 145(25), 161
Ringbom, A., 327(33), 332(33),
 339
Rinke, K., 447(5), 448
Roberts, E. J., 294(8), 305(8),
 311
Roberts, J., 176(8), 179
Robertson, E., 619(57), 628
Robie, R. A., 566(213), 580, 687(5),
 699
Robins, R. G., 409(31), 412,
 507(6), 515(6), 516(6), 517(6),
 518(6), 519(6), 520(6), 521(6),
 522(6), 524(6), 525
Robinson, R. A., 304(18,28),
 305(28), 308(28), 312, 407(14),
 412
Rocek, J., 482(13), 483
Rock, P. A., 305(31), 306(33),
 309(31), 312, 371(5), 381,
 408(23), 412
Romano, V., 475(6), 480(6), 483
Romberger, S. B., 282(57), 285
Rondinini, S., 717(3), 724
Roseveare, W. E., 316(15), 319
Ross, J. W., Jr., 539(14), 547
Rosseinsky, D. R., 44(13), 47
Rossini, F. D., 261(11), 264(11),
 264
Rossotti, F. J. C., 516(14), 517(28),
 525, 526
Rossotti, H. S., 516(14), 517(28),

[Rossotti, H. S.,]
  525, 526
Rotinian, A. L., 119(28), 125, 373(12), 381
Rouelle, F., 62(24), 66
Rourke, M. D., 316(14), 319
Rozenkevitch, N. A., 670(2), 670, 671(2a,2b), 671
Ruband, M., 383(6), 390
Ruben, S., 254(7), 256
Rubin, B., 540(5), 546
Ruediger, S., 767(29), 768(38), 783
Ruetschi, P., 230(24), 231(28), 235
Rumer, I. A., 623(75,76,78), 628, 670(2), 670, 671(2a,2b), 671, 672(1), 672
Rumjantsev, Yu. V., 242(19), 247
Russ, C. R., 321(2), 338
Russel, H., 144(21), 161
Russell, T. P., 280(50), 284
Ryan, J. L., 618(49,50,52), 662(50), 663(8), 627, 664, 665(3), 667, 669(5), 669, 670(4), 670, 671(3), 671, 672(2), 672
Rybalko, V. F., 676(3), 686
Rydberg, J., 645(2), 650
Ryhl, T., 343(17), 364
Ryn, J., 197(5), 200

Sadiq, M., 391(9), 407(9), 412
Sagi, S. R., 465(18), 469(18), 474
Salem, T. M., 217(5), 220
Salmon, J. F., 321(2), 338
Saltman, W., 237(2), 246
Sambucetti, C. J., 578(11), 580
Samhoun, K., 622(71), 628, 637(2), 639, 663(6), 664, 670(7), 670, 671(1a), 671, 672(3,4,5), 672, 673
Sammour, H. M., 181(9), 187
Samoilov, V. M., 387(10), 390
Sandell, E. B., 424(35), 427
Sannikov, Y. I., 517(26), 526
Saprykin, A. S., 660(1), 663(1), 664(2), 664, 667
Sargeson, A. M., 425(46), 427
Sargic, V., 319(18), 320
Sartori, G., 327(32), 332(32), 338
Sauerbrunn, R. D., 424(35), 427
Schäfer, H., 447(5), 448, 527(4),

[Schäfer, H.,]
  529(28,31,33), 531(16), 533(46), 535, 536, 537
Schaffer, G. E., Jr., 149(44), 162
Schlechter, L., 475(2), 480(2), 483
Schmal, N. G., 343(10), 364
Schmets, J., 540(8), 541(8), 546
Schmid, G. M., 319(20), 320
Schmidt, K., 615(48), 623(83), 627, 629, 660(4a,b), 664, 665(6), 667
Schmier, A., 435(12), 439
Schmulbach, C. D., 145(36), 162
Schnick, I., 239(4), 246
Schoepp, L., 517(23), 526
Schott, H. F., 114(17), 124, 579(15), 580
Schreiner, F., 765(11), 766(19), 768(44,47), 769(47,51), 770(47, 51,54), 771(51,54), 783, 784
Schrobilgen, G. J., 765(15), 783
Schrobilgen, R. J., 765(13), 783
Schroth, H., 517(24), 526
Schufle, A., 242(17), 247
Schuhmann, R., 172(4), 176(4), 177(4), 179
Schulte, F., 527(4), 535
Schultz, F. A., 473(39), 475
Schults, W. D., 619(55), 628
Schumacher, E., 768(37), 783
Schuman, R. H., 163(16), 171(16), 172
Schumb, W. C., 408(20), 412
Schumm, R. H., 40(2), 46, 94(1), 100(1), 105(1), 113(1), 117(1), 118(23), 122(1), 124, 125, 127(4), 137(4), 139, 159(59), 162, 249(1), 256, 257(6), 262(6), 264(6), 264, 266(11,12), 275(11, 12), 276(11), 277(11,12), 278(11), 279(11,12), 280(11), 281(11), 282(11), 283, 287(3, 12), 293, 391(3), 409(3), 411, 430(2,3), 439, 524(9), 525, 590(18), 620(19,69), 621(19,69), 622(19), 626, 628, 650(3), 650, 776(84), 785
Schurig, H., 437(21), 439
Schwabe, K., 276(23), 284, 658(5), 659
Schwarz, H. A., 60(20), 66
Schwarzenbach, G., 147(43), 162,

[Schwarzenbach, G.,]
  261(13), 265, 282(55), 285,
  329(22), 330(22), 338, 516(12,
  16), 517(12), 525
Schweigardt, F., 177(9), 179
Sealfeld, F. E., 163(4), 171
Segré, E., 91(1), 92
Selig, H., 766(22), 768(22,40,42,43,
  46), 783, 784
Selikhov, G. G., 772(62), 784
Semenov, G. A., 242(20), 247
Sen, D., 294(3), 309(3), 311
Senkowski, T., 430(5), 439
Sessa, P. A., 765(18), 766(18),
  783
Sevoyukov, N. N., 584(2), 585
Shafer, M. W., 466(23), 474
Shaffer G. E., Jr., 147(33), 162
Shakhova, Z. F., 503(18), 506
Shamsiev, A. S., 345(20), 364,
  388(14), 390
Shannon, R. D., 616(45), 617(45),
  627
Sharma, V. K., 409(33), 412
Sharpe, A. G., 371(8), 372(8),
  376(19), 381, 382, 430(6), 439
Sharpe, T. F., 233(33), 235
Shchukarev, S. A., 242(20), 247,
  532(45), 533(45), 537
Sheft, I., 768(39), 783
Sherman, R. H., 163(5), 171
Sherrill, M. S., 408(20), 412,
  619(56), 628
Shevtsova, N. A., 485(1), 505
Shibata, F. L. E., 700(7), 710(7),
  711
Shierman, G. R., 391(7), 411
Shilov, V. P., 660(1,3), 663(1),
  664(2a,9), 664, 667
Shorokhova, V. I., 176(6,7), 177(6,
  7), 179
Shrawder, J., Jr., 229(18), 235
Shults, W., 669(8), 669
Shuman, R. H., 441(8), 444, 475(5),
  479(5), 483, 486(3), 505
Shurayh, F. R., 410(29), 412
Sibbing, E., 531(16), 536
Sidarous, S. F., 467(30), 475
Siegel, S., 767(31), 769(52),
  775(31), 783, 784
Sikkeland, T., 673(1,2), 673, 674
Sillen, L. G., 261(13), 265,
  278(33), 284, 304(14), 312,

[Sillén, L. G.,]
  329(22), 330(22), 338, 391(2),
  411, 481(9), 483, 517(30), 526,
  548(2), 552, 563(4), 565
Silva, R. J., 672(3), 673(1,2,3),
  673, 674(1), 674
Silver, G. L., 655(2a), 659
Silverman, M. D., 416(18), 417(18),
  427
Silvestroni, P., 327(32), 332(32),
  338
Simakin, G. A., 654(8), 655,
  664(2b), 667, 669(1,7), 669,
  670
Simakov, B. V., 354(26,27), 364
Simon, A. C., 233(36), 235
Singh, K. P., 230(23,25), 231(23),
  232(23), 235
Sinke, G. C., 100(10), 124
Sjoblom, R., 660(4a,4b), 664,
  665(6), 667
Skibsted, L. H., 313(1), 316(1),
  317(13), 318(1), 319
Skilandat, H., 371(3), 381
Skinner, H. A., 287(10), 293,
  447(15), 448
Sklarchuck, J., 230(24), 231(24), 235
Skoog, D. A., 619(62,64), 621(64),
  628
Skuratov, S. M., 84(3), 90, 532(43),
  537
Sladky, F. O., 767(32,33,34), 783
Sloth, E. N., 778(89), 785
Small, L. Y., 502(11), 505
Smirnov, V. A., 266(8), 283
Smirnova, E. K., 533(47), 537
Smirnova, L. Y., 619(61), 620(61),
  628
Smith, D. F., 768(45), 773(71),
  778(71), 784
Smith, G. F., 619(53), 627
Smith, J. D., 180(1,5), 187
Smith, M. E., 535(38), 536
Smith, R. M., 261(10), 262(10), 264
Smith, W. E., 305(29), 308(29), 312
Smith, W. T., 294(11), 304(11), 311,
  440(1,2), 441(1,2), 442(2),
  443(1), 444
Sobkowski, J. B., 650(4,5), 650
Soiegusa, F., 277(27), 284
Sokolov, V. B., 765(12), 772(69),
  773(69), 783, 784
Sokolskii, D. V., 343(15), 344(15),

[Sokolskii, D. V.,] 364
Sorochan, R. I., 178(12), 179
Sorum, C. H., 104(14), 124
Souchay, D., 518(32), 519(32), 526
Souchay, P., 472(38), 475
Spencer, J. F., 278(35), 284
Speranskaya, E. F., 467(28,29), 474, 475
Spevak, V. N., 358(37,38), 360(38), 364, 365
Spice, J. E., 337(38), 339
Spinner. B., 453(3), 474, 530(9, 10), 535
Spitsyn, V. I., 355(33), 356(35), 359(39), 364, 365, 660(1,3), 664(2a), 664, 667, 670(2), 670, 671(2a,2b), 671, 672(1), 672
Spittler, T. M., 768(39), 773(73, 80), 774(73), 775(80), 777(80), 783, 784, 785
Spitz, R. D., 88(19), 91
Spohr, R., 765(8), 783
Spooner, R. C., 619(56), 628
Sprague, E. D., 447(14), 448, 449(3), 451
Srinivasan, K. V., 280(51), 284
Stanlinski, B., 541(13), 547
Steemers, T., 619(65), 628
Stehlik, B., 310(46), 312
Stein, G., 63(29), 66, 763(2), 782
Stein, L., 92(7), 92, 765(6), 780(6, 94,97,98), 781(6), 782, 785
Ste-Maire, J., 254(6), 256
Stemprok, M., 282(57), 285
Stender, V. V., 220(13), 220
Stephenson, C. C., 411(36), 412
Stevenson, D. P., 154(50), 162
Stevenson, J. N., 667(2), 669
Stock, A. E., 555(1), 565
Stokely, J. R., 619(55), 628, 669(6,8), 669, 670(3), 670, 672(3), 673, 674(1), 674
Stokes, R. H., 230(26), 235, 407(14), 412
Stone, R. E., 672(2), 673
Stone, W. E., 88(14), 91
Storms, E. K., 528(24), 536
Stotz, R. W., 580(1), 585
Stoughton, R. W., 294(11), 304(11), 311
Straumanis, M. E., 545(19), 547,

[Straumanis, M. E.,] 676(5), 686
Strehlow, H., 378(29), 382, 410(24), 412, 518(31), 526
Streng, A. G., 49(2), 65, 765(10), 767(25), 768(25), 783
Streng, L. V., 765(10), 767(25), 768(25), 783
Stubbs, M. F., 242(17), 247
Studier, M. H., 773(76), 774(76), 778(89), 784, 785
Stull, D. R., 44(12), 47, 100(10), 124, 266(13), 275(13), 283
Sturrock, P. E., 147(42), 162
Stuve, J. M., 447(7), 448(2), 448, 451
Sugar, J., 589(1), 626
Sukhotin, A. M., 466(22), 471(22), 474, 541(10), 544(10), 546
Sullivan, J., 654(6), 655
Sullivan, J. C., 417(24), 427, 615(48), 623(83), 627, 629, 651(3), 655, 660(4a,4b), 664, 665(6), 667
Summer, R. A., 228(14), 235
Sundermann, H., 472(37), 475
Supawan, A., 466(24), 474
Sutin, N., 372(9), 381, 416(6), 426
Suvorov, A. V., 532(45), 533(45), 537
Svec, H. J., 163(4), 171, 765(7), 783
Swaddle, T. W., 481(8), 483
Swanson, J. A., 266(9), 276(9), 280(9), 281(9), 283
Sweet, J. R., 378(27), 382
Sweeton, F. H., 391(6), 407(6), 411
Sweetser, S. B., 408(20), 412
Swendeman, N., 257(9), 261(9), 262(9), 264(9), 264
Swendeman Loud, N. A., 54(5), 66, 261(15), 262(15), 263(15), 264(15), 265, 540(7), 541(7), 546, 566(4), 578(4), 580, 710(8), 711, 716(4), 717, 723(4), 724(4), 724, 726, 761(11), 762
Swieter, D. S., 473(39), 475
Swift, E. H., 88(16), 91, 114(17), 124, 178(11), 179, 521(40), 524(44), 526
Swofford, H. S., 416(11), 426
Symons, M. C. R., 437(19), 439, 482(12), 483
Syverud, A. N., 486(7), 501(7), 505

Tagami, M., 343(9), 364
Taillon, R., 287(11), 293
Takacs, I., 278(32), 282(32), 284
Takei, T., 354(28), 364
Takhasi, K., 141(18), 146(18), 161
Talukdar, P. K., 279(43), 284, 435(11), 439
Tamamushi, R., 266(16), 283
Tanaka, N., 266(16), 283, 379(32), 382
Tanaka, S., 354(28), 364
Taniguchi, H., 305(26), 312
Taniguchi, S., 265(2,3), 283
Tarrant, J. R., 672(4), 673(3), 673, 674
Tarter, H. V., 218(10), 220
Taube, H., 417(23,26,27), 418(26, 27), 420(27), 427
Taubenest, R., 768(37), 783
Taulli, T. A., 145(34), 146(34,38), 162
Taylor, A. C., 228(13), 235
Taylor, N. H., 545(18), 547
Taylor, R. P., 141(17), 161
Taylor, W. C., 65(35), 66
Templeton, D. H., 343(4), 363
Thiel, A., 242(16), 247, 331(31), 332(31), 338
Thiele, R., 437(20), 439
Thirsk, H. R., 279(43), 284, 434(9), 435(9,10,11), 439
Thomas, F. G., 545(16), 547
Thomas, N. T., 539(15), 541(9), 545(15), 546, 547
Thompson, A., 530(6), 535
Thompson, R. C., 615(48), 627
Thornton, P., 548(4), 552
Tikhinskii, G. F., 676(3), 686
Tikhomirova, V. I., 57(10), 66
Tikhonov, V. I., 772(62), 784
Tilley, R. I., 313(2), 316(2), 318(2), 319
Timofeev, G. A., 664(2b), 667, 669(1,7), 669, 670
Titova, V. N., 389(22), 390
Tom, G. M., 417(26), 418(26), 420(26), 427
Tomilov, A. P., 139(1), 154(1), 161, 166(14), 167(14), 171
Tompson, A., 242(17), 247
Tomsicek, W. J., 279(47), 284, 471(34), 475
Toni, J. E., 380(36), 382

Toth, K. S., 672(3), 673, 674(1), 674
Towns, M. B., 304(21), 312
Trachtenberg, I., 254(8), 256
Trasatti, J., 16(2), 17(3), 18(2), 19(3), 37
Travers, J. G., 585(3), 585
Treadwell, W. D., 430(4), 439, 502(12), 505
Tremaine, P. R., 391(7,8), 407(8), 411, 412
Treumann, W. B., 257(3), 264
Tridot, G., 424(36), 427
Trippel, A. F., 767(28), 768(28), 783
Trofimov, T. I., 669(9), 669
Troitskaya, N. V., 75(3), 77
Tryson, G., 19(8), 37
Tsayun, G. P., 388(17), 389(22), 390
Tsigdinos, G. A., 453(6,8), 474
Tsvelodub, L. D., 344(18), 364
Tuch, D. G., 548(6), 553
Tungusiva, L. I., 541(10), 544(10), 546
Turnay, T. A., 103(12), 107(12), 124
Turner, D. R., 206(12), 206, 213(12), 213, 292(14), 293

Uchimura, K., 545(17), 547
Ummat. P. K., 265(4), 283
Unmack, A., 141(8), 161
Urbach, H. B., 57(14), 66
Ustinskii, B. Z., 336(53), 337(53), 339

Vaghi, E., 700(5), 711, 716(3), 717
Vagramian, A. T., 114(20), 125, 178(10), 179, 407(13), 412
Valentine, J. S., 61(22), 62(22), 66
Valla, E., 331(30), 332(30), 338
Vanden Bosche, E. G., 326(11), 329(11), 338
Vander Linden, J. G. M., 371(7), 381
Vander Sluis, K. L., 587(3,4), 626
Vanderzee, C. E., 141(7), 161, 266(9), 276(9), 280(9), 281(9), 283, 305(29), 308(29), 312
VanGaal, H. L. M., 371(7), 381

VanHeteren, W. J., 217(3), 220
Van Muylder, J., 540(8), 541(8), 546, 615(42), 621(42), 623(42), 627
Van Rysselberghe, P., 227(11), 230(11), 232(11), 238(11), 235
VanWazer, J. R., 145(36), 162
Varga, G. M., Jr., 502(16), 503(16), 506
Varga, L. P., 535(39), 536
Vas, S., 773(78), 775(78), 784
Vasiliev, V. P., 176(6,7), 177(6,7), 179, 181(7), 186(13), 187, 227(9), 235, 522(10), 553
Vasilkova, I. V., 533(47), 537
Vas'ko, A. T., 466(21), 474, 486(2), 505
Venkataratnam, G., 677(12), 686
Verbeek, F., 619(54), 627
Vestling, C. S., 88(18), 91
Vetter, K. J., 411(37), 412, 430(7), 439
Vijh, A. K., 213(11), 213
Vinal, G. W., 233(35), 235
Vincet, H., 159(57), 162
Virmani, Y., 447(15), 448
Visco, R. E., 242(7,9), 246
Vlček, A. A., 367(1), 371(1), 376(20), 377(20), 381, 382
Voigt, A. F., 265(1), 283
Von Massow, R., 391(7), 411
Von Roda, E., 436(16), 439
Vorob'ev, A. F., 84(3), 90
Vorob'eva, S. V., 164(10), 171
Vosburgh, W. C., 233(34), 235, 277(25), 279(48), 280(48), 284
Vrjosek, G. G., 577(10), 578(12), 580
Vulliner, P. A., 767(33), 783

Wadsworth, E., 619(58), 628
Wagman, D. D., 40(2), 46, 54(3), 57(3), 65, 94(1), 100(1), 105(1), 113(1), 117(1), 122(1), 124, 127(4), 137(4), 139, 159(59), 162, 163(16), 171(16), 172, 249(1), 256, 257(6), 261(11), 262(6), 264(11), 264, 266(11,12), 275(11,12), 276(11), 277(11,12), 278(11), 279(11,12), 280(11), 281(11), 282(11), 283, 287(3,12,13), 293, 391(3), 409(3), 411, 430(2,3), 439,

[Wagman, D. D.,] 441(8), 444, 475(5), 479(5), 483, 486(3,4), 501(4), 505, 507(9), 515(9), 525, 528(21), 529(21), 532(21), 533(21), 536, 590(18), 620(19,69), 621(19,69), 622(19), 626, 628, 650(3), 650, 677(9), 686, 776(84), 785
Wahl, A. C., 309(41), 312
Walker, L. C., 486(7), 501(7), 505
Wallace, R. M., 416(15), 427
Wallin, T., 242(15), 247
Warf, J. C., 287(9), 293
Warner, J. S., 343(11), 364
Warnqvist, B., 372(9,10), 373(13), 381
Warriner, J. P., 773(77), 774(77), 784
Washburn, E. W., 164(9), 171
Wasilewski, R. J., 534(37), 536
Watson, J. G., 577(5), 580
Watt, G. D., 377(22), 382
Watt, G. W., 343(4), 363
Watters, J. I., 145(29,31), 147(31, 42), 161, 162
Weakley, T. J. R., 453(7), 474, 503(19), 506
Weast, R. C., 140(4), 161
Weaver, B., 667(1,2), 669
Weber, J. H., 321(2), 338
Wechsberg, M., 767(33), 783
Weeks, E. J., 180(3), 187
Wehner, P., 416(14), 427
Weingartner, F., 242(14), 247
Weinstock, B., 763(1), 765(1), 766(1), 767(1), 769(1), 772(1), 782
Weis, G. S., 321(2), 338
Weksli, Z., 766(23), 783
Welford, G., 516(13), 518(13), 525
Weller, W. W., 529(35), 536
Wellman, H. B., 345(19), 364
Wells, C. F., 372(11), 381
Wendt, H., 378(29), 382, 410(24), 412, 518(31), 526
Westrum, E. F., Jr., 100(10), 124
White, J. C., 217(2), 220
Whitney, J. F., 535(38), 536
Whittemore, D. O., 408(21), 412
Widmer, M., 282(56), 285
Wiedemann, H. G., 343(12), 364
Wiers, B., 309(43), 312
Wijs, H. J., 329(19), 330(19), 338
Wild, J. F., 672(5), 673

Author Index

Wilkins, R. G., 375(17), 379(31), 382
Wilkinson, G., 49(1), 65, 200(1), 206, 207(1), 213, 321(1), 338, 377(25), 382, 383(3), 390, 453(5), 464(5), 474, 547(1), 548(1), 552
Williams, J. H., 577(6), 580
Williams, R. J., 616(46), 627
Williams, R. J. P., 377(23), 382
Williamson, S. M., 763(3), 765(3), 767(27), 773(72), 778(72), 782, 783, 784
Willihnganz, E., 233(36), 235
Willis, J. B., 385(9), 390, 424(45), 427
Wilson, E. G., 769(50), 784
Wilson, G. L., 529(29,32), 536
Withers, G., 529(30), 536
Wolcott, H. A., 616(47), 627
Wolf, E. F., 230(20), 235
Woolley, E. M., 287(2), 293
Woonter, N., 330(27), 338
Worrell, J., 528(23), 536
Worsley, I. G., 507(5), 515(5), 516(5), 517(5), 522(5), 524(5), 525, 527(1), 534(1), 535
Wynne-Jones, W. F. K., 230(23, 25), 231(23), 232(23), 235, 240(6), 246, 276(21), 283

Yablokova, I. E., 57(13), 66
Yacoubi, N., 623(81,82), 628
Yakovleva, V. S., 120(29), 125
Yalman, R. G., 374(14), 381
Yamamoto, H., 354(28), 364
Yeager, E., 57(11), 66
Yelin, R. E., 379(31), 382
Yost, D. M., 88(14), 91, 114(17), 124, 144(21), 144(21), 154(50, 51), 161, 162, 424(34), 427, 520(37), 526
Young, J. P., 622(71), 628
Young, R. C., 531(15), 536
Yui, N., 141(18), 146(18), 161
Zaichenko, V. N., 466(21), 474
Zapponi, P. P., 221(5), 234, 278(31), 284
Zarechanskaya, V. V., 407(13), 412
Zasshi, K. K., 75(2), 77
Zebreva, A. I., 266(6), 283
Zedner, J., 334(41,42), 337(41,42), 339
Zelenskii, M. I., 343(16), 364
Zemva, B., 767(35), 783
Zhamagortzyants, M. A., 407(13), 412
Zharkova, T. A., 278(41), 284
Zhdanov, S. I., 102(5), 103(5), 105(5), 108(5), 124
Zheligovskaya, N. N., 355(33), 356(35), 359(39), 363(42), 364, 365
Zielen, A., 654(6,7), 655
Zirin, M. H., 257(4), 264
Zirkin, M. H., 780(94), 785
Zivkovic, P., 319(18), 320
Zlotowski, I., 710(12), 711, 716(7), 717, 723(7), 724
Zolotavin, V. L., 517(26), 526
Zurc, J., 147(43), 162
Zurin, A. I., 336(51), 339
Zyka, J., 231(29), 235

Subject Index

Actinides, 631-674
  actinium, 637-639
  americium, 660-664
  berkelium, 667-669
  californium, 669-670
  curium, 664-667
  einsteinium, 670-671
  electronic structure of gaseous atoms of, 632
  fermium, 671-672
  lawrencium, 674
  mendelevium, 672-673
  neptunium, 651-655
  nobelium, 673-674
  oxidation states of, 633
  plutonium, 655-659
  protactinium, 639-645
  thorium, 639
  uranium, 645-650
Actinium, 637-639
  thermodynamic data on, 638
Alkali metals and their compounds, thermodynamic data on, 729-759
Aluminum, 566-580
  thermodynamic data on, 567-576
Americium, 660-664
  thermodynamic data on, 661-662
Antimony, 172-179
  thermodynamic data on, 173-175
Aqueous xenon compounds, 769-780
  electrode potentials, 781-782
Argon, ionization potential of, 764
Arsenic, 162-172
  arsenic acid-arsenious acid couple, 164-165

[Arsenic]
  arsenic-halogen compounds, 165-171
  arsenic-sulfur compounds, 165
  arsenious acid-arsenic couple, a63-164
  arsine, 163
  potential diagrams for, 171
  thermodynamic data on, 168-170
Astatine, 91-92
Atomic electron affinities, 32

Barium, 717-724
  thermodynamic data on, 718-722
Berkelium, 667-669
  thermodynamic data on, 668
Beryllium, 675-686
  thermodynamic data on, 678-685
Bismuth, 180-187
  thermodynamic data on, 182-184
Boron, 555-566
  thermodynamic data on, 556-562
Bromine, 78-83
  bromine-bromide couple, 78-79
  bromine redox couples, 79-82
  notes on bromine potentials, 82-83
  oxidation states, 78
  thermodynamic data on, 80-81

Cadmium, 257-265
  thermodynamic data on, 258-260
Calcium, 700-711
  thermodynamic data on, 702-709
Californium, 669-670
Carbon, 189-200
  oxidation states, 189

[Carbon]
  potentials of carbon componds with nitrogen and sulfur, 198-199
  potentials of carbon compounds with oxygen and hydrogen, 189-198
  thermodynamic data on, 190-194
Cerium, 593-595, 619-621
Cesium, 727-762
  thermodynamic data on, 755-758
Chlorine, 70-77
  chlorine-chloride couple, 70-74
  chlorine redox couples, 74-76
  notes on chlorine potentials, 76-77
  oxidation states, 70
  potential diagrams for, 77
  thermodynamic data on, 72-73
Chromium, 475-483
  thermodynamic data on, 476-479
Cobalt, 367-382
  thermodynamic data on, 368-370
Combination of standard electrode potentials, 6-7
Conditional potential, 8
Conventions for standard electrode potentials, 3-4
Copper, 287-293
  thermodynamic data on, 288-291
Curium, 664-667
  thermodynamic data on, 666

Diatomic gaseous elements, enthalpies of dissocation of, 33
Dysprosium, 606-607

Einsteinium, 670-671
Electrical potentials in real systems, 13-15
Electrodes, equilibrium at, 15-16
Energy, units of, 2-3
Enthalpies of dissocation of diatomic gaseous elements, 33
Enthalpies of sublimation of crystalline elements, 31
Equilibrium at an electrode, 15-16
Erbium, 609, 610
Europium, 601-602, 622-623

Fermium, 671-672
Fluorine, 67-69
  fluorine-fluoride couple, 67-69

[Fluorine]
  thermodynamic data on, 68
Formal potential, 8
Francium, 727-762
  thermodynamic data on, 758-759

Gadolinium, 603-604
Gallium, 237-239
  thermodynamic data on, 238-239
Galvanic cells, standard electrode potentials from, 7-8
Galvani potential difference, 16
Gas phase, standard electrode potentials in, 5-6
Germanium, 207-213
  thermodynamic data on, 208-210
Gold, 313-320
  thermodynamic data on, 314-315

Hafnium, 547-553
  thermodynamic data on, 551
Halogen compounds of selenium, 115
Halogen compounds of sulfur, 108
Halogen compounds of tellurium, 120
Halogens, 67-92
  astatine, 91-92
  bromine, 78-83
    bromine-bromide couple, 78-79
    bromine redox couples, 79-82
    notes on bromine potentials, 82-83
    oxidation states, 78
    thermodynamic data on, 80-81
  chlorine, 70-77
    chlorine-chloride couple, 70-74
    chlorine redox couples, 74-76
    notes on chlorine potentials, 76-77
    oxidation states, 70
    potential diagrams for, 77
    thermodynamic data on, 72-73
  fluorine, 67-69
    fluorine-fluoride couple, 67-69
    thermodynamic data on, 68
  iodine, 84-91
    iodine-iodide couple, 84
    oxidation states, 84
    thermodynamic data on, 85
    triiodide-iodide couple, 84
Helium, ionization potential of, 764
Holmium, 607-608
Hydrogen, 39-47
  diagrams for, 46

Subject Index

[Hydrogen]
  D_2/D^+ couple, 45
  H_2/H^+ couple, 41-43
  H_2/H^- couple, 43-45
  oxidation states, 41
  single ionic quantities, 45-46
  thermodynamic data on, 40
Hydrogen phosphides, 154
Hydrosulfurous acid, 104-105
Hypoiodite-iodide couple, 84-86
Hypophosphoric acid, 147-149
Hypophosphorous acid, 149-154

Indium, 240-242
  thermodynamic data on, 240-241
Inert gases, 763-785
  aqueous xenon compounds, 769-780
    electrode potentials for, 781-782
  ionization potential of, 764-765
  krypton, 765-766
  radon, 780-781
  xenon, 766-767
  xenon fluorides, 767-769
Iodine, 84-91
  iodine-iodide couple, 84
  oxidation states, 84
  thermodynamic data on, 85
  triiodide-iodide couple, 84
Ionization potentials of the elements, 24-27
Ions, thermodynamic properties of, 10-11
Iridium, 385-390
  thermodynamic data on, 383-387
Iron, 391-412
  oxidation states, 391
  potential diagrams for, 411
  standard electrode potentials, 407-411
  thermodynamic data on, 391-407

Joule, 2

Krypton, 765-766
  ionization potential of, 764

Lanthanum, 619
  thermodynamic data on, 592-593
Lanthanide elements, 587-629
  divalent and tetravalent lanthanides, 618

[Lanthanide elements]
  electronic structures of gaseous atoms, of, 632
  individual elements, 618-623
  oxidation states of, 633
  potential diagrams for, 623-625
  thermodynamic data on, 591-614
  trivalent ions, 590-615
Lattice energies of ionic crystals, 33-34
Lawrencium, 674
Lead, 220-234
  lead storage battery, 233
  oxidation states, 220-221
  thermodynamic data on, 222-226
Lithium, 727-762
  thermodynamic data on, 729-735
Lutetium, 613-614

Magnesium, 687-699
  thermodynamic data on, 688-697
Manganese, 429-439
  thermodynamic data on, 431-433
Mendelevium, 672-673
Mercury, 265-285
  chemistry of, 265-266
  mercurous-mercuric couple, 280
  mercury-mercuric ion couple, 281-282
  mercury-mercuric oxide couple, 281
  mercury-mercuric sulfide couple, 281-282
  mercury-mercurous halide couples, 266-277
  mercury-mercurous ion couple, 266
  mercury-mercurous salt couples, 277-280
  oxidation states, 265
  potential diagrams for, 282-283
  thermodynamic data on, 267-275
Molybdenum, 453-475
  Gibbs energy of molybdate and molybdic acid, 466
  oxidation states, 453-465
  potential diagrams for, 471-473
  reduction of molybdic acid to V, IV, and III states, 466-471
  thermodynamic data on, 454-463

Neodymium, 597-598
Neon, ionization potential of, 764

Neptunium, 651-655
 thermodynamic data on, 652-653
Nickel, 321-339
 thermodynamic data on, 322-325
Niobium, 526-534
 thermodynamic data on, 528-529
Nitrogen, 127-139
 electrode potentials of nitrogen
  compounds, 129-143
 oxidation states, 128
 potential diagrams for, 138, 139
 thermodynamic data on, 135-137
Nobelium, 673-674
Nobel gases, ionization potential of, 764

Osmium, 421-427
 thermodynamic data on, 422-423
Oxygen, 49-66
 irradiated solutions, 59-61
 oxidation states, 49
 oxygen atom-water couple, 56
 oxygen-peroxide couple, 56-57
 oxygen radical anion, 63-64
 oxygen-water couple, 54-56
 ozone-water couple, 58-59
 potential diagrams for, 65
 singlet oxygen, 64-65
 superoxide ion, 61-63
 thermodynamic data on, 50-53
Ozone-water couple, 58-59

Palladium, 339-345
 thermodynamic data on, 340-342
Periodate-iodate couple, 89
Peroxide-water couple, 57-58
Persulfates, 107-108
Phosphorus, 139-162
 hypophosphoric acid, 147-149
 hypophosphorous acid, 149-154
 oxidation states, 140
 phosphine and other hydrogen
  phosphides, 154
 phosphorus-carbon compounds, 159
 phosphorus-halogen compounds, 159
 phosphorus-nitrogen compounds, 159
 phosphorus-phosphoric acid
  couple, 140-147
 phorphorus-sulfur compounds, 159

[Phosphorus]
 potential diagrams for, 159-161
 thermodynamic data on, 155-158
 triphosphoric acid, 147
Physicochemical properties, relation
  between electrode potentials
  and, 22-37
Platinum, 345-365
 thermodynamic data on, 346-352
Plutonium, 655-659
 thermodynamic data on, 656-657
Polonium, 121-124
 thermodynamic data on, 122
Polysulfides, 101
Potassium, 727-762
 thermodynamic data on, 744-752
Praseodymium, 595-597, 621-622
Promethium, 599
Protactinium, 639-645
 thermodynamic data on, 643-644

Radium, 724-726
 thermodynamic data on, 725
Radon, 780-781
 ionization potential of, 764
Rhenium, 444-451
 thermodynamic data on, 445-447
Rhodium, 382-385
 thermodynamic data on, 384
Rubidium, 727-762
 thermodynamic data on, 752-754
Ruthenium, 413-421
 thermodynamic data on, 414-415

Samarium, 599-601
Scandium, 580-585
 thermodynamic data on, 581-583
Selenium, 110-115
 thermodynamic data on, 112-113
Silicon, 200-206
 oxidation states, 200
 silicon potentials, 204-206
 thermodynamic data on, 201-203
Silver, 294-312
 thermodynamic data on, 295-303
Single electrode potentials, 16-18
 calculation of, 18-19
 other physicochemical properties
  and, 22-37
 physical significance of, 19-22
Singlet oxygen, 64-65
Sodium, 727-762
 thermodynamic data on, 735-744

Subject Index 833

Solvation, thermodynamic properties of, 28-30
Strontium, 711-717
    thermodynamic data on, 712-715
Sulfoxylic acid, 104
Sulfur, 93-110
    halogen compounds of, 108
    hydrosulfurous acid, 104-105
    oxidation states, 93-94
    persulfates, 107-108
    polysulfides, 101
    potential diagrams for, 110, 111
    sulfate-sulfite couple, 102-103
    sulfoxylic acid, 104
    sulfur organic compounds, 108-110
    sulfur-sulfide couple, 94-101
    sulfur-sulfite couple, 101-102
    stepwise reduction of sulfur compounds, 110
    thermodynamic data on, 95-100
    thionic acids, 105-106
    thiosulfate and tetrathionate, 103-104
Sulfur organic compounds, 108-110
Superoxide ion, 61-63
Symbols, 1-2

Tantalum, 534-537
    thermodynamic data on, 532-533
Technetium, 439-444
    thermodynamic data on, 441
Tellurium, 115-121
    thermodynamic data on, 117
Terbium, 604-605
Tetrathionate, 103-104
Thallium, 242-246
Thermodynamic condition for electrode equilibrium, 15-16
Thermodynamic data
    on actinium, 638
    on alkali metals and their compounds, 729-759
    on aluminum, 567-576
    on americium, 661-662
    on antimony, 173-175
    on arsenic, 168-170
    on barium, 718-722
    on berkelium, 668
    on beryllium, 678-685
    on bismuth, 182-184
    on boron, 556-562

[Thermodynamic data]
    on bromine, 80-81
    on cadmium, 258-260
    on calcium, 702-709
    on carbon, 190-194
    on cesium, 755-758
    on chlorine, 72-73
    on chromium, 476-479
    on cobalt, 368-370
    on copper, 288-291
    on curium, 666
    on fluorine, 68
    on francium, 758-759
    on gallium, 238-239
    on germanium, 208-210
    on gold, 314-315
    on hafnium, 551
    on hydrogen, 40
    on indium, 240-241
    on iodine, 85
    on iridium, 386-387
    on iron, 391-407
    on lanthanum, 592-593
    on the lanthanide elements, 591-614
    on lead, 222-226
    on lithium, 729-735
    on magnesium, 688-697
    on manganese, 431-433
    on mercury, 267-275
    on molybdenum, 454-463
    on neptunium, 652-653
    on nickel, 322-325
    on niobium, 528-529
    on nitrogen, 135-137
    on osmium, 422-423
    on oxygen, 50-53
    on palladium, 340-342
    on phosphorus, 155-158
    on platinum, 346-352
    on plutonium, 656-657
    on polonium, 122
    on potassium, 744-752
    on protactinium, 643-644
    on radium, 725
    on rhenium, 445-447
    on rhodium, 384
    on rubidium, 752-754
    on ruthenium, 414-415
    on scandium, 581-583
    on selenium, 112-113
    on silicon, 201-203
    on silver, 295-303

[Thermodynamic data]
  on sodium, 735-744
  standard electrode potentials
    from, 9
  on strontium, 712-715
  on sulfur, 95-100
  on tantalum, 532-533
  on technetium, 441
  on tellurium, 117
  on thallium, 244-245
  on thorium, 640-641
  on tin, 214-216
  on titanium, 542-544
  on tungsten, 487-501
  on uranium, 646-649
  on vanadium, 508-515
  on xenon fluorides, 769, 770, 771
  on xenon trioxide, 776
  on yttrium, 591
  on zinc, 250-253
  on zirconium, 549-550
Thermodynamic properties of ions, 10-11
Thermodynamic properties of solvation, 28-30
Thionic acids, 105-106
Thiosulfate, 103-104
Thorium, 639
  thermodynamic data on, 640-641
Thulium, 609, 611
Tin, 213-220
  thermodynamic data on, 214-216
Titanium, 539-547
  thermodynamic data on, 542-544

Triiodide-iodide couple, 84
Triphosphoric acid, 147
Tungsten, 484-506
  normal and formal potentials, 486-504
  oxidation states, 484-486
  potential diagrams for, 505
  thermodynamic data on, 487-501

Uranium, 645-650
  thermodynamic data on, 646-649
Units of energy and potential, 2-3

Vanadium, 507-526
  thermodynamic data on, 508-515
Volta potential difference, 16

Xenon, 766-767
  ionization potential of, 764
  See also Aqueous xenon compounds
Xenon fluorides, 767-769
  thermodynamic data on, 769, 770, 771
Xenon trioxide, 776

Ytterbium, 611-612
Yttrium, 618-619
  thermodynamic data on, 591

Zinc, 249-257
  thermodynamic data on, 250-253
Zirconium, 547-553
  thermodynamic data on, 549-550

ROWAN UNIVERSITY LIBRARY

3 3001 00861 9040

QD 571 .S74 1985
Standard potentials in
aqueous solution

DATE DUE